SYSTEM IDENTIFICATION

Prentice Hall International
Series in Systems and Control Engineering

M.J. Grimble, Series Editor

BANKS, S.P., *Control Systems Engineering: modelling and simulation, control theory and microprocessor implementation*
BANKS, S.P., *Mathematical Theories of Nonlinear Systems*
BENNETT, S., *Real-time Computer Control: an introduction*
CEGRELL, T., *Power Systems Control*
COOK, P.A., *Nonlinear Dynamical Systems*
LUNZE, J., *Robust Multivariable Feedback Control*
PATTON, R., CLARK, R.N., FRANK, P.M. (editors), *Fault Diagnosis in Dynamic Systems*
SÖDERSTRÖM, T., STOICA, P., *System Identification*
WARWICK, K., *Control Systems: an introduction*

SYSTEM IDENTIFICATION

TORSTEN SÖDERSTRÖM
Automatic Control and Systems Analysis Group
Department of Technology, Uppsala University
Uppsala, Sweden

PETRE STOICA
Department of Automatic Control
Polytechnic Institute of Bucharest
Bucharest, Romania

PRENTICE HALL
NEW YORK   LONDON   TORONTO   SYDNEY   TOKYO

First published 1989 by
Prentice Hall International (UK) Ltd
Campus 400, Maylands Avenue
Hemel Hempstead
Hertfordshire HP2 7EZ
A division of
Simon & Schuster International Group

© 1989 Prentice Hall International (UK) Ltd

All rights reserved. No part of this publication may be reproduced, stored in a retrieval system, or transmitted, in any form or by any means, electronic, mechanical, photocopying, recording or otherwise, without the prior permission, in writing, from the publisher. For permission within the United States of America contact Prentice Hall Inc., Englewood Cliffs, NJ 07632.

Printed and bound in Great Britain at the
University Press, Cambridge.

Library of Congress Cataloging-in-Publication Data

Söderström, Torsten.
  System identification/Torsten Söderström and Petre Stoica.
  p.   cm. – (Prentice Hall international series in systems and control engineering)
  Bibliography: p.   Includes indexes.
  ISBN 0-13-127606-9
  I. System identification. I. Stoica, P. (Petre).
  1949–   II. Title III. series.
  QA402.S8933   1988
  003 – dc19   87-29265

---

4 5 6 7 98 97 96 95

ISBN 0-13-127606-9

To Marianne and Anca
and to our readers

CONTENTS

PREFACE AND ACKNOWLEDGMENTS	xii
GLOSSARY	xv
EXAMPLES	xviii
PROBLEMS	xxi

1 INTRODUCTION — 1

2 INTRODUCTORY EXAMPLES — 9

    2.1 The concepts $\mathscr{S}, \mathscr{M}, \mathscr{I}, \mathscr{X}$ — 9
    2.2 A basic example — 10
    2.3 Nonparametric methods — 10
    2.4 A parametric method — 12
    2.5 Bias, consistency and model approximation — 18
    2.6 A degenerate experimental condition — 23
    2.7 The influence of feedback — 25
    Summary and outlook — 28
    Problems — 29
    Bibliographical notes — 31

3 NONPARAMETRIC METHODS — 32

    3.1 Introduction — 32
    3.2 Transient analysis — 32
    3.3 Frequency analysis — 37
    3.4 Correlation analysis — 42
    3.5 Spectral analysis — 43
    Summary — 50
    Problems — 51
    Bibliographical notes — 54
    Appendices
        A3.1 Covariance functions, spectral densities and linear filtering — 55
        A3.2 Accuracy of correlation analysis — 58

4 LINEAR REGRESSION — 60

    4.1 The least squares estimate — 60
    4.2 Analysis of the least squares estimate — 65
    4.3 The best linear unbiased estimate — 67
    4.4 Determining the model dimension — 71
    4.5 Computational aspects — 74
    Summary — 77
    Problems — 77
    Bibliographical notes — 83
    Complements

C4.1 Best linear unbiased estimation under linear constraints ... 83
C4.2 Updating the parameter estimates in linear regression models ... 86
C4.3 Best linear unbiased estimates for linear regression models with possibly singular residual covariance matrix ... 88
C4.4 Asymptotically best consistent estimation of certain nonlinear regression parameters ... 91

5 INPUT SIGNALS ... 96

5.1 Some commonly used input signals ... 96
5.2 Spectral characteristics ... 100
5.3 Lowpass filtering ... 112
5.4 Persistent excitation ... 117
Summary ... 125
Problems ... 126
Bibliographical notes ... 129
Appendix
  A5.1 Spectral properties of periodic signals ... 129
Complements
  C5.1 Difference equation models with persistently exciting inputs ... 133
  C5.2 Condition number of the covariance matrix of filtered white noise ... 135
  C5.3 Pseudorandom binary sequences of maximum length ... 137

6 MODEL PARAMETRIZATIONS ... 146

6.1 Model classifications ... 146
6.2 A general model structure ... 148
6.3 Uniqueness properties ... 161
6.4 Identifiability ... 167
Summary ... 168
Problems ... 168
Bibliographical notes ... 171
Appendix
  A6.1 Spectral factorization ... 172
Complements
  C6.1 Uniqueness of the full polynomial form model ... 182
  C6.2 Uniqueness of the parametrization and the positive definiteness of the input–output covariance matrix ... 183

7 PREDICTION ERROR METHODS ... 185

7.1 The least squares method revisited ... 185
7.2 Description of prediction error methods ... 188
7.3 Optimal prediction ... 192
7.4 Relationships between prediction error methods and other identification methods ... 198
7.5 Theoretical analysis ... 202
7.6 Computational aspects ... 211
Summary ... 216
Problems ... 216
Bibliographical notes ... 226
Appendix
  A7.1 Covariance matrix of PEM estimates for multivariable systems ... 226

Contents ix

	Complements	
	C7.1 Approximation models depend on the loss function used in estimation	228
	C7.2 Multistep prediction of ARMA processes	229
	C7.3 Least squares estimation of the parameters of full polynomial form models	233
	C7.4 The generalized least squares method	236
	C7.5 The output error method	239
	C7.6 Unimodality of the PEM loss function for ARMA processes	247
	C7.7 Exact maximum likelihood estimation of AR and ARMA parameters	249
	C7.8 ML estimation from noisy input-output data	256

8 INSTRUMENTAL VARIABLE METHODS   260

    8.1 Description of instrumental variable methods   260
    8.2 Theoretical analysis   264
    8.3 Computational aspects   277
    Summary   280
    Problems   280
    Bibliographical notes   284
    Appendices
        A8.1 Covariance matrix of IV estimates   285
        A8.2 Comparison of optimal IV and prediction error estimates   286
    Complements
        C8.1 Yule–Walker equations   288
        C8.2 The Levinson–Durbin algorithm   292
        C8.3 A Levinson-type algorithm for solving nonsymmetric Yule–Walker systems of equations   298
        C8.4 Min-max optimal IV method   303
        C8.5 Optimally weighted extended IV method   305
        C8.6 The Whittle–Wiggins–Robinson algorithm   310

9 RECURSIVE IDENTIFICATION METHODS   320

    9.1 Introduction   320
    9.2 The recursive least squares method   321
    9.3 Real-time identification   324
    9.4 The recursive instrumental variable method   327
    9.5 The recursive prediction error method   328
    9.6 Theoretical analysis   334
    9.7 Practical aspects   348
    Summary   350
    Problems   351
    Bibliographical notes   357
    Complements
        C9.1 The recursive extended instrumental variable method   359
        C9.2 Fast least squares lattice algorithm for AR modeling   361
        C9.3 Fast least squares lattice algorithm for multivariate regression models   373

10 IDENTIFICATION OF SYSTEMS OPERATING IN CLOSED LOOP   381

    10.1 Introduction   381
    10.2 Identifiability considerations   382
    10.3 Direct identification   389

  10.4 Indirect identification   395
  10.5 Joint input–output identification   396
  10.6 Accuracy aspects   401
  Summary   406
  Problems   407
  Bibliographical notes   412
  Appendix
    A10.1 Analysis of the joint input–output identification   412
  Complement
    C10.1 Identifiability properties of the PEM applied to ARMAX systems operating under general linear feedback   416

11 MODEL VALIDATION AND MODEL STRUCTURE DETERMINATION   422

  11.1 Introduction   422
  11.2 Is a model flexible enough?   423
  11.3 Is a model too complex?   433
  11.4 The parsimony principle   438
  11.5 Comparison of model structures   440
  Summary   451
  Problems   451
  Bibliographical notes   456
  Appendices
    A11.1 Analysis of tests on covariance functions   457
    A11.2 Asymptotic distribution of the relative decrease in the criterion function   461
  Complement
    C11.1 A general form of the parsimony principle   464

12 SOME PRACTICAL ASPECTS   468

  12.1 Introduction   468
  12.2 Design of the experimental condition $\mathscr{X}$ 468
  12.3 Treating nonzero means and drifts in disturbances   474
  12.4 Determination of the model structure $\mathscr{M}$ 482
  12.5 Time delays   487
  12.6 Initial conditions   490
  12.7 Choice of the identification method $\mathscr{I}$ 493
  12.8 Local minima   493
  12.9 Robustness   495
  12.10 Model verification   499
  12.11 Software aspects   501
  12.12 Concluding remarks   502
  Problems   502
  Bibliographical notes   509

APPENDIX A SOME MATRIX RESULTS   511

  A.1 Partitioned matrices   511
  A.2 The least squares solution to linear equations, pseudoinverses and the singular value decomposition   518

A.3 The QR method 527
A.4 Matrix norms and numerical accuracy 532
A.5 Idempotent matrices 537
A.6 Sylvester matrices 540
A.7 Kronecker products 542
A.8 An optimization result for positive definite matrices 544
Bibliographical notes 546

APPENDIX B SOME RESULTS FROM PROBABILITY THEORY AND STATISTICS 547

B.1 Convergence of stochastic variables 547
B.2 The Gaussian and some related distributions 552
B.3 Maximum *a posteriori* and maximum likelihood parameter estimates 559
B.4 The Cramér–Rao lower bound 560
B.5 Minimum variance estimation 565
B.6 Conditional Gaussian distributions 567
B.7 The Kalman–Bucy filter 568
B.8 Asymptotic covariance matrices for sample correlation and covariance estimates 570
B.9 Accuracy of Monte Carlo analysis 576
Bibliographical notes 579

REFERENCES 580
ANSWERS AND FURTHER HINTS TO THE PROBLEMS 596
AUTHOR INDEX 606
SUBJECT INDEX 609

PREFACE AND ACKNOWLEDGMENTS

System identification is the field of mathematical modeling of systems from experimental data. It has acquired widespread applications in many areas. In control and systems engineering, system identification methods are used to get appropriate models for synthesis of a regulator, design of a prediction algorithm, or simulation. In signal processing applications (such as in communications, geophysical engineering and mechanical engineering), models obtained by system identification are used for spectral analysis, fault detection, pattern recognition, adaptive filtering, linear prediction and other purposes. System identification techniques are also successfully used in non-technical fields such as biology, environmental sciences and econometrics to develop models for increasing scientific knowledge on the identified object, or for prediction and control.

This book is aimed to be used for senior undergraduate and graduate level courses on system identification. It will provide the reader a profound understanding of the subject matter as well as the necessary background for performing research in the field. The book is primarily designed for classroom studies but can be used equally well for self-studies.

To reach its twofold goal of being both a basic and an advanced text on system identification, which addresses both the student and the researcher, the book is organized as follows. The chapters contain a *main text* that should fit the needs for graduate and advanced undergraduate courses. For most of the chapters some additional (often more detailed or more advanced) results are presented in extra sections called *complements*. In a short or undergraduate course many of the complements may be skipped. In other courses, such material can be included at the instructor's choice to provide a more profound treatment of specific methods or algorithmic aspects of implementation. Throughout the book, the important general results are included in solid boxes. In a few places, intermediate results that are essential to later developments, are included in dashed boxes. More complicated derivations or calculations are placed in *chapter appendices* that follow immediately the chapter text. Several general background results from linear algebra, matrix theory, probability theory and statistics are collected in the *general appendices* A and B at the end of the book. All chapters, except the first one, include *problems* to be dealt with as exercises for the reader. Some problems are illustrations of the results derived in the chapter and are rather simple, while others are aimed to give new results and insight and are often more complicated. The problem sections can thus provide appropriate homework exercises as well as challenges for more advanced readers. For each chapter, the simple problems are given before the more advanced ones. A separate *solutions manual* has been prepared which contains solutions to all the problems.

The book does not contain computer exercises. However, we find it very important that the students really apply some identification methods, preferably on real data. This will give a deeper understanding of the practical value of identification techniques that is hard to obtain from just reading a book. As we mention in Chapter 12, there are several good program packages available that are convenient to use.

Concerning the references in the text, our purpose has been to give some key references and hints for a further reading. Any attempt to cover the whole range of references would be an enormous, and perhaps not particularly useful, task.

We assume that the reader has a background corresponding to at least a senior-level academic experience in electrical engineering. This would include a basic knowledge of introductory probability theory and statistical estimation, time series analysis (or stochastic processes in discrete time), and models for dynamic systems. However, in the text and the appendices we include many of the necessary background results.

The text has been used, in a preliminary form, in several different ways. These include regular graduate and undergraduate courses, intensive courses for graduate students and for people working in industry, as well as for extra reading in graduate courses and for independent studies. The text has been tested in such various ways at Uppsala University, Polytechnic Institute of Bucharest, Lund Institute of Technology, Royal Institute of Technology, Stockholm, Yale University, and INTEC, Santa Fe, Argentina. The experience gained has been very useful when preparing the final text.

In writing the text we have been helped in various ways by several persons, whom we would like to sincerely thank.

We acknowledge the influence on our research work of our colleagues Professor Karl Johan Åström, Professor Pieter Eykhoff, Dr Ben Friedlander, Professor Lennart Ljung, Professor Arye Nehorai and Professor Mihai Tertişco who, directly or indirectly, have had a considerable impact on our writing.

The text has been read by a number of persons who have given many useful suggestions for improvements. In particular we would like to sincerely thank Professor Randy Moses, Professor Arye Nehorai, and Dr John Norton for many useful comments. We are also grateful to a number of students at Uppsala University, Polytechnic Institute of Bucharest, INTEC at Santa Fe, and Yale University, for several valuable proposals. Colleagues and students at Uppsala University have also given valuable assistance in the proof-reading phase.

The first inspiration for writing this book is due to Dr Greg Meira, who invited the first author to give a short graduate course at INTEC, Santa Fe, in 1983. The material produced for that course has since then been extended and revised by us jointly before reaching its present form.

The preparation of the text has been a task extended over a considerable period of time. The often cumbersome job of typing and correcting the text has been done with patience and perseverance by Ylva Johansson, Ingrid Ringård, Maria Dahlin, Helena Jansson, Ann-Cristin Lundquist and Lis Timner. We are most grateful to them for their excellent work carried out over the years with great skill.

Several of the figures were originally prepared by using the packages IDPAC (developed at Lund Institute of Technology) for some parameter estimations and BLAISE (developed at INRIA, France) for some of the general figures.

We have enjoyed the very pleasant collaboration with Prentice Hall International.

We would like to thank Professor Mike Grimble, Andrew Binnie, Allison King, Glen Murray, Ruth Freestone and Christopher Glennie for their permanent encouragement and support. Richard Shaw deserves special thanks for the many useful comments made on the presentation. We acknowledge his help with gratitude.

Torsten Söderström
Uppsala

Petre Stoica
Bucharest

GLOSSARY

Notations

$\mathscr{D}$	set of parameter vectors describing models with stable predictors
$D_T(\mathscr{S}, \mathscr{M})$	set of models $\mathscr{M}$ describing the true system $\mathscr{S}$
E	expectation operator
$e(t)$	white noise (a sequence of independent random variables)
$F(q^{-1})$	data prefilter
$G(q^{-1})$	transfer function operator
$\hat{G}(q^{-1})$	estimated transfer function operator
$H(q^{-1})$	noise shaping filter
$\hat{H}(q^{-1})$	estimated noise shaping filter
$\mathscr{I}$	identification method
I	identity matrix
I_n	$(n\|n)$ identity matrix
$k_i(t)$	reflection coefficient
log	natural logarithm
$\mathscr{M}$	model set, model structure
$\mathscr{M}(\theta)$	model corresponding to the parameter vector θ
$(m\|n)$	matrix dimension is m by n
$\mathscr{N}$	null space of a matrix
$\mathscr{N}(m, P)$	normal (Gaussian) distribution of mean value m and covariance matrix P
N	number of data points
n	model order
nu	number of inputs
ny	number of outputs
$n\theta$	dimension of parameter vector
O_n	$(n\|n)$ matrix with zero elements
$O(x)$	$O(x)/x$ is bounded when $x \to 0$
$p(x\|y)$	probability density function of x given y
$\mathscr{R}$	range (space) of a matrix
$\mathscr{R}^n$	Euclidean space
$\mathscr{S}$	true system
A^T	transpose of the matrix A
tr	trace (of a matrix)
t	time variable (integer-valued for discrete time models)
$u(t)$	input signal (vector of dimension nu)
V	loss function
vec(A)	a column vector formed by stacking the columns of the matrix A on top of each other
$\mathscr{X}$	experimental condition
$y(t)$	output signal (vector of dimension ny)
$\hat{y}(t\|t-1)$	optimal (one step) predictor
$z(t)$	vector of instrumental variables

Glossary

$\gamma(t)$	gain sequence
$\delta_{s,t}$	Kronecker delta ($= 1$ if $s = t$, else $= 0$)
$\delta(\tau)$	Dirac function
$\varepsilon(t, \theta)$	prediction error corresponding to the parameter vector θ
θ	parameter vector
$\hat{\theta}$	estimate of parameter vector
θ_0	true value of parameter vector
Λ	covariance matrix of innovations
λ^2	variance of white noise
λ	forgetting factor
σ	variance or standard deviation of white noise
$\phi(\omega)$	spectral density
$\phi_u(\omega)$	spectral density of the signal $u(t)$
$\phi_{yu}(\omega)$	cross-spectral density between the signals $y(t)$ and $u(t)$
$\varphi(t)$	vector formed by lagged input and output data
Φ	regressor matrix
$\chi^2(n)$	χ^2 distribution with n degrees of freedom
$\psi(t)$	negative gradient of the prediction error $\varepsilon(t, \theta)$ with respect to θ
ω	angular frequency

Abbreviations

ABCE	asymptotically best consistent estimator
adj	adjoint (or adjugate) of a matrix, $\mathrm{adj}(A) \triangleq A^{-1} \det A$
AIC	Akaike's information criterion
AR	autoregressive
AR(n)	AR of order n
ARIMA	autoregressive integrated moving average
ARMA	autoregressive moving average
ARMA(n_1, n_2)	ARMA where AR and MA parts have order n_1 and n_2, respectively
ARMAX	autoregressive moving average with exogenous variables
ARX	autoregressive with exogenous variables
BLUE	best linear unbiased estimator
CARIMA	controlled autoregressive integrated moving average
cov	covariance matrix
dim	dimension
deg	degree
ELS	extended least squares
FIR	finite impulse response
FFT	fast Fourier transform
FPE	final prediction error
GLS	generalized least squares
iid	independent and identically distributed
IV	instrumental variables
LDA	Levinson–Durbin algorithm
LIP	linear in the parameters
LMS	least mean square
LS	least squares
MA	moving average
MA(n)	MA of order n
MAP	maximum *a posteriori*

MFD	matrix fraction description
mgf	moment generating function
MIMO	multi input, multi output
ML	maximum likelihood
mse	mean square error
MVE	minimum variance estimator
ODE	ordinary differential equation
OEM	output error method
pdf	probability density function
pe	persistently exciting
PEM	prediction error method
PI	parameter identifiability
PLR	pseudolinear regression
PRBS	pseudorandom binary sequence
RIV	recursive instrumental variable
RLS	recursive least squares
RPEM	recursive prediction error method
SA	stochastic approximation
SI	system identifiability
SISO	single input, single output
SVD	singular value decomposition
var	variance
w.p.1	with probability one
w.r.t	with respect to
WWRA	Whittle–Wiggins–Robinson algorithm
YW	Yule–Walker

Notational conventions

$H^{-1}(q^{-1})$	$[H(q^{-1})]^{-1}$
$\varphi^T(t)$	$[\varphi(t)]^T$
A^{-T}	$[A^{-1}]^T$
$Q^{1/2}$	matrix square root of a positive definite matrix Q: $(Q^{1/2})^T Q^{1/2} = Q$
$Q^{T/2}$	$[Q^{1/2}]^T$
$\|x\|_Q^2$	$x^T Q x$ with Q a symmetric positive definite weighting matrix
$\xrightarrow{dist}$	convergence in distribution
$A \geq B$	the difference matrix $(A - B)$ is nonnegative definite (here A and B are nonnegative definite matrices)
$A > B$	the difference matrix $(A - B)$ is positive definite
$\triangleq$	defined as
$:=$	assignment operator
$\sim$	distributed as
$\otimes$	Kronecker product
$\oplus$	modulo 2 summation of binary variables
$\oplus$	direct sum of subspaces
Σ	modulo 2 summation of binary variables
V'	gradient of the loss function V
V''	Hessian (matrix of second order derivatives) of the loss function V

EXAMPLES

1.1	A stirred tank	1
1.2	An industrial robot	1
1.3	Aircraft dynamics	2
1.4	Effect of a drug	2
1.5	Modeling a stirred tank	4
2.1	Transient analysis	10
2.2	Correlation analysis	12
2.3	A PRBS as input	15
2.4	A step function as input	19
2.5	Prediction accuracy	21
2.6	An impulse as input	23
2.7	A feedback signal as input	26
2.8	A feedback signal and an additional setpoint as input	26
3.1	Step response of a first-order system	33
3.2	Step response of a damped oscillator	34
3.3	Nonideal impulse response	37
3.4	Some lag windows	46
3.5	Effect of lag window on frequency resolution	47
4.1	A polynomial trend	60
4.2	A weighted sum of exponentials	60
4.3	Truncated weighting function	61
4.4	Estimation of a constant	65
4.5	Estimation of a constant (continued from Example 4.4)	70
4.6	Sensitivity of the normal equations	76
5.1	A step function	97
5.2	A pseudorandom binary sequence	97
5.3	An autoregressive moving average sequence	97
5.4	A sum of sinusoids	98
5.5	Characterization of PRBS	102
5.6	Characterization of an ARMA process	103
5.7	Characterization of a sum of sinusoids	104
5.8	Comparison of a filtered PRBS and a filtered white noise	111
5.9	Standard filtering	113
5.10	Increasing the clock period	113
5.11	Decreasing the probability of level change	115
5.12	Spectral density interpretations	117
5.13	Order of persistent excitation	121
5.14	A step function	124
5.15	A PRBS	124
5.16	An ARMA process	125

		Examples	
5.17	A sum of sinusoids		125
C5.3.1	Influence of the feedback path on period of a PRBS		139
6.1	An ARMAX model		149
6.2	A general SISO model structure		152
6.3	The full polynomial form		154
6.4	Diagonal form of a multivariable system		155
6.5	A state space model		157
6.6	A Hammerstein model		160
6.7	Uniqueness properties for an ARMAX model		162
6.8	Nonuniqueness of a multivariable model		165
6.9	Nonuniqueness of general state space models		166
A6.1.1	Spectral factorization for an ARMA process		177
A6.1.2	Spectral factorization for a state space model		178
A6.1.3	Another spectral factorization for a state space model		179
A6.1.4	Spectral factorization for a nonstationary process		180
7.1	Criterion functions		190
7.2	The least squares method as a prediction error method		191
7.3	Prediction for a first-order ARMAX model		192
7.4	Prediction for a first-order ARMAX model, continued		196
7.5	The exact likelihood function for a first-order autoregressive process		199
7.6	Accuracy for a linear regression		207
7.7	Accuracy for a first-order ARMA process		207
7.8	Gradient calculation for an ARMAX model		213
7.9	Initial estimate for an ARMAX model		215
7.10	Initial estimates for increasing model structures		215
8.1	An IV variant for SISO systems		263
8.2	An IV variant for multivariable systems		264
8.3	The noise-free regressor vector		266
8.4	A possible parameter vector ϱ		267
8.5	Comparison of sample and asymptotic covariance matrix of IV estimates		270
8.6	Generic consistency		271
8.7	Model structures for SISO systems		275
8.8	IV estimates for nested structures		278
C8.3.1	Constraints on the nonsymmetric reflection coefficients and location of the zeros of $A_n(z)$		302
9.1	Recursive estimation of a constant		321
9.2	RPEM for an ARMAX model		331
9.3	PLR for an ARMAX model		333
9.4	Comparison of some recursive algorithms		334
9.5	Effect of the initial values		335
9.6	Effect of the forgetting factor		340
9.7	Convergence analysis of the RPEM		345
9.8	Convergence analysis of PLR for an ARMAX model		346
10.1	Application of spectral analysis		383
10.2	Application of spectral analysis in the multivariable case		384
10.3	IV method for closed loop systems		386
10.4	A first-order model		390

xx *Examples*

10.5	No external input		392
10.6	External input		393
10.7	Joint input–output identification of a first-order system		398
10.8	Accuracy for a first-order system		401
C10.1.1	Identifiability properties of an nth-order system with an mth-order regulator		417
C10.1.2	Identifiability properties of a minimum-phase system under minimum variance control		418
11.1	An autocorrelation test		423
11.2	A cross-correlation test		426
11.3	Testing changes of sign		428
11.4	Use of statistical tests on residuals		429
11.5	Statistical tests and common sense		431
11.6	Effects of overparametrizing an ARMAX system		433
11.7	The FPE criterion		443
11.8	Significance levels for FPE and AIC		445
11.9	Numerical comparisons of the F-test, FPE and AIC		446
11.10	Consistent model structure determination		450
A11.1.1	A singular P_r matrix due to overparametrization		459
12.1	Influence of the input signal on model performance		469
12.2	Effect of nonzero means		474
12.3	Model structure determination $\mathscr{S} \in \mathscr{M}$		482
12.4	Model structure determination $\mathscr{S} \notin \mathscr{M}$		482
12.5	Treating time delays for ARMAX models		489
12.6	Initial conditions for ARMAX models		491
12.7	Effect of outliers on an ARMAX model		494
12.8	Effect of outliers on an ARX model		495
12.9	On-line robustification		497
A.1	Decomposition of vectors		518
A.2	An orthogonal projector		538
B.1	Cramér-Rao lower bound for a linear regression with uncorrelated residuals		562
B.2	Cramér-Rao lower bound for a linear regression with correlated residuals		564
B.3	Variance of $\hat{r}_k$ for a first-order AR process		574
B.4	Variance of $\hat{\varrho}_1$ for a first-order AR process		575

PROBLEMS

2.1 Bias, variance and mean square error
2.2 Convergence rate for consistent estimators
2.3 Illustration of unbiasedness and consistency properties
2.4 Least squares estimates with white noise as input
2.5 Least squares estimates with a step function as input
2.6 Least squares estimates with a step function as input, continued
2.7 Conditions for a minimum
2.8 Weighting sequence and step response

3.1 Determination of time constant T from step responses
3.2 Analysis of the step response of a second-order damped oscillator
3.3 Determining amplitude and phase
3.4 The covariance function of a simple process
3.5 Some properties of the spectral density function
3.6 Parseval's formula
3.7 Correlation analysis with truncated weighting function
3.8 Accuracy of correlation analysis
3.9 Improved frequency analysis as a special case of spectral analysis
3.10 Step response analysis as a special case of spectral analysis
3.11 Determination of a parametric model from the impulse response

4.1 A linear trend model I
4.2 A linear trend model II
4.3 The loss function associated with the Markov estimate
4.4 Linear regression with missing data
4.5 Ill-conditioning of the normal equations associated with polynomial trend models
4.6 Fourier harmonic decomposition as a special case of regression analysis
4.7 Minimum points and normal equations when Φ does not have full rank
4.8 Optimal input design for gain estimation
4.9 Regularization of a linear system of equations
4.10 Conditions for least squares estimates to be BLUE
4.11 Comparison of the covariance matrices of the least squares and Markov estimates
4.12 The least squares estimate of the mean is BLUE asymptotically

5.1 Nonnegative definiteness of the sample covariance matrix
5.2 A rapid method for generating sinusoidal signals on a computer
5.3 Spectral density of the sum of two sinusoids
5.4 Admissible domain for ϱ_1 and ϱ_2 of a stationary process
5.5 Admissible domain for ϱ_1 and ϱ_2 for MA(2) and AR(2) processes
5.6 Spectral properties of a random wave
5.7 Spectral properties of a square wave
5.8 Simple conditions on the covariances of a moving average process
5.9 Weighting function estimation with PRBS
5.10 The cross-covariance matrix of two autoregressive processes obeys a Lyapunov equation

xxii Problems

6.1 Stability boundary for a second-order system
6.2 Spectral factorization
6.3 Further comments on the nonuniqueness of stochastic state space models
6.4 A state space representation of autoregressive processes
6.5 Uniqueness properties of ARARX models
6.6 Uniqueness properties of a state space model
6.7 Sampling a simple continuous time system

7.1 Optimal prediction of a nonminimum phase first-order MA model
7.2 Kalman gain for a first-order ARMA model
7.3 Prediction using exponential smoothing
7.4 Gradient calculation
7.5 Newton–Raphson minimization algorithm
7.6 Gauss–Newton minimization algorithm
7.7 Convergence rate for the Newton–Raphson and Gauss–Newton algorithms
7.8 Derivative of the determinant criterion
7.9 Optimal predictor for a state space model
7.10 Hessian of the loss function for an ARMAX model
7.11 Multistep prediction of an AR(1) process observed in noise
7.12 An asymptotically efficient two-step PEM
7.13 Frequency domain interpretation of approximate models
7.14 An indirect PEM
7.15 Consistency and uniform convergence
7.16 Accuracy of PEM for a first-order ARMAX system
7.17 Covariance matrix for the parameter estimates when the system does not belong to the model set
7.18 Estimation of the parameters of an AR process observed in noise
7.19 Whittle's formula for the Cramér–Rao lower bound
7.20 A sufficient condition for the stability of least squares input–output models
7.21 Accuracy of noise variance estimate
7.22 The Steiglitz–McBride method

8.1 An expression for the matrix R
8.2 Linear nonsingular transformations of instruments
8.3 Evaluation of the R matrix for a simple system
8.4 IV estimation of the transfer function parameters from noisy input–noisy output data
8.5 Generic consistency
8.6 Parameter estimates when the system does not belong to the model structure
8.7 Accuracy of a simple IV method applied to a first-order ARMAX system
8.8 Accuracy of an optimal IV method applied to a first-order ARMAX system
8.9 A necessary condition for consistency of IV methods
8.10 Sufficient conditions for consistency of IV methods; extension of Problem 8.3
8.11 The accuracy of extended IV method does not necessarily improve when the number of instruments increases
8.12 A weighting sequence coefficient-matching property of the IV method

9.1 Derivation of the real-time RLS algorithm
9.2 Influence of forgetting factors on consistency properties of parameter estimates
9.3 Effects of $P(t)$ becoming indefinite
9.4 Convergence properties and dependence on initial conditions of the RLS estimate

9.5   Updating the square root of $P(t)$
9.6   On the condition for global convergence of PLR
9.7   Updating the prediction error loss function
9.8   Analysis of a simple RLS algorithm
9.9   An alternative form for the RPEM
9.10 A RLS algorithm with a sliding window
9.11 On local convergence of the RPEM
9.12 On the condition for local convergence of PLR
9.13 Local convergence of the PLR algorithm for a first-order ARMA process
9.14 On the recursion for updating $P(t)$
9.15 One-step-ahead optimal input design for RLS identification
9.16 Illustration of the convergence rate for stochastic approximation algorithms

10.1 The estimates of parameters of 'nearly unidentifiable' systems have poor accuracy
10.2 On the use and accuracy of indirect identification
10.3 Persistent excitation of the external signal
10.4 On the use of the output error method for systems operating under feedback
10.5 Identifiability properties of the PEM applied to ARMAX systems operating under a minimum variance control feedback
10.6 Another optimal open loop solution to the input design problem of Example 10.8
10.7 On parametrizing the information matrix in Example 10.8
10.8 Optimal accuracy with bounded input variance
10.9 Optimal input design for weighting coefficient estimation
10.10 Maximal accuracy estimation with output power constraint may require closed loop experiments

11.1 On the use of the cross-correlation test for the least squares method
11.2 Identifiability results for ARX models
11.3 Variance of the prediction error when future and past experimental conditions differ
11.4 An assessment criterion for closed loop operation
11.5 Variance of the multi-step prediction error as an assessment criterion
11.6 Misspecification of the model structure and prediction ability
11.7 An illustration of the parsimony principle
11.8 The parsimony principle does not necessarily apply to nonhierarchical model structures
11.9 On testing cross-correlations between residuals and input
11.10 Extension of the prediction error formula (11.40) to the multivariable case

12.1 Step response of a simple sampled-data system
12.2 Optimal loss function
12.3 Least squares estimation with nonzero mean data
12.4 Comparison of approaches for treating nonzero mean data
12.5 Linear regression with nonzero mean data
12.6 Neglecting transients
12.7 Accuracy of PEM and hypothesis testing for an ARMA (1,1) process
12.8 A weighted recursive least squares method
12.9 The controller form state space realization of an ARMAX system
12.10 Gradient of the loss function with respect to initial values
12.11 Estimates of initial values are not consistent
12.12 Choice of the input signal for accurate estimation of static gain

12.13 Variance of an integrated process
12.14 An example of optimal choice of the sampling interval
12.15 Effects on parameter estimates of input variations during the sampling interval

Chapter 1
INTRODUCTION

The need for modeling dynamic systems

System identification is the field of modeling dynamic systems from experimental data. A dynamic system can be conceptually described as in Figure 1.1. The system is driven by input variables $u(t)$ and disturbances $v(t)$. The user can control $u(t)$ but not $v(t)$. In some signal processing applications the inputs may be missing. The output signals are variables which provide useful information about the system. For a *dynamic* system the control action at time t will influence the output at time instants $s > t$.

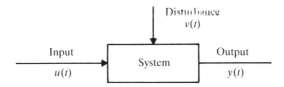

FIGURE 1.1 A dynamic system with input $u(t)$, output $y(t)$ and disturbance $v(t)$, where t denotes time.

The following examples of dynamic systems illustrate the need for mathematical models.

Example 1.1 *A stirred tank*
Consider the stirred tank shown in Figure 1.2, where two flows are mixed. The concentration in each of the flows can vary. The flows F_1 and F_2 can be controlled with valves. The signals $F_1(t)$ and $F_2(t)$ are the inputs to the system. The output flow $F(t)$ and the concentration $c(t)$ in the tank constitute the output variables. The input concentrations $c_1(t)$ and $c_2(t)$ cannot be controlled and are viewed as disturbances.
Suppose we want to design a regulator which acts on the flows $F_1(t)$ and $F_2(t)$ using the measurements of $F(t)$ and $c(t)$. The purpose of the regulator is to ensure that $F(t)$ or $c(t)$ remain as constant as possible even if the concentrations $c_1(t)$ and $c_2(t)$ vary considerably. For such a design we need some form of mathematical model which describes how the input, the output and the disturbances are related. ∎

Example 1.2 *An industrial robot*
An industrial robot can be seen as an advanced servo system. The robot arm has to perform certain movements, for example for welding at specific positions. It is then natural to regard the position of the robot arm as an output. The robot arm is controlled by electrical motors. The currents to these motors can be regarded as the control inputs. The movement of the robot can also be influenced by varying the load on the arm and by

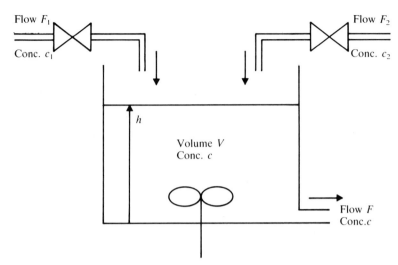

FIGURE 1.2 A stirred tank.

friction. Such variables are the disturbances. It is very important that the robot will move in a fast and reliable way to the desired positions without violating various geometrical constraints. In order to design an appropriate servo system it is of course necessary to have some model of how the behavior of the robot is influenced by the input and the disturbances. ∎

Example 1.3 *Aircraft dynamics*
An aircraft can be viewed as a complex dynamic system. Consider the problem of maintaining constant altitude and speed – these are the output variables. Elevator position and engine thrust are the inputs. The behavior of the airplane is also influenced by its load and by the atmospheric conditions. Such variables can be viewed as disturbances. In order to design an autopilot for keeping constant speed and course we need a model of how the aircraft's behavior is influenced by inputs and disturbances. The dynamic properties of an aircraft vary considerably, for example with speed and altitude, so identification methods will need to track these variations. ∎

Example 1.4 *Effect of a drug*
A medicine is generally required to produce an effect in a certain part of the body. If the drug is swallowed it will take some time before the drug passes the stomach and is absorbed in the intestines, and then some further time until it reaches the target organ, for example the liver or the heart. After some metabolism the concentration of the drug decreases and the waste products are secreted from the body. In order to understand what effect (and when) the drug has on the targeted organ and to design an appropriate schedule for taking the drug it is necessary to have some model that describes the properties of the drug dynamics. ∎

The above examples demonstrate the need for modeling dynamic systems both in technical and non-technical areas.

Many industrial processes, for example for production of paper, iron, glass or chemical compounds, must be controlled in order to run safely and efficiently. To design regulators, some type of model of the process is needed. The models can be of various types and degrees of sophistication. Sometimes it is sufficient to know the crossover frequency and the phase margin in a Bode plot. In other cases, such as the design of an optimal controller, the designer will need a much more detailed model which also describes the properties of the disturbances acting on the process.

In most applications of *signal processing* in forecasting, data communication, speech processing, radar, sonar and electrocardiogram analysis, the recorded data are filtered in some way and a good design of the filter should reflect the properties (such as high-pass characteristics, low-pass characteristics, existence of resonance frequencies, etc.) of the signal. To describe such spectral properties, a model of the signal is needed.

In many cases the primary aim of modeling is to aid in design. In other cases the knowledge of a model can itself be the purpose, as for example when describing the effect of a drug, as in Example 1.4. If the models can explain measured data satisfactorily then they might also be used to explain and understand the observed phenomena. In a more general sense modeling is used in many branches of science as an aid to describe and understand reality.

Sometimes it is interesting to model a technical system that does not exist, but may be constructed at some time in the future. Also in such a case the purpose of modeling is to gain insight into and knowledge of the dynamic behavior of the system. An example is a large space structure, where the dynamic behavior cannot be deduced by studying structures on earth, because of gravitational and atmospheric effects. Needless to say, for examples like this, the modeling must be based on theory and *a priori* knowledge, since experimental data are not available.

Types of model

Models of dynamic systems can be of many kinds, including the following:

- *Mental, intuitive or verbal models.* For example, this is the form of 'model' we use when driving a car ('turning the wheel causes the car to turn', 'pushing the brake decreases the speed', etc.)
- *Graphs and tables.* A Bode plot of a servo system is a typical example of a model in a graphical form. The step response, i.e. the output of a process excited with a step as input, is another type of model in graphical form. We will discuss the determination and use of such models in Chapters 2 and 3.
- *Mathematical models.* Although graphs may also be regarded as 'mathematical' models, here we confine this class of models to differential and difference equations. Such models are very well suited to the analysis, prediction and design of dynamic systems, regulators and filters. This is the type of model that will be predominantly discussed in the book. Chapter 6 presents various types of models and their properties from a system identification point of view.

4 Introduction

It should be stressed that although we speak generally about systems with inputs and outputs, the discussion here is to a large extent applicable also to time series analysis. In the latter case the system does not include any input signal. Signal models for times series can be useful in the design of spectral estimators, predictors, or filters that adapt to the signal properties.

Mathematical modeling and system identification

As discussed above, mathematical models of dynamic systems are useful in many areas and applications. Basically, there are two ways of constructing mathematical models:

- *Mathematical modeling.* This is an analytic approach. Basic laws from physics (such as Newton's laws and balance equations) are used to describe the dynamic behavior of a phenomenon or a process.
- *System identification.* This is an experimental approach. Some experiments are performed on the system; a model is then fitted to the recorded data by assigning suitable numerical values to its parameters.

Example 1.5 *Modeling a stirred tank*
Consider the stirred tank in Example 1.1. Assume that the liquid is incompressible, so that the density is constant; assume also that the mixing of the flows in the tank is very fast so that a homogeneous concentration c exists in the tank. To derive a mathematical model we will use balance equations of the form

net change = flow in − flow out

Applying this idea to the volume V in the tank,

$$\frac{dV}{dt} = F_1 + F_2 - F \qquad (1.1)$$

Applying the same idea to the dissolved substance,

$$\frac{d}{dt}(cV) = c_1 F_1 + c_2 F_2 - cF \qquad (1.2)$$

The model can be completed in several ways. The flow F may depend on the tank level h. This is certainly true if this flow is not controlled, for example by a valve. Ideally the flow in such a case is given by Torricelli's law:

$$F = a\sqrt{(2gh)} \qquad (1.3)$$

where a is the effective area of the flow and $g \approx 10$ m/sec^2. Equation (1.3) is an idealization and may not be accurate or even applicable. Finally, if the tank area A does not depend on the tank level h, then by simple geometry

$$V = Ah \qquad (1.4)$$

To summarize, equations (1.1) to (1.4) constitute a simple model of the stirred tank. The degree of validity of (1.3) is not obvious. The geometry of the tank is easy to measure, but the constant a in (1.3) is more difficult to determine. ∎

A comparison can be made of the two modeling approaches: mathematical modeling and system identification. In many cases the processes are so complex that it is not possible to obtain reasonable models using only physical insight (using first principles, e.g. balance equations). In such cases one is forced to use identification techniques. It often happens that a model based on physical insight contains a number of unknown parameters even if the structure is derived from physical laws. Identification methods can be applied to estimate the unknown parameters.

The models obtained by system identification have the following properties, in contrast to models based solely on mathematical modeling (i.e. physical insight):

- They have limited validity (they are valid for a certain working point, a certain type of input, a certain process, etc.).
- They give little physical insight, since in most cases the parameters of the model have no direct physical meaning. The parameters are used only as tools to give a good description of the system's overall behavior.
- They are relatively easy to construct and use.

Identification is not a foolproof methodology that can be used without interaction from the user. The reasons for this include:

- An appropriate model structure must be found. This can be a difficult problem, in particular if the dynamics of the system are nonlinear.
- There are certainly no 'perfect' data in real life. The fact that the recorded data are disturbed by noise must be taken into consideration.
- The process may vary with time, which can cause problems if an attempt is made to describe it with a time-invariant model.
- It may be difficult or impossible to measure some variables/signals that are of central importance for the model.

How system identification is applied

In general terms, an identification experiment is performed by exciting the system (using some sort of input signal such as a step, a sinusoid or a random signal) and observing its input and output over a time interval. These signals are normally recorded in a computer mass storage for subsequent 'information processing'. We then try to fit a parametric model of the process to the recorded input and output sequences. The first step is to determine an appropriate form of the model (typically a linear difference equation of a certain order). As a second step some statistically based method is used to estimate the unknown parameters of the model (such as the coefficients in the difference equation). In practice, the estimations of structure and parameters are often done iteratively. This means that a tentative structure is chosen and the corresponding parameters are estimated. The model obtained is then tested to see whether it is an appropriate representation of the system. If this is not the case, some more complex model structure must be considered, its parameters estimated, the new model validated, etc. The procedure is illustrated in Figure 1.3. Note that the 'restart' after the model validation gives an iterative scheme.

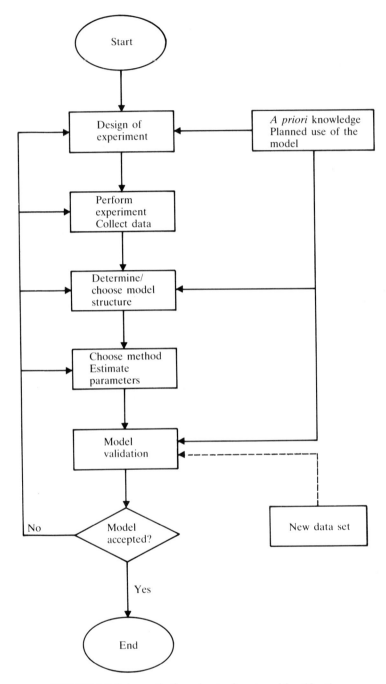

FIGURE 1.3 Schematic flowchart of system identification.

What this book contains

The following is a brief description of what the book contains (see also Summary and Outlook at the end of Chapter 2 for a more complete discussion):

- Chapter 2 presents some introductory examples of both nonparametric and parametric methods and some preliminary analysis. A more detailed description is given of how the book is organized.
- The book focuses on some central methods for identifying linear systems dynamics. This is the main theme in Chapters 3, 4 and 7–10. Both off-line and on-line identification methods are presented. Open loop as well as closed loop experiments are treated.
- A considerable amount of space is devoted to the problem of choosing a reasonable model structure and how to validate a model, i.e. to determine whether it can be regarded as an acceptable description of the process. Such aspects are discussed in Chapters 6 and 11.
- To some extent hints are given on how to design good identification experiments. This is dealt with in Chapters 5, 10 and 12. It is an important point: if the experiment is badly planned, the data will not be very useful (they may not contain much relevant information about the system dynamics, for example). Clearly, good models cannot be obtained from bad experiments.

What this book does not contain

It is only fair to point out the areas of system identification that are beyond the scope of this book:

- Identification of *distributed parameter systems* is not treated at all. For some survey papers in this field, see Polis (1982), Polis and Goodson (1976), Kubrusly (1977), Chavent (1979) and Banks *et al.* (1983).
- Identification of *nonlinear systems* is only marginally treated; see for example Example 6.6, where so-called Hammerstein models are treated. Some surveys of black box modeling of nonlinear systems have been given by Billings (1980), Mehra (1979), Haber and Keviczky (1976).
- Identification and *model approximation*. When the model structure used is not flexible enough to describe the true system dynamics, identification can be viewed as a form of model approximation or model reduction. Out of many available methods for model approximation, those based on partial realizations have given rise to much interest in the current literature. See, for example, Glover (1984) for a survey of such methods. Techniques for model approximation have close links to system identification, as described, for example, by Wahlberg (1985, 1986, 1987).
- Estimation of parameters in *continuous time models* is only discussed in parts of Chapter 2 and 3, and indirectly in the other chapters (see Example 6.5). In many cases, such as in the design of digital controllers, simulation, prediction, etc., it is sufficient to have a discrete time model. However, the parameters in a discrete time model most often have less physical sense than parameters of a continuous time model.
- *Frequency domain aspects* are only touched upon in the book (see Section 3.3 and

Examples 10.1, 10.2 and 12.1). There are several new results on how estimated models, even approximate ones, can be characterized and evaluated in the frequency domain; see Ljung (1985b, 1987).

Bibliographical notes

There are several books and other publications of general interest in the field of system identification. An old but still excellent survey paper was written by Åström and Eykhoff (1971). For further reading see the books by Ljung (1987), Norton (1986), Eykhoff (1974), Goodwin and Payne (1977), Kashyap and Rao (1976), Davis and Vinter (1985), Unbehauen and Rao (1987), Isermann (1987), Caines (1988), and Hannan and Deistler (1988). Advanced texts have been edited by Mehra and Lainiotis (1976), Eykhoff (1981), Hannan *et al.* (1985) and Leondes (1987). The text by Ljung and Söderström (1983) gives a comprehensive treatment of recursive identification methods, while the books by Söderström and Stoica (1983) and Young (1984) focus on the so-called instrumental variable identification methods. Other texts on system identification have been published by Mendel (1973), Hsia (1977) and Isermann (1974).

Several books have been published in the field of mathematical modeling. For some interesting treatments, see Aris (1978), Nicholson (1980), Wellstead (1979), Marcus-Roberts and Thompson (1976), Burghes and Borrie (1981), Burghes *et al.* (1982) and Blundell (1982).

The International Federation of Automatic Control (IFAC) has held a number of symposia on identification and system parameter estimation (Prague 1967, Prague 1970, the Hague 1973, Tbilisi 1976, Darmstadt 1979, Washington DC 1982, York 1985, Beijing 1988). In the proceedings of these conferences many papers have appeared. There are papers dealing with theory and others treating applications. The IFAC journal *Automatica* has published a special tutorial section on identification (Vol. 16, Sept. 1980) and a special issue on system identification (Vol. 18, Jan. 1981); and a new special issue is scheduled for Jan. 1990. Also, *IEEE Transactions on Automatic Control* has had a special issue on system identification and time-series analysis (Vol. 19. Dec. 1974).

Chapter 2
INTRODUCTORY EXAMPLES

2.1 The concepts $\mathcal{S}, \mathcal{M}, \mathcal{I}, \mathcal{X}$

This chapter introduces some basic concepts that will be valuable when describing and analyzing identification methods. The importance of these concepts will be demonstrated by some simple examples.

The result of an identification experiment will be influenced by (at least) the following four factors, which will be exemplified and discussed further both below and in subsequent chapters.

- *The system $\mathcal{S}$*. The physical reality that provides the experimental data will generally be referred to as the *process*. In order to perform a theoretical analysis of an identification it is necessary to introduce assumptions on the data. In such cases we will use the word *system* to denote a mathematical description of the process. In practice, where real data are used, the system is unknown and can even be an idealization. For simulated data, however, it is not only known but also used directly for the data generation in the computer. Note that to *apply* identification techniques it is not necessary to know the system. We will use the system concept only for investigation of how different identification methods behave under various circumstances. For that purpose the concept of a 'system' will be most useful.
- *The model structure $\mathcal{M}$*. Sometimes we will use *nonparametric models*. Such a model is described by a curve, function or table. A step response is an example. It is a curve that carries some information about the characteristic properties of a system. Impulse responses and frequency diagrams (Bode plots) are other examples of nonparametric models. However, in many cases it is relevant to deal with *parametric models*. Such models are characterized by a parameter vector, which we will denote by θ. The corresponding model will be denoted $\mathcal{M}(\theta)$. When θ is varied over some set of feasible values we obtain a model set (a set of models) or a model structure $\mathcal{M}$.
- *The identification method $\mathcal{I}$*. A large variety of identification methods have been proposed in the literature. Some important ones will be discussed later, especially in Chapters 7 and 8 and their complements. It is worth noting here that several proposed methods which are tied to different model structures, could and should be regarded as versions of the same approach, even if they are known under different names.
- *The experimental condition $\mathcal{X}$*. In general terms $\mathcal{X}$ is a description of how the identification experiment is carried out. This includes the selection and generation of the input signal, possible feedback loops in the process, the sampling interval, prefiltering of the data prior to estimation of the parameters, etc.

Before turning to some examples it should be mentioned that of the four concepts $\mathscr{S}$, $\mathscr{M}, \mathscr{I}, \mathscr{X}$, the system $\mathscr{S}$ must be regarded as fixed. It is 'given' in the sense that its properties cannot be changed by the user. The experimental condition $\mathscr{X}$ is determined when the data are collected from the process. It can often be influenced to some degree by the user. However, there may be restrictions – such as safety considerations or requirements of 'nearly normal' operations – that prevent a free choice of the experimental condition $\mathscr{X}$. Once the data are collected the user can still choose the identification method $\mathscr{I}$ and the model structure $\mathscr{M}$. Several choices of $\mathscr{I}$ and $\mathscr{M}$ can be tried on the same set of data until a satisfactory result is obtained.

2.2 A basic example

Throughout the chapter we will make repeated use of the two systems below, which are assumed to describe the generated data. These systems are given by

$$y(t) + a_0 y(t-1) = b_0 u(t-1) + e(t) + c_0 e(t-1) \tag{2.1}$$

where $e(t)$ is a sequence of independent and identically distributed random variables of zero mean and variance λ^2. Such a sequence will be referred to as *white noise*. Two different sets of parameter values will be used, namely

$$\begin{aligned}&\mathscr{S}_1: a_0 = -0.8 \quad b_0 = 1.0 \quad c_0 = 0.0 \quad \lambda = 1.0 \\ &\mathscr{S}_2: a_0 = -0.8 \quad b_0 = 1.0 \quad c_0 = -0.8 \quad \lambda = 1.0\end{aligned} \tag{2.2}$$

The system $\mathscr{S}_1$ can then be written as

$$\mathscr{S}_1: y(t) - 0.8 y(t-1) = 1.0 u(t-1) + e(t) \tag{2.3}$$

while $\mathscr{S}_2$ can be represented as

$$\mathscr{S}_2: \begin{aligned} x(t) - 0.8 x(t-1) &= 1.0 u(t-1) \\ y(t) &= x(t) + e(t) \end{aligned} \tag{2.4}$$

The white noise $e(t)$ thus enters into the systems in different ways. For the system $\mathscr{S}_1$ it appears as an equation disturbance, while for $\mathscr{S}_2$ it is added on the output (cf. Figure 2.1). Note that for $\mathscr{S}_2$ the signal $x(t)$ can be interpreted as the deterministic or noise-free output.

2.3 Nonparametric methods

In this section we will describe two nonparametric methods and apply them to the system $\mathscr{S}_1$.

Example 2.1 *Transient analysis*
A typical example of transient analysis is to let the input be a step and record the step response. This response will by itself give some characteristic properties (dominating

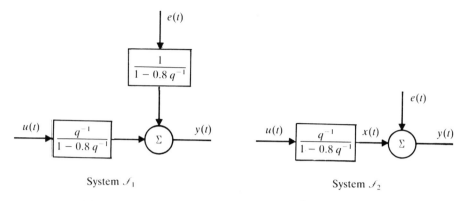

FIGURE 2.1 The systems $\mathscr{S}_1$ and $\mathscr{S}_2$. The symbol q^{-1} denotes the backward shift operator.

time constant, damping factor, static gain, etc.) of the process.

Figure 2.2 shows the result of applying a unit step to the system $\mathscr{S}_1$. Due to the high noise level it is very hard to deduce anything about the dynamic properties of the process. ∎

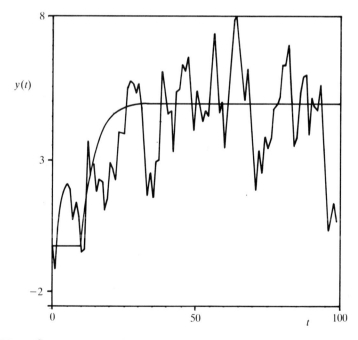

FIGURE 2.2 Step response of the system (jerky line). For comparison the true step response of the undisturbed system (smooth line) is shown. The step is applied at time $t = 10$.

12 Introductory examples

Example 2.2 *Correlation analysis*
A weighting function can be used to model the process:

$$y(t) = \sum_{k=0}^{\infty} h(k)u(t-k) + v(t) \qquad (2.5)$$

where $\{h(k)\}$ is the weighting function sequence (or weighting sequence, for short) and $v(t)$ is a disturbance term. Now let $u(t)$ be white noise, of zero mean and variance σ^2, which is independent of the disturbances. Multiplying (2.5) by $u(t-\tau)$ ($\tau > 0$) and taking expectation will then give

$$r_{yu}(\tau) \triangleq Ey(t)u(t-\tau) = \sum_{k=0}^{\infty} h(k) Eu(t-k)u(t-\tau) = \sigma^2 h(\tau) \qquad (2.6)$$

Based on this relation the weighting function coefficients $\{h(k)\}$ are estimated as

$$\hat{h}(\tau) = \frac{\dfrac{1}{N} \sum_{t=\tau+1}^{N} y(t)u(t-\tau)}{\dfrac{1}{N} \sum_{t=1}^{N} u^2(t)} \qquad (2.7)$$

where N denotes the number of data points. Figure 2.3 illustrates the result of simulating the system $\mathscr{S}_1$ with $u(t)$ as white noise with $\sigma = 1$, and the weighting function estimated as in (2.7).

The true weighting sequence of the system can be found to be

$$h(k) = 0.8^{k-1} \quad k \geq 1; \ h(0) = 0$$

The results obtained in Figure 2.3 would indicate some exponential decrease, although it is not easy to determine the parameter from the figure. To facilitate a comparison between the model and the true system the corresponding step responses are given in Figure 2.4.

It is clear from Figure 2.4 that the model obtained is not very accurate. In particular, its static gain (the stationary value of the response to a unit step) differs considerably from that of the true system. Nevertheless, it gives the correct magnitude of the time constant (or rise time) of the system. ∎

2.4 A parametric method

In this section the systems $\mathscr{S}_1$ and $\mathscr{S}_2$ are identified using a parametric method, namely the least squares method. In general, a parametric method can be characterized as a mapping from the recorded data to the estimated parameter vector.

Consider the model structure $\mathscr{M}$ given by the difference equation

$$y(t) + ay(t-1) = bu(t-1) + \varepsilon(t) \qquad (2.8)$$

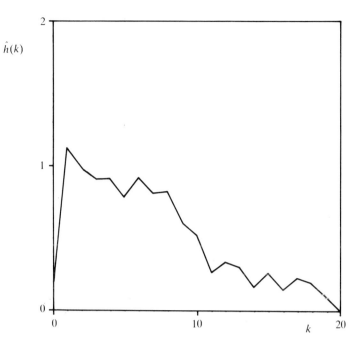

FIGURE 2.3 Estimated weighting function, Example 2.2.

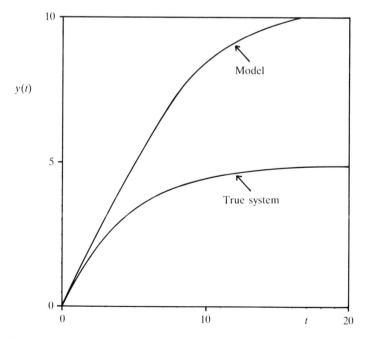

FIGURE 2.4 Step responses for the model obtained by correlation analysis and for the true (undisturbed) system.

14 Introductory examples

This model structure is a set of first-order linear discrete time models. The parameter vector is defined as

$$\theta = \begin{pmatrix} a \\ b \end{pmatrix} \tag{2.9}$$

In (2.8), $y(t)$ is the output signal at time t, $u(t)$ the input signal and $\varepsilon(t)$ a disturbance term. We will often call $\varepsilon(t)$ an equation error or a residual. The reason for including $\varepsilon(t)$ in the model (2.8) is that we can hardly hope that the model with $\varepsilon(t) \equiv 0$ (for all t) can give a perfect description of a real process. Therefore $\varepsilon(t)$ will describe the deviation in the data from a perfect (deterministic) first-order linear system. Using (2.8), (2.9), for a given data set $\{u(1), y(1), u(2), y(2), \ldots, u(N), y(N)\}$, $\{\varepsilon(t)\}$ can be regarded as functions of the parameter vector θ. To see this, simply rewrite (2.8) as

$$\varepsilon(t) = y(t) + ay(t-1) - bu(t-1) = y(t) - (-y(t-1) \ \ u(t-1))\theta \tag{2.10}$$

In the following chapters we will consider various generalizations of the simple model structure (2.8). Note that it is straightforward to extend it to an nth-order linear model simply by adding terms $a_i y(t-i)$, $b_i u(t-i)$ for $i = 2, \ldots, n$.

The identification method $\mathscr{I}$ should now be specified. In this chapter we will confine ourselves to the least squares method. Then the parameter vector is determined as the vector that minimizes the sum of squared equation errors. This means that we define the estimate by

$$\hat{\theta} = \arg \min_{\theta} V(\theta) \tag{2.11}$$

where the *loss function* $V(\theta)$ is given by

$$V(\theta) = \sum_{t=1}^{N} \varepsilon^2(t) \tag{2.12}$$

As can be seen from (2.10) the residual $\varepsilon(t)$ is a linear function of θ and thus $V(\theta)$ will be well defined for any value of θ.

For the simple model structure (2.8) it is easy to obtain an explicit expression of $V(\theta)$ as a function of θ. Denoting $\sum_{t=1}^{N}$ by Σ for short,

$$\begin{aligned} V(\theta) &= \Sigma[y(t) + ay(t-1) - bu(t-1)]^2 \\ &= [a^2 \Sigma y^2(t-1) + b^2 \Sigma u^2(t-1) - 2ab \Sigma y(t-1)u(t-1)] \\ &\quad + [2a \Sigma y(t)y(t-1) - 2b \Sigma y(t)u(t-1)] + \Sigma y^2(t) \end{aligned} \tag{2.13}$$

The estimate $\hat{\theta}$ is then obtained, according to (2.11), by minimizing (2.13). The minimum point is determined by setting the gradient of $V(\theta)$ to zero. This gives

$$0 = \frac{\partial V(\theta)}{\partial a} = 2[\hat{a} \Sigma y^2(t-1) - \hat{b} \Sigma y(t-1)u(t-1) + \Sigma y(t)y(t-1)]$$

$$0 = \frac{\partial V(\theta)}{\partial b} = 2[\hat{b} \Sigma u^2(t-1) - \hat{a} \Sigma y(t-1)u(t-1) - \Sigma y(t)u(t-1)] \tag{2.14}$$

or in matrix form

$$\begin{pmatrix} \Sigma y^2(t-1) & -\Sigma y(t-1)u(t-1) \\ -\Sigma y(t-1)u(t-1) & \Sigma u^2(t-1) \end{pmatrix} \begin{pmatrix} \hat{a} \\ \hat{b} \end{pmatrix}$$
$$= \begin{pmatrix} -\Sigma y(t)y(t-1) \\ \Sigma y(t)u(t-1) \end{pmatrix} \quad (2.15)$$

Note that (2.15) is a system of linear equations in the two unknowns $\hat{a}$ and $\hat{b}$. As we will see, there exists a unique solution under quite mild conditions.

In the following parameter estimates will be computed according to (2.15) for a number of cases. We will use simulated data generated in a computer but also theoretical analysis as a complement. In the analysis we will assume that the number of data points, N, is large. Then the following approximation can be made:

$$\frac{1}{N}\sum_{t=1}^{N} y^2(t-1) \approx Ey^2(t-1) \quad (2.16)$$

where E is the expectation operator for stochastic variables, and similarly for the other sums. It can be shown (see Lemma B.2 in Appendix B at the end of the book) that for all the cases treated here the left-hand side of (2.16) will converge, as N tends to infinity, to the right-hand side. The advantage of using expectations instead of sums is that in this way the analysis can deal with a deterministic problem, or more exactly a problem which does not depend on a particular realization of the data. For a deterministic signal the notation E will be used to mean

$$\lim_{N\to\infty} \frac{1}{N}\sum_{t=1}^{N}$$

Example 2.3 *A pseudorandom binary sequence as input*
The systems $\mathcal{S}_1$ and $\mathcal{S}_2$ were simulated generating 1 000 data points. The input signal was a PRBS (pseudorandom binary sequence). This signal shifts between two levels in a certain pattern such that its first- and second-order characteristics (mean value and covariance function) are quite similar to those of a white noise process. The PRBS and other signals are described in Chapter 5 and its complements. In the simulations the amplitude of $u(t)$ was $\sigma = 1.0$. The least squares estimates were computed according to (2.15). The results are given in Table 2.1. A part of the simulated data and the corresponding model outputs are depicted in Figure 2.5. The model output $x_m(t)$ is given by

$$x_m(t) + ax_m(t-1) = bu(t-1)$$

where $u(t)$ is the input used in the identification. One would expect that $x_m(t)$ should be close to the true *noise-free* output $x(t)$. The latter signal is, however, not known to the

TABLE 2.1 Parameter estimates for Example 2.3

Parameter	True value	Estimated value	
		System $\mathcal{S}_1$	System $\mathcal{S}_2$
a	-0.8	-0.795	-0.580
b	1.0	0.941	0.959

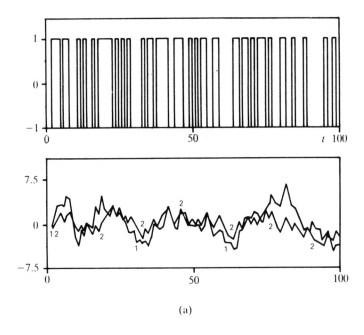

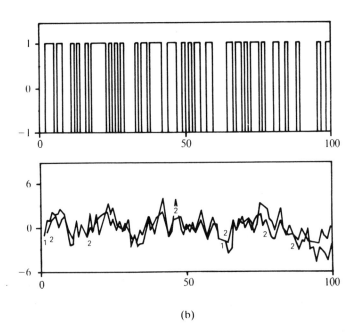

FIGURE 2.5 (a) Input (upper part), output (1, lower part) and model output (2, lower part) for Example 2.3, system $\mathscr{S}_1$. (b) Similarly for system $\mathscr{S}_2$.

user in the general case. Instead one can compare $x_m(t)$ to the 'measured' output signal $y(t)$. For a good model the signal $x_m(t)$ should explain all patterns in $y(t)$ that are due to the input. However, $x_m(t)$ should not be identical to $y(t)$ since the output also has a component that is caused by the disturbances. This stochastic component should not appear (or possibly only in an indirect fashion) in the model output $x_m(t)$.

Figure 2.6 shows the step responses of the true system and the estimated models. It is easily seen, in particular from Table 2.1 but also from Figure 2.6, that a good description is obtained of the system $\mathscr{S}_1$ while the system $\mathscr{S}_2$ is quite badly modeled. Let us now see if this result can be explained by a theoretical analysis. From (2.15) and (2.16) we obtain the following equations for the limiting estimates:

$$\begin{pmatrix} Ey^2(t) & -Ey(t)u(t) \\ -Ey(t)u(t) & Eu^2(t) \end{pmatrix} \begin{pmatrix} \hat{a} \\ \hat{b} \end{pmatrix} = \begin{pmatrix} -Ey(t)y(t-1) \\ Ey(t)u(t-1) \end{pmatrix} \quad (2.17)$$

Here we have used the fact that, by the stationarity assumption, $Ey^2(t-1) = Ey^2(t)$, etc. Now let $u(t)$ be a white noise process of zero mean and variance σ^2. This will be an accurate approximation of the PRBS used as far as the first- and second-order moments are concerned. See Section 5.2 for a discussion of this point. Then for the system (2.1), after some straightforward calculations,

$$Ey^2(t) = \frac{b_0^2 \sigma^2 + (1 + c_0^2 - 2a_0c_0)\lambda^2}{1 - a_0^2}$$

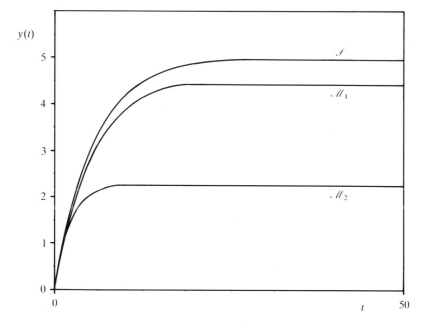

FIGURE 2.6 Step responses of the true system ($\mathscr{S}$), the estimated model of $\mathscr{S}_1$ ($\mathscr{M}_1$) and the estimated model of $\mathscr{S}_2$ ($\mathscr{M}_2$), Example 2.3.

$$Ey(t)u(t) = 0$$
$$Eu^2(t) = \sigma^2 \qquad (2.18a)$$
$$Ey(t)y(t-1) = \frac{-a_0 b_0^2 \sigma^2 + (c_0 - a_0)(1 - a_0 c_0)\lambda^2}{1 - a_0^2}$$
$$Ey(t)u(t-1) = b_0 \sigma^2$$

Using these results in (2.17) the following expressions are obtained for the limiting parameter estimates:

$$\hat{a} = a_0 + \frac{-c_0(1 - a_0^2)\lambda^2}{b_0^2 \sigma^2 + (1 + c_0^2 - 2a_0 c_0)\lambda^2} \qquad (2.18b)$$

$$\hat{b} = b_0$$

Thus for system $\mathscr{S}_1$, (2.2), where $c_0 = 0$,

$$\hat{a} = -0.8 \qquad \hat{b} = 1.0 \qquad (2.19)$$

which means that asymptotically (i.e. for large values of N) the parameter estimates will be close to the 'true values' a_0, b_0. This is well in accordance with what was observed in the simulations.

For the system $\mathscr{S}_2$ the corresponding results are

$$\hat{a} = \frac{-0.8\sigma^2}{\sigma^2 + 0.36\lambda^2} \approx -0.588 \qquad \hat{b} = 1.0 \qquad (2.20)$$

Thus in this case the estimate $\hat{a}$ will deviate from the true value. This was also obvious from the simulations; see Table 2.1 and Figure 2.6. (Compare (2.20) with the estimated values for $\mathscr{S}_2$ in Table 2.1.) The theoretical analysis shows that this was not due to 'bad luck' in the simulation nor to the data series being too short. No matter how many data pairs are used, there will, according to the theoretical expression (2.20), be a systematic deviation in the parameter estimate $\hat{a}$. ∎

2.5 Bias, consistency and model approximation

Following the example of the previous section it is appropriate to introduce a few definitions.

An estimate $\hat{\theta}$ is *biased* if its expected value deviates from the true value, i.e.

$$E\hat{\theta} \neq \theta_0 \qquad (2.21)$$

The difference $E\hat{\theta} - \theta_0$ is called *the bias*. If instead equality applies in (2.21), $\hat{\theta}$ is said to be *unbiased*.

In Example 2.3 it seems reasonable to believe that for large N the estimate $\hat{\theta}$ may be unbiased for the system $\mathscr{S}_1$ but that it is surely biased for the system $\mathscr{S}_2$.

Section 2.5   Bias, consistency and model approximation   19

We say that an estimate $\hat{\theta}$ is *consistent* if

$$\hat{\theta} \to \theta_0 \text{ as } N \to \infty \qquad (2.22)$$

Since $\hat{\theta}$ is a stochastic variable we must define in what sense the limit in (2.22) should be taken. A reasonable choice is 'limit with probability one'. We will generally use this alternative. Some convergence concepts for stochastic variables are reviewed in Appendix B (Section B.1).

The analysis carried out in Example 2.3 indicates that $\hat{\theta}$ is consistent for the system $\mathscr{S}_1$ but not for the system $\mathscr{S}_2$.

Loosely speaking, we say that a system is *identifiable* (in a given model set) if the corresponding parameter estimates are consistent. A more formal definition of identifiability is given in Section 6.4. Let us note that the identifiability properties of a given system will depend on the model structure $\mathscr{M}$, the identification method $\mathscr{I}$ and the experimental condition $\mathscr{X}$.

The following example demonstrates how the experimental condition can influence the result of an identification.

Example 2.4 *A step function as input*
The systems $\mathscr{S}_1$ and $\mathscr{S}_2$ were simulated, generating 1000 data points. This time the input was a unit step function. The least squares estimates were computed, and the numerical results are given in Table 2.2. Figure 2.7 shows the input, the output and model output.

TABLE 2.2 Parameter estimates for Example 2.4

Parameter	True value	Estimated value	
		System $\mathscr{S}_1$	System $\mathscr{S}_2$
a	−0.8	−0.788	−0.058
b	1.0	1.059	4.693

Again, a good model is obtained for system $\mathscr{S}_1$. For system $\mathscr{S}_2$ there is a considerable deviation from the true parameters. The estimates are also quite different from those in Example 2.3. In particular, now there is also a considerable deviation in the estimate $\hat{b}$.

For a theoretical analysis of the facts observed, equation (2.17) must be solved. For this purpose it is necessary to evaluate the different covariance elements which occur in (2.17). Let $u(t)$ be a step of size σ and introduce $S = b_0/(1 + a_0)$ as a notation for the static gain of the system. Then

$$Ey^2(t) = S^2\sigma^2 + \frac{(1 + c_0^2 - 2a_0c_0)\lambda^2}{1 - a_0^2}$$

$$Ey(t)u(t) = S\sigma^2$$

$$Eu^2(t) = \sigma^2 \qquad (2.23a)$$

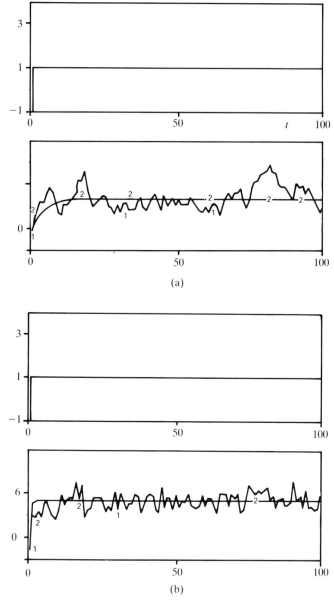

FIGURE 2.7 (a) Input (upper part), output (1, lower part) and model output (2, lower part) for Example 2.4, system $\mathscr{S}_1$. (b) Similarly for system $\mathscr{S}_2$.

$$Ey(t)\,y(t-1) = S^2\sigma^2 + \frac{(c_0 - a_0)\,(1 - a_0 c_0)\lambda^2}{1 - a_0^2}$$

$$Ey(t)u(t-1) = S\sigma^2$$

Using these results in (2.17) the following expressions for the parameter estimates are obtained:

$$\hat{a} = a_0 - \frac{c_0(1 - a_0^2)}{1 + c_0^2 - 2a_0c_0}$$

$$\hat{b} = b_0 - b_0c_0 \frac{1 - a_0}{1 + c_0^2 - 2a_0c_0} \qquad (2.23b)$$

Note that now both parameter estimates will in general deviate from the true parameter values. The deviations are independent of the size of the input step, σ, and they vanish if $c_0 = 0$. For system $\mathscr{S}_1$,

$$\hat{a} = -0.8 \qquad \hat{b} = 1.0 \qquad (2.24)$$

(as in Example 2.3), while for system $\mathscr{S}_2$ the result is

$$\hat{a} = 0.0 \qquad \hat{b} = \frac{b_0}{1 + a_0} = 5.0 \qquad (2.25)$$

which clearly deviates considerably from the true values. Note, though, that the static gain is estimated correctly, since from (2.23b)

$$\frac{\hat{b}}{1 + \hat{a}} = \frac{b_0}{1 + a_0} \qquad (2.26)$$

The theoretical results (2.24), (2.25) are quite similar to those based on the simulation; see Table 2.2. Note that in the noise-free case (when $\lambda^2 = 0$) a problem will be encountered. Then the matrix in (2.17) becomes

$$\sigma^2 \begin{pmatrix} S^2 & -S \\ -S & 1 \end{pmatrix}$$

Clearly this matrix is singular. Hence the system of equations (2.17) does not have a unique solution. In fact, the solutions in the noise-free case can be precisely characterized by

$$\frac{\hat{b}}{1 + \hat{a}} = S \qquad (2.27)$$

∎

We have seen in two examples that consistent estimates are obtained for system $\mathscr{S}_1$ while there is a systematic error in the parameter estimates for $\mathscr{S}_2$. The reason for this difference between $\mathscr{S}_1$ and $\mathscr{S}_2$ is that, even if both systems fit into the model structure $\mathscr{M}$ given by (2.8), only $\mathscr{S}_1$ will correspond to $\varepsilon(t)$ being white noise. A more detailed explanation for this behavior will be given in Chapter 7. (See also (2.34) below.) The models obtained for $\mathscr{S}_2$ can be seen as *approximations* of the true system. The approximation is clearly dependent on the experimental condition used. The following example presents some detailed calculations.

Example 2.5 *Prediction accuracy*
The model (2.8) will be used as a basis for prediction. Without knowledge about the

distribution or the covariance structure of $\varepsilon(t)$, a reasonable prediction of $y(t)$ given data up to (and including) time $t-1$ is given (see (2.8)) by

$$\hat{y}(t) = -ay(t-1) + bu(t-1) \tag{2.28}$$

This predictor will be justified in Section 7.3. The prediction error will, according to (2.1), satisfy

$$\begin{aligned}\tilde{y}(t) &= y(t) - \hat{y}(t) \\ &= (a - a_0)y(t-1) + (b_0 - b)u(t-1) + e(t) + c_0 e(t-1)\end{aligned} \tag{2.29}$$

The variance $W = E\tilde{y}^2(t)$ of the prediction error will be evaluated in a number of cases. For system $\mathcal{S}_2$ $c_0 = a_0$. First consider the true parameter values, i.e. $a = a_0$, $b = b_0$. Then the prediction error of $\mathcal{S}_2$ becomes

$$\tilde{y}(t) = e(t) + a_0 e(t-1)$$

and the prediction error variance

$$W_1 = \lambda^2(1 + a_0^2) \tag{2.30}$$

Note that the variance (2.30) is independent of the experimental condition. In the following we will assume that the prediction is used with $u(t)$ a step of size σ. Using the estimates (2.23), (2.25), in the stationary phase for system $\mathcal{S}_2$ the prediction error is

$$\begin{aligned}\tilde{y}(t) &= -a_0 y(t-1) + \left(b_0 - \frac{b_0}{1 + a_0}\right) u(t-1) + e(t) + a_0 e(t-1) \\ &= -a_0 \left[\frac{b_0}{1 + a_0} \sigma + e(t-1)\right] + \frac{a_0 b_0}{1 + a_0} \sigma + e(t) + a_0 e(t-1) \\ &= e(t)\end{aligned}$$

and thus

$$W_2 = \lambda^2 < W_1 \tag{2.31}$$

Note that here a better result is obtained than if the *true* values a_0 and b_0 were used! This is not unexpected since a and b are determined to make the sample variance $\Sigma\,\tilde{y}^2(t)/N$ as small as possible. In other words, the identification method uses a and b as vehicles to get a good prediction.

Note that it is crucial in the above calculation that the *same* experimental condition ($u(t)$ a step of size σ) is used in the identification and when evaluating the prediction. To see the importance of this fact, assume that the parameter estimates are determined from an experiment with $u(t)$ as white noise of variance $\bar{\sigma}^2$. Then (2.18), (2.20) with $c_0 = a_0$ give the parameter estimates. Now assume that the estimated model is used as a predictor for a step of size σ as input. Let S be the static gain, $S = b_0/(1 + a_0)$. From (2.29) we have

$$\begin{aligned}\tilde{y}(t) &= (a - a_0)y(t-1) + e(t) + a_0 e(t-1) \\ &= (a - a_0)[e(t-1) + S\sigma] + e(t) + a_0 e(t-1) \\ &= (a - a_0)S\sigma + e(t) + ae(t-1)\end{aligned}$$

Let r denote $b^2\tilde{\sigma}^2/(1 - a_0^2)\lambda^2$. Then from (2.18b),

$$a = a_0 - \frac{a_0}{r+1} = \frac{a_0 r}{r+1}$$

The expected value of $\bar{y}^2(t)$ becomes

$$\begin{aligned} W_3 &= \lambda^2(1 + a^2) + (a - a_0)^2 S^2 \sigma^2 \\ &= \lambda^2\left[1 + \left(\frac{a_0 r}{r+1}\right)^2\right] + \left(\frac{a_0}{r+1}\right)^2 S^2 \sigma^2 \end{aligned} \quad (2.32)$$

Clearly $W_3 > W_2$ always. Some straightforward calculations show that a value of W_3 can be obtained that is worse than W_1 (which corresponds to a predictor based on the true parameters a_0 and b_0). In fact $W_3 > W_1$ if and only if

$$S^2\sigma^2 > \lambda^2(2r + 1) \quad (2.33)$$

∎

In the following the discussion will be confined to system $\mathscr{S}_1$ only. In particular we will analyze the properties of the matrix appearing in (2.15). Assume that this matrix is 'well behaved'. Then there exists a unique solution. For system $\mathscr{S}_1$ this solution is asymptotically ($N \to \infty$) given by the true values a_0, b_0 since

$$\begin{aligned} &\frac{1}{N}\begin{pmatrix} \Sigma y^2(t-1) & -\Sigma y(t-1)u(t-1) \\ -\Sigma y(t-1)u(t-1) & \Sigma u^2(t-1) \end{pmatrix}\begin{pmatrix} a_0 \\ b_0 \end{pmatrix} \\ &\quad - \frac{1}{N}\begin{pmatrix} -\Sigma y(t)y(t-1) \\ \Sigma y(t)u(t-1) \end{pmatrix} \\ &= \frac{1}{N}\Sigma\begin{pmatrix} y(t-1) \\ -u(t-1) \end{pmatrix}[y(t) + a_0 y(t-1) - b_0 u(t-1)] \\ &= \frac{1}{N}\Sigma\begin{pmatrix} y(t-1) \\ -u(t-1) \end{pmatrix}e(t) \to E\begin{pmatrix} y(t-1) \\ -u(t-1) \end{pmatrix}e(t) = 0 \end{aligned} \quad (2.34)$$

The last equality follows since $e(t)$ is white noise and is hence independent of all past data.

2.6 A degenerate experimental condition

The examples in this and the following section investigate what happens when the square matrix appearing in (2.15) or (2.34) is not 'well behaved'.

Example 2.6 *An impulse as input*
The system $\mathscr{S}_1$ was simulated generating 1000 data points. The input was a unit impulse at $t = 1$. The least squares estimates were computed; the numerical results are given in Table 2.3. Figure 2.8 shows the input, the output and the model output.

TABLE 2.3 Parameter estimates for Example 2.6

Parameter	True value	Estimated value
a	−0.8	−0.796
b	1.0	2.950

This time it can be seen that the parameter a_0 is accurately estimated while the estimate of b_0 is poor. This inaccurate estimation of b_0 was expected since the input gives little contribution to the output. It is only through the influence of $u(\cdot)$ on $y(\cdot)$ that information about b_0 can be obtained. Concerning the parameter a_0, it will also describe the effect of the noise on the output. Since the noise is present in the data all the time, it is natural that a_0 is estimated much more accurately than b_0.

For a theoretical analysis consider (2.15) in the case where $u(t)$ is an impulse of magnitude σ at time $t = 1$. Using the notation

$$R_0 = \frac{1}{N} \sum_{t=1}^{N} y^2(t-1)$$

$$R_1 = \frac{1}{N} \sum_{t=1}^{N} y(t) y(t-1)$$

equation (2.15) can be transformed to

$$\begin{pmatrix} \hat{a} \\ \hat{b} \end{pmatrix} = \begin{pmatrix} NR_0 & -y(1)\sigma \\ -y(1)\sigma & \sigma^2 \end{pmatrix}^{-1} \begin{pmatrix} -NR_1 \\ y(2)\sigma \end{pmatrix}$$

$$= \frac{1}{R_0 - y^2(1)/N} \begin{pmatrix} -R_1 + y(1)y(2)/N \\ (-y(1)R_1 + y(2)R_0)/\sigma \end{pmatrix}$$

(2.35)

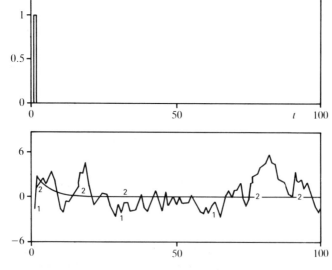

FIGURE 2.8 Input (upper part), output (1, lower part) and model output (2, lower part) for Example 2.6, system $\mathscr{S}_1$.

When N tends to infinity the contribution of $u(t)$ tends to zero and

$$R_0 \to \frac{\lambda^2}{1 - a_0^2} \qquad R_1 \to \frac{-a_0 \lambda^2}{1 - a_0^2} = -a_0 R_0$$

Thus the estimates $\hat{a}$ and $\hat{b}$ do converge. However, in contrast to what happened in Examples 2.3 and 2.4, the limits will depend on the realization (the recorded data), since they are given by

$$\begin{aligned} \hat{a} &= a_0 \\ \hat{b} &= (a_0 y(1) + y(2))/\sigma = b_0 + e(2)/\sigma \end{aligned} \qquad (2.36)$$

We see that $\hat{b}$ has a second term that makes it differ from the true value b_0. This deviation depends on the realization, so it will have quite different values depending on the actual data. In the present realization $e(2) = 1.957$, which according to (2.36) should give (asymptotically) $\hat{b} = 2.957$. This values agrees very well with the result given in Table 2.3.

The behavior of the least squares estimate observed above can be explained as follows. The case analyzed in this example is degenerate in two respects. First, the matrix in (2.35) multiplied by $1/N$ tends to a singular matrix as $N \to \infty$. However, in spite of this fact, it can be seen from (2.35) that the least squares estimate *uniquely* exists and can be computed for any value of N (possibly for $N \to \infty$). Second, and more important, in this case (2.34) is not valid since the sums involving the input signal do not tend to expectations ($u(t)$ being equal to zero almost all the time). Due to this fact, the least squares estimate converges to a stochastic variable rather than to a constant value (see the limiting estimate $\hat{b}$ in (2.36)). In particular, due to the combination of the two types of degenerency discussed above, the least squares method failed to provide consistent estimates. ∎

Examples 2.3 and 2.4 have shown that for system $\mathscr{S}_1$ consistent parameter estimates are obtained if the input is white noise or a step function (in the latter case it must be assumed that there is noise acting on the system so that $\lambda^2 > 0$; otherwise the system of equations (2.17) does not have a unique solution; see the discussion at the end of Example 2.4). If $u(t)$ is an impulse, however, the least squares method fails to give consistent estimates. The reason, in loose terms, is that an impulse function 'is equal to zero too often'. To guarantee consistency an input must be used that excites the process sufficiently. In technical terms the requirement is that the input signal be persistently exciting of order 1 (see Chapter 5 for a definition and discussions of this property).

2.7 The influence of feedback

It follows from the previous section that certain restrictions must be imposed on the input signal to guarantee that the matrix appearing in (2.15) is well behaved. The examples in this section illustrate what can happen when the input is determined through feedback from the output. The use of feedback might be necessary when making real identification experiments. The open loop system to be identified may be unstable, so without a stabilizing feedback it would be impossible to obtain anything but very short data series. Also safety requirements can be strong reasons for using a regulator in a feedback loop during the identification experiment.

Example 2.7 *A feedback signal as input*
Consider the model (2.8). Assume that the input is determined through a proportional feedback from the output,

$$u(t) = -ky(t) \tag{2.37}$$

Then the matrix in (2.15) becomes

$$\Sigma y^2(t-1) \begin{pmatrix} 1 & k \\ k & k^2 \end{pmatrix}$$

which is singular. As a consequence the least squares method cannot be applied. It can be seen in other ways that the system cannot be identified using the input (2.37). Assume that k is known (otherwise it is easily determined from measurements of $u(t)$ and $y(t)$ using (2.37)). Thus, only $\{y(t)\}$ carries information about the dynamics of the system. Using $\{u(t)\}$ cannot provide any more information. From (2.8) and (2.37),

$$\varepsilon(t) = y(t) + (a + bk)y(t-1) \tag{2.38}$$

This expression shows that only the linear combination $a + bk$ can be estimated from the data. All values of a and b that give the same value of $a + bk$ will also give the same sequence of residuals $\{\varepsilon(t)\}$ and the same value of the loss function. In particular, the loss function (2.12) will not have a unique mimimum. It is minimized along a straight line. In the asymptotic case ($N \to \infty$) this line is simply given by

$$\{\theta | a + bk = a_0 + b_0 k\}$$

Since there is a valley of minima, the Hessian (the matrix of second-order derivatives) must be singular. This matrix is precisely twice the matrix appearing in (2.15). This brings us back to the earlier observation that we cannot identify the parameters a and b using the input as given by (2.37). ∎

Based on Example 2.7 one may be led to believe that there is no chance of obtaining consistent parameter estimates if feedback must be used during the identification experiment. Fortunately, the situation is not so hopeless, as the following example demonstrates.

Example 2.8 *A feedback signal and an additional setpoint as input*
The system $\mathscr{S}_1$ was simulated generating 1000 data points. The input was a feedback from the output plus an additional signal,

$$u(t) = -ky(t) + r(t) \tag{2.39}$$

The signal $r(t)$ was generated as a PRBS of magnitude 0.5, while the feedback gain was chosen as $k = 0.5$. The least squares estimates were computed; the numerical values are given in Table 2.4. Figure 2.9 depicts the input, the output and the model output.

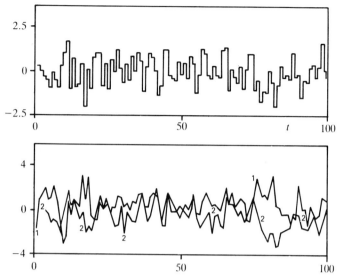

FIGURE 2.9 Input (upper part), output (1, lower part) and model output (2, lower part) for Example 2.8.

TABLE 2.4 Parameter estimates for Example 2.8

Parameter	True value	Estimated value
a	-0.8	-0.754
b	1.0	0.885

It can be seen that this time reasonable parameter estimates were obtained even though the data were generated using feedback.

To analyze this situation, first note that the consistency calculations (2.34) are still valid. It is thus sufficient to show that the matrix appearing in (2.15) is well behaved (in particular, nonsingular) for large N. For this purpose assume that $r(t)$ is white noise of zero mean and variance σ^2, which is independent of $e(s)$ for all t and s. (As already explained this is a close approximation to the PRBS used.) Then from (2.1) with $c_0 = 0$ and (2.39) (for convenience, we omit the index 0 of a and b), the following equations are obtained:

$$y(t) + (a + bk)y(t - 1) = br(t - 1) + e(t)$$
$$u(t) + (a + bk)u(t - 1) = r(t) + ar(t - 1) - ke(t)$$

This gives, after some calculations,

$$\begin{pmatrix} Ey^2(t) & -Ey(t)u(t) \\ -Ey(t)u(t) & Eu^2(t) \end{pmatrix} = \frac{1}{1 - (a + bk)^2}$$
$$\begin{pmatrix} (b^2\sigma^2 + \lambda^2) & -k(b^2\sigma^2 + \lambda^2) \\ -k(b^2\sigma^2 + \lambda^2) & k^2(b^2\sigma^2 + \lambda^2) + \{1 - (a + bk)^2\}\sigma^2 \end{pmatrix}$$

which is positive definite. Here we have assumed that the closed loop system is asymptotically stable so that $|a + bk| < 1$. ∎

Summary and outlook

The experiences gained so far and the conclusions that can be drawn may be summarized as follows.

1. When a *nonparametric method* such as transient analysis or correlation analysis is used, the result is easy to obtain but the derived model will be rather inaccurate. A *parametric method* was shown to give more accurate results.
2. When the least squares method is used, the consistency properties depend critically on how the noise enters into the system. This means that the inherent model structure may not be suitable unless the system has certain properties. This gives a requirement on the *system $\mathscr{S}$*. When that requirement is not satisfied, the estimated model will not be an exact representation of $\mathscr{S}$ even asymptotically (for $N \to \infty$). The model will only be an *approximation* of $\mathscr{S}$. The sense in which the model approximates the system is determined by the identification method used. Furthermore, the parameter values of the approximation model depend on the input signal (or, in more general terms, on the experimental condition).
3. As far as the *experimental condition $\mathscr{X}$* is concerned it is important that the input signal is *persistently exciting*. Roughly speaking, this implies that all modes of the system should be excited during the identification experiment.
4. When the experimental condition includes *feedback* from $y(t)$ to $u(t)$ it may not be possible to identify the system parameters even if the input is persistently exciting. On the other hand, when a persistently exciting reference signal is added to the system, the parameters may be estimated with reasonable accuracy.

Needless to say, the statements above have not been strictly proven. They merely rely on some simple first-order examples. However, the subsequent chapters will show how these conclusions can be proven to hold under much more general conditions.

The remaining chapters are organized in the following way.

Nonparametric methods are described in Chapter 3, where some methods other than those briefly introduced in Section 2.3 are analyzed. Nonparametric methods are sensitive to noise and do not give very accurate results. However, as they are easy to apply they are often useful means of deriving preliminary or crude models.

Chapter 4 treats *linear regression models*, confined to static models, that is models which do not include any dynamics. The extension to dynamic models is straightforward from a purely algorithmic point of view. The statistical properties of the parameter estimates in that case will be different, however, except in the special case of weighting function models. In particular, the analysis presented in Chapter 4 is crucially dependent on the assumption of static models. The extension to dynamic models is imbedded in a more general problem discussed in Chapter 7.

Chapter 5 is devoted to discussions of *input signals* and their properties relevant to system identification. In particular, the concept of persistent excitation is treated in some detail.

We have seen by studying the two simple systems $\mathscr{S}_1$ and $\mathscr{S}_2$ that the choice of model structure can be very important and that the model must match the system in a certain way. Otherwise it may not be possible to get consistent parameter estimates. Chapter 6 presents some general *model structures*, which describe general linear systems.

Chapters 7 and 8 discuss two different classes of important *identification methods*. Example 2.3 shows that the least squares method has some relation to minimization of the prediction error variance. In Chapter 7 this idea is extended to general model sets. The class of so-called prediction error identification methods is obtained in this way. Another class of identification methods is obtained by using instrumental variables techniques. We saw in (2.34) that the least squares method gives consistent estimates only for a certain type of disturbance acting on the system. The instrumental variable methods discussed in Chapter 8 can be seen as rather simple modifications of the least squares method (a linear system of equations similar to (2.15) is solved), but consistency can be guaranteed for a very general class of disturbances. The analysis carried out in Chapters 7 and 8 will deal with both the consistency and the asymptotic distribution of the parameter estimates.

In Chapter 9 *recursive identification methods* are introduced. For such methods the parameter estimates are modified every time a new input–output data pair becomes available. They are therefore perfectly suited to on-line or real-time applications. In particular, it is of interest to combine them with time-varying regulators or filters that depend on the current parameter vectors. In such a way one can design adaptive systems for control and filtering. It will be shown how the off-line identification methods introduced in previous chapters can be modified to recursive algorithms.

The role of the *experimental condition* in system identification is very important. A detailed discussion of this aspect is presented in Chapter 10. In particular, we investigate the conditions under which identifiability can be achieved when the system operates under feedback control during the identification experiment.

A very important phase in system identification is *model validation*. By this we mean different methods of determining if a model obtained by identification should be accepted as an appropriate description of the process or not. This is certainly a difficult problem. Chapter 11 provides some hints on how it can be tackled in practice. In particular, we discuss how to select between two or more competing model structures (which may, for example, correspond to different model orders).

It is sometimes claimed that system identification is more art than science. There are no foolproof methods that always and directly lead to a correct result. Instead, there are a number of theoretical results which are useful from a practical point of view. Even so, the user must combine the application of such a theory with common sense and intuition to get the most appropriate result. Chapter 12 should be seen in this light. There we discuss a number of practical issues and how the previously developed theory can help when dealing with *system identification in practice*.

Problems

Problem 2.1 *Bias, variance and mean square error*
Let $\hat{\theta}_i$, $i = 1, 2$ denote two estimates of the same scalar parameter θ. Assume, with N

denoting the number of data points, that

$$\text{bias}(\hat{\theta}_i) \triangleq E(\hat{\theta}_i) - \theta = \begin{cases} 1/N & \text{for } i = 1 \\ 0 & \text{for } i = 2 \end{cases}$$

$$\text{var}(\hat{\theta}_i) \triangleq E[\hat{\theta}_i - E(\hat{\theta}_i)]^2 = \begin{cases} 1/N & \text{for } i = 1 \\ 3/N & \text{for } i = 2 \end{cases}$$

where var $(\cdot)$ is the abbreviation for variance $(\cdot)$. The mean square error (mse) is defined as $E(\hat{\theta}_i - \theta)^2$. Which one of $\hat{\theta}_1$, $\hat{\theta}_2$ is the best estimate in terms of mse? Comment on the result.

Problem 2.2 *Convergence rates for consistent estimators*
For most consistent estimators of the parameters of *stationary* processes, the variance of the estimation error tends to zero as $1/N$ when $N \to \infty$ (N = the number of data points) (see, for example, Chapters 4, 7 and 8). For *nonstationary* processes, faster convergence rates may be expected. To see this, derive the variance of the least squares estimate of α in

$$y(t) = \alpha t + e(t) \qquad t = 1, 2, \ldots, N$$

where $e(t)$ is white noise with zero mean and variance λ^2.

Problem 2.3 *Illustration of unbiasedness and consistency properties*
Let $\{x_i\}_{i=1}^N$ be a sequence of independent and identically distributed Gaussian random variables with mean μ and variance σ. Both μ and σ are unknown. Consider the following estimate of μ:

$$\hat{\mu} = \frac{1}{N} \sum_{i=1}^N x_i$$

and the following two estimates for σ:

$$\hat{\sigma}_1 = \frac{1}{N} \sum_{i=1}^N (x_i - \hat{\mu})^2$$

$$\hat{\sigma}_2 = \frac{1}{N-1} \sum_{i=1}^N (x_i - \hat{\mu})^2$$

Determine the means and variances of the estimates $\hat{\mu}$, $\hat{\sigma}_1$ and $\hat{\sigma}_2$. Discuss their (un)biasedness and consistency properties. Compare $\hat{\sigma}_1$ and $\hat{\sigma}_2$ in terms of their mse's.
Hint. Lemma B.9 will be useful for calculating var$(\hat{\sigma}_i)$.

Remark. A generalization of the problem above is treated in Section B.9.

Problem 2.4 *Least squares estimates with white noise as input*
Verify the expressions (2.18a), (2.18b).

Problem 2.5 *Least squares estimates with a step function as input*
Verify the expressions (2.23a), (2.23b).

Problem 2.6 *Least squares estimates with a step function as input, continued*

(a) Verify the expression (2.26).
(b) Assume that the data are noise-free. Show that all solutions to the system of equations (2.17) are given by (2.27).

Problem 2.7 *Conditions for a minimum*

Show that the solution to (2.15) gives a minimum point of the loss function, and not another type of stationary point.

Problem 2.8 *Weighting sequence and step response*

Assume that the weighting sequence $\{h(k)\}_{k=0}^{\infty}$ of a system is given. Let $y(t)$ be the step response of the system. Show that $y(t)$ can be obtained by integrating the weighting sequence, in the following sense:

$$y(t) - y(t-1) = h(t)$$
$$y(-1) = 0$$

Bibliographical notes

The concepts of system, model structure, identification method and experimental condition have turned out to be valuable ways of describing the items that influence an identification result. These concepts have been described in Ljung (1976) and Gustavsson et al. (1977, 1981). A classical discussion along similar lines has been given by Zadeh (1962).

Chapter 3
NONPARAMETRIC METHODS

3.1 Introduction

This chapter describes some nonparametric methods for system identification. Such identification methods are characterized by the property that the resulting models are curves or functions, which are not necessarily parametrized by a finite-dimensional parameter vector. Two nonparametric methods were considered in Examples 2.1–2.2. The following methods will be discussed here:

- *Transient analysis.* The input is taken as a step or an impulse and the recorded output constitutes the model.
- *Frequency analysis.* The input is a sinusoid. For a linear system in steady state the output will also be sinusoidal. The change in amplitude and phase will give the frequency response for the used frequency.
- *Correlation analysis.* The input is white noise. A normalized cross-covariance function between output and input will provide an estimate of the weighting function.
- *Spectral analysis.* The frequency response can be estimated for arbitrary inputs by dividing the cross-spectrum between output and input to the input spectrum.

3.2 Transient analysis

With this approach the model used is either the step response or the impulse response of the system. The use of an impulse as input is common practice in certain applications, for example where the input is an injection of a substance, the future distribution of which is sought and traced with a sensor. This is typical in certain 'flow systems', for example in biomedical applications.

Step response

Sometimes it is of interest to fit a simple low-order model to a step response. This is illustrated in the following examples for some classes of first- and second-order systems, which are described using the transfer function model

$$Y(s) = G(s)U(s) \qquad (3.1)$$

where $Y(s)$ is the Laplace transform of the output signal $y(t)$, $U(s)$ is the Laplace transform of the input $u(t)$, and $G(s)$ is the transfer function of the system.

Example 3.1 *Step response of a first-order system*
Consider a system given by the transfer function

$$G(s) = \frac{K}{1 + sT}e^{-s\tau} \qquad (3.2a)$$

This means that the system is described by the first-order differential equation

$$T\frac{dy}{dt}(t) + y(t) = Ku(t - \tau) \qquad (3.2b)$$

Note that a time delay τ is included in the model. The step response of such a system is illustrated in Figure 3.1.

Figure 3.1 demonstrates a graphical method for determining the parameters K, T and τ from the step response. The gain K is given by the final value. By fitting the steepest tangent, T and τ can be obtained. The slope of this tangent is K/T, where T is the time constant. The tangent crosses the t axis at $t = \tau$, the time delay. ∎

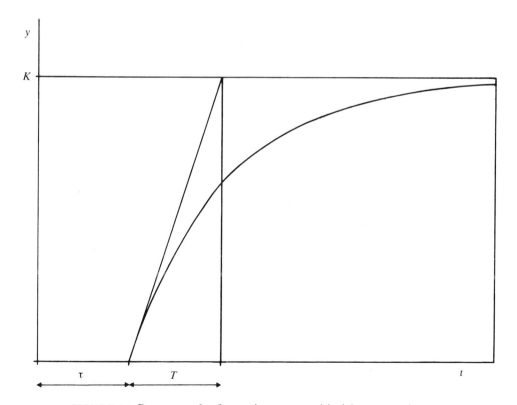

FIGURE 3.1 Response of a first-order system with delay to a unit step.

Example 3.2 *Step response of a damped oscillator*
Consider a second-order system given by

$$G(s) = \frac{K\omega_0^2}{s^2 + 2\zeta\omega_0 s + \omega_0^2} \qquad (3.3a)$$

In the time domain this system is described by

$$\frac{d^2y}{dt^2} + 2\zeta\omega_0 \frac{dy}{dt} + \omega_0^2 y = K\omega_0^2 u \qquad (3.3b)$$

Physically this equation describes a damped oscillator. After some calculations the step response is found to be

$$y(t) = K\left[1 - \frac{1}{\sqrt{(1-\zeta^2)}} e^{-\zeta\omega_0 t} \sin(\omega_0 \sqrt{(1-\zeta^2)} t + \tau)\right] \qquad (3.3c)$$

$$\tau = \arccos \zeta$$

This is illustrated in Figure 3.2.

It is obvious from Figure 3.2 how the relative damping ζ influences the character of the step response. The remaining two parameters, K and ω_0, merely act as scale factors. The gain K scales the amplitude axis while ω_0 scales the time axis. The three

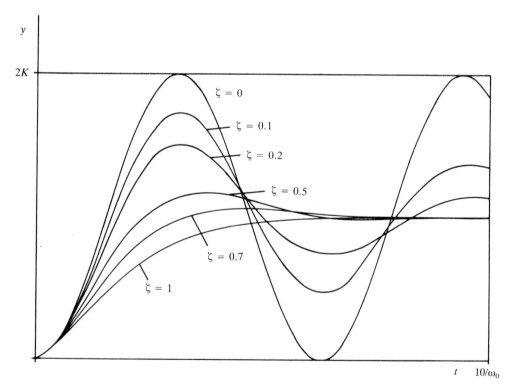

FIGURE 3.2 Response of a damped oscillator to a unit input step.

parameters of the model (3.3a), namely K, ζ and ω_0 could be determined by comparing the measured step response with Figure 3.2 and choosing the curve that is most similar to the recorded data. However, one can also proceed in a number of alternative ways. One possibility is to look at the local extrema (maxima and minima) of the step response. With some calculations it can be found from (3.3c) that they occur at times

$$t_k = k \frac{\pi}{\omega_0 \sqrt{(1 - \zeta^2)}} \qquad k = 1, 2, \ldots \tag{3.3d}$$

and that

$$y(t_k) = K[1 - (-1)^k M^k] \tag{3.3e}$$

where the overshoot M is given by

$$M = \exp[-\zeta\pi/\sqrt{(1 - \zeta^2)}] \tag{3.3f}$$

The relation (3.3f) between the overshoot M and the relative damping is illustrated in Figure 3.3.

The parameters K, ζ and ω_0 can be determined as follows (see Figure 3.4). The gain K is easily obtained as the final value (after convergence). The overshoot M can be

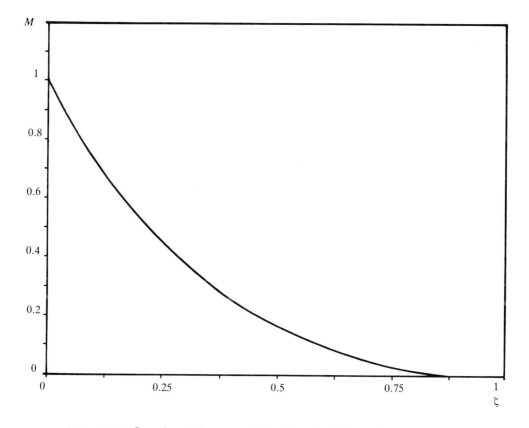

FIGURE 3.3 Overshoot M versus relative damping ζ for a damped oscillator.

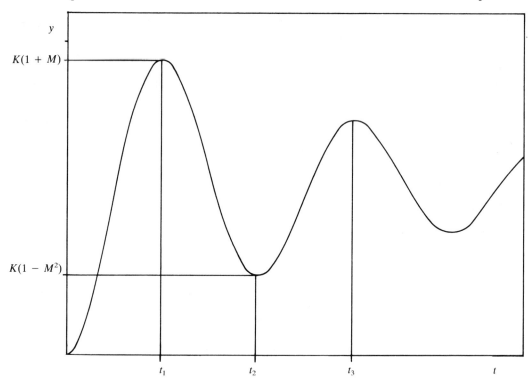

FIGURE 3.4 Determination of the parameters of a damped oscillator from the step response.

determined in several ways. One possibility is to use the first maximum. An alternative is to use several extrema and the fact (see (3.3e)) that the amplitude of the oscillations is reduced by a factor M for every half-period. Once M is determined, ζ can be derived from (3.3f):

$$\zeta = \frac{-\log M}{[\pi^2 + (\log M)^2]^{1/2}} \tag{3.3g}$$

From the step response the period T of the oscillations can also be determined. From (3.3d),

$$T = \frac{2\pi}{\omega_0 \sqrt{(1 - \zeta^2)}} \tag{3.3h}$$

Then ω_0 is given by

$$\omega_0 = \frac{2\pi}{T\sqrt{(1 - \zeta^2)}} = \frac{2}{T}[\pi^2 + (\log M)^2]^{1/2} \tag{3.3i}$$

∎

Impulse response

Theoretically, for an impulse response a Dirac function $\delta(t)$ is needed as input. Then the output will be equal to the weighting function $h(t)$ of the system. However, an ideal impulse cannot be realized in practice, and an approximate impulse must be used; for example

$$u(t) = \begin{cases} 1/\alpha & 0 \leq t < \alpha \\ 0 & \alpha \leq t \end{cases} \tag{3.4}$$

This input satisfies $\int u(t)dt = 1$ as the idealized impulse and should resemble it for sufficiently small values of the impulse length α.

Use of the approximate impulse (3.4) will give a distortion of the output, as can be seen from the following simple calculation:

$$y(t) = \int_0^\infty h(s)u(t-s)ds = \frac{1}{\alpha}\int_{\max(0,t-\alpha)}^t h(s)ds \approx h(t) \tag{3.5}$$

If the duration α of the impulse (3.4) is short compared to the time constants of interest, then the distortion introduced may be negligible. This fact is illustrated in the following example.

Example 3.3 *Nonideal impulse response*
Consider a damped oscillator with transfer function

$$G(s) = \frac{1}{s^2 + 0.4s + 1} \tag{3.6}$$

Figure 3.5 shows the weighting function and the responses to the approximate impulse (3.4) for various values of the impulse duration α. It can be seen that the (nonideal) impulse response deviates very little from the weighting function if α is small compared to the oscillation period. ∎

Problem 3.11 and Complement C7.5 contain a discussion of how a parametric model can be fitted to an estimated impulse response.

To summarize this section, we note that transient analysis is often simple to apply. It gives at least a first model which can be used to obtain rough estimates of the relative damping, the dominating time constants and a possible time delay. Therefore, transient analysis is a convenient way of deriving crude models. However, it is quite sensitive to noise.

3.3 Frequency analysis

For a discussion of frequency analysis it is convenient to use the continuous time model $Y(s) = G(s)U(s)$, see (3.1).

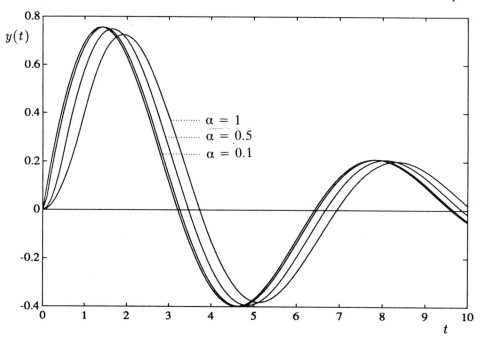

FIGURE 3.5 Weighting function and impulse responses for the damped oscillator (3.6) excited by the approximate impulses (3.4).

Basic frequency analysis

If the input signal is a sinusoid

$$u(t) = a\,\sin(\omega t) \qquad (3.7)$$

and the system is asymptotically stable, then in the steady state the output will become

$$y(t) = b\,\sin(\omega t + \varphi) \qquad (3.8)$$

where

$$b = a|G(i\omega)| \qquad (3.9a)$$
$$\varphi = \arg[G(i\omega)] \qquad (3.9b)$$

This can be proved as follows. Assume for convenience that the system is initially at rest. (The initial values will only give a transient effect, due to the assumption of stability.) Then the system can be represented using a weighting function $h(t)$ as follows:

$$y(t) = \int_0^t h(\tau) u(t-\tau) d\tau \qquad (3.10)$$

where $h(t)$ is the function whose Laplace transform equals $G(s)$. Now set

$$G_t(s) = \int_0^t h(\tau)e^{-s\tau}d\tau \tag{3.11}$$

Since

$$\sin \omega t = \frac{1}{2i}(e^{i\omega t} - e^{-i\omega t})$$

equations (3.7), (3.10), (3.11) give

$$y(t) = \frac{a}{2i}\int_0^t h(\tau)[e^{i\omega(t-\tau)} - e^{-i\omega(t-\tau)}]d\tau$$

$$= \frac{a}{2i}[e^{i\omega t}G_t(i\omega) - e^{-i\omega t}G_t(-i\omega)] \tag{3.12}$$

$$= \frac{a}{2i}|G_t(i\omega)|[e^{i\omega t}e^{i\arg G_t(i\omega)} - e^{-i\omega t}e^{-i\arg G_t(i\omega)}]$$

$$= a|G_t(i\omega)|\sin(\omega t + \arg G_t(i\omega))$$

When t tends to infinity, $G_t(i\omega)$ will tend to $G(i\omega)$. With this observation, the proof of (3.8), (3.9) is completed.

Note that normally the phase φ will be negative. By measuring the amplitudes a and b as well as the phase difference φ, the complex variable $G(i\omega)$ can be found from (3.9). If such a procedure is repeated for a number of frequencies then one can obtain a graphical representation of $G(i\omega)$ as a function of ω. Such Bode plots (or Nyquist plots or other equivalent representations) are well suited as tools for classical design of control systems.

The procedure outlined above is rather sensitive to disturbances. In practice it can seldom be used in such a simple form. This is not difficult to understand: assume that the true system can be described by

$$Y(s) = G(s)U(s) + E(s) \tag{3.13}$$

where $E(s)$ is the Laplace transform of some disturbance $e(t)$. Then instead of (3.8) we will have

$$y(t) = b\sin(\omega t + \varphi) + e(t) \tag{3.14}$$

and due to the presence of the noise it will be difficult to obtain an accurate estimate of the amplitude b and the phase difference φ.

Improved frequency analysis

There are ways to improve the basic frequency analysis method. This can be done by a correlation technique. The output is multiplied by $\sin \omega t$ and $\cos \omega t$ and the result is integrated over the interval $[0,T]$. This procedure is illustrated in Figure 3.6.

40 Nonparametric methods   Chapter 3

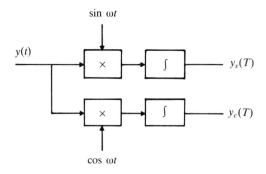

FIGURE 3.6 Improved frequency analysis.

For the improved frequency analysis, (3.14) yields

$$y_s(T) = \int_0^T y(t) \sin \omega t \, dt$$

$$= \int_0^T b \sin(\omega t + \varphi) \sin \omega t \, dt + \int_0^T e(t) \sin \omega t \, dt \qquad (3.15a)$$

$$= \frac{bT}{2} \cos \varphi - \frac{b}{2} \int_0^T \cos(2\omega t + \varphi) dt + \int_0^T e(t) \sin \omega t \, dt$$

$$y_c(T) = \int_0^T y(t) \cos \omega t \, dt$$

$$= \int_0^T b \sin(\omega t + \varphi) \cos \omega t \, dt + \int_0^T e(t) \cos \omega t \, dt \qquad (3.15b)$$

$$= \frac{bT}{2} \sin \varphi - \frac{b}{2} \int_0^T \sin(2\omega t + \varphi) dt + \int_0^T e(t) \cos \omega t \, dt$$

If the measurements are noise-free ($e(t) = 0$) and the integration time T is a multiple of the sinusoid period, say $T = k2\pi/\omega$, then

$$y_s(T) = \frac{bT}{2} \cos \varphi$$

$$y_c(T) = \frac{bT}{2} \sin \varphi \qquad (3.16)$$

From these relations it is easy to determine b and φ; then $|G(i\omega)|$ is calculated according to (3.9a). Note that (3.9) and (3.16) imply

$$y_s(T) = \frac{aT}{2} \text{Re } G(i\omega)$$

$$y_c(T) = \frac{aT}{2} \text{Im } G(i\omega)$$

(3.17)

which can also provide a useful form for describing $G(i\omega)$.

Sensitivity to noise

Intuitively, the approach (3.15)–(3.17) has better noise suppression properties than the basic frequency analysis method. The reason is that the effect of the noise is suppressed by the averaging inherent in (3.15).

A simplified analysis of the sensitivity to noise can be made as follows. Assume that $e(t)$ is a stationary process with covariance function $r_e(\tau)$. The variance of the output is then $r_e(0)$. The amplitude b can be difficult to estimate by inspection of the signals unless b^2 is much larger than $r_e(0)$. The variance of the term $y_s(T)$ can be analyzed as follows ($y_c(T)$ can be analyzed similarly). We have

$$\text{var}[y_s(T)] = E\left[\int_0^T e(t) \sin \omega t \, dt\right]^2$$

$$= E \int_0^T \int_0^T e(t_1) \sin \omega t_1 e(t_2) \sin \omega t_2 \, dt_1 dt_2 \qquad (3.18)$$

$$= \int_0^T \int_0^T r_e(t_1 - t_2) \sin \omega t_1 \sin \omega t_2 \, dt_1 dt_2$$

Assume that the noise covariance function $r_e(\tau)$ is bounded by an exponential function

$$|r_e(\tau)| \leq r_0 \exp(-\alpha|\tau|) \qquad (\alpha > 0) \qquad (3.19)$$

(For a stationary disturbance this is a weak assumption.) Then an upper bound for $\text{var}[y_s(T)]$ is derived as follows:

$$\text{var}[y_s(T)] \leq \int_0^T \int_0^T |r_e(t_1 - t_2)| \, dt_1 dt_2$$

$$= \int_0^T \left[\int_{-t_2}^{T-t_2} |r_e(\tau)| d\tau\right] dt_2$$

$$\leq \int_0^T \left[\int_{-\infty}^{\infty} |r_e(\tau)| d\tau\right] dt_2 \qquad (3.20)$$

$$\leq 2T \int_0^{\infty} r_0 \exp(-\alpha|\tau|) \, d\tau = \frac{2Tr_0}{\alpha}$$

Thus, the 'relative precision' of the improved frequency analysis method satisfies

$$\frac{\operatorname{var}[y_s(T)]}{\{E[y_s(T)]\}^2} \leq \frac{8}{\alpha \cos^2 \varphi} \frac{r_0}{b^2 T} \qquad (3.21)$$

This ratio decreases at least as fast as $1/T$ when T tends to infinity. For the basic frequency analysis without the correlation improvement, the relative precision is given by

$$\frac{\operatorname{var} y(t)}{\{E y(t)\}^2} = \frac{r_e(0)}{b^2 \sin^2(\omega t + \varphi)} \qquad (3.22)$$

which can be much larger than (3.21).

Commercial equipment is available for performing frequency analysis as in (3.15)–(3.17). The disadvantage of frequency analysis is that it often requires long experiment times. (Recall that for every frequency treated the system must be allowed to settle to the 'stationary phase' before the integrations are performed. For small ω it may be necessary to let the integration time T be very large.)

3.4 Correlation analysis

The form of model used in correlation analysis is

$$y(t) = \sum_{k=0}^{\infty} h(k) u(t-k) + v(t) \qquad (3.23)$$

or its continuous time counterpart. In (3.23) $\{h(k)\}$ is the weighting sequence and $v(t)$ is a disturbance term. Assume that the input is a stationary stochastic process which is independent of the disturbance. Then the following relation (called the Wiener–Hopf equation) holds for the covariance functions:

$$r_{yu}(\tau) = \sum_{k=0}^{\infty} h(k) r_u(\tau - k) \qquad (3.24)$$

where $r_{yu}(\tau) = E y(t + \tau) u(t)$ and $r_u(\tau) = E u(t + \tau) u(t)$ (see equation (A3.1.11) in Appendix A3.1 at the end of this chapter). The covariance functions in (3.24) can be estimated from the data as

$$\hat{r}_{yu}(\tau) = \frac{1}{N} \sum_{t=1-\min(\tau,0)}^{N-\max(\tau,0)} y(t + \tau) u(t) \qquad \tau = 0, \pm 1, \pm 2, \ldots$$

$$\hat{r}_u(\tau) = \frac{1}{N} \sum_{t=1}^{N-\tau} u(t + \tau) u(t) \qquad \hat{r}_u(-\tau) = \hat{r}_u(\tau) \qquad \tau = 0, 1, 2, \ldots \qquad (3.25)$$

Then an estimate $\{\hat{h}(k)\}$ of the weighting function $\{h(k)\}$ can be determined in principle by solving

$$\hat{r}_{yu}(\tau) = \sum_{k=0}^{\infty} \hat{h}(k)\hat{r}_u(\tau - k) \qquad (3.26)$$

This will in general give a linear system of infinite dimension. The problem is greatly simplified if we use white noise as input. Then it is known *a priori* that $r_u(\tau) = 0$ for $\tau \neq 0$. For this case (3.24) gives

$$h(k) = r_{yu}(k)/r_u(0) \qquad (3.27)$$

which is easy to estimate from the data using (3.25). In Appendix A3.2 the accuracy of this estimate is analyzed. Equipment exists for performing correlation analysis automatically in this way.

Another approach to the simplification of (3.26) is to consider a *truncated* weighting function. This will lead to a linear system of finite order. Assume that

$$h(k) = 0 \qquad k \geq M \qquad (3.28)$$

Such a model is often called a *finite impulse response* (FIR) model in signal processing applications. The integer M should be chosen to be large in comparison with the dominant time constants of the system. Then (3.28) will be a good approximation. Using (3.28), equation (3.26) becomes

$$\hat{r}_{yu}(\tau) = \sum_{k=0}^{M-1} \hat{h}(k)\hat{r}_u(\tau - k) \qquad (3.29)$$

Writing out this equation for $\tau = 0, 1, \ldots, M - 1$, the following linear system of equations is obtained:

$$\begin{pmatrix} \hat{r}_{yu}(0) \\ \vdots \\ \hat{r}_{yu}(M-1) \end{pmatrix} = \begin{pmatrix} \hat{r}_u(0) & \cdots & \hat{r}_u(M-1) \\ \hat{r}_u(1) & & \vdots \\ \vdots & & \vdots \\ \hat{r}_u(M-1) & \cdots & \hat{r}_u(0) \end{pmatrix} \begin{pmatrix} \hat{h}(0) \\ \vdots \\ \hat{h}(M-1) \end{pmatrix} \qquad (3.30)$$

Equation (3.29) can also be applied with more than M different values of τ giving rise to an overdetermined linear system of equations. The method of determining $\{h(k)\}$ from (3.30) is discussed in further detail in Chapter 4. The condition required in order to ensure that the system of equations (3.30) has a unique solution is derived in Section 5.4.

3.5 Spectral analysis

The final nonparametric method to be described is spectral analysis. As in the previous method, we start with the description (3.23), which implies (3.24). Taking discrete Fourier transforms, the following relation for the spectral densities can be derived from (3.24) (see Appendix A3.1):

44 *Nonparametric methods*   *Chapter 3*

$$\phi_{yu}(\omega) = H(e^{-i\omega})\phi_u(\omega) \qquad (3.31)$$

where

$$\phi_{yu}(\omega) = \frac{1}{2\pi} \sum_{\tau=-\infty}^{\infty} r_{yu}(\tau)e^{-i\tau\omega}$$

$$\phi_u(\omega) = \frac{1}{2\pi} \sum_{\tau=-\infty}^{\infty} r_u(\tau)e^{-i\tau\omega} \qquad (3.32)$$

$$H(e^{-i\omega}) = \sum_{k=0}^{\infty} h(k)e^{-ik\omega}$$

See (A3.1.12) for a derivation. Note that $\phi_{yu}(\omega)$ is complex valued.

Now the transfer function $H(e^{-i\omega})$ can be estimated from (3.31) as

$$\boxed{\hat{H}(e^{-i\omega}) = \hat{\phi}_{yu}(\omega)/\hat{\phi}_u(\omega)} \qquad (3.33)$$

To use (3.33) we must find a reasonable method for estimating the spectral densities. A straightforward approach would be to take

$$\hat{\phi}_{yu}(\omega) = \frac{1}{2\pi} \sum_{\tau=-N}^{N} \hat{r}_{yu}(\tau)e^{-i\tau\omega} \qquad (3.34)$$

(cf. (3.25), (3.32)), and similarly for $\hat{\phi}_u(\omega)$. The computations in (3.34) can be organized as follows. Using (3.25),

$$\hat{\phi}_{yu}(\omega) = \frac{1}{2\pi N} \sum_{\tau=-N}^{N} \sum_{t=1-\min(\tau,0)}^{N-\max(\tau,0)} y(t+\tau)u(t)e^{-i\tau\omega}$$

Next make the substitution $s = t + \tau$. Figure 3.7 illustrates how to derive the limits for the new summation index.

Since $e^{-i\tau\omega} = e^{-is\omega}e^{it\omega}$ we get

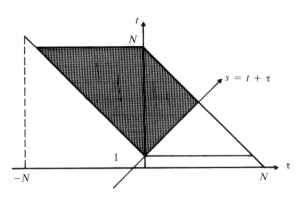

FIGURE 3.7 Change of summation variables. Summation is over the shaded area.

$$\hat{\phi}_{yu}(\omega) = \frac{1}{2\pi N} \sum_{s=1}^{N} \sum_{t=1}^{N} y(s)u(t)e^{-is\omega}e^{it\omega}$$

$$= \frac{1}{2\pi N} Y_N(\omega) U_N(-\omega) \tag{3.35}$$

where

$$Y_N(\omega) = \sum_{s=1}^{N} y(s)e^{-is\omega}$$
$$U_N(\omega) = \sum_{s=1}^{N} u(s)e^{-is\omega} \tag{3.36}$$

are the discrete Fourier transforms of the sequences $\{y(t)\}$ and $\{u(t)\}$, padded with zeros, respectively. For $\omega = 0, 2\pi/N, 4\pi/N, \ldots, \pi$ they can be computed efficiently using FFT (fast Fourier transform) algorithms.

In a similar way,

$$\hat{\phi}_u(\omega) = \frac{1}{2\pi N} U_N(\omega) U_N(-\omega) = \frac{1}{2\pi N} |U_N(\omega)|^2 \tag{3.37}$$

This estimate of the spectral density is called the *periodogram*. From (3.33), (3.35), (3.37), the estimate for the transfer function is

$$\hat{H}(e^{-i\omega}) = Y_N(\omega)/U_N(\omega) \tag{3.38}$$

This quantity is sometimes called the empirical transfer function estimate. See Ljung (1985b) for a discussion.

The foregoing approach to estimating the spectral densities and hence the transfer function will give poor results. For example, if $u(t)$ is a stochastic process, then the estimates (3.35), (3.37) do not converge (in the mean square sense) to the true spectrum as N, the number of data points, tends to infinity. In particular, the estimate $\hat{\phi}_u(\omega)$ will on average behave like $\phi_u(\omega)$, but its variance does not tend to zero as N tends to infinity. See Brillinger (1981) for a discussion of this point. One of the main reasons for this behavior is that the estimate $\hat{r}_{yu}(\tau)$ will be quite inaccurate for large values of τ (in which case only a few terms are used in (3.25)), but all covariance elements $\hat{r}_{yu}(\tau)$ are given the same weight in (3.34) regardless of their accuracy. Another more subtle reason may be explained as follows. In (3.34) $2N + 1$ terms are summed. Even if the estimation error of each term tended to zero as $N \to \infty$, there is no guarantee that the global estimation error of the sum also tends to zero. These difficulties may be overcome if the terms of (3.34) corresponding to large values of τ are weighted out. (The above discussion of the estimates $\hat{r}_{yu}(\tau)$ and $\hat{\phi}_{yu}(\omega)$ applies also to $\hat{r}_u(\tau)$ and $\hat{\phi}_u(\omega)$.) Thus, instead of (3.34) the following improved estimate of the cross-spectrum (and analogously for the auto-spectrum) is used:

$$\hat{\phi}_{yu}(\omega) = \frac{1}{2\pi} \sum_{\tau=-N}^{N} \hat{r}_{yu}(\tau) w(\tau) e^{-i\tau\omega} \tag{3.39}$$

where $w(\tau)$ is a so-called *lag window*. It should be equal to 1 for $\tau = 0$, decrease with increasing τ, and should be equal to 0 for large values of τ. ('Large values' refer to a certain proportion such as 5 or 10 percent of the number of data points, N). Several different forms of lag windows have been proposed in the literature; see Brillinger (1981).

Some simple windows are presented in the following example.

Example 3.4 *Some lag windows*
The following lag windows are often referred to in the literature:

$$w_1(\tau) = \begin{cases} 1 & |\tau| \leq M \\ 0 & |\tau| > M \end{cases} \tag{3.40a}$$

$$w_2(\tau) = \begin{cases} 1 - |\tau|/M & |\tau| \leq M \\ 0 & |\tau| > M \end{cases} \tag{3.40b}$$

$$w_3(\tau) = \begin{cases} \frac{1}{2}\left(1 + \cos\frac{\pi\tau}{M}\right) & |\tau| \leq M \\ 0 & |\tau| > M \end{cases} \tag{3.40c}$$

The window $w_1(\tau)$ is called rectangular, $w_2(\tau)$ is attributed to Bartlett, and $w_3(\tau)$ to Hamming and Tukey. These windows are depicted in Figure 3.8.

Note that all the windows vanish for $|\tau| > M$. If the parameter M is chosen to be sufficiently large, the periodogram will not be smoothed very much. On the other hand, a small M may mean that essential parts of the spectrum are smoothed out. It is not

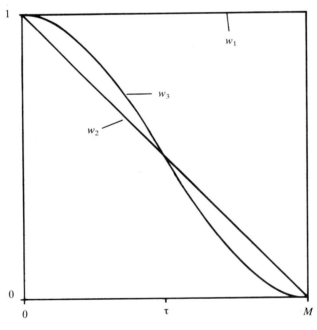

FIGURE 3.8 The lag windows $w_1(\tau)$, $w_2(\tau)$ and $w_3(\tau)$ given by (3.40).

Section 3.5　　　　　　　　　　　　　　　　　　　　　Spectral analysis　47

trivial to choose the parameter M. Roughly speaking M should be chosen according to the following two objectives:

1. M should be small compared to N (to reduce the erratic random fluctuations of the periodogram).
2. $|\hat{r}_u(\tau)| \ll \hat{r}_u(0)$ for $\tau \geq M$ (so as not to smooth out the parts of interest in the true spectrum). ∎

The use of a lag window is necessary to obtain a reasonable accuracy. On the other hand, the sharp peaks in the spectrum may be smeared out. It may therefore not be possible to separate adjacent peaks. Thus, use of a lag window will give a limited frequency resolution. The effect of a lag window on frequency resolution is illustrated in the following simple example.

Example 3.5 *Effect of lag window on frequency resolution*
Assume for simplicity that the input signal is a sinusoid, $u(t) = \sqrt{2} \sin \omega_0 t$, and that a rectangular window (3.40a) is used. (Note that when a sinusoid is used as input, lag windows are not necessary since $U_N(\omega)$ will behave as the true spectrum, i.e. will be much larger for the sinusoidal frequency than for all other arguments. However, we consider here such a case since it provides a simple illustration of the frequency resolution.) To emphasize the effect of the lag window, assume that the true covariance function is available. As shown in Example 5.7,

$$r_u(\tau) = \cos \omega_0 \tau \tag{3.41a}$$

Hence

$$\hat{\phi}_u(\omega) = \frac{1}{2\pi} \sum_{\tau=-M}^{M} \cos \omega_0 \tau \, e^{-i\tau\omega} \tag{3.41b}$$

The true spectrum is derived in Example 5.7 and is given by

$$\phi_u(\omega) = \frac{1}{2}[\delta(\omega - \omega_0) + \delta(\omega + \omega_0)] \tag{3.41c}$$

Thus the spectrum consists of two spikes at $\omega = \pm\omega_0$. Evaluating (3.41b),

$$\hat{\phi}_u(\omega) = \frac{1}{4\pi} \sum_{\tau=-M}^{M} [e^{i(\omega_0-\omega)\tau} + e^{-i(\omega_0+\omega)\tau}]$$

$$= \frac{1}{4\pi} \left(e^{-i(\omega_0-\omega)M} \frac{1 - e^{i(\omega_0-\omega)(2M+1)}}{1 - e^{i(\omega_0-\omega)}} + e^{i(\omega_0+\omega)M} \frac{1 - e^{-i(\omega_0+\omega)(2M+1)}}{1 - e^{-i(\omega_0+\omega)}} \right)$$

$$= \frac{1}{4\pi} \left(\frac{e^{-i(\omega_0-\omega)(M+1/2)} - e^{i(\omega_0-\omega)(M+1/2)}}{e^{-i(\omega_0-\omega)/2} - e^{i(\omega_0-\omega)/2}} \right.$$

$$\left. + \frac{e^{i(\omega_0+\omega)(M+1/2)} - e^{-i(\omega_0+\omega)(M+1/2)}}{e^{i(\omega_0+\omega)/2} - e^{-i(\omega_0+\omega)/2}} \right)$$

$$= \frac{1}{4\pi} \left(\frac{\sin(M + \frac{1}{2})(\omega_0 - \omega)}{\sin \frac{1}{2}(\omega_0 - \omega)} + \frac{\sin(M + \frac{1}{2})(\omega_0 + \omega)}{\sin \frac{1}{2}(\omega_0 + \omega)} \right) \tag{3.41d}$$

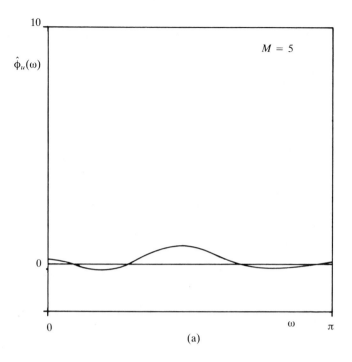

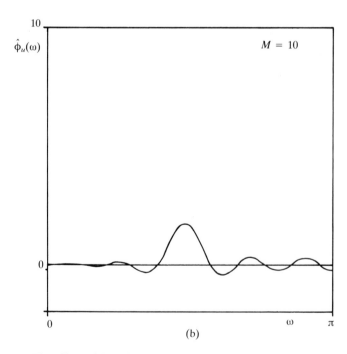

FIGURE 3.9 The effect of the width of a rectangular lag window on the windowed spectrum (3.41d), $\omega_0 = 1.5$.

Section 3.5 *Spectral analysis* 49

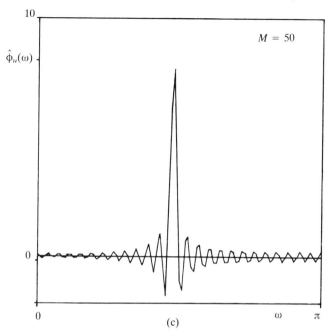

(c)

The windowed spectrum (3.41d) is depicted in Figure 3.9 for three different values of M. It can be seen that the peak is spread out (i.e. its width increases) as M decreases.
When M is large,

$$\hat{\phi}_u(\omega_0) \approx \frac{1}{4\pi} \frac{M + 1/2}{1/2} \approx \frac{M}{2\pi}$$

$$\hat{\phi}_u\left(\omega_0 + \frac{\pi}{2M}\right) \approx \frac{1}{4\pi} \frac{\sin(\pi/2)}{\sin(\pi/4M)} \approx \frac{1}{4\pi} \frac{1}{\pi/4M} = \frac{M}{\pi^2}$$

Hence an approximate expression for the width of the peak is

$$\Delta\omega = 2 \frac{\pi}{2M} = \frac{\pi}{M} \tag{3.41e}$$

Many signals can be described as a (finite or infinite) sum of superimposed sinusoids. According to the above calculations the true frequency content of the signal at $\omega = \omega_0$ appears in $\hat{\phi}_u(\omega)$ in the whole interval $(\omega_0 - \Delta\omega, \omega_0 + \Delta\omega)$. Hence peaks in the true spectrum that are separated by less than π/M are likely to be indistinguishable in the estimated spectrum. ∎

Spectral analysis is a versatile nonparametric method. There is no specific restriction on the input except that it must be uncorrelated with the disturbance. Spectral analysis has therefore become a popular method. The areas of application range from speech analysis and the study of mechanical vibrations to geophysical investigations, not to mention its use in the analysis and design of control systems.

Summary

- *Transient analysis* is easy to apply. It gives a step response or an impulse response (weighting function) as a model. It is very sensitive to noise and can only give a crude model.
- *Frequency analysis* is based on the use of sinusoids as inputs. It requires rather long identification experiments, especially if the correlation feature is included in order to reduce sensitivity to noise. The resulting model is a frequency response. It can be presented as a Bode plot or an equivalent representation of the transfer function.
- *Correlation analysis* is generally based on white noise as input. It gives the weighting function as a resulting model. It is rather insensitive to additive noise on the output signal.
- *Spectral analysis* can be applied with rather arbitrary inputs. The transfer function is obtained in the form of a Bode plot (or other equivalent form). To get a reasonably accurate estimate a lag window must be used. This leads to a limited frequency resolution.

As shown in Chapter 2, nonparametric methods are easy to apply but give only moderately accurate models. If high accuracy is needed a parametric method has to be used. In such cases nonparametric methods can be used to get a first crude model, which may give useful information on how to apply the parametric method.

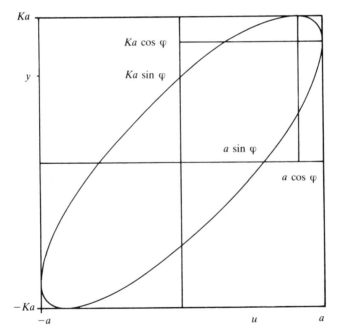

FIGURE 3.10 Input–output relation, Problem 3.3.

Problems

Problem 3.1 *Determination of time constant T from step response*
Prove the rule for determining T given in Figure 3.1.

Problem 3.2 *Analysis of the step response of a second-order damped oscillator*
Prove the relationships (3.3d)–(3.3f).

Problem 3.3 *Determining amplitude and phase*
One method of determining amplitude and phase changes, as needed for frequency analysis, is to plot $y(t)$ versus $u(t)$. If $u(t) = a \sin \omega t$, $y(t) = Ka \sin(\omega t + \varphi)$, show that the relationship depicted in the Figure 3.10 applies.

Problem 3.4 *The covariance function of a simple process*
Let $e(t)$ denote white noise of zero mean and variance λ^2. Consider the filtered white noise process

$$y(t) = \frac{1}{1 + aq^{-1}} e(t) \quad |a| < 1$$

where q^{-1} denotes the unit delay operator. Calculate the covariances $r_y(k) = Ey(t)y(t-k)$, $k = 0, 1, 2, \ldots$, of $y(t)$.

Problem 3.5 *Some properties of the spectral density function*
Let $\phi_u(\omega)$ denote the spectral density function of a stationary signal $u(t)$ (see (3.32)):

$$\phi_u(\omega) = \frac{1}{2\pi} \sum_{\tau=-\infty}^{\infty} r_u(\tau) e^{-i\omega\tau} \quad \omega \in [-\pi, \pi]$$

Assume that

$$\sum_{\tau=0}^{\infty} \tau |r_u(\tau)| < \infty$$

which guarantees the existence of $\phi_u(\omega)$. Show that $\phi_u(\omega)$ has the following properties:
(a) $\phi_u(\omega)$ is real valued and $\phi_u(-\omega) = \phi_u(\omega)$
(b) $\phi_u(\omega) \geq 0$ for all ω

Hint. Set $\varphi(t) = (u(t-1) \; \ldots \; u(t-n))^T$, $x = (1 \; e^{i\omega} \; \ldots \; e^{i(n-1)\omega})^T$. Then show that

$$0 \leq \frac{1}{n} E|\varphi^T(t)x|^2$$

$$= \frac{1}{n} \sum_{\tau=-n}^{n} (n - |\tau|) r_u(\tau) e^{i\omega\tau}$$

and find out what happens when n tends to infinity.

52 Nonparametric methods — Chapter 3

Problem 3.6 *Parseval's formula*
Let

$$H(q^{-1}) = \sum_{k=0}^{\infty} H_k q^{-k} \quad (m|n)$$

$$G(q^{-1}) = \sum_{k=0}^{\infty} G_k q^{-k} \quad (p|n)$$

be two stable matrix transfer functions, and let $e(t)$ be a white noise of zero mean and covariance matrix Λ $(n|n)$. Show that

$$\frac{1}{2\pi} \int_{-\pi}^{\pi} H(e^{i\omega}) \Lambda G^{\mathrm{T}}(e^{-i\omega}) d\omega = \sum_{k=0}^{\infty} H_k \Lambda G_k^{\mathrm{T}}$$

$$= E[H(q^{-1})e(t)][G(q^{-1})e(t)]^{\mathrm{T}}$$

The first equality above is called Parseval's formula. The second equality provides an 'interpretation' of the terms occurring in the first equality.

Problem 3.7 *Correlation analysis with truncated weighting function*
Consider equation (3.30) as a means of applying correlation analysis. Assume that the covariance estimates $\hat{r}_{yu}(\cdot)$ and $\hat{r}_u(\cdot)$ are without errors.

(a) Let the input be zero-mean white noise. Show that, regardless of the choice of M, the weighting function estimate is exact in the sense that

$$\hat{h}(k) = h(k) \quad k = 0, \ldots, M-1$$

(b) To show that the result of (a) does not apply for an arbitrary input, consider the input $u(t)$ given by

$$u(t) - \alpha u(t-1) = v(t) \quad |\alpha| < 1$$

where $v(t)$ is white noise of zero mean and variance σ^2, and the first-order system

$$y(t) + ay(t-1) = bu(t-1) \quad |a| < 1$$

Show that

$$h(0) = 0$$
$$h(k) = b(-a)^{k-1} \quad k \geq 1$$
$$\hat{h}(k) = h(k) \quad k = 0, \ldots, M-2$$
$$\hat{h}(M-1) = \frac{h(M-1)}{(1 + a\alpha)}$$

Hint

$$\begin{pmatrix} 1 & \alpha & \cdots & \alpha^{M-1} \\ \alpha & 1 & & \\ \vdots & & & \\ \vdots & & & \\ \alpha^{M-1} & & \cdots & 1 \end{pmatrix}^{-1} = \frac{1}{1-\alpha^2} \begin{pmatrix} 1 & -\alpha & & & \\ -\alpha & 1+\alpha^2 & \ddots & & 0 \\ & \ddots & \ddots & \ddots & \\ & 0 & \ddots & 1+\alpha^2 & -\alpha \\ & & & -\alpha & 1 \end{pmatrix}$$

(This result can be verified by direct calculations. It can be derived, for example, from (C7.7.18).)

Problem 3.8 *Accuracy of correlation analysis*
Consider the noise-free first-order system

$$y(t) + ay(t-1) = bu(t-1) \qquad |a| < 1$$

Assume that correlation analysis is applied in the case where $u(t)$ is zero-mean white noise. Evaluate the variances of the estimates $\hat{h}(k)$, $k = 1, 2, \ldots$ using the results of Appendix A3.2.

Problem 3.9 *Improved frequency analysis as a special case of spectral analysis*
Show that if the input is a sinusoid of the form

$$u(t) = a \sin \tilde{\omega} t \qquad \tilde{\omega} = \frac{2\pi}{N} n, \, n \in [0, N-1]$$

(N = number of data points)

the spectral analysis (3.33)–(3.36) reduces to the discrete time counterpart of the improved frequency analysis (3.15)–(3.17). More specifically, show that

$$\operatorname{Re}\hat{H}(e^{i\tilde{\omega}}) = \frac{2}{a} \frac{1}{N} \sum_{t=1}^{N} y(t) \sin \tilde{\omega} t$$

$$\operatorname{Im}\hat{H}(e^{i\tilde{\omega}}) = \frac{2}{a} \frac{1}{N} \sum_{t=1}^{N} y(t) \cos \tilde{\omega} t$$

Problem 3.10 *Step response analysis as a special case of spectral analysis*
Let $\{y(t)\}_{t=0}^{N}$ denote the response of a discrete time linear system with transfer function $H(q^{-1}) = \sum_{k=0}^{\infty} h(k) q^{-k}$ to a step signal $u(t)$ of amplitude a. Assume that $y(t) = 0$ for $t < 0$ and $y(t) \approx$ constant for $t > N$. Justify the following estimate of the system transfer function:

$$\hat{H}(q^{-1}) = \sum_{k=0}^{N} \hat{h}(k) q^{-k}; \quad \hat{h}(k) = \frac{y(k) - y(k-1)}{a} \qquad k = 0, \ldots, N$$

and show that it is approximately equal to the estimate provided by the spectral analysis.

Problem 3.11 *Determination of a parametric model from the impulse response*
Assume that we know (or have estimated) an impulse response $\{h(k)\}_{k=1}^{\infty}$. Consider an nth-order parametric model of the form

$$A(q^{-1})y(t) = B(q^{-1})u(t) \qquad \text{(i)}$$

which is to be determined from the impulse response $\{h(k)\}$, where

$$A(q^{-1}) = 1 + a_1 q^{-1} + \ldots + a_n q^{-n}$$
$$B(q^{-1}) = b_1 q^{-1} + \ldots + b_n q^{-n}$$

One possibility to determine $\{a_i, b_j\}$ from $\{h(k)\}$ is to require that the model (i) has a weighting function that coincides with the given sequence for $k = 1, \ldots, 2n$.

(a) Set $H(q^{-1}) = \sum_{k=1}^{\infty} h(k) q^{-k}$. Show that the above procedure can be described in a polynomial formulation as

$$B(q^{-1}) = A(q^{-1})H(q^{-1}) + O(q^{-2n-1}) \qquad \text{(ii)}$$

and that (ii) is equivalent to the following linear system of equations:

$$\begin{pmatrix} 0 & & & 0 & \vdots & 1 & & 0 \\ -h(1) & 0 & & & \vdots & & \ddots & \\ -h(2) & -h(1) & & & \vdots & 0 & & 1 \\ \vdots & & \ddots & & \vdots & \multicolumn{3}{c}{\text{-----------}} \\ \vdots & & & -h(1) & \vdots & & 0 & \\ \vdots & & & \vdots & \vdots & & & \\ -h(2n-1) & \ldots & & -h(n) & \vdots & & & \end{pmatrix} \begin{pmatrix} a_1 \\ \vdots \\ a_n \\ b_1 \\ \vdots \\ b_n \end{pmatrix} = \begin{pmatrix} h(1) \\ \vdots \\ \vdots \\ \vdots \\ h(2n) \end{pmatrix} \qquad \text{(iii)}$$

Also derive (iii) directly from the difference equation (i), using the fact that $\{h(k)\}$ is the impulse response of the system.

(b) Assume that $\{h(k)\}$ is the noise-free impulse response of an nth-order system

$$A_0(q^{-1})y(t) = B_0(q^{-1})u(t)$$

where A_0, B_0 are coprime. Show that the above procedure gives a perfect model in the sense that $A(q^{-1}) = A_0(q^{-1})$, $B(q^{-1}) = B_0(q^{-1})$.

Bibliographical notes

Eykhoff (1974) and the tutorial papers by Rake (1980, 1987) and Glover (1987) give some general and more detailed results on nonparametric identification methods.

Some different ways of determining a parametric model from a step response have been given by Schwarze (1964).

Frequency analysis has been analyzed thoroughly by Åström (1975), while Davies (1970) gives a further treatment of correlation analysis.

The book by Jenkins and Watts (1969) is still a standard text on spectral analysis. For more recent references in this area, see Brillinger (1981), Priestley (1982), Ljung (1985b, 1987), Hannan (1970), Wellstead (1981), and Bendat and Piersol (1980). Kay (1988) also presents many parametric methods for spectral analysis. The FFT algorithm for efficiently computing the discrete Fourier transforms is due to Cooley and Tukey (1965). See also Bergland (1969) for a tutorial description.

Appendix A3.1
Covariance functions, spectral densities and linear filtering

Let $u(t)$ be an nu-dimensional stationary stochastic process. Assume that its mean value is m_u and its covariance function is

$$r_u(\tau) = E[u(t + \tau) - m_u][u(t) - m_u]^T \tag{A3.1.1}$$

Its spectral density is then, by definition,

$$\phi_u(\omega) \triangleq \frac{1}{2\pi} \sum_{\tau=-\infty}^{\infty} r_u(\tau) e^{-i\tau\omega} \tag{A3.1.2}$$

The inverse relation to (A3.1.2) describes how the covariance function can be found from the spectral density. This relation is given by

$$r_u(\tau) = \int_{-\pi}^{\pi} \phi_u(\omega) e^{i\tau\omega} d\omega \tag{A3.1.3}$$

As a verification, the right-hand side of (A3.1.3) can be evaluated using (A3.1.2), giving

$$\int_{-\pi}^{\pi} \frac{1}{2\pi} \sum_{\tau'=-\infty}^{\infty} r_u(\tau') e^{-i\tau'\omega} e^{i\tau\omega} d\omega = \frac{1}{2\pi} \sum_{\tau'=-\infty}^{\infty} r_u(\tau') \int_{-\pi}^{\pi} e^{i(\tau-\tau')\omega} d\omega$$

$$= \sum_{\tau'=-\infty}^{\infty} r_u(\tau') \delta_{\tau,\tau'} = r_u(\tau)$$

which proves the relation (A3.1.3).

Now consider a linear filtering of $u(t)$, that is

$$y(t) = \sum_{k=0}^{\infty} h(k) u(t - k) \tag{A3.1.4}$$

where $y(t)$ is an ny-dimensional signal and $\{h(k)\}$ a sequence of $(ny|nu)$-dimensional matrices. We assume that the filter in (A3.1.4) is stable, which implies that $\|h(k)\| \to 0$ as $k \to \infty$.

Under the given conditions the signal $y(t)$ is stationary. The aim of this appendix is to derive its mean value m_y, covariance function $r_y(\tau)$ and spectral density $\phi_y(\omega)$; and in addition the cross-covariance function $r_{yu}(\tau)$ and the cross-spectral density $\phi_{yu}(\omega)$. It will be convenient to introduce the filter, or transfer function operator

56 Nonparametric methods   Chapter 3

$$H(q^{-1}) = \sum_{k=0}^{\infty} h(k)q^{-k} \tag{A3.1.5}$$

where q^{-1} is the backward shift operator. Using $H(q^{-1})$, the filtering (A3.1.4) can be rewritten as

$$y(t) = H(q^{-1})u(t) \tag{A3.1.6}$$

The mean value of $y(t)$ is easily found from (A3.1.4):

$$m_y = Ey(t) = E \sum_{k=0}^{\infty} h(k)u(t-k) = \sum_{k=0}^{\infty} h(k)m_u = H(1)m_u \tag{A3.1.7}$$

Note that $H(1)$ can be interpreted as the static (dc) gain of the filter.

Now consider how the deviations from the mean values $\tilde{y}(t) \triangleq y(t) - m_y$, $\tilde{u}(t) \triangleq u(t) - m_u$ are related. From (A3.1.4) and (A3.1.7),

$$\tilde{y}(t) = \sum_{k=0}^{\infty} h(k)u(t-k) - \sum_{k=0}^{\infty} h(k)m_u = \sum_{k=0}^{\infty} h(k)[u(t-k) - m_u]$$

$$= \sum_{k=0}^{\infty} h(k)\tilde{u}(t-k) = H(q^{-1})\tilde{u}(t) \tag{A3.1.8}$$

Thus $(\tilde{u}(t), \tilde{y}(t))$ are related in the same way as $(u(t), y(t))$. When analyzing the covariance functions, strictly speaking we should deal with $\tilde{u}(t), \tilde{y}(t)$. For simplicity we drop the $\tilde{}$ notation. This means formally that $u(t)$ is assumed to have zero mean. Note, however, that the following results are true also for $m_u \neq 0$.

Consider first the covariance function of $y(t)$. Some straightforward calculations give

$$r_y(\tau) = Ey(t+\tau)y^T(t)$$

$$= \sum_{j=0}^{\infty} \sum_{k=0}^{\infty} h(j)Eu(t+\tau-j)u^T(t-k)h^T(k)$$

$$= \sum_{j=0}^{\infty} \sum_{k=0}^{\infty} h(j)r_u(\tau - j + k)h^T(k) \tag{A3.1.9}$$

In most situations this relation is not very useful, but its counterpart for the spectral densities has an attractive form. Applying the definition (A3.1.2),

$$\phi_y(\omega) = \frac{1}{2\pi} \sum_{\tau=-\infty}^{\infty} r_y(\tau)e^{-i\tau\omega}$$

$$= \frac{1}{2\pi} \sum_{\tau=-\infty}^{\infty} \sum_{j=0}^{\infty} \sum_{k=0}^{\infty} h(j)e^{-ij\omega}r_u(\tau - j + k)e^{-i(\tau-j+k)\omega}h^T(k)e^{ik\omega}$$

$$= \frac{1}{2\pi} \sum_{j=0}^{\infty} \sum_{k=0}^{\infty} h(j)e^{-ij\omega}\left[\sum_{\tau'=-\infty}^{\infty} r_u(\tau')e^{-i\tau'\omega}\right]h^T(k)e^{ik\omega}$$

$$= \left[\sum_{j=0}^{\infty} h(j)e^{-ij\omega}\right]\phi_u(\omega)\left[\sum_{k=0}^{\infty} h^T(k)e^{ik\omega}\right]$$

or

$$\phi_y(\omega) = H(e^{-i\omega})\phi_u(\omega)H^T(e^{i\omega}) \qquad (A3.1.10)$$

This is a useful relation. It describes how the frequency content of the output depends on the input spectral density $\phi_u(\omega)$ and on the transfer function $H(e^{i\omega})$. For example, suppose the system has a weakly damped resonance frequency ω_0. Then $|H(e^{i\omega_0})|$ will be large and so will $\phi_y(\omega_0)$ (assuming $\phi_u(\omega_0) \neq 0$).

Next consider the cross-covariance function. For this case

$$\begin{aligned}
r_{yu}(\tau) &= Ey(t+\tau)u^T(t) \\
&= \sum_{j=0}^{\infty} h(j)Eu(t+\tau-j)u^T(t) \\
&= \sum_{j=0}^{\infty} h(j)r_u(\tau-j)
\end{aligned} \qquad (A3.1.11)$$

For the cross-spectral density, some simple calculations give

$$\begin{aligned}
\phi_{yu}(\omega) &= \frac{1}{2\pi} \sum_{\tau=-\infty}^{\infty} r_{yu}(\tau)e^{-i\tau\omega} \\
&= \frac{1}{2\pi} \sum_{\tau=-\infty}^{\infty} \sum_{j=0}^{\infty} h(j)e^{-ij\omega}r_u(\tau-j)e^{-i(\tau-j)\omega} \\
&= \sum_{j=0}^{\infty} h(j)e^{-ij\omega}\left[\frac{1}{2\pi}\sum_{\tau'=-\infty}^{\infty} r_u(\tau')e^{-i\tau'\omega}\right]
\end{aligned}$$

or

$$\phi_{yu}(\omega) = H(e^{-i\omega})\phi_u(\omega) \qquad (A3.1.12)$$

The results of this appendix were derived for stationary processes. Ljung (1985c) has shown that they remain valid, with appropriate interpretations, for quasi-stationary signals. Such signals are stochastic processes with deterministic components. In analogy with (A3.1.1), mean and covariance functions are then defined as

$$m_u = \lim_{N\to\infty} \frac{1}{N}\sum_{t=1}^{N} Eu(t) \qquad (A3.1.13a)$$

$$r_u(\tau) = \lim_{N\to\infty} \frac{1}{N}\sum_{t=1}^{N} E(u(t+\tau)-m_u)(u(t)-m_u)^T \qquad (A3.1.13b)$$

assuming the limits above exist. Once the covariance function is defined, the spectral density can be introduced as in (A3.1.2). As mentioned above, the general results (A3.1.10) and (A3.1.12) for linear filtering also hold for quasi-stationary signals.

Appendix A3.2
Accuracy of correlation analysis

Consider the system

$$y(t) = \sum_{k=0}^{\infty} h(k)u(t - k) + v(t) \tag{A3.2.1}$$

where $v(t)$ is a stationary stochastic process, independent of the input signal. In Section 3.4 the following estimate for the weighting function was derived:

$$\hat{h}(k) = \frac{\hat{r}_{yu}(k)}{\hat{r}_u(0)} = \frac{\frac{1}{N}\sum_{t=1}^{N} y(t + k)u(t)}{\frac{1}{N}\sum_{t=1}^{N} u^2(t)} \quad k = 0, 1, \ldots \tag{A3.2.2}$$

assuming that $u(t)$ is white noise of zero mean and variance σ^2.

The accuracy of the estimate (A3.2.2) can be determined using ergodic theory (see Section B.1 of Appendix B). We find $\hat{r}_{yu}(k) \to r_{yu}(k)$, $\hat{r}_u(0) \to r_u(0)$ and hence $\hat{h}(k) \to h(k)$ as N tends to infinity. To examine the deviation $\hat{h}(k) - h(k)$ for a finite but large N it is necessary to find the accuracy of the covariance estimates $\hat{r}_{yu}(k)$ and $\hat{r}_u(0)$. This can be done as in Section B.8, where results on the accuracy of sample autocovariance functions are derived. However, here we choose a more direct way. First note that

$$\hat{h}(k) - h(k) = \frac{1}{\hat{r}_u(0)} [\hat{r}_{yu}(k) - h(k)\hat{r}_u(0)]$$

$$\approx \frac{1}{r_u(0)} \frac{1}{N} \left[\sum_{t=1}^{N} \{y(t + k) - h(k)u(t)\} u(t) \right]$$

$$= \frac{1}{\sigma^2} \frac{1}{N} \sum_{t=1}^{N} \left\{ \sum_{\substack{i=0 \\ i \neq k}}^{\infty} h(i)u(t + k - i) + v(t + k) \right\} u(t) \tag{A3.2.3}$$

The covariance $P_{\mu\nu}$ between $\{\hat{h}(\mu) - h(\mu)\}$ and $\{\hat{h}(\nu) - h(\nu)\}$ is calculated as follows:

$$P_{\mu\nu} \approx \frac{1}{N^2\sigma^4} E\left[\sum_{t=1}^{N} \left\{ \sum_{\substack{i=0 \\ i \neq \mu}}^{\infty} h(i)u(t + \mu - i) + v(t + \mu) \right\} u(t) \right]$$

$$\times \left[\sum_{s=1}^{N} \left\{ \sum_{\substack{j=0 \\ j \neq \nu}}^{\infty} h(j)u(s + \nu - j) + v(s + \nu) \right\} u(s) \right]$$

$$= \frac{1}{N^2\sigma^4} \sum_{t=1}^{N} \sum_{s=1}^{N} \sum_{\substack{i=0 \\ i \neq \mu}}^{\infty} \sum_{\substack{j=0 \\ j \neq \nu}}^{\infty} h(i)h(j)E[u(t + \mu - i)\,u(t)u(s + \nu - j)u(s)]$$

$$+ \frac{1}{N^2\sigma^4} \sum_{t=1}^{N} \sum_{s=1}^{N} Ev(t + \mu)v(s + \nu) Eu(t)u(s) \tag{A3.2.4}$$

where use has been made of the fact that u and v are uncorrelated. The second term in (A3.2.4) is easily found to be

$$\frac{1}{N\sigma^2} r_v(\mu - v)$$

To evaluate the first term, note that for a white noise process

$$Eu(t + \mu - i)u(t)u(s + v - j)u(s) = \sigma^4 \delta_{\mu-i,0}\delta_{v-j,0}$$
$$+ \sigma^4 \delta_{t+\mu-i,s+v-j}\delta_{t,s} + \sigma^4 \delta_{t+\mu-i,s}\delta_{t,s+v-j}$$
$$+ \{Eu^4(t) - 3\sigma^4\}\delta_{\mu-i,0}\delta_{v-j,0}\delta_{t,s}$$

For convenience, set $h(i) = 0$ for $i < 0$. Inserting the above expression into (A3.2.4) we now find

$$P_{\mu v} \approx \frac{r_v(\mu - v)}{N\sigma^2} + \frac{1}{N} \sum_{\substack{i=0 \\ i\neq\mu}}^{\infty} \sum_{\substack{j=0 \\ j\neq v}}^{\infty} h(i)h(j)\delta_{\mu-i,v-j} + \frac{1}{N^2} \sum_{\substack{\tau=-N \\ \tau\neq 0}}^{N} (N - |\tau|)h(\tau + \mu)h(v - \tau)$$

$$= \frac{r_v(\mu - v)}{N\sigma^2} + \frac{1}{N} \sum_{\substack{i=0 \\ i\neq\mu}}^{\infty} h(i)h(i - \mu + v)$$

$$+ \frac{1}{N^2} \sum_{\substack{\tau=-\mu \\ \tau\neq 0}}^{v} (N - |\tau|)h(\tau + \mu)h(v - \tau)$$

$$\approx \frac{r_v(\mu - v)}{N\sigma^2} + \frac{1}{N} \sum_{i=0}^{\infty} h(i)h(i + |v - \mu|)$$

$$+ \frac{1}{N} \sum_{\tau=-\mu}^{v} h(\tau + \mu)h(v - \tau) - \frac{2}{N} h(\mu)h(v) \quad (A3.2.5)$$

Note that the covariance element $P_{\mu v}$ will *not* vanish in the noise-free case ($v(t) \equiv 0$). In contrast, the variances–covariances of the estimation errors associated with the parametric methods studied in Chapters 7 and 8, vanish in the noise-free case. Further note that the covariance elements $P_{\mu v}$ approach zero when N tends to infinity. This is in contrast to spectral analysis (the counterpart of correlation analysis in the frequency domain), which does not give consistent estimates of H.

Chapter 4
LINEAR REGRESSION

4.1 The least squares estimate

This chapter presents a discussion and analysis of the concept of linear regression. This is indeed a very common concept in statistics. Its origin can be traced back to Gauss (1809), who used such a technique for calculating orbits of the planets.

The linear regression is the simplest type of *parametric* model. The corresponding model structure can be written as

$$y(t) = \varphi^T(t)\theta \tag{4.1}$$

where $y(t)$ is a measurable quantity, $\varphi(t)$ is an n-vector of known quantities and θ is an n-vector of unknown parameters. The elements of the vector $\varphi(t)$ are often called *regression variables* or *regressors* while $y(t)$ is called the *regressed variable*. We will call θ the *parameter vector*. The variable t takes integer values. Sometimes t denotes a time variable but this is not necessarily the case.

It is straightforward to extend the model (4.1) to the multivariable case, and then

$$y(t) = \Phi^T(t)\theta \tag{4.2}$$

where $y(t)$ is a p-vector, $\Phi(t)$ an $(n|p)$ matrix and θ an n-vector. Least squares estimation of the parameters in multivariable models of the form (4.2) will be treated in some detail in Complement C7.3.

Example 4.1 *A polynomial trend*
Suppose the model is of the form

$$y(t) = a_0 + a_1 t + \ldots + a_r t^r$$

with unknown coefficients $a_0, \ldots, a_r$. This can be written in the form (4.1) by defining

$$\varphi(t) = (1 \quad t \quad \ldots \quad t^r)^T$$
$$\theta = (a_0 \quad a_1 \quad \ldots \quad a_r)^T$$

Such a model can be used to describe a trend in a time series. The integer r is typically taken as 0 or 1. When $r = 0$ only the mean value is described by the model. ■

Example 4.2 *A weighted sum of exponentials*
In the analysis of transient signals a suitable model is

$$y(t) = b_1 e^{-k_1 t} + b_2 e^{-k_2 t} + \ldots + b_n e^{-k_n t}$$

It is assumed here that $k_1, \ldots, k_n$ (the inverse time constants) are known but that the weights $b_1, \ldots, b_n$ are unknown. Then a model of the form (4.1) is obtained by setting

$$\varphi(t) = (e^{-k_1 t} \quad \ldots \quad e^{-k_n t})^T$$
$$\theta = (b_1 \quad \ldots \quad b_n)^T$$

■

Example 4.3 *Truncated weighting function*
A truncated weighting function model was described in Section 3.4. Such a model is given by

$$y(t) = h_0 u(t) + h_1 u(t-1) + \ldots + h_{M-1} u(t - M + 1)$$

The input signal $u(t)$ is recorded during the experiment and can hence be considered as known. In this case

$$\varphi(t) = (u(t) \quad u(t-1) \quad \ldots \quad u(t - M + 1))^T$$
$$\theta = (h_0 \quad \ldots \quad h_{M-1})^T$$

This type of model will often require many parameters in order to give an accurate description of the dynamics (M typically being of the order 20–50; in certain signal processing applications it may be several hundreds or even thousands). Nevertheless, it is quite simple conceptually and fits directly into the framework discussed in this chapter. ■

The problem is to find an estimate $\hat{\theta}$ of the parameter vector θ from measurements $y(1)$, $\varphi(1), \ldots, y(N), \varphi(N)$. Given these measurements, a system of linear equations is obtained, namely

$$y(1) = \varphi^T(1)\theta$$
$$y(2) = \varphi^T(2)\theta$$
$$\vdots$$
$$y(N) = \varphi^T(N)\theta$$

This can be written in matrix notation as

$$Y = \Phi\theta \tag{4.3}$$

where

$$Y = \begin{pmatrix} y(1) \\ \vdots \\ y(N) \end{pmatrix}, \quad \text{an } (N|1) \text{ vector} \tag{4.4a}$$

$$\Phi = \begin{pmatrix} \varphi^T(1) \\ \vdots \\ \varphi^T(N) \end{pmatrix}, \quad \text{an } (N|n) \text{ matrix} \tag{4.4b}$$

One way to find θ from (4.3) would of course be to choose the number of measurements, N, to be equal to n. Then Φ becomes a square matrix. If this matrix is nonsingular the

62 *Linear regression* Chapter 4

linear system of equations, (4.3), could easily be solved for θ. In practice, however, noise, disturbances and model misfit are good reasons for using a number of data greater than *n*. With the additional data it should be possible to get an improved estimate. When $N > n$ the linear system of equations, (4.3), becomes overdetermined. An exact solution will then in general not exist.

Now introduce the equation errors as

$$\varepsilon(t) = y(t) - \varphi^T(t)\theta \tag{4.5}$$

and stack these in a vector ε defined similarly to (4.4a):

$$\varepsilon = \begin{pmatrix} \varepsilon(1) \\ \vdots \\ \varepsilon(N) \end{pmatrix}$$

In the statistical literature the equation errors are often called *residuals*. The *least squares estimate* of θ is defined as the vector $\hat{\theta}$ that minimizes the loss function

$$V(\theta) = \frac{1}{2}\sum_{t=1}^{N}\varepsilon^2(t) = \frac{1}{2}\varepsilon^T\varepsilon = \frac{1}{2}\|\varepsilon\|^2 \tag{4.6}$$

where $\|\cdot\|$ denotes the Euclidean vector norm. According to (4.5) the equation error $\varepsilon(t)$ is a linear function of the parameter vector θ.

The solution of the optimization problem stated above is given in the following lemma.

Lemma 4.1
Consider the loss function $V(\theta)$ given by (4.5), (4.6). Suppose that the matrix $\Phi^T\Phi$ is positive definite. Then $V(\theta)$ has a unique minimum point given by

$$\hat{\theta} = (\Phi^T\Phi)^{-1}\Phi^T Y \tag{4.7}$$

The corresponding minimal value of $V(\theta)$ is

$$\min_{\theta} V(\theta) = V(\hat{\theta}) = \frac{1}{2}[Y^T Y - Y^T \Phi(\Phi^T\Phi)^{-1}\Phi^T Y] \tag{4.8}$$

Proof. Using (4.3), (4.5) and (4.6) an explicit expression for the loss function $V(\theta)$ can be obtained. The point is to see that $V(\theta)$ as a function of θ has quadratic, linear and constant terms. Therefore it is possible to use the technique of completing the squares. We have

$$\varepsilon = Y - \Phi\theta$$

and

$$V(\theta) = \frac{1}{2}[Y - \Phi\theta]^T[Y - \Phi\theta]$$

$$= \frac{1}{2}[\theta^T\Phi^T\Phi\theta - \theta^T\Phi^T Y - Y^T\Phi\theta + Y^T Y]$$

Hence

$$V(\theta) = \frac{1}{2}[\theta - (\Phi^T\Phi)^{-1}\Phi^T Y]^T(\Phi^T\Phi)[\theta - (\Phi^T\Phi)^{-1}\Phi^T Y]$$

$$+ \frac{1}{2}[Y^T Y - Y^T\Phi(\Phi^T\Phi)^{-1}\Phi^T Y]$$

The second term does not depend on θ. Since $\Phi^T\Phi$ by assumption is positive definite, the first term is always greater than or equal to zero. Thus $V(\theta)$ can be minimized by setting the first term to zero. This gives (4.7), as required. The minimal value (4.8) of the loss function then follows directly. ∎

Remark 1. The matrix $\Phi^T\Phi$ is by construction always nonnegative definite. When it is singular (only positive semidefinite) the above calculation is not valid. In that case one can instead evaluate the gradient of the loss function. Setting the gradient to zero,

$$0 = \frac{dV(\theta)}{d\theta} = -Y^T\Phi + \theta^T(\Phi^T\Phi)$$

which can be written as

$$(\Phi^T\Phi)\theta = \Phi^T Y \tag{4.9}$$

When $\Phi^T\Phi$ is singular, Φ does not have full rank (i.e. rank $\Phi < n$). In such a case, equation (4.9) will have infinitely many solutions. They span a subspace which describes a valley of minimum points of $V(\theta)$. Note, however, that if the experiment and the model structure are well chosen then Φ will have full rank. ∎

Remark 2. Some basic results from linear algebra (on overdetermined linear systems of equations) are given in Appendix A (see Lemmas A.7 to A.15). In particular, the least squares solutions are characterized and the so-called pseudoinverse is introduced; this replaces the usual inverse when the matrix is rectangular or does not have full rank. In particular, when Φ is $(N|n)$ and of rank n then the matrix $(\Phi^T\Phi)^{-1}\Phi^T$ which appears in (4.7) is the pseudoinverse of Φ (see Lemma A.11). ∎

Remark 3. The form (4.7) of the least squares estimate can be rewritten in the equivalent form

$$\hat{\theta} = \left[\sum_{t=1}^{N}\varphi(t)\varphi^T(t)\right]^{-1}\left[\sum_{t=1}^{N}\varphi(t)y(t)\right] \tag{4.10}$$

In many cases $\varphi(t)$ is known as a function of t. Then (4.10) might be easier to implement than (4.7) since the matrix Φ of large dimension is not needed in (4.10). Also the form (4.10) is the starting point in deriving several recursive estimates which will be presented in Chapter 9. Note, however, that for a sound numerical implementation, neither (4.7) nor (4.10) should be used, as will be explained in Section 4.5. Both these forms are quite sensitive to rounding errors. ∎

Remark 4. The least-squares problem can be given the following geometrical interpretation. Let the column vectors of Φ be denoted by $\Phi_1, \ldots, \Phi_n$. These vectors belong to $\mathcal{R}^N$. The problem is to seek a linear combination of $\Phi_1, \ldots, \Phi_n$ such that it approximates Y as closely as possible. The best approximation in a least squares sense is given by the orthogonal projection of Y onto the subspace spanned by $\Phi_1, \ldots, \Phi_n$ (see Figure 4.1). Let the orthogonal projection be denoted Y^*. Then it is required that

$$\Phi_i^T(Y - Y^*) = 0 \quad i = 1, \ldots, n$$

$$Y^* = \sum_{j=1}^{n} \Phi_j \theta_j$$

for some weights θ_j to be determined. This gives

$$\Phi_i^T Y = \sum_{j=1}^{n} \Phi_i^T \Phi_j \theta_j \quad i = 1, \ldots, n$$

In matrix form this becomes

$$\begin{pmatrix} \Phi_1^T\Phi_1 & \cdots & \Phi_1^T\Phi_n \\ \vdots & & \vdots \\ \Phi_n^T\Phi_1 & \cdots & \Phi_n^T\Phi_n \end{pmatrix} \begin{pmatrix} \theta_1 \\ \vdots \\ \theta_n \end{pmatrix} = \begin{pmatrix} \Phi_1^T Y \\ \vdots \\ \Phi_n^T Y \end{pmatrix}$$

which is precisely (4.9). ∎

The following example illustrates the least squares solution (4.7).

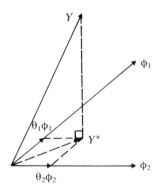

FIGURE 4.1 Geometrical illustration of the least squares solution for the case $N = 3$, $n = 2$.

Example 4.4 *Estimation of a constant*
Assume that the model (4.1) is

$$y(t) = b$$

This means that a constant is to be estimated from a number of noisy measurements. In this case

$$\varphi(t) = 1 \quad \theta = b$$

$$\Phi = \begin{pmatrix} 1 \\ \vdots \\ 1 \end{pmatrix}$$

and (4.7) becomes

$$\hat{\theta} = \frac{1}{N}[y(1) + \ldots + y(N)]$$

This expression is simply the arithmetic mean of all the measurements. ∎

4.2 Analysis of the least squares estimate

We now analyze the statistical properties of the estimate (4.7). In doing so some assumptions must be imposed on the data. Therefore assume that the data satisfy

$$y(t) = \varphi^T(t)\theta_0 + e(t) \tag{4.11a}$$

where θ_0 is called the true parameter vector. Assume further that $e(t)$ is a stochastic variable with zero mean and variance λ^2. In matrix form equation (4.11a) is written as

$$Y = \Phi\theta_0 + e \tag{4.11b}$$

where

$$e = \begin{pmatrix} e(1) \\ \vdots \\ e(N) \end{pmatrix}$$

Lemma 4.2
Consider the estimate (4.7). Assume that the data obey (4.11) with $e(t)$ *white noise* of zero mean and variance λ^2. Then the following properties hold:

(i) $\hat{\theta}$ is an unbiased estimate of θ_0.
(ii) The covariance matrix of $\hat{\theta}$ is given by

$$\text{cov}(\hat{\theta}) = \lambda^2(\Phi^T\Phi)^{-1} \tag{4.12}$$

(iii) An unbiased estimate of λ^2 is given by

$$s^2 = 2V(\hat{\theta})/(N - n) \qquad (4.13)$$

Proof. Equations (4.7) and (4.11) give

$$\hat{\theta} = (\Phi^T\Phi)^{-1}\Phi^T\{\Phi\theta_0 + e\} = \theta_0 + (\Phi^T\Phi)^{-1}\Phi^T e$$

and hence

$$E\hat{\theta} = \theta_0 + (\Phi^T\Phi)^{-1}\Phi^T Ee = \theta_0$$

which proves (i).

To prove (ii) note that the assumption of white noise implies $Eee^T = \lambda^2 I$. Then

$$\begin{aligned} E(\hat{\theta} - \theta_0)(\hat{\theta} - \theta_0)^T &= E[(\Phi^T\Phi)^{-1}\Phi^T e][(\Phi^T\Phi)^{-1}\Phi^T e]^T \\ &= (\Phi^T\Phi)^{-1}\Phi^T Eee^T \Phi(\Phi^T\Phi)^{-1} \\ &= (\Phi^T\Phi)^{-1}\Phi^T \lambda^2 I \Phi(\Phi^T\Phi)^{-1} \\ &= \lambda^2(\Phi^T\Phi)^{-1} \end{aligned}$$

which proves (ii).

The minimal value $V(\hat{\theta})$ of the loss function can be written, according to (4.8) and (4.11), as

$$\begin{aligned} V(\hat{\theta}) &= \frac{1}{2}[\Phi\theta_0 + e]^T[I - \Phi(\Phi^T\Phi)^{-1}\Phi^T][\Phi\theta_0 + e] \\ &= \frac{1}{2}e^T[I - \Phi(\Phi^T\Phi)^{-1}\Phi^T]e \end{aligned}$$

Consider the estimate s^2 given by (4.13). Its mean value can be evaluated as

$$\begin{aligned} Es^2 &= 2EV(\hat{\theta})/(N - n) = E\text{tr}\{e^T[I_N - \Phi(\Phi^T\Phi)^{-1}\Phi^T]e\}/(N - n) \\ &= E\text{tr}\{[I_N - \Phi(\Phi^T\Phi)^{-1}\Phi^T]ee^T\}/(N - n) \\ &= \text{tr}[I_N - \Phi(\Phi^T\Phi)^{-1}\Phi^T]\lambda^2 I_N/(N - n) \\ &= [\text{tr}I_N - \text{tr}\{\Phi(\Phi^T\Phi)^{-1}\Phi^T\}]\lambda^2/(N - n) \\ &= [\text{tr}I_N - \text{tr}\{(\Phi^T\Phi)^{-1}(\Phi^T\Phi)\}]\lambda^2/(N - n) \\ &= [\text{tr}I_N - \text{tr}I_n]\lambda^2/(N - n) = [N - n]\lambda^2/(N - n) = \lambda^2 \end{aligned}$$

In the calculations above I_k denotes the identity matrix of dimension k. We used the fact that for matrices A and B of compatible dimensions $\text{tr}(AB) = \text{tr}(BA)$. The calculations show that s^2 is an unbiased estimate of λ^2. ∎

Remark 1. Note that it is essential in the proof that Φ is a deterministic matrix. When taking expectations it is then only necessary to take e into consideration. ∎

Remark 2. In Appendix B (Lemma B.15) it is shown that for every unbiased estimate $\hat{\theta}$ there is a lower bound, P_{CR}, on the covariance matrix of $\hat{\theta}$. This means that

$$\text{cov}(\hat{\theta}) \geq P_{CR} \tag{4.14}$$

Example B.1 analyzes the least squares estimate (4.7) as well as the estimate s^2, (4.13), under the assumption that the data are Gaussian distributed and satisfy (4.11). It is shown there that $\text{cov}(\hat{\theta})$ attains the lower bound, while $\text{var}(s^2)$ does so only asymptotically, i.e. for a very large number of data points. ∎

4.3 The best linear unbiased estimate

In Lemma 4.2 it was assumed that the disturbance $e(t)$ in (4.11) was white noise, which means that $e(t)$ and $e(s)$ are uncorrelated for all $t \neq s$. Now consider what will happen when this assumption is relaxed. Assume that (4.11) holds and that

$$Eee^T = R \tag{4.15}$$

where R is a positive definite matrix. Looking at the proof of the lemma, it can be seen that $\hat{\theta}$ is still an unbiased estimate of θ_0. However, the covariance matrix of $\hat{\theta}$ becomes

$$\text{cov}(\hat{\theta}) = (\Phi^T\Phi)^{-1}\Phi^T R \Phi(\Phi^T\Phi)^{-1} \tag{4.16}$$

Next we will extend the class of identification methods and consider general linear estimates of θ_0. By a *linear* estimate we mean that $\hat{\theta}$ is a linear function of the data vector Y. Such estimates can be written as

$$\hat{\theta} = Z^T Y \tag{4.17}$$

where Z is an $(N|n)$ matrix which does not depend on Y. The least squares estimate (4.7) is a special case obtained by taking

$$Z = \Phi(\Phi^T\Phi)^{-1}$$

We shall see how to choose Z so that the corresponding estimate is unbiased and has a minimal covariance matrix. The result is known as the *best linear unbiased estimate* (BLUE) and also as the *Markov estimate*. It is given in the following lemma.

Lemma 4.3
Consider the estimate (4.17). Assume that the data satisfy (4.11), (4.15). Let

$$Z^* = R^{-1}\Phi(\Phi^T R^{-1}\Phi)^{-1} \tag{4.18}$$

Then the estimate given by (4.17), (4.18) is an unbiased estimate of θ_0. Furthermore, its covariance matrix is minimal in the sense that the inequality

> $$\mathrm{cov}_{Z^*}(\hat{\theta}) = (\Phi^T R^{-1} \Phi)^{-1} \leq \mathrm{cov}_Z(\hat{\theta}) \qquad (4.19)$$

holds for all unbiased linear estimates. ($P_1 \leq P_2$ means that $P_2 - P_1$ is nonnegative definite).

Proof. The requirement on unbiasedness gives

$$\theta_0 = E\hat{\theta} = EZ^T(\Phi\theta_0 + e) = Z^T\Phi\theta_0$$

so, since θ_0 is arbitrary,

$$Z^T\Phi = I \qquad (4.20)$$

In particular, note that the choice (4.18) of Z satisfies this condition.

The covariance matrix of the estimate (4.17) is given in the general case by

$$\mathrm{cov}_Z(\hat{\theta}) = E(Z^T Y - \theta_0)(Z^T Y - \theta_0)^T = Z^T R Z \qquad (4.21)$$

In particular, if $Z = Z^*$ as given by (4.18), then

$$\mathrm{cov}_{Z^*}(\hat{\theta}) = (\Phi^T R^{-1} \Phi)^{-1} \Phi^T R^{-1} R R^{-1} \Phi (\Phi^T R^{-1} \Phi)^{-1} = (\Phi^T R^{-1} \Phi)^{-1} \qquad (4.22)$$

which proves the equality in (4.19). It remains to show the inequality in (4.19). For illustration, we will give four different proofs based on different methodologies.

Proof A. Let $\tilde{\theta}$ denote the estimation error $\hat{\theta} - \theta_0$ for a general estimate (subject to (4.20)); and let $\tilde{\theta}^*$ denote the error which corresponds to the choice (4.18) of Z. Then

$$\mathrm{cov}_Z(\hat{\theta}) = E\tilde{\theta}\tilde{\theta}^T = E[\tilde{\theta} - \tilde{\theta}^*][\tilde{\theta} - \tilde{\theta}^*]^T + E\tilde{\theta}\tilde{\theta}^{*T} + E\tilde{\theta}^*\tilde{\theta}^T - E\tilde{\theta}^*\tilde{\theta}^{*T} \qquad (4.23)$$

However, we already know that

$$E\tilde{\theta}^*\tilde{\theta}^{*T} = (\Phi^T R^{-1} \Phi)^{-1}$$

and it follows that

$$E\tilde{\theta}\tilde{\theta}^{*T} = EZ^T ee^T Z^* = Z^T R Z^* = Z^T R R^{-1} \Phi (\Phi^T R^{-1} \Phi)^{-1}$$
$$= Z^T \Phi (\Phi^T R^{-1} \Phi)^{-1} = (\Phi^T R^{-1} \Phi)^{-1} = E\tilde{\theta}^*\tilde{\theta}^{*T}$$
$$= [E\tilde{\theta}^*\tilde{\theta}^T]^T$$

Note that we have used the constraint (4.20) in these calculations. Now (4.23) gives

$$\mathrm{cov}_Z(\hat{\theta}) = E[\tilde{\theta} - \tilde{\theta}^*][\tilde{\theta} - \tilde{\theta}^*]^T + (\Phi^T R^{-1} \Phi)^{-1} \geq (\Phi^T R^{-1} \Phi)^{-1}$$

which proves (4.19).

Proof B. The matrix

$$\begin{pmatrix} Z^T R Z & Z^T R Z^* \\ Z^{*T} R Z & Z^{*T} R Z^* \end{pmatrix} = \begin{pmatrix} Z^T \\ Z^{*T} \end{pmatrix} R (Z \ Z^*)$$

is obviously nonnegative definite. Then in particular

Section 4.3 The best linear unbiased estimate 69

$$Z^T R Z - (Z^T R Z^*)(Z^{*T} R Z^*)^{-1}(Z^{*T} R Z) \geq 0 \qquad (4.24)$$

(see Lemma A.3 in Appendix A). However,

$$Z^T R Z^* = Z^T R R^{-1} \Phi (\Phi^T R^{-1} \Phi)^{-1} = Z^T \Phi (\Phi^T R^{-1} \Phi)^{-1} = (\Phi^T R^{-1} \Phi)^{-1}$$

$$Z^{*T} R Z = (Z^T R Z^*)^T = (\Phi^T R^{-1} \Phi)^{-1}$$

Since this holds for all Z (subject to (4.20)), it is also true in particular that

$$Z^{*T} R Z^* = (\Phi^T R^{-1} \Phi)^{-1}$$

(cf. (4.22)). Now, (4.21), (4.24) give

$$\text{cov}_Z(\hat{\theta}) = Z^T R Z \geq (\Phi^T R^{-1} \Phi)^{-1} = \text{cov}_{Z^*}(\hat{\theta})$$

which proves (4.19).

Proof C. Making use of (4.20),

$$\text{cov}_Z(\hat{\theta}) - \text{cov}_{Z^*}(\hat{\theta}) = Z^T R Z - (\Phi^T R^{-1} \Phi)^{-1}$$
$$= Z^T R Z - Z^T \Phi (\Phi^T R^{-1} \Phi)^{-1} \Phi^T Z \qquad (4.25)$$
$$= Z^T [R - \Phi (\Phi^T R^{-1} \Phi)^{-1} \Phi^T] Z$$

However, the matrix in square brackets in this last expression can be written as

$$R - \Phi (\Phi^T R^{-1} \Phi)^{-1} \Phi^T = [R - \Phi (\Phi^T R^{-1} \Phi)^{-1} \Phi^T] R^{-1} [R - \Phi (\Phi^T R^{-1} \Phi)^{-1} \Phi^T]$$

and it is hence nonnegative definite. It then follows easily from (4.25) that

$$\text{cov}_Z(\hat{\theta}) - \text{cov}_{Z^*}(\hat{\theta}) \geq 0$$

which is the required result.

Proof D. Let α be an arbitrary n-dimensional vector. Finding the best linear unbiased estimate $\hat{\theta}$ by minimizing the covariance matrix (4.21) is equivalent to minimizing $\alpha^T Z^T R Z \alpha$ subject to (4.20). This constrained optimization problem is solved using the method of Lagrange multipliers. The Lagrangian for this problem is

$$L(Z, \Lambda) = \alpha^T Z^T R Z \alpha + \text{tr}\{\Lambda (Z^T \Phi - I)\}$$

where the $(n|n)$ matrix Λ represents the multipliers. Using the definition

$$\left[\frac{\partial L}{\partial Z}\right]_{ij} = \frac{\partial L}{\partial Z_{ij}}$$

for the derivative of a scalar function with respect to a matrix, the following result can be obtained:

$$0 = \frac{\partial L}{\partial Z} = 2 R Z \alpha \alpha^T + \Phi \Lambda \qquad (4.26)$$

This equation is to be solved together with (4.20). Multiplying (4.26) on the left by $\Phi^T R^{-1}$,

$$0 = 2\Phi^T Z \alpha \alpha^T + (\Phi^T R^{-1} \Phi)\Lambda$$

In view of (4.20), this gives

$$\Lambda = -2(\Phi^T R^{-1} \Phi)^{-1} \alpha \alpha^T$$

Substituting this in (4.26),

$$Z\alpha\alpha^T = -\frac{1}{2} R^{-1} \Phi \Lambda = R^{-1} \Phi (\Phi^T R^{-1} \Phi)^{-1} \alpha \alpha^T$$

Thus

$$[Z - R^{-1} \Phi (\Phi^T R^{-1} \Phi)^{-1}] \alpha \alpha^T = 0$$

for *all* α. This in turn implies that

$$Z = R^{-1} \Phi (\Phi^T R^{-1} \Phi)^{-1}$$

Thus to minimize the covariance matrix of an unbiased estimate we must choose Z as in (4.18). ∎

Remark 1. Suppose that $R = \lambda^2 I$. Then $Z^* = \Phi(\Phi^T \Phi)^{-1}$, which leads to the simple least squares estimate (4.7). In the case where $e(t)$ is white noise the least squares estimate is therefore the best linear unbiased estimate. ∎

Remark 2. One may ask whether there are *nonlinear* estimates with better accuracy than the BLUE. It is shown in Example B.2 that if the disturbances $e(t)$ have a Gaussian distribution, then the *linear* estimate given by (4.17), (4.18) is the best one among all nonlinear unbiased estimates. If $e(t)$ is not Gaussian distributed then there may exist nonlinear estimates which are more accurate than the BLUE. ∎

Remark 3. The result of the lemma can be slightly generalized as follows. Let $\hat{\theta}$ be the BLUE of θ_0, and let A be a constant matrix. Then the BLUE of $A\theta_0$ is $A\hat{\theta}$. This can be shown by calculations analogous to those of the above proof. Note that equation (4.20) will have to be modified to $Z^T \Phi = A$. ∎

Remark 4. In the complements to this chapter we give several extensions to the above lemma. In Complement C4.1 we consider the BLUE when a linear constraint of the form $A\theta_0 = b$ is imposed. The case when the residual covariance matrix R may be singular is dealt with in Complement C4.3. Complement C4.4 contains some extensions to an important class of nonlinear models. ∎

We now illustrate the BLUE (4.17), (4.18) by means of a simple example.

Example 4.5 *Estimation of a constant (continued from Example 4.4)*
Let the model be

$$y(t) = b$$

Assume that the measurement errors are independent but that they may have different variances, so

$$y(t) = b_0 + e(t) \qquad Ee^2(t) = \lambda_t^2$$

Then

$$\Phi = \begin{pmatrix} 1 \\ \vdots \\ 1 \end{pmatrix} \qquad R = \begin{pmatrix} \lambda_1^2 & & & 0 \\ & \lambda_2^2 & & \\ & & \ddots & \\ 0 & & & \lambda_N^2 \end{pmatrix}$$

Thus the BLUE of b_0 is

$$\hat{\theta} = \frac{1}{\sum_{j=1}^{N}(1/\lambda_j^2)} \sum_{i=1}^{N}(1/\lambda_i^2) y(i)$$

This is a weighted arithmetic mean of the observations. Note that the weight of $y(i)$, i.e.

$$\frac{1}{\sum_{j=1}^{N}(1/\lambda_j^2)} \cdot 1/\lambda_i^2$$

is small if this measurement is inaccurate (λ_i large) and vice versa. ∎

4.4 Determining the model dimension

A heuristic discussion

The treatment so far has investigated some statistical properties of the least squares and other linear estimates of the regression parameters. The discussion now turns to the choice of an appropriate model dimension. Consider the situation where there is a sequence of model structures of increasing dimension. For Example 4.1 this would simply mean that r is increased. With more free parameters in a model structure, a better fit will be obtained to the observed data. The important thing here is to investigate whether or not the improvement in the fit is significant. Consider first an ideal case. Assume that the data are noise-free or $N = \infty$, and that there is a model structure, say $\mathcal{M}^*$, such that, with suitable parameter values, it can describe the system exactly. Then the relationship of loss function to the model structure will be as shown in Figure 4.2.

In this ideal case the loss $V(\hat{\theta})$ will remain constant as long as $\mathcal{M}$ is 'at least as large as' $\mathcal{M}^*$. Note that for $N \to \infty$ and $\mathcal{M} \supset \mathcal{M}^*$ we have $2V(\hat{\theta})/N \to Ee^2(t)$. In the

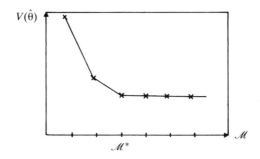

FIGURE 4.2 Minimal value of the loss function versus the model structure. Ideal case (noise-free data or $N \to \infty$). The model structure $\mathcal{M}^*$ corresponds to the true system.

practical case, however, when the data are noisy and $N < \infty$, the situation is somewhat different. Then the minimal loss $V(\hat{\theta})$ will decrease slowly with increasing $\mathcal{M}$, as illustrated in Figure 4.3.

The problem is thus to decide whether the decrease $\Delta V = V_1 - V_2$ is 'small' or not (see Figure 4.3). If it is small then the model structure $\mathcal{M}_1$ should be chosen; otherwise the larger model $\mathcal{M}_2$ should be preferred. For a normalized test quantity it seems reasonable to consider $\Delta V = (V_1 - V_2)/V_2$. Further, it can be argued that if the true system can be described within the smaller model structure $\mathcal{M}_1$, then the decrease ΔV should tend to zero as the number of data points, N, tends to infinity. One would therefore expect $N(V_1 - V_2)/V_2$ to be an appropriate test quantity. For 'small' values of this quantity, the structure $\mathcal{M}_1$ should be selected, while otherwise $\mathcal{M}_2$ should be chosen.

Statistical analysis

To get a more precise picture of how to apply the test introduced above, a statistical analysis is needed. This is given in the following lemma.

Lemma 4.4
Consider two model structures $\mathcal{M}_1$ and $\mathcal{M}_2$ given by

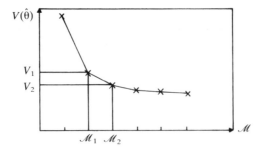

FIGURE 4.3 Minimal value of the loss function versus the model structure. Realistic case (noisy data and $N < \infty$).

Section 4.4   Determining the model dimension   73

$$\mathcal{M}_1: y(t) = \varphi_1^T(t)\theta_1 \tag{4.27}$$

$$\mathcal{M}_2: y(t) = \varphi_2^T(t)\theta_2 \tag{4.28}$$

where $n_1 = \dim \theta_1 < n_2 = \dim \theta_2$ and

$$\varphi_2(t) = \begin{pmatrix} \varphi_1(t) \\ \tilde{\varphi}_2(t) \end{pmatrix}$$

$$\theta_2 = \begin{pmatrix} \theta_1 \\ \tilde{\theta}_2 \end{pmatrix}$$

Assume that the data satisfy

$$\mathcal{S}: y(t) = \varphi_1^T(t)\theta_0 + e(t) \tag{4.29}$$

where $\{e(t)\}$ is a sequence of independent $\mathcal{N}(0, \lambda^2)$ distributed random variables. Let V_i denote the minimal value of the loss function as given by (4.8) for the model structure $\mathcal{M}_i$, $i = 1, 2$. Then the following results are true:

(i) $\dfrac{2V_2}{\lambda^2} \sim \chi^2(N - n_2)$

(ii) $\dfrac{2(V_1 - V_2)}{\lambda^2} \sim \chi^2(n_2 - n_1)$

(iii) $V_1 - V_2$ and V_2 are independent random variables

Proof. The proof relies heavily on Lemma A.29, which is rather technical. As in the proof of Lemma 4.2 we have

$$V_1 = \tfrac{1}{2} e^T P_1 e \qquad V_2 = \tfrac{1}{2} e^T P_2 e$$

where

$$P_1 = I - \Phi_1(\Phi_1^T\Phi_1)^{-1}\Phi_1^T$$

$$P_2 = I - \Phi_2(\Phi_2^T\Phi_2)^{-1}\Phi_2^T$$

We now apply Lemma A.29 (where F_1 corresponds to Φ_1 and F corresponds to Φ_2). Set $\bar{e} = Ue/\lambda$, where U is a specific orthogonal matrix which simultaneously diagonalizes P_2 and $P_1 - P_2$. Since U is orthogonal we have $\bar{e} \sim \mathcal{N}(0, I)$.

Next note that

$$2V_2/\lambda^2 = e^T P_2 e/\lambda^2 = \bar{e}^T \begin{pmatrix} I_{N-n_2} & 0 \\ 0 & 0_{n_2} \end{pmatrix} \bar{e}$$

$$2(V_1 - V_2)/\lambda^2 = e^T(P_1 - P_2)e/\lambda^2 = \bar{e}^T \begin{pmatrix} 0_{N-n_2} & 0 & 0 \\ 0 & I_{n_2-n_1} & 0 \\ 0 & 0 & 0_{n_1} \end{pmatrix} \bar{e}$$

The results now follow by invoking Lemma B.13. ∎

Remark. The $\chi^2(n)$ distribution and some of its properties are described in Section B.2 of Appendix B. ∎

Corollary. The quantity

$$f = \frac{V_1 - V_2}{n_2 - n_1} \frac{N - n_2}{V_2} \qquad (4.30)$$

is distributed $F(n_2 - n_1, N - n_2)$.

Proof. The result is immediate by definition of the F-distribution; see Section B.2. ∎

The variable f in (4.30) can be used to construct a statistical test for comparing the model structures $\mathcal{M}_1$ and $\mathcal{M}_2$. For each model structure the parameter estimates and the minimal value of the loss function can be determined. In this way the test quantity f can easily be computed. If N is large compared to n_2 the $\chi^2(n_2 - n_1)$ distribution can be used instead of $F(n_2 - n_1, N - n_2)$ (see Lemma B.14). Thus, if f is smaller than $\chi^2_\alpha(n_2 - n_1)$ then we can accept $\mathcal{M}_1$ at a significance level of α. Here $\chi^2_\alpha(n_2 - n_1)$ is the test threshold defined as follows. If x is a random variable which is $\chi^2(n_2 - n_1)$ distributed and $\alpha \in (0, 1)$ is a given number, then by definition $P(x > \chi^2_\alpha(n_2 - n_1)) = \alpha$. As a rough rule of thumb, $\mathcal{M}_1$ should be accepted when $f < (n_2 - n_1) + \sqrt{[8(n_2 - n_1)]}$ and rejected otherwise (see the end of Section B.2). This corresponds approximately to $\alpha = 0.05$. Thus, if $f \geq (n_2 - n_1) + \sqrt{[8(n_2 - n_1)]}$, the larger model structure $\mathcal{M}_2$ should be regarded as more appropriate.

It is not easy to make a strict analysis of how to select α, for the following reason. Note that

$$\alpha = P(H_0 \text{ is rejected when } H_0 \text{ is true})$$

where H_0 is the so-called null hypothesis,

H_0: The smaller model structure $\mathcal{M}_1$ is adequate

This observation offers some guidelines for choosing α. (A common choice is $\alpha = 0.05$.) However, it is not possible to calculate the risk for the other type of error:

$$P(H_0 \text{ is accepted when } H_0 \text{ is false})$$

which corresponds to a given value of α.

The determination of the model dimension is examined further in Chapter 11, where additional specific results are derived.

4.5 Computational aspects

This section considers some aspects of the numerical computation of the least squares estimate (4.7). The following topics are covered:

- Solution of the normal equations.
- Orthogonal triangularization.
- A recursive algorithm.

Normal equations

The first approach is to compute $\Phi^T\Phi$ and $\Phi^T Y$ and then solve the so-called normal equations (cf. (4.9)):

$$(\Phi^T\Phi)\theta = (\Phi^T Y) \tag{4.31}$$

This is indeed a straightforward approach but it is sensitive to numerical rounding errors. (See Example 4.6 below.)

Orthogonal triangularization

The second approach, orthogonal triangularization, is also known as the *QR method*. The basic idea is the following: Instead of directly 'solving' the original overdetermined linear system of equations

$$\Phi\theta = Y \tag{4.32}$$

it is multiplied on the left by an orthogonal matrix Q to give

$$Q\Phi\theta = QY \tag{4.33}$$

This will not affect the loss function (4.6), since

$$\|QY - Q\Phi\theta\|^2 = \|Q(Y - \Phi\theta)\|^2 = (Y - \Phi\theta)^T Q^T Q (Y - \Phi\theta)$$
$$= (Y - \Phi\theta)^T (Y - \Phi\theta) = \|Y - \Phi\theta\|^2$$

Suppose now that the orthogonal matrix Q can be chosen so that $Q\Phi$ has a 'convenient' form. In Appendix A, Lemmas A.16–A.20, it is shown how Q can be constructed to make $Q\Phi$ upper triangular. This means that

$$Q\Phi = \begin{pmatrix} R \\ 0 \end{pmatrix} \quad QY = \begin{pmatrix} z_1 \\ z_2 \end{pmatrix} \tag{4.34}$$

where R is a square, upper triangular matrix. The loss function then becomes

$$V(\theta) = \|Q\Phi\theta - QY\|^2 = \|R\theta - z_1\|^2 + \|z_2\|^2 \tag{4.35}$$

Assuming that R is nonsingular (which is equivalent to Φ being of full rank, and also to $\Phi^T\Phi$ being nonsingular), it is easy to see that $V(\theta)$ is minimized for θ given by

$$R\theta = z_1 \tag{4.36}$$

The minimum value is

$$\min_\theta V(\theta) = \|z_2\|^2 = z_2^T z_2 \tag{4.37}$$

76   Linear regression   Chapter 4

It is an easy task to solve the linear system (4.36) since R is a triangular matrix.

The QR method requires approximately twice as many computations as a direct solution to the normal equations. Its advantage is that it is much less sensitive to rounding errors. Assume that the relative errors in the data are of magnitude δ and that the precision in the computation is η. Then in order to avoid unreasonable errors in the result we should require $\eta < \delta^2$ for the normal equations, whereas $\eta < \delta$ is sufficient for the QR method. A further discussion of this point is given in Appendix A (see Section A.4).

The following example illustrates that the normal equations are more sensitive to rounding errors than the QR method.

Example 4.6 *Sensitivity of the normal equations*
Consider the system

$$\begin{pmatrix} 3 & 3-\delta \\ 4 & 4+\delta \end{pmatrix} x = \begin{pmatrix} -1 \\ 1 \end{pmatrix} \tag{4.38}$$

where δ is a small number. Since the column vectors of the matrix in (4.38) are almost parallel, one can expect that the solution is sensitive to small changes in the coefficients. The exact solution is

$$x = \begin{pmatrix} -1/\delta \\ 1/\delta \end{pmatrix} \tag{4.39}$$

The normal equations are easily found to be

$$\begin{pmatrix} 25 & 25+\delta \\ 25+\delta & 25+2\delta+2\delta^2 \end{pmatrix} x = \begin{pmatrix} 1 \\ 1+2\delta \end{pmatrix} \tag{4.40}$$

If a QR method is applied to (4.38) with Q constructed as a Householder transformation (see Lemma A.18), then

$$Q = \frac{1}{5}\begin{pmatrix} 3 & 4 \\ 4 & -3 \end{pmatrix}$$

Applying the QR method we get the following triangular system

$$\begin{pmatrix} 5 & 5+\delta/5 \\ 0 & -7\delta/5 \end{pmatrix} x = \begin{pmatrix} 1/5 \\ -7/5 \end{pmatrix} \tag{4.41}$$

Assume now that the equations are solved on a computer using finite arithmetic and that due to truncation errors $\delta^2 = 0$. The 'QR equations' (4.41) are then not affected and the solution is still given by (4.39). However, after truncation the normal equations (4.40) become

$$\begin{pmatrix} 25 & 25+\delta \\ 25+\delta & 25+2\delta \end{pmatrix} x = \begin{pmatrix} 1 \\ 1+2\delta \end{pmatrix} \tag{4.42}$$

The solution to (4.42) is readily found to be

$$x = \begin{pmatrix} 49/\delta + 2 \\ -49/\delta \end{pmatrix} \quad (4.43)$$

Note that this solution differs considerably from the true solution (4.39) to the original problem. ∎

Recursive algorithm

A third approach is to use a recursive algorithm. The mathematical description and analysis of such algorithms are given in Chapter 9. The idea is to rewrite the estimate (4.7) in the following form:

$$\hat{\theta}(t) = \hat{\theta}(t-1) + K(t)[y(t) - \varphi^T(t)\hat{\theta}(t-1)] \quad (4.44)$$

Here $\hat{\theta}(t)$ denotes the estimate based on t equations (t rows in Φ). The term $y(t) - \varphi^T(t)\hat{\theta}(t-1)$ describes how well the 'measurement' $y(t)$ can be explained by using the parameter vector $\hat{\theta}(t-1)$ obtained from the previous data. Further, in (4.44), $K(t)$ is a gain vector. Complement C4.2 shows among other things how a linear regression model can be updated when new information becomes available. This is conceptually and algebraically very close to a recursive algorithm.

Summary

Linear regression models have been defined and the least squares estimates of their unknown parameters were derived (Lemma 4.1). The statistical properties of the least squares parameter estimates were examined (Lemma 4.2). The estimates were shown to be unbiased and an expression for their covariance matrix was given. It was a crucial assumption that the regression variables were deterministic and known functions. Next the analysis was extended to general linear unbiased estimates. In particular Lemma 4.3 derived the estimate in this class with the smallest covariance matrix of the parameter estimates. Finally, Lemma 4.4 provided a systematic way to determine which one of a number of competitive model structures should be chosen. It was also pointed out that orthogonal triangularization (the QR method) is a numerically sound way to compute a least squares estimate.

Problems

Problem 4.1 *A linear trend model I*
Consider the linear regression model

$$y(t) = a + bt + \varepsilon(t)$$

Find the least squares estimates of a and b. Treat the following cases.

78   Linear regression   Chapter 4

(a) First consider the following situations:
   (i) The data are $y(1), y(2), \ldots, y(N)$. Set
   $$S_0 = \sum_{t=1}^{N} y(t) \qquad S_1 = \sum_{t=1}^{N} ty(t)$$
   (ii) The data are $y(-N), y(-N+1), \ldots, y(N)$. Set
   $$S_0 = \sum_{t=-N}^{N} y(t) \qquad S_1 = \sum_{t=-N}^{N} ty(t)$$
   Hint. $\sum_{t=1}^{N} t = N(N+1)/2$; $\sum_{t=1}^{N} t^2 = N(N+1)(2N+1)/6$.

(b) Next suppose that the parameter a is first estimated as
   $$\hat{a} = \frac{1}{N} S_0 \quad \text{(case (i))}$$
   $$\hat{a} = \frac{1}{2N+1} S_0 \quad \text{(case (ii))}$$
   Then b is estimated (by the least squares method) in the model structure
   $$y(t) - \hat{a} = bt + \varepsilon(t)$$
   What will $\hat{b}$ become? Compare with (a).

In parts (c)–(e) consider the situation in (a) (i) and assume that the data satisfy
$$y(t) = a + bt + e(t)$$
where $e(t)$ is white noise of variance λ^2.

(c) Find the variance of the quantity $s(t) = \hat{a} + \hat{b}t$. What is the value of this variance for $t = 1$ and for $t = N$? What is its minimal value with respect to t?

(d) Write the covariance matrix of $\hat{\theta} = (\hat{a} \ \hat{b})^T$ in the form
$$P = \text{cov}(\hat{\theta}) = \begin{pmatrix} \sigma_1^2 & \varrho\sigma_1\sigma_2 \\ \varrho\sigma_1\sigma_2 & \sigma_2^2 \end{pmatrix}$$
Find the asymptotic value of the *correlation coefficient* ϱ when N tends to infinity.

(e) Introduce the *concentration ellipsoid*
$$(\hat{\theta} - \theta)^T P^{-1} (\hat{\theta} - \theta) = \zeta$$
(Roughly, vectors inside the ellipsoid are likely while vectors outside are unlikely, provided ζ is given a value $\sim n$. In fact,
$$E(\hat{\theta} - \theta)^T P^{-1}(\hat{\theta} - \theta) = \text{tr } P^{-1} E(\hat{\theta} - \theta)(\hat{\theta} - \theta)^T$$
$$= \text{tr } P^{-1}\{(\theta_0 - \theta)(\theta_0 - \theta)^T + P\}$$
$$= n + (\theta_0 - \theta)^T P^{-1}(\theta_0 - \theta) \approx n, \text{ if } \theta_0 - \theta \text{ small enough.}$$
If $\hat{\theta}$ is Gaussian distributed, $\hat{\theta} \sim \mathcal{N}(\theta_0, P)$, then $(\hat{\theta} - \theta_0)^T P^{-1}(\hat{\theta} - \theta_0) \sim \chi^2(n)$, see Lemma B.12). Find and plot the concentration ellipsoids when $\lambda^2 = 0.1$, $\zeta = 2$ and the two cases (i) $N = 3$, (ii) $N = 8$.

Problem 4.2 *A linear trend model II*
Consider the linear regression model

$$\mathcal{M}_1: y(t) = a + bt + \varepsilon_1(t)$$

Assume that the aim is to estimate the parameter b. One alternative is of course to use linear regression to estimate both a and b.

Another alternative is to work with differenced data. For this purpose introduce

$$z(t) = y(t) - y(t-1)$$

Then $\mathcal{M}_1$ gives the new model structure

$$\mathcal{M}_2: z(t) = b + \varepsilon_2(t)$$

(where $\varepsilon_2(t) = \varepsilon_1(t) - \varepsilon_1(t-1)$). Linear regression can be applied to $\mathcal{M}_2$ for estimating b only (treating $\varepsilon_2(t)$ as the equation error).

Compare the variances of the estimate of b obtained in the two cases. Assume that the data are collected at times $t = 1, \ldots, N$ and obey

$$\mathcal{S}: y(t) = a_0 + b_0 t + e(t)$$

where $e(t)$ is white noise of zero mean and unit variance.

Problem 4.3 *The loss function associated with the Markov estimate*
Show that the Markov estimate (4.17), (4.18) minimizes the following criterion:

$$V(\theta) = e^T R^{-1} e \quad e = Y - \Phi\theta$$

Problem 4.4 *Linear regression with missing data*
Consider the regression equation

$$Y = \Phi\theta + e$$

where

$$Y = \begin{pmatrix} y_1 \\ y_2 \end{pmatrix} \quad \Phi = \begin{pmatrix} \Phi_1 \\ \Phi_2 \end{pmatrix} \quad e = \begin{pmatrix} e_1 \\ e_2 \end{pmatrix}$$

and

$$E e_1 = 0 \quad E e_1 e_1^T = I$$

Assume that y_1, Φ_1 and Φ_2 are available but y_2 is missing. Derive the least squares estimates of θ and y_2 defined as

$$\hat{\theta}, \hat{y}_2 = \arg\min_{\theta, y_2} (Y - \Phi\theta)^T (Y - \Phi\theta)$$

Problem 4.5 *Ill-conditioning of the normal equations associated with polynomial trend models*
Consider the normal equations (4.31) corresponding to the polynomial trend model of Example 4.1. Show that these equations are ill-conditioned (see Section A.4 of Appendix A). More exactly, show that the condition number of the matrix $(\Phi^T \Phi)$ satisfies

$$\text{cond}(\Phi^T\Phi) \geq O(N^{2r}/(2r+1)) \quad \text{for } N \text{ large}$$

where r denotes the polynomial degree.

Hint. Use the relations

$$\lambda_{\max}(A) \geq \max_i (a_{ii})$$

$$\lambda_{\min}(A) \leq \min_i (a_{ii})$$

which hold for a symmetric matrix $A = [a_{ij}]$.

Problem 4.6 *Fourier harmonic decomposition as a special case of regression analysis*
Consider a regression model of the form (4.1) where

$$\theta = (a_1 \quad \ldots \quad a_n \quad b_1 \quad \ldots \quad b_n)^T$$

$$\varphi(t) = (\cos \omega_1 t \quad \ldots \quad \cos \omega_n t \quad \sin \omega_1 t \quad \ldots \quad \sin \omega_n t)^T$$

$$\omega_i = \frac{2\pi}{N} i \quad N = \text{number of data points}$$

$$n \leq \left[\frac{N}{2}\right] \quad \text{(the integer part of } N/2\text{)}$$

Such a regression model is a possible way of presenting the following harmonic decomposition model:

$$y(t) = \sum_{k=1}^{n} (a_k \cos \omega_k t + b_k \sin \omega_k t) \quad t = 1, \ldots, N$$

Show that the least squares estimates of $\{a_k\}$ and $\{b_k\}$ are equal to the Fourier coefficients

$$a_k = \frac{2}{N} \sum_{t=1}^{N} y(t) \cos \omega_k t$$
$$b_k = \frac{2}{N} \sum_{t=1}^{N} y(t) \sin \omega_k t \quad (k = 1, \ldots, n)$$

Hint. First show that the following equalities hold:

$$\sum_{t=1}^{N} \cos \omega_k t \cos \omega_p t = \frac{N}{2} \delta_{k,p}$$

$$\sum_{t=1}^{N} \sin \omega_k t \sin \omega_p t = \frac{N}{2} \delta_{k,p}$$

$$\sum_{t=1}^{N} \cos \omega_k t \sin \omega_p t = 0$$

Problem 4.7 *Minimum points and normal equations when Φ does not have full rank*
As a simple illustration of the case when Φ does not have full rank consider

$$\Phi = \begin{pmatrix} 1 & 1 \\ \vdots & \vdots \\ 1 & 1 \end{pmatrix}$$

Find all the minimum points of the loss function $V(\theta) = \|Y - \Phi\theta\|^2$. Compare them with the set of solutions to the normal equations

$$(\Phi^T\Phi)\theta = \Phi^T Y$$

Also calculate the pseudoinverse of Φ (see Section A.2) and the minimum norm solution to the normal equations.

Problem 4.8 *Optimal input design for gain estimation*
Consider the following straight-line (or static gain) equation:

$$y(t) = \theta u(t) + e(t) \qquad Ee(t)e(s) = \lambda^2 \delta_{t,s}$$

which is the simplest case of a regression. The parameter θ is estimated from N data points $\{y(t), u(t)\}$ using the least squares (LS) method. The variance of the LS estimate $\hat{\theta}$ of θ is given (see Lemma 4.2) by

$$\sigma^2 \triangleq \mathrm{var}(\hat{\theta}) = \lambda^2 \bigg/ \sum_{t=1}^{N} u^2(t)$$

Determine the optimal sequence $\{u(1), \ldots, u(N)\}$ which minimizes σ^2 under the constraint $|u(t)| \leq \beta$. Comment on the solution obtained.

Problem 4.9 *Regularization of a linear system of equations*
When a linear system of equations is almost singular it may be regularized to make it less sensitive to numerical errors. This means that the identity matrix multiplied by a small number is added to the original system matrix. Hence, instead of solving

$$Ax = b \tag{i}$$

the modified system of equations

$$(A + \delta I)x = b \tag{ii}$$

is solved.

Assume that the matrix A is symmetric and positive semidefinite (and hence singular). Then it can be factorized as

$$A = BB^T \quad \text{where } B \text{ is } (n|p), \quad \text{rank } B = p < n \tag{iii}$$

To guarantee that a solution exists it must be assumed that

$$b = Bd \tag{iv}$$

for some $(p|1)$ vector d.
Derive and compare the following solutions:

(a) The minimum-norm least squares solution, $x_1 = A^\dagger b$, where $A^\dagger$ denotes the pseudo-inverse of A.
  Hint. Use properties of pseudoinverses; see Section A.2 of Appendix A.

(b) Writing the equations as $B(B^T x - d) = 0$ we may choose to drop B and get the system $B^T x = d$, which is underdetermined. Find the minimum norm least squares solution of this system, $x_2 = (B^T)^\dagger d$.

(c) The regularized solution is $x_3 = (A + \delta I)^{-1} b$. What happens when δ tends to zero?

Problem 4.10 *Conditions for least squares estimates to be BLUE*

For the linear regression model

$$y = \Phi\theta + e \qquad Ee = 0 \qquad Eee^T = R > 0$$

$$\text{rank } \Phi = \dim \theta$$

show that (i) and (ii) below are two equivalent sets of necessary and sufficient conditions for the least squares estimate of θ to be the best linear unbiased estimate (BLUE).

$$\Phi^T R^{-1}[I - \Phi(\Phi^T \Phi)^{-1}\Phi^T] = 0 \tag{i}$$

$$R\Phi = \Phi F \quad \text{for some nonsingular matrix } F \tag{ii}$$

Give an example of a covariance matrix R which satisfies (i), or equivalently (ii).

Hint. For a nontrivial example of R satisfying (i), consider $R = I + \alpha\varphi\varphi^T$ where φ is any column of Φ and α is such that $R > 0$.

Problem 4.11 *Comparison of the covariance matrices of the least squares and Markov estimates*

Consider the linear regression model of Problem 4.10. Let P_{LS} and P_M denote the covariance matrices of the least squares and the Markov estimates respectively of the parameter vector θ (see (4.16) and (4.19)). It follows from Lemma 4.3 that

$$P_{LS} \geq P_M$$

Provide a simple direct proof of this inequality. Use that proof to obtain condition (i) of Problem 4.10 which is necessary and sufficient for the least squares estimate (LSE) to be BLUE.

Hint. Use some calculations similar to proof B of Lemma 4.3.

Problem 4.12 *The least squares estimate of the mean is BLUE asymptotically*

The least squares estimate of a in

$$y(t) = a + e(t) \qquad t = 1, \ldots, N$$

where the vector of measurement errors $[e(1) \quad \ldots \quad e(N)]^T$ has zero mean and covariance R, has variance given by

$$P_{LS} = (\Phi^T \Phi)^{-1} \Phi^T R \Phi (\Phi^T \Phi)^{-1} \qquad \Phi^T = [1 \quad \ldots \quad 1]$$

(see (4.16)). The variance of the BLUE is

$$P_{BLUE} = (\Phi^T R^{-1} \Phi)^{-1}$$

Let the process $\{e(t)\}$ be stationary with spectral density $\phi_e(\omega) > 0$ for $\omega \in (-\pi, \pi)$ and covariance function $r(k) = Ee(t)e(t - k)$ decaying exponentially to zero as $k \to \infty$. Then show that

$$\lim_{N\to\infty} NP_{\text{LS}} = \lim_{N\to\infty} NP_{\text{BLUE}} = 2\pi\phi_e(0)$$

Hint. For evaluating $\lim_{N\to\infty} NP_{\text{BLUE}}$, first use the fact that $\lambda_{\min}(R) > 0$ (see e.g. Complement C5.2) to establish that

$$\|R^{-1}\|_\infty = \max_k \sum_{p=1}^N |(R^{-1})_{kp}| \leq C\sqrt{N}$$

for some constant C.

Remark. The result above follows from a more general result first presented by Grenander and Rosenblatt (1956). The latter result states, essentially, that the least squares estimate of the regression parameters is asymptotically (for $N \to \infty$) BLUE provided the spectrum of the residuals is constant over all the elements of the 'spectrum' of the regression functions.

Bibliographical notes

There are many books in the statistical literature which treat linear regression. One reason for this is that linear regressions constitute a 'simple' yet fundamental class of models. See Rao (1973), Dhrymes (1978), Draper and Smith (1981), Weisberg (1985), Åström (1968), and Wetherill *et al.* (1986) for further reading. Nonlinear regression is treated by Jennrich (1969) and Ratkowsky (1983), for example. The testing of statistical hypotheses is treated in depth by Lehmann (1986).

For the differentiation (4.26), see for example Rogers (1980), who also gives a number of related results on differentiation with respect to a matrix.

Complement C4.1
Best linear unbiased estimation under linear constraints

Consider the regression equation

$$Y = \Phi\theta + e \qquad Ee = 0 \qquad Ee e^T = R > 0$$

and the following linear restrictions on the unknown parameter vector θ:

$$A\theta = b \qquad A \text{ of dimension } (m|n\theta)$$

$$\text{rank } A = m < n\theta$$

The problem is to find the BLUE of θ. We present two approaches to solve this problem.

Approach A

This approach is due to Dhrymes (1978), Rao (1973) and others.
The Lagrangian function associated with the above problem is

84 *Linear regression* Chapter 4

$$F(\theta, \alpha) = \tfrac{1}{2}(Y - \Phi\theta)^T R^{-1}(Y - \Phi\theta) + \alpha^T(A\theta - b)$$

(see Problem 4.3) where α denotes the vector of Lagrange multipliers. By equating the derivatives of F with respect to θ and α to zero, we get

$$-\Phi^T R^{-1}(Y - \Phi\theta) + A^T\alpha = 0 \qquad \text{(C4.1.1)}$$

$$A\theta = b \qquad \text{(C4.1.2)}$$

Let $Q \triangleq (\Phi^T R^{-1} \Phi)^{-1}$. Multiplication of (C4.1.1) on the left by AQ gives

$$AQA^T\alpha = AQ\Phi^T R^{-1} Y - A\theta = A\hat{\theta} - b$$

where $\hat{\theta} = Q\Phi^T R^{-1} Y$ is the unconstrained BLUE (see (4.17), (4.18)). Thus, since AQA^T is positive definite,

$$\alpha = (AQA^T)^{-1}(A\hat{\theta} - b)$$

which inserted into (C4.1.1) gives the constrained BLUE

$$\tilde{\theta} = Q\Phi^T R^{-1} Y - QA^T\alpha$$

or

$$\boxed{\tilde{\theta} = \hat{\theta} - QA^T(AQA^T)^{-1}(A\hat{\theta} - b)} \qquad \text{(C4.1.3)}$$

Approach B

The problem of determining $\tilde{\theta}$ can be stated in the following way: find the BLUE of θ in the following regression equation with singular residual covariance matrix

$$\begin{pmatrix} Y \\ b \end{pmatrix} = \begin{pmatrix} \Phi \\ A \end{pmatrix} \theta + \begin{pmatrix} e \\ 0 \end{pmatrix} \qquad \text{(C4.1.4)}$$

The covariance matrix of the residuals is

$$\begin{pmatrix} R & 0 \\ 0 & \varepsilon I \end{pmatrix}$$

with $\varepsilon = 0$. First take $\varepsilon > 0$ (which makes the matrix nonsingular) and then apply the standard BLUE result (4.18). Next suppose that ε tends to zero. Note that it would also be possible to apply the results for a BLUE with singular residual covariance matrix as developed in Complement C4.3. The BLUE for $\varepsilon > 0$ is given by

$$\theta_\varepsilon = \left\{ (\Phi^T \; A^T) \begin{pmatrix} R^{-1} & 0 \\ 0 & \tfrac{1}{\varepsilon}I \end{pmatrix} \begin{pmatrix} \Phi \\ A \end{pmatrix} \right\}^{-1} \left\{ (\Phi^T \; A^T) \begin{pmatrix} R^{-1} & 0 \\ 0 & \tfrac{1}{\varepsilon}I \end{pmatrix} \begin{pmatrix} Y \\ b \end{pmatrix} \right\}$$

Provided the limit exists, we can write

$$\tilde{\theta} = \lim_{\varepsilon \to 0} \theta_\varepsilon$$

With $Q = (\Phi^T R^{-1} \Phi)^{-1}$ the expression for θ_ε becomes

$$\theta_\varepsilon = \left(Q^{-1} + \frac{1}{\varepsilon} A^T A\right)^{-1} \left(\Phi^T R^{-1} Y + \frac{1}{\varepsilon} A^T b\right)$$

Next note that

$$\left(Q^{-1} + \frac{1}{\varepsilon} A^T A\right)^{-1} = Q - Q \frac{1}{\varepsilon} A^T \left(I + \frac{1}{\varepsilon} A Q A^T\right)^{-1} A Q$$

(see Lemma A.1). Thus

$$\theta_\varepsilon = \hat\theta - Q A^T(\varepsilon I + A Q A^T)^{-1} A \hat\theta + \frac{1}{\varepsilon} Q A^T b$$

$$- \frac{1}{\varepsilon} Q A^T \left(I + \frac{1}{\varepsilon} A Q A^T\right)^{-1} A Q \frac{1}{\varepsilon} A^T b$$

$$= \hat\theta - Q A^T(\varepsilon I + A Q A^T)^{-1} A \hat\theta$$

$$+ \frac{1}{\varepsilon} Q A^T \left(I + \frac{1}{\varepsilon} A Q A^T\right)^{-1} \left[\left(I + \frac{1}{\varepsilon} A Q A^T\right) - \frac{1}{\varepsilon} A Q A^T\right] b$$

$$= \hat\theta - Q A^T(\varepsilon I + A Q A^T)^{-1}(A\hat\theta - b)$$

$$\to \hat\theta - Q A^T(A Q A^T)^{-1}(A\hat\theta - b) \quad \text{as} \quad \varepsilon \to 0.$$

This is precisely the result (C4.1.3).

The covariance matrix of the constrained BLUE $\tilde\theta$ can be readily evaluated. From

$$\tilde\theta - \theta = (\hat\theta - \theta) - Q A^T (A Q A^T)^{-1} A (\hat\theta - \theta)$$

$$= [I - Q A^T (A Q A^T)^{-1} A](\hat\theta - \theta)$$

it follows (recall that $\text{cov}(\hat\theta - \theta) = Q$) that

$$\text{cov}(\tilde\theta - \theta) = [I - Q A^T (A Q A^T)^{-1} A] Q [I - A^T (A Q A^T)^{-1} A Q]$$

$$\boxed{\text{cov}(\tilde\theta - \theta) = Q - Q A^T (A Q A^T)^{-1} A Q}$$

Note that

$$\text{cov}(\tilde\theta - \theta) \leq \text{cov}(\hat\theta - \theta)$$

which can be seen as an illustration of the 'parsimony principle' (see Complement C11.1). That is to say, taking into account the constraints on θ leads to more accurate estimates. Finally note that $\tilde\theta$ obeys the constraints

$$A\tilde\theta = A\hat\theta - A Q A^T (A Q A^T)^{-1}(A\hat\theta - b) = b$$

as expected. As a consequence the matrix $\text{cov}(\tilde\theta - \theta)$ must be singular. This is readily seen since $A \, \text{cov}(\tilde\theta - \theta) = 0$.

Complement C4.2
Updating the parameter estimates in linear regression models

New measurements

Consider the regression equation

$$\tilde{Y} = \tilde{\Phi}\theta + \varepsilon \tag{C4.2.1}$$

where

$$\tilde{Y} = \begin{pmatrix} Y \\ \bar{\theta} \end{pmatrix} \begin{matrix} \}ny \\ \}n\theta \end{matrix} \qquad \tilde{\Phi} = \begin{pmatrix} \Phi \\ I \end{pmatrix} \begin{matrix} \}ny \\ \}n\theta \end{matrix} \qquad \varepsilon = \begin{pmatrix} \varepsilon_1 \\ \varepsilon_2 \end{pmatrix} \begin{matrix} \}ny \\ \}n\theta \end{matrix}$$

and

$$E\varepsilon = 0 \qquad E\varepsilon\varepsilon^T = \begin{pmatrix} S & 0 \\ 0 & P \end{pmatrix} \begin{matrix} \}ny \\ \}n\theta \end{matrix} > 0$$

The equation $\bar{\theta} = \theta + \varepsilon_2$ (the last block of (C4.2.1)) can be viewed as 'information' obtained in some previous estimation stage. The remaining equation, viz. $Y = \Phi\theta + \varepsilon_1$, is the information provided by a new set of measurements. The problem is to find the BLUE of θ by properly updating $\bar{\theta}$.

The BLUE of θ is given by

$$\begin{aligned}
\hat{\theta} &= \left[\tilde{\Phi}^T \begin{pmatrix} S^{-1} & 0 \\ 0 & P^{-1} \end{pmatrix} \tilde{\Phi} \right]^{-1} \left[\tilde{\Phi}^T \begin{pmatrix} S^{-1} & 0 \\ 0 & P^{-1} \end{pmatrix} \tilde{Y} \right] \\
&= (\Phi^T S^{-1} \Phi + P^{-1})^{-1} (\Phi^T S^{-1} Y + P^{-1}\bar{\theta}) \\
&= \bar{\theta} + (\Phi^T S^{-1} \Phi + P^{-1})^{-1} \Phi^T S^{-1} [Y - \Phi\bar{\theta}]
\end{aligned} \tag{C4.2.2}$$

(see (4.18)). Since

$$(\Phi^T S^{-1} \Phi + P^{-1})^{-1} = P - P\Phi^T (S + \Phi P \Phi^T)^{-1} \Phi P$$

(cf. Lemma A.1), it follows that

$$\begin{aligned}
(\Phi^T S^{-1} \Phi + P^{-1})^{-1} \Phi^T S^{-1} &= P\Phi^T (S + \Phi P \Phi^T)^{-1} [(S + \Phi P \Phi^T) \\
&\quad - \Phi P \Phi^T] S^{-1} \\
&= P\Phi^T (S + \Phi P \Phi^T)^{-1}
\end{aligned} \tag{C4.2.3}$$

Inserting this into (C4.2.2) gives

$$\boxed{\hat{\theta} = \bar{\theta} + P\Phi^T (S + \Phi P \Phi^T)^{-1} (Y - \Phi\bar{\theta})} \tag{C4.2.4}$$

This expression for $\hat{\theta}$ is computationally more advantageous than (C4.2.2) since in general $ny < n\theta$ (quite often $ny = 1$). Thus the matrix inverse to be computed in (C4.2.4) in general has a smaller dimension than that in (C4.2.2).

The covariance matrix of $\hat{\theta}$ is given (cf. (4.19)) by

$$\hat{P} = \left[\tilde{\Phi}^T \begin{pmatrix} S^{-1} & 0 \\ 0 & P^{-1} \end{pmatrix} \tilde{\Phi} \right]^{-1} = (\Phi^T S^{-1} \Phi + P^{-1})^{-1}$$

which can be rewritten as in the preceding derivation to give

$$\hat{P} = P - P\Phi^T(S + \Phi P \Phi^T)^{-1} \Phi P \tag{C4.2.5}$$

Note that it follows explicitly that $\hat{P} \leq P$. This means that the accuracy of $\hat{\theta}$ is better than that of $\bar{\theta}$, which is very natural. The use of the additional information carried by Y should indeed improve the accuracy.

Decreasing dimension

Consider the regression model

$$Y = \Phi\theta + \varepsilon \qquad E\varepsilon = 0 \qquad E\varepsilon\varepsilon^T = R > 0 \tag{C4.2.6}$$

for which the BLUE of θ,

$$\hat{\theta} = Q\Phi^T R^{-1} Y \qquad Q \triangleq (\Phi^T R^{-1} \Phi)^{-1}$$

has been computed. In some situations it may be required to compute the BLUE of the parameters of a *reduced-dimension* regression, by 'updating' or modifying $\hat{\theta}$. Specifically, assume that after computing $\hat{\theta}$ it was realized that the regressor corresponding to the last (say) column of Φ does not influence Y. The problem is to determine the BLUE of the parameters in a regression where the last column of Φ was eliminated. Furthermore this should be done in a numerically efficient manner, presumably by 'updating' $\hat{\theta}$.

This problem can be stated equivalently as follows: find the BLUE $\tilde{\theta}$ of θ in (C4.2.6) under the constraint

$$A\theta = b \qquad A = (0 \quad \ldots \quad 0 \quad 1) \qquad b = 0$$

Making use of (C4.1.3),

$$\tilde{\theta} = [I - QA^T(AQA^T)^{-1}A]\hat{\theta} \tag{C4.2.7}$$

Let $(\psi_1 \ldots \psi_{n\theta})^T$ denote the last column of Q, and let $\hat{\theta}_i$ be the ith component of $\hat{\theta}$. Then, from (C4.2.7),

$$\tilde{\theta} = \begin{pmatrix} 1 & 0 & \ldots & 0 & -\psi_1/\psi_{n\theta} \\ & \ddots & & \vdots & \vdots \\ & & & 1 & -\psi_{n\theta-1}/\psi_{n\theta} \\ & & & 0 & 0 \end{pmatrix} \hat{\theta} = \begin{pmatrix} \hat{\theta}_1 - \frac{\psi_1}{\psi_{n\theta}}\hat{\theta}_{n\theta} \\ \vdots \\ \hat{\theta}_{n\theta-1} - \frac{\psi_{n\theta-1}}{\psi_{n\theta}}\hat{\theta}_{n\theta} \\ 0 \end{pmatrix} \tag{C4.2.8}$$

The BLUE of the reduced-order regression parameters are given by the first ($n\theta - 1$) components of the above vector.

Note that the result (C4.2.8) is closely related to the Levinson–Durbin recursion discussed in Complement C8.2. That recursion corresponds to a special structure of the model (more exactly, of the Φ matrix), which leads to very simple expressions for $\{\psi_i\}$.

If more components of $\hat{\theta}$ than the last one should be constrained to zero, the above procedure can be repeated. The variables ψ_j must of course then be redefined for each reduction in the dimension. One can alternatively proceed by taking

$$A = \begin{bmatrix} 0 & I \end{bmatrix} \quad b = 0$$

where A is $(m|n\theta)$ if the last m components of θ should be constrained to zero. By partitioning Q and $\hat{\theta}$ as

$$Q = \begin{pmatrix} Q_{11} & Q_{12} \\ Q_{12}^T & Q_{22} \end{pmatrix} \begin{matrix} \}n\theta - m \\ \}m \end{matrix} \qquad \hat{\theta} = \begin{pmatrix} \hat{\theta}_1 \\ \hat{\theta}_2 \end{pmatrix} \begin{matrix} \}n\theta - m \\ \}m \end{matrix}$$

equation (C4.2.7) becomes

$$\tilde{\theta} = \begin{pmatrix} \hat{\theta}_1 \\ \hat{\theta}_2 \end{pmatrix} - \begin{pmatrix} Q_{12} \\ Q_{22} \end{pmatrix} Q_{22}^{-1} \hat{\theta}_2$$

or

$$\tilde{\theta} = \begin{pmatrix} \hat{\theta}_1 - Q_{12} Q_{22}^{-1} \hat{\theta}_2 \\ 0 \end{pmatrix} \tag{C4.2.9}$$

Complement C4.3
Best linear unbiased estimates for linear regression models with possibly singular residual covariance matrix

Consider the regression model

$$Y = \Phi\theta + e \qquad Eee^T = R \tag{C4.3.1}$$

(see (4.11), (4.15)). It is assumed that

$$\text{rank } \Phi = \dim \theta = n\theta \tag{C4.3.2}$$

However, no assumption is made on rank R. This represents a substantial generalization over the case treated previously where it was assumed that $R > 0$. Note that assumption (C4.3.2) is needed for unbiased estimates of θ to exist. If (C4.3.2) is relaxed then unbiased estimates exist only for *certain* linear combinations of the unknown parameter vector θ (see Werner, 1985, and the references therein for treatment of the case where neither rank Φ nor rank R are constrained in any way).

The main result of this complement is the following.

Lemma C4.3.1
Under the assumptions introduced above the BLUE of θ is given by

$$\theta^* = [\Phi^T(R + \Phi\Phi^T)^\dagger \Phi]^{-1} \Phi^T(R + \Phi\Phi^T)^\dagger Y \tag{C4.3.3}$$

where $A^\dagger$ denotes the pseudoinverse of the matrix A. (see Section A.2 or Rao (1973) for definition, properties and computation of the pseudoinverse.)

Proof. First it has to be shown that the inverse appearing in (C4.3.3) exists. Let

$$\widetilde{R} = R + \Phi\Phi^T \tag{C4.3.4}$$

and for any $n\theta$-vector α let

$$\beta = \Phi\alpha \tag{C4.3.5}$$

We have to show that

$$\alpha^T \Phi^T \widetilde{R}^\dagger \Phi \alpha = 0 \tag{C4.3.6}$$

implies $\alpha = 0$. The following series of implications can be readily verified:

$$(C4.3.6) \Rightarrow \beta^T \widetilde{R}^\dagger \beta = 0 \Rightarrow \beta^T \widetilde{R}^\dagger = 0 \Rightarrow \beta^T \widetilde{R}^\dagger \widetilde{R} = 0 \Rightarrow \widetilde{R}^\dagger \widetilde{R} \beta = 0 \Rightarrow \widetilde{R}\beta = 0$$

$$\Rightarrow \beta^T(R + \Phi\Phi^T)\beta = 0 \Rightarrow \Phi^T\beta = 0 \Rightarrow (\Phi^T\Phi)\alpha = 0 \Rightarrow \alpha = 0$$

which proves that only $\alpha = 0$ satisfies (C4.3.6) and thus that $\Phi^T\widetilde{R}^\dagger\Phi$ is a positive definite matrix. Here we have repeatedly used the properties of pseudoinverses given in Lemma A.15.

Next observe that θ^* is an unbiased estimate of θ. It remains to prove that θ^* has the smallest covariance matrix in the class of linear unbiased estimators of θ:

$$\hat{\theta} = Z^T Y$$

where by unbiasedness $Z^T\Phi = I$ (see (4.17), (4.20)). The covariance matrix of θ^* is given by

$$\text{cov}(\theta^*) = (\Phi^T\widetilde{R}^\dagger\Phi)^{-1}\Phi^T\widetilde{R}^\dagger R \widetilde{R}^\dagger \Phi (\Phi^T\widetilde{R}^\dagger\Phi)^{-1}$$

$$= (\Phi^T\widetilde{R}^\dagger\Phi)^{-1}\Phi^T\widetilde{R}^\dagger(\widetilde{R} - \Phi\Phi^T)\widetilde{R}^\dagger\Phi(\Phi^T\widetilde{R}^\dagger\Phi)^{-1}$$

$$= (\Phi^T\widetilde{R}^\dagger\Phi)^{-1} - I$$

Since $\text{cov}(\hat{\theta}) = Z^T R Z$ (see (4.21)), it remains only to prove that

$$Z^T R Z + I - (\Phi^T \widetilde{R}^\dagger \Phi)^{-1} \geq 0 \tag{C4.3.7}$$

Using the constraint $Z^T \Phi = I$ on the matrix Z, the inequality above can be rewritten as

$$Z^T R Z + Z^T \Phi [I - (\Phi^T \widetilde{R}^\dagger \Phi)^{-1}] \Phi^T Z = Z^T [\widetilde{R} - \Phi(\Phi^T \widetilde{R}^\dagger \Phi)^{-1} \Phi^T] Z \geq 0 \tag{C4.3.8}$$

The following identity can be readily verified:

$$\tilde{R} - \Phi(\Phi^T\tilde{R}^\dagger\Phi)^{-1}\Phi^T = [\tilde{R} - \Phi(\Phi^T\tilde{R}^\dagger\Phi)^{-1}\Phi^T]\tilde{R}^\dagger[\tilde{R} - \Phi(\Phi^T\tilde{R}^\dagger\Phi)^{-1}\Phi^T]$$
$$- (I - \tilde{R}\tilde{R}^\dagger)\Phi(\Phi^T\tilde{R}^\dagger\Phi)^{-1}\Phi^T \qquad \text{(C4.3.9)}$$
$$- \Phi(\Phi^T\tilde{R}^\dagger\Phi)^{-1}\Phi^T(I - \tilde{R}^\dagger\tilde{R})$$

Next we show that

$$(I - \tilde{R}\tilde{R}^\dagger)\Phi = 0 \qquad \text{(C4.3.10)}$$

Let α be an $n\theta$-vector and let

$$\beta = (I - \tilde{R}\tilde{R}^\dagger)\Phi\alpha = (I - \tilde{R}^\dagger\tilde{R})\Phi\alpha \qquad \text{(C4.3.11)}$$

In (C4.3.11) we have used Lemma A.15. It follows from (C4.3.11) that

$$\tilde{R}\beta = 0 \qquad \text{(C4.3.12)}$$

which due to (C4.3.4) implies that

$$\beta^T\Phi = 0 \qquad \text{(C4.3.13)}$$

Premultiplying (C4.3.11) by β^T and using (C4.3.12), (C4.3.13), we get

$$\beta^T\beta = [\beta^T\Phi - \beta^T\tilde{R}\tilde{R}^\dagger\Phi]\alpha = 0$$

Thus $\beta = 0$, and since α is arbitrary, (C4.3.10) follows. Combining (C4.3.9) and (C4.3.10) the following factorization for the right-hand side of (C4.3.8) is obtained:

$$Z^T[\tilde{R} - \Phi(\Phi^T\tilde{R}^\dagger\Phi)^{-1}\Phi^T]Z = Z^T[\tilde{R} - \Phi(\Phi^T\tilde{R}^\dagger\Phi)^{-1}\Phi^T]\tilde{R}^\dagger$$
$$\times [\tilde{R} - \Phi(\Phi^T\tilde{R}^\dagger\Phi)^{-1}\Phi^T]Z \qquad \text{(C4.3.14)}$$

Since $\tilde{R}^\dagger \geq 0$ it can be concluded that (C4.3.8) or equivalently (C4.3.7) is true. ∎

It is instructive to check that (C4.3.3) reduces to the standard Gauss–Markov estimate, (4.17), (4.18), when $R > 0$. For $R > 0$, a straightforward application of Lemma A.1 gives

$$\Phi^T(R + \Phi\Phi^T)^{-1} = \Phi^T[R^{-1} - R^{-1}\Phi(I + \Phi^TR^{-1}\Phi)^{-1}\Phi^TR^{-1}]$$
$$= (I + \Phi^TR^{-1}\Phi - \Phi^TR^{-1}\Phi)(I + \Phi^TR^{-1}\Phi)^{-1}\Phi^TR^{-1}$$
$$= (I + \Phi^TR^{-1}\Phi)^{-1}\Phi^TR^{-1}$$

Thus

$$\theta^* = (\Phi^TR^{-1}\Phi)^{-1}(I + \Phi^TR^{-1}\Phi)(I + \Phi^TR^{-1}\Phi)^{-1}\Phi^TR^{-1}Y$$
$$= (\Phi^TR^{-1}\Phi)^{-1}\Phi^TR^{-1}Y$$

which is the 'standard' Markov estimate given by (4.17), (4.18).

In some applications the matrix R is nonsingular but ill-conditioned. The estimation of sinusoidal frequencies from noisy data by using the optimally weighted overdetermined Yule–Walker method (see Stoica, Friedlander and Söderström (1986) for details) is such an application. In such cases the BLUE could be computed using the standard formula (4.17), (4.18). However, the matrix $R + \Phi\Phi^T$ may be better conditioned to inversion than R and thus the (theoretically equivalent) formula (C4.3.3) may be a better choice even in this case.

Complement C4.4
Asymptotically best consistent estimation of certain nonlinear regression parameters

Earlier in the chapter several results for linear regressions were given. In this complement some of these results will be generalized to certain nonlinear models. Consider the following special nonlinear regression model:

$$Y_N = g(\theta_0) + e_N \tag{C4.4.1}$$

where Y_N and e_N are M-vectors, $g(\cdot)$ is a vector-valued differentiable function, and θ_0 is the vector of unknown parameters to be estimated. Both Y and e depend on the integer N. This dependence will be illustrated below by means of an example. It is assumed that the function $g(\cdot)$ admits a left inverse. The covariance matrix of e_N is not known for $N < \infty$. However, it is known that

$$\lim_{N \to \infty} NEe_N e_N^T = R(\theta_0) \tag{C4.4.2}$$

Note that R is allowed to depend on θ_0. It follows from (C4.4.1) and (C4.4.2) that the difference $[Y_N - g(\theta_0)]$ tends to zero as $1/\sqrt{N}$ when $N \to \infty$. Thus Y_N is a consistent estimate of $g(\theta_0)$; it is sometimes called a root-N consistent estimate. Note that it is assumption (C4.4.2) which makes the nonlinear regression model (C4.4.1) 'special'.

Let

$$\tilde{\theta} = f(Y_N) \tag{C4.4.3}$$

where $f(\cdot)$ is a differentiable function, denote a general consistent (for $N \to \infty$) estimate of θ_0. The objective is to find a consistent estimate of θ_0, which is asymptotically best in the following sense. Let $\hat{\theta}$ denote the asymptotically best consistent estimate (ABCE) of θ_0, assuming such a $\hat{\theta}$ exists, and let

$$P_M(\hat{\theta}) = \lim_{N \to \infty} NE(\hat{\theta} - \theta_0)(\hat{\theta} - \theta_0)^T$$

be its asymptotic covariance matrix (as N tends to infinity). Then

$$P_M(\tilde{\theta}) \geq P_M(\hat{\theta}) \tag{C4.4.4}$$

for any other consistent estimate $\tilde{\theta}$ as defined by (C4.4.3).

An example

As an illustration of the above problem formulation consider a stationary process $x(t)$ whose second-order properties are completely characterized by some parameter vector θ_0. An estimate of θ_0 has to be made from observations $x(1), \ldots, x(N)$. In a first stage indirect information about θ_0 is obtained by calculating the sample covariances

$$\hat{r}_k = \frac{1}{N} \sum_{t=1}^{N-k} x(t)x(t+k) \qquad k = 0, \ldots, M-1$$

92 *Linear regression* **Chapter 4**

Under weak conditions the vector $Y_N = (\hat{r}_0 \ \ldots \ \hat{r}_{M-1})^T$ is a root $- N$ consistent estimate of the vector of theoretical covariances $g(\theta_0) = (r_0 \ \ldots \ r_{M-1})^T$. Furthermore, an expression for the asymptotic covariance matrix $R(\theta_0)$ of the estimation error $\sqrt{N}[Y_N - g(\theta_0)]$ is available (see Bartlett, 1946, 1966; Stoica et al., 1985c; and also Appendix B.8). Therefore (C4.4.1) and (C4.4.2) hold.

In a second stage θ_0 is determined from Y_N. In particular, it may be required to calculate the ABCE of θ_0. Note that the first few sample covariances $\{\hat{r}_0, \ldots, \hat{r}_{M-1}\}$ where $M \ll N$, may contain almost all the 'information' about θ_0 in the initial sample $(x(1), \ldots, x(N))$. Thus it may be advantageous from the computational standpoint to base the estimation on $\{\hat{r}_0, \ldots, \hat{r}_{M-1}\}$ rather than on the raw data $\{x(1), \ldots, x(N)\}$. See Porat and Friedlander (1985, 1986), Stoica et al. (1985c) for details.

After this brief digression we return to the nonlinear regression problem at the beginning of this complement. In the following a number of interesting results pertaining to this problem are derived.

A lower bound on the covariance matrix of any consistent estimator of θ_0

The assumed consistency of $\tilde{\theta}$, (C4.4.3), imposes some restriction on the function $f(\cdot)$. To see this, note first that the continuity of $f(\cdot)$ and the fact that $e_N \to 0$ as $N \to \infty$ imply

$$\lim_{N \to \infty} \tilde{\theta} = f(g(\theta_0))$$

Since $\tilde{\theta}$ is a consistent estimate it follows that

$$f(g(\theta_0)) = \theta_0 \qquad (C4.4.5)$$

As this must hold for an arbitrary θ_0, it follows that $f(\cdot)$ is a left inverse of $g(\cdot)$. Moreover (C4.4.5) implies

$$\left.\frac{\partial f}{\partial g}\right|_{g=g(\theta_0)} \left.\frac{\partial g}{\partial \theta}\right|_{\theta=\theta_0} = I \qquad (C4.4.6)$$

With

$$F = \left.\frac{\partial f(g)}{\partial g}\right|_{g=g(\theta_0)} \quad \text{an } (n\theta|M) \text{ matrix}$$

$$G = \left.\frac{\partial g}{\partial \theta}\right|_{\theta=\theta_0} \quad \text{an } (M|n\theta) \text{ matrix} \qquad (C4.4.7)$$

the condition (C4.4.6) can be written more compactly as

$$FG = I \qquad (C4.4.8)$$

Next we derive the asymptotic covariance matrix (for $N \to \infty$) of $\tilde{\theta}$ in (C4.4.3). A Taylor series expansion of $\theta = f(Y_N)$ around $\theta_0 = f(g(\theta_0))$ gives

$$\tilde{\theta} = \theta_0 + F[Y_N - g(\theta_0)] + O(|e_N|^2)$$

which gives

$$\sqrt{N}(\tilde{\theta} - \theta_0) = F\sqrt{N}e_N + O(1/\sqrt{N})$$

Therefore

$$P_M(\tilde{\theta}) = \lim_{N \to \infty} NE(\tilde{\theta} - \theta_0)(\tilde{\theta} - \theta_0)^T = FRF^T$$

where for brevity the argument θ_0 of R has been omitted.

We are now in a position to state the main result of this subsection. Let

$$P_M^0 = (G^T R^{-1} G)^{-1} \quad (C4.4.9)$$

Then

$$P_M(\tilde{\theta}) = FRF^T \geq P_M^0 \quad (C4.4.10)$$

To prove (C4.4.10) recall that F and G are related by (C4.4.8). Thus (C4.4.10) is equivalent to

$$FRF^T \geq FG(G^T R^{-1} G)^{-1} G^T F^T$$

However, this is exactly the type of inequality encountered in (4.25). In the following it is shown that the lower bound P_M^0 is achievable. This means that there exists a consistent estimate of θ_0, whose asymptotic covariance matrix is equal to P_M^0.

An ABC estimate

Define

$$\hat{\theta} = \arg \min_{\theta} V(\theta) \quad (C4.4.11a)$$

where

$$V(\theta) = \frac{1}{2}[Y_N - g(\theta)]^T R^{-1}(\theta)[Y_N - g(\theta)] \quad (C4.4.11b)$$

The asymptotic properties of $\hat{\theta}$ (for $N \to \infty$) can be established as follows. A simple Taylor series expansion gives

$$0 = V'(\hat{\theta})^T = V'(\theta_0)^T + V''(\theta_0)(\hat{\theta} - \theta_0) + O(\|\hat{\theta} - \theta_0\|^2) \quad (C4.4.12)$$

Using the fact that $e_N = O(1/\sqrt{N})$, one can write

$$V'(\theta_0)^T = -G^T R^{-1} e_N + \frac{1}{2} \begin{pmatrix} e_N^T \dfrac{\partial R^{-1}(\theta)}{\partial \theta_1} e_N \\ \vdots \\ e_N^T \dfrac{\partial R^{-1}(\theta)}{\partial \theta_{n\theta}} e_N \end{pmatrix}_{\theta=\theta_0} \quad (C4.4.13a)$$

$$= -G^T R^{-1} e_N + O(1/N)$$

and
$$V''(\theta_0) = G^T R^{-1} G + O(1/\sqrt{N}) \tag{C4.4.13b}$$
It follows from (C4.4.12), (C4.4.13) that
$$(\hat{\theta} - \theta_0) = O(1/\sqrt{N})$$
and that
$$\sqrt{N}(\hat{\theta} - \theta_0) = (G^T R^{-1} G)^{-1} G^T R^{-1} \sqrt{N} e_N + O(1/\sqrt{N}) \tag{C4.4.14}$$
assuming the inverse exists. From (C4.4.14) it follows that
$$P_M(\hat{\theta}) = (G^T R^{-1} G)^{-1} = P_M^0 \tag{C4.4.15}$$
which concludes the proof that $\hat{\theta}$ is an ABCE of θ_0.

It is worth noting that replacement of $R(\theta)$ in (C4.4.11) by a root-N consistent estimate of $R(\theta_0)$ does not change the asymptotic properties of the parameter estimate. That is to say, the estimate

$$\hat{\hat{\theta}} = \arg\min_{\theta}\left\{\frac{1}{2}[Y_N - g(\theta)]^T \hat{R}^{-1} [Y_N - g(\theta)]\right\}$$

where $\hat{R} - R(\theta_0) = O(1/\sqrt{N})$, is asymptotically equal to $\hat{\theta}$. Indeed, calculations similar to those made above when analyzing $\hat{\theta}$ show that
$$(\hat{\hat{\theta}} - \theta_0) = (G^T R^{-1} G)^{-1} G^T R^{-1} e_N + O(1/N) = (\hat{\theta} - \theta_0) + O(1/N)$$
Note that $\hat{\hat{\theta}}$ is more convenient computationally than $\hat{\theta}$, since the matrix $\hat{R}$ does not need to be re-computed and inverted at each iteration of a minimization algorithm, as does $R(\theta)$ in (C4.4.11). However, $\hat{\hat{\theta}}$ can be used only if a consistent estimate $\hat{R}$ of R is available.

Remark. In the case of the nonlinear regression model (C4.4.1), (C4.4.2) it is more convenient to consider consistent rather than unbiased estimates. Despite this difference, observe the strong analogies between the results of the theory of unbiased estimation of the linear regression parameters and the results introduced above (for example, compare (4.20) with (C4.4.6), (C4.4.8); Lemma 4.3 with (C4.4.9), (C4.4.10); etc).

Some properties of P_M^0

Under weak conditions the matrices $\{P_M^0\}$ form a sequence of nonincreasing positive definite matrices, as M increases. To be more specific, assume that the model (C4.4.1) is extended by adding one new equation. The ABCE of θ_0 in the extended regression has the following asymptotic (for $N \to \infty$) covariance matrix (see (C4.4.15)):

$$P_{M+1}^0 = (G_{M+1}^T R_{M+1}^{-1} G_{M+1})^{-1}$$

where

$$G_{M+1} = \begin{pmatrix} G_M \\ \psi \end{pmatrix}$$

$$R_{M+1} = \begin{pmatrix} R_M & \beta \\ \beta^T & \alpha \end{pmatrix}$$

showing explicitly the dependence of G, R and P^0 on M. The exact expressions for the scalar α and the vectors ψ and β (which correspond to the new equation added to (C4.4.1)) are of no interest for the present discussion.

Using the nested structure of G and R and Lemma A.2, the following relationship can be derived:

$$\begin{aligned}(P^0_{M+1})^{-1} &= (G_M^T \quad \psi^T)\left\{\begin{pmatrix} I \\ 0 \end{pmatrix} R_M^{-1}(I \quad 0) \right. \\ &\quad \left. + \begin{pmatrix} -R_M^{-1}\beta \\ 1 \end{pmatrix}(-\beta^T R_M^{-1} \quad 1)/\gamma\right\}\begin{pmatrix} G_M \\ \psi \end{pmatrix} \\ &= (P^0_M)^{-1} + [\psi^T - G_M^T R_M^{-1}\beta][\psi^T - G_M^T R_M^{-1}\beta]^T/\gamma\end{aligned} \quad (C4.4.16)$$

where

$$\gamma = \alpha - \beta^T R_M^{-1}\beta$$

Since $R_{M+1} > 0$ implies $\gamma > 0$ (see Lemma A.2), it readily follows from (C4.4.16) that

$$P^0_M \geq P^0_{M+1} \quad (C4.4.17)$$

which completes the proof of the claim introduced at the beginning of this subsection. Thus by adding new equations to (C4.4.1) the asymptotic accuracy of the ABCE increases, which is an intuitively pleasing property. Furthermore, it follows from (C4.4.17) that P^0_M must have a limit as $M \to \infty$. It would be interesting to be able to evaluate the limiting matrix

$$P^0_\infty = \lim_{M \to \infty} P^0_M$$

However, this seems possible only in specific cases where more structure is introduced into the problem.

The problem of estimating the parameters of a finite-order model of a stationary process from sample covariances was touched on previously. For this case it was proved in Porat and Friedlander (1985, 1986), Stoica *et al.* (1985c) that $P^0_\infty = P_{CR}$; where P_{CR} denotes the Cramér–Rao lower bound on the covariance matrix of any consistent estimate of θ_0 (see Section B.4). Thus, in this case the estimate $\hat{\theta}$ is not only asymptotically (for $N \to \infty$) the best estimate of θ_0 based on a fixed number M of sample covariances, but when both *N and M* tend to infinity it is also the most accurate possible estimate.

Chapter 5
INPUT SIGNALS

5.1 Some commonly used input signals

The input signal used in an identification experiment can have a significant influence on the resulting parameter estimates. Several examples in Chapter 2 illustrated this fact, and further examples are given in Chapters 10 and 12. This chapter presents some types of input which are often used in practice. The following types of input signal will be described and analyzed:

- Step function.
- Pseudorandom binary sequence.
- Autoregressive moving average process.
- Sum of sinusoids.

Examples of these input signals are given in this section. Their spectral properties are described in Section 5.2. In Section 5.3 it is shown how the inputs can be modified in various ways in order to give them a low-frequency character. Section 5.4 demonstrates that the input spectrum must satisfy certain properties in order to guarantee that the system can be identified. This will lead to the concept of persistent excitation.

Some practical aspects on the choice of input are discussed in Section 12.2. In some situations the choice of input is imposed by the type of identification method employed. For instance, transient analysis requires a step or an impulse as input, while correlation analysis generally uses a pseudorandom binary sequence as input signal. In other situations, however, the input may be chosen in many different ways and the problem of choosing it becomes an important aspect of designing system identification experiments.

We generally assume that the system to be identified is a sampled data system. This implies that the input and output data are recorded in discrete time. In most cases we will use discrete time models to describe the system. In reality the input will be a continuous time signal. During the sampling intervals it may be held constant by sending it through a sample and hold circuit. Note that this is the normal way of generating inputs in digital control. In other situations, however, the system may operate with a continuous time controller or the input signal may not be under the investigator's control (the so-called 'normal operation' mode). In such situations the input signal cannot in general be restored between the sampling instants. For the forthcoming analysis it will be sufficient to describe the input and its properties in discrete time. We will not be concerned with the behavior of the input during the sampling intervals.

With a few exceptions we will deal with linear models only. It will then be sufficient to characterize signals in terms of first- and second-order moments (mean value and

Section 5.1   *Commonly used input signals* 97

covariance function). Note however, that two signals can have equal mean and covariance function but still have drastically different realizations. As an illustration think of a white random variable distributed as $\mathcal{N}(0, 1)$, and another white random variable that equals 1 with probability 0.5 and -1 with probability 0.5. Both variables will have zero mean and unit variance, although their realizations (outcomes) will look quite different.

There follow some examples of typical input signals.

Example 5.1 *A step function*
A step function is given by

$$u(t) = \begin{cases} 0 & t < 0 \\ u_0 & t \geq 0 \end{cases} \tag{5.1}$$

The user has to choose only the amplitude u_0. For systems with a large signal-to-noise ratio, an input step can give valuable information about the dynamics. Rise time, overshoot and static gain are directly related to the step response. Also the major time constants and a possible resonance can be at least crudely estimated from a step response. ∎

Example 5.2 *A pseudorandom binary sequence*
A pseudorandom binary sequence (PRBS) is a signal that shifts between two levels in a certain fashion. It can be generated by using shift registers for realizing a finite state system, (see Complement C5.3; Davies, 1970; or Eykhoff, 1974), and is a periodic signal. In most cases the period is chosen to be of the same order as the number of samples in the experiment, or larger. PRBS was used in Example 2.3 and is illustrated in Figure 2.5.

When applying a PRBS, the user must select the two levels, the period and the clock period. The clock period is the minimal number of sampling intervals after which the sequence is allowed to shift. In Example 2.3 the clock period was equal to one sampling interval. ∎

Example 5.3 *An autoregressive moving average sequence*
There are many ways of generating pseudorandom numbers on a computer (see, for example, Rubinstein, 1981; or Morgan, 1984, for a description). Let $\{e(t)\}$ be a pseudorandom sequence which is similar to white noise in the sense that

$$\frac{1}{N} \sum_{t=1}^{N} e(t)e(t + \tau) \to 0 \quad \text{as } N \to \infty \quad (\tau \neq 0) \tag{5.2}$$

This relation is to hold for τ at least as large as the dominating time constant of the unknown system. From the sequence $\{e(t)\}$ a rather general input $u(t)$ can be obtained by linear filtering as follows:

$$u(t) + c_1 u(t - 1) + \ldots + c_m u(t - m) = e(t) + d_1 e(t - 1) + \ldots \\ + d_m e(t - m) \tag{5.3}$$

Signals such as $u(t)$ given by (5.3) are often called ARMA (autoregressive moving average) processes. When all $c_i = 0$ it is called an MA (moving average) process, while for an AR (autoregressive) process all $d_i = 0$. Occasionally the notation ARMA (m_1, m_2) is used, where m_1 and m_2 denote the number of c_i- and d_i-coefficients, respectively. ARMA models are discussed in some detail in Chapter 6 as a way of modeling time series.

With this approach the user has to select the filter parameters m, $\{c_j\}$, $\{d_j\}$ and the random generator for $e(t)$. The latter includes the distribution of $e(t)$ which often is taken as Gaussian or rectangular, but other choices are possible.

The filtering (5.3) can be written as

$$C(q^{-1})u(t) = D(q^{-1})e(t) \tag{5.4a}$$

or

$$u(t) = \frac{D(q^{-1})}{C(q^{-1})} e(t) \tag{5.4b}$$

where q^{-1} is the backward shift operator ($q^{-1}e(t) = e(t-1)$, etc.) and

$$\begin{aligned} C(q^{-1}) &= 1 + c_1 q^{-1} + \ldots + c_m q^{-m} \\ D(q^{-1}) &= 1 + d_1 q^{-1} + \ldots + d_m q^{-m} \end{aligned} \tag{5.4c}$$

The filter parameters should be chosen so that the polynomials $C(z)$ and $D(z)$ have all zeros outside the unit circle. The requirement on $C(z)$ guarantees that $u(t)$ is a stationary signal. It follows from spectral factorization theory (see Appendix A6.1) that the requirement on $D(z)$ does not impose any constraints. There will always exist an equivalent representation such that $D(z)$ has all zeros on or outside the unit circle, as long as only the spectral density of the signal is being considered. The above requirement on $D(z)$ will be most useful in a somewhat different context, when deriving optimal predictors (Section 7.3).

Different choices of the filter parameters $\{c_i, d_j\}$ lead to input signals with various frequency contents and various shapes of time realizations. Simulations of three different ARMA processes are illustrated in Figure 5.1. (The continuous curves shown are obtained by linear interpolation.)

The curves (a) and (b) show a rather periodic pattern. The 'resonance' is more pronounced in (b), which is explained by the fact that in (b) $C(z)$ has zeros closer to the unit circle than in (a). The curve for (c) has quite a different pattern. It is rather irregular and different values seem little correlated unless they lie very close together.

■

Example 5.4 *A sum of sinusoids*
In this class of input signals $u(t)$ is given by

$$u(t) = \sum_{j=1}^{m} a_j \sin(\omega_j t + \varphi_j) \tag{5.5a}$$

where the angular frequencies $\{\omega_j\}$ are distinct,

$$0 \leq \omega_1 < \omega_2 < \ldots < \omega_m \leq \pi \tag{5.5b}$$

For a sum of sinusoids the user has to choose the amplitudes $\{a_i\}$, the frequencies $\{\omega_i\}$ and the phases $\{\varphi_i\}$.

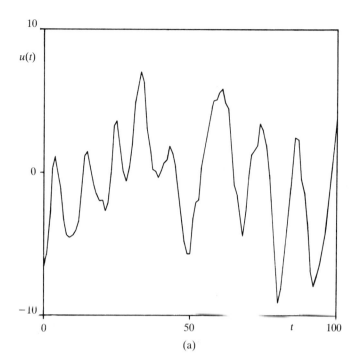

(a)

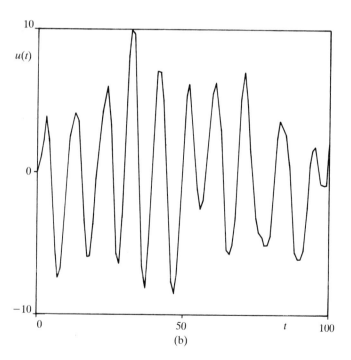

(b)

FIGURE 5.1 Simulation of three ARMA processes.
(a) $C(q^{-1}) = 1 - 1.5q^{-1} + 0.7q^{-2}$, $D(q^{-1}) = 1$.
(b) $C(q^{-1}) = 1 - 1.5q^{-1} + 0.9q^{-2}$, $D(q^{-1}) = 1$.
(c) $C(q^{-1}) = 1$, $D(q^{-1}) = 1 - 1.5q^{-1} + 0.7q^{-2}$.

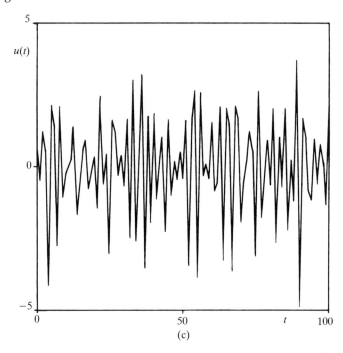

(c)

A term with $\omega_1 = 0$ corresponds to a constant $a_1 \sin \varphi_1$. A term with $\omega_m = \pi$ will oscillate with a period of two sampling intervals. Indeed, for $\omega_m = \pi$,

$$a_m \sin(\omega_m(t+1) + \varphi_m) = -a_m \sin(\omega_m t + \varphi_m)$$

Figure 5.2 illustrates a sum of two sinusoids. Both continuous time and sampled signals passed through a zero-order holding device are shown. ∎

5.2 Spectral characteristics

In many cases it is sufficient to describe a signal by its first- and second-order moments (i.e. the mean value and the covariance function). When dealing with linear systems and quadratic criteria, the corresponding identification methods can be analyzed using first- and second-order properties only as will be seen in the chapters to follow.

For a stationary stochastic process $y(t)$ the mean m and the covariance function $r(\tau)$ are defined as

$$m \triangleq Ey(t)$$
$$r(\tau) \triangleq E[y(t+\tau) - m][y(t) - m]^T \tag{5.6a}$$

where E denotes the expectation operator.

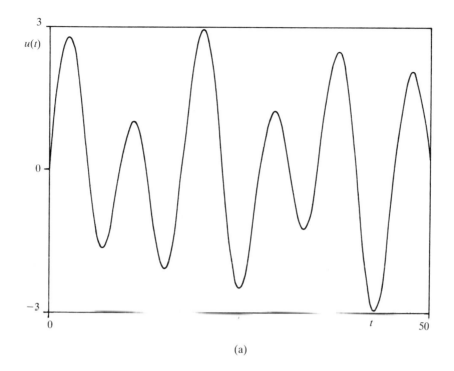

(a)

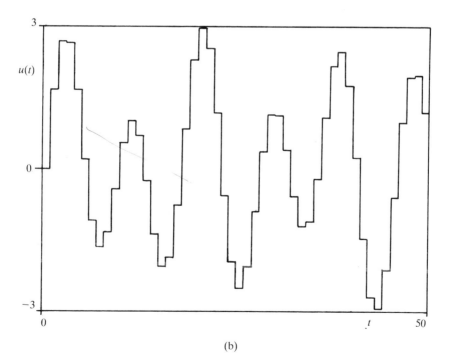

(b)

FIGURE 5.2 A sum of two sinusoids ($a_1 = 1$, $a_2 = 2$, $\omega_1 = 0.4$, $\omega_2 = 0.7$, $\varphi_1 = \varphi_2 = 0$). (a) The continuous time signal. (b) Its sampled form.

For a deterministic signal the corresponding definitions are obtained by substituting for E the limit of a normalized sum:

$$m \triangleq \lim_{N \to \infty} \frac{1}{N} \sum_{t=1}^{N} y(t)$$

$$r(\tau) \triangleq \lim_{N \to \infty} \frac{1}{N} \sum_{t=1}^{N} [y(t+\tau) - m][y(t) - m]^T \quad (5.6b)$$

assuming that the limits exist. (See also Appendix A5.1.)

Note that for many stochastic processes, the definitions (5.6a) and (5.6b) of m and $r(\tau)$ are equivalent. Such processes are called ergodic (see Definition B.2 in Appendix B).

For stochastic processes and for deterministic signals the spectral density $\phi(\omega)$ can be defined as the discrete Fourier transform of the covariance function:

$$\phi(\omega) \triangleq \frac{1}{2\pi} \sum_{\tau = -\infty}^{\infty} r(\tau) e^{-i\tau\omega} \quad (5.7)$$

(see also (A3.1.2)).

The function $\phi(\omega)$ is defined for ω in the interval $(-\pi, \pi)$. It is not difficult to see from (5.7) that $\phi(\omega)$ is a nonnegative definite Hermitian matrix (i.e. $\phi^*(\omega) = \phi(\omega)$ where $*$ denotes transpose complex conjugate). In particular, this implies that its diagonal elements are real valued and nonnegative even functions, $\phi_{jj}(\omega) = \phi_{jj}(-\omega) \geq 0$ (see also Problem 3.5). The inverse transformation to (5.7) is given by

$$r(\tau) = \int_{-\pi}^{\pi} \phi(\omega) e^{i\tau\omega} d\omega \quad (5.8)$$

(cf. (A3.1.3)).

The spectral density will have drastically different structures for periodic signals than for aperiodic signals. (See Appendix A5.1 for an analysis of periodic signals.)

The examples that follow examine the covariance functions and the spectral densities of some of the signals introduced at the beginning of this chapter.

Example 5.5 *Characterization of a PRBS*

Let $u(t)$ be a PRBS that shifts between the values a and $-a$, and let its period be M. Its covariance function can be shown to be (see complement C5.3)

$$r_u(\tau) = \begin{cases} a^2 & \tau = 0, \pm M, \pm 2M, \ldots \\ -a^2/M & \text{elsewhere} \end{cases} \quad (5.9a)$$

The spectral density of the signal can be computed using the formulas derived in Appendix A5.1; the result is

$$\phi_u(\omega) = \sum_{k=0}^{M-1} C_k \delta\left(\omega - 2\pi \frac{k}{M}\right) \tag{5.9b}$$

The coefficients $\{C_k\}$ are given by (A5.1.15). Using (5.9a),

$$C_0 = \frac{1}{M} \sum_{\tau=0}^{M-1} r_u(\tau) = \frac{1}{M}\left(a^2 - (M-1)\frac{a^2}{M}\right) = \frac{a^2}{M^2} \tag{5.9c}$$

and for $k > 0$ (using the convention $\alpha \triangleq e^{i2\pi/M}$)

$$\begin{aligned}
C_k &= \frac{1}{M}\left(a^2 - \sum_{\tau=1}^{M-1} \alpha^{-\tau k} a^2/M\right) \\
&= \frac{a^2}{M^2}\left(M - \alpha^{-k}\frac{1 - \alpha^{-(M-1)k}}{1 - \alpha^{-k}}\right) \\
&= \frac{a^2}{M^2}\left(M - \frac{\alpha^{-k} - \alpha^{-M}}{1 - \alpha^{-k}}\right) = \frac{a^2}{M^2}(M+1)
\end{aligned} \tag{5.9d}$$

(Observe that $\alpha^{-M} = 1$.) Combining (5.9b) (5.9d) the spectral density is obtained as

$$\boxed{\phi_u(\omega) = \frac{a^2}{M^2}\left[\delta(\omega) + (M+1)\sum_{k=1}^{M-1}\delta\left(\omega - 2\pi\frac{k}{M}\right)\right]} \tag{5.9e}$$

The spectral properties of a PRBS are investigated further in Example 5.8. ∎

Example 5.6 *Characterization of an ARMA process*
In the case of a white noise process $u(t)$,

$$r_u(\tau) = \lambda^2 \delta_{\tau,0} \tag{5.10a}$$

where $\delta_{\tau,0}$ is Kronecker's delta ($\delta_{\tau,0} = 1$ if $\tau = 0$, $\delta_{\tau,0} = 0$ if $\tau \neq 0$). The spectral density is easily found from the definition (5.7). It turns out to be constant

$$\phi_u(\omega) = \frac{\lambda^2}{2\pi} \tag{5.10b}$$

Next consider a filtered signal $u(t)$ as in (5.3). Applying (A3.1.10), the spectral density in this case is

$$\boxed{\phi_u(\omega) = \frac{\lambda^2}{2\pi}\frac{D(e^{i\omega})D(e^{-i\omega})}{C(e^{i\omega})C(e^{-i\omega})} = \frac{\lambda^2}{2\pi}\left|\frac{D(e^{i\omega})}{C(e^{i\omega})}\right|^2} \tag{5.10c}$$

From this expression it is obvious that the use of the polynomial coefficients $\{c_i, d_j\}$ will considerably affect the frequency content of the signal $u(t)$. If the polynomial $C(z)$ has zeros close to the unit circle, say close to $e^{i\omega_0}$ and $e^{-i\omega_0}$, then $\phi_u(\omega)$ will have large resonance peaks close to the frequency $\omega = \omega_0$. This means that the frequency $\omega = \omega_0$ is strongly emphasized in the signal. Similarly, if the polynomial $D(z)$ has zeros close to

$e^{i\omega_0}$ and $e^{-i\omega_0}$ then $\phi_u(\omega_0)$ will be small and the frequency component of $u(t)$ corresponding to $\omega = \omega_0$ will be negligible. Figures 5.3 and 5.4 illustrate how the polynomial coefficients $\{c_i, d_j\}$ affect the covariance function and the spectral density (5.10c), respectively.

Figures 5.3 and 5.4 illustrate clearly that the signals in cases (a) and (b) are resonant (with an angular frequency of about 0.43 and 0.66 respectively). The resonance is sharper in (b). Note how these frequencies also show up in the simulations (see Figure 5.1a, b). Figures 5.3c, 5.4c illustrate the high-frequency character of the signal in case (c). ∎

Example 5.7 *Characterization of a sum of sinusoids*
Consider a sum of sinusoids as in Example 5.4,

$$u(t) = \sum_{j=1}^{m} a_j \sin(\omega_j t + \varphi_j) \tag{5.11a}$$

with

$$0 \leq \omega_1 < \omega_2 \ldots < \omega_m \leq \pi \tag{5.11b}$$

To analyze the spectral properties of $u(t)$ consider first the quantity

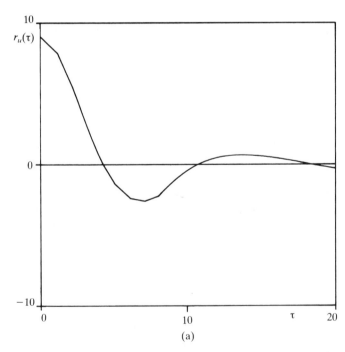

FIGURE 5.3 The covariance function of some ARMA processes. (a) $\lambda^2 = 1$, $C(z) = 1 - 1.5z + 0.7z^2$, $D(z) = 1$. (b) $\lambda^2 = 1$, $C(z) = 1 - 1.5z + 0.9z^2$, $D(z) = 1$. (c) $\lambda^2 = 1$, $C(z) = 1$, $D(z) = 1 - 1.5z + 0.7z^2$.

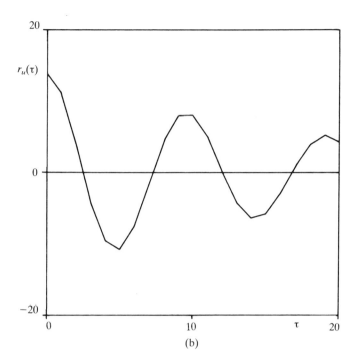

(b)

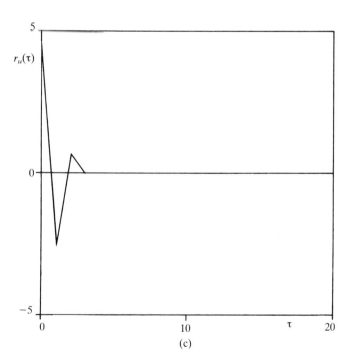

(c)

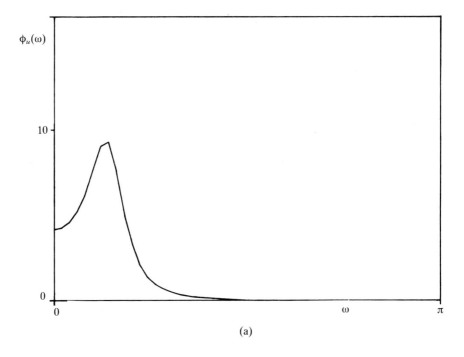

(a)

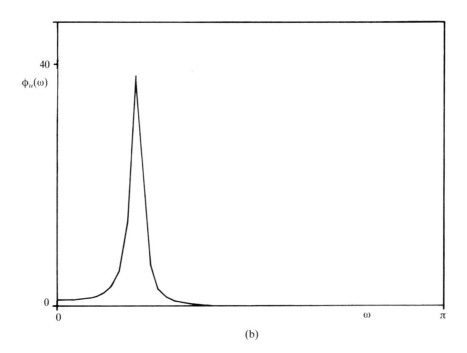

(b)

FIGURE 5.4 The spectral densities of some ARMA processes. (a) $\lambda^2 = 1$, $C(z) = 1 - 1.5z + 0.7z^2$, $D(z) = 1$. (b) $\lambda^2 = 1$, $C(z) = 1 - 1.5z + 0.9z^2$, $D(z) = 1$. (c) $\lambda^2 = 1$, $C(z) = 1$, $D(z) = 1 - 1.5z + 0.7z^2$.

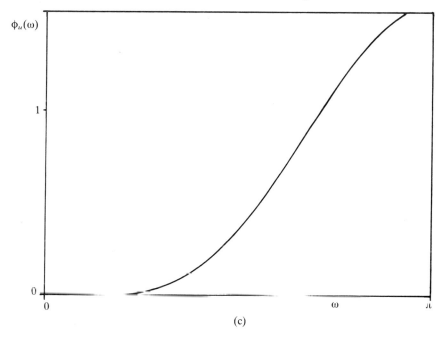

(c)

$$S_N - \frac{1}{N} \sum_{t=1}^{N} \sin(\omega t + \varphi) \qquad (5.12a)$$

If ω is a multiple of 2π then $S_N = \sin \varphi$. For all other arguments,

$$S_N = \frac{1}{N} \operatorname{Im} \sum_{t=1}^{N} e^{i(\omega t + \varphi)} = \frac{1}{N} \operatorname{Im}\left\{e^{i\varphi} e^{i\omega} \frac{1 - e^{i\omega N}}{1 - e^{i\omega}}\right\}$$

and

$$|S_N| \leq \frac{1}{N} \frac{|1 - e^{i\omega N}|}{|1 - e^{i\omega}|} \leq \frac{1}{N} \frac{2}{|e^{i\omega/2} - e^{-i\omega/2}|} = \frac{1}{N} \frac{1}{|\sin \omega/2|}$$

Hence

$$S_N \to \begin{cases} \sin \varphi & \text{if } \omega = 2\pi k \text{ (} k \text{ an integer)} \\ 0 & \text{elsewhere} \end{cases} \quad \text{as } N \to \infty \qquad (5.12b)$$

Since $u(t)$ is a deterministic function the definitions (5.6b) of m and $r(\tau)$ must be applied. From (5.12a, b), the mean value is

$$m = \begin{cases} a_1 \sin \varphi_1 & \text{if } \omega_1 = 0 \\ 0 & \text{if } \omega_1 \neq 0 \end{cases} \qquad (5.12c)$$

In the case $\omega_1 = 0$ this is precisely the first term in (5.11a). Thus the signal $u(t) - m$ contains only strictly positive frequencies. To simplify the notation in the following calculations $u(t) - m$ will be replaced by $u(t)$ and it is assumed that $\omega_1 > 0$. Then

$$r(\tau) = \lim_{N \to \infty} \frac{1}{N} \sum_{t=1}^{N} u(t+\tau)u(t)$$

$$= \sum_{j=1}^{m} \sum_{k=1}^{m} a_j a_k \lim_{N \to \infty} \frac{1}{N} \sum_{t=1}^{N} \sin(\omega_j t + \omega_j \tau + \varphi_j) \sin(\omega_k t + \varphi_k)$$

$$= \sum_{j=1}^{m} \sum_{k=1}^{m} a_j a_k \lim_{N \to \infty} \frac{1}{2N} \sum_{t=1}^{N} \{\cos(\omega_j t + \omega_j \tau + \varphi_j - \omega_k t - \varphi_k)$$

$$- \cos(\omega_j t + \omega_j \tau + \varphi_j + \omega_k t + \varphi_k)\}$$

Application of (5.12a, b) produces

$$\lim_{N \to \infty} \frac{1}{N} \sum_{t=1}^{N} \cos(\omega_j t + \omega_j \tau + \varphi_j - \omega_k t - \varphi_k)$$

$$= \lim_{N \to \infty} \frac{1}{N} \sum_{t=1}^{N} \sin\left((\omega_k - \omega_j)t + \frac{\pi}{2} - \omega_j \tau - \varphi_j + \varphi_k\right)$$

$$= \sin\left(\frac{\pi}{2} - \omega_j \tau - \varphi_j + \varphi_k\right) \delta_{j,k} = \cos(\omega_j \tau) \delta_{j,k}$$

If $\omega_m < \pi$ or $j + k < 2m$,

$$\lim_{N \to \infty} \frac{1}{N} \sum_{t=1}^{N} \cos(\omega_j t + \omega_k t + \omega_j \tau + \varphi_j + \varphi_k)$$

$$= \lim_{N \to \infty} \frac{1}{N} \sum_{t=1}^{N} \sin\left(-(\omega_j + \omega_k)t + \frac{\pi}{2} - \omega_j \tau - \varphi_j - \varphi_k\right) = 0$$

while for $\omega_m = \pi$, $j = k = m$,

$$\lim_{N \to \infty} \frac{1}{N} \sum_{t=1}^{N} \cos(\omega_j t + \omega_k t + \omega_j \tau + \varphi_j + \varphi_k)$$

$$= \lim_{N \to \infty} \frac{1}{N} \sum_{t=1}^{N} \sin\left(-2\pi t + \frac{\pi}{2} - \omega_m \tau - 2\varphi_m\right)$$

$$= \cos(\pi \tau + 2\varphi_m)$$

From these calculations the covariance function can be written as

$$r(\tau) = \sum_{j=1}^{m} C_j \cos(\omega_j \tau)$$

$$C_j = \frac{a_j^2}{2}$$

(5.12d)

Section 5.2 *Spectral characteristics* 109

If $\omega_m = \pi$ the weight C_m should be modified. The contribution of the last sinusoid to $r(\tau)$ is then, according to the previous calculation,

$$\frac{a_m^2}{2} \cos(\omega_m \tau) - \frac{a_m^2}{2} \cos(\omega_m \tau + 2\varphi_m)$$

$$= \frac{a_m^2}{2} [\cos(\pi \tau) - \cos(\pi \tau) \cos(2\varphi_m) + \sin(\pi \tau) \sin(2\varphi_m)]$$

$$= \frac{a_m^2}{2} \cos \pi\tau [1 - \cos(2\varphi_m)]$$

$$= a_m^2 \sin^2(\varphi_m) \cos(\pi\tau)$$

Thus for this particular case,

$$C_m = a_m^2 \sin^2 \varphi_m \tag{5.12e}$$

Note that (5.12e) is fairly natural: considering only one sinusoid with $a = 1$, $\omega = \pi$ we get $y(1) = -\sin \varphi_m$, $y(2) = \sin \varphi_m$, and in general $y(t) = (-1)^t \sin \varphi_m$. This gives $r_y(\tau) = \sin^2 \varphi_m (-1)^\tau$, which is perfectly consistent with (5.12d, e).

It remains to find the spectral density corresponding to the covariance function (5.12d). The procedure is similar to the calculations in Appendix A5.1. An expression for $\phi(\omega)$ is postulated and then it is verified by proving that (5.8) holds. We therefore claim that the spectral density corresponding to (5.12d) is

$$\boxed{\phi(\omega) = \sum_{j=1}^{m} \frac{C_j}{2} [\delta(\omega - \omega_j) + \delta(\omega + \omega_j)] \tag{5.12f}}$$

Substituting (5.12f) in (5.8),

$$r(\tau) = \int_{-\pi}^{\pi} \sum_{j=1}^{m} \frac{C_j}{2} [\delta(\omega - \omega_j) + \delta(\omega + \omega_j)] e^{i\tau\omega} d\omega$$

$$= \sum_{j=1}^{m} \frac{C_j}{2} [e^{i\tau\omega_j} + e^{-i\tau\omega_j}] = \sum_{j=1}^{m} C_j \cos(\omega_j \tau)$$

which is precisely (5.12d).

As a numerical illustration, Figure 5.5 illustrates the covariance function $r(\tau)$ for the sum of sinusoids depicted in Figure 5.2. ∎

Similarities between PRBS and white noise

In the informal analysis of Chapter 2 a PRBS was approximated by white noise. The validity of such an approximation is now examined.

First note that only second-order properties (the covariance functions) will be compared here. The distribution functions can be quite different. A PRBS has a two-

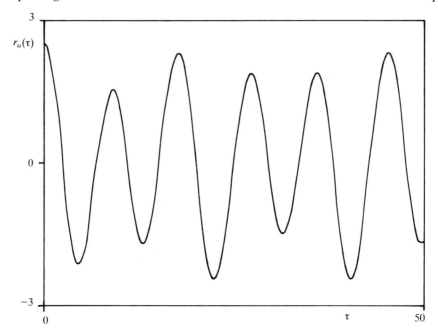

FIGURE 5.5 Covariance function for the sum of two sinusoids $u(t) = \sin 0.4t + 2 \sin 0.7t$, Example 5.4.

point distribution (since it can take two distinct values only) while the noise in most cases has a continuous probability density function.

A comparison of the covariance functions, which are given by (5.9a) and (5.10a), shows that they are very similar for moderate values of τ provided $\lambda^2 = a^2$ and M is large. It is therefore justifiable to say that a PRBS has similar properties to a white noise. Since a PRBS is periodic, the spectral densities look quite different. The density (5.9e) is a weighted sum of Dirac functions whereas (5.10b) is a constant function. However, it is not so relevant to compare spectral densities as such: what matters are the covariances between various filtered input signals (cf. Chapters 2, 7 and 8). Let

$$x_1(t) = G_1(q^{-1})u(t) \quad x_2(t) = G_2(q^{-1})u(t)$$

where $G_1(q^{-1})$ and $G_2(q^{-1})$ are two asymptotically stable filters. Then the covariance between $x_1(t)$ and $x_2(t)$ can be calculated to be

$$Ex_1(t)x_2(t) = \int_{-\pi}^{\pi} G_1(e^{-i\omega})G_2(e^{i\omega})\phi_u(\omega)d\omega$$

(cf. Problem 3.6; equation (5.8); and Appendix A3.1). Hence it is relevant to consider integrals of the form

$$I = \int_{-\pi}^{\pi} f(\omega)\phi(\omega)d\omega$$

where $f(\omega)$ is a continuous function of ω, and $\phi(\omega)$ is the spectral density of a PRBS or a white noise sequence. Using the spectral density (5.10b) the integral becomes

Section 5.2   Spectral characteristics   111

$$I_1 = \frac{a^2}{2\pi} \int_{-\pi}^{\pi} f(\omega)d\omega = \frac{a^2}{2\pi} \int_0^{2\pi} f(\omega)d\omega \tag{5.13a}$$

Using the spectral density (5.9e) the integral becomes

$$I_2 = \frac{a^2}{M^2}\left[f(0) + (M+1)\sum_{k=1}^{M-1} f\left(2\pi\frac{k}{M}\right)\right] \tag{5.13b}$$

Now assume that the integral in (5.13a) is approximated by a Riemann sum. Then the interval $(0, 2\pi)$ is divided into M subintervals, each of length $2\pi/M$, and the integral is replaced by a sum in the following way:

$$I_1 \approx \frac{a^2}{2\pi}\frac{2\pi}{M}\sum_{k=0}^{M-1} f\left(2\pi\frac{k}{M}\right) \triangleq I_3 \tag{5.13c}$$

The approximation error tends to zero as M tends to infinity. It is easy to see that I_2 and I_3 are very similar. In fact,

$$I_2 - I_3 = \frac{a^2}{M^2}\left[(1-M)f(0) + \sum_{k=1}^{M-1} f\left(2\pi\frac{k}{M}\right)\right] \tag{5.13d}$$

which is of order $O(1/M)$.

The following example illustrates numerically the difference between the covariance functions of a filtered PRBS and a filtered white noise.

Example 5.8 *Comparison of a filtered PRBS and a filtered white noise*
Let $u_1(t)$ be white noise of zero mean and variance λ^2 and define $y_1(t)$ as

$$y_1(t) - ay_1(t-1) = u_1(t) \tag{5.14a}$$

The covariance function of $y_1(t)$ is given by

$$r_1(\tau) = \frac{\lambda^2}{1-a^2}a^\tau \quad (\tau \geq 0) \tag{5.14b}$$

Let $u_2(t)$ be a PRBS of period M and amplitude λ and set

$$y_2(t) - ay_2(t-1) = u_2(t) \tag{5.15a}$$

The covariance function of $y_2(t)$ is calculated as follows. From (5.8), (5.9e) (see also Appendix A3.1),

$$r_2(\tau) = \int_{-\pi}^{\pi} \phi_{y_2}(\omega)e^{i\tau\omega}d\omega$$

$$= \text{Re} \int_{-\pi}^{\pi} \phi_{u_2}(\omega)\left|\frac{1}{1-ae^{i\omega}}\right|^2 e^{i\tau\omega}d\omega$$

$$= \frac{\lambda^2}{M^2}\int_0^{2\pi}\left[\delta(\omega) + (M+1)\sum_{k=1}^{M-1}\delta\left(\omega - 2\pi\frac{k}{M}\right)\right] \tag{5.15b}$$

$$\times \frac{1}{1+a^2-2a\cos\omega}\cos(\tau\omega)d\omega$$

$$= \frac{\lambda^2}{M^2}\left[\frac{1}{(1-a)^2} + (M+1)\sum_{k=1}^{M-1}\frac{1}{1+a^2-2a\cos(2\pi k/M)}\cos\left(\tau 2\pi\frac{k}{M}\right)\right]$$

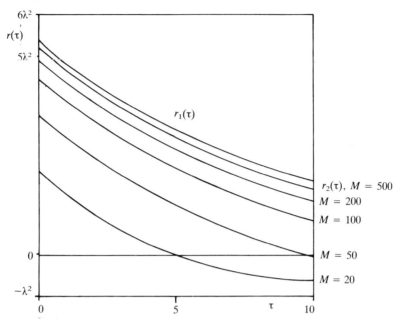

FIGURE 5.6 Plots of $r_1(\tau)$ and $r_2(\tau)$ versus τ. The filter parameter is $a = 0.9$; $r_2(\tau)$ is plotted for $M = 500, 200, 100, 50$ and 20.

Figure 5.6 shows the plots of the covariance functions $r_1(\tau)$ and $r_2(\tau)$ versus τ, for different values of M. It can be seen from the figure that $r_1(\tau)$ and $r_2(\tau)$ are very close for large values of M. This is in accordance with the previous general discussion showing that PRBS and white noise may have similar properties. Note also that for small values of M there are significant differences between $r_1(\tau)$ and $r_2(\tau)$. ∎

5.3 Lowpass filtering

The frequency content of an input signal and its distribution over the interval $[0, \pi]$ can be of considerable importance in identification. In most cases the input should emphasize the low-frequency properties of the system and hence it should have a rich content of low frequencies. This section examines some ways of modifying a white noise process (which has a constant spectral density) into a low-frequency signal. In this way it should be possible to emphasize the low-frequency properties of the system during the identification experiment. Example 12.1 will show how the frequency content of the input can influence the accuracy of the identified model in different frequency ranges. The modifications which are aimed at emphasizing the low frequencies can be applied to signals other than white noise, but the discussion here is limited to white noise in order to keep the analysis simple. The different approaches to be illustrated in the following examples are:

Section 5.3 **Lowpass filtering**

- Standard filtering.
- Increasing the clock period.
- Decreasing the probability of level change.

Example 5.9 *Standard filtering*
This approach was explained in Examples 5.3 and 5.6. The user has to choose the filter parameters $\{c_i, d_j\}$. It was seen in Example 5.6 how these parameters can be chosen to obtain various frequency properties. In particular, $u(t)$ will be a low-frequency signal if the polynomial $C(z)$ has zeros close to $z = 1$. Then, for small ω, $|C(e^{i\omega})|$ will be small and $\phi(\omega)$ large (see (5.10c)). ∎

Example 5.10 *Increasing the clock period*
Let $e(t)$ be a white noise process. Then the input $u(t)$ is obtained from $e(t)$ by keeping the same value for N steps. In this case N is called the increased clock period. This means that

$$u(t) = e\left(\left[\frac{t-1}{N}\right] + 1\right) \quad t = 1, 2, \ldots \tag{5.16a}$$

where $[x]$ = integer part of x. This approach is illustrated in Figure 5.7.

Assume that $e(t)$ has zero mean and unit variance. The signal $u(t)$ will not be stationary. By construction it holds that $u(t)$ and $u(s)$ are uncorrelated if $|t - s| \geq N$. Therefore let $\tau = t - s \in [0, N-1]$. Note that $Eu(t + \tau)u(t)$ is a periodic function of t, with period N.

The covariance function will be evaluated using (5.6b), which gives

$$r(\tau) = \lim_{M \to \infty} \frac{1}{M} \sum_{t=1}^{M} u(t + \tau)u(t)$$

Note that this definition of the covariance function is relevant to the asymptotic behavior of most identification methods. This is why (5.6b) is used here even if $M(t)$ is not a deterministic signal.

Now set $M = pN$ and

$$v_\tau(s) = \frac{1}{N} \sum_{k=1}^{N} u(sN - N + k + \tau)u(sN - N + k)$$

Then

$$r(\tau) = \lim_{p \to \infty} \frac{1}{pN} \sum_{s=1}^{P} \sum_{k=1}^{N} u(sN - N + k + \tau)u(sN - N + k)$$

$$= \lim_{p \to \infty} \frac{1}{p} \sum_{s=1}^{P} v_\tau(s)$$

Now $v_\tau(s_1)$ and $v_\tau(s_2)$ are uncorrelated for $|s_1 - s_2| > 2$, since they then are formed from the input of disjoint intervals. Hence Lemma B.1 can be applied, and

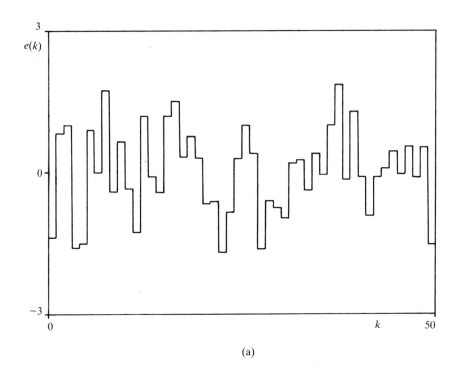

(a)

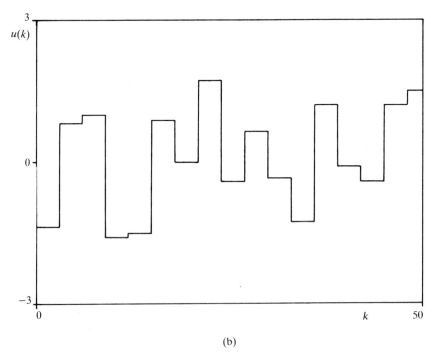

(b)

FIGURE 5.7 The effect of increasing the clock period. (a) $\{e(k)\}$. (b) $\{u(k)\}$; $N = 3$.

$$r(\tau) = Ev_\tau(s) = \frac{1}{N} \sum_{k=1}^{N} Eu(sN - N + k + \tau)u(sN - N + k)$$

$$= \frac{1}{N} \sum_{k=1}^{N} Eu(k + \tau)u(k)$$

$$= \frac{1}{N} \left[\sum_{k=1}^{N-\tau} Eu(k + \tau)u(k) + \sum_{k=N-\tau+1}^{N} Eu(k + \tau)u(k) \right]$$

$$= \frac{1}{N} \left[\sum_{k=1}^{N-\tau} Ee(1)e(1) + \sum_{k=N-\tau+1}^{N} Ee(2)e(1) \right]$$

$$= \frac{N - \tau}{N} \tag{5.16b}$$

The covariance function (5.16b) can also be obtained by filtering the white noise $e(t)$ through the filter

$$H(q^{-1}) = \frac{1}{\sqrt{N}}(1 + q^{-1} + \ldots + q^{-N+1}) = \frac{1}{\sqrt{N}} \frac{1 - q^{-N}}{1 - q^{-1}} \tag{5.16c}$$

This can be shown using (A3.1.9). For the present case we have

$$h(j) = \begin{cases} 1/\sqrt{N} & j = 0, \ldots, N - 1 \\ 0 & \text{elsewhere} \end{cases}$$

Then (A3.1.9) gives

$$r_u(\tau) = \sum_{j=0}^{\infty} \sum_{k=0}^{\infty} h(j)h(k)r_e(\tau - j + k)$$

$$= \sum_{j=0}^{N-1} \sum_{k=0}^{N-1} \frac{1}{N} \delta_{\tau-j+k,0}$$

$$= \frac{1}{N} \sum_{k=0}^{N-1-\tau} 1 = \frac{N - \tau}{N}$$

which coincides with (5.16b). ∎

Remark. There are some advantages of using a signal with increased clock period over a white noise filtered by (5.16c). In the first case the input will be constant over long periods of time. This means that in the recorded data, prior to any parameter estimation, we may directly see the beginning of the step response. This can be valuable information *per se*. The measured data will thus approximately contain transient analysis. Secondly, a continuously varying input signal will cause more wearing of the actuators. ∎

Example 5.11 *Decreasing the probability of level change*
Let $e(t)$ be a white noise process with zero mean and variance λ^2. Define $u(t)$ as

116   *Input signals*   *Chapter 5*

$$u(t) = \begin{cases} u(t-1) & \text{with probability } \alpha \\ e(t) & \text{with probability } 1 - \alpha \end{cases} \qquad (5.17a)$$

The probability described by (5.17a) is independent of the state in previous time steps.

It is intuitively clear that if α is chosen close to 1, the input will remain constant over long intervals and hence be of a low-frequency character. Figure 5.8 illustrates such a signal.

When evaluating the covariance function $r_u(\tau)$ note that there are two random sequences involved in the generation of $u(t)$. The white noise $\{e(t)\}$ is one sequence, while the other sequence accounts for occurrence of changes. These two sequences are independent. We have

$$\begin{aligned} u(t+\tau) &= e(t_1) \quad \text{for some } t_1 \leq t + \tau \\ u(t) &= e(t_2) \quad \text{for some } t_2 \leq t \end{aligned} \qquad (5.17b)$$

Thus for $\tau \geq 0$,

$$\begin{aligned} r_u(\tau) &= Eu(t+\tau)u(t) = Ee(t_1)e(t_2) \\ &= E[e(t_1)e(t_2)|t_1 \neq t_2] P(\text{at least one change has occurred in the interval } [t, t+\tau] \\ &\quad + E[e(t_1)e(t_2)|t_1 = t_2] P(\text{no change has occurred in the interval } [t, t+\tau]) \\ &= 0 \times (1 - \alpha^\tau) + \lambda^2 \alpha^\tau \\ &= \lambda^2 \alpha^\tau \end{aligned} \qquad (5.17c)$$

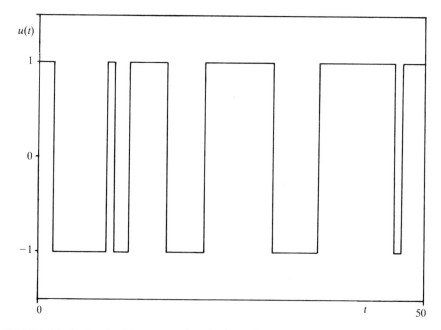

FIGURE 5.8 A signal $u(t)$ generated as in (5.17a). The white noise $\{e(t)\}$ has a two-point distribution ($P(e(t) = 1) = P(e(t) = -1) = 0.5$). The parameter $\alpha = 0.8$.

This covariance function would also be obtained by filtering $e(t)$ with a first-order filter

$$H(q^{-1}) = \frac{(1 - \alpha^2)^{1/2}}{1 - \alpha q^{-1}} \qquad (5.17d)$$

To see this, note that the weighting sequence of this filter is given by

$$h(j) = (1 - \alpha^2)^{1/2} \alpha^j \qquad j \geq 0$$

Then from (A3.1.8),

$$r_u(\tau) = \sum_{j=0}^{\infty} \sum_{k=0}^{\infty} h(j) h(k) r_e(\tau - j + k)$$

$$= \lambda^2 \sum_{k=0}^{\infty} h(k) h(k + \tau)$$

$$= \lambda^2 (1 - \alpha^2) \sum_{k=0}^{\infty} \alpha^{2k+\tau} = \lambda^2 \alpha^\tau$$

which coincides with (5.17c). ∎

These examples have shown different methods to increase the low-frequency content of a signal. For the methods based on an increased clock period and a decreased probability of change we have also given 'equivalent filter interpretations'. These methods will give signals whose spectral properties could also be obtained by standard lowpass filtering using the 'equivalent filters'.

The next example illustrates how the spectral densities are modified.

Example 5.12 *Spectral density interpretations*
Consider first the covariance function (5.16b). Since it can be associated with the filter (5.16c), the corresponding spectral density is readily found to be

$$\phi_1(\omega) = \frac{1}{2\pi} \left| \frac{1}{\sqrt{N}} \frac{1 - e^{-i\omega N}}{1 - e^{-i\omega}} \right|^2 = \frac{1}{2\pi} \frac{1}{N} \frac{2 - 2\cos N\omega}{2 - 2\cos \omega} = \frac{1}{2\pi} \frac{1}{N} \frac{1 - \cos N\omega}{1 - \cos \omega}$$

Next consider the covariance function (5.17c) with $\lambda^2 = 1$. From (5.17d) the spectral density is found to be

$$\phi_2(\omega) = \frac{1}{2\pi} \frac{1 - \alpha^2}{|1 - \alpha e^{i\omega}|^2} = \frac{1}{2\pi} \frac{1 - \alpha^2}{1 + \alpha^2 - 2\alpha \cos \omega}$$

These spectral densities are illustrated in Figure 5.9, where it is clearly seen how the frequency content is concentrated at low frequencies. It can also be seen how the low-frequency character is emphasized when N is increased or α approaches 1. ∎

5.4 Persistent excitation

Consider the estimate (3.30) of a truncated weighting function. Then, asymptotically (for $N \to \infty$), the coefficients $\{h(k)\}_{k=0}^{M-1}$ are determined as the solution to

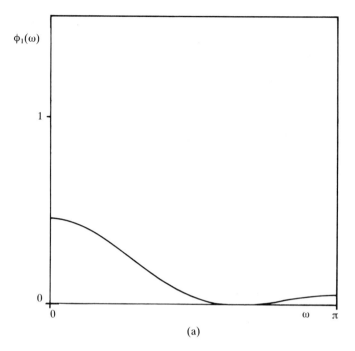

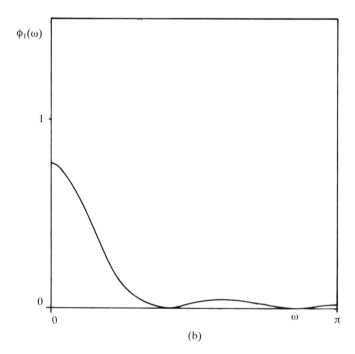

FIGURE 5.9 (a) Illustration of $\phi_1(\omega)$, $N = 3$, Example 5.12. (b) Similarly for $N = 5$. (c) Similarly for $N = 10$. (d) Illustration of $\phi_2(\omega)$, for $\lambda^2 = 1$, $\alpha = 0.3$, 0.7 and 0.9, Example 5.12.

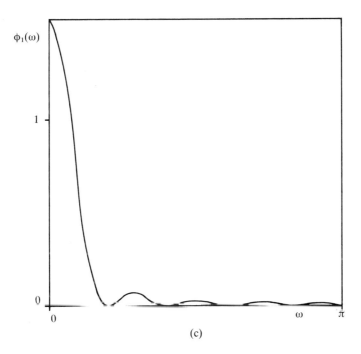

(c)

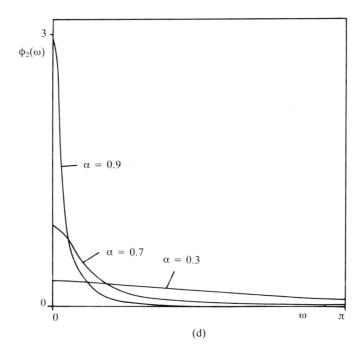

(d)

$$\begin{pmatrix} r_u(0) & \cdots & r_u(M-1) \\ \vdots & \ddots & \vdots \\ r_u(M-1) & \cdots & r_u(0) \end{pmatrix} \begin{pmatrix} h(0) \\ \vdots \\ h(M-1) \end{pmatrix} = \begin{pmatrix} r_{yu}(0) \\ \vdots \\ r_{yu}(M-1) \end{pmatrix} \qquad (5.18)$$

For a unique solution to exist the matrix appearing in (5.18) must be nonsingular. This leads to the concept of persistent excitation.

Definition 5.1
A signal $u(t)$ is said to be persistently exciting (pe) of order n if:

(i) the following limit exists:

$$r_u(\tau) = \lim_{N \to \infty} \frac{1}{N} \sum_{t=1}^{N} u(t+\tau) u^T(t); \qquad (5.19)$$

and

(ii) the matrix

$$R_u(n) = \begin{pmatrix} r_u(0) & r_u(1) & \cdots & r_u(n-1) \\ r_u(-1) & r_u(0) & & \vdots \\ \vdots & & \ddots & \\ r_u(1-n) & \cdots & & r_u(0) \end{pmatrix} \qquad (5.20)$$

is positive definite. ■

Remark 1. As noted after (5.6), many stationary stochastic processes are ergodic. In such cases one can substitute

$$\lim_{N \to \infty} \frac{1}{N} \sum_{t=1}^{N}$$

in (5.19) by the expectation operator E. Then the matrix $R_u(n)$ is the usual covariance matrix (supposing for simplicity that $u(t)$ has zero mean). ■

Remark 2. In the context of adaptive control an alternative definition of pe is used (see e.g. Anderson, 1982; Goodwin and Sin, 1984; Bai and Sastry, 1985. See also Lai and Wei, 1986.) There $u(t)$ is said to be persistently exciting of order n if for all t there is an integer m such that

$$\varrho_1 I > \sum_{k=t}^{t+m} \varphi(k) \varphi^T(k) > \varrho_2 I \qquad (5.21)$$

where $\varrho_1, \varrho_2 > 0$ and the vector $\varphi(t)$ is given by

$$\varphi(t) = (u^T(t-1) \ \cdots \ u^T(t-n))^T$$

To see the close relation between Definition 5.1 and (5.21), note that the matrix (5.20) can be written as

$$R_u(n) = \lim_{N \to \infty} \frac{1}{N} \sum_{t=1}^{N} \varphi(t)\varphi^T(t) \tag{5.22}$$

∎

As illustration of the concept of persistent excitation, the following example considers some simple input signals.

Example 5.13 *Order of persistent excitation*
Let $u(t)$ be white noise, say of zero mean and variance σ^2. Then $r_u(\tau) = \sigma^2 \delta_{0,\tau}$ and $R_u(n) = \sigma^2 I_n$, which always is positive definite. Thus white noise is persistently exciting of all orders.

Next, let $u(t)$ be a step function of magnitude σ. Then $r_u(\tau) = \sigma^2$ for all τ. Hence $R_u(n)$ is nonsingular if and only if $n = 1$. Thus a step function is pe of order 1 only.

Finally, let $u(t)$ be an impulse: $u(t) = 1$ for $t = 0$, and 0 otherwise. This gives $r_u(\tau) = 0$ for all τ and $R_u(n) = 0$. This signal is not pe of any order. ∎

Remark 1. Sometimes one includes a term corresponding to an estimated mean value in (5.19) for the definition of pe signals. This means that (5.19) is replaced by

$$m_u = \lim_{N \to \infty} \frac{1}{N} \sum_{t=1}^{N} u(t)$$

$$r_u(\tau) = \lim_{N \to \infty} \frac{1}{N} \sum_{t=1}^{N} [u(t+\tau) - m_u][u^T(t) - m_u^T] \tag{5.23}$$

In this way the order of persistent excitation can be decreased by at most one. This can be seen to be the case since

$$R_u(n)^{\text{new}} = R_u(n)^{\text{old}} - mm^T$$

$$m^T = (m_u^T \quad \ldots \quad m_u^T)$$

and therefore rank $R_u(n)^{\text{new}} \geq$ rank $R_u(n)^{\text{old}} - 1$. With this convention a white noise is still pe of any order. However, for a step function we now get $r_u(\tau) = 0$ all τ. Then $R_u(n)$ becomes singular for all n. Thus a step function is not pe of any order with this alternative definition. ∎

Remark 2. The concept of persistent excitation was introduced here using the truncated weighting function model. However, the use of this concept is not limited to the weighting function estimation problem. As will be seen in subsequent chapters, a necessary condition for consistent estimation of an nth-order linear system is that the input signal is persistently exciting of order $2n$. Some detailed calculations can be found in Example 11.6. In some cases, notably when the least squares method is applied, it is sufficient (and necessary) to use an input that is pe of order n. This result explains why consistent parameter estimates were not obtained in Example 2.6 when the input was an impulse. (See also Complement C5.1.) ∎

Remark 3. The statements of Remark 2 are applicable to *consistent* estimation in *noisy* systems. For noise-free systems, it is not necessary that the input is persistently exciting. Consider for example an nth-order linear system initially at rest. Assume that an impulse is applied, and the impulse response is recorded. From the first $2n$ (nonzero) values of the impulse it is possible to find the system parameters (cf. Problem 3.11). Hence the system can be identified even though the input is not persistently exciting. The reason is that noise-free systems can be identified from a finite number of data points ($N < \infty$) whereas persistent excitation concerns the input properties for $N \to \infty$ (which are relevant to the analysis of the consistency of the parameter estimates in noisy systems). ∎

Remark 4. Sometimes it would be valuable to have a more detailed concept than the order of persistent excitation. Complement C5.2 discusses the condition number of $R_u(n)$ as a more detailed measure of the persistency properties of an input signal, and presents some results tied to ARMA processes. ∎

In what follows some properties of persistently exciting signals are presented. Such an analysis was originally undertaken by Ljung (1971), while an extension to multivariable signals was made by Stoica (1981a). To simplify the proofs we restrict them to stationary ergodic, stochastic processes, but similar results hold for deterministic signals; see Ljung (1971) for more details.

Property 1
Let $u(t)$ be a multivariable ergodic process of dimension nu. Assume that its spectral density matrix is positive definite in at least n distinct frequencies (within the interval $(-\pi, \pi)$). Then $u(t)$ is persistently exciting of order n.

Proof. Let $g = (g_1^T \ldots g_n^T)^T$ be an arbitrary $n \times nu$-vector and set $G(q^{-1}) = \sum_{i=1}^n g_i q^{-i}$. Consider the equation

$$0 = g^T R_u(n) g = E[G^T(q^{-1})u(t)][G^T(q^{-1})u(t)]^T$$
$$= \int_{-\pi}^{\pi} G^T(e^{-i\omega}) \phi_u(\omega) G(e^{i\omega}) d\omega$$

where $\phi_u(\omega)$ is the spectral density matrix of $u(t)$. Since $\phi_u(\omega)$ is nonnegative definite,

$$G^T(e^{-i\omega}) \phi_u(\omega) G(e^{i\omega}) \equiv 0$$

Thus $G(e^{-i\omega})$ is equal to zero in n distinct frequencies. However, since $G(z)$ is a (vector) polynomial of degree $n - 1$ only, this implies $g = 0$. Thus the matrix $R_u(n)$ is positive definite and $u(t)$ is persistently exciting of order n. ∎

Property 2
An ARMA process is persistently exciting of any finite order.

Proof. The assertion follows immediately from Property 1 since the spectral density matrix of an ARMA process is positive definite for almost all frequencies in $(-\pi, \pi)$. ∎

For *scalar* processes the condition of Property 1 is also *necessary* for $u(t)$ to be pe of order n (see Property 3 below). This is not true in the multivariable case, as shown by Stoica (1981a).

Property 3

Let $u(t)$ be a scalar signal that is persistently exciting of order n. Then its spectral density is nonzero in at least n frequencies.

Proof. The proof is by contradiction. From the calculation of Property 1,

$$0 = g^T R_u(n) g \Leftrightarrow \phi_u(\omega)|G(e^{i\omega})|^2 \equiv 0$$

Assume that the spectral density is nonzero in at most $n - 1$ frequencies. Then we can choose the polynomial $G(z)$ (of degree $n - 1$) to vanish where $\phi_u(\omega)$ is nonzero. This means that there is a nonzero vector g such that $g^T R_u(n) g = 0$. Hence u is not pe of order n. This is a contradiction and so the result is proved. ∎

A scalar signal that is pe of order n but not of order $n + 1$ has a spectral density which is nonzero in precisely n distinct points (in the interval $(-\pi, \pi)$). It can be generated as a sum of $[n/2 + 1]$ sinusoids (cf. Example 5.17).

Property 4

Let $u(t)$ be a multivariable ergodic signal with spectral density matrix $\phi_u(\omega)$. Assume that $\phi_u(\omega)$ is positive definite for at least n distinct frequencies. Let $H(q^{-1})$ be an $(nu|nu)$ asymptotically stable linear filter and assume that $\det[H(z)]$ has no zero on the unit circle. Then the filtered signal $y(t) = H(q^{-1})u(t)$ is persistently exciting of order n.

Proof. Since

$$\phi_y(\omega) = H(e^{-i\omega})\phi_u(\omega)H^T(e^{i\omega})$$

the result is an immediate consequence of Property 1. ∎

The preceding result can be somewhat strengthened for scalar signals.

Property 5

Let $u(t)$ be a scalar signal that is persistently exciting of order n. Assume that $H(q^{-1})$ is an asymptotically stable filter with k zeros on the unit circle. Then the filtered signal $y(t) = H(q^{-1})u(t)$ is persistently exciting of order m with $n - k \leq m \leq n$.

Proof. We have

$$\phi_y(\omega) = \phi_u(\omega)|H(e^{i\omega})|^2$$

Now $\phi_u(\omega)$ is nonzero in n points; $|H(e^{i\omega})|^2$ is zero in precisely k points; hence $\phi_y(\omega)$ is nonzero in m points, where $n - k \leq m \leq n$. The result now follows from Property 1. ∎

Note that the exact value of m depends on the location of the zeros of $|H(e^{i\omega})|^2$ and

124 Input signals   Chapter 5

whether $\phi_u(\omega)$ is nonzero for these frequencies. If in particular $H(q^{-1})$ has no zeros on the unit circle then $u(t)$ and $H(q^{-1})u(t)$ are pe of the same order.

Property 6
Let $u(t)$ be a stationary process and let

$$z(t) = \sum_{i=1}^{n} H_i u(t-i)$$

Then $Ez(t)z^T(t) = 0$ implies $H_i = 0$, $i = 1, \ldots, n$ if and only if $u(t)$ is persistently exciting of order n.

Proof. Set

$$\bar{H} = (H_1 \ \ldots \ H_n)^T \qquad \varphi(t) = (u^T(t-1) \ \ldots \ u^T(t-n))^T$$

Then $z(t) = \bar{H}^T\varphi(t)$ and

$$0 = Ez(t)z^T(t) = E\bar{H}^T\varphi(t)\varphi^T(t)\bar{H} = \bar{H}^T[E\varphi(t)\varphi^T(t)]\bar{H}$$
$$= \bar{H}^T R_u(n)\bar{H}$$

Thus $Ez(t)z^T(t) = 0$ implies $\bar{H} = 0$ if and only if $R_u(n)$ is positive definite. However, this condition is the same as $u(t)$ being pe of order n. ∎

The following examples analyze the input signals introduced in Section 5.1 with respect to their persistent excitation properties.

Example 5.14 *A step function*
As already discussed in Example 5.13, a step function is pe of order 1, but of no greater order. ∎

Example 5.15 *A PRBS*
We consider a PRBS of period M. Let h be an arbitrary non-zero n-vector, where $n \leq M$. Set

$$e = (1 \ \ 1 \ \ \ldots \ \ 1)^T \qquad (n|1)$$

The covariance function of the PRBS is given by (5.9a). Therefore, for $h \neq 0$

$$h^T R_u(n) h = h^T \begin{pmatrix} a^2 & \ldots & -a^2/M \\ -a^2/M & . & \vdots \\ \vdots & . & \\ -a^2/M & \ldots & a^2 \end{pmatrix} h$$

$$= h^T\left[\left(a^2 + \frac{a^2}{M}\right)I - \frac{a^2}{M}ee^T\right]h = a^2\left(1 + \frac{1}{M}\right)h^T h - \frac{a^2}{M}(h^T e)^2$$

$$\geq a^2\left(1 + \frac{1}{M}\right)h^T h - \frac{a^2}{M}h^T h e^T e = a^2 h^T h\left[1 + \frac{1}{M} - \frac{n}{M}\right] > 0$$

The inequality follows from the Cauchy–Schwartz inequality.
In addition,

$$R_u(M+1) = \begin{pmatrix} a^2 & -a^2/M & \cdots & a^2 \\ -a^2/M & a^2 & & -a^2/M \\ \vdots & & \ddots & \vdots \\ -a^2/M & & & \cdot \\ a^2 & -a^2/M & \cdots & a^2 \end{pmatrix}$$

Since the first and the last row are equal, the matrix $R_u(M+1)$ is singular. Hence a PRBS with period M is pe of order equal to but not greater than M. ∎

Remark. A general periodic signal of period M can be persistently exciting of at most order M. This can be realized as follows. From (A5.1.8) it may be concluded that the spectral density is positive in at most M distinct frequencies. Then it follows from Properties 1 and 3 that the signal is pe of an order n that cannot exceed M. For the specific case of a PRBS we know from Example 5.5 that the spectral density is positive for exactly M distinct frequencies. Then Property 1 implies that it is pe of order M. ∎

Example 5.16 *An ARMA process*
Consider the ARMA process $u(t)$ introduced in Example 5.3. It follows from Property 2 that $u(t)$ is pe of any finite order. ∎

Example 5.17 *A sum of sinusoids*
Consider the signal

$$u(t) = \sum_{j=1}^{m} a_j \sin(\omega_j t + \varphi_j)$$

$$0 \leq \omega_1 < \omega_2 \ldots < \omega_m \leq \pi$$

which was introduced in Example 5.4. The spectral density was given in (5.12f) as

$$\phi_u(\omega) = \sum_{j=1}^{m} \frac{C_j}{2} [\delta(\omega - \omega_j) + \delta(\omega + \omega_j)]$$

Thus the spectral density is nonzero (in the interval $(-\pi, \pi]$) in exactly n points, where

$$n = \begin{cases} 2m & \text{if } 0 < \omega_1, \omega_m < \pi \\ 2m - 1 & \text{if } 0 = \omega_1 \text{ or } \omega_m = \pi \\ 2m - 2 & \text{if } 0 - \omega_1 \text{ and } \omega_m - \pi \end{cases} \quad (5.24)$$

It then follows from Properties 1 and 3 that $u(t)$ is pe of order n, as given by (5.24), but not of any greater order. ∎

Summary

Section 5.1 introduced some typical input signals that often are used in identification experiments. These included PRBS and ARMA processes. In Section 5.2 they were characterized in terms of the covariance function and spectral density.

Section 5.3 described several ways of implementing lowpass filtering. This is of interest for shaping low-frequency inputs. Such inputs are useful when the low frequencies in a model to be estimated are of particular interest.

Section 5.4 introduced the concept of persistent excitation, which is fundamental when analyzing parameter identifiability. A signal is persistently exciting (pe) of order n if its covariance matrix of order n is positive definite. In the frequency domain this condition is equivalent to requiring that the spectral density of the signal is nonzero in at least n points. It was shown that an ARMA process is pe of infinite order, while a sum of sinusoids is pe only of a finite order (in most cases equal to twice the number of sinusoids). A PRBS with period M is pe of order M, while a step function is pe of order one only.

Problems

Problem 5.1 *Nonnegative definiteness of the sample covariance matrix*
The following are two commonly used estimates of the covariance function of a stationary process:

$$R_k = \frac{1}{N} \sum_{t=1}^{N-k} y(t) y^T(t+k) \qquad R_{-k} = R_k^T \qquad k \geq 0$$

and

$$\tilde{R}_k = \frac{1}{N-k} \sum_{t=1}^{N-k} y(t) y^T(t+k) \qquad \tilde{R}_{-k} = \tilde{R}_k^T \qquad k \geq 0$$

where $\{y(1), \ldots, y(N)\}$ is the sample of observations.

(a) Show that the sample covariance matrix $[\tilde{R}_{i-j}]$ is not necessarily nonnegative definite.
(b) Let $R_k = 0$, $|k| \geq N$. Then show that the sample covariance matrix of any dimension $[R_{i-j}]$ is nonnegative definite.

Problem 5.2 *A rapid method for generating sinusoidal signals on a computer*
The sinusoidal signal

$$u(t) = a \sin(\bar{\omega} t + \varphi) \qquad t = 1, 2, \ldots$$

obeys the following recurrence equation

$$u(t) - (2 \cos \bar{\omega}) u(t-1) + u(t-2) = 0 \qquad \text{(i)}$$

Show this in two ways: (a) using simple trigonometric equalities; and (b) using the spectral properties of sinusoidal signals and the formula for the transfer of spectral densities through linear systems. Use the property above to conceive a computationally efficient method for generating sinusoidal signals on a computer. Implement this method and study empirically its computational efficiency and the propagation of numerical errors compared to standard procedures.

Problems 127

Problem 5.3 *Spectral density of the sum of two sinusoids*
Consider the signal

$$u(t) = a_1 \sin \omega_1 t + a_2 \sin \omega_2 t$$

where $\omega_j = 2\pi k_j/M$ ($j = 1, 2$) and k_1, k_2 are integers in the interval $[1, M-1]$. Derive the spectral density of $u(t)$ using (5.12d), (A5.1.8), (A5.1.15). Compare with (5.12f).

Problem 5.4 *Admissible domain for ϱ_1 and ϱ_2 of a stationary process*
Let $\{r_k\}$ denote the covariance at lag k of a stationary process and let $\varrho_k = r_k/r_0$ be the kth correlation coefficient. Derive the admissible domain for ϱ_1, ϱ_2.
Hint. The correlation matrix $[\varrho_{i-j}]$ must be nonnegative definite.

Problem 5.5 *Admissible domain of ϱ_1 and ϱ_2 for MA(2) and AR(2) processes*
Which is the domain spanned by the first two correlations ϱ_1, ϱ_2 of a MA(2) process? What is the corresponding domain for an AR(2) process? Which one of these two domains is the largest?

Problem 5.6 *Spectral properties of a random wave*
Consider a random wave $u(t)$ generated as follows:

$$u(t) = \pm a$$

$$u(t) = \begin{cases} u(t-1) & \text{with probability } 1-\alpha \\ -u(t-1) & \text{with probability } \alpha \end{cases}$$

where $0 < \alpha < 1$. The probability of change at time t is independent of what happened at previous time instants.

(a) Derive the covariance function.
   Hint. Use the ideas of Example 5.11.
(b) Determine the spectral density. Also show that the signal has low-frequency character ($\phi(\omega)$ decreases with increasing ω) if and only if $\alpha \leq 0.5$.

Problem 5.7 *Spectral properties of a square wave*
Consider a square wave of period $2n$, defined by

$$u(t) = 1 \quad t = 0, \ldots, n-1$$
$$u(t + n) = -u(t) \quad \text{all } t$$

(a) Derive its covariance function.
(b) Show that the spectral density is given by

$$\phi_u(\omega) = \sum_{j=1}^{n} \frac{2}{n^2} \frac{1}{1 - \cos \pi \frac{2j-1}{n}} \delta\left(\omega - \frac{\pi}{n}(2j-1)\right)$$

Hint. $\sum_{j=0}^{n-1} jz^{j-1} = \frac{d}{dz} \sum_{j=0}^{n-1} z^j = -n\frac{z^{n-1}}{(1-z)} + \frac{1-z^n}{(1-z)^2}$.

(c) Of what order will $u(t)$ be persistently exciting?

Problem 5.8 *Simple conditions on the covariances of a moving average process*
Let $\{r_k\}_{k=0}^n$ denote some real scalars, with $r_0 > 0$. Show that
(a) $\{r_k\}$ are the covariances of a nth-order MA process if

$$\sum_{k=1}^n |r_k| \leq \frac{1}{2} r_0 \quad \text{(sufficient condition)}$$

(b) $\{r_k\}$ are the covariances of a nth-order MA process only if

$$\sum_{k=1}^n |r_k| \leq \frac{n}{2} r_0 \quad \text{(necessary condition)}$$

Problem 5.9 *Weighting function estimation with PRBS*
Derive a closed-form formula for the weighting coefficient estimates (5.18) when the input is a PRBS of amplitudes ± 1 and period $N \geq M$. Show that

$$\hat{h}(k) = \frac{N}{(N+1)(N-M+1)} \left[\sum_{\substack{i=0 \\ i \neq k}}^{M-1} \hat{r}_{yu}(i) + (N - M + 2)\hat{r}_{yu}(k) \right]$$

If N is much larger than M show that this can be simplified to $\hat{h}(k) \approx \hat{r}_{yu}(k)$. Next observe that for $N = M$, the formula above reduces to $\hat{h}(k) \approx \hat{r}_{yu}(k) + \sum_{i=0}^{N-1} \hat{r}_{yu}(i)$. This might appear to be a contradiction of the fact that for large N the covariance matrix of a PRBS with unit amplitude is approximately equal to the identity matrix. Resolve this contradiction.

Problem 5.10 *The cross-covariance matrix of two autoregressive processes obeys a Lyapunov equation*
Consider the following two stationary AR processes:

$$(1 + a_1 q^{-1} + \ldots + a_n q^{-n}) y(t) = e(t)$$
$$(1 + b_1 q^{-1} + \ldots + b_m q^{-m}) x(t) = e(t)$$
$$Ee(t)e(s) = \delta_{t,s}$$

Let

$$R = E \begin{pmatrix} y(t-1) \\ \vdots \\ y(t-n) \end{pmatrix} \begin{pmatrix} x(t-1) & \ldots & x(t-m) \end{pmatrix} \quad \text{(cross-covariance matrix)}$$

$$A = \begin{pmatrix} -a_1 & \ldots & -a_{n-1} & -a_n \\ 1 & & & 0 \\ & \ddots & 0 & \vdots \\ 0 & & 1 & 0 \end{pmatrix}$$

$$B = \begin{pmatrix} -b_1 & \ldots & -b_{m-1} & -b_m \\ 1 & & & 0 \\ & \ddots & 0 & \vdots \\ 0 & & 1 & 0 \end{pmatrix}$$

Show that

$$R - ARB^T = \mu_n \mu_m^T$$

where μ_n is the first nth-order unit vector; that is $\mu_n = (1 \ 0 \ \ldots \ 0)^T$.

Bibliographical notes

Brillinger (1981), Hannan (1970), Anderson (1971), Oppenheim and Willsky (1983), and Aoki (1987) are good general sources on time series analysis.

PRBS and other pseudorandom signals are described and analyzed in detail by Davies (1970). See also Verbruggen (1975).

The concept of persistent excitation originates from Åström and Bohlin (1965). A further analysis of this concept was carried out by Ljung (1971) and Stoica (1981a).

Digital lowpass filtering is treated, for example, by Oppenheim and Schafer (1975), and Rabiner and Gold (1975).

Appendix A5.1
Spectral properties of periodic signals

Let $u(t)$ be a deterministic signal that is periodic with period M, i.e.

$$u(t) = u(t - M) \quad \text{all } t \tag{A5.1.1}$$

The mean value m and the covariance function $r(\tau)$ are then defined as

$$m \triangleq \lim_{N \to \infty} \frac{1}{N} \sum_{t=1}^{N} u(t) = \frac{1}{M} \sum_{t=1}^{M} u(t)$$

$$r(\tau) \triangleq \lim_{N \to \infty} \frac{1}{N} \sum_{t=1}^{N} [u(t + \tau) - m][u(t) - m]^T \tag{A5.1.2}$$

$$= \frac{1}{M} \sum_{t=1}^{M} [u(t + \tau) - m][u(t) - m]^T$$

The expressions for the limits in the definitions of m and $r(\tau)$ can be readily established. For general signals (deterministic or stochastic) it holds that

$$r(-\tau) = r^T(\tau) \tag{A5.1.3}$$

In addition, for periodic signals we evidently have

$$r(M + \tau) = r(\tau) \tag{A5.1.4}$$

and hence

$$r(M - k) = r^T(k) \tag{A5.1.5}$$

130   Input signals   Chapter 5

The general relations (A3.1.2), (A3.1.3) between a covariance function and the associated spectral density apply also in this case. However, when $r(\tau)$ is a periodic function, the spectral density will no longer be a smooth function of ω. Instead it will consist of a number of weighted Dirac pulses. This means that only certain frequencies are present in the signal.

For periodic signals, an alternative to the definition of spectral (A3.1.2)

$$\phi(\omega) = \frac{1}{2\pi} \sum_{\tau=-\infty}^{\infty} r(\tau) e^{-i\tau\omega}$$

is given by the discrete Fourier transform (DFT)

$$\bar{\phi}_n = \sum_{\tau=0}^{M-1} r(\tau) e^{-i2\pi\tau n/M} \qquad n = 0, \ldots, M-1 \qquad (A5.1.6)$$

Then instead of the relation (A3.1.3),

$$r(\tau) = \int_{-\pi}^{\pi} e^{i\tau\omega} \phi(\omega) d\omega$$

we have

$$r(\tau) = \frac{1}{M} \sum_{n=0}^{M-1} \bar{\phi}_n e^{i2\pi\tau n/M} \qquad (A5.1.7)$$

We will, however, keep the original definition of a spectral density. For a treatment of the DFT, see Oppenheim and Schafer (1975), and Čižek (1986). The relation between $\phi(\omega)$ and $\{\bar{\phi}_n\}$ will become clear later (see (A5.1.8), (A5.1.15) and the subsequent discussion).

For periodic signals it will be convenient to define the spectral density over the interval $[0, 2\pi]$ rather than $[-\pi, \pi]$. This does not introduce any restriction since $\phi(\omega)$ is by definition a periodic (even) function with period 2π.

For periodic signals it is not very convenient to evaluate the spectral density from (A3.1.2). An alternative approach is to try the structure

$$\phi(\omega) = \sum_{k=0}^{M-1} C_k \delta\left(\omega - 2\pi \frac{k}{M}\right) \qquad (A5.1.8)$$

where C_k are matrix coefficients to be determined. It will be shown that with certain values of $\{C_k\}$ the equation (A3.1.3) is satisfied by (A5.1.8).

First note that substitution of (A5.1.8) in the relation (A3.1.3) gives

$$r(\tau) = \int_0^{2\pi} \sum_{k=0}^{M-1} C_k \delta\left(\omega - 2\pi \frac{k}{M}\right) e^{i\tau\omega} d\omega$$

$$= \sum_{k=0}^{M-1} C_k \alpha^{k\tau} \qquad (A5.1.9)$$

where
$$\alpha \triangleq e^{i2\pi/M} \tag{A5.1.10}$$

Note that
$$\alpha \neq 1, \alpha^M = 1 \tag{A5.1.11}$$

It follows directly from (A5.1.11) that the right-hand side of (A5.1.9) is periodic with period M. It is therefore sufficient to verify (A5.1.9) for the values $\tau = 0, \ldots, M - 1$. For this purpose introduce the $(M|M)$ matrix

$$U = \frac{1}{\sqrt{M}} \begin{pmatrix} 1 & 1 & 1 & \cdots & 1 \\ 1 & \alpha & \alpha^2 & & \alpha^{M-1} \\ 1 & \alpha^2 & \alpha^4 & & \cdot \\ \vdots & \vdots & \vdots & & \vdots \\ 1 & \alpha^{M-1} & \alpha^{2(M-1)} & \cdots & \alpha^{(M-1)^2} \end{pmatrix} \tag{A5.1.12}$$

Then note that (A5.1.9) for $\tau = 0, \ldots, M - 1$ can be written in matrix form as

$$\begin{pmatrix} r(0) \\ r(1) \\ \vdots \\ r(M-1) \end{pmatrix} = (\sqrt{M} U \otimes I) \begin{pmatrix} C_0 \\ \vdots \\ C_{M-1} \end{pmatrix} \tag{A5.1.13}$$

where $\otimes$ denotes Kronecker product (see Section A.7 of Appendix A). The matrix U can be viewed as a Vandermonde matrix and it is therefore easy to prove that it is nonsingular. This means that $\{C_k\}_{k=0}^{M-1}$ can be derived uniquely from $\{r(\tau)\}_{\tau=0}^{M-1}$. In fact U turns out to be unitary, i.e.
$$U^{-1} = U^* \tag{A5.1.14}$$

where U^* denotes the complex conjugate transpose of U. To prove (A5.1.14) evaluate the elements of UU^*. When doing this it is convenient to let the indexes vary from 0 to $M - 1$. Since $U_{jk} = \alpha^{jk}/\sqrt{M}$,

$$(UU^*)_{jk} = \sum_{p=0}^{M-1} U_{jp} U_{pk}^* = \frac{1}{M} \sum_{p=0}^{M-1} \alpha^{jp} \alpha^{-pk}$$

$$= \frac{1}{M} \sum_{p=0}^{M-1} \alpha^{p(j-k)}$$

Hence
$$(UU^*)_{jj} = \frac{1}{M} \sum_{p=0}^{M-1} 1 = 1$$

and for $j \neq k$
$$(UU^*)_{jk} = \frac{1}{M} \frac{1 - \alpha^{M(j-k)}}{1 - \alpha^{j-k}} = 0$$

These calculations prove that $UU^* = I$, which is equivalent to (A5.1.14).

Using (A5.1.12)–(A5.1.14) an expression can be derived for the coefficients $\{C_k\}$ in the spectral density. From (A5.1.14) and Lemma A.32,

$$(U \otimes I)^{-1} = U^* \otimes I$$

Thus

$$C_k = \frac{1}{\sqrt{M}}[U_{k0}^* r(0) + U_{k1}^* r(1) + \ldots + U_{k,M-1}^* r(M-1)]$$

$$= \frac{1}{M}[r(0) + \alpha^{-k} r(1) + \ldots + \alpha^{-k(M-1)} r(M-1)]$$

or

$$C_k = \frac{1}{M} \sum_{j=0}^{M-1} \alpha^{-kj} r(j) \tag{A5.1.15}$$

Note that from (A5.1.6) and (A5.1.15) it follows that

$$C_k = \frac{1}{M} \bar{\phi}_k$$

Hence the weights of the Dirac pulses in the spectral density are equal to the normalized DFT of the covariance function.

In the remaining part of this appendix it is shown that the matrix C_k is Hermitian and nonnegative definite. Introduce the matrix

$$R = \begin{pmatrix} r(0) & r(1) & \ldots & r(M-1) \\ r(-1) & r(0) & & \\ \vdots & & \ddots & \\ r(1-M) & & \ldots & r(0) \end{pmatrix}$$

$$= \lim_{N \to \infty} \frac{1}{N} \sum_{t=1}^{N} \left\{ \begin{pmatrix} u(t-1) - m \\ \vdots \\ u(t-M) - m \end{pmatrix} \begin{pmatrix} u^T(t-1) - m^T & \ldots & u^T(t-M) - m^T \end{pmatrix} \right\}$$

which by construction is nonnegative definite. Further, introduce the complex-valued matrix

$$a_k \triangleq \begin{pmatrix} 1 & \alpha^k & \alpha^{2k} & \ldots & \alpha^{(M-1)k} \end{pmatrix} \otimes I$$

The quadratic form $a_k R a_k^*$ is a Hermitian and nonnegative definite matrix by construction. Then (A5.1.5), (A5.1.10), (A5.1.11) give

$$a_k R a_k^* = r(0)M + \sum_{\tau=1}^{M-1} (M - \tau) r(\tau) \alpha^{-k\tau} + \sum_{\tau=1}^{M-1} (M - \tau) r(-\tau) \alpha^{k\tau}$$

$$= r(0)M + \sum_{\tau=1}^{M-1} (M - \tau) r(\tau) \alpha^{-k\tau} + \sum_{l=1}^{M-1} lr(l) \alpha^{k(M-l)}$$

$$= M \left[r(0) + \sum_{\tau=1}^{M-1} r(\tau) \alpha^{-k\tau} \right] = M^2 C_k$$

This calculation shows that the C_k, for $k = 0, \ldots, M - 1$, are Hermitian and nonnegative definite matrices. Finally observe that if $u(t)$ is a persistently exciting signal of order M, then R is positive definite. Since the matrices a_k have full row rank, it follows that in such a case $\{C_k\}$ are positive definite matrices.

Complement C5.1
Difference equation models with persistently exciting inputs

The persistent excitation property has been introduced in relation to the LS estimation of weighting function (or all-zeros) models. It is shown in this complement that the pe property is relevant to difference equation (or poles–zeros) models as well. The output $y(t)$ of a linear, asymptotically stable, rational filter $B(q^{-1})/A(q^{-1})$ with input $u(t)$ can be written as

$$A(q^{-1}) y(t) = B(q^{-1}) u(t) + \varepsilon(t) \tag{C5.1.1}$$

where

$$A(q^{-1}) = 1 + a_1 q^{-1} + \ldots + a_{na} q^{-na}$$
$$B(q^{-1}) = b_1 q^{-1} + \ldots + b_{nb} q^{-nb}$$

and where $\varepsilon(t)$ denotes some random disturbance. With

$$\varphi(t) \triangleq (-y(t-1) \quad \ldots \quad -y(t-na) \quad u(t-1) \quad \ldots \quad u(t-nb))^T$$
$$\theta \triangleq (a_1 \quad \ldots \quad a_{na} \quad b_1 \quad \ldots \quad b_{nb})^T$$

the difference equation (C5.1.1) can be written as a linear regression

$$y(t) = \varphi^T(t) \theta + \varepsilon(t) \tag{C5.1.2}$$

Similarly to the case of all-zeros models, the existence of the least squares estimate $\hat{\theta}$ will be asymptotically equivalent to the nonsingularity (positive definiteness) of the covariance matrix

$$R = E\varphi(t)\varphi^T(t)$$

As shown below, the condition $R > 0$ is intimately related to the persistent excitation property of $u(t)$. The cases $\varepsilon(t) \equiv 0$ and $\varepsilon(t) \neq 0$ will be considered separately.

Noise-free output

For $\varepsilon(t) \equiv 0$ one can write

$$\varphi(t) = \begin{matrix} na\left\{\vphantom{\begin{pmatrix}1\\1\\1\\1\end{pmatrix}}\right. \\ nb\left\{\vphantom{\begin{pmatrix}1\\1\\1\\1\end{pmatrix}}\right. \end{matrix} \begin{pmatrix} -b_1 & \cdots & -b_{nb} & 0 & & \\ & \ddots & & \ddots & & \\ 0 & & -b_1 & \cdots & -b_{nb} \\ 1 & a_1 & \cdots & a_{na} & 0 & \\ & \ddots & & & \ddots & \\ 0 & & 1 & a_1 & \cdots & a_{na} \end{pmatrix} \begin{pmatrix} \dfrac{1}{A(q^{-1})}u(t-1) \\ \vdots \\ \dfrac{1}{A(q^{-1})}u(t-na-nb) \end{pmatrix}$$

$$\triangleq \mathscr{S}(-B, A)\tilde{\varphi}(t)$$

Thus

$$R = \mathscr{S}(-B, A) E\tilde{\varphi}(t)\tilde{\varphi}^{\mathrm{T}}(t)\mathscr{S}^{\mathrm{T}}(-B, A) \triangleq \mathscr{S}(-B, A)\tilde{R}\mathscr{S}^{\mathrm{T}}(-B, A)$$

which shows that

$$\{R > 0\} \Leftrightarrow \{\mathscr{S} \text{ nonsingular and } \tilde{R} > 0\}$$

Now $\mathscr{S}(-B, A)$ is the Sylvester matrix associated with the polynomials $-B(z)$ and $A(z)$ (see Definition A.8). It is nonsingular if and only if $A(z)$ and $B(z)$ are coprime (see Lemma A.30).

Regarding the matrix $\tilde{R}$, it is positive definite if and only if $u(t)$ is pe of order $(na + nb)$; see Property 5 in Section 5.4. Thus

$$\{R > 0\} \Leftrightarrow \{u(t) \text{ is pe of order } (na + nb),$$
$$\text{and } A(z), B(z) \text{ are coprime}\} \qquad \text{(C5.1.3)}$$

Noisy output

Let

$$\bar{\varepsilon}(t) = -\frac{1}{A(q^{-1})}\varepsilon(t) \qquad x(t) = \frac{B(q^{-1})}{A(q^{-1})}u(t)$$

and assume that $Eu(t)\varepsilon(s) = 0$ for all t and s. Then one can write

$$R = E \begin{pmatrix} x(t-1) \\ \vdots \\ x(t-na) \\ u(t-1) \\ \vdots \\ u(t-nb) \end{pmatrix} \cdot (x(t-1) \quad \cdots \quad x(t-na) \quad u(t-1) \quad \cdots \quad u(t-nb))$$

$$+ E \begin{pmatrix} \bar{\varepsilon}(t-1) \\ \vdots \\ \bar{\varepsilon}(t-na) \\ 0 \\ \vdots \\ 0 \end{pmatrix} \begin{pmatrix} \bar{\varepsilon}(t-1) & \cdots & \bar{\varepsilon}(t-na) & 0 & \cdots & 0 \end{pmatrix}$$

$$\triangleq \begin{pmatrix} \tilde{A} & B \\ B^T & C \end{pmatrix} + \begin{pmatrix} \bar{A} & 0 \\ 0 & 0 \end{pmatrix} \tag{C5.1.4}$$

Clearly the condition $C > 0$ is necessary for $R > 0$. Under the assumption that $\bar{A} > 0$, it is also sufficient. Indeed, it follows from Lemma A.3 that if $C > 0$ then

$$\tilde{A} - BC^{-1}B^T \geq 0$$

and

$$\text{rank } R = nb + \text{rank } (\tilde{A} + \bar{A} - BC^{-1}B^T)$$

Thus, assuming $\bar{A} > 0$,

$$\text{rank } R = na + nb$$

The conclusion is that under weak conditions on the noise $\varepsilon(t)$ (more precisely $\bar{A} > 0$), the following equivalence holds:

$$\boxed{\{R > 0\} \Leftrightarrow \{u(t) \text{ is pe of order } (nb)\}} \tag{C5.1.5}$$

Complement C5.2
Condition number of the covariance matrix of filtered white noise

As shown in Properties 2 and 5 of Section 5.4, filtered white noise is a persistently exciting signal of any order. Thus, the matrix

$$R_m = E \begin{pmatrix} u(t-1) \\ \vdots \\ u(t-m) \end{pmatrix} \begin{pmatrix} u(t-1) & \cdots & u(t-m) \end{pmatrix} \qquad u(t) = H(q^{-1})e(t) \tag{C5.2.1}$$

where $Ee(t)e(s) = \delta_{t,s}$, and $H(q^{-1})$ is an asymptotically stable filter, is positive definite for any m. From a practical standpoint it would be more useful to have information on the condition number of R_m (cond (R_m)) rather than to just know that $R_m > 0$. In the following a simple upper bound on

$$\text{cond}(R_m) \triangleq \lambda_{\max}(R_m)/\lambda_{\min}(R_m) \tag{C5.2.2}$$

is derived (see Stoica et al., 1985c; Grenander and Szegö, 1958).

First recall that for a symmetric matrix R the following relationships hold:

$$\lambda_{\min} = \inf_x \frac{x^T R x}{x^T x}$$

$$\lambda_{\max} = \sup_x \frac{x^T R x}{x^T x}$$

Since the matrices R_{m+1} and R_m are nested, i.e.

$$R_{m+1} = \begin{pmatrix} R_m & * \\ * & * \end{pmatrix}$$

it follows that

$$\lambda_{\min}(R_{m+1}) \leq \lambda_{\min}(R_m)$$
$$\lambda_{\max}(R_{m+1}) \geq \lambda_{\max}(R_m)$$

Thus

$$\text{cond}(R_{m+1}) \geq \text{cond}(R_m) \tag{C5.2.3}$$

Next we prove that

$$\sigma_{\min} \triangleq \lim_{m \to \infty} \lambda_{\min}(R_m) = \inf_\omega |H(e^{i\omega})|^2 \tag{C5.2.4}$$

$$\sigma_{\max} \triangleq \lim_{m \to \infty} \lambda_{\max}(R_m) = \sup_\omega |H(e^{i\omega})|^2 \tag{C5.2.5}$$

Let ϱ be a real number, and consider the matrix $R_m - \varrho I$. The (k, p) element of this matrix is given by

$$\frac{1}{2\pi} \int_{-\pi}^{\pi} |H(e^{i\omega})|^2 e^{i\omega(k-p)} d\omega - \varrho \delta_{k,p} = \frac{1}{2\pi} \int_{-\pi}^{\pi} [|H(e^{i\omega})|^2 - \varrho] e^{i\omega(k-p)} d\omega \tag{C5.2.6}$$

If ϱ satisfies

$$|H(e^{i\omega})|^2 - \varrho \geq 0 \quad \text{for } \omega \in (-\pi, \pi) \tag{C5.2.7}$$

then it follows from (C5.2.6) that $R_m - \varrho I$ is the covariance matrix of a process with the spectral density function equal to the left-hand side of (C5.2.7). Thus

$$R_m \geq \varrho I \quad \text{for all } m \tag{C5.2.8}$$

If (C5.2.7) does not hold, then (C5.2.8) cannot be true. Now $\sigma_{\min}$ is uniquely defined by the following two conditions:

$$R_m \geq \sigma_{\min} I \quad \text{for } all \; m$$

and

$$R_m \geq (\sigma_{\min} + \varepsilon) I \quad \text{some } \varepsilon > 0, \text{ cannot hold for all } m$$

From the above discussion it readily follows that $\sigma_{\min}$ is given by (C5.2.4). The proof of (C5.2.5) is similar and is therefore omitted.

The result above implies that

$$\operatorname{cond}(R_m) \leq \sup_{\omega} |H(e^{i\omega})|^2 / \inf_{\omega} |H(e^{i\omega})|^2 \qquad (C5.2.9)$$

with equality holding in the limit as $m \to \infty$. Thus, if $H(q^{-1})$ has zeros on or close to the unit circle, then the matrix R_m is expected to be ill-conditioned for large m.

Complement C5.3
Pseudorandom binary sequences of maximum length

Pseudorandom binary sequences (PRBS) are two-state signals which may be generated, for example, by using a shift register of order n as depicted in Figure C5.3.1.

The register state variables are fed with 1 or 0. Every initial state vector is allowed except the all-zero state. When the clock pulse is applied, the value of the kth state is transferred to the $(k + 1)$th state and a new value is introduced into the first state through the feedback path.

The feedback coefficients, $a_1, \ldots, a_n$, are either 0 or 1. The modulo-two addition, denoted by $\oplus$ in the figure, is defined in Table C5.3.1. The system operates in discrete time. The clock period is equal to the sampling time.

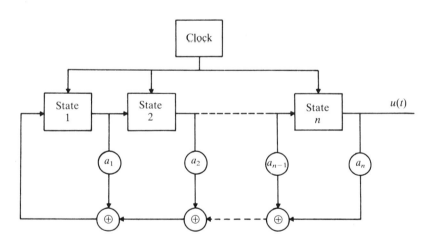

FIGURE C5.3.1 Shift register with modulo-two feedback path.

TABLE C5.3.1 Modulo-two addition of two binary variables

u_1	u_2	$u_1 \oplus u_2$
0	0	0
1	0	1
0	1	1
1	1	0

The system shown in Figure C5.3.1 can be represented in state space form as

$$x(t+1) = \begin{pmatrix} a_1 & \cdots & & a_n \\ 1 & & & 0 \\ & \ddots & & \\ 0 & & 1 & 0 \end{pmatrix} x(t) \qquad (C5.3.1)$$

$$u(t) = (0 \quad \cdots \quad 0 \quad 1)x(t)$$

where all additions must be carried out modulo-two. Note the similarity with 'standard' state space models. In fact, with appropriate interpretations, it is possible to use several 'standard' results for state space models for the model (C5.3.1). Models of the above form where the state vector can take only a finite number of values are often called finite state systems. For a general treatment of such systems, refer to Zadeh and Polak (1969), Golomb (1967), Peterson and Weldon (1972).

It follows from the foregoing discussion that the shift register of Figure C5.3.1 will generate a sequence of ones and zeros. This is called a pseudorandom binary sequence. The name may be explained as follows. A PRBS is a purely deterministic signal: given the initial state of the register, its future states can be computed exactly. However, its correlation function resembles the correlation function of white random noise. This is true at least for certain types of PRBS called maximum length PRBS (ml PRBS). For this reason, this type of sequence is called 'pseudorandom'. Of course, the sequence is called 'binary' since it contains two states only.

The PRBS introduced above takes the values 0 and 1. To generate a signal that shifts between the values a and b, simply take

$$y(t) = a + (b - a)u(t) \qquad (C5.3.2)$$

In particular, if $b = -a$ (for symmetry) and $b = 1$ (for simplicity; it will anyhow act as a scaling factor only), then

$$y(t) = -1 + 2u(t) \qquad (C5.3.3)$$

This complement shows how to generate an m.l. PRBS by using the shift register of Figure C5.3.1. We also establish some important properties of ml PRBS and in particular justify the claim introduced above that the correlation properties of ml PRBS resemble those of white noise.

Maximum length PRBS

For a shift register with n states there is a possible total of 2^n different state vectors composed of ones and zeros. Thus the maximum period of a sequence generated using such a system is 2^n. In fact, 2^n is an upper bound which cannot be attained. The reason is that occurrence of the all-zero state must be prevented. If such a state occurred then the state vector would remain zero for all future times. It follows that the maximum possible period is

$$M = 2^n - 1 \qquad (C5.3.4)$$

where

n = order (number of stages) of the shift register

A PRBS of period equal to M is called a maximum length PRBS. Whether or not an nth-order system will generate an m.l. PRBS depends on its feedback path. This is illustrated in the following example.

Example C5.3.1 *Influence of the feedback path on the period of a PRBS*
Let the initial state of a three-stage ($n = 3$) shift register be $(1 \ 0 \ 0)^T$. For the following feedback paths the corresponding sequences of generated states will be determined.

(a) Feedback from states 1 and 2, i.e. $a_1 = 1$, $a_2 = 1$, $a_3 = 0$. The corresponding sequence of state vectors

$$\begin{pmatrix}1\\0\\0\end{pmatrix}, \begin{pmatrix}1\\1\\0\end{pmatrix}, \begin{pmatrix}0\\1\\1\end{pmatrix}, \begin{pmatrix}1\\0\\1\end{pmatrix}, \begin{pmatrix}1\\1\\0\end{pmatrix}$$

has period equal to 3.

(b) Feedback from states 1 and 3, i.e. $a_1 = 1$, $a_2 = 0$, $a_3 = 1$. The corresponding sequence of state vectors

$$\begin{pmatrix}1\\0\\0\end{pmatrix}, \begin{pmatrix}1\\1\\0\end{pmatrix}, \begin{pmatrix}1\\1\\1\end{pmatrix}, \begin{pmatrix}0\\1\\1\end{pmatrix}, \begin{pmatrix}1\\0\\1\end{pmatrix}, \begin{pmatrix}0\\1\\0\end{pmatrix}, \begin{pmatrix}0\\0\\1\end{pmatrix}, \begin{pmatrix}1\\0\\0\end{pmatrix}$$

has the maximum period $M = 2^3 - 1 = 7$.

(c) Feedback from states 2 and 3, i.e. $a_1 = 0$, $a_2 = 1$, $a_3 = 1$. The corresponding sequences of state vectors

$$\begin{pmatrix}1\\0\\0\end{pmatrix}, \begin{pmatrix}0\\1\\0\end{pmatrix}, \begin{pmatrix}1\\0\\1\end{pmatrix}, \begin{pmatrix}1\\1\\0\end{pmatrix}, \begin{pmatrix}1\\1\\1\end{pmatrix}, \begin{pmatrix}0\\1\\1\end{pmatrix}, \begin{pmatrix}0\\0\\1\end{pmatrix}, \begin{pmatrix}1\\0\\0\end{pmatrix}$$

has again the maximum possible period $M = 7$.

(d) Feedback from all states, i.e. $a_1 = 1$, $a_2 = 1$, $a_3 = 1$. The corresponding sequence of the state vectors

$$\begin{pmatrix}1\\0\\0\end{pmatrix}, \begin{pmatrix}1\\1\\0\end{pmatrix}, \begin{pmatrix}0\\1\\1\end{pmatrix}, \begin{pmatrix}0\\0\\1\end{pmatrix}, \begin{pmatrix}1\\0\\0\end{pmatrix}$$

has period equal to 4. ■

There are at least two reasons for concentrating on *maximum length* (ml) PRBS:

- The correlation function of an ml PRBS resembles that of a white noise (see below). This property is not guaranteed for nonmaximum length PRBSs.
- As shown in Section 5.4, a periodic signal is persistently exciting of an order which cannot exceed the value of its period. Since persistent excitation is a vital condition for identifiability, a long period will give more flexibility in this respect.

The problem of choosing the feedback path of a shift register to give an ml PRBS is discussed in the following.

Let

$$A(q^{-1}) = 1 \oplus a_1 q^{-1} \oplus a_2 q^{-2} \oplus \ldots \oplus a_n q^{-n} \qquad \text{(C5.3.5)}$$

where q^{-1} is the unit delay operator. The PRBS $u(t)$ generated using (C5.3.1) obeys the following homogeneous equation:

$$A(q^{-1})u(t) = 0 \qquad \text{(C5.3.6)}$$

This can be shown as follows. From (C5.3.1),

$$x_n(t + j) = x_{n-j}(t) \qquad j = 1, \ldots, (n-1)$$

and

$$\begin{aligned} A(q^{-1})u(t) &= x_n(t) \oplus a_1 x_n(t-1) \oplus \ldots \oplus a_n x_n(t-n) \\ &= x_1(t-n+1) \oplus a_1 x_1(t-n) \oplus \ldots \oplus a_n x_n(t-n) = 0 \end{aligned}$$

The problem to study is the choice of the feedback coefficients $\{a_i\}$ such that the equation (C5.3.6) has no solution $u(t)$ with period smaller than $2^n - 1$. A necessary and sufficient condition on $\{a_i\}$ for this property to hold is provided by the following lemma.

Lemma C5.3.1

The homogeneous recursive equation (C5.3.6) has only solutions of period $2^n - 1$, if and only if the following two conditions are satisfied:

(i) The binary polynomial $A(q^{-1})$ is irreducible [i.e. there do not exist any two polynomials $A_1(q^{-1})$ and $A_2(q^{-1})$ with binary coefficients such that $A(q^{-1}) = A_1(q^{-1})A_2(q^{-1})$].
(ii) $A(q^{-1})$ is a factor of $1 \oplus q^{-M}$ but is not a factor of $1 \oplus q^{-p}$ for any $p < M = 2^n - 1$.

Proof. We first show that condition (i) is necessary. For this purpose assume that $A(q^{-1})$ is not irreducible. Then there are binary polynomials $A_1(q^{-1})$ and $A_2(q^{-1})$ with

$$\deg A_1 = n_1 > 0 \qquad \deg A_2 = n_2 > 0 \qquad \text{and } n_1 + n_2 = n$$

such that the following factorization holds:

$$A(q^{-1}) = A_1(q^{-1})A_2(q^{-1})$$

Let $u_1(t)$ and $u_2(t)$ be solutions to the equations

$$\begin{aligned} A_1(q^{-1})u_1(t) &= 0 \\ A_2(q^{-1})u_2(t) &= 0 \end{aligned} \qquad \text{(C5.3.7)}$$

Then

$$u(t) = u_1(t) \oplus u_2(t) \qquad \text{(C5.3.8)}$$

is a solution to (C5.3.6). Now, according to the above discussion, the maximum period of

$u_i(t)$ is $2^{n_i} - 1$ ($i = 1, 2$). Thus, the maximum possible period of $u(t)$, (C5.3.8), is at most equal to $(2^{n_1} - 1)(2^{n_2} - 1)$, which is strictly less than $2^n - 1$. This establishes the necessity of condition (i).

We next show that if $A(q^{-1})$ is irreducible then the equation (C5.3.6) has solutions of period p if and only if $A(q^{-1})$ is a factor of $1 \oplus q^{-p}$. The 'if' part is immediate. If there exists a binary polynomial $P(q^{-1})$ such that

$$A(q^{-1})P(q^{-1}) = 1 \oplus q^{-p}$$

then it follows that

$$(1 \oplus q^{-p})u(t) = P(q^{-1})[A(q^{-1})u(t)] = 0$$

which implies that $u(t)$ has period p.

The 'only if' part is more difficult to prove. Introduce the following 'z-transform' of the binary sequence $u(t)$

$$U(z) = \sum_{t=0}^{\infty} u(t)z^t \qquad (C5.3.9)$$

where Σ denotes modulo-two summation. Let

$$P(z) = u(0) \oplus u(1)z \oplus \ldots \oplus u(p-1)z^{p-1}$$

Due to the assumption that the period of $u(t)$ is equal to p, we can write

$$U(z) = P(z) \oplus z^p P(z) \oplus z^{2p} P(z) \oplus \ldots = P(z)[1 \oplus z^p \oplus z^{2p} \oplus \ldots]$$

Hence

$$U(z)[1 \oplus z^p] = P(z) \qquad (C5.3.10)$$

On the other hand, we also have (cf. (C5.3.6))

$$U(z) = \sum_{t=0}^{\infty} \left[\sum_{i=1}^{n} a_i u(t-i) \right] z^t = \sum_{i=1}^{n} a_i z^i \sum_{t=0}^{\infty} u(t-i)z^{t-i}$$

$$= \sum_{i=1}^{n} a_i z^i \left[\sum_{t=-i}^{-1} u(t)z^t \oplus U(z) \right] = Q(z) \oplus \left[\sum_{i=1}^{n} a_i z^i \right] U(z) \qquad (C5.3.11)$$

where

$$Q(z) \triangleq \sum_{i=1}^{n} a_i \sum_{t=-i}^{-1} u(t)z^{t+i} \qquad (C5.3.12)$$

is a binary polynomial of degree $(n-1)$, the coefficients of which depend on the initial values $u(-1), \ldots, u(-n)$ and the feedback coefficients of the shift register. It follows that

$$U(z) = Q(z) \oplus [A(z) \oplus 1]U(z)$$
$$= Q(z) \oplus A(z)U(z) \oplus U(z)$$

and hence

$$Q(z) = A(z)U(z) \qquad (C5.3.13)$$

Now, (C5.3.10) and (C5.3.13) give

$$A(z)P(z) = Q(z)[1 \oplus z^p] \qquad (C5.3.14)$$

Since $A(z)$ is assumed to be irreducible, it must be a factor of either $Q(z)$ or $1 \oplus z^p$. However, the degree of $Q(z)$ is smaller than the degree of $A(z)$. Hence $A(z)$ cannot be a factor of $Q(z)$. Thus, $A(z)$ is a factor of $1 \oplus z^p$, which was the required result.

The necessity of condition (ii) and the sufficiency of conditions (i) and (ii) follow immediately from this last result. ∎

Tables of polynomials satisfying the conditions of the lemma are available in the literature. For example, Davies (1970) gives a table of all the polynomials which satisfy the conditions (i) and (ii), for n from 3 to 10. Table C5.3.2 lists those polynomials with a minimum number of nonzero coefficients which satisfy the conditions of the lemma, for n ranging from 3 to 10. (Note that for some values of n there are more than two such polynomials; see Davies (1970).)

TABLE C5.3.2 Polynomials $A(z)$ satisfying the conditions (i) and (ii) of Lemma C5.3.1 ($n = 3, \ldots, 10$)

n	$A(z)$	
3	$1 \oplus z \oplus z^3$	$1 \oplus z^2 \oplus z^3$
4	$1 \oplus z \oplus z^4$	$1 \oplus z^3 \oplus z^4$
5	$1 \oplus z^2 \oplus z^5$	$1 \oplus z^3 \oplus z^5$
6	$1 \oplus z \oplus z^6$	$1 \oplus z^5 \oplus z^6$
7	$1 \oplus z \oplus z^7$	$1 \oplus z^3 \oplus z^7$
8	$1 \oplus z \oplus z^2 \oplus z^7 \oplus z^8$	$1 \oplus z \oplus z^6 \oplus z^7 \oplus z^8$
9	$1 \oplus z^4 \oplus z^9$	$1 \oplus z^5 \oplus z^9$
10	$1 \oplus z^3 \oplus z^{10}$	$1 \oplus z^7 \oplus z^{10}$

Remark. In the light of Lemma C5.3.1, the period lengths encountered in Example C5.3.1 can be understood more clearly. Case (a) corresponds to the feedback polynomial

$$A(q^{-1}) = 1 \oplus q^{-1} \oplus q^{-2}$$

which does not satisfy condition (ii) of the lemma since

$$A(q^{-1})(1 \oplus q^{-1}) = 1 \oplus q^{-3}$$

Case (b) corresponds to

$$A(q^{-1}) = 1 \oplus q^{-1} \oplus q^{-3}$$

given in Table C5.3.2. Case (c) corresponds to

$$A(q^{-1}) = 1 \oplus q^{-2} \oplus q^{-3}$$

given in Table C5.3.2. Case (d) corresponds to

$$A(q^{-1}) = 1 \oplus q^{-1} \oplus q^{-2} \oplus q^{-3}$$

which does not satisfy condition (ii) since

$$A(q^{-1})(1 \oplus q^{-1}) = 1 \oplus q^{-4}$$

■

Covariance function of maximum length PRBS

The ml PRBS has a number of properties, which are described by Davies (1970). Here we will evaluate its mean and covariance function. To do so we will use two specific properties, denoted by P1 and P2 below.

Property P1
If $u(t)$ is an ml PRBS of period $M = 2^n - 1$, then within one period it contains $(M + 1)/2 = 2^{n-1}$ ones and $(M - 1)/2 = 2^{n-1} - 1$ zeros. ■

Property P1 is fairly obvious. During one period the state vector, $x(t)$, will take all possible values, except the zero vector. Out of the 2^n possible state vectors, half of them, i.e. 2^{n-1}, will contain a one in the last position (giving $u(t) = 1$).

Property P2
Let $u(t)$ be an ml PRBS of period M. Then for $k = 1, \ldots, M - 1$,

$$u(t) \oplus u(t - k) = u(t - l) \tag{C5.3.15}$$

for some $l \in [1, M - 1]$ that depends on k. ■

Property P2 can be realized as follows. According to (C5.3.6), $A(q^{-1})u(t) = 0$ for all initial values. Conversely, if $A(q^{-1})u(t) = 0$ then $u(t)$ is an ml PRBS which is the solution to (C5.3.6) for some initial values. Now

$$A(q^{-1})[u(t) \oplus u(t - k)] = [A(q^{-1})u(t)] \oplus [A(q^{-1})u(t - k)] = 0$$

Hence $u(t) \oplus u(t - k)$ is an ml PRBS, corresponding to some initial values. Since all possible state vectors appear during one period of $u(t)$, the relation (C5.3.15) follows.

One further simple property is required, which is valid for binary variables.

Property P3
If x and y are binary variables, then

$$xy = \frac{1}{2}[x + y - (x \oplus y)] \tag{C5.3.16}$$

■

Property P3 is easy to verify by direct computation of all possible cases; see Table C5.3.3.

TABLE C5.3.3 Verification of property P3

x	y	$x \oplus y$	$\frac{1}{2}[x + y - (x \oplus y)]$	xy
0	0	0	0	0
0	1	1	0	0
1	0	1	0	0
1	1	0	1	1

The mean and the covariance function of an ml PRBS can now be evaluated. Let the period be $M = 2^n - 1$. According to Appendix A5.1,

$$m = \frac{1}{M} \sum_{t=1}^{M} u(t)$$

$$r(\tau) = \frac{1}{M} \sum_{t=1}^{M} [u(t + \tau) - m][u(t) - m] \tag{C5.3.17}$$

and $r(\tau)$ is periodic with period M.

Property P1 gives

$$m = \frac{1}{M}\left(\frac{M + 1}{2}\right) = \frac{1}{2} + \frac{1}{2M} \tag{C5.3.18}$$

The mean value is hence slightly greater than 0.5.

To evaluate the covariance function note first that

$$r(0) = \frac{1}{M} \sum_{t=1}^{M} [u(t) - m]^2 = \frac{1}{M} \sum_{t=1}^{M} u^2(t) - m^2$$

$$= \frac{1}{M} \sum_{t=1}^{M} u(t) - m^2 = m - m^2 = m(1 - m) \tag{C5.3.19}$$

$$= \frac{M + 1}{2M} \frac{M - 1}{2M} = \frac{M^2 - 1}{4M^2}$$

The variance is thus slightly less than 0.25. Next note that for $\tau = 1, \ldots, M - 1$, properties P1–P3 give

$$r(\tau) = \frac{1}{M} \sum_{t=1}^{M} [u(t + \tau) - m][u(t) - m]$$

$$= \frac{1}{M} \sum_{t=1}^{M} u(t + \tau)u(t) - m^2$$

$$= \frac{1}{2M} \sum_{t=1}^{M} [u(t + \tau) + u(t) - \{u(t + \tau) \oplus u(t)\}] - m^2 \tag{C5.3.20}$$

$$= m - \frac{1}{2M} \sum_{t=1}^{M} u(t + \tau - l) - m^2$$

$$= \frac{m}{2} - m^2 = \frac{m}{2}(1 - 2m) = -\frac{M + 1}{4M^2}$$

Concerning the signal $y(t) = -1 + 2u(t)$ from (C5.3.3), the following results are obtained:

$$m_y = -1 + 2m_u = \frac{1}{M} \approx 0 \qquad \text{(C5.3.21a)}$$

$$r_y(0) = 4r_u(0) = 1 - \frac{1}{M^2} \approx 1 \qquad \text{(C5.3.21b)}$$

$$r_y(\tau) = 4r_{\tilde{u}}(\tau) = -\frac{1}{M} - \frac{1}{M^2} \approx -\frac{1}{M} \quad \tau = 1, \ldots, M-1 \qquad \text{(C5.3.21c)}$$

The approximations in (C5.3.21) are obtained by neglecting the slight difference of the mean from zero.

It can be seen that for M sufficiently large the covariance function of $y(t)$ resembles that of white noise of unit variance. The analogy between ml PRBS and white noise is further discussed in Section 5.2.

Due to their easy generation and their convenient properties the ml PRBSs have been used widely in system identification. The PRBS resembles white noise as far as the spectral properties are concerned. Furthermore, by various forms of linear filtering (increasing the clock period to several sampling intervals being one way of lowpass filtering; see Section 5.3), it is possible to generate a signal with prescribed spectral properties. These facts have made the PRBS a convenient probing signal for many identification methods.

The PRBS has also some specific properties that are advantageous for nonparametric methods. The covariance matrix $[r(i-j)]$ corresponding to an ml PRBS can be inverted analytically (see Problem 5.9). Also, calculation of the cross-correlation of a PRBS with another signal needs only addition operations (multiplications are not needed). These facts make the ml PRBS very convenient for nonparametric identification of weighting function models (see, for example, Section 3.4 and Problem 5.9).

Chapter 6
MODEL PARAMETRIZATIONS

6.1 Model classifications

This chapter examines the role played by the model structure $\mathcal{M}$ in system identification, and presents both a general description of linear models and a number of examples. In this section, some general comments are made on classification of models. A distinction can be made between:

- *'Intuitive' or 'mental' models.*
- *Mathematical models.* These could also include models given in the form of graphs or tables.

Models can be classified in other ways. For example:

- *Analog models*, which are based on analogies between processes in different areas. For example, a mechanical and an electrical oscillator can be described by the same second-order linear differential equation, but the coefficients will have different physical interpretations. Analog computers are based on such principles: differential equations constituting a model of some system are solved by using an 'analog' or 'equivalent' electrical network. The voltages at various points in this network are recorded as functions of time and give the solution to the differential equations.
- *Physical models*, which are mostly laboratory-scale units that have the same essential characteristics as the (full-scale) processes they model.

In system science mathematical models are useful because they can provide a description of a physical phenomenon or a process, and can be used as a tool for the design of a regulator or a filter.

Mathematical models can be derived in two ways:

- *Modeling*, which refers to derivation of models from basic laws in physics, economics, etc. One often uses fundamental balance equations, for example net accumulation = input flow − output flow, which can be applied to a range of variables, such as energy, mass, money, etc.
- *Identification*, which refers to the determination of dynamic models from experimental data. It includes the set-up of the identification experiment, i.e. data acquisition, the determination of a suitable form of the model as well as of its parameters, and a validation of the model.

These methods have already been discussed briefly in Chapter 1.

Classification

Mathematical models of dynamic systems can be classified in various ways. Such models describe how the effect of an input signal will influence the system behavior at subsequent times. In contrast, for static models, which were exemplified in Chapter 4, there is no 'memory'. Hence the effect of an 'input variable' is only instantaneous. The ways of classifying dynamic models include the following:

- *Single input, single output (SISO) models—multivariable models.* SISO models refer to processes where a description is given of the influence of *one* input on *one* output. When more variables are involved a multivariable model results. Most of the theory in this book will hold for multivariable models, although mostly SISO models will be used for illustration. It should be noted that multi-input, single output (MISO) models are in most cases as easy to derive as SISO models, whereas multi-input, multi-output (MIMO) models are more difficult to determine.
- *Linear models—nonlinear models.* A model is linear if the output depends linearly on the input and possible disturbances; otherwise it is nonlinear. With a few exceptions, only linear models will be discussed here.
- *Parametric models—nonparametric models.* A parametric model is described by a set of parameters. Some simple parametric models were introduced in Chapter 2, and we will concentrate on such models in the following. Chapter 3 provided some examples of nonparametric models, which can consist of a function or a graph.
- *Time invariant models—time varying models.* Time invariant models are certainly the more common. For time varying models special identification methods are needed. In such cases where a model has parameters that change with time, one often speaks about tracking or real-time identification when estimating the parameters.
- *Time domain models—frequency domain models.* Typical examples of time domain models are differential and difference equations, while a spectral density and a Bode plot are examples of frequency domain models. The major part of this book deals with time domain models.
- *Discrete time models—continuous time models.* A discrete time model describes the relation between inputs and outputs at discrete time points. It will be assumed that these points are equidistant and the time between two points will be used as time unit. Therefore the time t will take values 1, 2, 3 ... for discrete time models, which is the dominating class discussed here. Note that a continuous time *model*, such as a differential equation, can very well be fitted to discrete time *data*!
- *Lumped models—distributed parameter models.* Lumped models are described by or based on a *finite* number of ordinary differential or difference equations. If the number of equations is *infinite* or the model is based on *partial* differential equations, then it is called a distributed parameter model. The treatment here is confined to lumped models.
- *Deterministic models—stochastic models.* For a deterministic model the output can be exactly calculated as soon as the input signal is known. In contrast, a stochastic model contains random terms that make such an exact calculation impossible. The random terms can be seen as a description of disturbances. This book concentrates mainly on stochastic models.

Note that the term 'linear models' above refers to the way in which $y(t)$ depends on past data. Another concept concerns models that are *linear in the parameters* θ to be estimated (sometimes abbreviated to LIP). Then $y(t)$ depends linearly on θ. The identification methods to be discussed in Chapter 8 are restricted to such models. The linear regression model (4.1),

$$y(t) = \varphi^T(t)\theta$$

is both linear (since $y(t)$ depends linearly on $\varphi(t)$) and linear in the parameters (since $y(t)$ depends linearly on θ). Note however that we can allow $\varphi(t)$ to depend on the measured data in a nonlinear fashion. Example 6.6 will illustrate such a case.

The choice of which type of model to use is highly problem-dependent. Chapter 11 discusses some means of choosing an appropriate model within a given type.

6.2 A general model structure

The general form of model structure that will be used here is the following:

$$\boxed{\begin{aligned}\mathcal{M}(\theta):\; & y(t) = G(q^{-1}; \theta)u(t) + H(q^{-1}; \theta)e(t) \\ & Ee(t)e^T(s) = \Lambda(\theta)\delta_{t,s}\end{aligned}} \quad (6.1)$$

In (6.1), $y(t)$ is the *ny*-dimensional output at time t and $u(t)$ the *nu*-dimensional input. Further, $e(t)$ is a sequence of independent and identically distributed (iid) random variables with zero mean. Such a sequence is referred to as *white noise*. The reason for this name is that the corresponding spectral density is constant over the whole frequency range (see Chapter 5). The analogy can be made with white light, which contains 'all' frequencies. In (6.1) $G(q^{-1}; \theta)$ is an $(ny|nu)$-dimensional filter and $H(q^{-1}; \theta)$ an $(ny|ny)$-dimensional filter. The argument q^{-1} denotes the backward shift operator, so $q^{-1}u(t) = u(t-1)$, etc. In most cases the filters $G(q^{-1}; \theta)$ and $H(q^{-1}; \theta)$ will be of finite order. Then they are rational functions of q^{-1}. The model (6.1) is depicted in Figure 6.1.

The filters $G(q^{-1}; \theta)$ and $H(q^{-1}; \theta)$ as well as the noise covariance matrix $\Lambda(\theta)$ are functions of the parameter vector θ. Often θ (which we assume to be $n\theta$-dimensional) is restricted to lie in a subset of $\mathcal{R}^{n\theta}$. This set is given by

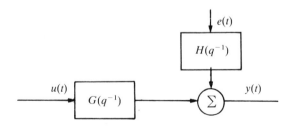

FIGURE 6.1 Block diagram of the model structure (6.1).

$$\mathcal{D} = \{\theta | H^{-1}(q^{-1}; \theta) \text{ and } H^{-1}(q^{-1}; \theta)G(q^{-1}; \theta) \text{ are asymptotically stable,} \\ G(0; \theta) = 0, H(0; \theta) = I, \Lambda(\theta) \text{ is nonnegative definite}\} \quad (6.2)$$

The reasons for these restrictions in the definition of $\mathcal{D}$ will become clear in the next chapter, where it will be shown that when θ belongs to $\mathcal{D}$ there is a simple form for the optimal prediction of $y(t)$ given $y(t-1), u(t-1), y(t-2), u(t-2), \ldots$

Note that for the moment $H(q^{-1}; \theta)$ is not restricted to be asymptotically stable. In later chapters it will be necessary to do so occasionally in order to impose stationarity of the data. Models with unstable $H(q^{-1}; \theta)$ can be useful for describing drift in the data as will be illustrated in Example 12.2. Since the disturbance term in such cases is not stationary, it is assumed to start at $t = 0$. Stationary disturbances, which correspond to an asymptotically stable $H(q^{-1}; \theta)$ filter can be assumed to start at $t = -\infty$.

For stationary disturbances with rational spectral densities it is a consequence of the spectral factorization theorem (see for example Appendix A6.1; Anderson and Moore, 1979; Åström, 1970) that they can be modeled within the restrictions given by (6.2). Spectral factorization can also be applied to nonstationary disturbances as illustrated in Example A6.1.4.

Equation (6.1) describes a general linear model. The following examples describe typical model structures by specifying the parametrization. That is to say, they specify how $G(q^{-1}; \theta)$, $H(q^{-1}; \theta)$ and $\Lambda(\theta)$ depend on the parameter vector θ.

Example 6.1 *An ARMAX model*

Let $y(t)$ and $u(t)$ be scalar signals and consider the model structure

$$A(q^{-1})y(t) = B(q^{-1})u(t) + C(q^{-1})e(t) \quad (6.3)$$

where

$$A(q^{-1}) = 1 + a_1 q^{-1} + \ldots + a_{na} q^{-na}$$
$$B(q^{-1}) = b_1 q^{-1} + \ldots + b_{nb} q^{-nb} \quad (6.4)$$
$$C(q^{-1}) = 1 + c_1 q^{-1} + \ldots + c_{nc} q^{-nc}$$

The parameter vector is taken as

$$\theta = (a_1 \ \ldots \ a_{na} \ b_1 \ \ldots \ b_{nb} \ c_1 \ \ldots \ c_{nc})^T \quad (6.5)$$

The model (6.3) can be written explicitly as the difference equation

$$y(t) + a_1 y(t-1) + \ldots + a_{na} y(t-na) = b_1 u(t-1) + \ldots + b_{nb} u(t-nb) \\ + e(t) + c_1 e(t-1) + \ldots + c_{nc} e(t-nc) \quad (6.6)$$

but the form (6.3) using the polynomial formalism will be more convenient.

The parameter vector could be complemented with the noise variance

150 *Model parametrizations* Chapter 6

$$\lambda^2 = Ee^2(t) \tag{6.7}$$

so that

$$\theta' = (\theta^T \ \lambda^2)^T$$

is the new parameter vector, of dimension $na + nb + nc + 1$.

The model (6.3) is called an ARMAX model, which is short for an ARMA model (autoregressive moving average) with an exogenous signal (i.e. a control variable $u(t)$). Figure 6.2 gives block diagrams of the model (6.3).

To see how this relates to (6.1), note that (6.3) can be rewritten as

$$y(t) = \frac{B(q^{-1})}{A(q^{-1})} u(t) + \frac{C(q^{-1})}{A(q^{-1})} e(t)$$

Thus for the model structure (6.3),

$$G(q^{-1}; \theta) = \frac{B(q^{-1})}{A(q^{-1})}$$

$$H(q^{-1}; \theta) = \frac{C(q^{-1})}{A(q^{-1})} \tag{6.8}$$

$$\Lambda(\theta) = \lambda^2$$

The set $\mathcal{D}$ is given by

$$\mathcal{D} = \{\theta | \text{ The polynomial } C(z) \text{ has all zeros outside the unit circle}\}$$

(a)
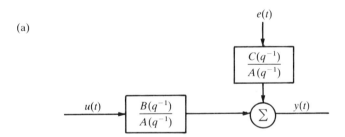

(b)
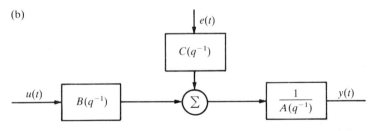

FIGURE 6.2 Equivalent block diagrams of an ARMAX model.

A more standard formulation of the requirement $\theta \in \mathcal{D}$ is that the *reciprocal* polynomial

$$C^*(z) = z^{nc} + c_1 z^{nc-1} + \ldots + c_{nc} = z^{nc} C(z^{-1})$$

has all zeros *inside* the unit circle.

There are several important special cases of (6.3):

- An autoregressive (AR) model is obtained when $nb = nc = 0$. (Then a pure time series is modeled, i.e. no input signal is assumed to be present.) For this case

$$A(q^{-1})y(t) = e(t)$$
$$\theta = (a_1 \ldots a_{na})^{\mathrm{T}} \tag{6.9}$$

- A moving average (MA) model is obtained when $na = nb = 0$. Then

$$y(t) = C(q^{-1})e(t)$$
$$\theta = (c_1 \ldots c_{nc})^{\mathrm{T}} \tag{6.10}$$

- An autoregressive moving average (ARMA) model is obtained when $nb = 0$. Then

$$A(q^{-1})y(t) = C(q^{-1})e(t)$$
$$\theta = (a_1 \ldots a_{na} \; c_1 \ldots c_{nc})^{\mathrm{T}} \tag{6.11}$$

When $A(q^{-1})$ is constrained to contain the factor $1 - q^{-1}$ the model is called autoregressive integrated moving average (ARIMA). Such models are useful for describing drifting disturbances and are further discussed in Section 12.3.

- A finite impulse response (FIR) model is obtained when $na = nc = 0$. It can also be called a truncated weighting function model. Then

$$y(t) = B(q^{-1})u(t) + e(t)$$
$$\theta = (b_1 \ldots b_{nb})^{\mathrm{T}} \tag{6.12}$$

- Another special case is when $nc = 0$. The model structure then becomes

$$A(q^{-1})y(t) = B(q^{-1})u(t) + e(t)$$
$$\theta = (a_1 \ldots a_{na} \; b_1 \ldots b_{nb})^{\mathrm{T}} \tag{6.13a}$$

This is sometimes called an ARX (controlled autoregressive) model. This structure can in some sense be viewed as a linear regression. To see this, rewrite the model (6.13a) as

$$y(t) = \varphi^{\mathrm{T}}(t)\theta + e(t) \tag{6.13b}$$

where

$$\varphi(t) = (-y(t-1) \ldots -y(t-na) \; u(t-1) \ldots u(t-nb))^{\mathrm{T}} \tag{6.13c}$$

Note though that here the regressors (the elements of $\varphi(t)$) are not deterministic functions. This means that the *analysis* carried out in Chapter 4 for linear regression models cannot be applied to (6.13). ∎

152 Model parametrizations   Chapter 6

Example 6.2 *A general SISO model structure*
The ARMAX structure (6.3) is in a sense quite general. Any linear, finite-order system with stationary disturbances having a rational spectral density can be described by an ARMAX model. (Then the integers na, nb and nc as well as θ and λ^2 have, of course, to take certain appropriate values according to context (see below).) However, there are alternative ways to parametrize a linear finite-order system. To describe a general case consider the equation

$$A(q^{-1})y(t) = \frac{B(q^{-1})}{F(q^{-1})}u(t) + \frac{C(q^{-1})}{D(q^{-1})}e(t) \qquad Ee^2(t) = \lambda^2 \qquad (6.14)$$

with $A(q^{-1})$, $B(q^{-1})$, $C(q^{-1})$ as in (6.4), and

$$D(q^{-1}) = 1 + d_1 q^{-1} + \ldots + d_{nd} q^{-nd}$$
$$F(q^{-1}) = 1 + f_1 q^{-1} + \ldots + f_{nf} q^{-nf} \qquad (6.15)$$

The parameter vector is now

$$\theta = (a_1 \ldots a_{na} \; b_1 \ldots b_{nb} \; c_1 \ldots c_{nc} \; d_1 \ldots d_{nd} \; f_1 \ldots f_{nf})^T \qquad (6.16)$$

Block diagrams of the general model (6.14) are given in Figure 6.3.

It should be stressed that it is seldom necessary to use the model structure (6.14) in its *general* form. On the contrary, for practical use one often restricts it by setting one or more polynomials to unity. For example, choosing $nd = nf = 0$ (i.e. $D(q^{-1}) \equiv F(q^{-1}) \equiv 1$) produces the ARMAX model (6.3). The value of the form (6.14) lies in its generality. It includes a number of important forms as special cases. This makes it possible to describe and analyze several cases simultaneously.

Clearly, (6.14) gives

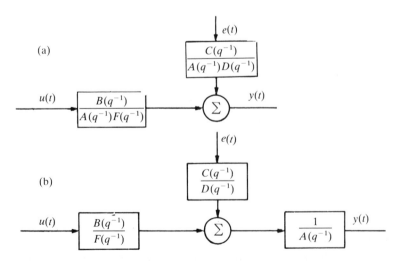

FIGURE 6.3 Equivalent block diagrams of the general SISO model (6.14).

$$G(q^{-1}; \theta) = \frac{B(q^{-1})}{A(q^{-1})F(q^{-1})}$$

$$H(q^{-1}; \theta) = \frac{C(q^{-1})}{A(q^{-1})D(q^{-1})} \qquad (6.17)$$

$$\Lambda(\theta) = \lambda^2$$

The set $\mathcal{D}$ becomes

$$\mathcal{D} = \{\theta | \text{ The polynomials } C(z) \text{ and } F(z) \text{ have all zeros outside the unit circle}\}$$

Typical special cases of (6.14) include the following:

- If $nd = nf = 0$ the result is the ARMAX model

$$A(q^{-1})y(t) = B(q^{-1})u(t) + C(q^{-1})e(t) \qquad (6.18)$$

and its variants as discussed in Example 6.1.
- If $na = nc = nd = 0$ then

$$y(t) = \frac{B(q^{-1})}{F(q^{-1})} u(t) + e(t) \qquad (6.19)$$

In this case $H(q^{-1}; \theta) = 1$. Equation (6.19) is sometimes referred to as an output error structure, since it implies that

$$e(t) = y(t) - \frac{B(q^{-1})}{F(q^{-1})} u(t) \qquad (6.20)$$

is the output error, i.e. the difference between the measurable output $y(t)$ and the model output $B(q^{-1})/F(q^{-1})u(t)$.

- If $na = 0$ then

$$y(t) = \frac{B(q^{-1})}{F(q^{-1})} u(t) + \frac{C(q^{-1})}{D(q^{-1})} e(t) \qquad (6.21)$$

A particular property of this structure is that $G(q^{-1}; \theta)$ and $H(q^{-1}; \theta)$ have no common parameter; see (6.17). This can have a beneficial influence on the identification result in certain cases. It will be possible to estimate $G(q^{-1}; \theta)$ consistently even if the parametrization of $H(q^{-1}; \theta)$ is not appropriate (see Chapter 7 for details).

It is instructive to consider how different versions of the general model structure (6.14) are related. To describe an arbitrary linear system with given filters $G(q^{-1})$ and $H(q^{-1})$, the relations (6.17) must be satisfied, in which case it is necessary that the model comprises at least three polynomials. The following types of models are all equivalent in the sense that they 'solve' (6.17) with respect to the model polynomials.

- An ARMAX model implies that $D(q^{-1}) = F(q^{-1}) = 1$. Hence $G(q^{-1}; \theta)$ and $H(q^{-1}; \theta)$ have the same denominator. This may be relevant if the disturbances enter the process early (i.e. close to the input). If the disturbances enter late there will often

be common factors in $A(q^{-1})$ and $B(q^{-1})$ as well as in $A(q^{-1})$ and $C(q^{-1})$. For example, if the disturbances enter as white measurement noise then $H(q^{-1}; \theta) = 1$, which gives $A(q^{-1}) = C(q^{-1})$. In general $A(q^{-1})$ must be chosen as the (least) common denominator of $G(q^{-1}; \theta)$ and $H(q^{-1}; \theta)$.
- The model (6.21) implies that $A(q^{-1}) = 1$. Hence the filters $G(q^{-1}; \theta)$ and $H(q^{-1}; \theta)$ are parametrized with different parameters. This model might require fewer parameters than an ARMAX model. However, if $G(q^{-1})$ and $H(q^{-1})$ do have common poles, it is an advantage to use this fact in the model description. Such poles will also be estimated more accurately if an ARMAX model is used.
- The general form (6.14) of the model makes it possible to allow $G(q^{-1}; \theta)$ and $H(q^{-1}; \theta)$ to have partly the same poles.

Another structure that has gained some popularity is the so-called ARARX structure, for which $C(q^{-1}) = F(q^{-1}) = 1$. The name ARARX refers to the fact that the disturbance is modeled as an AR process and the system dynamics as an ARX model. This model is not as general as those above, since in this case

$$H(q^{-1}; \theta) \equiv \frac{1}{A(q^{-1})D(q^{-1})}$$

This identity may not have any exact solution. Instead a polynomial $D(q^{-1})$ of high degree may be needed to get a good approximate solution. The advantage of this structure is that it leads to simple estimation algorithms, such as the generalized least squares method (see Complement C7.4). It is also well suited for the optimal instrumental variable method (see Example 8.7 and the subsequent discussion). ∎

The next two examples describe ways of generalizing the structure (6.13) to the multivariable case.

Example 6.3 *The full polynomial form*
For a linear multivariable system consider the model

$$A(q^{-1})y(t) = B(q^{-1})u(t) + e(t) \tag{6.22}$$

where now $A(q^{-1})$ and $B(q^{-1})$ are the following matrix polynomials of dimension $(ny|ny)$ and $(ny|nu)$ respectively:

$$\begin{aligned} A(q^{-1}) &= I + A_1 q^{-1} + \ldots + A_{na} q^{-na} \\ B(q^{-1}) &= B_1 q^{-1} + \ldots + B_{nb} q^{-nb} \end{aligned} \tag{6.23}$$

Assume that all elements of the matrices $A_1, \ldots, A_{na}, B_1, \ldots, B_{nb}$ are unknown. They will thus be included in the parameter vector. Then the model (6.22) can alternatively be written as

$$y(t) = \phi^T(t)\theta + e(t) \tag{6.24}$$

$$\phi^{\mathrm{T}}(t) = \begin{pmatrix} \varphi^{\mathrm{T}}(t) & & 0 \\ & \ddots & \\ 0 & & \varphi^{\mathrm{T}}(t) \end{pmatrix} \quad (6.25\mathrm{a})$$

$$\varphi^{\mathrm{T}}(t) = (-y^{\mathrm{T}}(t-1) \ \ldots \ -y^{\mathrm{T}}(t-na) \ u^{\mathrm{T}}(t-1) \ \ldots \ u^{\mathrm{T}}(t-nb)) \quad (6.25\mathrm{b})$$

$$\theta = \begin{pmatrix} \theta^1 \\ \vdots \\ \theta^{ny} \end{pmatrix} \quad (6.25\mathrm{c})$$

$$\begin{pmatrix} (\theta^1)^{\mathrm{T}} \\ \vdots \\ (\theta^{ny})^{\mathrm{T}} \end{pmatrix} = (A_1 \ \ldots \ A_{na} \ B_1 \ \ldots \ B_{nb}) \quad (6.25\mathrm{d})$$

The set $\mathcal{D}$ associated with (6.22) is the whole space $\mathcal{R}^{n\theta}$.
For an illustration of (6.25) consider the case $ny = 2$, $nu = 3$, $na = 1$, $nb = 2$ and set

$$A_1 = \begin{pmatrix} a^{11} & a^{12} \\ a^{21} & a^{22} \end{pmatrix} \quad B_1 = \begin{pmatrix} b_1^{11} & b_1^{12} & b_1^{13} \\ b_1^{21} & b_1^{22} & b_1^{23} \end{pmatrix} \quad B_2 = \begin{pmatrix} b_2^{11} & b_2^{12} & b_2^{13} \\ b_2^{21} & b_2^{22} & b_2^{23} \end{pmatrix}$$

Then

$$\varphi^{\mathrm{T}}(t) = (-y_1(t-1) \ -y_2(t-1) \ u_1(t-1) \ u_2(t-1) \ u_3(t-1) \ u_1(t-2) \\ u_2(t-2) \ u_3(t-2))$$

$$\theta^1 = (a^{11} \ a^{12} \ b_1^{11} \ b_1^{12} \ b_1^{13} \ b_2^{11} \ b_2^{12} \ b_2^{13})^{\mathrm{T}}$$

$$\theta^2 = (a^{21} \ a^{22} \ b_1^{21} \ b_1^{22} \ b_1^{23} \ b_2^{21} \ b_2^{22} \ b_2^{23})^{\mathrm{T}} \quad \blacksquare$$

The full polynomial form (6.24) gives a natural and straightforward extension of the scalar model (6.13). Identification methods for estimating the parameters in the full polynomial form will look simple. However, this structure also has some drawbacks. The full polynomial form is not a canonical parametrization. Roughly speaking, this means that there are linear systems of the form (6.22) whose transfer function matrix $A^{-1}(q^{-1})B(q^{-1})$ cannot be uniquely parametrized by θ. Such systems cannot be identified using the full polynomial model, as should be intuitively clear. Note, however, that these 'problematic' systems are rare, so that the full polynomial form can be used to represent uniquely almost all linear systems of the form (6.22). The important problem of the uniqueness of the parametrization is discussed in the next section. A detailed analysis of the full polynomial model is provided in Complement C6.1.

Example 6.4 *Diagonal form of a multivariable system*
This model can be seen as another generalization of (6.13) to the multivariable case. Here the $(ny|ny)$ matrix $A(q^{-1})$ is assumed to be diagonal. More specifically,

$$A(q^{-1}; \theta)y(t) = B(q^{-1}; \theta)u(t) + e(t) \tag{6.26a}$$

$$A(z; \theta) = \begin{pmatrix} a_1(z; \theta) & & 0 \\ & \ddots & \\ 0 & & a_{ny}(z; \theta) \end{pmatrix} \tag{6.26b}$$

where

$$a_i(z; \theta) = 1 + a_{1,i}z + \ldots + a_{nai,i}z^{nai}$$

are scalar polynomials. Further,

$$B(z; \theta) = \begin{pmatrix} B_1(z; \theta) \\ \vdots \\ B_{ny}(z; \theta) \end{pmatrix} \tag{6.26c}$$

where $B_i(z; \theta)$ is the following row-vector polynomial of dimension nu:

$$B_i(z; \theta) = b_{1,i}^T z + \ldots b_{nbi,i}^T z^{nbi} \quad (b_{k,i} \text{ are } (nu|1) \text{ vectors})$$

The integer-valued parameters nai and nbi, $i = 1, \ldots, ny$ are the structural indices of the parametrization. The model can be written as

$$y(t) = \phi^T(t)\theta + e(t) \tag{6.27a}$$

where the parameter vector θ and the matrix $\phi(t)$ are given by

$$\theta = \begin{pmatrix} \theta^1 \\ \vdots \\ \theta^{ny} \end{pmatrix} \quad \phi(t) = \begin{pmatrix} \varphi_1(t) & & 0 \\ & \ddots & \\ 0 & & \varphi_{ny}(t) \end{pmatrix} \tag{6.27b}$$

$$\theta^i = (a_{1,i} \ldots a_{nai,i} \ b_{1,i}^T \ldots b_{nbi,i}^T)^T \tag{6.27c}$$

$$\varphi_i(t) = (-y_i(t-1) \ldots -y_i(t-nai) \ u^T(t-1) \ldots u^T(t-nbi))^T \tag{6.27d}$$

The diagonal form model is a canonical parametrization of a linear model of the type of (6.22) (see, for example, Kashyap and Rao, 1976). However, compared to the full polynomial model, it has some drawbacks. First, estimation of the parameters in full polynomial form models may lead to simpler algorithms than those associated with diagonal models (see Chapter 8). Second, the structure of the diagonal model (6.27) clearly is more complicated than that of the full polynomial model (6.25). For the model (6.27) $2ny$ structure indices have to be determined while the model (6.25) needs determination of only two structure indices (the degrees na and nb). Determination of a large number of structural parameters, which in practical applications is most often done by scanning all the combinations of the structure parameters that are thought to be possible, may lead to a prohibitive computational burden. Third, the model (6.27) may often contain more unknown parameters than (6.25), as the following simple example demonstrates: consider a linear system given by (6.22) with $ny = nu$ and $na = nb = 1$; then the model (6.24), (6.25) will contain $2ny^2$ parameters, while the diagonal model (6.26), (6.27) will most often contain about ny^3 parameters. ∎

For both the full polynomial form and the diagonal form models the number of parameters used can be larger than necessary. If the number of parameters in a model is reduced, the user may gain two things. One is that the numerical computation required to obtain the parameter estimates will in general terms be simpler since the problem is of a lower dimension. The other advantage, which will be demonstrated later, is that using fewer free parameters will in a certain sense lead to a more accurate model (see (11.40)).

The following example presents a model for multivariable systems where the internal structure is of importance.

Example 6.5 *A state space model*
Consider a linear stochastic model in the state space form

$$\begin{aligned} x(t+1) &= A(\theta)x(t) + B(\theta)u(t) + v(t) \\ y(t) &= C(\theta)x(t) + \bar{e}(t) \end{aligned} \quad (6.28)$$

Here $v(t)$ and $\bar{e}(t)$ are (multivariate) white noise sequences with zero means and the following covariances:

$$\begin{aligned} Ev(t)v^T(s) &= R_1(\theta)\delta_{t,s} \\ Ev(t)\bar{e}^T(s) &= R_{12}(\theta)\delta_{t,s} \\ E\bar{e}(t)\bar{e}^T(s) &= R_2(\theta)\delta_{t,s} \end{aligned} \quad (6.29)$$

The matrices $A(\theta)$, $B(\theta)$, $C(\theta)$, $R_1(\theta)$, $R_{12}(\theta)$, $R_2(\theta)$ can depend on the parameter vector in different ways. A typical case is when a model of the form (6.28) is partially known but some of its matrix elements remain to be estimated. No particular parametrization will be specified here.

Next consider how the model (6.28) should be transformed into the general form (6.1) introduced at the beginning of this chapter. The transfer function $G(q^{-1}; \theta)$ is easily found: it can be seen that for the model (6.28) the influence of the input $u(t)$ on the output $y(t)$ is characterized by the transfer function

$$G(q^{-1}; \theta) = C(\theta)[qI - A(\theta)]^{-1}B(\theta) \quad (6.30)$$

To find the filter $H(q^{-1}; \theta)$ and the covariance matrix $\Lambda(\theta)$ is a bit more complicated. Equation (6.28) must be transformed into the so-called innovation form. Note that in (6.1) the only noise source is $e(t)$, which has the same dimension as the output $y(t)$. However, in (6.28) there are two noise sources acting on $y(t)$: the 'process noise' $v(t)$ and the 'measurement noise' $\bar{e}(t)$. This problem can be solved using spectral factorization, as explained in Appendix A6.1. We must first solve the Riccati equation

$$\begin{aligned} P(\theta) = A(\theta)P(\theta)A^T(\theta) + R_1(\theta) - [A(\theta)P(\theta)C^T(\theta) + R_{12}(\theta)] \\ \times [C(\theta)P(\theta)C^T(\theta) + R_2(\theta)]^{-1}[C(\theta)P(\theta)A^T(\theta) + R_{12}^T(\theta)] \end{aligned} \quad (6.31)$$

taking the symmetric positive definite solution, and then compute the Kalman gain

$$K(\theta) = [A(\theta)P(\theta)C^T(\theta) + R_{12}(\theta)][C(\theta)P(\theta)C^T(\theta) + R_2(\theta)]^{-1} \quad (6.32)$$

It is obvious that both P and K computed from (6.31) and (6.32) will depend on the parameter vector θ. The system (6.28) can then be described, using the one-step predictions $\hat{x}(t+1|t)$ as state variables, as follows:

$$\begin{aligned}\hat{x}(t+1|t) &= A(\theta)\hat{x}(t|t-1) + B(\theta)u(t) + K(\theta)\bar{y}(t) \\ y(t) &= C(\theta)\hat{x}(t|t-1) + \bar{y}(t)\end{aligned} \qquad (6.33)$$

where

$$\bar{y}(t) = y(t) - C(\theta)\hat{x}(t|t-1) \qquad (6.34)$$

is the output innovation at time t. The innovation has a covariance matrix equal to

$$\text{cov}[\bar{y}(t)] = C(\theta)P(\theta)C^T(\theta) + R_2(\theta) \qquad (6.35)$$

The innovations $\bar{y}(t)$ play the role of $e(t)$ in (6.1). Hence from (6.33), (6.35), $H(q^{-1}; \theta)$ and $\Lambda(\theta)$ are given by

$$\begin{aligned}H(q^{-1}; \theta) &= I + C(\theta)[qI - A(\theta)]^{-1}K(\theta) \\ \Lambda(\theta) &= C(\theta)P(\theta)C^T(\theta) + R_2(\theta)\end{aligned} \qquad (6.36)$$

Now consider the set $\mathcal{D}$, in (6.2). For this case (6.33), (6.34) give

$$\begin{aligned}\hat{x}(t+1|t) &= [A(\theta) - K(\theta)C(\theta)]\hat{x}(t|t-1) + B(\theta)u(t) + K(\theta)y(t) \\ \bar{y}(t) &= -C(\theta)\hat{x}(t|t-1) + y(t)\end{aligned}$$

Comparing this with

$$\bar{y}(t) = e(t) = H^{-1}(q^{-1}; \theta)y(t) - H^{-1}(q^{-1}; \theta)G(q^{-1}; \theta)u(t)$$

derived from (6.1), it follows that

$$\begin{aligned}H^{-1}(q^{-1}; \theta) &= I - C(\theta)[qI - A(\theta) + K(\theta)C(\theta)]^{-1}K(\theta) \\ H^{-1}(q^{-1}; \theta)G(q^{-1}; \theta) &= C(\theta)[qI - A(\theta) + K(\theta)C(\theta)]^{-1}B(\theta)\end{aligned}$$

The poles of $H^{-1}(q^{-1}; \theta)$ and $H^{-1}(q^{-1}; \theta)G(q^{-1}; \theta)$ are therefore given by the eigenvalues of the matrix $A(\theta) - K(\theta)C(\theta)$. Since the positive definite solution of the Riccati equation (6.31) was selected, all these eigenvalues lie inside the unit circle (see Appendix A6.1; and Anderson and Moore, 1979). Hence the restriction $\theta \in \mathcal{D}$, (6.2), does not introduce any further constraints here: this restriction is automatically handled as a by-product.

Continuous time models

An important special case of the discrete time state space model (6.28) occurs by sampling a parametrized continuous time model. To be strict one should use a stochastic differential equation driven by a Wiener process, see Åström (1970). Such a linear continuous-time stochastic model can be written as

$$dx(t) = F(\theta)x(t)dt + G(\theta)u(t)dt + dw(t)$$
$$Edw(t)dw(t)^T = R(\theta)dt \qquad (6.37a)$$

where $w(t)$ is a Wiener process.

An informal way to represent (6.37a) is to use a (continuous time) white noise process $\tilde{w}(t)$ (corresponding to the derivative of the Wiener process $w(t)$). Then the model can be written as

$$\dot{x}(t) = F(\theta)x(t) + G(\theta)u(t) + \tilde{w}(t)$$
$$E\tilde{w}(t)\tilde{w}^T(s) = R(\theta)\delta(t-s) \qquad (6.37b)$$

where $\delta(t)$ is the Dirac delta-function. Equation (6.37b) can be complemented with

$$y(t) = C(\theta)x(t) + \bar{e}(t) \qquad E\bar{e}(t)\bar{e}^T(s) = R_2(\theta)\delta_{t,s} \qquad (6.38)$$

showing how certain linear combinations of the state vector are measured in discrete time.

Note that model parametrizations based on physical insight can often be more easily described by the continuous time model (6.37) than by the discrete time model (6.28). The reason is that many basic physical laws which can be used in modeling are given in the form of differential equations. As a simple illustration let $x_1(t)$ be the position and $x_2(t)$ the velocity of some moving object. Then trivially $\dot{x}_1(t) = x_2(t)$, which must be one of the equations of (6.37). Therefore physical laws will allow certain 'structural information' to be included in the continuous time model.

The solution of (6.37) is given by (see, for example, Kailath, 1980; Åström, 1970)

$$x(t) = e^{F(\theta)(t-t_0)}x(t_0) + \int_{t_0}^{t} e^{F(\theta)(t-s)}G(\theta)u(s)ds + \int_{t_0}^{t} e^{F(\theta)(t-s)}\tilde{w}(s)ds \qquad (6.39a)$$

where $x(t_0)$ denotes the initial condition. The last term should be written more formally as

$$\int_{t_0}^{t} e^{F(\theta)(t-s)}dw(s) \qquad (6.39b)$$

Next, assume that the input signal is kept *constant* over the sampling intervals. Then the sampling of (6.37), (6.38) will give a discrete time model of the form (6.28), (6.29) with

$$A(\theta) = e^{F(\theta)h}$$
$$B(\theta) = \int_0^h e^{F(\theta)s}G(\theta)ds$$
$$R_1(\theta) = \int_0^h e^{F(\theta)s}R(\theta)e^{F^T(\theta)s}ds \qquad (6.40)$$
$$R_{12}(\theta) = 0$$

where h denotes the sampling interval and it is assumed that the process noise $w(t)$ (or $\tilde{w}(t)$) and the measurement noise $\bar{e}(s)$ are independent for all t and s.

To verify (6.40) set $t_0 = t$ and $t = t + h$ in (6.39a). Then

$$x(t + h) = e^{F(\theta)h}x(t) + \left[\int_t^{t+h} e^{F(\theta)(t+h-s)}G(\theta)ds\right]u(t) + v(t)$$

where the process noise $v(t)$ is defined as

$$v(t) = \int_t^{t+h} e^{F(\theta)(t+h-s)}\tilde{w}(s)ds$$

The sequence $\{v(t)\}$ is uncorrelated (white noise). The covariance matrix of $v(t)$ is given by

$$Ev(t)v^T(t) = E\int_t^{t+h}\int_t^{t+h} e^{F(\theta)(t+h-s')}\tilde{w}(s')\tilde{w}^T(s'')e^{F^T(\theta)(t+h-s'')}ds'ds''$$

$$= \int_t^{t+h}\int_t^{t+h} e^{F(\theta)(t+h-s')}R(\theta)\delta(s' - s'')e^{F^T(\theta)(t+h-s'')}ds'ds''$$

$$= \int_t^{t+h} e^{F(\theta)(t+h-s)}R(\theta)e^{F^T(\theta)(t+h-s)}ds$$

$$= \int_0^h e^{F(\theta)s}R(\theta)e^{F^T(\theta)s}ds \qquad \blacksquare$$

So far only linear models have been considered. Nonlinear dynamics can of course appear in many ways. Sometimes the nonlinearity appears only as a nonlinear transformation of the signals involved. Such cases can be incorporated directly into the previous framework simply by redefining the signals. This is illustrated in the following example.

Example 6.6 *A Hammerstein model*
Consider the scalar model

$$A(q^{-1})y(t) = B_1(q^{-1})u(t) + B_2(q^{-1})u^2(t) + \ldots + B_m(q^{-1})u^m(t) + e(t) \qquad (6.41)$$

The relationship between u and y is clearly nonlinear. By defining a new artificial input

$$\bar{u}(t) = \begin{pmatrix} u(t) \\ u^2(t) \\ \vdots \\ u^m(t) \end{pmatrix}$$

and setting

$$\bar{B}(q^{-1}) = (B_1(q^{-1}) \quad B_2(q^{-1}) \quad \ldots \quad B_m(q^{-1}))$$

the model (6.41) can be written as a multi-input, single-output equation

$$A(q^{-1})y(t) = \bar{B}(q^{-1})\bar{u}(t) + e(t) \qquad (6.42)$$

Section 6.3 *Uniqueness properties* 161

However, this is a standard *linear* model if $\bar{u}(t)$ is now regarded as the input signal. One can then use all the powerful tools applicable to linear models to estimate the parameters, evaluate the properties of the estimates, etc. ∎

6.3 Uniqueness properties

So far in this chapter, a general model structure for linear systems has been described as well as a number of typical special cases of the structure (6.1). Chapter 11 will return to the question of how to determine an appropriate model structure from a given set of data.

Choosing a class of model structures

In general terms it is first necessary to choose a *class of model structures* and then make an appropriate choice within this class. The general SISO model (6.14) is one example of a class of model structures; the state space models (6.28) and the nonlinear difference equations (6.41) correspond to other classes. For the class (6.14), the choice of model structure consists of selecting the polynomial degrees na, nb, nc, nd and nf. For a multivariable model such as the state space equation (6.28), both the structure parameter(s) (in the case of (6.28), the dimension of the state vector) *and* the parametrization itself (i.e. the way in which θ enters into the matrices $A(\theta)$, $B(\theta)$, etc.) should be chosen.

The choice of a class of model structures to a large extent should be made according to the aim of the modeling (that is to say, the model set should be chosen that best fits the final purpose). At this stage it is sufficient to note that there are indeed many other factors that influence the selection of a model structure class. Four of the most important factors are:

- *Flexibility.* It should be possible to use the model structure to describe most of the different system dynamics that can be expected in the application. Both the number of free parameters and the way they enter into the model are important.
- *Parsimony.* The model structure should be *parsimonious*. This means that the model should contain the smallest number of free parameters required to represent the true system adequately.
- *Algorithm complexity.* Some identification methods such as the prediction error method (PEM) (see Chapter 7) can be applied to a variety of model structures. However, the form of structure selected can considerably influence the amount of computation needed.
- *Properties of the criterion function.* The asymptotic properties of PEM estimates depend crucially on the criterion function. The existence of local (i.e. nonglobal) minima as well as nonunique global minima is very much dependent on the model structure used.

Some detailed discussion of the factors above can be found in Chapters 7, 8 and 11. Specifically, parsimony is discussed in Complement C11.1; computational complexity in

Sections 7.6, 8.3 and Complements C7.3, C7.4, C8.2 and C8.6; properties of the criterion function in Complements C7.3, C7.5 and C7.6; and the subject of flexibility is touched on in several examples in Chapters 7 and 11. See also Ljung and Söderström (1983) for a further discussion of these factors and their role in system identification.

General uniqueness considerations

There is one aspect related to parsimony that should be analyzed here. This concerns the problem of adequately and uniquely describing a given system within a certain model structure. To formalize such a problem one must of course introduce some assumptions on the true system (i.e. the mechanism that produces the data $u(1), y(1), u(2), y(2), \ldots$). It should be stressed that such assumptions are needed for the *analysis* only. The *application* of identification techniques is not dependent on the validity of such assumptions.

Assume that the true system $\mathscr{S}$ is linear, discrete time, and that its disturbances have rational spectral density. Then it can be described as

$$y(t) = G_s(q^{-1})u(t) + H_s(q^{-1})e_s(t)$$
$$Ee_s(t)e_s^T(t') = \Lambda_s \delta_{t,t'}$$
(6.43)

Introduce the set

$$D_T(\mathscr{S}, \mathscr{M}) = \{\theta |\ G_s(q^{-1}) \equiv G(q^{-1}; \theta),\ H_s(q^{-1}) \equiv H(q^{-1}; \theta),$$
$$\Lambda_s = \Lambda(\theta)\}$$
(6.44)

The set $D_T(\mathscr{S}, \mathscr{M})$ consists of those parameter vectors for which the model structure $\mathscr{M}$ gives a perfect description of the true system $\mathscr{S}$. Three situations can occur:

- The set $D_T(\mathscr{S}, \mathscr{M})$ may be empty. Then no perfect description of the system can be obtained in $\mathscr{M}$, no matter how the parameter vector is chosen. One can say that the model structure has too few parameters to describe the system adequately. This is called *underparametrization*.
- The set $D_T(\mathscr{S}, \mathscr{M})$ may consist of one point. This will then be denoted by θ_0. This is the ideal case; θ_0 is called the true parameter vector.
- The set $D_T(\mathscr{S}, \mathscr{M})$ may consist of several points. Then there are several models within the model set that give a perfect description of the system. This situation is sometimes referred to as *overparametrization*. In such a case one can expect that numerical problems may occur when the parameter estimates are sought. This is certainly the case when the points of $D_T(\mathscr{S}, \mathscr{M})$ are not isolated. In many cases, such as those illustrated in Example 6.7, the set $D_T(\mathscr{S}, \mathscr{M})$ will in fact be a connected subset (or even a linear subspace).

ARMAX models

Example 6.7 *Uniqueness properties for an ARMAX model*
Consider a (scalar) ARMAX model, (6.3)–(6.4),

Section 6.3   Uniqueness properties   163

$$A(q^{-1})y(t) = B(q^{-1})u(t) + C(q^{-1})e(t) \quad Ee^2(t) = \lambda^2 \tag{6.45}$$

Let the true system be given by

$$A_s(q^{-1})y(t) = B_s(q^{-1})u(t) + C_s(q^{-1})e_s(t) \quad Ee_s^2(t) = \lambda_s^2 \tag{6.46}$$

with

$$\begin{aligned}
A_s(q^{-1}) &= 1 + a_1^s q^{-1} + \ldots + a_{na_s}^s q^{-na_s} \\
B_s(q^{-1}) &= b_1^s q^{-1} + \ldots + b_{nb_s}^s q^{-nb_s} \\
C_s(q^{-1}) &= 1 + c_1^s q^{-1} + \ldots + c_{nc_s}^s q^{-nc_s}
\end{aligned} \tag{6.47}$$

Assume that the polynomials A_s, B_s and C_s are coprime, i.e. that there is no common factor to all three polynomials.

The identities in (6.44) defining the set $D_T(\mathcal{S}, \mathcal{M})$ now become

$$\frac{B_s(q^{-1})}{A_s(q^{-1})} \equiv \frac{B(q^{-1})}{A(q^{-1})} \quad \frac{C_s(q^{-1})}{A_s(q^{-1})} \equiv \frac{C(q^{-1})}{A(q^{-1})} \quad \lambda_s^2 = \lambda^2 \tag{6.48}$$

For (6.48) to have a solution it is necessary that

$$na \geq na_s \quad nb \geq nb_s \quad nc \geq nc_s \tag{6.49}$$

This means that every model polynomial should have at least as large a degree as the corresponding system polynomial. The inequalities in (6.49) can be summarized as

$$n^* \triangleq \min(na - na_s, nb - nb_s, nc - nc_s) \geq 0 \tag{6.50}$$

One must now find the solution of (6.48) with respect to the coefficients of A, B, C and λ^2. Assume that (6.50) holds. Trivially, $\lambda^2 = \lambda_s^2$. To continue, let us first discuss the more simple case of a pure ARMA process. Then $nb = nb_s = 0$ and the first identity in (6.48) can be dispensed with. In the second identity note that A_s and C_s have no common factor and further that both sides must have the same poles and zeros. These observations imply

$$\begin{aligned}
A(q^{-1}) &= A_s(q^{-1})D(q^{-1}) \\
C(q^{-1}) &= C_s(q^{-1})D(q^{-1})
\end{aligned} \tag{6.51}$$

where

$$D(q^{-1}) = 1 + d_1 q^{-1} + \ldots + d_{nd} q^{-nd}$$

has arbitrary coefficients. The degree nd is given by

$$nd = \deg D = \min(na - na_s, nc - nc_s) = n^*$$

Thus if $n^* > 0$ there are infinitely many solutions to (6.48) (obtained by varying $d_1 \ldots d_{n^*}$). On the other hand, if $n^* = 0$ then $D(q^{-1}) \equiv 1$ and (6.48) has a unique solution. The condition $n^* = 0$ means that at least one of the polynomials $A(q^{-1})$ and $C(q^{-1})$ has the same degree as the corresponding polynomial of the true system.

Next examine the somewhat more complicated case of a general ARMAX structure. Assume that $A_s(q^{-1})$ and $C_s(q^{-1})$ have exactly nl ($nl \geq 0$) common zeros. Then there exist unique polynomials $A_0(q^{-1})$, $C_0(q^{-1})$ and $L(q^{-1})$, such that

$$L(q^{-1}) = 1 + l_1 q^{-1} + \ldots + l_{nl} q^{-nl}$$
$$A_s(q^{-1}) \equiv A_0(q^{-1}) L(q^{-1})$$
$$C_s(q^{-1}) \equiv C_0(q^{-1}) L(q^{-1}) \qquad (6.52)$$
$$A_0(q^{-1}), C_0(q^{-1}) \text{ coprime}$$
$$B_s(q^{-1}), L(q^{-1}) \text{ coprime}$$

The second identity in (6.48) gives

$$\frac{C_0(q^{-1})}{A_0(q^{-1})} \equiv \frac{C(q^{-1})}{A(q^{-1})}$$

Since $A_0(q^{-1})$ and $C_0(q^{-1})$ are coprime, it follows that

$$A(q^{-1}) \equiv A_0(q^{-1}) M(q^{-1})$$
$$C(q^{-1}) \equiv C_0(q^{-1}) M(q^{-1}) \qquad (6.53)$$

where

$$M(q^{-1}) = 1 + m_1 q^{-1} + \ldots + m_{nm} q^{-nm}$$

has arbitrary coefficients and degree

$$\deg M = nm = \min(na - \deg A_0,\ nc - \deg C_0)$$
$$= nl + \min(na - na_s,\ nc - nc_s) \qquad (6.54)$$

Using (6.52), (6.53) in the first identity of (6.48) gives

$$\frac{B_s(q^{-1})}{A_0(q^{-1}) L(q^{-1})} \equiv \frac{B(q^{-1})}{A_0(q^{-1}) M(q^{-1})} \qquad (6.55)$$

Now cancel the factor $A_0(q^{-1})$. Since $B_s(q^{-1})$ and $L(q^{-1})$ are coprime (cf. (6.52)), it follows from (6.55) that

$$B(q^{-1}) \equiv B_s(q^{-1}) D(q^{-1})$$
$$M(q^{-1}) \equiv L(q^{-1}) D(q^{-1}) \qquad (6.56)$$

where

$$D(q^{-1}) = 1 + d_1 q^{-1} + \ldots + d_{nd} q^{-nd} \qquad (6.57a)$$

has arbitrary coefficients. Its degree is given by

$$\deg D = nd = \min(nb - nb_s,\ nm - nl) = n^* \qquad (6.57b)$$

Further (6.52), (6.53) and (6.56) together give the general solution to (6.48):

$$A(q^{-1}) = A_s(q^{-1}) D(q^{-1})$$
$$B(q^{-1}) = B_s(q^{-1}) D(q^{-1}) \qquad (6.58)$$
$$C(q^{-1}) = C_s(q^{-1}) D(q^{-1})$$

Section 6.3 *Uniqueness properties* 165

where $D(q^{-1})$ must have all zeros outside the unit circle (cf. (6.2)) but is otherwise arbitrary.

To summarize, for an ARMAX model:

- If $n^* < 0$ there is no point in the set $D_T(\mathscr{S}, \mathscr{M})$.
- If $n^* = 0$ there is exactly one point in the set $D_T(\mathscr{S}, \mathscr{M})$.
- If $n^* > 0$ there are infinitely many points in $D_T(\mathscr{S}, \mathscr{M})$. (More exactly, $D_T(\mathscr{S}, \mathscr{M})$ is a linear subspace of dimension n^*.) ∎

The following paragraphs present some results on the uniqueness properties of other model structures.

Diagonal form and full polynomial form

Consider first the diagonal form (6.26)–(6.27) for a multivariable system, and concentrate on the deterministic part only. The transfer function operator from $u(t)$ to $y(t)$ is given by

$$G(q^{-1}; \theta) = A^{-1}(q^{-1}; \theta) B(q^{-1}; \theta)$$

$$= \begin{pmatrix} B_1(q^{-1}; \theta)/a_1(q^{-1}; \theta) \\ \vdots \\ B_{ny}(q^{-1}; \theta)/a_{ny}(q^{-1}; \theta) \end{pmatrix} \tag{6.59a}$$

The condition $n^* \geq 0$, (6.50), can now be generalized to

$$n_i^* \triangleq \min(nai - nai_s, nbi - nbi_s) \geq 0 \quad i = 1, \ldots, ny \tag{6.59b}$$

which is necessary for the existence of a parameter vector θ_0 that gives the true transfer function. The condition required for θ_0 to be unique is

$$n_i^* = 0 \quad i = 1, \ldots, ny \tag{6.59c}$$

For the full polynomial form (6.22), it is often but not always possible to guarantee uniqueness. Exact conditions for uniqueness are derived in Complement C6.1. The following example demonstrates that nonuniqueness may easily occur for multivariable models (not necessarily of the full polynomial form).

Example 6.8 *Nonuniqueness of a multivariable model*
Consider a deterministic multivariable system with $ny = 2$, $nu = 1$ given by

$$y(t) - \begin{pmatrix} \alpha & 2 - \alpha \\ \alpha & 2 - \alpha \end{pmatrix} y(t-1) = \begin{pmatrix} 1 \\ 1 \end{pmatrix} u(t-1) \tag{6.60a}$$

where α is a parameter. The corresponding transfer function operator is given by

$$G(q^{-1}) = \begin{pmatrix} q - \alpha & -2 + \alpha \\ -\alpha & q - 2 + \alpha \end{pmatrix}^{-1} \begin{pmatrix} 1 \\ 1 \end{pmatrix} = \frac{q^{-1}}{1 - 2q^{-1}} \begin{pmatrix} 1 \\ 1 \end{pmatrix} \tag{6.60b}$$

which is independent of α. Hence all models of the form (6.60a), for any value of α, will give the same transfer function operator $G(q^{-1})$. In particular, this means that the system cannot be uniquely represented by a first-order full polynomial form model. Note that this conclusion follows immediately from the result (C6.1.3) in Complement C6.1.

■

State space models

The next example considers state space models without any specific parametrization, and shows that if all matrix elements are assumed unknown, then uniqueness cannot hold. (See also Problem 6.3.)

Example 6.9 *Nonuniqueness of general state space models*
Consider the multivariable model

$$\begin{aligned} x(t+1) &= Ax(t) + Bu(t) + v(t) \\ y(t) &= Cx(t) + e(t) \end{aligned} \tag{6.61a}$$

where $\{v(t)\}$ and $\{e(t)\}$ are mutually independent white noise sequences with zero means and covariance matrices R_1 and R_2, respectively. Assume that all the matrix elements are free to vary. Consider also a second model

$$\begin{aligned} \bar{x}(t+1) &= \bar{A}\bar{x}(t) + \bar{B}u(t) + \bar{v}(t) \\ y(t) &= \bar{C}\bar{x}(t) + e(t) \\ E\bar{v}(t)\bar{v}^T(s) &= \bar{R}_1\delta_{t,s} \quad Ee(t)e^T(s) = R_2\delta_{t,s} \\ E\bar{v}(t)e^T(s) &= 0 \end{aligned} \tag{6.61b}$$

where

$$\bar{A} = QAQ^{-1} \quad \bar{B} = QB \quad \bar{C} = CQ^{-1}$$
$$\bar{R} = QR_1Q^T$$

and Q is an arbitrary nonsingular matrix. The models above are equivalent in the sense that they have the same transfer function from u to y and their outputs have the same second-order properties. To see this, first calculate the transfer function operator from $u(t)$ to $y(t)$ of the model (6.61b).

$$\begin{aligned} G(q^{-1}) &= \bar{C}[qI - \bar{A}]^{-1}\bar{B} \\ &= CQ^{-1}[qI - QAQ^{-1}]^{-1}QB \\ &= C[Q^{-1}\{qI - QAQ^{-1}\}Q]^{-1}B \\ &= C[qI - A]^{-1}B \end{aligned}$$

This transfer function is independent of the matrix Q. To analyze the influence of the stochastic terms on $y(t)$, it is more convenient in this case to examine the spectral density

$\phi_y(\omega)$ rather than to explicitly derive $H(q^{-1})$ and Λ in (6.1). Once $\phi_y(\omega)$ is known, $H(q^{-1})$ and Λ are uniquely given by the spectral factorization theorem. To evaluate $\phi_y(\omega)$ for the model (6.61b), set $u(t) \equiv 0$. Then

$$\begin{aligned}2\pi\phi_y(\omega) &= \bar{C}[e^{i\omega}I - \bar{A}]^{-1}\bar{R}_1[e^{-i\omega}I - \bar{A}^T]^{-1}\bar{C}^T + R_2 \\ &= CQ^{-1}[e^{i\omega}I - \bar{A}]^{-1}QR_1Q^T[e^{-i\omega}I - \bar{A}^T]^{-1}Q^{-T}C^T + R_2 \\ &= C[Q^{-1}(e^{i\omega}I - \bar{A})Q]^{-1}R_1[Q^T(e^{-i\omega}I - \bar{A}^T)Q^{-T}]^{-1}C^T + R_2 \\ &= C[e^{i\omega}I - A]^{-1}R_1[e^{-i\omega}I - A^T]^{-1}C^T + R_2\end{aligned}$$

Thus the spectral density (and hence $H(q^{-1})$ and Λ) are also independent of Q. Since Q can be chosen as an arbitrary nonsingular matrix, it follows that the model is not unique in the sense that $D_T(\mathscr{S}, \mathscr{M})$ consists of an infinity of points. To get a unique model it is necessary to impose restrictions of some form on the matrix elements. This is a topic that belongs to the field of canonical forms. See the bibliographical notes at the end of this chapter for some references. ∎

6.4 Identifiability

The concept of identifiability can be introduced in a number of ways, but the following is convenient for the present purposes.

When an identification method $\mathscr{I}$ is applied to a parametric model structure $\mathscr{M}$ the resulting estimate is denoted by $\hat{\theta}(N; \mathscr{S}, \mathscr{M}, \mathscr{I}, \mathscr{X})$. Clearly the estimate will depend not only on $\mathscr{I}$ and $\mathscr{M}$ but also on the number of data points N, the true system $\mathscr{S}$, and the experimental condition $\mathscr{X}$.

The system $\mathscr{S}$ is said to be *system identifiable under* $\mathscr{M}$, $\mathscr{I}$ *and* $\mathscr{X}$, abbreviated SI($\mathscr{M}, \mathscr{I}, \mathscr{X}$), if

$$\hat{\theta}(N; \mathscr{S}, \mathscr{M}, \mathscr{I}, \mathscr{X}) \to D_T(\mathscr{S}, \mathscr{M}) \quad \text{as } N \to \infty \tag{6.62}$$

(with probability one). For SI($\mathscr{M}, \mathscr{I}, \mathscr{X}$) it is in particular required that the set $D_T(\mathscr{S}, \mathscr{M})$ (introduced in (6.44)) is nonempty. If it contains more than one point then (6.62) must be interpreted as

$$\lim_{N \to \infty} \inf_{\theta \in D_T(\mathscr{S}, \mathscr{M})} \| \hat{\theta}(N; \mathscr{S}, \mathscr{M}, \mathscr{I}, \mathscr{X}) - \theta \| = 0 \tag{6.63}$$

The meaning of (6.63) is that the shortest distance between the estimate $\hat{\theta}$ and the set $D_T(\mathscr{S}, \mathscr{M})$ of all parameter vectors describing $G(q^{-1})$ and $H(q^{-1})$ exactly, tends to zero as the number of data points tends to infinity.

We say that the system $\mathscr{S}$ is *parameter identifiable under* $\mathscr{M}$, $\mathscr{I}$ *and* $\mathscr{X}$, abbreviated PI($\mathscr{M}, \mathscr{I}, \mathscr{X}$), if it is SI($\mathscr{M}, \mathscr{I}, \mathscr{X}$) and $D_T(\mathscr{S}, \mathscr{M})$ consists of exactly one point. This is the ideal case. If the system is PI($\mathscr{M}, \mathscr{I}, \mathscr{X}$) then the parameter estimate $\hat{\theta}$ will be unique for large values of N and also consistent (i.e. $\hat{\theta}$ converges to the true value, as given by the definition of $D_T(\mathscr{S}, \mathscr{M})$).

Here the concept of identifiability has been separated into two parts. The convergence of the parameter estimate $\hat{\theta}$ to the set $D_T(\mathscr{S}, \mathscr{M})$ (i.e. the system identifiability) is a

168 Model parametrizations   Chapter 6

property that basically depends on the identification method $\mathscr{I}$. This is a most desirable property and should hold for as general experimental conditions $\mathscr{X}$ as possible. It is then 'only' the model parametrization or model structure $\mathscr{M}$ that determines whether the system is also parameter identifiable. It is of course desirable to choose the model structure so that the set $D_T(\mathscr{I}, \mathscr{M})$ has precisely one point. Some practical aspects on this problem are given in Chapter 11.

Summary

In this chapter various aspects of the choice of model structure have been discussed. Sections 6.1 and 6.2 described various ways of classifying model structures, and a general form of model structure for linear multivariable systems was given. Some restrictions on the model parameters were noted. Section 6.2 also gave examples of the way in which some well-known model structures can be seen as special cases of the general form (6.1). Section 6.3 discussed the important problem of model structure uniqueness (i.e. the property of a model set to represent uniquely an arbitrary given system). The case of ARMAX models was analyzed in detail. Finally, identifiability concepts were introduced in Section 6.4.

Problems

Problem 6.1 *Stability boundary for a second-order system*
Consider a second-order AR model
$$y(t) + a_1 y(t-1) + a_2 y(t-2) = e(t)$$
Derive and plot the area in the (a_1, a_2)-plane for which the model is asymptotically stable.

Problem 6.2 *Spectral factorization*
Consider a stationary stochastic process $y(t)$ with spectral density
$$\phi_y(\omega) = \frac{1}{2\pi} \frac{5 - 4\cos\omega}{8.2 - 8\cos\omega}$$
Show that this process can be represented as a first-order ARMA process and derive its parameters.

Problem 6.3 *Further comments on the nonuniqueness of stochastic state space models*
Consider the following stochastic state space model (cf. (6.61)):
$$\begin{aligned} x(t+1) &= Ax(t) + Bv(t) \\ y(t) &= Cx(t) + e(t) \end{aligned} \tag{i}$$
where

$$E\begin{pmatrix}v(t)\\e(t)\end{pmatrix}(v^T(s)\ e^T(s)) = R\delta_{t,s}$$

It was shown in Example 6.9 that the representation (i) of the stochastic process $y(t)$ is not unique unless the matrices A, B and C are canonically parametrized.

There is also another type of nonuniqueness of (i) as a second-order representation of $y(t)$, induced by an 'inappropriate' parametrization of R (the matrices A, B and C being canonically parametrized or even given). To see this, consider (i) with

$$A = 1/2 \quad B = 1 \quad C = 1$$

$$R = \begin{pmatrix} 10 - \frac{3}{4}\varrho & \frac{1}{2}\varrho \\ \frac{1}{2}\varrho & \varrho \end{pmatrix} \quad \varrho \in (0, 10) \tag{ii}$$

Note that the matrix R is positive definite as required. Determine the spectral density function of $y(t)$ corresponding to (i), (ii). Since this function does not depend on ϱ, conclude that $y(t)$ is not uniquely identifiable in the representation (i), (ii) from second-order data.

Remark. This exercise is patterned after Anderson and Moore (1979), where more details on representing stationary second-order processes by state space models may be found.

Problem 6.4 *A state space representation of autoregressive processes*
Consider an autoregressive process $y(t)$ given (see (6.9)) by

$$y(t) + a_1 y(t-1) + \ldots + a_n y(t-n) = e(t)$$
$$A(z) = 1 + a_1 z + \ldots + a_n z^n \neq 0 \quad \text{for } |z| \leq 1$$

Derive a state space representation of $y(t)$

$$x(t+1) = Ax(t) + Be(t+1)$$
$$y(t) = Cx(t) \tag{i}$$

with the matrix A in the following companion form:

$$A = \begin{pmatrix} -a_1 & \cdots & -a_{n-1} & -a_n \\ 1 & & 0 & 0 \\ & \ddots & & \vdots \\ 0 & & 1 & 0 \end{pmatrix}$$

Discuss the identifiability properties of the representation. In particular, is it uniquely identifiable from second-order data?

Problem 6.5 *Uniqueness properties of ARARX models*
Consider the set $D_T(\mathcal{S}, \mathcal{M})$ for the system

$$A_s(q^{-1})y(t) = B_s(q^{-1})u(t) + \frac{1}{C_s(q^{-1})}e(t)$$

and the model structure

$$A(q^{-1})y(t) = B(q^{-1})u(t) + \frac{1}{C(q^{-1})}e(t)$$

(which we call ARARX), where

$$A_s(q^{-1}) = 1 + a_1^s q^{-1} + \ldots + a_{na_s}^s q^{-na_s}$$
$$B_s(q^{-1}) = b_1^s q^{-1} + \ldots + b_{nb_s}^s q^{-nb_s}$$
$$C_s(q^{-1}) = 1 + c_1^s q^{-1} + \ldots + c_{nc_s}^s q^{-nc_s}$$
$$A(q^{-1}) = 1 + a_1 q^{-1} + \ldots + a_{na} q^{-na}$$
$$B(q^{-1}) = b_1 q^{-1} + \ldots + b_{nb} q^{-nb}$$
$$C(q^{-1}) = 1 + c_1 q^{-1} + \ldots + c_{nc} q^{-nc}$$

(a) Derive sufficient conditions for $D_T(\mathcal{S}, \mathcal{M})$ to consist of exactly one point.
(b) Give an example where $D_T(\mathcal{S}, \mathcal{M})$ contains more than one point.

Remark. Note that in the case of overparametrization the set D_T consists of a *finite* number of *isolated* points. In contrast, for overparameterized ARMAX models D_T is a connected set (in fact a subspace) (see Example 6.7). A detailed study of the topic of this problem is contained in Stoica and Söderström (1982e).

Problem 6.6 *Uniqueness properties of a state space model*
Consider the state space model

$$x(t+1) = \begin{pmatrix} a_{11} & a_{12} \\ a_{12} & a_{22} \end{pmatrix} x(t) + \begin{pmatrix} b \\ 0 \end{pmatrix} u(t)$$

$$\theta = (a_{11} \;\; a_{12} \;\; a_{22} \;\; b)^T$$

Examine the uniqueness properties for the following cases:

(a) The first state variable is measured.
(b) The second state variable is measured.
(c) Both state variables are measured.

Assume in all cases that the true system is included in the model set.

Problem 6.7 *Sampling a simple continuous time system*
Consider the following continuous time system:

$$\dot{x} = \begin{pmatrix} 0 & K \\ 0 & 0 \end{pmatrix} x + \begin{pmatrix} K \\ 0 \end{pmatrix} u + \begin{pmatrix} 0 \\ 1 \end{pmatrix} v \quad Ev(t)v(s) = r\delta(t-s)$$

$$y = (1 \;\; 0) x$$

(This can be written as

$$Y(s) = \frac{K}{s}U(s) + \frac{K}{s^2}V(s).)$$

Assume that the gain K is unknown and is to be estimated from *discrete time* measurements

$$y(t) = (1 \quad 0)x(t) \quad t = 1, \ldots, N$$

Find the discrete time description (6.1) of the system, assuming the input is constant over the sampling intervals.

Bibliographical notes

Various ways of parametrizing models are described in the survey paper by Hajdasinski et al. (1982). Ljung and Söderström (1983) give general comments on the choice of parametrization.

The ARMAX model (6.3) has been used frequently in identification since the seminal paper by Åström and Bohlin (1965) appeared. The 'X' in 'ARMAX' refers to the exogenous (control) variable $u(t)$, according to the terminology used in econometrics. The general SISO model (6.14) is discussed, e.g., by Ljung and Söderström (1983).

Diagonal right matrix fraction description models are sometimes of interest. Identification algorithms for such models were considered by Nehorai and Morf (1984).

For a discussion of state space models such as (6.28), (6.29) see, e.g., Ljung and Söderström (1983). For the role of spectral factorization, the Riccati equation and the Kalman filter in this context, see Anderson and Moore (1979) or Åström (1970). The latter reference also describes continuous time stochastic models of the form (6.37) and sampling thereof. Efficient numerical algorithms for performing spectral factorization in polynomial form have been given by Wilson (1969) and Kučera (1979).

Many special model structures for multivariable systems have been proposed in the literature. They can be regarded as ways of parametrizing (6.1) or (6.28) viewed as black-box models (i.e. without using a priori knowledge (physical insight)). For some examples of this kind see Rissanen (1974), Kailath (1980), Guidorzi (1975, 1981), Hannan (1976), Ljung and Rissanen (1976), van Overbeek and Ljung (1982), Gevers and Wertz (1984, 1987a, 1987b), Janssen (1987a, b), Corrêa and Glover (1984a, 1984b), Stoica (1983). Gevers (1986), Hannan and Kavalieris (1984), and Deistler (1986) give rather detailed descriptions for ARMA models, while Corrêa and Glover (1987) discuss specific parametrizations for instrumental variable estimation.

Specific compartmental models, which are frequently used in biomedical applications, have been analyzed for example by Godfrey (1983), Walter (1982), Godfrey and DiStefano (1985), and Walter (1987).

Leontaritis and Billings (1985) as well as Carrol and Ruppert (1984) deal with various forms of nonlinear models.

Nguyen and Wood (1982) give a survey of various results on identifiability.

Appendix A6.1
Spectral factorization

This appendix examines the so-called *spectral factorization problem*. The following lemma is required.

Lemma A6.1
Consider the function

$$f(z) = \sum_{k=-n}^{n} f_k z^k \qquad f_k = f_{-k} \text{ (real-valued)} \qquad f_n \neq 0 \qquad \text{(A6.1.1a)}$$

Assume that

$$f(e^{i\omega}) \geq 0 \qquad \text{all } \omega \qquad \text{(A6.1.1b)}$$

The following results hold true:

(i) If $\bar{z}$ is a zero of $f(z)$, so is $\bar{z}^{-1}$.
(ii) There is a polynomial

$$g(z) = z^n + g_1 z^{n-1} + \ldots + g_n$$

with real-valued coefficients and all zeros inside or on the unit circle, and a (real-valued) positive constant C, such that

$$f(z) = C g(z) g(z^{-1}) \qquad \text{all } z \qquad \text{(A6.1.2)}$$

(iii) If $f(e^{i\omega}) > 0$ for all ω, then the polynomial $g(z)$ can be chosen with all zeros strictly inside the unit circle.

Proof. Part (i) follows by direct verification:

$$f(\bar{z}^{-1}) = \sum_{k=-n}^{n} f_k \bar{z}^{-k} = \sum_{k=-n}^{n} f_{-k} \bar{z}^{-k} = \sum_{k=-n}^{n} f_k \bar{z}^k = f(\bar{z}) = 0$$

As a consequence $f(z)$ can be written as

$$f(z) = h(z) h(z^{-1}) k(z)$$

where $h(z)$ has all zeros strictly inside the unit circle (and hence $h(z^{-1})$ has all zeros strictly outside the unit circle) and $k(z)$ all zeros on the unit circle. Hence

$$f(e^{i\omega}) = h(e^{i\omega}) h(e^{-i\omega}) k(e^{i\omega}) = |h(e^{i\omega})|^2 k(e^{i\omega})$$

From (A6.1.1b), $k(e^{i\omega}) \geq 0$ for all ω. Therefore all zeros of $k(z)$ must have an even multiplicity. (If $\bar{z} = e^{i\varphi}$ were a zero with odd multiplicity then $k(e^{i\omega})$ would shift sign at $\omega = \varphi$.) As a consequence, the zeros of $f(z)$ can be written as $\bar{z}_1, \bar{z}_2, \ldots, \bar{z}_n, \bar{z}_1^{-1}, \ldots, \bar{z}_n^{-1}$, where $0 < |\bar{z}_i| \leq 1$ for $i = 1, \ldots, n$. Set

$$g(z) = \prod_{i=1}^{n} (z - \bar{z}_i) \qquad C = (-1)^n f_n \bigg/ \prod_{j=1}^{n} \bar{z}_j$$

Appendix A6.1

which gives

$$Cg(z)g(z^{-1}) = (-1)^n \frac{f_n}{\prod_{j=1}^n \bar{z}_j} \prod_{i=1}^n (z - \bar{z}_i) \prod_{k=1}^n (z^{-1} - \bar{z}_k)$$

$$= (-1)^n \frac{f_n}{\prod_{j=1}^n \bar{z}_j} \prod_{i=1}^n (z - \bar{z}_i) \prod_{k=1}^n \{-z^{-1}\bar{z}_k(z - \bar{z}_k^{-1})\}$$

$$= f_n \prod_{i=1}^n (z - \bar{z}_i) z^{-n} \prod_{k=1}^n (z - \bar{z}_k^{-1}) = f(z)$$

which proves the relation (A6.1.2). To complete part (ii), it remains to prove that g_i is real-valued and that C is positive. First note that any complex-valued $\bar{z}_i$ with $|\bar{z}_i| \leq 1$ appears in a complex-conjugated pair. Hence $g(z)$ has real-valued coefficients. That the constant C is positive follows easily from

$$0 \leq f(e^{i\omega}) = C|g(e^{i\omega})|^2$$

To prove part (iii) note that

$$0 < f(e^{i\omega}) = C|g(e^{i\omega})|^2 \quad \text{all } \omega$$

which precludes that $g(z)$ can have any zero on the unit circle. ■

The lemma can be applied to *rational* spectral densities. Such a spectral density is a rational function (a ratio between two polynomials) of $\cos \omega$ (or equivalently of $e^{i\omega}$). It can hence be written as

$$\phi(\omega) = \frac{\sum_{k=-m}^{m} \beta_k e^{ik\omega}}{\sum_{j=-n}^{n} \alpha_j e^{ij\omega}} \quad (\beta_{-k} = \beta_k, \alpha_{-j} = \alpha_j \text{ are real-valued}) \tag{A6.1.3}$$

By varying the integers m and n and the coefficients $\{\alpha_j\}$, $\{\beta_k\}$ a large set of spectral densities can be obtained. This was illustrated to some extent in Example 5.6. According to a theorem by Weierstrass (see Pearson, 1974), any continuous function can be approximated arbitrarily closely by a polynomial if the polynomial degree is chosen large enough. This gives a further justification of using (A6.1.3) for describing a spectral density. By applying the lemma to the rational density (A6.1.3) twice (to the numerator and the denominator) it is found that $\phi(\omega)$ can be factorized as

$$\phi(\omega) = \frac{\lambda^2}{2\pi} \frac{C(e^{i\omega})C(e^{-i\omega})}{A(e^{i\omega})A(e^{-i\omega})} \tag{A6.1.4}$$

where λ is a real-valued number, and $A(z)$, $C(z)$ are polynomials

$$A(z) = 1 + a_1 z + \ldots + a_n z^n$$
$$C(z) = 1 + c_1 z + \ldots + c_m z^m$$

with all zeros outside, or for $C(z)$ possibly on the unit circle. The polynomial $A(z)$ cannot have zeros on the unit circle since that would make $\phi(\omega)$ infinite. If $\phi(\omega) > 0$ for all ω then $C(z)$ can be chosen with all zeros strictly outside the unit circle.

The consequence of (A6.1.4) is that as far as the second-order properties are concerned (i.e. the spectral density), the signal can be described as an ARMA process

$$A(q^{-1})y(t) = C(q^{-1})e(t) \quad Ee^2(t) = \lambda^2 \qquad (A6.1.5)$$

Indeed the process (A6.1.5) has the spectral density given by (A6.1.4) (cf. (A3.1.10)). In practice the signal whose spectral density is $\phi(\omega)$ may be caused by a number of interacting noise sources. However it is convenient to use the ARMA model (A6.1.5) to describe its spectral density (or equivalently its covariance function). For many identification problems it is relevant to characterize the involved signals by their second-order properties only.

Remark. The polynomial $C(z)$ could have been chosen with some zeros inside the unit circle. However, the choice made above will turn out to be the most suitable one when deriving optimal predictors (see Section 7.3). ∎

The result (A6.1.4) was derived for a scalar signal. For the multivariable case the following result holds. Let the rational spectral density $\phi(\omega)$ be nonsingular (det $\phi(\omega) \neq 0$) for all frequencies. Then there are a unique rational filter $H(q^{-1})$ and a positive definite matrix Λ such that

$$\phi(\omega) = \frac{1}{2\pi} H(e^{-i\omega}) \Lambda H^T(e^{i\omega}) \qquad (A6.1.6a)$$

$H(q^{-1})$ and $H^{-1}(q^{-1})$ are asymptotically stable $\qquad$ (A6.1.6b)

$H(0) = I \qquad (A6.1.6c)$

(see, e.g., Anderson and Moore, 1979). The consequence of this result is that the signal, whose spectral density is $\phi(\omega)$, can be described by the model

$$y(t) = H(q^{-1})e(t) \quad Ee(t)e^T(s) = \Lambda \delta_{t,s} \qquad (A6.1.7)$$

(cf. (A3.1.10)).

Next consider how the filter $H(q^{-1})$ and the covariance matrix Λ can be found for a system given in state space form; in other words, how to solve the spectral factorization problem for a state space system. Consider the model

$$\begin{aligned} x(t+1) &= Ax(t) + v(t) \\ y(t) &= Cx(t) + e(t) \end{aligned} \qquad (A6.1.8)$$

where $v(t)$ and $e(t)$ are mutually independent white noise sequences with zero means and covariance matrices R_1 and R_2, respectively. The matrix A is assumed to have all eigenvalues inside the unit circle. Let P be a positive definite matrix that satisfies the algebraic Riccati equation

$$P = APA^T + R_1 - APC^T(CPC^T + R_2)^{-1}CPA^T \tag{A6.1.9}$$

Set

$$K = APC^T(CPC^T + R_2)^{-1}$$
$$H(q^{-1}) = I + C[qI - A]^{-1}K \tag{A6.1.10}$$
$$\Lambda = CPC^T + R_2$$

This solves the factorization problem (A6.1.6) (as will be shown below). It is obvious that $H(0) = I$ and that $H(q^{-1})$ is asymptotically stable. To show (A6.1.6a), note from (A3.1.10) that

$$2\pi\phi(\omega) = R_2 + C(e^{i\omega}I - A)^{-1}R_1(e^{-i\omega}I - A^T)^{-1}C^T$$

By the construction of $H(q^{-1})$,

$$H(e^{-i\omega})\Lambda H^T(e^{i\omega}) = [I + C(e^{i\omega}I - A)^{-1}K]\Lambda[I + K^T(e^{-i\omega}I - A^T)^{-1}C^T]$$
$$= \Lambda + C(e^{i\omega}I - A)^{-1}APC^T + CPA^T(e^{-i\omega}I - A^T)^{-1}C^T$$
$$+ C(e^{i\omega}I - A)^{-1}APC^T(CPC^T + R_2)^{-1}CPA^T$$
$$\times (e^{-i\omega}I - A^T)^{-1}C^T$$
$$= CPC^T + R_2 + C(e^{i\omega}I - A)^{-1}APC^T$$
$$+ CPA^T(e^{-i\omega}I - A^T)^{-1}C^T$$
$$+ C(e^{i\omega}I - A)^{-1}(APA^T + R_1 - P)(e^{-i\omega}I - A^T)^{-1}C^T$$
$$= 2\pi\phi(\omega) + C(e^{i\omega}I - A)^{-1}[(e^{i\omega}I - A)P(e^{-i\omega}I - A^T)$$
$$+ AP(e^{-i\omega}I - A^T) + (e^{i\omega}I - A)PA^T$$
$$+ APA^T - P](e^{-i\omega}I - A^T)^{-1}C^T$$
$$= 2\pi\phi(\omega)$$

Hence (A6.1.6a) holds for $H(q^{-1})$ and Λ given by (A6.1.10). It remains to examine the stability properties of $H^{-1}(q^{-1})$. Using the matrix inversion lemma (Lemma A.1 in Appendix A), equation (A6.1.10) gives

$$H^{-1}(q^{-1}) = [I + C(qI - A)^{-1}K]^{-1}$$
$$= I - C[(qI - A) + KC]^{-1}K \tag{A6.1.11}$$
$$= I - C[qI - (A - KC)]^{-1}K$$

Hence the stability properties of $H^{-1}(q^{-1})$ are completely determined by the location of the eigenvalues of $A - KC$. To examine the location of these eigenvalues one can study the stability of the system

$$x(t + 1) = (A - KC)^T x(t) \tag{A6.1.12}$$

since $A - KC$ and $(A - KC)^T$ have the same set of eigenvalues.

In the study of (A6.1.12) the following candidate Lyapunov function will be used (recall that P was assumed to be positive definite):

$$V(x) = x^T P x$$

For this function,
$$V(x(t+1)) - V(x(t)) = x^T(t)[(A - KC)P(A - KC)^T - P]x(t) \quad (A6.1.13)$$

The matrix in square brackets can be rewritten (using (A6.1.9)) as
$$\begin{aligned}(A - KC)P(A - KC)^T - P &= (APA^T - P) + (-KCPA^T - APC^TK^T + KCPC^TK^T)\\ &= [-R_1 + K(CPC^T + R_2)K^T]\\ &\quad - 2K(CPC^T + R_2)K^T + KCPC^TK^T\\ &= -R_1 - KR_2K^T\end{aligned}$$

which is nonpositive definite. Hence $H^{-1}(q^{-1})$ is at least stable, and possibly asymptotically stable, cf. Remark 2 below.

Remark 1. The foregoing results have a very close connection to the (stationary) optimal state predictor. This predictor is given by the Kalman filter (see Section B.7), and has the structure

$$\begin{aligned}\hat{x}(t+1|t) &= A\hat{x}(t|t-1) + K\tilde{y}(t)\\ y(t) &= C\hat{x}(t|t-1) + \tilde{y}(t)\end{aligned} \quad (A6.1.14)$$

where $\tilde{y}(t)$ is the output innovation at time t. The covariance matrix of $\tilde{y}(t)$ is given by

$$\text{cov}[\tilde{y}(t)] = CPC^T + R_2$$

From (A6.1.14) it follows that
$$\begin{aligned}y(t) &= \tilde{y}(t) + C(qI - A)^{-1}K\tilde{y}(t)\\ &= H(q^{-1})\tilde{y}(t)\end{aligned}$$

One can also 'invert' (A6.1.14) by considering $y(t)$ as input and $\tilde{y}(t)$ as output. In this way,

$$\begin{aligned}\hat{x}(t+1|t) &= A\hat{x}(t|t-1) + K[y(t) - C\hat{x}(t|t-1)]\\ \tilde{y}(t) &= y(t) - C\hat{x}(t|t-1)\end{aligned}$$

or
$$\begin{aligned}\hat{x}(t+1|t) &= (A - KC)\hat{x}(t|t-1) + Ky(t)\\ \tilde{y}(t) &= -C\hat{x}(t|t-1) + y(t)\end{aligned} \quad (A6.1.15)$$

from which the expression (A6.1.11) for $H^{-1}(q^{-1})$ easily follows. Thus by solving the spectral factorization problem associated with (A6.1.8), the innovation form (or equivalently the stationary Kalman filter) of the system (A6.1.8) is obtained implicitly. ∎

Remark 2. $H^{-1}(q^{-1})$ will be *asymptotically* stable under weak conditions. In case $H^{-1}(q^{-1})$ is only stable (having poles *on* the unit circle), then $A - KC$ has eigenvalues on the unit circle (see A.6.1.11). This implies that det $H(q^{-1})$ has a zero on the unit circle. To see this, note that by Corollary 1 of Lemma A.5,

$$\det H(e^{-i\omega}) = \det[I + C(e^{i\omega}I - A)^{-1}K]$$
$$= \det[I + KC(e^{i\omega}I - A)^{-1}]$$
$$= \det[e^{i\omega}I - A + KC] \times \det[(e^{i\omega}I - A)^{-1}]$$

Hence by (A6.1.6a) the spectral density matrix $\phi(\omega)$ is singular at some ω. Conversely, if $\phi(\omega)$ is singular for some frequencies, the matrix $A - KC$ will have some eigenvalues on the unit circle and $H^{-1}(q^{-1})$ will not be asymptotically stable. ∎

Remark 3. The Riccati equation (A6.1.9) has been treated extensively in the literature. See for example Kučera (1972), Anderson and Moore (1979), Goodwin and Sin (1984) and de Sousa et al. (1986). The equation has at most one symmetric positive definite solution. In some 'degenerate' cases there may be no positive definite solution, but a positive semidefinite solution exists and is the one of interest (see Example A6.1.2 for an illustration). Some general sufficient conditions for existence of a positive definite solution are as follows. Factorize R_1 as $R_1 = BB^T$. Then it is sufficient to require:

(i) $R_2 > 0$;
(ii) (A, B) controllable (i.e. rank $(B \ AB \ \ldots \ A^{n-1}B) = n = \dim A$).

Some weaker but more technical necessary and sufficient conditions are given by de Sousa et al. (1986). ∎

Remark 4. If $v(t)$ and $e(t)$ are correlated so that

$$Ev(t)e^T(s) = R_{12}\delta_{t,s}$$

the results remain valid if the Riccati equation (A6.1.9) and the gain vector K (A6.1.10) are changed to, see Anderson and Moore (1979),

$$P = APA^T + R_1 - (APC^T + R_{12})(CPC^T + R_2)^{-1}(CPA^T + R_{12}^T)$$
$$K = (APC^T + R_{12})(CPC^T + R_2)^{-1} \quad (A6.1.16)$$

∎

Example A6.1.1 *Spectral factorization for an ARMA process*
Consider the ARMA (1, 1) process

$$y(t) + ay(t - 1) = e(t) + ce(t - 1) \quad |a| < 1, c \neq 0$$
$$Ee(t)e(s) = \delta_{t,s} \quad (A6.1.17a)$$

The spectral density of $y(t)$ is easily found to be

$$\phi(\omega) = \frac{1}{2\pi}\left|\frac{1 + ce^{i\omega}}{1 + ae^{i\omega}}\right|^2 = \frac{1}{2\pi}\frac{1 + c^2 + 2c\cos\omega}{1 + a^2 + 2a\cos\omega} \quad (A6.1.17b)$$

(cf. A3.1.10). If $|c| \leq 1$ the representation (A6.1.17a) is the one described in (A6.1.6). For completeness consider also the case $|c| > 1$. The numerator of the spectral density is equal to $f(z) = cz + (1 + c^2) + cz^{-1} = (1 + cz)(1 + cz^{-1})$. Its zeros are $\bar{z}_1 = -1/c$ and

178 *Model parametrizations*   *Chapter 6*

$\bar{z}_1^{-1} = -c$. It can be rewritten as $f(z) = c^2(z + 1/c)(z^{-1} + 1/c)$. To summarize, the following is the invertible and stable cofactor of the spectral density (A6.1.17b):

Case (i): $|c| \leq 1$ $H(q^{-1}) = \dfrac{1 + cq^{-1}}{1 + aq^{-1}}$ $\Lambda = 1$

Case (ii): $|c| \geq 1$ $H(q^{-1}) = \dfrac{1 + (1/c)q^{-1}}{1 + aq^{-1}}$ $\Lambda = c^2$

(A6.1.17c)

∎

Example A6.1.2 *Spectral factorization for a state space model*
The system (A6.1.17a) can be represented in state space form as

$$x(t+1) = \begin{pmatrix} -a & 1 \\ 0 & 0 \end{pmatrix} x(t) + \begin{pmatrix} 1 \\ c \end{pmatrix} e(t+1)$$

$$y(t) = (1 \;\; 0)x(t)$$

(A6.1.18a)

Here

$$R_1 = \begin{pmatrix} 1 & c \\ c & c^2 \end{pmatrix} \quad\quad R_2 = 0$$

Since the second row of A is zero it follows that any solution of the algebraic Riccati equation must have the structure

$$P = \begin{pmatrix} 1 + \alpha & c \\ c & c^2 \end{pmatrix}$$

(A6.1.18b)

where α is a scalar quantity to be determined. Inserting (A6.1.18b) in (A6.1.9) and evaluating the 1,1-element gives

$$1 + \alpha = (c - a)^2 + a^2\alpha + 1 - \dfrac{(c - a - a\alpha)^2}{1 + \alpha}$$

(A6.1.18c)

which is rewritten as

$$(\alpha + 1)[\alpha(1 - a^2) - (c - a)^2] + [-a\alpha + (c - a)]^2 = 0$$

$$\alpha^2 + \alpha[-(c - a)^2 + (1 - a^2) - 2a(c - a)] = 0$$

$$\alpha(\alpha + 1 - c^2) = 0$$

This equation has two solutions, $\alpha_1 = 0$ and $\alpha_2 = c^2 - 1$. We will discuss the choice of solution shortly. First note that the gain K is given by

$$K = \begin{pmatrix} 1 \\ 0 \end{pmatrix} \dfrac{c - a - a\alpha}{1 + \alpha}$$

(A6.1.18d)

(see (A6.1.10)). Therefore

$$H(q^{-1}) = 1 + (1 \ 0)\begin{pmatrix} q+a & -1 \\ 0 & q \end{pmatrix}^{-1} K = \frac{1 + \left(\frac{c}{1+\alpha}q^{-1}\right)}{1 + aq^{-1}} \quad \text{(A6.1.18e)}$$

$$\Lambda = 1 + \alpha \quad \text{(A6.1.18f)}$$

The remaining discussion considers separately the two cases introduced in Example A6.1.1.

Case (i). Assume $|c| \leq 1$. Then the solution $\alpha_2 = c^2 - 1$ can be ruled out since it corresponds to

$$P = \begin{pmatrix} c^2 & c \\ c & c^2 \end{pmatrix}$$

which is indefinite (except for $|c| = 1$, for which $\alpha_1 = \alpha_2$). Hence

$$P = \begin{pmatrix} 1 & c \\ c & c^2 \end{pmatrix}$$

$$H(q^{-1}) = \frac{1 + cq^{-1}}{1 + aq^{-1}} \quad \Lambda = 1 \quad \text{(A6.1.18g)}$$

Case (ii). Assume $|c| \geq 1$. Then

$$P_1 = \begin{pmatrix} 1 & c \\ c & c^2 \end{pmatrix} \quad P_2 = \begin{pmatrix} c^2 & c \\ c & c^2 \end{pmatrix}$$

with P_2 positive definite, whereas P_1 is only positive semidefinite. There are two ways to conclude that α_2 is the required solution in this case. First, it is P_2 that is the positive definite solution to (A6.1.9). Note also that P_2 is the largest nonnegative definite solution in the sense $P_2 - P_1 \geq 0$. Second, the filter $H^{-1}(q^{-1})$ will be stable for $\alpha = \alpha_2$ but unstable for $\alpha = \alpha_1$. Inserting $\alpha_2 = c^2 - 1$ into (A6.1.18e, f) gives

$$H(q^{-1}) = \frac{1 + (1/c)q^{-1}}{1 + aq^{-1}} \quad \Lambda = c^2 \quad \text{(A6.1.18h)}$$

As expected, the results (A6.1.18g, h) coincide with the previously derived (A6.1.17c).

■

Example A6.1.3 *Another spectral factorization for a state space model*
An alternative way of representing the system (A6.1.17a) in state space form is

$$\begin{aligned} x(t+1) &= -ax(t) + (c-a)e(t) \\ y(t) &= x(t) + e(t) \end{aligned} \quad \text{(A6.1.19a)}$$

In this case

$$R_1 = (c-a)^2 \quad R_{12} = (c-a) \quad R_2 = 1$$

The Riccati equation (A6.1.16) reduces to

180   *Model parametrizations*   Chapter 6

$$p = a^2 p + (c-a)^2 - \frac{(-ap+c-a)^2}{p+1} \tag{A6.1.19b}$$

Note that this equation is equivalent to (A6.1.18c). The solutions are hence $p_1 = 0$ (applicable for $|c| \le 1$) and $p_2 = c^2 - 1$ (applicable for $|c| > 1$). The gain K becomes, according to (A6.1.16),

$$K = \frac{-ap + c - a}{p+1}$$

Then

$$H(q^{-1}) = \frac{1 + (K+a)q^{-1}}{1 + aq^{-1}} = \frac{1 + \dfrac{c}{p+1} q^{-1}}{1 + aq^{-1}} \qquad \Lambda = p + 1 \tag{A6.1.19c}$$

which leads to the same solutions (A6.1.17c) as before.   ∎

Example A6.1.4 *Spectral factorization for a nonstationary process*
Let $x(t)$ be a drift, which can be described as

$$x(t+1) = x(t) + v(t) \qquad Ev(t)v(s) = \lambda_v^2 \delta_{t,s} \tag{A6.1.20a}$$

Assume that $x(t)$ is measured with some observation noise which is independent of $v(t)$

$$y(t) = x(t) + e(t) \qquad Ee(t)e(s) = \lambda_e^2 \delta_{t,s} \tag{A6.1.20b}$$

Since the system (A6.1.20a) has a pole *on* the unit circle, $y(t)$ is not a stationary process. However, after differentiating it becomes stationary. Introduce

$$z(t) = (1 - q^{-1})y(t) = v(t-1) + e(t) - e(t-1) \tag{A6.1.20c}$$

The right-hand side of (A6.1.20c) is certainly stationary and has a covariance function that vanishes for lags larger than 1. Hence $z(t)$ can be described by an equivalent MA(1) model:

$$z(t) = (1 + cq^{-1})\varepsilon(t) \qquad E\varepsilon(t)\varepsilon(s) = \lambda_\varepsilon^2 \delta_{t,s} \tag{A6.1.20d}$$

Comparison of the covariance functions of (A6.1.20c) and (A6.1.20d) gives

$$\begin{aligned}(1 + c^2)\lambda_\varepsilon^2 &= \lambda_v^2 + 2\lambda_e^2 \\ c\lambda_\varepsilon^2 &= -\lambda_e^2\end{aligned} \tag{A6.1.20e}$$

These equations have a unique solution satisfying $|c| < 1$. This solution is readily found to be

$$\begin{aligned}c &= -1 - \frac{\beta}{2} + \left(\beta + \frac{\beta^2}{4}\right)^{1/2} \\ \lambda_\varepsilon^2 &= \lambda_e^2 \left[1 + \frac{\beta}{2} + \left(\beta + \frac{\beta^2}{4}\right)^{1/2}\right]\end{aligned} \tag{A6.1.20f}$$

where $\beta = \lambda_v^2/\lambda_e^2$. In conclusion, the measurements can be described equivalently by

$$y(t) = H(q^{-1})\varepsilon(t)$$

where $H(q^{-1}) = (1 + cq^{-1})/(1 - q^{-1})$ has an asymptotically stable inverse. ∎

Wilson (1969) and Kučera (1979) have given efficient algorithms for performing spectral factorization. To describe such an algorithm it is sufficient to consider (A6.1.1a), (A6.1.2). Rescale the problem by setting $C = 1$ and let g_0 be free. Then the problem is to solve

$$f(z) = g(z)g(z^{-1}) \tag{A6.1.21a}$$

where

$$\begin{aligned} f(z) &= f_0 + f_1(z + z^{-1}) + \ldots + f_n(z^n + z^{-n}) \\ g(z) &= g_0 z^n + g_1 z^{n-1} + \ldots + g_n \end{aligned} \tag{A6.1.21b}$$

for the unknowns $\{g_i\}$. Note that to factorize an ARMA spectrum, two problems of the above type must be solved. The solution is required for which $g(z)$ has all zeros inside or on the unit circle. Identifying the coefficients in (A6.1.21a),

$$f_k = \sum_{i=0}^{n-k} g_i g_{i+k} \qquad k = 0, \ldots, n \tag{A6.1.21c}$$

This is a nonlinear system of equations with $\{g_i\}$ as unknowns, which can be rewritten in a more compact form as follows:

$$\begin{aligned} \bar{f} &\triangleq \begin{pmatrix} f_0 \\ \vdots \\ f_n \end{pmatrix} = \begin{pmatrix} g_0 & g_1 & \cdots & g_n \\ & \ddots & \ddots & \vdots \\ & & \ddots & g_1 \\ 0 & & & g_0 \end{pmatrix} \begin{pmatrix} g_0 \\ g_1 \\ \vdots \\ g_n \end{pmatrix} \triangleq T(g)g \\ &= \begin{pmatrix} g_0 & \cdots & g_{n-1} & g_n \\ \vdots & \ddots & \ddots & \\ g_{n-1} & \ddots & & \\ g_n & & 0 & \end{pmatrix} \begin{pmatrix} g_0 \\ g_1 \\ \vdots \\ g_n \end{pmatrix} \triangleq H(g)g \\ &= \frac{1}{2}[T(g) + H(g)]g \end{aligned} \tag{A6.1.22}$$

Observe that the matrices T and H introduced in (A6.1.22) are Toeplitz and Hankel, respectively. (T is Toeplitz because T_{ij} depends only on $i - j$. H is Hankel because H_{ij} depends only on $i + j$.) The nonlinear equation (A6.1.22) can be solved using the Newton–Raphson algorithm. The derivative of the right-hand side of (A6.1.22), with respect to g, can readily be seen to be $[T(g) + H(g)]$. Thus the basic iteration of the Newton–Raphson algorithm for solving (A6.1.22) is as follows:

$$g^{k+1} = g^k + [T(g^k) + H(g^k)]^{-1}\left\{\bar{f} - \frac{1}{2}[T(g^k) + H(g^k)]g^k\right\}$$
$$= \frac{1}{2}g^k + [T(g^k) + H(g^k)]^{-1}\bar{f}$$
(A6.1.23)

where the superscript k denotes the iteration number.

Merchant and Parks (1982) give some fast algorithms for the inversion of the Toeplitz-plus-Hankel matrices like that in (A6.1.23), whose use may be indicated whenever n is large. Kučera (1979) describes a version of the algorithm (A6.1.23) in which explicit matrix inversion is not needed. The Newton–Raphson recursion (A6.1.23) was first derived in Wilson (1969), where it was also shown that it has the following properties:

- If $g^k(z)$ has all zeros inside the unit circle, then so does $g^{k+1}(z)$.
- The recursion is globally convergent to the solution of (A6.1.21a) with $g(z)$ having all zeros inside the unit circle, provided $g^0(z)$ satisfies this stability condition (which can be easily realized). Furthermore, close to the solution, the rate of convergence is quadratic.

Complement C6.1
Uniqueness of the full polynomial form model

Let $G(z)$ denote a transfer function matrix. The full polynomial form model (6.22), (6.23) corresponds to the following parametrization of $G(z)$:

$$G(z) = A^{-1}(z)B(z) \qquad (ny|nu)$$
$$A(z) = I + A_1 z + \ldots + A_{na} z^{na} \qquad (ny|ny) \qquad \text{(C6.1.1)}$$
$$B(z) = B_1 z + \ldots + B_{nb} z^{nb} \qquad (ny|nu)$$

where all the elements of the matrix coefficients $\{A_i, B_j\}$ are assumed to be unknown. The parametrization (C6.1.1) is the so-called full matrix fraction description (FMFD), studied by Hannan (1969, 1970, 1976), Stoica (1983), Söderström and Stoica (1983). Here we analyze the uniqueness properties of the FMFD. That is to say, we delineate the transfer function matrices $G(z)$ which, for properly chosen (na, nb), can be uniquely represented by a FMFD.

Lemma C6.1.1
Let the strictly proper transfer function matrix $G(z)$ be represented in the form (C6.1.1) (for any such $G(z)$ there exists an infinity of representations of the form (C6.1.1) having various degrees (na, nb); see Kashyap and Rao (1976), Kailath (1980)). Then (C6.1.1) is unique in the class of matrix fraction descriptions (MFD) of degrees (na, nb) if and only if

$$A(z), B(z) \text{ are left coprime} \qquad \text{(C6.1.2)}$$
$$\text{rank }(A_{na} \quad B_{nb}) = ny \qquad \text{(C6.1.3)}$$

Proof. Assume that the conditions (C6.1.2), (C6.1.3) hold. Let $\tilde{A}^{-1}(z)\tilde{B}(z)$ be another MFD of degrees (na, nb) of $G(z)$. Then

$$\tilde{A}(z) = L(z)A(z)$$
$$\tilde{B}(z) = L(z)B(z) \tag{C6.1.4}$$

for some polynomial $L(z)$ (Kailath, 1980). Now $A(0) = \tilde{A}(0) = I$ which implies $L(0) = I$. Furthermore, (C6.1.3) implies $L(z) \equiv I$. For if $L(z) = I + L_1 z$, for example, then $\deg \tilde{A} = na$ and $\deg \tilde{B} = nb$ if and only if

$$L_1(A_{na} \quad B_{nb}) = 0 \tag{C6.1.5}$$

But this implies $L_1 = 0$ in view of (C6.1.3). Thus the sufficiency of (C6.1.2), (C6.1.3) is proved.

The necessity of (C6.1.2) is obvious. For if $A(z)$ and $B(z)$ are not left coprime then there exists a polynomial $K(z)$ of degree $nk \geq 1$ and polynomials $\bar{A}(z)$, $\bar{B}(z)$ such that

$$A(z) = K(z)\bar{A}(z)$$
$$B(z) = K(z)\bar{B}(z) \tag{C6.1.6}$$

Replacing $K(z)$ in (C6.1.6) by any other polynomial of degree not greater than nk leads to another MFD (na, nb) of $G(z)$. Concerning the necessity of (C6.1.3), set $L(z) = I + L_1 z$ in (C6.1.4), where $L_1 \neq 0$ satisfies (C6.1.5). Then (C6.1.4) is another MFD (na, nb) of $G(z)$, in addition to $\{A(z), B(z)\}$. ∎

There are $G(z)$ matrices which do not satisfy the conditions (C6.1.2), (C6.1.3) (see Example 6.8). However, these conditions are satisfied by almost all strictly proper $G(z)$ matrices. This is so since the condition rank $(A_{na} \quad B_{nb}) < ny$ imposes some nontrivial restrictions on the matrices A_{na}, B_{nb}, which may hold for some, but only for a 'few' systems. Note, however, that if the matrix $(A_{na} \quad B_{nb})$ is almost rank-deficient, then use of the FMFD for identification purposes may lead to ill-conditioned numerical problems.

Complement C6.2
Uniqueness of the parametrization and the positive definiteness of the input–output covariance matrix

This complement extends the result of Complement C5.1 concerning noise-free output, to the multivariable case. For the sake of clarity we study the full polynomial form model only. More general parametrizations of the coefficient matrices in (6.22), (6.23) can be analyzed in the same way (see Söderström and Stoica, 1983). Consider equation (6.22) in the noise-free case ($e(t) \equiv 0$). It can be rewritten as

$$y(t) = \phi^T(t)\theta \tag{C6.2.1}$$

where $\phi(t)$ and θ are defined in (6.25). As in the scalar case, the existence of the least squares estimate of θ in (C6.2.1) will be asymptotically equivalent to the positive definiteness of the covariance matrix

184 *Model parametrizations*   Chapter 6

$$R = E\phi(t)\phi^T(t)$$

It is shown below that the condition $R > 0$ is intimately related to the uniqueness of the full polynomial form.

Lemma C6.2.1
Assume that the input signal $u(t)$ is persistently exciting of a sufficiently high order. Then the matrix R is positive definite if and only if there is no other pair of polynomials, say $\bar{A}(z)$, $\bar{B}(z)$, having the same form as $A(z)$, $B(z)$, respectively, and such that $\bar{A}^{-1}(z)\bar{B}(z) = A^{-1}(z)B(z)$.

Proof. Let $\tilde{\theta}$ be a vector of the same dimension as θ, and consider the equation

$$\tilde{\theta}^T R \tilde{\theta} = 0 \tag{C6.2.2}$$

with $\tilde{\theta}$ as unknown. The following equivalences can be readily verified ($\tilde{A}(z)$ and $\tilde{B}(z)$ being the polynomials constructed from $\tilde{\theta}$ exactly in the way $A(z)$ and $B(z)$ are obtained from θ):

$$(C6.2.2) \Leftrightarrow \phi^T(t)\tilde{\theta} = 0 \text{ all } t \Leftrightarrow y(t) = \phi^T(t)(\theta - \tilde{\theta}) \text{ all } t$$

$$\Leftrightarrow y(t) = [A(q^{-1}) + I - \tilde{A}(q^{-1})]^{-1}[B(q^{-1}) - \tilde{B}(q^{-1})]u(t) \text{ all } t$$

$$\Leftrightarrow \{A^{-1}(q^{-1})B(q^{-1}) - [A(q^{-1}) + I - \tilde{A}(q^{-1})]^{-1}[B(q^{-1}) - \tilde{B}(q^{-1})]\}u(t)$$

$$\equiv 0 \text{ all } t \tag{C6.2.3}$$

Since $u(t)$ is assumed to be a persistently exciting signal, (C6.2.3) is equivalent to

$$A^{-1}(z)B(z) \equiv [A(z) + I - \tilde{A}(z)]^{-1}[B(z) - \tilde{B}(z)] \tag{C6.2.4}$$

Hence the matrix R is positive definite if and only if the only solution of (C6.2.4) is $\tilde{A}(z) - I \equiv 0$, $\tilde{B}(z) \equiv 0$ (i.e. $\tilde{\theta} = 0$). ■

Chapter 7
PREDICTION ERROR METHODS

7.1 The least squares method revisited

In Chapter 4 the least squares method was applied to static linear regression models. This section considers how linear regressions can be extended to cover dynamic models. The statistical analysis carried out in Chapter 4 will no longer be valid.

When using linear regressions in the dynamic case, models of the following form are considered (cf. (6.12), (6.13)):

$$A(q^{-1})y(t) = B(q^{-1})u(t) + \varepsilon(t) \tag{7.1a}$$

where

$$A(q^{-1}) = 1 + a_1 q^{-1} + \ldots + a_{na} q^{-na}$$
$$B(q^{-1}) = b_1 q^{-1} + \ldots + b_{nb} q^{-nb} \tag{7.1b}$$

In (7.1a) the term $\varepsilon(t)$ denotes the equation error. As noted in Chapter 6 the model (7.1) can be equivalently expressed as

$$y(t) = \varphi^T(t)\theta + \varepsilon(t) \tag{7.2a}$$

where

$$\varphi^T(t) = (-y(t-1) \ldots -y(t-na) \quad u(t-1) \ldots u(t-nb))$$
$$\theta = (a_1 \ldots a_{na} \quad b_1 \ldots b_{nb})^T \tag{7.2b}$$

The model (7.2) has exactly the same form as considered in Chapter 4. Hence, it is already known that the parameter vector which minimizes the sum of squared equation errors,

$$V_N(\theta) = \frac{1}{N} \sum_{t=1}^{N} \varepsilon^2(t) \tag{7.3}$$

is given by

$$\hat{\theta} = \left[\frac{1}{N} \sum_{t=1}^{N} \varphi(t)\varphi^T(t)\right]^{-1} \left[\frac{1}{N} \sum_{t=1}^{N} \varphi(t)y(t)\right] \tag{7.4}$$

186 Prediction error methods

This identification method is known as *the least squares (LS) method*. The name 'equation error method' also appears in the literature. The reason is, of course, that $\varepsilon(t)$, whose sample variance is minimized, appears as an equation error in the model (7.1).

Note that all the discussions about algorithms for computing $\hat{\theta}$ (see Section 4.5) will remain valid. The results derived there depend only on the 'algebraic structure' of the estimate (7.4). For the statistical properties, though, it is of crucial importance whether $\varphi(t)$ is an a priori given quantity (as for static models considered in Chapter 4), or whether it is a realization of a stochastic process (as for the model (7.1)). The reason why this difference is important is that for the dynamic models, when taking expectations of various quantities as in Section 4.2, it is no longer possible to treat Φ as a constant matrix.

Analysis

Consider the least squares estimate (7.4) applied to the model (7.1), (7.2). Assume that the data obey

$$A_0(q^{-1})y(t) = B_0(q^{-1})u(t) + v(t) \qquad (7.5a)$$

or equivalently

$$y(t) = \varphi^T(t)\theta_0 + v(t) \qquad (7.5b)$$

Here θ_0 is called the true parameter vector. Assume that $v(t)$ is a stationary stochastic process.

If the estimate $\hat{\theta}$ in (7.4) is 'good', it should be close to the true parameter vector θ_0. To examine if this is the case, an expression is derived for the estimation error

$$\begin{aligned}\hat{\theta} - \theta_0 &= \left[\frac{1}{N}\sum_{t=1}^{N}\varphi(t)\varphi^T(t)\right]^{-1}\left[\frac{1}{N}\sum_{t=1}^{N}\varphi(t)y(t)\right.\\ &\quad \left.- \left\{\frac{1}{N}\sum_{t=1}^{N}\varphi(t)\varphi^T(t)\right\}\theta_0\right]\\ &= \left[\frac{1}{N}\sum_{t=1}^{N}\varphi(t)\varphi^T(t)\right]^{-1}\left[\frac{1}{N}\sum_{t=1}^{N}\varphi(t)v(t)\right]\end{aligned} \qquad (7.6)$$

Under weak conditions (see Lemma B.2) the sums in (7.6) tend to the corresponding expected values as the number of data points, N, tends to infinity. Hence $\hat{\theta}$ is consistent (that is, $\hat{\theta}$ tends to θ_0 as N tends to infinity) if

$$E\varphi(t)\varphi^T(t) \text{ is nonsingular} \qquad (7.7a)$$

$$E\varphi(t)v(t) = 0 \qquad (7.7b)$$

Condition (7.7a) is satisfied in most cases. There are a few exceptions:

- The input is not persistently exciting of order *nb*.
- The data are completely noise-free ($v(t) \equiv 0$) and the model order is chosen too high (which implies that $A_0(q^{-1})$ and $B_0(q^{-1})$ have common factors).
- The input $u(t)$ is generated by a linear low-order feedback from the output.

Explanations as to why (7.7a) does not hold under these special circumstances are given in Complements C6.1, C6.2 and in Chapter 10.

Unlike (7.7a), condition (7.7b) is in most cases *not* satisfied. An important exception is when $v(t)$ is white noise (a sequence of independent random variables). In such a case $v(t)$ will be uncorrelated with all past data and in particular with $\varphi(t)$. However, when $v(t)$ is not white, it will normally be correlated with past outputs, since $y(t)$ depends (through (7.5a)) on $v(s)$ for $s \leq t$. Hence (7.7b) will not hold.

Recall that in Chapter 2 it was examplified in several cases that the consistent estimation of θ requires $v(t)$ to be white noise. Examples were also given of some of the exceptions for which condition (7.7a) does not hold.

Modifications

The least squares method is certainly simple to use. As shown above, it gives consistent parameter estimates only under rather restrictive conditions. In some cases the lack of consistency may be tolerable. If the signal-to-noise ratio is large, the bias will be small. If a regulator design is to be based on the identified model, some bias can in general be acceptable. This is because a reasonable regulator should make the closed loop system insensitive to parameter variations in the open loop part.

In other situations, however, it can be of considerable importance to have consistent parameter estimates. In this and the following chapter, two different ways are given of modifying the LS method so that consistent estimates can be obtained under less restrictive conditions. The modifications are:

- Minimization of the prediction error for other 'more detailed' model structures. This idea leads to the class of prediction error methods to be dealt with in this chapter.
- Modification of the normal equations associated with the least squares estimate. This idea leads to the class of instrumental variable methods, which are described in Chapter 8.

It is appropriate here to comment on the prediction error approach and why the LS method is a special case of this approach. Neglecting the equation error $\varepsilon(t)$ in the model (7.1a), one can predict the output at time t as

$$\hat{y}(t) = -a_1 y(t-1) - \ldots - a_{na} y(t-na) + b_1 u(t-1)$$
$$+ \ldots + b_{nb} u(t-nb) \quad (7.8a)$$
$$= \varphi^T(t)\theta$$

Hence

$$\varepsilon(t) = y(t) - \hat{y}(t) \quad (7.8b)$$

can be interpreted as a prediction error. Therefore, the LS method determines the parameter vector which makes the sum of squared prediction errors as small as possible. Note that the predictor (7.8a) is constructed in a rather *ad hoc* manner. It is not claimed to have any generally valid statistical properties, such as mean square optimality.

7.2 Description of prediction error methods

A model obtained by identification can be used in many ways, depending on the purpose of modeling. In many applications the model is used for prediction. Note that this is often inherent when the model is to be used as a basis for control system synthesis. Most systems are stochastic, which means that the output at time t cannot be determined exactly from data available at time $t - 1$. It is thus valuable to know at time $t - 1$ what the output of the process is likely to be at time t in order to take an appropriate control action, i.e. to determine the input $u(t - 1)$.

It therefore makes sense to determine the model parameter vector θ so that the prediction error

$$\varepsilon(t, \theta) = y(t) - \hat{y}(t|t - 1; \theta) \tag{7.9}$$

is small. In (7.9), $\hat{y}(t|t - 1; \theta)$ denotes a prediction of $y(t)$ given the data up to and including time $t - 1$ (i.e. $y(t - 1)$, $u(t - 1)$, $y(t - 2)$, $u(t - 2)$, ...) and based on the model parameter vector θ.

To formalize this idea, consider the general model structure introduced in (6.1):

$$\begin{gathered} y(t) = G(q^{-1}; \theta)u(t) + H(q^{-1}; \theta)e(t) \\ Ee(t)e^T(s) = \Lambda(\theta)\delta_{t,s} \end{gathered} \tag{7.10}$$

Assume that $G(0; \theta) = 0$, i.e. that the model has at least one pure delay from input to output. As a general linear predictor, consider

$$\hat{y}(t|t - 1; \theta) = L_1(q^{-1}; \theta)y(t) + L_2(q^{-1}; \theta)u(t) \tag{7.11a}$$

which is a function of past data only if the predictor filters $L_1(q^{-1}; \theta)$ and $L_2(q^{-1}; \theta)$ are constrained by

$$L_1(0; \theta) = 0 \quad L_2(0; \theta) = 0 \tag{7.11b}$$

The predictor (7.11) can be constructed in various ways for any given model (7.10). Once the model and the predictor are given, the prediction errors are computed as in (7.9). The parameter estimate $\hat{\theta}$ is then chosen to make the prediction errors $\varepsilon(1, \theta)$, ..., $\varepsilon(N, \theta)$ small.

To define a prediction error method the user has to make the following choices:

- *Choice of model structure.* This concerns the parametrization of $G(q^{-1}; \theta)$, $H(q^{-1}; \theta)$ and $\Lambda(\theta)$ in (7.10) as functions of θ.
- *Choice of predictor.* This concerns the filters $L_1(q^{-1}; \theta)$ and $L_2(q^{-1}; \theta)$ in (7.11), once the model is specified.
- *Choice of criterion.* This concerns a scalar-valued function of all the prediction errors $\varepsilon(1, \theta)$, ..., $\varepsilon(N, \theta)$, which will assess the performance of the predictor used; this criterion is to be minimized with respect to θ to choose the 'best' predictor in the class considered.

The choice of model structure was discussed in Chapter 6. In Chapter 11, which deals with model validation, it is shown how an appropriate model parametrization can be determined once a parameter estimation has been performed.

The predictor filters $L_1(q^{-1}; \theta)$ and $L_2(q^{-1}; \theta)$ can in principle be chosen in many ways. The most common way is to let (7.11a) be the *optimal mean square predictor*. This means that the filters are chosen so that under the given model assumptions the prediction errors have as small a variance as possible. In Section 7.3 the optimal predictors are derived for some general model structures. The use of optimal predictors is often assumed without being explicitly stated in the prediction error method. Under certain regularity conditions such optimal predictors will give optimal properties of the parameter estimates so obtained, as will be shown in this chapter. Note, though, that the predictor can also be defined in an *ad hoc* nonprobabilistic sense. Problem 7.3 gives an illustration. When the predictor is defined from deterministic considerations, it is reasonable to let the weighting sequences associated to $L_1(q^{-1}; \theta)$ and $L_2(q^{-1}; \theta)$ have fast decaying coefficients to make the influence of erroneous initial conditions insignificant. The filters should also be chosen so that imperfections in the measured data are well damped.

The criterion which maps the sequence of prediction errors into a scalar can be chosen in many ways. Here the following class of criteria is adopted. Define the sample covariance matrix

$$R_N(\theta) = \frac{1}{N} \sum_{t=1}^{N} \varepsilon(t, \theta)\varepsilon^T(t, \theta) \qquad (7.12)$$

where N denotes the number of data points. If the system has one output only ($ny = 1$) then $\varepsilon(t, \theta)$ is a scalar and so is $R_N(\theta)$. In such a case $R_N(\theta)$ can be taken as a criterion to be minimized. In the multivariable case, $R_N(\theta)$ is a positive definite matrix. Then the criterion

$$V_N(\theta) = h(R_N(\theta)) \qquad (7.13)$$

is chosen, where $h(Q)$ is a scalar-valued function defined on the set of positive definite matrices Q, which must satisfy certain conditions. $V_N(\theta)$ is frequently called a *loss function*. Note that the number of data points, N, is used as a subscript for convenience only. The requirement on the function $h(Q)$ is that it must be monotonically increasing. More specifically, let Q be positive definite and ΔQ nonnegative definite. Then it is required that

$$h(Q + \Delta Q) \geq h(Q) \qquad (7.14)$$

and that equality holds only for $\Delta Q = 0$.

The following example illustrates some possibilities for the choice of $h(Q)$.

Example 7.1 *Criterion functions*
One possible choice of $h(Q)$ is

$$h_1(Q) = \operatorname{tr} SQ \tag{7.15a}$$

where S is a (symmetric) positive definite weighting matrix. Then $S = GG^T$ for some nonsingular matrix G, which implies that

$$h_1(Q + \Delta Q) - h_1(Q) = \operatorname{tr} S\Delta Q = \operatorname{tr} GG^T \Delta Q = \operatorname{tr} G^T \Delta Q G \geq 0$$

Thus the condition (7.14) is satisfied. Note also that if ΔQ is nonzero then the inequality will be strict.

Another possibility for $h(Q)$ is

$$h_2(Q) = \det Q \tag{7.15b}$$

Let $Q = GG^T$, where G is nonsingular since Q is positive definite. Then

$$h_2(Q + \Delta Q) - h_2(Q) = \det[Q(I + Q^{-1}\Delta Q)] - \det Q$$
$$= \det Q[\det(I + G^{-T}G^{-1}\Delta Q) - 1]$$
$$= \det Q[\det(I + G^{-1}\Delta Q G^{-T}) - 1]$$

The above calculations have made use of the relation

$$\det(I + AB) = \det(I + BA)$$

(see Corollary 1 to Lemma A.5). Let $G^{-1} \Delta Q G^{-T}$ have eigenvalues $\lambda_1, \ldots, \lambda_{ny}$. Since this matrix is symmetric and nonnegative definite, $\lambda_i \geq 0$. As the determinant of a matrix is equal to the product of its eigenvalues, it follows that

$$h_2(Q + \Delta Q) - h_2(Q) = \det Q \left[\prod_{i=1}^{ny} (1 + \lambda_i) - 1 \right] \geq 0$$

Equality will hold if and only if $\lambda_i = 0$ for all i. This is precisely the case when $\Delta Q = 0$.

∎

Remark. It should be noted that the criterion can be chosen in other ways. A more general form of the loss function is, for example,

$$V_N(\theta) = \frac{1}{N} \sum_{t=1}^{N} l(t, \theta, \varepsilon(t, \theta)) \tag{7.16}$$

where the scalar-valued function $l(t, \theta, \varepsilon)$ must satisfy some regularity conditions. It is also possible to apply the prediction error approach to nonlinear models. The only requirement is, naturally enough, that the models provide a way of computing the prediction errors $\varepsilon(t, \theta)$ from the data. In this chapter the treatment is limited to linear models and the criterion (7.13). ∎

Some further comments may be made on the criteria of Example 7.1:

- The choices (7.15a) and (7.15b) require practically the same amount of computation when applied off-line, since the main computational burden is to find $R_N(\theta)$. However, the choice (7.15a) is more convenient when deriving recursive (on-line) algorithms as will be shown in Chapter 9.
- The choice (7.15b) gives optimal accuracy of the parameter estimates under weak conditions. The criterion (7.15a) will do so only if $S = \Lambda^{-1}$. (See Section 7.5 for proofs of these assertions.) Since Λ is very seldom known in practice, this choice is not useful in practical situations.
- The choice (7.15b) will be shown later to be optimal for Gaussian distributed disturbances. The criterion (7.16) can be made robust to outliers (abnormal data) by appropriate choice of the function $l(\cdot, \cdot, \cdot)$. (To give a robust parameter estimate, $l(\cdot, \cdot, \cdot)$ should increase with ε at a rate that is less than quadratic.)

The following example illustrates the choices of model structure, predictor and criterion in a simple case.

Example 7.2 *The least squares method as a prediction error method*
Consider the least squares method described in Section 7.1. The model structure is then given by (7.1a). The predictor is given by (7.8a):

$$\hat{y}(t) = [1 - A(q^{-1})]y(t) + B(q^{-1})u(t)$$

Hence, in this case

$$L_1(q^{-1}; \theta) = 1 - A(q^{-1})$$
$$L_2(q^{-1}; \theta) = B(q^{-1})$$

Note that the condition (7.11b) is satisfied. Finally, the criterion is given by (7.3). ∎

To summarize, the prediction error method (PEM) can be described as follows:
- Choose a model structure of the form (7.10) and a predictor of the form (7.11).
- Select a criterion function $h(Q)$; see (7.13).
- Determine the parameter estimate $\hat{\theta}$ as the (global) minimum point of the loss function $h(R_N(\theta))$:

$$\hat{\theta} = \arg\min_{\theta} h(R_N(\theta)) \qquad (7.17)$$

To evaluate the loss function at any value of θ, the prediction errors $\{\varepsilon(t, \theta)\}_{t=1}^{N}$ are determined from (7.9), (7.11). Then the sample covariance matrix $R_N(\theta)$ is evaluated according to (7.12).

Figure 7.1 provides an illustration of the basic principle of the prediction error method.

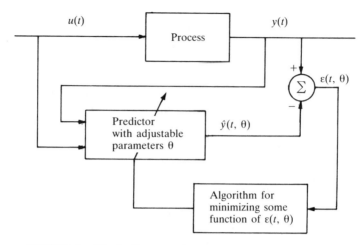

FIGURE 7.1 Block diagram of the prediction error method.

7.3 Optimal prediction

An integral part of a prediction error method is the calculation of the predictors for a given model. In most cases the optimal predictor, which minimizes the variance of the prediction error, is used. In this section the optimal predictors are derived for some general model structures.

To illustrate the main ideas in some detail, consider first a simple example.

Example 7.3 *Prediction for a first-order ARMAX model*
Consider the model

$$y(t) + ay(t-1) = bu(t-1) + e(t) + ce(t-1) \qquad (7.18)$$

where $e(t)$ is zero mean white noise of variance λ^2. The parameter vector is

$$\theta = (a \quad b \quad c)^T$$

Assume that $u(t)$ and $e(s)$ are independent for $t < s$. Hence the model allows feedback from $y(\cdot)$ to $u(\cdot)$. The output at time t satisfies

$$y(t) = [-ay(t-1) + bu(t-1) + ce(t-1)] + [e(t)] \qquad (7.19)$$

The two terms on the right-hand side of (7.19) are independent, since $e(t)$ is white noise. If $y^*(t)$ is an arbitrary prediction of $y(t)$ (based on data up to time $t-1$) it therefore follows that

$$E[y(t) - y^*(t)]^2 = E[-ay(t-1) + bu(t-1) \\ + ce(t-1) - y^*(t)]^2 + \lambda^2 \geq \lambda^2 \qquad (7.20)$$

This gives a lower bound, namely λ^2, on the prediction error variance. An *optimal*

Section 7.3 *Optimal prediction* 193

predictor $\hat{y}(t|t-1;\theta)$ is one which minimizes this variance. Equality in (7.20) is achieved for

$$\hat{y}(t|t-1;\theta) = -ay(t-1) + bu(t-1) + ce(t-1) \tag{7.21}$$

The problem with (7.21) is, of course, that it cannot be used as it stands since the term $e(t-1)$ is not measurable. However, $e(t-1)$ can be reconstructed from the measurable data $y(t-1)$, $u(t-1)$, $y(t-2)$, $u(t-2)$, ..., as shown below. Substituting (7.18) in (7.21),

$$\begin{aligned}\hat{y}(t|t-1;\theta) &= -ay(t-1) + bu(t-1) + c[y(t-1) + ay(t-2) - bu(t-2) \\ &\quad - ce(t-2)] \\ &= (c-a)y(t-1) + acy(t-2) + bu(t-1) - bcu(t-2) \\ &\quad - c^2[y(t-2) + ay(t-3) - bu(t-3) - ce(t-3)] \\ &= (c-a)y(t-1) + (ac-c^2)y(t-2) - ac^2 y(t-3) \\ &\quad + bu(t-1) - bcu(t-2) + bc^2 u(t-3) + c^3 e(t-3) \end{aligned} \tag{7.22}$$

$$= \ldots$$

$$= \sum_{i=1}^{t-1}(c-a)(-c)^{i-1}y(t-i) - a(-c)^{t-1}y(0)$$

$$+ b\sum_{i=1}^{t}(-c)^{i-1}u(t-i) - (-c)^{t}e(0)$$

Under the assumption that $|c| < 1$, which by definition is true for $\theta \in \mathcal{D}$ (see Example 6.1), the last term in (7.22) can be neglected for large t. It will only have a decaying transient effect. In order to get a realizable predictor this term will be neglected. Then

$$\hat{y}(t|t-1;\theta) = \sum_{i=1}^{t-1}(c-a)(-c)^{i-1}y(t-i) - a(-c)^{t-1}y(0)$$

$$+ b\sum_{i=1}^{t}(-c)^{i-1}u(t-i) \tag{7.23}$$

The expression (7.23) is, however, not well suited to practical implementation. To derive a more convenient expression, note that (7.23) implies

$$\hat{y}(t|t-1;\theta) + c\hat{y}(t-1|t-2;\theta) = (c-a)y(t-1) + bu(t-1) \tag{7.24}$$

which gives a simple recursion for computation of the optimal prediction. The prediction error $\varepsilon(t,\theta)$, (7.9), will obey a similar recursion. From (7.9) and (7.24) it follows that

$$\begin{aligned}\varepsilon(t,\theta) + c\varepsilon(t-1,\theta) &= y(t) + cy(t-1) - [(c-a)y(t-1) + bu(t-1)] \\ &= y(t) + ay(t-1) - bu(t-1) \end{aligned} \tag{7.25}$$

The recursion (7.25) needs an initial value $\varepsilon(0,\theta)$. This value is, however, unknown. To overcome this difficulty, $\varepsilon(0,\theta)$ is in most cases taken as zero. Since $|c| < 1$ by assumption, the effect of $\varepsilon(0, \theta)$ will only be a decaying transient: it will not give any significant contribution to $\varepsilon(t, \theta)$ when t is sufficiently large. Note that setting $\varepsilon(0, \theta) = 0$ in (7.25) is equivalent to using $\hat{y}(0|-1; \theta) = y(0)$ to initialize the predictor recursion (7.24). Further discussions on the choice of the initial value $\varepsilon(0, \theta)$ are given in Complement C7.7 and Section 12.6.

Note the similarity between the model (7.18) and the equation (7.25) for the prediction error $\varepsilon(t, \theta)$: $e(t)$ is simply replaced by $\varepsilon(t, \theta)$. This is how to proceed in practice when *calculating* the prediction error. The foregoing analysis has *derived* the expression for the prediction error. The calculations above can be performed in a more compact way using the polynomial formulation. The output of the model (7.18) can then be written as

$$y(t) = \frac{bq^{-1}}{1 + aq^{-1}} u(t) + \frac{1 + cq^{-1}}{1 + aq^{-1}} e(t)$$

$$= \left[\frac{bq^{-1}}{1 + aq^{-1}} u(t) + \frac{(c-a)q^{-1}}{1 + aq^{-1}} e(t) \right] + [e(t)]$$

$$= \left[\frac{bq^{-1}}{1 + aq^{-1}} u(t) + \frac{(c-a)q^{-1}}{1 + aq^{-1}} \frac{1}{1 + cq^{-1}} \{(1 + aq^{-1})y(t) - bq^{-1}u(t)\} \right] + [e(t)]$$

$$= \left[\frac{bq^{-1}}{(1 + aq^{-1})(1 + cq^{-1})} \{(1 + cq^{-1}) - (c-a)q^{-1}\} u(t) + \frac{(c-a)q^{-1}}{1 + cq^{-1}} y(t) \right] + [e(t)]$$

$$= \left[\frac{bq^{-1}}{1 + cq^{-1}} u(t) + \frac{(c-a)q^{-1}}{1 + cq^{-1}} y(t) \right] + [e(t)]$$

Thus

$$\hat{y}(t|t-1; \theta) = \frac{bq^{-1}}{1 + cq^{-1}} u(t) + \frac{(c-a)q^{-1}}{1 + cq^{-1}} y(t) \tag{7.26}$$

which is just another form of (7.24). When working with filters in this way it is assumed that data are available from the infinite past. This means that there is no transient effect due to an initial value. Since data in practice are available only from time $t = 1$ onwards, the form (7.26) of the predictor implicitly introduces a transient effect. In (7.22) this transient effect appeared as the last term. Note that even if the polynomial formalism gives a quick and elegant derivation of the optimal predictor, for practical implementations a difference equation form like (7.24) must be used. Of course, the result (7.26) can easily be reformulated as a difference equation, thus leading to a convenient form for implementation. ∎

Next consider the general linear model

Section 7.3
Optimal prediction 195

$$y(t) = G(q^{-1}; \theta)u(t) + H(q^{-1}; \theta)e(t)$$
$$Ee(t)e^T(s) = \Lambda(\theta)\delta_{t,s} \quad (7.27)$$

which was introduced in (6.1); see also (7.10). Assume that $G(0; \theta) = 0$, $H(0; \theta) = I$ and that $H^{-1}(q^{-1}; \theta)$ and $H^{-1}(q^{-1}; \theta)G(q^{-1}; \theta)$ are asymptotically stable (i.e. $\theta \in \mathcal{D}$, (6.2)). Assume also that $u(t)$ and $e(s)$ are uncorrelated for $t < s$. This condition holds if either the system operates in open loop with disturbances uncorrelated with the input, or the input is determined by causal feedback.

The optimal predictor is easily found from the following calculations:

$$y(t) = G(q^{-1}; \theta)u(t) + \{H(q^{-1}; \theta) - I\}e(t) + e(t)$$
$$= [G(q^{-1}; \theta)u(t) + \{H(q^{-1}; \theta) - I\}H^{-1}(q^{-1}; \theta)\{y(t)$$
$$- G(q^{-1}; \theta)u(t)\}] + e(t) \quad (7.28a)$$
$$= [H^{-1}(q^{-1}; \theta)G(q^{-1}; \theta)u(t) + \{I - H^{-1}(q^{-1}; \theta)\}y(t)] + e(t)$$
$$\triangleq z(t) + e(t)$$

Note that $z(t)$ and $e(t)$ are uncorrelated. Let $y^*(t)$ be an arbitrary predictor of $y(t)$ based on data up to time $t - 1$. Then the following inequality holds for the prediction error covariance matrix:

$$E[y(t) - y^*(t)][y(t) - y^*(t)]^T$$
$$= E[z(t) + e(t) - y^*(t)][z(t) + e(t) - y^*(t)]^T \quad (7.28b)$$
$$= E[z(t) - y^*(t)][z(t) - y^*(t)]^T + \Lambda(\theta) \geq \Lambda(\theta)$$

Hence $z(t)$ is the optimal mean square predictor, and $e(t)$ the prediction error. This can be written as

$$\hat{y}(t|t - 1; \theta) = H^{-1}(q^{-1}; \theta)G(q^{-1}; \theta)u(t) + \{I - H^{-1}(q^{-1}; \theta)\}y(t)$$
$$\varepsilon(t, \theta) = e(t) = H^{-1}(q^{-1}; \theta)\{y(t) - G(q^{-1}; \theta)u(t)\} \quad (7.29)$$

Note that the assumption $G(0; \theta) = 0$ means that the predictor $\hat{y}(t|t - 1; \theta)$ depends only on previous inputs (i.e. $u(t - 1)$, $u(t - 2)$, ...) and not on $u(t)$. Similarly, since $H(0; \theta) = I$ and hence $H^{-1}(0, \theta) = I$, the predictor does not depend on $y(t)$ but only on former output values $y(t - 1)$, $y(t - 2)$ ...

Further, note that use of the set $\mathcal{D}$, introduced in (6.2), means that θ is restricted to those values for which the predictor (7.29) is asymptotically stable. This was in fact the reason for introducing the set $\mathcal{D}$.

For the particular case treated in Example 7.3,

$$G(q^{-1}; \theta) = \frac{bq^{-1}}{1 + aq^{-1}} \quad H(q^{-1}; \theta) = \frac{1 + cq^{-1}}{1 + aq^{-1}}$$

Then (7.29) gives

$$\hat{y}(t|t-1;\theta) = \frac{1+aq^{-1}}{1+cq^{-1}} \frac{bq^{-1}}{1+aq^{-1}} u(t) + \left[1 - \frac{1+aq^{-1}}{1+cq^{-1}}\right] y(t)$$

$$= \frac{bq^{-1}}{1+cq^{-1}} u(t) + \frac{(c-a)q^{-1}}{1+cq^{-1}} y(t)$$

which is identical to (7.26).

Next consider optimal prediction for systems given in the state space form (6.28):

$$\begin{aligned} x(t+1) &= A(\theta)x(t) + B(\theta)u(t) + v(t) \\ y(t) &= C(\theta)x(t) + e(t) \end{aligned} \quad (7.30)$$

where $v(t)$ and $e(t)$ are mutually uncorrelated white noise sequences with zero means and covariance matrices $R_1(\theta)$ and $R_2(\theta)$, respectively. The optimal one-step predictor of $y(t)$ is given by the Kalman filter (see Section B.7 in Appendix B; cf. also (6.33)),

$$\begin{aligned} \hat{x}(t+1|t) &= A(\theta)\hat{x}(t|t-1) + B(\theta)u(t) + K(\theta)[y(t) - C(\theta)\hat{x}(t|t-1)] \\ \hat{y}(t|t-1) &= C(\theta)\hat{x}(t|t-1) \end{aligned} \quad (7.31)$$

where the gain $K(\theta)$ is given by

$$K(\theta) = A(\theta)P(\theta)C^T(\theta)[C(\theta)P(\theta)C^T(\theta) + R_2(\theta)]^{-1} \quad (7.32a)$$

and where $P(\theta)$ is the solution of the following algebraic Riccati equation:

$$P(\theta) = A(\theta)P(\theta)A^T(\theta) + R_1(\theta) - K(\theta)C(\theta)P(\theta)A^T(\theta) \quad (7.32b)$$

This predictor is mean square optimal if the disturbances are Gaussian distributed. For other distributions it is the *optimal linear predictor*. (This is also true for (7.29).)

Remark. As noted in Appendix A6.1 for the state space model (7.30), there are strong connections between spectral factorization and optimal prediction. In particular, the factorization of the spectral density matrix of the disturbance term of $y(t)$, makes it possible to write the model (7.30) (which has two noise sources) in the form of (7.27), for which the optimal predictor is easily derived. ∎

As an illustration of the equivalence between the above two methods for finding the optimal predictor, the next example reconsiders the ARMAX model of Example 7.3 but uses the results for the state space model (7.30) to derive the predictor.

Example 7.4 *Prediction for a first-order ARMAX model, continued*
Consider the model (7.18)

$$\begin{aligned} y(t) + ay(t-1) &= bu(t-1) + e(t) + ce(t-1) \\ |c| < 1 \quad Ee(t)e(s) &= \lambda^2 \delta_{t,s} \end{aligned} \quad (7.33)$$

which is represented in state space form as

Section 7.3 — Optimal prediction

$$x(t+1) = \begin{pmatrix} -a & 1 \\ 0 & 0 \end{pmatrix} x(t) + \begin{pmatrix} b \\ 0 \end{pmatrix} u(t) + \begin{pmatrix} 1 \\ c \end{pmatrix} e(t+1)$$

$$y(t) = (1\ \ 0)x(t)$$

(7.34)

(cf. (A6.1.18a)). As in Example A6.1.2 the solution of the Riccati equation has the form

$$P = \lambda^2 \begin{pmatrix} 1+\alpha & c \\ c & c^2 \end{pmatrix}$$

where the scalar α satisfies

$$\alpha = (c-a)^2 + a^2\alpha - \frac{(c-a-a\alpha)^2}{1+\alpha}$$

This gives $\alpha = 0$ (since $|c| < 1$). The gain vector K is found to be

$$K = \begin{pmatrix} 1 \\ 0 \end{pmatrix} \frac{c-a-a\alpha}{1+\alpha} = \begin{pmatrix} c-a \\ 0 \end{pmatrix}$$

(7.35)

According to (7.31) the one-step optimal predictor of the output will be

$$\hat{x}(t+1|t) = \begin{pmatrix} -a & 1 \\ 0 & 0 \end{pmatrix} \hat{x}(t|t-1) + \begin{pmatrix} b \\ 0 \end{pmatrix} u(t)$$

$$+ \begin{pmatrix} c-a \\ 0 \end{pmatrix} \{y(t) - (1\ \ 0)\hat{x}(t|t-1)\}$$

(7.36)

$$\hat{y}(t|t-1) = (1\ \ 0)\hat{x}(t|t-1)$$

This can be written in standard state space form as

$$\hat{x}(t+1|t) = \begin{pmatrix} -c & 1 \\ 0 & 0 \end{pmatrix} \hat{x}(t|t-1) + \begin{pmatrix} b \\ 0 \end{pmatrix} u(t) + \begin{pmatrix} c-a \\ 0 \end{pmatrix} y(t)$$

$$\hat{y}(t|t-1) = (1\ \ 0)\hat{x}(t|t-1)$$

and it follows that

$$\hat{y}(t|t-1) = (1\ \ 0) \begin{pmatrix} q+c & -1 \\ 0 & q \end{pmatrix}^{-1} \begin{pmatrix} bu(t) + (c-a)y(t) \\ 0 \end{pmatrix}$$

$$= \frac{1}{q+c}[bu(t) + (c-a)y(t)]$$

$$= \frac{bq^{-1}}{1+cq^{-1}} u(t) + \frac{(c-a)q^{-1}}{1+cq^{-1}} y(t)$$

This is precisely the result (7.26) obtained previously. ∎

We have now seen how to compute the optimal predictors and the prediction errors for general linear models. It should be stressed that it is a model assumption only that $e(t)$ in (7.27) is white noise. This is used to derive the form of the predictor. We can compute and apply the predictor (7.29) even if this model assumption is not satisfied by the data.

Thus the model assumption should be regarded as a tool to construct the predictor (7.29). Note, however, that if $e(t)$ is not white but correlated, the predictor (7.29) will no longer be optimal. The same discussion applies to the disturbances in the state space model (7.30). Complement C7.2 describes optimal *multistep* prediction of ARMA processes.

7.4 Relationships between prediction error methods and other identification methods

It was shown in Section 7.1 that the *least squares method* is a special case of the prediction error method (PEM). There are other cases for which the prediction error method is known under other names:

- For the model structure

$$A(q^{-1})y(t) = B(q^{-1})u(t) + D^{-1}(q^{-1})e(t) \tag{7.37}$$

the PEM is sometimes called the *generalized least squares* (GLS) *method*, although GLS originally was associated with a certain numerical minimization procedure. Various specific results for the GLS method are given in Complement C7.4.

- Consider the model structure

$$y(t) = G(q^{-1}; \theta)u(t) + e(t) \tag{7.38}$$

Then the prediction error calculated according to (7.29) becomes

$$\varepsilon(t, \theta) = y(t) - G(q^{-1}; \theta)u(t)$$

which is the difference between the measured output $y(t)$ and the 'noise-free model output' $G(q^{-1}; \theta)u(t)$. In such a case the PEM is often called an *output error method* (OEM). This method is analyzed in Complement C7.5.

The maximum likelihood method

There is a further important issue to mention in connection with PEMs, namely the relation to the maximum likelihood (ML) method.

For this purpose introduce the further assumption that the noise in the model (7.10) is Gaussian distributed. The maximum likelihood estimate of θ is obtained by maximizing the likelihood function, i.e. the probability density function (pdf) of the observations conditioned on the parameter vector θ (see Section B.3 of Appendix B). Now there is a 1–1 transformation between $\{y(t)\}$ and $\{e(t)\}$ as given by (7.10) if the effect of initial conditions is neglected; see below for details. Therefore it is equally valid to use the pdf of the disturbances. Using the expression for the multivariable Gaussian distribution function (see (B.12)), it is found that the likelihood function is given by

$$L(\theta) = \frac{1}{(2\pi)^{Nn_y/2}[\det \Lambda(\theta)]^{N/2}} \exp\left[-\frac{1}{2} \sum_{t=1}^{N} \varepsilon^T(t, \theta)\Lambda^{-1}(\theta)\varepsilon(t, \theta)\right] \tag{7.39}$$

Taking natural logarithms of both sides of (7.39) the following expression for the log-likelihood is obtained:

$$\log L(\theta) = -\frac{1}{2}\sum_{t=1}^{N} \varepsilon^T(t,\theta)\Lambda^{-1}(\theta)\varepsilon(t,\theta) - \frac{N}{2}\log \det \Lambda(\theta) + \text{constant} \quad (7.40)$$

Strictly speaking, $L(\theta)$ in (7.39) is not the *exact* likelihood function. The reason is that the transformation of $\{y(t)\}$ to $\{\varepsilon(t,\theta)\}$ using (7.29) has neglected the transient effect due to the initial values. $L(\theta)$ in (7.39) should be called the *L*-function conditioned on the initial values, or the conditional *L*-function. Complement C7.7 presents a derivation of the exact *L*-function for ARMA models, in which case the initial values are treated in an appropriate way. The exact likelihood is more complicated to evaluate than the conditional version introduced here. However, when the number of data points is large the difference between the conditional and the exact *L*-functions is small and hence the corresponding estimates are very similar. This is illustrated in the following example.

Example 7.5 *The exact likelihood function for a first-order autoregressive process*
Consider the first-order autoregressive model

$$y(t) + ay(t-1) = e(t) \quad |a| < 1 \quad (7.41)$$

where $e(t)$ is Gaussian white noise of zero mean and variance λ^2. Set $\theta = (a \ \lambda)^T$ and evaluate the likelihood function in the following way. Using Bayes' rule,

$$p(y(1), \ldots, y(N)) = p(y(N), \ldots, y(2)|y(1))p(y(1))$$
$$= p(y(N), \ldots, y(3)|y(2), y(1))p(y(2)|y(1))p(y(1))$$
$$= \left[\prod_{k=2}^{N} p(y(k)|y(k-1), \ldots, y(1))\right]p(y(1))$$

For $k \geq 2$,

$$p(y(k)|y(k-1), \ldots, y(1)) = p(e(k))$$
$$= \frac{1}{\sqrt{(2\pi)}\lambda}\exp\left[-\frac{1}{2\lambda^2}(y(k) + ay(k-1))^2\right]$$

and

$$p(y(1)) = \frac{1}{\sqrt{(2\pi)}\sigma}\exp\left[-\frac{1}{2\sigma^2}y^2(1)\right]$$

where $\sigma^2 = Ey^2(t) = \lambda^2/(1-a^2)$.

Hence the exact likelihood function is given by

$$L_e(\theta) = p(y(1) \ldots y(N)|\theta)$$
$$= \left\{\prod_{k=2}^{N} \frac{1}{\sqrt{(2\pi)}\lambda}\exp\left[-\frac{1}{2\lambda^2}(y(k)\right.\right.$$

$$+\ ay(k-1))^2 \Big] \Big\} \frac{1}{\sqrt{(2\pi)}\sigma} \exp\Big[-\frac{1}{2\sigma^2}y^2(1)\Big]$$

The log-likelihood function becomes

$$\log L_e(\theta) = \sum_{k=2}^{N} \log\Big\{\frac{1}{\sqrt{(2\pi)}\lambda}\exp\Big[-\frac{1}{2\lambda^2}(y(k)+ay(k-1))^2\Big]\Big\}$$

$$+ \log\Big\{\frac{1}{\sqrt{(2\pi)}\sigma}\exp\Big[-\frac{1}{2\sigma^2}y^2(1)\Big]\Big\}$$

$$= -\frac{N}{2}\log 2\pi - (N-1)\log\lambda - \frac{1}{2\lambda^2}\sum_{k=2}^{N}\{y(k)+ay(k-1)\}^2$$

$$-\frac{1}{2}\log\frac{\lambda^2}{1-a^2} - \frac{1-a^2}{2\lambda^2}y^2(1)$$

For comparison, consider the conditional likelihood function. For the model (7.41),

$$\varepsilon(t) = y(t) + ay(t-1)$$

The first prediction error, $\varepsilon(1)$, is not defined unless $y(0)$ is specified: so set it here to $\varepsilon(1) = y(1)$, which corresponds to the assumption $y(0) = 0$. The conditional log-likelihood function, denoted by $\log L(\theta)$, is given by

$$\log L(\theta) = -\frac{N}{2}\log 2\pi - N\log\lambda - \frac{1}{2\lambda^2}\sum_{k=2}^{N}\{y(k)+ay(k-1)\}^2 - \frac{1}{2\lambda^2}y^2(1)$$

(cf. (7.40)). Note that $\log L_e(\theta)$ and $\log L(\theta)$ both are of order N for large N. However, the difference $\log L_e(\theta) - \log L(\theta)$ is only of order $O(1)$ as N tends to infinity. Hence the exact and the conditional estimates are very close for a large amount of data. With some further calculations it can in fact be shown that

$$\hat{\theta}_{\text{exact}} = \hat{\theta}_{\text{cond}} + O\Big(\frac{1}{N}\Big)$$

In this case the conditional estimate $\hat{\theta}_{\text{cond}}$ will be the LS estimate (7.4) and is hence easy to compute. The estimate $\hat{\theta}_{\text{exact}}$ can be found as the solution to a third-order equation, see, e.g., Anderson (1971). ∎

Returning to the general case, assume that all the elements of Λ are to be estimated and that $\varepsilon(t, \theta)$ and $\Lambda(\theta)$ have no common parameters. For notational convenience we will assume that the unknown parameters in Λ are in addition to those in θ and will therefore drop the argument θ in $\Lambda(\theta)$.

For simplicity consider first the *scalar* case ($ny = 1$, which implies that $\varepsilon(t, \theta)$ and Λ are scalars). The ML estimates $\hat{\theta}$, $\hat{\Lambda}$ are obtained as the maximizing elements of $L(\theta, \Lambda)$. First maximize with respect to Λ the expression

$$\log L(\theta, \Lambda) = -\frac{1}{\Lambda}\frac{N}{2}\frac{1}{N}\sum_{t=1}^{N}\varepsilon^2(t, \theta) - \frac{N}{2}\log\Lambda + \text{constant}$$

Section 7.4 Relationships with other methods

$$= -\frac{N}{2}[R_N(\theta)/\Lambda + \log \Lambda] + \text{constant} \quad (7.42)$$

Straightforward differentiation gives

$$\frac{\partial \log L(\theta, \Lambda)}{\partial \Lambda} = -\frac{N}{2}\left(-\frac{R_N(\theta)}{\Lambda^2} + \frac{1}{\Lambda}\right)$$

$$\frac{\partial^2 \log L(\theta, \Lambda)}{\partial \Lambda^2} = \frac{N}{2}\left(-\frac{2R_N(\theta)}{\Lambda^3} + \frac{1}{\Lambda^2}\right)$$

There is only one stationary point, namely

$$\Lambda = R_N(\theta)$$

Furthermore, the second-order derivative is negative at this point. Hence it is a maximum. The estimate of Λ is thus found to be

$$\hat{\Lambda} = R_N(\theta) \quad (7.43)$$

where θ is to be replaced by its optimal value, which is yet to be determined. Inserting (7.43) into (7.42) gives

$$\log L(\theta, \hat{\Lambda}) = -\frac{N}{2}\log R_N(\theta) + \text{constant}$$

so $\hat{\theta}$ is obtained by minimizing $R_N(\theta)$. The minimum point will be the estimate $\hat{\theta}$ and the minimal value $R_N(\hat{\theta})$ will become the estimate $\hat{\Lambda}$. So the prediction error estimate can be interpreted as the maximum likelihood estimate provided the disturbances are Gaussian distributed.

Next consider the *multivariable* case (dim $y(t) = ny \geq 1$). A similar result as above for scalar systems holds, but the calculations are a little bit more complicated. In this case

$$\log L(\theta, \Lambda) = -\frac{N}{2}[\text{tr } R_N(\theta)\Lambda^{-1} + \log \det \Lambda] + \text{constant} \quad (7.44)$$

The statement that $L(\theta, \Lambda)$ is maximized with respect to Λ for $\Lambda = R_N(\theta)$ is equivalent to

$$\text{tr } R\Lambda^{-1} + \log \det \Lambda \geq \text{tr } I + \log \det R \Leftrightarrow$$

$$\text{tr } R\Lambda^{-1} + \log[\det \Lambda/\det R] \geq ny \Leftrightarrow \quad (7.45)$$

$$\text{tr } R\Lambda^{-1} - \log \det R\Lambda^{-1} \geq ny$$

Next write R as $R = GG^T$ and set $Y = G^T\Lambda^{-1}G$. The matrix Y is symmetric and positive definite, with eigenvalues $\lambda_1 \ldots \lambda_{ny}$ ($\lambda_i > 0$). Now (7.45) is equivalent to

$$\text{tr } GG^T\Lambda^{-1} - \log \det GG^T\Lambda^{-1} \geq ny \Leftrightarrow$$

$$\text{tr } G^T\Lambda^{-1}G - \log \det G^T\Lambda^{-1}G \geq ny \Leftrightarrow$$

$$\text{tr } Y - \log \det Y \geq ny \Leftrightarrow$$

$$\sum_{i=1}^{ny} \lambda_i - \log \prod_{i=1}^{ny} \lambda_i \geq ny \Leftrightarrow$$

$$\sum_{i=1}^{ny} [\lambda_i - \log \lambda_i - 1] \geq 0$$

Since for any positive λ, $\log \lambda \leq \lambda - 1$, the above inequality is obviously true.
Hence the likelihood function is maximized with respect to Λ for

$$\hat{\Lambda} = R_N(\theta) \tag{7.46}$$

where θ is to be replaced by its value $\hat{\theta}$ which maximizes (7.44). It can be seen from (7.44) that $\hat{\theta}$ is determined as the minimizing element of

$$V(\theta) = \det R_N(\theta) \tag{7.47}$$

This means that here one uses the function $h_2(Q)$ of Example 7.1.

The above analysis has demonstrated an interesting relation between the PEM and the ML method. If it is assumed that the disturbances in the model are Gaussian distributed, then the ML method becomes a prediction error method corresponding to the loss function (7.47). In fact, for this reason, prediction error methods have often been known as ML methods.

7.5 Theoretical analysis

An analysis is given here of the estimates described in the foregoing sections. In particular, the limiting properties of the estimated parameters as the number of data points tends to infinity will be examined. In the following, $\hat{\theta}_N$ denotes the parameter estimate based on N data points. Thus $\hat{\theta}_N$ is a minimum point of $V_N(\theta)$.

Basic assumptions

A1. The data $\{u(t), y(t)\}$ are stationary processes.
A2. The input is persistently exciting.
A3. The Hessian $V_N''(\theta)$ is nonsingular at least locally around the minimum points of $V_N(\theta)$.
A4. The filters $G(q^{-1}; \theta)$ and $H(q^{-1}; \theta)$ are smooth (differentiable) functions of the parameter vector θ.

Assumption A3 is weak. For models that are not overparametrized, it is a consequence of A1, A2, A4. Note that A3 is further examined in Example 11.6 for ARMAX models. The other assumptions A1, A2 and A4 are also fairly weak.

Section 7.5 *Theoretical analysis* 203

For part of the analysis the following additional assumption will be needed:

A5. The set $D_T(\mathcal{S}, \mathcal{M})$ introduced in (6.44) consists of precisely one point.

Note that we do not assume that the system operates in open loop. In fact, as will be seen in Chapter 10, one can often apply prediction error methods with success even if the data are collected from a closed loop experiment.

Asymptotic estimates

When N tends to infinity, the sample covariances converge to the corresponding expected values according to the ergodicity theory for stationary signals (see Lemma B.2; and Hannan, 1970). Then since the function $h(\cdot)$ is assumed to be continuous, it follows that

$$V_N(\theta) = h(R_N(\theta)) \to h(R_\infty(\theta)) \triangleq V_\infty(\theta), \text{ as } N \to \infty \qquad (7.48)$$

where

$$R_\infty = E\varepsilon(t, \theta)\varepsilon^T(t, \theta) \qquad (7.49)$$

It is in fact possible to show that the convergence in (7.48) is uniform on compact (i.e. closed and bounded) sets in the θ-space (see Ljung, 1978).

If the convergence in (7.48) is uniform, it follows that $\hat{\theta}_N$ *converges to a minimium point of* $V_\infty(\theta)$ (cf. Problem 7.15). Denote such a minimum by θ^*. This is an important result. Note, in particular, that Assumption A5 has not been used so far, so the stated result does not require the model structure to be large enough to cover the true system.

If the set $D_T(\mathcal{S}, \mathcal{M})$ is empty then an approximate prediction model is obtained. The approximation is in fact most reasonable. The parameter vector θ^* is by definition such that the prediction error $\varepsilon(t, \theta)$ has as small a variance as possible. Examples were given in Chapter 2 (see Examples 2.3, 2.4) of how such an approximation will depend on the experimental condition. It is shown in Complement C7.1 that in the multivariable case θ^* will also depend on the chosen criterion.

Consistency analysis

Next assume that the set $D_T(\mathcal{S}, \mathcal{M})$ is nonempty. Let θ_0 be an arbitrary element of $D_T(\mathcal{S}, \mathcal{M})$. This means that the true system satisfies

$$y(t) = G(q^{-1}; \theta_0)u(t) + H(q^{-1}; \theta_0)e(t) \qquad Ee(t)e^T(t) = \Lambda(\theta_0) \qquad (7.50)$$

where $e(t)$ is white noise. If the set $D_T(\mathcal{S}, \mathcal{M})$ has only one point (Assumption A5) then we can call θ_0 *the* true parameter vector. Next analyze the 'minimum' points of $R_\infty(\theta)$. It follows from (7.29), (7.50) that

$$\begin{aligned}\varepsilon(t, \theta) &= H^{-1}(q^{-1}; \theta)[G(q^{-1}; \theta_0)u(t) + H(q^{-1}; \theta_0)e(t) - G(q^{-1}; \theta)u(t)] \\ &= H^{-1}(q^{-1}; \theta)[G(q^{-1}; \theta_0) - G(q^{-1}; \theta)]u(t) \\ &\quad + H^{-1}(q^{-1}; \theta)H(q^{-1}; \theta_0)e(t)\end{aligned} \qquad (7.51)$$

Since $G(0, \theta) = 0$, $H(0, \theta) = H^{-1}(0, \theta) = I$ for all θ,

$\varepsilon(t, \theta) = e(t) + $ a term independent of $e(t)$

Thus

$$R_\infty(\theta) = E\varepsilon(t, \theta)\varepsilon^T(t, \theta) \geq Ee(t)e^T(t) = \Lambda(\theta_0) \tag{7.52}$$

This is independent of the way the input is generated. It is only necessary to assume that any possible feedback from $y(\cdot)$ to $u(\cdot)$ is causal. Since (7.52) gives a lower bound that is attained for $\theta = \theta_0$ it follows that $\theta^* = \theta_0$ is a possible limiting estimate. Note that the calculations above have shown that θ_0 'minimizes' $R_\infty(\theta)$, but they did not establish that no other minimizers of R_∞ exist. In the following it is proved that for systems operating in open loop, all the minimizers of R_∞ are the points θ_0 of D_T. The case of closed loop systems is more complicated (in such a case there may also be 'false' minimizers not belonging to D_T), and is studied in Chapter 10.

Assume that the system operates in open loop so that $u(t)$ and $e(s)$ are independent for all t and s. Then $R_\infty(\theta) = \Lambda(\theta_0)$ implies (cf. (7.51)):

$$H^{-1}(q^{-1}; \theta)[G(q^{-1}; \theta_0) - G(q^{-1}; \theta)]u(t) \equiv 0$$

$$H^{-1}(q^{-1}; \theta)H(q^{-1}; \theta_0) \equiv I$$

The second relation gives

$$H(q^{-1}; \theta) \equiv H(q^{-1}; \theta_0)$$

Using Assumption A2 and Property 5 of persistently exciting signals (see Section 5.4) one can conclude from the first identity that

$$G(q^{-1}; \theta) \equiv G(q^{-1}; \theta_0)$$

Thus we have $\theta \in D_T(\mathcal{S}, \mathcal{M})$.

The above result shows that under weak conditions *the PEM estimate $\hat{\theta}_N$ is consistent*. Note that under the general assumptions A1–A4 the system is system identifiable (see Section 6.4). If A5 is also satisfied, *the system is parameter identifiable*. This is essentially the same as saying that $\hat{\theta}_N$ is consistent.

Approximation

There is an important instance of approximation that should be pointed out. 'Approximation' here means that the model structure is not rich enough to include the true system. Assume that the system is given by

$$y(t) = G_s(q^{-1})u(t) + H_s(q^{-1})e(t) \tag{7.53}$$

and that the model structure is

$$y(t) = G(q^{-1}; \theta_1)u(t) + H(q^{-1}; \theta_2)e(t) \qquad Ee(t)e^T(t) = \Lambda(\theta_3) \tag{7.54}$$

Here the parameter vector has been split into three parts, θ_1, θ_2 and θ_3. It is crucial for what follows that G and H in (7.54) have different parameters. Further assume that there is a parameter vector θ_{10} such that

$$G_s(q^{-1}) \equiv G(q^{-1}; \theta_{10}) \tag{7.55}$$

Let the input $u(t)$ and the disturbance $e(s)$ in (7.53) be independent for all t and s. This is a reasonable assumption if the system operates in open loop. Then the two terms of (7.51) are uncorrelated. Hence

$$\begin{aligned} R_\infty(\theta) &= E\varepsilon(t,\theta)\varepsilon^T(t,\theta) \\ &\geq E[H^{-1}(q^{-1};\theta_2)H_s(q^{-1})e(t)][H^{-1}(q^{-1};\theta_2)H_s(q^{-1})e(t)]^T \end{aligned} \tag{7.56}$$

Equality holds in (7.56) if $\theta_1 = \theta_{10}$. This means that the limiting estimate θ^* is such that $\theta_1^* = \theta_{10}$. In other words, asymptotically a perfect description of the transfer function $G_s(q^{-1})$ is obtained even though the noise filter H may not be adequately parametrized. Recall that G and H have different parameters and that the system operates in open loop. Output error models are special cases where these assumptions are applicable, since $H(q^{-1};\theta_2) \equiv I$ for these models.

Asymptotic distribution of the parameter estimates

Following this discussion of the limit of $\hat{\theta}_N$, this subsection examines the limiting distribution. The estimate $\hat{\theta}_N$ will be shown to be asymptotically Gaussian distributed.

The estimate $\hat{\theta}_N$ is a minimum point of the loss function $V_N(\theta)$. In the following it will be assumed that the set $D_T(\mathcal{S}, \mathcal{M})$ has exactly one point, i.e. there exists a unique true vector θ_0 (Assumption A5). A Taylor series expansion of $V_N'(\hat{\theta}_N)^T$ around θ_0 retaining only the first two terms gives

$$\begin{aligned} 0 = V_N'(\hat{\theta}_N)^T &\approx V_N'(\theta_0)^T + V_N''(\theta_0)(\hat{\theta}_N - \theta_0) \\ &\approx V_N'(\theta_0)^T + V_\infty''(\theta_0)(\hat{\theta}_N - \theta_0) \end{aligned} \tag{7.57}$$

The second approximation follows since $V_N''(\theta_0) \to V_\infty''(\theta_0)$ with probability 1 as $N \to \infty$. Here V' denotes the gradient of V, and V'' the Hessian. The approximation in (7.57) has an error that tends to zero faster than $\|\hat{\theta}_N - \theta_0\|$. Since $\hat{\theta}_N$ converges to θ_0 as N tends to infinity, for large N the dominating term in the estimation error $\hat{\theta}_N - \theta_0$ can be written as

$$\sqrt{N}(\hat{\theta}_N - \theta_0) \approx -[V_\infty''(\theta_0)]^{-1}[\sqrt{N}V_N'(\theta_0)^T] \tag{7.58}$$

The matrix $V_\infty''(\theta_0)$ is deterministic. It is nonsingular under very general conditions when $D_T(\mathcal{S}, \mathcal{M})$ consists of one single point θ_0. However, the vector $\sqrt{N}V_N'(\theta_0)^T$ is a random variable. It will be shown in the following, using Lemma B.3, that it is asymptotically Gaussian distributed with zero mean and a covariance matrix denoted by P_0. Then Lemma B.4 and (7.58) give

$$\sqrt{N}(\hat{\theta}_N - \theta_0) \xrightarrow{dist} \mathcal{N}(0, P) \tag{7.59}$$

with

$$P = [V_\infty''(\theta_0)]^{-1} P_0 [V_\infty''(\theta_0)]^{-1} \tag{7.60}$$

The matrices in (7.60) must be evaluated. First consider *scalar* systems (dim $y(t) = 1$). For this case

$$V_N(0) = \frac{1}{N} \sum_{t=1}^{N} \varepsilon^2(t, \theta) \qquad V_\infty(\theta) = E\varepsilon^2(t, \theta) \tag{7.61}$$

Introduce the notation

$$\psi(t, \theta) = -\left(\frac{\partial \varepsilon(t, \theta)}{\partial \theta}\right)^T \tag{7.62}$$

which is an $(n\theta|1)$ vector. By straightforward differentiation

$$V'_N(\theta) = -\frac{2}{N} \sum_{t=1}^{N} \varepsilon(t, \theta)\psi^T(t, \theta)$$

$$V''_N(\theta) = \frac{2}{N} \sum_{t=1}^{N} \psi(t, \theta)\psi^T(t, \theta) + \frac{2}{N} \sum_{t=1}^{N} \varepsilon(t, \theta) \frac{\partial^2}{\partial \theta^2} \varepsilon(t, \theta) \tag{7.63}$$

For the model structures considered the first- and second-order derivatives of the prediction error, i.e. $\psi(t, \theta)$ and $\partial^2 \varepsilon(t, \theta)/\partial \theta^2$, will depend on the data up to time $t - 1$. They will therefore be independent of $e(t) \equiv \varepsilon(t, \theta_0)$. Thus

$$V'_N(\theta_0) = -\frac{2}{N} \sum_{t=1}^{N} e(t)\psi^T(t, \theta_0) \tag{7.64a}$$

$$V''_\infty(\theta_0) = 2E\psi(t, \theta_0)\psi^T(t, \theta_0) \tag{7.64b}$$

Since $e(t)$ and $\psi(t, \theta_0)$ are uncorrelated, the result (7.59) follows from Lemmas B.3, B.4. The matrix P_0 can be found from

$$P_0 = \lim_{N \to \infty} EN V'_N(\theta_0)^T V'_N(\theta_0)$$

$$= \lim_{N \to \infty} \frac{4}{N} \sum_{t=1}^{N} \sum_{s=1}^{N} Ee(t)\psi(t, \theta_0)e(s)\psi^T(s, \theta_0)$$

To evaluate this limit, note that $e(t)$ is white noise. Hence $e(t)$ is independent of $e(s)$ for all $s \neq t$ and also independent of $\psi(s, \theta_0)$ for $s \leq t$. Therefore

$$P_0 = \lim_{N \to \infty} \left\{ \frac{4}{N} \sum_{t=1}^{N} \sum_{s=1}^{t-1} Ee(t)E\psi(t, \theta_0)e(s)\psi^T(s, \theta_0) \right.$$

$$+ \frac{4}{N} \sum_{t=1}^{N} \sum_{s=t+1}^{N} Ee(t)\psi(t, \theta_0)\psi^T(s, \theta_0)Ee(s)$$

$$\left. + \frac{4}{N} \sum_{t=1}^{N} \sum_{s=t}^{t} Ee(t)\psi(t, \theta_0)e(s)\psi^T(s, \theta_0) \right\}$$

$$= \lim_{N \to \infty} \frac{4}{N} \sum_{t=1}^{N} Ee^2(t)E\psi(t, \theta_0)\psi^T(t, \theta_0)$$

$$= 4\Lambda E\psi(t, \theta_0)\psi^T(t, \theta_0) \tag{7.65}$$

Finally, from (7.60), (7.64b), (7.65) the following expression for the asymptotic normalized covariance matrix is obtained:

$$P = \Lambda[E\psi(t, \theta_0)\psi^T(t, \theta_0)]^{-1} \qquad (7.66)$$

Note that a reasonable estimate of P can be found as

$$\hat{P} = \hat{\Lambda}\left[\frac{1}{N}\sum_{t=1}^{N} \psi(t, \hat{\theta}_N)\psi^T(t, \hat{\theta}_N)\right]^{-1} \qquad (7.67)$$

This means that the accuracy of $\hat{\theta}_N$ can be estimated from the data.

The result (7.66) is now illustrated by means of two simple examples before discussing how it can be extended to multivariable systems.

Example 7.6 *Accuracy for a linear regression*
Assume that the model structure is

$$A(q^{-1})y(t) = B(q^{-1})u(t) + e(t)$$

which can be written in the form

$$y(t) = \varphi^T(t)\theta + e(t)$$

(cf. (6.12), (6.13), (7.1)). Since $\varepsilon(t, \theta) = y(t) - \varphi^T(t)\theta$, it follows that

$$\psi(t, \theta) = -\left(\frac{\partial \varepsilon(t, \theta)}{\partial \theta}\right)^T = \varphi(t)$$

Thus from (7.66)

$$P = \Lambda[E\varphi(t)\varphi^T(t)]^{-1} \qquad (7.68)$$

It is interesting to compare (7.68) with the corresponding result for the static case (see Lemma 4.2). For the static case, with the present notation, for a *finite* data length:
(a) $\hat{\theta}$ is unbiased
(b) $\sqrt{N}(\hat{\theta} - \theta_0)$ is Gaussian distributed $\mathcal{N}(0, P)$ where

$$P = \Lambda\left(\frac{1}{N}\sum_{t=1}^{N}\varphi(t)\varphi^T(t)\right)^{-1} \qquad (7.69)$$

In the dynamic case these results do not hold exactly for a finite N. Instead the following asymptotic results hold:

(a) $\hat{\theta}$ is consistent
(b) $\sqrt{N}(\hat{\theta} - \theta_0)$ is asymptotically Gaussian distributed $\mathcal{N}(0, P)$ where

$$P = \Lambda[E\varphi(t)\varphi^T(t)]^{-1}$$

■

Example 7.7 *Accuracy for a first-order ARMA process*
Let the model structure be

$$y(t) + ay(t-1) = e(t) + ce(t-1) \qquad Ee^2(t) = \Lambda \qquad \theta = (a \ \ c)^T \qquad (7.70)$$

Then

$$\varepsilon(t, \theta) = \frac{1 + aq^{-1}}{1 + cq^{-1}} y(t)$$

$$\frac{\partial \varepsilon}{\partial a}(t, \theta) = \frac{q^{-1}}{1 + cq^{-1}} y(t)$$

$$\frac{\partial \varepsilon}{\partial c}(t, \theta) = -\frac{1 + aq^{-1}}{(1 + cq^{-1})^2} q^{-1} y(t) = -\frac{q^{-1}}{1 + cq^{-1}} \varepsilon(t, \theta)$$

Thus

$$P = \Lambda \begin{pmatrix} Ey^F(t-1)^2 & -Ey^F(t-1)\varepsilon^F(t-1) \\ -Ey^F(t-1)\varepsilon^F(t-1) & E\varepsilon^F(t-1)^2 \end{pmatrix}^{-1}_{\theta=\theta_0}$$

where

$$y^F(t) = \frac{1}{1 + cq^{-1}} y(t) \qquad \varepsilon^F(t) = \frac{1}{1 + cq^{-1}} \varepsilon(t, \theta)$$

Since (7.70) is assumed to give a true description of the system,

$$y^F(t) = \frac{1}{1 + aq^{-1}} e(t) \qquad \varepsilon^F(t) = \frac{1}{1 + cq^{-1}} e(t)$$

It is therefore found after some straightforward calculations that

$$P = \Lambda \begin{pmatrix} \Lambda/(1-a^2) & -\Lambda/(1-ac) \\ -\Lambda/(1-ac) & \Lambda/(1-c^2) \end{pmatrix}^{-1}$$

$$= \frac{1}{(c-a)^2} \begin{pmatrix} (1-a^2)(1-ac)^2 & (1-a^2)(1-ac)(1-c^2) \\ (1-a^2)(1-ac)(1-c^2) & (1-ac)^2(1-c^2) \end{pmatrix} \quad (7.71)$$

The matrix P is independent of the noise variance Λ. Note from (7.71) that the covariance elements all increase without bound when c approaches a. Observe that when $c = a$, i.e. when the true system is a white noise process, the model (7.70) is overparametrized. In such a situation one cannot expect convergence of the estimate $\hat{\theta}_N$ to a certain point. For that case the asymptotic loss function $V_\infty(\theta)$ will not have a unique minimum since all θ satisfying $a = c$ will minimize $V_\infty(\theta)$. ∎

So far the scalar case ($ny = 1$) has been discussed in some detail. For the *multivariable* case ($ny \geq 1$) the calculations are a bit more involved. In Appendix A7.1 it is shown that

$$P = [E\psi(t)H\psi^T(t)]^{-1}[E\psi(t)H\Lambda H\psi^T(t)][E\psi(t)H\psi^T(t)]^{-1} \quad (7.72)$$

Here $\psi(t)$ denotes

$$\psi(t) = -\left(\frac{\partial \varepsilon(t, \theta)}{\partial \theta}\right)^T\bigg|_{\theta=\theta_0} \quad (7.73)$$

(cf. (7.62)). Now $\psi(t)$ is a matrix of dimension $(n\theta|ny)$. Further, in (7.72) H is a matrix of dimension $(ny|ny)$ defined by

$$H = \left.\frac{\partial h(Q)}{\partial Q}\right|_{Q=\Lambda} \tag{7.74}$$

Recall that $h(Q)$ defines the loss function (see (7.15)). The matrix H is constant. The derivative in (7.74) should be interpreted as

$$H_{ij} = \frac{\partial h(Q)}{\partial Q_{ij}}$$

Optimal accuracy

It is also shown in Appendix A7.1 that there is a lower bound on P:

$$P \geq [E\psi(t)\Lambda^{-1}\psi^T(t)]^{-1} \tag{7.75}$$

with equality for

$$H = \Lambda^{-1} \tag{7.76}$$

or a scalar multiple thereof.

This result can be used to choose the function $h(Q)$ so as to maximize the accuracy. If $h(Q) = h_1(Q) = \text{tr } SQ$, it follows from (7.74) that $H = S$ and hence the optimal accuracy for this choice of $h(Q)$ can be obtained by setting $S = \Lambda^{-1}$. However, this alternative is not realistic since it requires knowledge of the noise covariance matrix Λ.

Next consider the case $h(Q) = h_2(Q) = \det Q$. This choice gives $H = \text{adj}(Q) = Q^{-1} \det Q$. When evaluated at $Q = \Lambda$ it follows from (7.76) that equality is obtained in (7.75). Moreover, this choice of $h(Q)$ is perfectly realizable. The only drawback is that it requires somewhat more calculations for evaluation of the loss function. The advantage is that optimal accuracy is obtained. Recall that this choice of $h(Q)$ corresponds to the ML method under the Gaussian hypothesis.

Underparametrized models

Most of the previous results are stated for the case when the true system belongs to the considered model structure (i.e. $\mathscr{S} \in \mathscr{M}$, or equivalently $D_T(\mathscr{S}, \mathscr{M})$ is not empty). Assume for a moment that the system is more complex than the model structure. As stated after (7.48), the parameter estimates will converge to a minimum point of the asymptotic loss function

$$\hat{\theta}_N \to \theta^* \triangleq \arg\min_\theta V_\infty(\theta), \quad N \to \infty \tag{7.77}$$

The estimates will still be asymptotically Gaussian distributed

$$\sqrt{N}(\hat{\theta}_N - \theta^*) \xrightarrow{\text{dist}} \mathcal{N}(0, P) \tag{7.78a}$$

$$P = [V''_\infty(\theta^*)]^{-1} [\lim_{N \to \infty} NE\{V'_N(\theta^*)\}^T\{V'_N(\theta^*)\}][V''_\infty(\theta^*)]^{-1} \tag{7.78b}$$

(cf. (7.59), (7.60)). To evaluate the covariance matrix P given by (7.78b) it is necessary to use the properties of the data as well as the model structure.

Statistical efficiency

It will now be shown that for Gaussian distributed disturbances the optimal PEM (then identical to the ML method) is asymptotically statistically efficient. This means (cf. Section B.4 of Appendix B) that the covariance matrix of the optimal PEM is equal to the Cramér–Rao lower bound. Note that due to the general consistency properties of PEM the 'bias factor' $\partial \gamma(\theta)/\partial \theta$ in the Cramér–Rao lower bound (B.31) is asymptotically equal to the identity matrix. Thus the Cramér–Rao lower bound formula (B.24) for unbiased estimators asymptotically applies also to consistent (PEM or other) estimators.

That the ML estimate is statistically efficient for Gaussian distributed data is a classical result for independent observations (see, for example, Cramér, 1946). In the present case, $\{y(t)\}$ are dependent. However, the innovations are independent and there is a linear transformation between the output measurements and the innovations. Hence the statistical efficiency of the ML estimate in our case is rather natural.

The log-likelihood function (for Gaussian distributed disturbances) is given by (7.44)

$$\log L(\theta, \Lambda) = -\frac{1}{2} \sum_{t=1}^{N} \varepsilon^T(t, \theta) \Lambda^{-1} \varepsilon(t, \theta) - \frac{N}{2} \log \det \Lambda + \text{constant}$$

Assuming that $\varepsilon(t, \theta)$ and Λ are independently parametrized and that all the elements of Λ are unknown,

$$\frac{\partial \log L(\theta, \Lambda)}{\partial \theta} = \sum_{t=1}^{N} \varepsilon^T(t, \theta) \Lambda^{-1} \psi^T(t, \theta) \tag{7.79a}$$

$$\frac{\partial \log L(\theta, \Lambda)}{\partial \Lambda_{ij}} = \frac{1}{2} \sum_{t=1}^{N} \varepsilon^T(t, \theta) \Lambda^{-1} e_i e_j^T \Lambda^{-1} \varepsilon(t, \theta) - \frac{N}{2} \frac{1}{\det \Lambda} [\text{adj}(\Lambda)]_{ij}$$

$$= \frac{N}{2} [\Lambda^{-1} R_N(\theta) \Lambda^{-1}]_{ij} - \frac{N}{2} [\Lambda^{-1}]_{ij} \tag{7.79b}$$

Here e_i denotes the ith unit vector.

The Fisher information matrix J (see (B.29)), is given by

$$J = E \left(\begin{array}{c} \left(\frac{\partial \log L(\theta, \Lambda)}{\partial \theta}\right)^T \\ \left(\frac{\partial \log L(\theta, \Lambda)}{\partial \Lambda}\right)^T \end{array} \right) \left(\frac{\partial \log L(\theta, \Lambda)}{\partial \theta} \quad \frac{\partial \log L(\theta, \Lambda)}{\partial \Lambda} \right) \bigg|_{\theta=\theta_0, \Lambda=\Lambda_0} \tag{7.80}$$

where the derivative with respect to Λ is expressed in an informal way. It follows from (7.79) that the information matrix is block diagonal, since $\varepsilon(t, \theta_0)$ and $\psi(t, \theta_0)$ are mutually uncorrelated and $\{\varepsilon(t, \theta_0)\}$ is an uncorrelated sequence. This can be shown as follows:

$$2E\left(\frac{\partial \log L(\theta, \Lambda)}{\partial \theta}\right)^T \left(\frac{\partial \log L(\theta, \Lambda)}{\partial \Lambda_{ij}}\right)\bigg|_{\theta=\theta_0, \Lambda=\Lambda_0}$$

$$= \sum_{t=1}^{N} \sum_{s=1}^{N} E\psi(t, \theta_0)\Lambda_0^{-1} e(t) e^T(s) \Lambda_0^{-1} e_i e_j^T \Lambda_0^{-1} e(s)$$

$$= \sum_{t=1}^{N} \sum_{s=1}^{t-1} E\psi(t, \theta_0)\Lambda_0^{-1} \underbrace{[Ee(t)]}_{=0} e^T(s) \Lambda_0^{-1} e_i e_j^T \Lambda_0^{-1} e(s)$$

$$+ \sum_{t=1}^{N} \sum_{s=t}^{N} \underbrace{E[\psi(t, \theta_0)]}_{=0} \Lambda_0^{-1} Ee(t) e^T(s) \Lambda_0^{-1} e_i e_j^T \Lambda_0^{-1} e(s) = 0$$

We also have

$$E\left(\frac{\partial \log L(\theta, \Lambda)}{\partial \theta}\right)^T \left(\frac{\partial \log L(\theta, \Lambda)}{\partial \theta}\right)\bigg|_{\theta=\theta_0, \Lambda=\Lambda_0}$$

$$= E\sum_{t=1}^{N} \sum_{s=1}^{N} \psi(t, \theta_0)\Lambda^{-1} e(t) e^T(s) \Lambda^{-1} \psi^T(s, \theta_0)$$

$$= \sum_{t=1}^{N} \sum_{s=1}^{t-1} E\psi(t, \theta_0)\Lambda^{-1} \underbrace{[Ee(t)]}_{=0} e^T(s) \Lambda^{-1} \psi^T(s, \theta_0) \quad (7.81)$$

$$+ \sum_{t=1}^{N} \sum_{s=t+1}^{N} E\psi(t, \theta_0)\Lambda^{-1} e(t) \underbrace{[Ee^T(s)]}_{=0} \Lambda^{-1} \psi^T(s, \theta_0)$$

$$+ \sum_{t=1}^{N} E\psi(t, \theta_0)\Lambda^{-1}[Ee(t)e^T(t)]\Lambda^{-1} \psi^T(t, \theta_0)$$

$$= NE\psi(t, \theta_0)\Lambda^{-1} \psi^T(t, \theta_0)$$

By comparing with (7.59), (7.72) and (7.75) it can now be seen that under the Gaussian assumption *the PEM estimate $\hat{\theta}_N$ is asymptotically statistically efficient if the criterion $h(\cdot)$ is chosen so that $H = \Lambda^{-1}$* (or a positive scalar times Λ^{-1}). Note that this condition is always satisfied in the scalar case ($ny = 1$).

7.6 Computational aspects

In the special case where $\varepsilon(t, \theta)$ depends linearly on θ the minimization of $V_N(\theta)$ can be done analytically. This case corresponds to linear regression. Some implementation issues for linear regressions were discussed in Section 4.5. See also Lawson and Hanson (1974) and Complement C7.3 for further aspects on how to organize the computation of the least squares estimate of linear regression model parameters.

Optimization algorithms

In most cases the minimum of $V_N(\theta)$ cannot be found analytically. For such cases the minimization must be performed using a numerical search routine. A commonly used method is the *Newton–Raphson algorithm*:

$$\hat{\theta}^{(k+1)} = \hat{\theta}^{(k)} - \alpha_k [V_N''(\hat{\theta}^{(k)})]^{-1} V_N'(\hat{\theta}^{(k)})^T \qquad (7.82)$$

Here $\hat{\theta}^{(k)}$ denotes the kth iteration point in the search. The sequence of scalars α_k in (7.82) is used to control the step length. Note that in a strict theoretical sense, the Newton–Raphson algorithm corresponds to (7.82) with $\alpha_k = 1$ (see Problem 7.5). However, in practice a variable step length is often necessary. For instance, to improve the convergence of (7.82) one may choose α_k so that

$$\alpha_k = \arg\min_{\alpha} V_N(\hat{\theta}^{(k)} - \alpha [V_N''(\hat{\theta}^{(k)})]^{-1} V_N'(\hat{\theta}^{(k)})^T) \qquad (7.83)$$

Note that α_k can also be used to guarantee that $\hat{\theta}^{(k+1)} \in \mathscr{D}$ for all k, as required.

The derivatives of $V(\theta)$ can be found in (7.63), (A7.1.3), (A7.1.4). In general,

$$V_N'(\theta) = -\frac{2}{N} \sum_{t=1}^{N} \varepsilon^T(t, \theta) H \psi^T(t, \theta) \qquad (7.84)$$

and

$$V_N''(\theta) = \frac{2}{N} \sum_{t=1}^{N} \psi(t, \theta) H \psi^T(t, \theta) - \frac{2}{N} \sum_{t=1}^{N} \varepsilon^T(t, \theta) \left(\frac{\partial H}{\partial \theta}\right)^T \psi^T(t, \theta)$$
$$- \frac{2}{N} \sum_{t=1}^{N} \varepsilon^T(t, \theta) H \frac{\partial}{\partial \theta} \psi^T(t, \theta) \qquad (7.85)$$

Here $\partial H/\partial \theta$ and $\partial \psi/\partial \theta$ are written in an informal way. In the scalar case (i.e. $ny = 1$) explicit expressions of these factors are easily found (see (7.63)).

At the global minimum point $\varepsilon(t, \theta)$ becomes asymptotically (as N tends to infinity) white noise which is independent of $\psi(t, \theta)$. Then

$$V_N''(\theta_0) \approx \frac{2}{N} \sum_{t=1}^{N} \psi(t, \theta_0) H \psi^T(t, \theta_0) \qquad (7.86)$$

(cf. (A7.1.4)). It is appealing to neglect the second and third terms in (7.85) when using the algorithm (7.82). There are two reasons for this:

- The approximate $V_N''(\theta)$ in (7.86) is by construction guaranteed to be positive definite. Therefore the loss function will decrease in every iteration if α_k is appropriately chosen (cf. Problem 7.7).
- The computations are simpler since $\partial H/\partial \theta$ and $\partial \psi(t, \theta)/\partial \theta$ need not be evaluated.

The algorithm obtained in this way can be written as

$$\hat{\theta}^{(k+1)} = \hat{\theta}^{(k)} + \alpha_k \left[\sum_{t=1}^{N} \psi(t, \hat{\theta}^{(k)}) H \psi^T(t, \hat{\theta}^{(k)}) \right]^{-1}$$
$$\times \left[\sum_{t=1}^{N} \psi(t, \hat{\theta}^{(k)}) H \varepsilon(t, \hat{\theta}^{(k)}) \right] \quad (7.87)$$

This is called the *Gauss–Newton algorithm*. When N is large the two algorithms (7.82) and (7.87) will behave quite similarly if $\hat{\theta}^{(k)}$ is close to the minimum point. For any N the local convergence of the Newton–Raphson algorithm is quadratic, i.e. when $\hat{\theta}^{(k)}$ is close to $\hat{\theta}$ (the minimum point) then $\|\hat{\theta}^{(k+1)} - \hat{\theta}\|$ is of the same magnitude as $\|\hat{\theta}^{(k)} - \hat{\theta}\|^2$. This means roughly that the number of significant digits in the estimate $\hat{\theta}^{(k)}$ is doubled in every iteration. The Gauss–Newton algorithm will give linear convergence in general. It will be quadratically convergent when the last two terms in (7.85) are zero. In practice these terms are small but nonzero; then the convergence will be linear but fast (so-called superlinear convergence; cf. Problem 7.7).

It should be mentioned here that an interesting alternative to (7.82) is to apply a recursive algorithm (see Chapter 9) to the data a couple of times. The initial values for one run of the algorithm on the data are then formed from the final values of the previous run. This will often give considerably faster convergence and will thus lead to a saving of computer time.

Evaluation of gradients

For some model structures the gradient $\psi(t, \theta)$ can be computed efficiently. The following example shows how this can be done for an nth-order ARMAX model.

Example 7.8 *Gradient calculation for an ARMAX model*
Consider the model

$$A(q^{-1})y(t) = B(q^{-1})u(t) + C(q^{-1})e(t) \quad (7.88)$$

The prediction errors $\varepsilon(t, \theta)$ are then given by

$$C(q^{-1})\varepsilon(t, \theta) = A(q^{-1})y(t) - B(q^{-1})u(t) \quad (7.89)$$

First differentiate this equation with respect to a_i to get

$$C(q^{-1}) \frac{\partial \varepsilon(t, \theta)}{\partial a_i} = y(t - i)$$

Thus

$$\frac{\partial \varepsilon(t, \theta)}{\partial a_i} = \frac{1}{C(q^{-1})} y(t - i) \quad (7.90a)$$

Therefore to determine $\{\partial \varepsilon(t, \theta)/\partial a_i\}$, $i = 1, \ldots, n$, it is sufficient to compute

$[1/C(q^{-1})]y(t)$. This means that only *one* filtering procedure is necessary for all the elements $\partial\varepsilon(t, \theta)/\partial a_i$, $i = 1, \ldots, n$.

Differentiation of (7.89) with respect to b_i ($1 \le i \le n$) gives similarly

$$C(q^{-1}) \frac{\partial \varepsilon(t, \theta)}{\partial b_i} = -u(t - i) \tag{7.90b}$$

The derivatives with respect to c_i ($1 \le i \le n$) are found from

$$\varepsilon(t - i, \theta) + C(q^{-1}) \frac{\partial \varepsilon(t, \theta)}{\partial c_i} = 0$$

which gives

$$\frac{\partial \varepsilon(t, \theta)}{\partial c_i} = -\frac{1}{C(q^{-1})} \varepsilon(t - i, \theta) \tag{7.90c}$$

To summarize, for the model structure (7.88) the vector $\psi(t, \theta)$ can be efficiently computed as follows. First compute the filtered signals

$$y^F(t) = \frac{1}{C(q^{-1})} y(t) \qquad u^F(t) = \frac{1}{C(q^{-1})} u(t) \qquad \varepsilon^F(t) = \frac{1}{C(q^{-1})} \varepsilon(t)$$

Then $\psi(t, \theta)$ is given by

$$\psi(t, \theta) = -(y^F(t-1) \ldots y^F(t-n) \quad -u^F(t-1) \ldots -u^F(t-n) \quad -\varepsilon^F(t-1) \ldots -\varepsilon^F(t-n))^T$$

However, there is often no practical use for the numerical value of $\psi(t, \theta)$ as such. (Recursive algorithms are an important exception, see Chapter 9.) Instead it is important to compute quantities such as

$$V'_N(\theta) = -\frac{2}{N} \sum_{t=1}^{N} \varepsilon(t, \theta) \psi^T(t, \theta)$$

As an example, the first n elements of $V'_N(\theta)$ are easily computed from the signals $\varepsilon(t, \theta)$ and $y^F(t)$ to be

$$[V'_N(\theta)]_i = \frac{2}{N} \sum_{t=1}^{N} \varepsilon(t, \theta) y^F(t - i) \qquad i = 1, \ldots, n$$

The remaining elements can be computed using similar expressions. ∎

Initial estimate $\hat{\theta}^{(0)}$

To start the Newton–Raphson (7.82) or the Gauss-Newton (7.87) algorithm an initial estimate $\hat{\theta}^{(0)}$ is required. The choice of initial value will often significantly influence the number of iterations needed to find the minimum point, and is a highly problem-dependent question. In some cases there might be *a priori* information, which then of course should be taken into account. For some model structures there are special methods that can be applied for finding an appropriate initial estimate. Two examples follow.

Example 7.9 *Initial estimate for an ARMAX model*
Consider an ARMAX model

$$A(q^{-1})y(t) = B(q^{-1})u(t) + C(q^{-1})e(t) \tag{7.91a}$$

where for simplicity it is assumed that all the polynomials are of degree n. The model can also be written formally as a linear regression:

$$y(t) = \varphi^T(t)\theta + e(t) \tag{7.91b}$$

with

$$\varphi^T(t) = (-y(t-1) \ldots -y(t-n) \quad u(t-1) \ldots u(t-n) \quad e(t-1) \ldots e(t-n))$$
$$\theta = (a_1 \ldots a_n \quad b_1 \ldots b_n \quad c_1 \ldots c_n)^T \tag{7.91c}$$

Assume for a moment that $\varphi(t)$ is known. Then θ can be estimated by using a standard LS estimate

$$\hat{\theta}^{(0)} = \left[\sum_{t=1}^{N} \varphi(t)\varphi^T(t)\right]^{-1} \left[\sum_{t=1}^{N} \varphi(t)y(t)\right] \tag{7.92}$$

To make (7.92) realizable, $e(t-i)$ in (7.91c) must be replaced by some estimate $\hat{e}(t-i)$, $i = 1, \ldots, n$. Such estimates can be found using a linear regression model

$$\bar{A}(q^{-1})y(t) = \bar{B}(q^{-1})u(t) + e(t) \tag{7.93}$$

of *high* order, say $\bar{n}$. The prediction errors obtained from the estimated model (7.93) are then used as estimates of $e(t-i)$ in $\psi(t)$. For this approach to be successful one must have

$$\bar{A}(q^{-1}) \approx \frac{A(q^{-1})}{C(q^{-1})} \quad \bar{B}(q^{-1}) \approx \frac{B(q^{-1})}{C(q^{-1})} \tag{7.94}$$

These approximations will be valid if $\bar{n}$ is sufficiently large. The closer the zeros of $C(z)$ are located to the unit circle, the larger is the necessary value of $\bar{n}$. ∎

Example 7.10 *Initial estimates for increasing model structures*
When applying an identification method the user normally starts with a small model structure (a low-order model) and estimates its parameters. The model structure is successively increased until a reasonable model is obtained. In such situations the 'previous' models may be used to provide appropriate initial values for the parameters of higher-order models.

To make the discussion more concrete, consider a scalar ARMA model

$$A(q^{-1})y(t) = C(q^{-1})e(t) \tag{7.95a}$$

where

$$A(q^{-1}) = 1 + a_1 q^{-1} + \ldots + a_n q^{-n}$$
$$C(q^{-1}) = 1 + c_1 q^{-1} + \ldots + c_n q^{-n} \tag{7.95b}$$

Suppose the parameters of a model of order n have been estimated, but this model is found to be unsatisfactory when assessing its 'performance'. Then one would like to try an ARMA model of order $n + 1$. Let the parameter vectors be denoted by

$$\hat{\theta}^n = (\hat{a}_1 \ldots \hat{a}_n \ \hat{c}_1 \ldots \hat{c}_n)^T$$
$$\theta^{n+1} = (a_1 \ldots a_{n+1} \ c_1 \ldots c_{n+1})^T \qquad (7.96)$$

for the two model structures. A very rough initial value for θ^{n+1} would be $\theta^{n+1} = 0$. However, the vector

$$\bar{\theta}^{n+1} = (\hat{a}_1 \ldots \hat{a}_n \ 0 \ \hat{c}_1 \ldots \hat{c}_n \ 0)^T \qquad (7.97)$$

will give a better fit (a smaller value of the criterion). It is in fact the best value for the nth-order ARMA parameters. Note that the class of nth-order models can be viewed as a subset of the $(n + 1)$th-order model set. The initial value (7.97) for the optimization algorithm will be the best that can be selected using the information available from $\hat{\theta}^n$. In particular, observe that if the true ARMA order is n then $\bar{\theta}^{n+1}$ given by (7.97) will be very close to the optimal value for the parameters of the $(n + 1)$th-order model. ∎

Summary

In Section 7.1 it was shown how linear regression techniques can be used for identification of dynamic models. It was demonstrated that the parameter estimates so obtained are consistent only under restrictive conditions. Section 7.2 introduced prediction error methods as a generalization of the least squares method. The parameter estimate is determined as the minimizing vector of a suitable scalar-valued function of the sample covariance matrix of the prediction errors. Section 7.3 gave a description of optimal predictors. In particular, it was shown how to compute the optimal prediction errors for a general model structure. In Section 7.4 the relationship of the PEM to other identification methods was discussed. In particular, it was shown that the PEM can be interpreted as a maximum likelihood method for Gaussian distributed disturbances.

The PEM parameter estimates were analyzed in Section 7.5, where it was seen that, under weak assumptions, they are consistent and asymptotically Gaussian distributed. Explicit expressions were given for the asymptotic covariance matrix of the estimates and it was demonstrated that, under the Gaussian hypothesis, the PEM estimates are statistically efficient (i.e. have the minimum possible variance).

Finally, in Section 7.6 several computational aspects on implementing prediction error methods were presented. A numerical search method must, in general, be used. In particular various Newton methods were described, including efficient ways for evaluating the criterion gradient and choosing the initial value.

Problems

Problem 7.1 *Optimal prediction of a nonminimum phase first-order MA model*
Consider the process given by

$$y(t) = e(t) + 2e(t - 1)$$

where $e(t)$ is white noise of zero mean and unit variance. Derive the optimal mean square one-step predictor and find the variance of the prediction error.

Problem 7.2 *Kalman gain for a first-order ARMA model*
Verify (7.35).

Problem 7.3 *Prediction using exponential smoothing*
One way of predicting a signal $y(t)$ is to use a so-called exponential smoothing. This can be described as

$$\hat{y}(t + 1|t) = \frac{1 - \alpha}{1 - \alpha q^{-1}} y(t)$$

where α is a parameter, $0 < \alpha < 1$.

(a) Assume that $y(t) = m$ for all t. Show that in stationarity the predicted value will be equal to the exact value m.
(b) For which ARMA model is exponential smoothing an optimal predictor?

Problem 7.4 *Gradient calculation*
Consider the model structure

$$y(t) = \frac{B(q^{-1})}{F(q^{-1})} u(t) + \frac{C(q^{-1})}{D(q^{-1})} e(t)$$

with the parameter vector

$$\theta = (b_1 \ \ldots \ b_{nb} \ \ c_1 \ \ldots \ c_{nc} \ \ d_1 \ \ldots \ d_{nd} \ \ f_1 \ \ldots \ f_{nf})^{\mathrm{T}}$$

Derive the gradient $\partial \varepsilon(t, \theta)/\partial \theta$ of the prediction error.

Problem 7.5 *Newton–Raphson minimization algorithm*
Let $V(\theta)$ be an analytic function of θ. An iterative algorithm for minimizing $V(\theta)$ with respect to θ can be obtained in the following way. Let $\theta^{(k)}$ denote the vector θ at the kth iteration. Take $\theta^{(k+1)}$ as the minimum point of the quadratic approximation of $V(\theta)$ around $\theta^{(k)}$. Show that this principle leads to the algorithm (7.82) with $\alpha_k = 1$.

Problem 7.6 *Gauss–Newton minimization algorithm*
The Gauss–Newton algorithm for minimizing sum-of-squares functions of the form

$$V(\theta) = \sum_{t=1}^{N} \varepsilon^2(t, \theta)$$

where $\varepsilon(\cdot, \theta)$ is a differentiable function of the parameter vector θ, can be obtained from the Newton–Raphson algorithm by making a certain approximation on the Hessian matrix (see (7.82)–(7.86)). It can also be obtained by 'quasilinearization' and in fact it is sometimes called the quasilinearization minimization method. To be more exact, let $\theta^{(k)}$

denote the vector θ obtained at iteration k, and let $\tilde{\varepsilon}(t, \theta)$ denote the linear approximation of $\varepsilon(t, \theta)$ around $\theta^{(k)}$:

$$\tilde{\varepsilon}(t, \theta) = \varepsilon(t, \theta^{(k)}) - \psi^T(t, \theta^{(k)}) (\theta - \theta^{(k)})$$

where

$$\psi(t, \theta) = -\left(\frac{\partial \varepsilon(t, \theta)}{\partial \theta}\right)^T$$

Determine

$$\theta^{(k+1)} = \arg \min_\theta \sum_{t=1}^N \tilde{\varepsilon}^2(t, \theta) \qquad (i)$$

and show that the recursion so obtained is precisely the Gauss–Newton algorithm.

Problem 7.7 *Convergence rate for the Newton–Raphson and Gauss–Newton algorithms*
Consider the algorithms

$$A_1: x^{(k+1)} = x^{(k)} - [V''(x^{(k)})]^{-1}[V'(x^{(k)})]^T$$
$$A_2: x^{(k+1)} = x^{(k)} - S[V'(x^{(k)})]^T$$

for minimization of the function $V(x)$. The matrix S is positive definite.

(a) Introduce a positive scalar α in the algorithm A_2 for controlling the step length

$$A_2': x^{(k+1)} = x^{(k)} - \alpha S[V'(x^{(k)})]^T$$

Show that this algorithm gives a decreasing sequence of function values ($V(x^{(k+1)}) \leq V(x^{(k)})$) if α is chosen small enough.

(b) Apply the algorithms to the quadratic function $V(x) = \frac{1}{2}x^T A x + x^T b + c$, where A is a positive definite matrix. The minimum point of $V(x)$ is $x^* = -A^{-1}b$. Show that for A_1,

$$x^{(1)} = x^*$$

(Hence, A_1 converges in one step for quadratic functions.) Show that for A_2,

$$[x^{(k+1)} - x^*] = [I - SA][x^{(k)} - x^*]$$

(Assuming $I - SA$ has all eigenvalues strictly inside the unit circle $x^{(k)}$ will converge, and the convergence will be linear. If in particular $S = A^{-1} + Q$ and Q is small, then the convergence will be fast (superlinear).)

Problem 7.8 *Derivative of the determinant criterion*
Consider the function $h(Q) = \det Q$. Show that its derivative is

$$H = \frac{\partial h(Q)}{\partial Q} = (\det Q) Q^{-1}$$

Problem 7.9 *Optimal predictor for a state space model*
Consider a state space model of the form (7.30). Derive the mean square optimal predictor and present it in the form (7.11).

(a) Use the result (7.31).
(b) Use (6.33) to rewrite the state space form into the input–output form (7.27). Then use (7.29) to find the optimal filters $L_1(q^{-1}; \theta)$ and $L_2(q^{-1}; \theta)$.

Problem 7.10 *Hessian of the loss function for an ARMAX model*
Consider the loss function

$$V(\theta) = \sum_{t=1}^{N} \varepsilon^2(t, \theta)$$

for the model

$$A(q^{-1})y(t) = B(q^{-1})u(t) + C(q^{-1})\varepsilon(t, \theta)$$

Derive the Hessian (the matrix of second-order derivatives) of the loss function.

Problem 7.11 *Multistep prediction of an AR(1) process observed in noise*
Let the process $y(t)$ be defined as

$$y(t) = \frac{1}{1 + aq^{-1}} e(t) + \varepsilon(t)$$

where $|a| < 1$, $Ee(t)e(s) = \lambda_e^2 \delta_{t,s}$, $E\varepsilon(t)\varepsilon(s) = \lambda_\varepsilon^2 \delta_{t,s}$ and $Ee(t)\varepsilon(s) = 0$ for all t, s. Assume that $e(t)$ and $\varepsilon(t)$ are Gaussian distributed. Derive the mean-square optimal k-step predictor of $y(t + k)$. What happens if the Gaussian hypothesis is relaxed?

Problem 7.12 *An asymptotically efficient two-step PEM*
Let $\varepsilon(t, \theta)$ denote the prediction errors at time instant t of a general PE model with parameters θ. Assume that $\varepsilon(t, \theta)$ is an ergodic process for any admissible value of θ and that it is an analytic function of θ. Also, assume that there exists a parameter vector θ_0 (the 'true' parameter vector) such that $\{\varepsilon(t, \theta_0)\}$ is a white noise sequence. Consider the following two-step procedure:

S1: Determine a consistent estimate $\tilde{\theta}$ of θ_0

S2: $\hat{\theta} = \tilde{\theta} + \left[\sum_{t=1}^{N} \psi^T(t, \tilde{\theta})\psi(t, \tilde{\theta}) \right]^{-1} \left[\sum_{t=1}^{N} \psi^T(t, \tilde{\theta})\varepsilon(t, \tilde{\theta}) \right]$

where N denotes the number of data points, and

$$\psi(t, \theta) = -\frac{\partial \varepsilon(t, \theta)}{\partial \theta}$$

Assume that $\tilde{\theta}$ has a covariance matrix that tends to zero as $1/N$, as N tends to infinity. Show that the asymptotic covariance matrix of $\hat{\theta}$ is given by (7.66). (This means that the

Gauss–Newton algorithm S2 initialized in $\bar{\theta}$ converges in one iteration to the prediction error estimate.)

Problem 7.13 *Frequency domain interpretation of approximate models*
Consider the system

$$y(t) = G(q^{-1})u(t) + e(t)$$

where the noise $e(t)$ is not correlated with the input signal $u(t)$. Suppose the purpose of identification is the estimation of the transfer function $G(q^{-1})$, which may be of infinite order. For simplicity, a rational approximation of $G(q^{-1})$ of the form $B(q^{-1})/A(q^{-1})$, where $A(q^{-1})$ and $B(q^{-1})$ are polynomials, is sought. The coefficients of $B(q^{-1})$ and $A(q^{-1})$ are estimated by using the least squares method as well as the output error method. Interpret the two approximate models so obtained in the frequency domain, assuming that the number of data points N tends to infinity and that $u(t)$ and $e(t)$ are stationary processes.

Problem 7.14 *An indirect PEM*
Consider the system

$$y(t) = b_0 u(t-1) + \frac{1}{1 + a_0 q^{-1}} e(t)$$

where $u(t)$ and $e(t)$ are mutually independent white noise sequences with zero means and variances σ^2 and λ^2, respectively.

(a) Consider two ways of identifying the system.
   Case (i): The model structure is given by

   $$\mathcal{M}_1: \; y(t) = bu(t-1) + \frac{1}{1 + aq^{-1}} e(t)$$

   $$\theta_1 = (a \; b)^T$$

   and a prediction error method is used.
   Case (ii): The model structure is given by

   $$\mathcal{M}_2: \; y(t) + ay(t-1) = b_1 u(t-1) + b_2 u(t-2) + e(t)$$

   $$\theta_2 = (a \; b_1 \; b_2)^T$$

   and the least squares method is used.
   Determine the asymptotic variances of the estimates. Compare the variances of $\hat{a}$ and $\hat{b}$ obtained by using $\mathcal{M}_1$ with the variances of $\hat{a}$ and $\hat{b}_1$ obtained by using $\mathcal{M}_2$.

(b) Case (i) gives better accuracy but requires much more computation than case (ii). One can therefore think of the following approach. First compute $\hat{\theta}_2$ as in case (ii). As a second step the parameters of $\mathcal{M}_1$ are estimated from

   $$f(\hat{\theta}_1^*) = \hat{\theta}_2$$

where (compare $\mathcal{M}_1$ and $\mathcal{M}_2$)

$$f(\theta_1) = \begin{pmatrix} a \\ b \\ ab \end{pmatrix}$$

Since this is an overdetermined system (3 equations, 2 unknowns), an exact solution is in general not possible. To overcome this difficulty the estimate $\hat{\theta}_1^*$ can be defined as the minimum point of

$$V(\theta_1) = [\hat{\theta}_2 - f(\theta_1)]^T Q [\hat{\theta}_2 - f(\theta_1)]$$

where Q is a positive definite weighting matrix. (Note that $V(\theta_1)$ does not depend explicitly on the data, so the associated minimization problem should require much less computations than that of case (i).)

Show that the asymptotic covariance matrix of $\hat{\theta}_1^*$ is given by

$$P_{\hat{\theta}_1^*} = (F^T Q F)^{-1} F^T Q P_{\hat{\theta}_2} Q F (F^T Q F)^{-1}$$

where $P_{\hat{\theta}_2}$ is the covariance matrix of $\hat{\theta}_2$ and

$$F = \left. \frac{df(\theta_1)}{d\theta_1} \right|_{\theta_1 = \theta_{10}}$$

Hint. Let θ_{10}, θ_{20} denote the true parameter vectors. Then by a Taylor series expansion $\hat{\theta}_2 - f(\hat{\theta}_1^*) \approx (\hat{\theta}_2 - \theta_{20}) - F(\hat{\theta}_1^* - \theta_{10})$

$$\left. \frac{1}{2} \frac{dV(\theta_1)}{d\theta_1} \right|_{\theta_1 = \hat{\theta}_1^*} \approx -[\hat{\theta}_2 - f(\hat{\theta}_1^*)]^T Q F$$

From these equations an asymptotically valid expression for $\hat{\theta}_1^* - \theta_{10}$ can be obtained.

(c) Show that the covariance matrix $P_{\hat{\theta}_1^*}$ is minimized with respect to Q by the choice $Q = P_{\hat{\theta}_2}^{-1}$ in the sense that

$$P_{\hat{\theta}_1^*} \geq (F^T P_{\hat{\theta}_2}^{-1} F)^{-1} = P_{\hat{\theta}_1^*} | Q = P_{\hat{\theta}_2}^{-1}$$

Evaluate the right-hand side explicitly for the system above. Compare with the covariance matrix for $\hat{\theta}_1$.

Remark. The choice $Q = P_{\hat{\theta}_2}$ is not realistic in practice. Instead $Q = \hat{P}_{\hat{\theta}_2}$ can be used. ($\hat{P}_{\hat{\theta}_2}$ is an estimate of $P_{\hat{\theta}_2}$ obtained by replacing E by $1/N \sum_{t=1}^{N}$ in the expression for $P_{\hat{\theta}_2}$.) With this weighting matrix, $\hat{\theta}_1^* = \hat{\theta}_1$ as shown in Complement C7.4.

Problem 7.15 *Consistency and uniform convergence*

Assume that $V_N(\theta)$ converges *uniformly* to $V_\infty(\theta)$ (as N tends to infinity) in a compact set Ω, and that $V_\infty(\theta)$ is continuous and has a unique global minimum point θ^* in Ω. Prove that for any global minimum point $\hat{\theta}$ of $V_N(\theta)$ in Ω, it holds that $\hat{\theta} \to \theta^*$ as N tends to infinity.

Hint. Let $\varepsilon > 0$ be arbitrary and set $\Omega^* = \{\theta; \| \theta - \theta^* \| < \varepsilon\}$. Next choose a number

δ so that $\inf_{\Omega-\Omega^*} V_\infty(\theta) \geq V_\infty(\theta^*) + \delta$. Then choose N_0 such that $|V_N(\theta) - V_\infty(\theta)| < \delta/3$ for all $N \geq N_0$ and all $\theta \in \Omega$. Deduce that $\hat{\theta} \in \Omega^*$.

Problem 7.16 *Accuracy of PEM for a first-order ARMAX system*
Consider the system

$$y(t) + ay(t-1) = bu(t-1) + e(t) + ce(t-1)$$

where $e(t)$ is white noise with zero mean and variance λ^2. Assume that $u(t)$ is white noise of zero mean and variance σ^2, and independent of the $e(t)$, and that the system is identified using a PEM. Show that the normalized covariance matrix of the parameter estimates is given by

$$P = \lambda^2 \begin{pmatrix} \dfrac{b^2\sigma^2(1+ac)}{(1-a^2)(1-ac)(1-c^2)} + \dfrac{\lambda^2}{1-a^2} & \dfrac{bc\sigma^2}{(1-ac)(1-c^2)} & -\dfrac{\lambda^2}{1-ac} \\ \dfrac{bc\sigma^2}{(1-ac)(1-c^2)} & \dfrac{\sigma^2}{1-c^2} & 0 \\ -\dfrac{\lambda^2}{1-ac} & 0 & \dfrac{\lambda^2}{1-c^2} \end{pmatrix}^{-1}$$

$$= \dfrac{(1-a^2)(1-ac)^2(1-c^2)}{\sigma^2[b^2\sigma^2 + \lambda^2(c-a)^2]}$$

$$\times \begin{pmatrix} \dfrac{\sigma^2\lambda^2}{1-c^2} & -\dfrac{bc\sigma^2\lambda^2}{(1-ac)(1-c^2)} & \dfrac{\sigma^2\lambda^2}{(1-ac)} \\ -\dfrac{bc\sigma^2\lambda^2}{(1-ac)(1-c^2)} & \dfrac{b^2\sigma^2\lambda^2(1-a^2c^2) + \lambda^4(c-a)^2(1-c^2)}{(1-a^2)(1-ac)^2(1-c^2)} & -\dfrac{bc\sigma^2\lambda^2}{(1-ac)^2} \\ \dfrac{\sigma^2\lambda^2}{(1-ac)} & -\dfrac{bc\sigma^2\lambda^2}{(1-ac)^2} & \dfrac{\sigma^2\{b^2\sigma^2 + \lambda^2(1-ac)^2\}}{(1-a^2)(1-ac)^2} \end{pmatrix}$$

Problem 7.17 *Covariance matrix for the parameter estimates when the system does not belong to the model set*
Let $y(t)$ be a stationary Gaussian process with covariance function $r_k = Ey(t)y(t-k)$ and assume that r_k tends exponentially to zero as k tends to infinity (which means that there are constants $C > 0$, $0 < \alpha < 1$, such that $|r_k| < C\alpha^{|k|}$). Assume that a first-order AR model

$$y(t) + ay(t-1) = e(t)$$

is fitted to the measured data $\{y(1), \ldots, y(N)\}$.

(a) Show that

$$\sqrt{N}(\hat{a} - a^*) \xrightarrow{\text{dist}} \mathcal{N}(0, P) \quad \text{as} \quad N \to \infty$$

where $a^* = -r_1/r_0$ and

$$P = \sum_{\tau=-\infty}^{\infty} \left[\left(1 + 2\dfrac{r_1^2}{r_0^2}\right)r_\tau^2 - 4\dfrac{r_1}{r_0}r_\tau r_{\tau+1} + r_{\tau-1}r_{\tau+1}\right]\bigg/r_0^2$$

Hint. Use Lemma B.9.

(b) Show how a^* and P simplify if the process $y(t)$ is a first-order AR process.

Remark. For an extension of the result above to general (arbitrary order) AR models, see Stoica, Nehorai and Kay (1987).

Problem 7.18 *Estimation of the parameters of an AR process observed in noise*
Consider an nth-order AR process that is observed with noise,

$$A_0(q^{-1})x(t) = v(t)$$
$$y(t) = x(t) + e(t)$$

$v(t)$, $e(t)$ being mutually independent white noise sequences of zero mean and variances λ_v^2, λ_e^2.

(a) How can a model for $y(t)$ be obtained using a prediction error method?
(b) Assume that a prediction error method is applied using the model structure

$$A(q^{-1})x(t) = v(t)$$
$$y(t) = x(t) + e(t)$$
$$\theta = (a_1 \ldots a_n \; \lambda_v^2 \; \lambda_e^2)^T$$

Show how to compute the prediction errors $\varepsilon(t, \theta)$ from the data. Also show that the criterion

$$V(\theta) = \sum_{t=1}^{N} \varepsilon^2(t, \theta)$$

has many global minima. What can be done to solve this problem and estimate θ uniquely?

Remark. More details on this type of estimation problem and its solution can be found in Nehorai and Stoica (1988).

Problem 7.19 *Whittle's formula for the Cramér–Rao lower bound*
Let $y(t)$ be a Gaussian stationary process with continuous spectral density $\phi(\omega)$ completely defined by the vector θ of (unknown) parameters. Then the Cramér–Rao lower bound (CRLB) for any consistent estimate of θ is given by Whittle's formula:

$$P_w = \left\{ \frac{1}{4\pi i} \oint \frac{1}{\phi^2(z)} \left(\frac{\partial \phi(z)}{\partial \theta} \right)^T \left(\frac{\partial \phi(z)}{\partial \theta} \right) \frac{dz}{z} \right\}^{-1}$$

Let $y(t)$ be a scalar ARMA process. Then an alternative expression for the CRLB is given by (7.66). Prove the equivalence between the expression above and (7.66).

Remark. The above formula for the covariance matrix corresponding to the CRLB is due to Whittle (1953).

Problem 7.20 *A sufficient condition for the stability of least squares input–output models*
Let $y(t)$ and $u(t)$ denote the (ny-dimensional) output and (nu-dimensional) input, respectively, of a multivariable system. The only assumption made about the system is that $\{u(t)\}$ and $\{y(t)\}$ are stationary signals. A full polynomial form model (see 6.22),

$$\hat{A}(q^{-1})y(t) = \hat{B}(q^{-1})u(t) + \varepsilon(t)$$
$$\hat{A}(q^{-1}) = I + \hat{A}_1 q^{-1} + \ldots + \hat{A}_n q^{-n}$$
$$\hat{B}(q^{-1}) = \hat{B}_1 q^{-1} + \ldots + \hat{B}_n q^{-n}$$

is fitted to input–output data from the system, by the least squares (LS) method. This means that the unknown matrix coefficients $\{\hat{A}_i, \hat{B}_j\}$ are determined such that asymptotically (i.e. for an infinite number of data points)

$$E\|\hat{A}(q^{-1})y(t) - \hat{B}(q^{-1})u(t)\|^2 = \min \tag{i}$$

where $\|\cdot\|$ denotes the Euclidean norm. Since the model is only an approximation of the identified system, one cannot speak about the consistency or related properties of the estimates $\{\hat{A}_i, \hat{B}_j\}$. However, one can speak about the stability properties of the model, as for some applications (such as controller design, simulation and prediction) it may be important that the model is stable, i.e.

$$\det[\hat{A}(z)] \neq 0 \quad \text{for } |z| \leq 1 \tag{ii}$$

The LS models are *not* necessarily stable. Their stability properties will depend on the system identified and the characteristics of its input $u(t)$ (see Söderström and Stoica, 1981b). Show that a sufficient condition for the LS model (i) to be stable is that $u(t)$ is a *white signal*. Assume that the system is causal, i.e. $y(t)$ depends only on the past values of the input $u(t)$, $u(t-1)$, etc.

Hint. A simple proof of stability can be obtained using the result of Complement C8.6 which states that the matrix polynomial

$$P(q^{-1}) = I + P_1 q^{-1} + \ldots + P_n q^{-n} \quad (m|m)$$

given by

$$E\|P(q^{-1})x(t)\|^2 = \min$$

where $x(t)$ is any (m-dimensional) stationary signal, is stable (i.e. satisfies a condition similar to (ii)).

Problem 7.21 *Accuracy of noise variance estimate*
Consider prediction error estimation in the single output model

$$y(t) = G(q^{-1}, \theta)u(t) + H(q^{-1}, \theta)e(t)$$

Assume that the system belongs to the model set and that the true innovations $\{e(t)\}$ form a white noise process with zero mean, variance λ^2 and fourth-order moment $\mu = Ee^4(t)$. Consider the following estimate $\hat{\lambda}^2$ of λ^2 (cf. (7.43)):

$$\hat{\lambda}^2 = V_N(\hat{\theta}_N) = \frac{1}{N} \sum_{t=1}^{N} \varepsilon^2(t, \hat{\theta}_N)$$

Show that

$$\lim_{N \to \infty} NE(\hat{\lambda}^2 - \lambda^2) = -\lambda^2 \dim \theta$$

$$\lim_{N \to \infty} NE(\hat{\lambda}^2 - \lambda^2)^2 = \mu - \lambda^4$$

Hint. Write $\hat{\lambda}^2 - \lambda^2 = V_N(\hat{\theta}_N) - V_N(\theta_0) + 1/N \sum_{t=1}^{N} e^2(t) - \lambda^2$ and make a Taylor series expansion of $V_N(\theta_0)$ around $\hat{\theta}_N$.

Remark. For Gaussian distributed noise, $\mu = 3\lambda^4$. Hence in that case

$$\lim_{N \to \infty} NE(\hat{\lambda}^2 - \lambda^2)^2 = 2\lambda^4$$

Problem 7.22 *The Steiglitz-McBride method*
Consider the output error model

$$y(t) = \frac{B(q^{-1})}{A(q^{-1})} u(t) + \varepsilon(t)$$

where deg A = deg B = n. The following iterative for determining $A(q^{-1})$ and $B(q^{-1})$ is based on successive linear least squares fits

$$(\hat{A}_{k+1}, \hat{B}_{k+1}) = \arg \min_{(A, B)} \sum_{t=1}^{N} \left[A(q^{-1}) \left\{ \frac{1}{\hat{A}_k(q^{-1})} y(t) \right\} - B(q^{-1}) \left\{ \frac{1}{\hat{A}_k(q^{-1})} u(t) \right\} \right]^2 \quad \text{(i)}$$

Assume that the data satisfy

$$y(t) = \frac{B_0(q^{-1})}{A_0(q^{-1})} u(t) + v(t)$$

where $A_0(q^{-1})$ and $B_0(q^{-1})$ are coprime and of degree n, $u(t)$ is persistently exciting of order $2n$ and $v(t)$ is a stationary disturbance that is independent of the input. Consider the asymptotic case where N, the number of data, tends to infinity.

(a) Assume that $v(t)$ is white noise. Show that the only possible stationary solution of (i) is given by $A(q^{-1}) = A_0(q^{-1})$, $B(q^{-1}) = B_0(q^{-1})$.
(b) Assume that $v(t)$ is correlated noise. Show that $A(q^{-1}) = A_0(q^{-1})$, $B(q^{-1}) = B_0(q^{-1})$ is generally not a possible stationary solution to (i).

Remark. The method was proposed by Steiglitz and McBride (1965). For a deeper analysis of its properties see Stoica and Söderström (1981b), Söderström and Stoica (1988). Note that it is a consequence of (b) that the method does not converge to minimum points of the output error loss function.

Bibliographical notes

(*Sections 7.2, 7.4*). The ML approach for Gaussian distributed disturbances was first proposed for system identification by Åström and Bohlin (1965). As mentioned in the text, the ML method can be seen as a prediction error identification method. For a further description of PEMs, see Ljung (1976, 1978), Caines (1976), Åström (1980), Box and Jenkins (1976). The generalized least squares method (see (7.37)) was proposed by Clarke (1967) and analyzed by Söderström (1974). For a more formal derivation of the accuracy result (7.59), (7.60), see Caines and Ljung (1976), Ljung and Caines (1979).

The output error method (i.e. a PEM applied to a model with $H(q^{-1}; \theta) \equiv 1$) has been analyzed by Kabaila (1983), Söderström and Stoica (1982). For further treatments of the results (7.77), (7.78) for underparametrized models, see Ljung (1978), Ljung and Glover (1981). The properties of underparametrized models, such as bias and variance of the transfer function estimate $G(e^{i\omega}; \hat{\theta}_N)$ for fixed ω, have been examined by Ljung (1985a, b), Wahlberg and Ljung (1986).

(*Section 7.3*). Detailed derivations of the optimal prediction for ARMAX and state space models are given by Åström (1970), and Anderson and Moore (1979).

(*Section 7.6*). For some general results on numerical search routines for optimization, see Dennis and Schnabel (1983) and Gill *et al.* (1981). The gradient of the loss function can be computed efficiently by using the adjoint system; see Hill (1985) and van Zee and Bosgra (1982). The possibility of computing the prediction error estimate by applying a recursive algorithm a number of passes over the data has been described by Ljung (1982); see also Young (1984), Solbrand *et al.* (1985). The idea of approximately estimating the prediction errors by fitting a high-order linear regression (cf. Example 7.9) is developed, for example, by Mayne and Firoozan (1982) (see also Stoica, Söderström, Ahlén and Solbrand (1984, 1985)).

Appendix A7.1
Covariance matrix of PEM estimates for multivariable systems

It follows from (7.58) and Lemmas B.3 and B.4 that the covariance matrix P is given by

$$P = [V''_\infty(\theta_0)]^{-1} [\lim_{N \to \infty} NE\{V'_N(\theta_0)\}^T V'_N(\theta_0)][V''_\infty(\theta_0)]^{-1} \qquad (A7.1.1)$$

Hence the primary goal is to find $V'_N(\theta_0)$ and $V''_\infty(\theta_0)$ for a general multivariable model structure and a general loss function

$$V_N(\theta) = h(R_N(\theta)) \qquad R_N(\theta) = \frac{1}{N} \sum_{t=1}^{N} \varepsilon(t, \theta) \varepsilon^T(t, \theta) \qquad (A7.1.2)$$

Using the notation

$$\varepsilon_i(t) = \frac{\partial}{\partial \theta_i} \varepsilon(t, \theta) \qquad \varepsilon_{ij}(t) = \frac{\partial^2}{\partial \theta_i \partial \theta_j} \varepsilon(t, \theta)$$

$$H_N(\theta) = \frac{\partial}{\partial Q} h(Q)|_{Q=R_N(\theta)}$$

$$H = H_\infty(\theta_0)$$

$$\psi(t) = -(\varepsilon_1(t) \ldots \varepsilon_{n\theta}(t))^T_{\theta=\theta_0} = -\left(\frac{\partial \varepsilon(t)}{\partial \theta}\right)^T_{\theta=\theta_0}$$

straightforward differentiation gives

$$\frac{\partial}{\partial \theta_i} V_N(\theta) = \sum_{j,k} \frac{\partial h}{\partial R_{N_{jk}}} \frac{\partial R_{N_{jk}}}{\partial \theta_i}$$

$$= \mathrm{tr}\left(H_N(\theta) \frac{1}{N} \sum_{t=1}^{N} \{\varepsilon_i(t)\varepsilon^T(t) + \varepsilon(t)\varepsilon_i^T(t)\}\right)$$

$$= \frac{2}{N} \sum_{t=1}^{N} \varepsilon^T(t) H_N(\theta) \varepsilon_i(t)$$

$$\frac{\partial^2}{\partial \theta_i \partial \theta_j} V_\infty(\theta) = 2 \frac{\partial}{\partial \theta_j} E\varepsilon^T(t) H_\infty(\theta) \varepsilon_i(t)$$

$$= 2\left[E\varepsilon_j^T(t) H_\infty(\theta) \varepsilon_i(t) + E\varepsilon^T(t) \frac{\partial H_\infty(\theta)}{\partial \theta_j} \varepsilon_i(t) + E\varepsilon^T(t) H_\infty(\theta) \varepsilon_{ij}(t) \right]$$

Thus

$$V'_N(\theta)_{\theta=\theta_0} \approx -\frac{2}{N} \sum_{t=1}^{N} e^T(t) H \psi^T(t) \quad (A7.1.3)$$

$$V''_\infty(\theta)_{\theta=\theta_0} = 2[E\psi(t) H \psi^T(t)] \quad (A7.1.4)$$

The expression for the Hessian follows since $\varepsilon(t)_{\theta=\theta_0} = e(t)$ is uncorrelated with $\varepsilon_i(t)$ and $\varepsilon_{ij}(t)$. The central matrix in (A7.1.1) is thus given by

$$P_0 = \lim_{N \to \infty} N \frac{4}{N^2} E \sum_{t=1}^{N} \psi(t) H e(t) \sum_{s=1}^{N} e^T(s) H \psi^T(s)$$

It can be evaluated as in the scalar case (see the calculations leading to (7.65)). The result is

$$P_0 = 4[E\psi(t) H \Lambda H \psi^T(t)] \quad (A7.1.5)$$

Now, (A7.1.1), (A7.1.4), and (A7.1.5) give (7.72).

To show (7.75), proceed as follows. Clearly the matrix

$$E\begin{pmatrix} \psi(t)H\Lambda H\psi^T(t) & \psi(t)H\psi^T(t) \\ \psi(t)H\psi^T(t) & \psi(t)\Lambda^{-1}\psi^T(t) \end{pmatrix} = E\begin{pmatrix} \psi(t)H\Lambda \\ \psi(t) \end{pmatrix} \Lambda^{-1}(\Lambda H \psi^T(t) \quad \psi^T(t))$$

is nonnegative definite. It then follows from Lemma A.3 that

$$E\psi(t)\Lambda^{-1}\psi^T(t) - [E\psi(t)H\psi^T(t)][E\psi(t)H\Lambda H\psi^T(t)]^{-1}[E\psi(t)H\psi^T(t)] \geq 0 \quad (A7.1.6)$$

This inequality is easily rewritten as

$$E\psi(t)\Lambda^{-1}\psi^T(t) \geq P^{-1}$$

which in turn is equivalent to (7.75). It is trivial to verify that (7.76) gives equality in (7.75).

Complement C7.1
Approximation models depend on the loss function used in estimation

In Section 2.4 and 7.5 it has been demonstrated that if the system which generated the data cannot be described exactly in the model class, then the estimated model may depend drastically on the experimental condition even asymptotically (i.e. for $N \to \infty$, N the number of data points).

For multivariable systems an additional problem occurs. Let $\varepsilon(t)$ denote the residuals of the model with parameters θ, with dim $\varepsilon(t) > 1$. Let the parameters θ be determined by minimizing a criterion of the form (7.13). When the system does not belong to the model class considered, the loss function $h(\cdot)$ used in estimation will significantly influence the estimated model even for $N \to \infty$. To show this we present a simple example patterned after Caines (1978).

Consider the system

$$\mathcal{S}: y(t) = e(t) \qquad t = 1, 2, \ldots$$

$$Ee(t)e^T(s) = \begin{pmatrix} 1 & 0 \\ 0 & 1 \end{pmatrix} \delta_{t,s}$$

and the model class

$$\mathcal{M}: y(t) = \begin{pmatrix} 2\theta & 0 \\ 0 & \sqrt{(1-\theta)} \end{pmatrix} y(t-1) + \varepsilon(t)$$

Clearly $\mathcal{S} \notin \mathcal{M}$. That is to say, there is no θ such that $\mathcal{S} \equiv \mathcal{M}(\theta)$. For $\theta \in (0, \frac{1}{2})$ the model is asymptotically stable and as a consequence $\varepsilon(t)$ is an ergodic process. Thus

$$\lim_{N \to \infty} \frac{1}{N} \sum_{t=1}^{N} \varepsilon(t)\varepsilon^T(t) = E\varepsilon(t)\varepsilon^T(t) \triangleq Q$$

Simple calculations give

$$Q = E\left\{e(t) - \begin{pmatrix} 2\theta & 0 \\ 0 & \sqrt{1-\theta} \end{pmatrix} e(t-1)\right\}\left\{e(t) - \begin{pmatrix} 2\theta & 0 \\ 0 & \sqrt{(1-\theta)} \end{pmatrix} e(t-1)\right\}^T$$

$$= I + \begin{pmatrix} 4\theta^2 & 0 \\ 0 & 1-\theta \end{pmatrix}$$

Therefore

$$h_1(Q) \triangleq \text{tr } Q = 3 - \theta + 4\theta^2$$

$$h_2(Q) \triangleq \det Q = (1 + 4\theta^2)(2 - \theta) = 2 - \theta + 8\theta^2 - 4\theta^3$$

and

$$\frac{\partial h_1}{\partial \theta} = 8\theta - 1$$

$$\frac{\partial h_2}{\partial \theta} = -12\theta^2 + 16\theta - 1$$

which give

$$\hat{\theta}_1 \triangleq \arg\min h_1(\theta) = 0.125$$

$$\hat{\theta}_2 \triangleq \arg\min h_2(\theta) = (4 - \sqrt{13})/6 \simeq 0.066$$

Thus, when $\mathscr{S} \notin \mathscr{M}$ different loss functions lead to different parameter estimates, even asymptotically.

For some general results pertaining to the influence of the loss function used on the outcomes of an approximate PEM identification, see Söderström and Stoica (1981a).

Complement C7.2
Multistep prediction of ARMA processes

Consider the ARMA model

$$A(q^{-1})y(t) = C(q^{-1})e(t) \tag{C7.2.1}$$

where

$$A(q^{-1}) = 1 + a_1 q^{-1} + \ldots + a_{na} q^{-na}$$

$$C(q^{-1}) = 1 + c_1 q^{-1} + \ldots + c_{nc} q^{-nc}$$

and where $\{e(t)\}$, $t = 0, \pm 1, \ldots$ is a sequence of uncorrelated and identically distributed Gaussian random variables with zero mean and variance λ^2. According to the spectral factorization theorem (See Appendix A6.1) one can assume without introducing any restriction that $A(z)C(z)$ has all zeros outside the unit circle:

$$A(z)C(z) = 0 \Rightarrow |z| > 1$$

One can also assume that $A(z)$ and $C(z)$ have no common zeros. The parameters $\{a_i, c_j\}$ are supposed to be given. Let Y^t denote the information available at the time instant t:

$$Y^t = \{y(t), y(t-1), \ldots\}$$

The optimal k-step predictor

The problem is to determine the mean square optimal k-step prediction of $y(t + k)$; i.e. an estimate $\hat{y}(t + k|t)$ of $y(t + k)$, which is a function of Y^t and is such that the variance of the prediction error

$$y(t + k) - \hat{y}(t + k|t)$$

is minimized. Introduce the following two polynomials:

$$F(z) = 1 + f_1 z + \ldots + f_{k-1} z^{k-1}$$

$$G(z) = g_0 + g_1 z + \ldots + g_{l-1} z^{l-1} \qquad l = \max(na - 1, nc - k)$$

through

$$C(z) \equiv F(z)A(z) + z^k G(z) \qquad (C7.2.2)$$

Due to the assumption that $A(z)$ and $C(z)$ are coprime polynomials, this identity uniquely defines F and G. (If $l < 0$ take $G(z) \equiv 0$.) Inserting (C7.2.2) into (C7.2.1) gives

$$y(t + k) = F(q^{-1})e(t + k) + \frac{G(q^{-1})}{C(q^{-1})} y(t)$$

The first term in the right-hand side of the above relation is independent of Y^t. Thus

$$E[y(t + k) - \hat{y}(t + k|t)]^2 = E\left(\frac{G(q^{-1})}{C(q^{-1})} y(t) - \hat{y}(t + k|t)\right)^2 \qquad (C7.2.3)$$
$$+ E[F(q^{-1})e(t + k)]^2$$

which shows that the optimal k-step predictor is given by

$$\hat{y}(t + k|t) = \frac{G(q^{-1})}{C(q^{-1})} y(t) \qquad (C7.2.4)$$

The prediction error is an MA($k - 1$) process

$$y(t + k) - \hat{y}(t + k|t) = F(q^{-1})e(t + k) \qquad (C7.2.5)$$

with variance

$$(1 + f_1^2 + \ldots + f_{k-1}^2)\lambda^2$$

Note that $e(t)$ is the error of the optimal one-step predictor.

Concerning the assumptions under which the optimal prediction is given by (C7.2.4), it is apparent from (C7.2.3) that the two terms

$$\frac{G(q^{-1})}{C(q^{-1})} y(t) - \hat{y}(t + k|t) \qquad \text{and} \qquad F(q^{-1})e(t + k) \qquad (C7.2.6)$$

must be uncorrelated for (C7.2.4) to hold. This condition is satisfied in the following situations:

- Assume that $\{e(t)\}$ is a sequence of *independent* (not only uncorrelated) variables. Note that this is the case if the sequence of uncorrelated random variables $\{e(t)\}$ is Gaussian distributed (see e.g. Appendix B.6). Then the two terms in (C7.2.6) will be independent and the calculation (C7.2.3) will hold.

- Assume that the prediction is constrained to be a *linear* function of Y^t. Then the two terms in (C7.2.6) will be uncorrelated and (C7.2.3), (C7.2.4) will hold. It follows that for any distribution of the data the mean square optimal *linear* k-step prediction is given by (C7.2.4).

In the following, a computationally efficient method for solving the polynomial equation (C7.2.2) for $k = 1, 2, \ldots$ will be presented. To emphasize that $F(z)$ and $G(z)$ as defined by (C7.2.2) depend on k, denote them by $F_k(z)$ and $G_k(z)$. It is important to observe that $F_k(z)$ consists of the first k terms in the series expansion of $C(z)/A(z)$ around $z = 0$. Thus, the following recursion must hold:

$$F_k(z) = F_{k-1}(z) + f_{k-1}z^{k-1} \qquad \text{(C7.2.7)}$$

From (C7.2.7), (C7.2.2) for k and $k - 1$, it follows that

$$A(z)F_{k-1}(z) + z^{(k-1)}G_{k-1}(z) \equiv A(z)[F_{k-1}(z) + f_{k-1}z^{(k-1)}] + z^k G_k(z)$$

which reduces to

$$A(z)f_{k-1} + zG_k(z) = G_{k-1}(z) \qquad \text{(C7.2.8)}$$

Assume for simplicity that $na = nc = n$ and let

$$G_k(z) = g_k^0 + g_k^1 z + \ldots + g_k^{n-1}z^{(n-1)}$$

By equating the different powers of z in (C7.2.8) the following formulas for computing f_{k-1}, $\{g_k^i\}_{i=0}^{n-1}$ for $k = 2, 3, \ldots$ are obtained:

$$\boxed{\begin{aligned} f_{k-1} &= g_{k-1}^0 \\ g_k^i &= g_{k-1}^{i+1} - a_{i+1}f_{k-1} \qquad i = 0, \ldots, n - 1 \end{aligned}} \qquad \text{(C7.2.9)}$$

Here $g_{k-1}^n = 0$. The initial values (corresponding to $k = 1$) for (C7.2.9) are easily found from (C7.2.2) to be

$$f_0 = 1$$
$$g_1^i = c_{i+1} - a_{i+1} \qquad i = 0, \ldots, n - 1$$

Using the method presented above for calculating $\hat{y}(t + k|t)$ one can develop k-step PEMs for estimating the ARMA parameters; i.e. PEMs which determine the parameter estimates by minimizing the sample variance of the k-step prediction errors:

$$\varepsilon_k(t) \triangleq y(t + k) - \hat{y}(t + k|t) \qquad t = 1, 2, \ldots \qquad \text{(C7.2.10)}$$

(see Åström, 1980; Stoica and Söderström, 1984; Stoica and Nehorai, 1987a). Recall that in the main part of this chapter only *one-step* PEMs were discussed.

Multistep optimal prediction

In some applications it may be desirable to determine the parameter estimates by minimizing a suitable function of the prediction errors $\{\varepsilon_k(t)\}$ for several (consecutive) values of k. For example, one may want to minimize the following loss function:

$$\sum_{k=1}^{m}\left[\frac{1}{N}\sum_{t=1}^{N-k}\varepsilon_k^2(t)\right] \tag{C7.2.11}$$

In other words, the parameters are estimated by using what may be called a *multi-step* PEM. An efficient method for computing the prediction errors $\{\varepsilon_k(t)\}_{k=1}^m$ (for $t = 1, \ldots, N - k$) needed in (C7.2.11) runs as follows (Stoica and Nehorai, 1987a):
Observe that (C7.2.7) implies

$$\varepsilon_k(t) \triangleq F_k(q^{-1})e(t + k) = \varepsilon_{k-1}(t + 1) + f_{k-1}\varepsilon_1(t) \quad (k \geq 2) \tag{C7.2.12}$$

where by definition $\varepsilon_1(t) = e(t + 1)$. Thus $\varepsilon_1(t)$ must first be computed as

$$\varepsilon_1(t) = \frac{A(q^{-1})}{C(q^{-1})} y(t + 1) \quad t = 1, 2, \ldots$$

Next (C7.2.12) is used to determine the other prediction errors needed in (C7.2.11).

In the context of multistep prediction of ARMA process, it is interesting to note that the predictions $\hat{y}(t + k|t)$, $\hat{y}(t + k - 1|t)$, etc., can be shown to obey a certain recursion. Thus, it follows from Lemma B.16 of Appendix B that the optimal mean-square k-step prediction of $y(t + k)$ under the Gaussian hypothesis is given by

$$\hat{y}(t + k|t) = E[y(t + k)|Y^t]$$

Taking conditional expectation $E[\cdot|Y^t]$ in the ARMA equation (C7.2.1),

$$\boxed{\begin{aligned}&\hat{y}(t + k|t) + a_1\hat{y}(t + k - 1|t) + \ldots + a_{na}\hat{y}(t + k - na|t) \\ &= c_k e(t) + \ldots + c_{nc} e(t + k - nc)\end{aligned}} \tag{C7.2.13}$$

where

$$\hat{y}(t + i|t) = \begin{cases} \hat{y}(t + i|t) & \text{for } i > 0 \\ y(t + i) & \text{for } i \leq 0 \end{cases}$$

and where it is assumed that $k \leq nc$. If $k > nc$ then the right-hand side of (C7.2.13) is zero. Note that for AR processes, the right-hand side of (C7.2.13) is zero for all $k \geq 1$. Thus for pure AR processes, (C7.2.13) becomes a simple and quite intuitive multistep prediction formula.

The relationship (C7.2.13) is important from a theoretical standpoint. It shows that for an ARMA process the predictor space, i.e. the space generated by the vectors

$$\begin{pmatrix} \hat{y}(t + 1|t) \\ t = 1, 2, \ldots \end{pmatrix}, \begin{pmatrix} \hat{y}(t + 2|t) \\ t = 1, 2, \ldots \end{pmatrix}, \text{ etc.}$$

is finite dimensional. More exactly, it has dimension not larger than nc. (See Akaike, 1981, for an application of this result to the problem of stochastic realization.) The recursive-in-k prediction formula (C7.2.13) also has practical applications. For instance, it is useful in on-line prediction applications of ARMA models (see Holst, 1977).

Complement C7.3
Least squares estimation of the parameters of full polynomial form models

Consider the following full polynomial form model of a MIMO system (see Example 6.3):

$$y(t) + A_1 y(t-1) + \ldots + A_{na} y(t-na) = B_1 u(t-1) + \ldots \\ + B_{nb} u(t-nb) + \varepsilon(t) \tag{C7.3.1}$$

where $\dim y = ny$, $\dim u = nu$, and where *all* entries of the matrix coefficients $\{A_i, B_j\}$ are assumed unknown. It can be readily verified that (C7.3.1) can be rewritten as

$$y(t) = \Theta^T \varphi(t) + \varepsilon(t) \tag{C7.3.2}$$

where

$$\Theta^T = (A_1 \ldots A_{na} \; B_1 \ldots B_{nb}) \qquad (ny | (na \cdot ny + nb \cdot nu))$$
$$\varphi^T(t) = (-y^T(t-1) \ldots -y^T(t-na) \; u^T(t-1) \ldots u^T(t-nb))$$
$$((na \cdot ny + nb \cdot nu) | 1)$$

Note that (C7.3.2) is an alternative form to (6.24), (6.25).

As shown in Lemma A.34, for matrices A, B of compatible dimensions,

$$\text{vec}(AB) = (I \otimes A) \text{vec } B = (B^T \otimes I) \text{vec } A \tag{C7.3.3}$$

where for some $m|n$ matrix H with the ith column denoted by h_i, $\text{vec } H = (h_1^T \ldots h_n^T)^T$. Using this result, (C7.3.2) can be rewritten in the form of a multivariate linear regression:

$$y(t) = \text{vec}[\varphi^T(t) \Theta] + \varepsilon(t) \\ = \phi^T(t) \theta + \varepsilon(t) \tag{C7.3.4}$$

where (see also (6.24), (6.25))

$$\phi^T(t) = I \otimes \varphi^T(t) \qquad (ny | ny(na \cdot ny + nb \cdot nu))$$
$$\theta = \text{vec } \Theta \qquad (ny(na \cdot ny + nb \cdot nu) | 1)$$

The problem to be discussed is how to determine the LS estimates of $\{A_i, B_j\}$ from N pairs of measurements of $u(t)$ and $y(t)$.

First consider the model (C7.3.2). Introduce the sample covariance matrix of the residuals

$$Q = \sum_{t=1}^{N} \varepsilon(t) \varepsilon^T(t)$$

and the notation

234 Prediction error methods   Chapter 7

$$R = \sum_{t=1}^{N} \varphi(t)\varphi^{T}(t)$$

$$\Gamma = \sum_{t=1}^{N} \varphi(t) y^{T}(t)$$

Next note that

$$\begin{aligned}
Q &= \sum_{t=1}^{N} [y(t) - \Theta^{T}\varphi(t)][y(t) - \Theta^{T}\varphi(t)]^{T} \\
&= \sum_{t=1}^{N} y(t) y^{T}(t) - \Theta^{T}\Gamma - \Gamma^{T}\Theta + \Theta^{T} R \Theta \qquad \text{(C7.3.5)} \\
&= [\Theta - R^{-1}\Gamma]^{T} R [\Theta - R^{-1}\Gamma] + \sum_{t=1}^{N} y(t) y^{T}(t) - \Gamma^{T} R^{-1}\Gamma
\end{aligned}$$

Since the matrix R is positive definite, and the second and third terms in (C7.3.5) do not depend on Θ it follows that

$$Q \geq Q|_{\Theta = \hat{\Theta}}$$

where

$$\hat{\Theta} = R^{-1}\Gamma \qquad \text{(C7.3.6)}$$

The LS estimate (C7.3.6) is usually derived by minimizing tr Q. However, as shown above it 'minimizes' the whole sample covariance matrix Q. As a consequence of this strong property, $\hat{\Theta}$ will minimize any nondecreasing function of Q, such as det Q or tr WQ, with W any positive definite weighting matrix. Note that even though such a function may be strongly nonlinear in Θ, it can be shown that (C7.3.6) is its only stationary point (Söderström and Stoica, 1980).

Next consider the model (C7.3.4). The LS estimate of θ in this model is usually defined as

$$\hat{\theta} = \arg\min_{\theta} \sum_{t=1}^{N} \varepsilon^{T}(t) W \varepsilon(t)$$

where W is some symmetric positive definite weighting matrix. Introduce

$$\tilde{R} = \sum_{t=1}^{N} \phi(t) W \phi^{T}(t)$$

$$\tilde{\Gamma} = \sum_{t=1}^{N} \phi(t) W y(t)$$

Then

$$\sum_{t=1}^{N} \varepsilon^{T}(t) W \varepsilon(t) = \sum_{t=1}^{N} [y(t) - \varphi^{T}(t)\theta]^{T} W [y(t) - \varphi^{T}(t)\theta]$$

$$= \sum_{t=1}^{N} y^{T}(t) W y(t) - \tilde{\Gamma}^{T}\theta - \theta^{T}\tilde{\Gamma} + \theta^{T}\tilde{R}\theta$$

$$= [\theta - \tilde{R}^{-1}\tilde{\Gamma}]^{T} \tilde{R} [\theta - \tilde{R}^{-1}\tilde{\Gamma}] + \sum_{t=1}^{N} y^{T}(t) W y(t) - \tilde{\Gamma}^{T}\tilde{R}^{-1}\tilde{\Gamma}$$

Since $\tilde{R}$ is a positive definite matrix it is readily concluded from the above equation that

$$\hat{\theta} = \tilde{R}^{-1}\tilde{\Gamma} \tag{C7.3.7}$$

The LS estimate (C7.3.7) seems to depend on W. However, this should *not* be so. The two models (C7.3.2) and (C7.3.4) are just two *equivalent* parametrizations of the same model (C7.3.1). Also, the loss function minimized by $\hat{\theta}$ can be written as tr WQ and hence is in the class of functions minimized by Θ. Thus $\hat{\theta}$ must be the minimizer of the whole covariance matrix Q, and indeed

$$\hat{\theta} = \text{vec } \hat{\Theta} \tag{C7.3.8}$$

This can be proved algebraically. Note that by using the Kronecker product (see Definition A.10 and Lemma A.34) we can write

$$\hat{\theta} = \tilde{R}^{-1}\tilde{\Gamma} = \left\{ \sum_{t=1}^{N} [I \otimes \varphi(t)][W \otimes 1][I \otimes \varphi^{T}(t)] \right\}^{-1} \left\{ \sum_{t=1}^{N} [I \otimes \varphi(t)][W \otimes 1] y(t) \right\}$$

$$= \left\{ W \otimes \sum_{t=1}^{N} \varphi(t)\varphi^{T}(t) \right\}^{-1} \left\{ \sum_{t=1}^{N} [W \otimes \varphi(t)] y(t) \right\}$$

$$= \sum_{t=1}^{N} [W^{-1} \otimes R^{-1}][W \otimes \varphi(t)] y(t) = \sum_{t=1}^{N} [I \otimes R^{-1}\varphi(t)] y(t) \tag{C7.3.9}$$

Now, it follows from (C7.3.3) and (C7.3.6) that

$$\text{vec } \hat{\Theta} = \text{vec } (R^{-1}\Gamma) = (I \otimes R^{-1}) \text{ vec } \Gamma$$

$$= (I \otimes R^{-1}) \sum_{t=1}^{N} \text{vec } [\varphi(t) y^{T}(t)]$$

$$= \sum_{t=1}^{N} [I \otimes R^{-1}][I \otimes \varphi(t)] y(t)$$

$$= \sum_{t=1}^{N} [I \otimes R^{-1}\varphi(t)] y(t) \tag{C7.3.10}$$

Comparison of (C7.3.9) and (C7.3.10) completes the proof.

The conclusion is that there is no reason for using the estimate (C7.3.7), which is much more complicated computationally than the *equivalent* estimate (C7.3.6). (Note that the matrix R has dimension $na \cdot ny + nb \cdot nu$ while $\tilde{R}$ has dimension $ny \cdot (na \cdot ny + nb \cdot nu)$.)

Finally, note that if some elements of the matrix coefficients $\{A_i, B_j\}$ are known, then the model (C7.3.4) and the least squares estimate of its parameters can be easily adapted to use this information, while the model (C7.3.2) cannot be used in such a case.

Complement C7.4
The generalized least squares method

Consider the following model of a MIMO system:

$$A(q^{-1})y(t) = B(q^{-1})u(t) + v(t)$$
$$D(q^{-1})v(t) = \varepsilon(t) \tag{C7.4.1}$$

where

$$\begin{aligned} A(q^{-1}) &= I + A_1 q^{-1} + \ldots + A_{na} q^{-na} & (ny|ny) \\ B(q^{-1}) &= B_1 q^{-1} + \ldots + B_{nb} q^{-nb} & (ny|nu) \\ D(q^{-1}) &= I + D_1 q^{-1} + \ldots + D_{nd} q^{-nd} & (ny|ny) \end{aligned} \tag{C7.4.2}$$

Thus the equation errors $\{v(t)\}$ are modeled as an autoregression. All the elements of the matrix coefficients $\{A_i, B_j, D_k\}$ are assumed to be unknown. The problem is to determine the PE estimates $\{\hat{A}_i, \hat{B}_j, \hat{D}_k\}$ of the matrix coefficients. Define these estimates as

$$\{\hat{A}_i, \hat{B}_j, \hat{D}_k\} = \arg \min_{\{A_i, B_j, D_k\}} \operatorname{tr} Q(A, B, D)$$

where

$$Q = \sum_{t=1}^{N} \varepsilon(t, A, B, D) \varepsilon^{\mathrm{T}}(t, A, B, D)$$

and where N denotes the number of data points, and the dependence of the prediction errors ε on $\{A_i, B_j, D_k\}$ was stressed by notation. The minimization can be performed in various ways.

A relaxation algorithm

The prediction errors $\varepsilon(t)$ of the model under study have a special feature. For given D they are linear in A and B, and vice versa. This feature can be exploited to obtain a simple relaxation algorithm for minimizing tr Q. Specifically, the algorithm consists of iterating the following two steps until convergence.

S1: For given $\hat{D}$ determine

$$\hat{A}, \hat{B} = \arg\min_{A,B} \text{tr } Q(A, B, \hat{D})$$

S2: For given $\hat{A}$ and $\hat{B}$ determine

$$\hat{D} = \arg\min_{D} \text{tr } Q(\hat{A}, \hat{B}, D)$$

The results of Complement C7.3 can be used to obtain explicit expressions for $\hat{A}$, $\hat{B}$ and $\hat{D}$.

For step S1 let

$$\theta = \text{vec}(A_1 \ldots A_{na} \; B_1 \ldots B_{nb})^T$$
$$\varphi(t) = (-y^T(t-1) \ldots -y^T(t-na) \; u^T(t-1) \ldots u^T(t-nb))^T$$
$$\phi^T(t) = I \otimes \varphi^T(t)$$

Then

$$\varepsilon(t, A, B, \hat{D}) = \hat{D}(q^{-1})[A(q^{-1})y(t) - B(q^{-1})u(t)]$$
$$= \hat{D}(q^{-1})[y(t) - \phi^T(t)\theta] = y_F(t) - \phi_F^T(t)\theta$$

where

$$y_F(t) = \hat{D}(q^{-1})y(t)$$
$$\phi_F^T(t) = \hat{D}(q^{-1})\phi^T(t)$$

The minimizer of tr $Q(A, B, \hat{D})$ now follows from (C7.3.7):

$$\hat{\theta} = \left[\sum_{t=1}^{N} \phi_F(t)\phi_F^T(t)\right]^{-1} \left[\sum_{t=1}^{N} \phi_F(t)y_F(t)\right] \tag{C7.4.3}$$

For step S2 introduce

$$\hat{v}(t) = \hat{A}(q^{-1})y(t) - \hat{B}(q^{-1})u(t)$$
$$\psi(t) = -(\hat{v}^T(t-1) \ldots \hat{v}^T(t-nd))^T$$
$$\Gamma = (D_1 \ldots D_{nd})$$

One can then write

$$\varepsilon(t, \hat{A}, \hat{B}, D) = D(q^{-1})[\hat{A}(q^{-1})y(t) - \hat{B}(q^{-1})u(t)]$$
$$= D(q^{-1})\hat{v}(t) = \hat{v}(t) - \Gamma\psi(t)$$

It now follows from (C7.3.6) that the minimizer of tr $Q(\hat{A}, \hat{B}, D)$ is given by

$$\hat{\Gamma} = \left[\sum_{t=1}^{N} \hat{v}(t)\psi^T(t)\right]\left[\sum_{t=1}^{N} \psi(t)\psi^T(t)\right]^{-1} \tag{C7.4.4}$$

238   Prediction error methods   Chapter 7

Thus each step of the algorithm amounts to solving an LS problem. This gave the name of generalized LS (GLS) to the algorithm. It was introduced for scalar systems by Clarke (1967). Extensions to MIMO systems have been presented by Goodwin and Payne (1977) and Keviczky and Bányász (1976), but those appear to be more complicated than the algorithm above. Note that since relaxation algorithms converge linearly close to the minimum, the convergence rate of the GLS procedure may be slow. A faster algorithm is discussed below. Finally note that the GLS procedure does not need any special parametrization of the matrices $\{A_i, B_j, D_k\}$. If, however, some knowledge of the elements of A, B or D is available, this can be accommodated in the algorithm with minor modifications.

An indirect GLS procedure

Consider the model (C7.4.1) in the scalar case ($ny = nu = 1$)

$$A(q^{-1})y(t) = B(q^{-1})u(t) + \frac{1}{D(q^{-1})}\varepsilon(t) \tag{C7.4.5}$$

Introduce the following notation:

$$F(q^{-1}) = A(q^{-1})D(q^{-1}) = 1 + f_1 q^{-1} + \ldots + f_{nf} q^{-nf} \quad (nf = na + nd)$$
$$G(q^{-1}) = B(q^{-1})D(q^{-1}) = g_1 q^{-1} + \ldots + g_{ng} q^{-ng} \quad (ng = nb + nd)$$
$$\alpha = (f_1 \ldots f_{nf} \; g_1 \ldots g_{ng})^T \tag{C7.4.6}$$
$$\varphi(t) = (-y(t-1) \ldots -y(t-nf) \; u(t-1) \ldots u(t-ng))^T$$

Then one can write

$$\varepsilon_\alpha(t, \alpha) \triangleq F(q^{-1})y(t) - G(q^{-1})u(t) = y(t) - \varphi^T(t)\alpha$$

The least squares estimate (LSE) of α,

$$\hat\alpha = \arg\min_\alpha \sum_{t=1}^N \varepsilon_\alpha^2(t, \alpha)$$

satisfies

$$\sum_{t=1}^N \varepsilon_\alpha(t, \alpha) \frac{\partial \varepsilon_\alpha(t, \alpha)}{\partial \alpha}\bigg|_{\alpha=\hat\alpha}^T = -\sum_{t=1}^N \varepsilon_\alpha(t, \hat\alpha)\varphi(t) = 0 \tag{C7.4.7}$$

and is given by

$$\hat\alpha = \left[\sum_{t=1}^N \varphi(t)\varphi^T(t)\right]^{-1}\left[\sum_{t=1}^N \varphi(t)y(t)\right]$$

With θ denoting the vector which contains the (unknown) coefficients of $A(q^{-1})$, $B(q^{-1})$ and $D(q^{-1})$, let $\alpha(\theta)$ denote the function induced by (C7.4.6). For example, let $na = nb = nd = 1$. Then the function $\alpha(\theta)$ is given by

$$\alpha(\theta) = (a_1 + d_1 \; a_1 d_1 \; b_1 \; b_1 d_1)^T$$

Next note that the residuals $\varepsilon(t)$ of (C7.4.5) can be written as

$$\varepsilon(t) = \varepsilon_\alpha(t, \alpha(\theta)) = y(t) - \varphi^T(t)\alpha(\theta)$$
$$= y(t) - \varphi^T(t)\hat{\alpha} + \varphi^T(t)[\hat{\alpha} - \alpha(\theta)]$$
$$= \varepsilon_\alpha(t, \hat{\alpha}) + \varphi^T(t)[\hat{\alpha} - \alpha(\theta)]$$

Thus

$$\sum_{t=1}^{N} \varepsilon^2(t) = \sum_{t=1}^{N} \varepsilon_\alpha^2(t, \hat{\alpha}) + [\hat{\alpha} - \alpha(\theta)]^T P [\hat{\alpha} - \alpha(\theta)]$$

$$+ 2[\hat{\alpha} - \alpha(\theta)]^T \sum_{t=1}^{N} \varepsilon_\alpha(t, \hat{\alpha})\varphi(t) \tag{C7.4.8}$$

where

$$P = \sum_{t=1}^{N} \varphi(t)\varphi^T(t)$$

The third term in (C7.4.8) is equal to zero (see (C7.4.7)), and the first one does not depend on θ. Thus the PE estimate of θ which minimizes the left-hand side of (C7.4.8) can alternatively be obtained as

$$\hat{\theta} = \arg \min_{\theta} [\hat{\alpha} - \alpha(\theta)]^T P [\hat{\alpha} - \alpha(\theta)] \tag{C7.4.9}$$

Note that the data are used only in the calculation of $\hat{\alpha}$. The matrix P follows as a byproduct of that calculation. The loss function in (C7.4.9) does not depend explicitly on the data. Thus it may be expected that the 'indirect' GLS procedure (C7.4.9) is more efficient computationally than the relaxation algorithm previously described. Practical experience with (C7.4.9) has shown that it may lead to considerable computer time savings (Söderström, 1975c; Stoica, 1976). The basic ideas of the indirect PEM described above may be applied also to more general model structures (Stoica and Söderström, 1987).

Complement C7.5
The output error method

Consider the following SISO system:

$$y(t) = G^0(q^{-1})u(t) + H^0(q^{-1})e(t) \qquad Ee(t)e(s) = \lambda^2 \delta_{t,s}$$

where $G^0(q^{-1})$ is stable, and $H^0(q^{-1})$ is stable and invertible. Furthermore, let

$$G^0(q^{-1}) = \frac{B^0(q^{-1})}{A^0(q^{-1})} = \frac{b_1^0 q^{-1} + \ldots + b_{nb}^0 q^{-nb}}{1 + a_1^0 q^{-1} + \ldots + a_{na}^0 q^{-na}}$$

240 Prediction error methods Chapter 7

and
$$\theta_0 = (a_1^0 \ldots a_{na}^0 \; b_1^0 \ldots b_{nb}^0)^T$$

Assume that $B^0(\cdot)$ and $A^0(\cdot)$ have no common zeros, and that the system operates in open loop so that $Eu(t)e(s) = 0$ for all t and s.

The output error (OE) estimate of θ_0 is by definition

$$\hat{\theta} = \arg\min_\theta \sum_{t=1}^N \varepsilon^2(t, \theta) \qquad \varepsilon(t, \theta) = y(t) - \frac{B(q^{-1})}{A(q^{-1})} u(t) \qquad \text{(C7.5.1)}$$

where N is the number of data points, and θ, B and A are defined similarly to θ_0, B^0 and A^0. For simplicity, na and nb are assumed to be known. In the following we derive the asymptotic distribution of $\hat{\theta}$, compare the covariance matrix of this distribution with that corresponding to the PEM (which also provides an estimate of the noise dynamics), and present sufficient conditions for the unimodality of the loss function in (C7.5.1).

Asymptotic distribution

First note that $\hat{\theta}$ will converge to θ_0 as N tends to infinity (cf. (7.53)–(7.56)).
Further it follows from (7.78) that

$$\sqrt{N}(\hat{\theta} - \theta_0) \xrightarrow[N\to\infty]{\text{dist}} \mathcal{N}(0, P_{\text{OEM}}) \qquad \text{(C7.5.2)}$$

It remains to evaluate the covariance matrix P_{OEM} given by (7.78b):

$$P_{\text{OEM}} = [V''_\infty(\theta_0)]^{-1} [\lim_{N\to\infty} NE\{V'_N(\theta)\}|_{\theta=\theta_0}^T \{V'_N(\theta)\}|_{\theta=\theta_0}][V''_\infty(\theta_0)]^{-1} \qquad \text{(C7.5.3)}$$

where
$$V_N(\theta) = \frac{1}{N} \sum_{t=1}^N \varepsilon^2(t, \theta)$$

Direct differentiation gives

$$\{V'_N(\theta)\}^T|_{\theta=\theta_0} = \frac{2}{N} \sum_{t=1}^N \varepsilon(t, \theta_0)\psi(t)$$

$$V''_\infty(\theta_0) = 2E\left[\psi(t)\psi^T(t) + \varepsilon(t, \theta_0) \frac{\partial^2 \varepsilon(t, \theta)}{\partial \theta^2}\bigg|_{\theta=\theta_0}\right]$$

where
$$\psi(t) = \left[\frac{\partial}{\partial \theta} \varepsilon(t, \theta)\right]^T\bigg|_{\theta=\theta_0}$$
$$= \left(\frac{B^0(q^{-1})}{A^0(q^{-1})^2} u(t-1) \ldots \frac{B^0(q^{-1})}{A^0(q^{-1})^2} u(t-na) \; \frac{-1}{A^0(q^{-1})} u(t-1) \ldots \frac{-1}{A^0(q^{-1})} u(t-nb)\right)^T$$

Since $\varepsilon(t, \theta_0) = H^0(q^{-1})e(t)$ is independent of $\dfrac{\partial^2 \varepsilon(t, \theta)}{\partial \theta^2}\bigg|_{\theta=\theta_0}$ it follows that

$$V''_\infty(\theta_0) = 2E[\psi(t)\psi^T(t)] \tag{C7.5.4}$$

Next note that $e(t)$ and $\psi(s)$ are independent for all t and s. Using the notation

$$H^0(q^{-1}) = \sum_{i=0}^{\infty} h_i q^{-i} \qquad r(\tau) = E[H^0(q^{-1})e(t)][H^0(q^{-1})e(t+\tau)]$$

one can write

$$\lim_{N\to\infty} NE\{V'_N(\theta)\}|_{\theta=\theta_0}^T \{V'_N(\theta)\}|_{\theta=\theta_0}$$

$$= \lim_{N\to\infty} \frac{4}{N} E \sum_{p=1}^{N} H^0(q^{-1})e(p)\psi(p) \sum_{t=1}^{N} H^0(q^{-1})e(t)\psi^T(t)$$

$$= \lim_{N\to\infty} \frac{4}{N} \sum_{p=1}^{N}\sum_{t=1}^{N} [E\psi(p)\psi^T(t)][EH^0(q^{-1})e(p)H^0(q^{-1})e(t)] \tag{C7.5.5}$$

$$= \lim_{N\to\infty} \frac{4}{N} \sum_{\tau=-N}^{N} (N-|\tau|)r(\tau)E\psi(t)\psi^T(t+\tau)$$

The term containing $|\tau|$ in (C7.5.5) tends to zero as N tends to infinity (see Appendix A8.1 for a similar discussion). The remaining term becomes (set $h_i = 0$ for $i < 0$)

$$4 \sum_{\tau=-\infty}^{\infty} r(\tau) E\psi(t)\psi^T(t+\tau)$$

$$= 4 \sum_{\tau=-\infty}^{\infty} \sum_{i=-\infty}^{\infty} \lambda^2 h_i h_{i+\tau} E\psi(t-i-\tau)\psi^T(t+\tau-i-\tau) \tag{C7.5.6}$$

$$= 4\lambda^2 \sum_{i=-\infty}^{\infty} \sum_{j=-\infty}^{\infty} E[h_j\psi(t-j)][h_i\psi^T(t-i)]$$

$$= 4\lambda^2 E[H^0(q^{-1})\psi(t)][H^0(q^{-1})\psi(t)]^T$$

It follows from (C7.5.3)–(C7.5.6) that

$$\boxed{P_{\text{OEM}} = \lambda^2 [E\psi(t)\psi^T(t)]^{-1}[EH^0(q^{-1})\psi(t)H^0(q^{-1})\psi^T(t)][E\psi(t)\psi^T(t)]^{-1}} \tag{C7.5.7}$$

Comparison with the PEM

Let α denote an additional vector needed to parametrize the noise shaping filter $H^0(q^{-1})$. Assume that α and θ have no common parameters. The asymptotic covariance matrix of the PE estimates of θ and α is given by

$$\text{cov}\left\{\begin{pmatrix}\hat{\theta}\\\hat{\alpha}\end{pmatrix}\right\} = \lambda^2 \left\{ E\begin{pmatrix}\frac{\partial\tilde{\varepsilon}(t)^T}{\partial\theta}\\\frac{\partial\tilde{\varepsilon}(t)^T}{\partial\alpha}\end{pmatrix}\bigg|_{\theta_0,\alpha_0} \left(\frac{\partial\tilde{\varepsilon}(t)}{\partial\theta} \quad \frac{\partial\tilde{\varepsilon}(t)}{\partial\alpha}\right)\bigg|_{\theta_0,\alpha_0} \right\}^{-1} \tag{C7.5.8}$$

(see (7.49)), where

$$\tilde{\varepsilon}(t) = H^{-1}(q^{-1}; \alpha)[y(t) - G(q^{-1}; \theta)u(t)]$$

The random variables $\left.\dfrac{\partial \tilde{\varepsilon}(t)}{\partial \alpha}\right|_{\theta_0, \alpha_0}$ and $\left.\dfrac{\partial \tilde{\varepsilon}(t)}{\partial \theta}\right|_{\theta_0, \alpha_0}$ clearly are uncorrelated. Thus (C7.5.8) is block-diagonal, with the block corresponding to θ given by

$$P_{\text{PEM}} = \lambda^2 \left\{ E \frac{1}{H^0(q^{-1})} \psi(t) \frac{1}{H^0(q^{-1})} \psi^T(t) \right\}^{-1} \quad \text{(C7.5.9)}$$

$\left(\text{notice that } \partial \tilde{\varepsilon}(t)/\partial \theta|_{\theta_0, \alpha_0} = \dfrac{1}{H^0(q^{-1})} \psi^T(t)\right)$

Next compare the accuracies of the OEM and PEM. One may expect that

$$P_{\text{OEM}} \geq P_{\text{PEM}} \quad \text{(C7.5.10)}$$

The reason for this expectation is as follows. For Gaussian distributed data P_{PEM} is equal to the Cramér–Rao lower bound on the covariance matrix of any consistent estimator of θ. Thus (C7.5.10) must hold in that case. Now the matrices in (C7.5.10) depend on the second-order properties of the data and noise, and not on their distributions. Therefore the inequality (C7.5.10) must continue to hold for other distributions.

A simple algebraic proof of (C7.5.10) is as follows. Let $\phi_\psi(\omega)$ denote the spectral density matrix of $\psi(t)$. The matrix

$$E \begin{pmatrix} \dfrac{1}{H^0(q^{-1})} \psi(t) \dfrac{1}{H^0(q^{-1})} \psi^T(t) & \psi(t)\psi^T(t) \\ \psi(t)\psi^T(t) & H^0(q^{-1})\psi(t) H^0(q^{-1})\psi^T(t) \end{pmatrix}$$

$$= \int_{-\pi}^{\pi} \begin{pmatrix} \dfrac{1}{|H^0(e^{i\omega})|^2} \phi_\psi(\omega) & \phi_\psi(\omega) \\ \phi_\psi(\omega) & |H^0(e^{i\omega})|^2 \phi_\psi(\omega) \end{pmatrix} d\omega$$

clearly is nonnegative definite. Then it follows from Lemma A.3 that

$$E \frac{1}{H^0(q^{-1})} \psi(t) \frac{1}{H^0(q^{-1})} \psi^T(t) = \lambda^2 P_{\text{PEM}}^{-1}$$

$$\geq [E\psi(t)\psi^T(t)][EH^0(q^{-1})\psi(t)H^0(q^{-1})\psi^T(t)]^{-1}$$

$$\times [E\psi(t)\psi^T(t)]$$

$$= \lambda^2 P_{\text{OEM}}^{-1}$$

which concludes the proof of (C7.5.10).

Interesting enough, P_{OEM} may be equal to P_{PEM}. It was shown by Kabaila (1983) that for the specific input

$$u(t) = \sum_{k=1}^{n} \alpha_k \sin(\omega_k t + \varphi_k) \qquad \begin{array}{l} \alpha_k, \varphi_k \in R; \omega_k \in (0, \pi) \\ n = (na + nb)/2 \end{array} \qquad \text{(C7.5.11)}$$

it holds that

$$P_{\text{OEM}} = P_{\text{PEM}} \qquad \text{(C7.5.12)}$$

Observe that in (C7.5.11) we implicitly assumed that $na + nb$ is even. This was only done to simplify to some extent the following discussion. For $na + nb$ odd, the result (C7.5.12) continues to hold provided we allow $\omega_k = \pi$ in (C7.5.11) such that the input still has a spectral density of exactly $na + nb$ lines.

The result above may be viewed as a special case of a more general result presented in Grenander and Rosenblatt (1956) and mentioned in Problem 4.12. (The result of Grenander and Rosenblatt deals with linear regressions. Note, however, that P_{OEM} and P_{PEM} can be viewed as the *asymptotic* covariance matrices of the LSE and the Markov estimate of the parameters of a regression model with $\psi(t)$ as the regressor vector at time instant t and $H^0(q^{-1})$ as the noise shaping filter.)

There follows a simple proof of (C7.5.12) for the input (C7.5.11). The spectral density of (C7.5.11) is given by

$$\phi_u(\omega) = \sum_{k=1}^{n} \frac{\alpha_k^2}{4} [\delta(\omega - \omega_k) + \delta(\omega + \omega_k)]$$

(cf. (5.12f)). Let

$$F(q^{-1}) = \left(\frac{B^0(q^{-1})}{A^0(q^{-1})^2} q^{-1} \quad \cdots \quad \frac{B^0(q^{-1})}{A^0(q^{-1})^2} q^{-na} \quad \frac{-q^{-1}}{A^0(q^{-1})} \quad \cdots \quad \frac{-q^{-nb}}{A^0(q^{-1})} \right)^{\text{T}}$$

Then

$$\psi(t) = F(q^{-1})u(t)$$

and thus

$$EH^0(q^{-1})\psi(t) \cdot H^0(q^{-1})\psi^{\text{T}}(t)$$

$$= \int_{-\pi}^{\pi} |H^0(e^{i\omega})|^2 F(e^{i\omega}) F^{\text{T}}(e^{-i\omega}) \phi_u(\omega) d\omega$$

$$= \sum_{k=1}^{n} \alpha_k^2 |H^0(e^{i\omega_k})|^2 \{\text{Re}[F(e^{i\omega_k}) F^{\text{T}}(e^{-i\omega_k})]\}/2$$

$$= \sum_{k=1}^{n} |H^0(e^{i\omega_k})|^2 [g_k g_k^{\text{T}} + f_k f_k^{\text{T}}]$$

$$= (g_1 \; f_1 \; \cdots \; g_n \; f_n) \begin{pmatrix} |H^0(e^{i\omega_1})|^2 I_2 & & 0 \\ & \cdot & \\ & & \cdot \\ 0 & & |H^0(e^{i\omega_n})|^2 I_2 \end{pmatrix} \begin{pmatrix} g_1^T \\ f_1^T \\ \vdots \\ g_n^T \\ f_n^T \end{pmatrix} \triangleq MHM^T \quad \text{(C7.5.13)}$$

where g_k and f_k denote the real and imaginary parts, respectively, of $\alpha_k F(e^{i\omega_k})/\sqrt{2}$. Similarly, one can show that

$$E \frac{1}{H^0(q^{-1})} \psi(t) \frac{1}{H^0(q^{-1})} \psi^T(t) = MH^{-1}M^T$$

and, by specializing (C7.5.13) to $H^0(z) \equiv 1$, that

$$E\psi(t)\psi^T(t) = MM^T$$

Since A_0 and B_0 are coprime and the input (C7.5.11) is pe of order $na + nb$, the matrix MHM^T and hence M must be nonsingular (see Complement C5.1). Thus

$$P_{OEM} = \lambda^2 M^{-T}M^{-1}MHM^TM^{-T}M^{-1} = \lambda^2 M^{-T}HM^{-1}$$

and

$$P_{PEM} = \lambda^2(MH^{-1}M^T)^{-1} = \lambda^2 M^{-T}HM^{-1}$$

which proves (C7.5.12) for the specific input given by (C7.5.11).

The above result is quite interesting from a theoretical standpoint. However, its practical relevance appears limited. The (optimal) number of parameters $na + nb$ of the model is usually determined by processing the data measured; normally it is not known *a priori*. However, the condition that the input (C7.5.11) is a sum of exactly n sinusoids, appears to be essential for (C7.5.12) to hold.

Conditions for unimodality

If the loss function in (C7.5.1) is not unimodal then inherent difficulties in solving the optimization problem (C7.5.1) may occur. For $N < \infty$ the loss function is a random variable whose shape depends on the realization. Consider the asymptotic (for $N \to \infty$) loss function

$$W(\theta) = E\left[y(t) - \frac{B(q^{-1})}{A(q^{-1})} u(t)\right]^2$$

$$= E\left\{\left[\frac{B^0(q^{-1})}{A^0(q^{-1})} - \frac{B(q^{-1})}{A(q^{-1})}\right] u(t)\right\}^2 + E[H^0(q^{-1})e(t)]^2$$

The second term above does not depend on θ. To find the stationary points of $W(\theta)$, differentiate the first term with respect to a_i and b_j. This gives

$$E\left[\frac{B(q^{-1})}{A^2(q^{-1})}u(t-i)\right]\left[\frac{B^0(q^{-1})}{A^0(q^{-1})} - \frac{B(q^{-1})}{A(q^{-1})}\right]u(t) = 0 \quad i = 1, \ldots, na \quad \text{(C7.5.14a)}$$

$$E\left[\frac{1}{A(q^{-1})}u(t-j)\right]\left[\frac{B^0(q^{-1})}{A^0(q^{-1})} - \frac{B(q^{-1})}{A(q^{-1})}\right]u(t) = 0 \quad j = 1, \ldots, nb \quad \text{(C7.5.14b)}$$

For simplicity consider the case $na = nb = n$.

In the following it will be shown that the only solution of (C7.5.14) is $A = A^0$ and $B = B^0$ provided $u(t)$ is white noise. Clearly the result is relevant for the OEM problem (C7.5.1), at least for large N. It is also relevant for a different problem, as explained below.

Let $\hat{g}_k$, $k = 1, \ldots, N$ denote some consistent estimate of the system's weighting sequence, obtained by one of the many available techniques (for example, by using the least squares method; see the end of Section 3.4). A parametric model $B(q^{-1})/A(q^{-1})$ of the system transfer function may then be determined by minimizing (w.r.t. $\{a_i, b_j\}$) the criterion

$$V = \sum_{k=1}^{N} \{\hat{g}_k - g_k(a_i, b_j)\}^2$$

where $g_k(a_i, b_j)$ is the weighting function sequence of $B(q^{-1})/A(q^{-1})$. For large N, $\hat{g}_k$ is

$$V \approx \sum_{k=1}^{\infty} \{g_k^0 - g_k(a_i, b_j)\}^2 = E\left\{\left[\frac{B^0(q^{-1})}{A^0(q^{-1})} - \frac{B(q^{-1})}{A(q^{-1})}\right]u(t)\right\}^2$$

where $Eu(t)u(s) = \delta_{t,s}$. Therefore, the result stated above is relevant also for the problem of rational approximation of an estimated weighting sequence.

To prove that (C7.5.14) has a unique solution, it will be shown first that there is no solution with $B(z) \equiv 0$. If $B(z) \equiv 0$ then (C7.5.14b) implies that

$$E \frac{1}{A(q^{-1})}u(t-j) \frac{B^0(q^{-1})}{A^0(q^{-1})}u(t) = 0 \quad j = 1, \ldots, n \quad \text{(C7.5.15)}$$

Let $\bar{u}(t) = \frac{B^0(q^{-1})}{A^0(q^{-1})}u(t)$. Then

$$E \frac{1}{A(q^{-1})}u(t-n-1)\bar{u}(t)$$

$$= E \frac{1}{A(q^{-1})}u(t-n-1)[B^0(q^{-1})u(t) - a_1^0\bar{u}(t-1) - \ldots - a_n^0\bar{u}(t-n)]$$

$$= E\left[\frac{1}{A(q^{-1})}u(t-n-1)B^0(q^{-1})u(t)\right] - a_1^0 E \frac{1}{A(q^{-1})}u(t-n)\bar{u}(t)$$

$$- \ldots - a_n^0 E \frac{1}{A(q^{-1})}u(t-1)\bar{u}(t)$$

$$= 0$$

The first term is zero since the input $u(t)$ is white noise. The other terms are also zero, by (C7.5.15). Similarly,

$$E \frac{1}{A(q^{-1})} u(t-k)\bar{u}(t) = 0 \quad \text{all } k \geq 1$$

Thus

$$E\left[\frac{B^0(q^{-1})}{A^0(q^{-1})} u(t)\right]^2 = E\left[\frac{B^0(q^{-1})A(q^{-1})}{A^0(q^{-1})} \frac{1}{A(q^{-1})} u(t)\right][\bar{u}(t)] = 0$$

which implies $B^0(z) \equiv 0$; but this is a contradiction.

Next consider nondegenerate solutions with $B(z) \neq 0$. Since the polynomials A, B need not be coprime, it is convenient to write them in factorized form as

$$A(z) = \bar{A}(z)L(z)$$
$$B(z) = \bar{B}(z)L(z)$$

(C7.5.16)

where $\bar{A}(z)$ and $\bar{B}(z)$ are coprime, and

$$L(z) = 1 + l_1 z + \ldots + l_m z^m \quad m \in [0, n-1]$$

Since the polynomial L (including its degree) is arbitrary, any pair of polynomials A, B can be written as above. Inserting (C7.5.16) into (C7.5.14) gives

$$n \left\{ \begin{pmatrix} 0 & \bar{b}_1 & \ldots & \bar{b}_{n-m} & & & 0 \\ & \ddots & & & \ddots & & \\ 0 & 0 & \bar{b}_1 & \ldots & \bar{b}_{n-m} & & \\ 1 & \bar{a}_1 & \ldots & \bar{a}_{n-m} & 0 & & \\ & \ddots & & & \ddots & & \\ 0 & 1 & \bar{a}_1 & \ldots & \bar{a}_{n-m} & & \end{pmatrix} \right. E \frac{1}{A(q^{-1})\bar{A}(q^{-1})} \begin{pmatrix} u(t-1) \\ \vdots \\ \vdots \\ \vdots \\ u(t-2n+m) \end{pmatrix} \bar{u}(t) = 0$$

$$\underbrace{\phantom{\begin{pmatrix} 0 & \bar{b}_1 & \ldots \end{pmatrix}}}_{\mathcal{S}(\bar{B}, \bar{A})}$$

where

$$\bar{u}(t) = \frac{\bar{A}(q^{-1})B^0(q^{-1}) - A^0(q^{-1})\bar{B}(q^{-1})}{A^0(q^{-1})\bar{A}(q^{-1})} u(t)$$

Since $\bar{A}$ and $\bar{B}$ have no common zeros, the Sylvester matrix $\mathcal{S}(\bar{B}, \bar{A})$, which has dimension $(2n|(2n-m))$, has full rank (see Lemma A.30 and its corollaries). Hence

$$E \frac{1}{\bar{A}(q^{-1})A(q^{-1})} u(t-i)\bar{u}(t) = 0 \quad i = 1, \ldots, 2n - m$$

Introduce the notation

$$H(z) = \bar{A}(z)B^0(z) - A^0(z)\bar{B}(z) \quad (\deg H = 2n - m)$$
$$G(z) = A^0(z)\bar{A}(z) \triangleq 1 + g_1 z + \ldots + g_{2n-m} z^{2n-m}$$

Next note that

$$E \frac{1}{\bar{A}(q^{-1})A(q^{-1})} u(t - 2n + m - 1)\bar{u}(t)$$

$$= E \frac{1}{\bar{A}(q^{-1})A(q^{-1})} u(t - 2n + m - 1)[H(q^{-1})u(t)$$

$$- g_1\bar{u}(t - 1) - g_{2n-m}\bar{u}(t - 2n + m)]$$

$$= 0$$

In a similar manner it can be shown that

$$E \frac{1}{\bar{A}(q^{-1})A(q^{-1})} u(t - i) \frac{H(q^{-1})}{G(q^{-1})} u(t) = 0 \quad \text{all } i \geq 1$$

which implies that

$$E \left[\frac{H(q^{-1})}{G(q^{-1})} u(t)\right]^2 = E\left[\frac{H(q^{-1})\bar{A}(q^{-1})A(q^{-1})}{G(q^{-1})} \frac{1}{\bar{A}(q^{-1})A(q^{-1})} u(t)\right]$$

$$\times \left[\frac{H(q^{-1})}{G(q^{-1})} u(t)\right]$$

$$= 0$$

Thus, $H(z) \equiv 0$, or equivalently $\bar{A}(z)B^0(z) \equiv A^0(z)\bar{B}(z)$. Since $A^0(z)$ and $B^0(z)$ have no common zeros, it is concluded that $A(z) \equiv A^0(z)$ and $B(z) \equiv B^0(z)$.

The above result on the unimodality of the (asymptotic) loss function associated with the OEM can be extended slightly. More exactly, it can be shown that the result continues to hold for a class of ARMA input signals of 'sufficiently low order' (compared to the order of the identified system) (see Söderström and Stoica, 1982).

In the circuits and systems literature the result shown above is sometimes called 'Stearns' conjecture' (see, for example, Widrow and Stearns, 1985, and the references therein). In that context the result is relevant to the design of rational infinite impulse response filters.

Complement C7.6
Unimodality of the PEM loss function for ARMA processes

Consider an ARMA process $y(t)$ given by

$$A^0(q^{-1})y(t) = C^0(q^{-1})e(t) \quad Ee(t)e(s) = \lambda^2 \delta_{t,s}$$

where

$$A^0(q^{-1}) = 1 + a_1^0 q^{-1} + \ldots + a_{na}^0 q^{-na}$$
$$C^0(q^{-1}) = 1 + c_1^0 q^{-1} + \ldots + c_{nc}^0 q^{-nc}$$

248 *Prediction error methods* *Chapter 7*

According to the spectral factorization theorem (see Appendix A6.1), there is no restriction in assuming that $A^0(z)$ and $C^0(z)$ have no common roots and that $A^0(z)C^0(z) \neq 0$ for $|z| \leq 1$. The prediction error (PE) estimates of the (unknown) coefficients of $A^0(q^{-1})$ and $C^0(q^{-1})$ are by definition

$$\hat{\theta} = \arg \min_{\theta} \frac{1}{N} \sum_{t=1}^{N} \varepsilon^2(t, \theta) \qquad \varepsilon(t, \theta) = \frac{A(q^{-1})}{C(q^{-1})} y(t) \tag{C7.6.1}$$

where N denotes the number of data points, $A(q^{-1})$ and $C(q^{-1})$ are polynomials defined similarly to $A^0(q^{-1})$ and $C^0(q^{-1})$, and θ denotes the unknown parameter vector

$$\theta = (a_1 \ldots a_{na} \quad c_1 \ldots c_{nc})^{\mathrm{T}}$$

For simplicity we assume that *na* and *nc* are known. The more general case of unknown *na* and *nc* can be treated similarly (Åström and Söderström, 1974; Stoica and Söderström, 1982a).

In the following it will be shown that for N sufficiently large, the loss function in (C7.6.1) is unimodal. The proof of this neat property follows the same lines as that presented in Complement C7.5 for OEMs. When N tends to infinity the loss function in (C7.6.1) tends to

$$W(\theta) \triangleq E\varepsilon^2(t, \theta)$$

The stationary points of $W(\theta)$ are the solutions of the following equations:

$$\begin{aligned}E\varepsilon(t, \theta) \frac{1}{A(q^{-1})} \varepsilon(t - i, \theta) = 0 \qquad & i = 1, \ldots, na \\ E\varepsilon(t, \theta) \frac{1}{C(q^{-1})} \varepsilon(t - j, \theta) = 0 \qquad & j = 1, \ldots, nc\end{aligned} \tag{C7.6.2}$$

Since the solutions $A(q^{-1})$ and $C(q^{-1})$ of (C7.6.2) are not necessarily coprime polynomials, it is convenient to write them in factorized form as

$$\begin{aligned} A(z) &= \bar{A}(z)D(z) \\ C(z) &= \bar{C}(z)D(z) \end{aligned} \tag{C7.6.3}$$

where $\bar{A}(z)$ and $\bar{C}(z)$ are coprime, and

$$D(z) = 1 + d_1 z + \ldots + d_{nd} z^{nd} \qquad nd \in [0, \min(na, nc)]$$

Note that $D(z) \equiv 1$ for coprime polynomials $A(z)$, $C(z)$.

Using (C7.6.3), (C7.6.2) can be rewritten as

$$\underbrace{\begin{pmatrix} 1 & \bar{c}_1 \ldots \bar{c}_{nc-nd} & & \\ 0 & \ddots & \ddots & 0 \\ & & 1 & \bar{c}_1 \ldots \bar{c}_{nc-nd} \\ 1 & \bar{a}_1 \ldots \bar{a}_{na-nd} & & \\ 0 & \ddots & \ddots & 0 \\ & & 1 & \bar{a}_1 \ldots \bar{a}_{na-nd} \end{pmatrix}}_{\mathscr{S}(\bar{C}, \bar{A})} \left. \begin{matrix} \}na \\ \\ \}nc \end{matrix} \right. E \left\{ \begin{pmatrix} \frac{1}{\bar{A}(q^{-1})\bar{C}(q^{-1})} \varepsilon(t-1, \theta) \\ \vdots \\ \frac{1}{\bar{A}(q^{-1})\bar{C}(q^{-1})} \varepsilon(t-na-nc+nd, \theta) \end{pmatrix} \varepsilon(t, \theta) \right\} y = 0 \tag{C7.6.4}$$

The Sylvester $\mathscr{S}(\bar{C}, \bar{A})$ $(na + nc)|(na + nc - nd)$ matrix has full rank (see Lemma A.30 and its corollaries). Thus, (C7.6.4) implies

$$E \frac{1}{\bar{A}(q^{-1})C(q^{-1})} \varepsilon(t - i, \theta)\varepsilon(t, \theta) = 0 \quad i = 1, \ldots, na + nc - nd \quad \text{(C7.6.5)}$$

Introduce

$$G(z) \triangleq \bar{C}(z)A^0(z) = 1 + g_1 z + \ldots + g_{ng} z^{ng} \quad ng \triangleq na + nc - nd$$

With this notation, it follows from (C7.6.5) and (C7.6.1) that

$$\varepsilon(t, \theta) = \frac{A(q^{-1})}{C(q^{-1})} y(t) = \frac{\bar{A}(q^{-1})C^0(q^{-1})}{G(q^{-1})} e(t) \quad \text{(C7.6.6)}$$

and

$$E \frac{1}{\bar{A}(q^{-1})C(q^{-1})} \varepsilon(t - ng - 1, \theta)\varepsilon(t, \theta)$$

$$= E \frac{1}{\bar{A}(q^{-1})C(q^{-1})} \varepsilon(t - ng - 1, \theta)[\bar{A}(q^{-1})C^0(q^{-1})e(t) - g_1\varepsilon(t - 1, \theta)$$

$$\ldots - g_{ng}\varepsilon(t - ng, \theta)]$$

$$= E \frac{1}{\bar{A}(q^{-1})C(q^{-1})} \varepsilon(t - ng - 1, \theta)\bar{A}(q^{-1})C^0(q^{-1})e(t) = 0$$

One can show in the same manner that

$$E \frac{1}{\bar{A}(q^{-1})C(q^{-1})} \varepsilon(t - k, \theta)\varepsilon(t, \theta) = 0 \quad \text{for all } k \geq 1$$

Then it readily follows that

$$E\varepsilon(t - k, \theta)\varepsilon(t, \theta) = E\left\{\bar{A}(q^{-1})C(q^{-1})\left[\frac{1}{\bar{A}(q^{-1})C(q^{-1})} \varepsilon(t - k, \theta)\right]\varepsilon(t, \theta)\right\}$$

$$= 0 \quad \text{for all } k \geq 1$$

Thus $\varepsilon(t, \theta)$, with θ satisfying (C7.6.2), is a sequence of independent random variables. In view of (C7.6.6) this is possible if and only if

$$\bar{A}(z)C^0(z) \equiv A^0(z)\bar{C}(z) \quad \text{(C7.6.7)}$$

Since $A^0(z)$ and $C^0(z)$ are coprime polynomials, the only solution to (C7.6.7) is $A(z) = A^0(z)$, $C(z) = C^0(z)$, which concludes the proof.

Complement C7.7
Exact maximum likelihood estimation of AR and ARMA parameters

The first part of this complement deals with ARMA processes. The second part will specialize to AR processes, for which the analysis will be much simplified.

ARMA processes

Consider an ARMA process $y(t)$ given by

$$A(q^{-1})y(t) = C(q^{-1})e(t) \qquad Ee(t)e(s) = \lambda^2 \delta_{t,s} \qquad \text{(C7.7.1)}$$

where

$$A(q^{-1}) = 1 + a_1 q^{-1} + \ldots + a_{na} q^{-na}$$
$$C(q^{-1}) = 1 + c_1 q^{-1} + \ldots + c_{nc} q^{-nc}$$

The unknown parameters λ and

$$\theta = (a_1 \ldots a_{na} \; c_1 \ldots c_{nc})^{\mathrm{T}}$$

are to be estimated from a sample of observations

$$Y \triangleq (y(1) \ldots y(N))^{\mathrm{T}}$$

It is assumed that the polynomials $A(z)$ and $C(z)$ have no common roots and that $A(z)C(z) \neq 0$ for $|z| \leq 1$. According to the spectral factorization theorem (see Appendix A6.1), this assumption does not introduce any restriction. It is also assumed that $e(t)$ is Gaussian distributed. Since $y(t)$ depends linearly on the white noise sequence it follows that the data vector Y is also Gaussian distributed. Thus, the conditional probability function $p(Y|\theta, \lambda)$ is given by

$$p(Y|\theta, \lambda) = (2\pi\lambda^2)^{-N/2} (\det \Omega)^{-1/2} \exp\left(-\frac{1}{2\lambda^2} Y^{\mathrm{T}} \Omega^{-1} Y\right) \qquad \text{(C7.7.2)}$$

where

$$\Omega \triangleq \frac{1}{\lambda^2} \mathrm{cov}(Y|\theta)$$

The maximum likelihood estimates (MLE) of θ and λ are obtained by maximizing (C7.7.2). A variety of search procedures can be used for this purpose. They all require the evaluation of $p(Y|\theta, \lambda)$ for given θ and λ. As shown below, the evaluation of the covariance matrix Ω can be made with a modest computational effort. A direct evaluation of $p(Y|\theta, \lambda)$ would, however, require $O(N^2)$ arithmetic operations if the Toeplitz structure of Ω (i.e. Ω_{ij} depends only on $|i - j|$) is exploited, and $O(N^3)$ operations otherwise. This could be a prohibitive computational burden for most applications. Fortunately, there are fast procedures for evaluating $p(Y|\theta, \lambda)$, which need only $O(N)$ arithmetic operations. The aim of this complement is to describe the procedure of Ansley (1979), which is not only quite simple conceptually but seems also to be one of the most efficient numerical procedures for evaluating $p(Y|\theta, \lambda)$. Other fast procedures are presented in Dent (1977), Gardner et al. (1980), Gueguen and Scharf (1980), Ljung and Box (1979), Newbold (1974), and Dugré et al. (1986).

Let

$$z(t) = \begin{cases} y(t) & 1 \leq t \leq m \\ A(q^{-1})y(t) & m+1 \leq t \leq N \end{cases} \quad m = \max(na, nc) \quad \text{(C7.7.3)}$$

$$Z = (z(1) \ \ldots \ z(N))^{\mathrm{T}}$$

and let

$$\Sigma = \frac{1}{\lambda^2} \mathrm{cov}(Z|\theta)$$

For the transformation (C7.7.3) it follows that

$$\frac{\partial Z}{\partial Y} = \begin{pmatrix} 1 & & & & & \\ & \ddots & & & & \\ 0 & & 1 & & 0 & \\ a_{na} & \ldots & a_1 & 1 & & \\ 0 & & & \ddots & & \\ & & & a_{na} & \ldots & a_1 & 1 \end{pmatrix}$$

Hence the Jacobian of (C7.7.3) (i.e. the determinant of the matrix $\partial Z/\partial Y$) is equal to one. It therefore follows that

$$p(Y|\theta, \lambda) = (2\pi\lambda^2)^{-N/2}(\det \Sigma)^{-1/2} \exp\left(-\frac{1}{2\lambda^2} Z^{\mathrm{T}} \Sigma^{-1} Z\right) \quad \text{(C7.7.4)}$$

Next consider the evaluation of Σ for given θ and λ. Let $s \geq 0$. Then:

(a) For $1 \leq t \leq m, t+s \leq m$,
$$Ez(t)z(t+s) = Ey(t)y(t+s) \triangleq r_s \quad \text{(C7.7.5a)}$$

(b) For $1 \leq t \leq m, t+s > m$,
$$Ez(t)z(t+s) = Ey(t)C(q^{-1})e(t+s) \triangleq \alpha_s \quad \text{(C7.7.5b)}$$

(c) For $t > m$,
$$Ez(t)z(t+s) = EC(q^{-1})e(t)C(q^{-1})e(t+s)$$
$$= \begin{cases} (c_0 c_s + \ldots + c_{nc-s}c_{nc})\lambda^2 & s \leq nc \\ 0 & s > nc \end{cases} \quad \text{(C7.7.5c)}$$

The covariances $\{r_s\}$ of $y(t)$ can be evaluated as follows. Let

$$\varrho_k = Ey(t)e(t-k)$$

Multiplying (C7.7.1) by $y(t-k)$ and taking expectations,

$$r_k + a_1 r_{k-1} + \ldots + a_{na} r_{k-na} = c_k \varrho_0 + \ldots + c_{nc} \varrho_{nc-k} \quad \text{(C7.7.6)}$$

where it is assumed that $0 \leq k \leq nc$. Further, from (C7.7.5b),

$$\alpha_k = Ey(t-k)C(q^{-1})e(t) = c_k\varrho_0 + \ldots + c_{nc}\varrho_{nc-k} \quad (k \geq 0) \tag{C7.7.7a}$$

Clearly $\alpha_k = 0$ for $k > nc$. Now, multiplying (C7.7.1) by $e(t-j)$ and taking expectations produces

$$\varrho_j + a_1\varrho_{j-1} + \ldots + a_{na}\varrho_{j-na} = \begin{cases} 0 & j > nc \\ \lambda^2 c_j & 0 \leq j \leq nc \end{cases} \tag{C7.7.7b}$$

Since $\varrho_j = 0$ for $j < 0$, the sequence $\{\varrho_j\}$ can be obtained easily from (C7.7.7b). Note that to compute the sequence $\{\alpha_k\}$ requires only $\{\varrho_j\}_{j=0}^{nc}$. Thus the following procedure can be used to evaluate $\{r_s\}$, for given θ and λ:

(a) Determine $\{\varrho_j\}$, $j \in [0, nc]$ ($\varrho_j = 0$, $j < 0$) from (C7.7.7b).
    Evaluate $\{\alpha_k\}$, $k \in [0, nc]$ ($\alpha_k = 0$, $k > nc$) from (C7.7.7a).
(b) Solve the following linear system (cf. (C7.7.6)) to obtain $\{r_k\}$, $k \in [0, na]$:

$$\left\{ \begin{pmatrix} 1 & & & 0 \\ a_1 & 1 & & 0 \\ \vdots & & \ddots & \\ a_{na} & a_{na-1} & \ldots & 1 \end{pmatrix} + \begin{pmatrix} 0 & a_1 & \ldots & a_{na} \\ 0 & a_2 & & a_{na} \\ \vdots & & \ddots & \\ 0 & a_{na} & & 0 \\ 0 & & & \end{pmatrix} \right\} \begin{pmatrix} r_0 \\ \cdot \\ \cdot \\ \cdot \\ r_{na} \end{pmatrix} = \begin{pmatrix} \alpha_0 \\ \cdot \\ \cdot \\ \cdot \\ \alpha_{na} \end{pmatrix}$$

The other covariances $\{r_k\}$, $k > na$, can then be obtained from (C7.7.6).

Note that the sequence $\{\alpha_k\}$ is obtained as a by-product of the procedure above. Thus there is a complete procedure for evaluation of Σ.

Next note that Σ is a *banded* matrix, with the band width equal to $2m$. Now the idea behind the transformation (C7.7.3) can be easily understood. Since in general $m \ll N$, the lower triangular Cholesky factor L of Σ,

$$\Sigma = LL^{\mathrm{T}} \tag{C7.7.8}$$

is also a banded matrix with the band width equal to m, and it can be determined in $O(N)$ arithmetic operations (see Lemma A.6 and its proof; see also Friedlander, 1983). Let the $(N|1)$ vector e be defined by

$$Le = Z \tag{C7.7.9}$$

Computation of e can be done in $O(N)$ operations. Evaluation of $\det L$ also needs $O(N)$ operations.

Inserting (C7.7.8) and (C7.7.9) in (C7.7.4) produces

$$p(Y|\theta, \lambda) = (2\pi\lambda^2)^{-N/2}(\det L)^{-1} \exp\left(-\frac{1}{2\lambda^2} e^{\mathrm{T}} e\right)$$

Thus the log-likelihood function is given by

$$L(\theta, \lambda) \triangleq \log p(Y|\theta, \lambda) = \mathrm{const} - \frac{N}{2} \log \lambda^2 - \log(\det L) - \frac{1}{2\lambda^2} e^{\mathrm{T}} e \tag{C7.7.10}$$

It is not difficult to see that Σ, and hence the matrix L, do not depend on λ. Differentiation of $L(\theta, \lambda)$ with respect to λ^2 then gives

$$\frac{\partial L(\theta, \lambda)}{\partial \lambda^2} = -\frac{N}{2}\frac{1}{\lambda^2} + \frac{1}{2\lambda^4} e^T e$$

Thus the MLE of λ^2 is

$$\hat{\lambda}^2 = \frac{1}{N} e^T e \tag{C7.7.11}$$

Inserting $\hat{\lambda}^2$ into (C7.7.10) gives the concentrated log-likelihood function

$$\max_{\lambda^2} L(\theta, \lambda) = \text{const} - \frac{N}{2} \log\left(\frac{1}{N} e^T e\right) - \log(\det L)$$

$$= \text{const} - \frac{N}{2} \log\left(\frac{1}{N} e^T e (\det L)^{2/N}\right)$$

Thus the exact MLE of θ is given by

$$\hat{\theta} = \arg\min_{\theta} \{\bar{e}^T \bar{e}\} \tag{C7.7.12}$$

where $\bar{e} = (\det L)^{1/N} e$. A procedure for computing $\bar{e}$ in $O(N)$ arithmetic operations was described above. A variety of numerical minimization algorithms (based on loss function evaluations only) can be used to solve the optimization problem (C7.7.12).

AR processes

For AR processes $C(z) \equiv 1$, and the previous procedure for exact evaluation of the likelihood function simplifies considerably. Indeed, for $nc = 0$ the matrix Σ becomes

$$\Sigma = \begin{pmatrix} R & 0 \\ 0 & I \end{pmatrix}$$

where

$$R = \frac{1}{\lambda^2} \text{cov}(\bar{y}|\theta) \quad \bar{y} = (y(1) \ldots y(na))^T \quad \theta = (a_1 \ldots a_{na})^T$$

Thus for AR processes the likelihood function is given (cf. (C7.7.4)) by

$$p(Y|\theta, \lambda) = (2\pi\lambda^2)^{-N/2} (\det R)^{-1/2} \exp\left\{-\frac{1}{2\lambda^2}\left(\bar{y}^T R^{-1} \bar{y} + \sum_{t=na+1}^{N} [A(q^{-1})y(t)]^2\right)\right\}$$

Both R^{-1} and det R can be computed in $O(na^2)$ arithmetic operations from λ and θ by using the Levinson–Durbin algorithm (see, e.g., Complement C8.2).

Note that it is possible to derive a closed form expression for R^{-1} as a function of θ. At the end of this complement a proof of this interesting result is given, patterned after Godolphin and Unwin (1983) (see also Gohberg and Heinig, 1974; Kailath, Vieira and Morf, 1978).

Let

$$A = \begin{pmatrix} 1 & a_1 & \cdots & a_{na-1} \\ & \ddots & \ddots & \vdots \\ & & \ddots & a_1 \\ 0 & & & 1 \end{pmatrix} \quad B = \begin{pmatrix} a_{na} & a_{na-1} & \cdots & a_1 \\ & \ddots & \ddots & \vdots \\ & & \ddots & a_{na-1} \\ 0 & & & a_{na} \end{pmatrix}$$

Also introduce $\{\alpha_k\}$ through

$$\sum_{i=0}^{\infty} \alpha_i z^i \triangleq \frac{1}{A(z)} \tag{C7.7.13}$$

with $\alpha_0 = 1$ and $\alpha_k = 0$ for $k < 0$. This definition is slightly different from (C7.7.5b) (α_i in (C7.7.13) is equal to α_{-i} in (C7.7.5b)). Similarly, introduce $a_0 = 1$ and $a_k = 0$ for $k < 0$ or $k > na$. Multiplying both sides of

$$A(q^{-1})y(t) = e(t)$$

by $y(t - k)$, $k \geq 1$, and $y(t + k)$, $k \geq 0$, and taking expectations,

$$\begin{aligned} r_k + a_1 r_{k-1} + \ldots + a_{na} r_{k-na} &= 0 & k \geq 1 \\ r_k + a_1 r_{k+1} + \ldots + a_{na} r_{k+na} &= \alpha_k \lambda^2 & k \geq 0 \end{aligned} \tag{C7.7.14}$$

Next note that (C7.7.13) implies that

$$1 \equiv \left(\sum_{i=0}^{na} a_i z^i \right) \left(\sum_{j=0}^{\infty} \alpha_j z^j \right) = \sum_{j=0}^{\infty} \sum_{i=0}^{na} a_i \alpha_j z^{i+j}$$

$$= \sum_{k=0}^{\infty} \left(\sum_{i=0}^{na} a_i \alpha_{k-i} \right) z^k$$

Thus

$$\sum_{i=0}^{na} a_i \alpha_{k-i} = 0 \quad \text{for } k \geq 1 \tag{C7.7.15}$$

From (C7.7.15) it follows that

$$A^{-1} = \begin{pmatrix} 1 & \alpha_1 & \cdots & \alpha_{na-1} \\ & \ddots & \ddots & \vdots \\ & & \ddots & \alpha_1 \\ 0 & & & 1 \end{pmatrix} \tag{C7.7.16}$$

To see this note that the i, j element of the matrix AA^{-1}, with A^{-1} given by (C7.7.16), is (for $i, j = 1, \ldots, na$)

$$\sum_{k=1}^{na} a_{k-i}\alpha_{j-k} = \sum_{p=1-i}^{na-i} a_p \alpha_{j-i-p} = \sum_{p=0}^{na} a_p \alpha_{(j-i)-p}$$

$$= \begin{cases} 0 & \text{for } j - i < 0 \text{ since } \alpha_k = 0 \text{ for } k < 0 \\ 1 & \text{for } i = j \\ 0 & \text{for } j - i > 0 \text{ by (C7.7.15)} \end{cases}$$

which proves that the matrix in (C7.7.16) is really the inverse of A. Let

$$\tilde{R} \triangleq \frac{1}{\lambda^2} \begin{pmatrix} r_{na} & r_{na+1} & \cdots & r_{2na-1} \\ r_{na-1} & r_{na} & \cdots & r_{2na-2} \\ \vdots & & & \\ r_1 & r_2 & \cdots & r_{na} \end{pmatrix}$$

It will be shown that

$$RA^T + \tilde{R}B = A^{-1} \tag{C7.7.17a}$$

$$RB^T + \tilde{R}A = 0 \tag{C7.7.17b}$$

Evaluating the i, j element of the left-hand side in (C7.7.17a) gives

$$\lambda^2[RA^T + \tilde{R}B]_{ij} = \sum_{k=1}^{na} r_{i-k}a_{k-j} + \sum_{k=1}^{na} r_{na+k-i}a_{na+k-j}$$

$$= \sum_{k=j}^{na} r_{i-k}a_{k-j} + \sum_{k=1}^{j} r_{na+k-i}a_{na+k-j}$$

$$= \sum_{p=0}^{na-j} r_{i-p-j}a_p + \sum_{p=na-j+1}^{na} r_{i-p-j}a_p$$

$$= \sum_{p=0}^{na} r_{i-p-j}a_p$$

If $j \geq i$ this is equal to

$$\sum_{p=0}^{na} r_{p-i+j}a_p = \lambda^2 \alpha_{j-i} = \lambda^2[A^{-1}]_{ij}$$

according to (C7.7.14). Similarly, for $i > j$ the first part of (C7.7.14) gives

$$\sum_{p=0}^{na} r_{i-p-j}a_p = 0 = \lambda^2[A^{-1}]_{ij}$$

Hence (C7.7.17a) is verified. The proof of (C7.7.17b) proceeds in a similar way. The i, j element of the left-hand side is given by

$$\lambda^2[RB^T + \tilde{R}A]_{ij} = \sum_{k=1}^{na} r_{i-k}a_{na+j-k} + \sum_{k=1}^{na} r_{na+k-i}a_{j-k}$$

$$= \sum_{k=j}^{na} r_{i-k}a_{na+j-k} + \sum_{k=1}^{j} r_{na+k-i}a_{j-k}$$

$$= \sum_{p=j}^{na} r_{i-na-j+p}a_p + \sum_{p=0}^{j-1} r_{na-i+j-p}a_p$$

$$= \sum_{p=0}^{na} r_{na-i+j-p}a_p$$

$$= 0$$

The last equality follows from (C7.7.14) since $na - i + j \geq 1$. This proves (C7.7.17b). The relations (C7.7.17) give easily

$$RA^T - RB^TA^{-1}B = A^{-1}$$

or

$$R(A^TA - B^TA^{-1}BA) = I$$

It is easy to show that the matrices A and B commute. Hence it may be concluded from the above equation that

$$R^{-1} = (A^TA - B^TB) \qquad (C7.7.18)$$

The above expression for R^{-1} is sometimes called the Gohberg–Heinig–Semencul formula.

Complement C7.8
ML estimation from noisy input–output data

Consider the system shown in Figure C7.8.1; it can be described by the equations

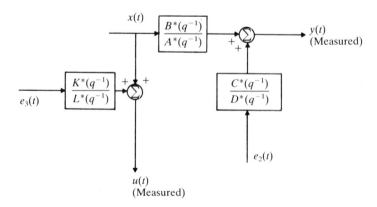

FIGURE C7.8.1 A system with noise-corrupted input and output signals.

$$y(t) = \frac{B^*(q^{-1})}{A^*(q^{-1})} x(t) + \frac{C^*(q^{-1})}{D^*(q^{-1})} e_2(t)$$
$$u(t) = x(t) + \frac{K^*(q^{-1})}{L^*(q^{-1})} e_3(t)$$
(C7.8.1)

where $A^*(q^{-1})$ etc. are polynomials in the unit delay operator q^{-1}, and

$$Ee_i(t)e_i(s) = \lambda_i^2 \delta_{t,s} \quad i = 2, 3$$
$$Ee_2(t)e_3(s) = 0 \quad \text{for all } t, s$$

The special feature of the system (C7.8.1) is that both the output *and* the input measurements are corrupted by noise. The noise corrupting the input may be due, for example, to the errors made by a measurement device. The problem of identifying systems from noise-corrupted input–output data is often called *errors-in-variables problem*.

In order to estimate the parameters of the system, the noise-free input $x(t)$ must be parametrized. Assume that $x(t)$ is an ARMA process,

$$x(t) = \frac{G^*(q^{-1})}{H^*(q^{-1})} e_1(t)$$
(C7.8.2)

where G^* and H^* are polynomials in q^{-1}, and

$$Ee_1(t)e_1(s) = \lambda_1^2 \delta_{t,s}$$
$$Ee_1(t)e_i(s) = 0 \quad i = 2, 3, \text{ for all } t, s$$

The input signals occurring in the normal operating of many processes can be well described by ARMA models. Assume also that all the signals in the system are stationary, and that the description (C7.8.1) of the system is minimal.

Let θ denote the vector of unknown parameters

$$\theta = (\text{coefficients of } \{A^*, B^*, C^*, D^*, G^*, H^*, K^*, L^*\}; \lambda_1, \lambda_2, \lambda_3)^T$$

and let U and Y denote the available measurements of the input and output, respectively,

$$U = (u(1) \ldots u(N))^T$$
$$Y = (y(1) \ldots y(N))^T$$

The MLE of θ is obtained by maximizing the conditional probability of the data

$$p(Y, U|\theta)$$
(C7.8.3)

with respect to θ. In general, an algorithm for evaluating (C7.8.3) will be needed.

The system equations (C7.8.1), (C7.8.2) can be written as

$$A(q^{-1})y(t) = C(q^{-1})e_1(t) + D(q^{-1})e_2(t)$$
$$B(q^{-1})u(t) = G(q^{-1})e_1(t) + H(q^{-1})e_3(t)$$
(C7.8.4)

The coefficients of A, B, etc., can be determined from θ in a straightforward way. Let na, etc., denote the degree of $A(q^{-1})$, etc., and let

$$m = \max(na, nb, nc, nd, ng, nh)$$

258 *Prediction error methods* **Chapter 7**

The results of Complement C7.7 could be used to derive an 'exact algorithm' for the evaluation of (C7.8.3). The algorithm so obtained would have a simple basic structure but the detailed expressions involved would be rather cumbersome. In the following, a much simpler approximate algorithm is presented. The larger N, the more exact the result obtained with this approximate algorithm will be.

Let

$$\tilde{z}(t) = \begin{pmatrix} A(q^{-1})y(t) \\ B(q^{-1})u(t) \end{pmatrix} \tag{C7.8.5}$$

Clearly $\tilde{z}$ is a bivariate MA process of order not greater than m (see (C7.8.4)). Thus, the covariance matrix

$$\tilde{\Sigma} = E \begin{pmatrix} \tilde{z}(1) \\ \tilde{z}(2) \\ \vdots \end{pmatrix} (\tilde{z}^T(1) \quad \tilde{z}^T(2) \quad \ldots)$$

is banded, with the band width equal to $4m + 1$. Now, since $\tilde{z}(t)$ is an MA(m) process it follows from the spectral factorization theorem that there exists a unique (2|2) polynomial matrix $S(q^{-1})$, with $S(0) = I$, deg $S = m$, and det $S(z) \neq 0$ for $|z| \leq 1$, such that

$$\tilde{z}(t) = S(q^{-1})\varepsilon(t) \tag{C7.8.6}$$

where

$$E\varepsilon(t)\varepsilon^T(s) = \Lambda\delta_{t,s} \quad \Lambda > 0$$

(see, e.g., Anderson and Moore, 1979).

From (C7.8.5), (C7.8.6) it follows that the prediction errors (or the innovations) of the bivariate process $(y^T(t) \quad u^T(t))^T$ are given by

$$\varepsilon(t) = S^{-1}(q^{-1}) \begin{pmatrix} A(q^{-1})y(t) \\ B(q^{-1})u(t) \end{pmatrix} \tag{C7.8.7}$$

The initial conditions needed to start (C7.8.7) may be chosen to be zero. Next, the likelihood function can be evaluated approximately (cf. (7.39)) as

$$p(Y, U|\theta) \approx (2\pi)^{-N}(\det \Lambda)^{-N/2} \exp\left\{-\frac{1}{2}\sum_{t=1}^{N} \varepsilon^T(t)\Lambda^{-1}\varepsilon(t)\right\}$$

There remains the problem of determining $S(q^{-1})$ and Λ for a given θ. A numerically efficient procedure for achieving this task is the Cholesky factorization of the infinite-dimensional covariance matrix $\tilde{\Sigma}$.

First note that $\tilde{\Sigma}$ can be readily evaluated as a function of θ. For $s \geq 0$,

$E\tilde{z}(t)\tilde{z}^T(t+s)$

$$= \begin{pmatrix} \lambda_1^2(c_0 c_s + \ldots + c_{nc-s} c_{nc}) + \lambda_2^2(d_0 d_s + \ldots + d_{nd-s} d_{nd}) & \lambda_1^2(c_0 g_s + \ldots + c_{ng-s} g_{ng}) \\ \lambda_1^2(g_0 c_s + \ldots + g_{nc-s} c_{nc}) & \lambda_1^2(g_0 g_s + \ldots + g_{ng-s} g_{ng}) + \lambda_3^2(h_0 h_s + \ldots + h_{nh-s} h_{nh}) \end{pmatrix}$$

where $c_k = 0$ for $k < 0$ and $k > nc$, and similarly for the other coefficients. A numerically efficient algorithm for computation of the Cholesky factorization of $\tilde{\Sigma}$,

$$\tilde{\Sigma} = LL^T$$

is presented by Friedlander (1983); see also Lemma A.6 and its proof. L is a lower triangular banded matrix. The (2|2) block entries in the rows of L converge to the coefficients of the MA description (C7.8.6). The number of block rows which need to be computed until convergence is obtained is in general much smaller than N. Computation of a block row of L requires $O(m)$ arithmetic operations. Note that since $S(q^{-1})$ obtained in this way is guaranteed to be invertible (i.e. det $S(z) \neq 0$ for $|z| \leq 1$), the vector θ need not be constrained in the course of optimization. Finally note that the time-invariant innovation representation (C7.8.7) of the process $(y^T(t) \; u^T(t))^T$ can also be found by using the stationary Kalman filter (see Söderström, 1981).

This complement has shown how to evaluate the criterion (C7.8.3). To find the parameter estimates one can use any standard optimization algorithm based on criterion evaluations only. See Dennis and Schnabel (1983), and Gill *et al.* (1981) for examples of such algorithms.

For a further analysis of how to identify systems when the input is corrupted by noise, see Anderson and Deistler (1984), Anderson (1985), and Stoica and Nehorai (1987c).

Chapter 8
INSTRUMENTAL VARIABLE METHODS

8.1 Description of instrumental variable methods

The least squares method was introduced for static models in Section 4.1 and for dynamic models in Section 7.1. It is easy to apply but has a substantial drawback: the parameter estimates are consistent only under restrictive conditions. It was mentioned in Section 7.1 that the LS method could be modified in different ways to overcome this drawback. This chapter presents the modification leading to the class of instrumental variable (IV) methods. The idea is to modify the normal equations.

This section, which is devoted to a general description of the IV methods, is organized as follows. First the model structure is introduced. Next, for completeness, a brief review is given of the LS method. Then various classes of IV methods are defined and it is shown how they can be viewed as generalizations of the LS method.

Model structure

The IV method is used to estimate the system dynamics (the transfer function from the input $u(t)$ to the output $y(t)$). In this context the model structure (6.24) will be used:

$$y(t) = \phi^T(t)\theta + \varepsilon(t) \qquad (8.1)$$

where $y(t)$ is the ny-dimensional output at time t, $\phi^T(t)$ is an $(ny|n\theta)$ dimensional matrix whose elements are delayed input and output components, θ is a $n\theta$-dimensional parameter vector and $\varepsilon(t)$ is the equation error. It was shown in Chapter 6 that the model

$$A(q^{-1})y(t) = B(q^{-1})u(t) + \varepsilon(t) \qquad (8.2)$$

where $A(q^{-1})$, $B(q^{-1})$ are the polynomial matrices

$$\begin{aligned}A(q^{-1}) &= I + A_1 q^{-1} + \ldots + A_{na} q^{-na} \\ B(q^{-1}) &= B_1 q^{-1} + \ldots + B_{nb} q^{-nb}\end{aligned} \qquad (8.3)$$

can be rewritten in the form (8.1). If all entries of matrix coefficients $\{A_i, B_j\}$ are unknown, then the so-called full polynomial form is obtained (see Example 6.3). In this case

Section 8.1   *Description of instrumental variable methods*   261

$$\phi^T(t) = \begin{pmatrix} \varphi^T(t) & & 0 \\ & \ddots & \\ 0 & & \varphi^T(t) \end{pmatrix} \tag{8.4a}$$

$$\varphi^T(t) = (-y^T(t-1) \ldots -y^T(t-na) \; u^T(t-1) \ldots u^T(t-nb)) \tag{8.4b}$$

$$\theta = \begin{pmatrix} \theta^1 \\ \vdots \\ \theta^{ny} \end{pmatrix} \quad \begin{pmatrix} (\theta^1)^T \\ \vdots \\ (\theta^{ny})^T \end{pmatrix} = (A_1 \ldots A_{na} \; B_1 \ldots B_{nb}) \tag{8.5}$$

In most of the analysis which follows it is assumed that the true system is given by

$$y(t) = \phi^T(t)\theta_0 + v(t) \tag{8.6}$$

where $v(t)$ is a stochastic disturbance term and θ_0 is the vector of 'true' parameters.

The LS method revisited

Recalling (7.4), the simple least squares (LS) estimate of θ_0 is given by

$$\hat{\theta} = \left[\sum_{t=1}^{N} \phi(t)\phi^T(t)\right]^{-1} \left[\sum_{t=1}^{N} \phi(t)y(t)\right] \tag{8.7}$$

Using the true system description (8.6), the difference between the estimate $\hat{\theta}$ and the true value θ_0 can be determined (cf. (7.6)):

$$\hat{\theta} - \theta_0 = \left[\frac{1}{N}\sum_{t=1}^{N} \phi(t)\phi^T(t)\right]^{-1} \left[\frac{1}{N}\sum_{t=1}^{N} \phi(t)v(t)\right] \tag{8.8}$$

When N tends to infinity this becomes

$$\hat{\theta} - \theta_0 = [E\phi(t)\phi^T(t)]^{-1}[E\phi(t)v(t)] \tag{8.9}$$

Thus the LS estimate $\hat{\theta}$, (8.7), will have an asymptotic bias (or expressed in another way, it will not be consistent) unless

$$E\phi(t)v(t) = 0 \tag{8.10}$$

However, noting that $\phi(t)$ depends on the output and thus implicitly on past values of $v(t)$ through (8.6), it is seen that (8.10) is quite restrictive. In fact, it can be shown that in general (8.10) is satisfied if and essentially only if $v(t)$ is white noise. This was illustrated in Examples 2.3 and 2.4. This disadvantage of the LS estimate can be seen as the motivation for introducing the instrumental variable method.

Basic IV methods

The idea underlying the introduction of the IV estimates of the parameter vector θ_0 may be explained in several ways. A simple version is as follows. Assume that $Z(t)$ is a

$(nz|ny)$ matrix, the entries of which are signals *uncorrelated* with the disturbance $v(t)$. Then one may try to estimate the parameter vector θ of the model (8.1) by exploiting this property, which means that θ is required to satisfy the following system of linear equations:

$$\frac{1}{N}\sum_{t=1}^{N} Z(t)\varepsilon(t) = \frac{1}{N}\sum_{t=1}^{N} Z(t)[y(t) - \phi^T(t)\theta] = 0 \qquad (8.11)$$

If $nz = n\theta$, then (8.11) gives rise to the so-called *basic* IV estimate of θ:

$$\boxed{\hat{\theta} = \left[\sum_{t=1}^{N} Z(t)\phi^T(t)\right]^{-1}\left[\sum_{t=1}^{N} Z(t)y(t)\right]} \qquad (8.12)$$

where the inverse is assumed to exist. The elements of the matrix $Z(t)$ are usually called the *instruments*. They can be chosen in different ways (as exemplified below) subject to certain conditions guaranteeing the consistency of the estimate (8.12). These conditions will be specified later. Evidently the IV estimate (8.12) is a generalization of the LS estimate (8.7): for $Z(t) = \phi(t)$, (8.12) reduces to (8.7).

Extended IV methods

The *extended* IV estimates of θ_0 are obtained by generalizing (8.12) in two directions. Such IV estimation methods allow for an augmented $Z(t)$ matrix (i.e. one can have $nz \geq n\theta$) as well as a prefiltering of the data. The extended IV estimate is given by

$$\boxed{\hat{\theta} = \arg\min_{\theta} \left\|\left[\sum_{t=1}^{N} Z(t)F(q^{-1})\phi^T(t)\right]\theta - \left[\sum_{t=1}^{N} Z(t)F(q^{-1})y(t)\right]\right\|_Q^2} \qquad (8.13)$$

Here $Z(t)$ is the IV matrix of dimension $(nz|ny)$ $(nz \geq n\theta)$, $F(q^{-1})$ is an $(ny|ny)$ dimensional, asymptotically stable (pre-)filter and $\|x\|_Q^2 = x^T Q x$, where Q is a positive definite weighting matrix. When $F(q^{-1}) \equiv I$, $nz = n\theta$, $(Q = I)$, the basic IV estimate (8.12) is obtained. Note that the estimate (8.13) is the weighted least squares solution of an overdetermined linear system of equations. (There are nz equations and $n\theta$ unknowns.) The solution is readily found to be

$$\hat{\theta} = (R_N^T Q R_N)^{-1} R_N^T Q r_N \qquad (8.14a)$$

where

$$R_N = \frac{1}{N}\sum_{t=1}^{N} Z(t)F(q^{-1})\phi^T(t) \qquad (8.14b)$$

$$r_N = \frac{1}{N}\sum_{t=1}^{N} Z(t)F(q^{-1})y(t) \qquad (8.14c)$$

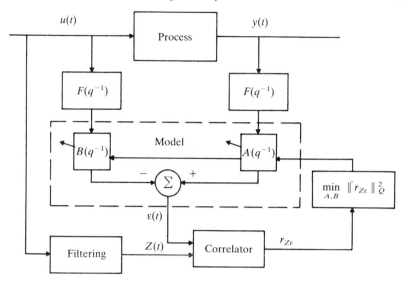

FIGURE 8.1 Block diagram of the extended instrumental variable method. $r_{Z\varepsilon}$ denotes the sample correlation between $Z(t)$ and $\varepsilon(t)$.

This form of the solution is merely of theoretical interest. It is not suitable for implementation. Appropriate numerical algorithms for implementing the extended IV estimate (8.13) are discussed in Sections 8.3 and A.3.

Figure 8.1 is a schematic illustration of the basic principle of the extended IVM.

Choice of instruments

The following two examples illustrate some simple possibilities for choosing the IV matrix $Z(t)$. A discussion of the conditions which these choices must satisfy is deferred to a later stage, where the idea underlying the construction of $Z(t)$ in the following examples also will be clarified. In Complement C8.1 it is described how the IV techniques can be used also for pure time series (with no input present) and how they relate to the so-called Yule–Walker approach for estimating the parameters of ARMA models.

Example 8.1 *An IV variant for SISO systems*

One possibility for choosing the instruments for a SISO system is the following:

$$Z(t) = (-\eta(t-1) \ldots -\eta(t-na) \ u(t-1) \ldots u(t-nb))^T \tag{8.15a}$$

where the signal $\eta(t)$ is obtained by filtering the input,

$$C(q^{-1})\eta(t) = D(q^{-1})u(t) \tag{8.15b}$$

The coefficients of the polynomials C and D can be chosen in many ways. One special choice is to let C and D be *a priori* estimates of A and B, respectively. Another special

264 *Instrumental variable methods*   *Chapter 8*

case is where $C(q^{-1}) = 1$, $D(q^{-1}) = -q^{-nb}$. Then $Z(t)$ becomes, after a reordering of the elements

$$Z(t) = (u(t-1) \ldots u(t-na-nb))^{\mathrm{T}} \tag{8.15c}$$

Note that a reordering or, more generally, *a nonsingular linear transformation of the instruments $Z(t)$ has no influence on the corresponding basic IV estimate*. Indeed, it can easily be seen from (8.12) that a change of instruments from $Z(t)$ to $TZ(t)$, where T is a nonsingular matrix, will not change the parameter estimate $\hat{\theta}$ (cf. Problem 8.2). ■

Example 8.2 *An IV variant for multivariable systems*
One possible choice of the instrumental variables for an MIMO system in the full polynomial form is the following (cf. (8.4)):

$$Z(t) = \begin{pmatrix} z(t) & & 0 \\ & \ddots & \\ 0 & & z(t) \end{pmatrix} \tag{8.16a}$$

with

$$z^{\mathrm{T}}(t) = K(q^{-1})(u^{\mathrm{T}}(t-1) \ldots u^{\mathrm{T}}(t-m)) \tag{8.16b}$$

In (8.16b), $K(q^{-1})$ is a scalar filter, and m is an integer satisfying $m \times nu \geq na \times ny + nb \times nu$. ■

8.2 Theoretical analysis

Assumptions

In order to discuss the properties of the IV estimate $\hat{\theta}$, (8.13), when the number of data points tends to infinity, it is necessary to introduce a number of assumptions about the true system (8.6), the model structure (8.1), the IV method and the experimental conditions under which the data are collected. These assumptions will be considered to be valid throughout the analysis.

    A1: The system is strictly causal and asymptotically stable.
    A2: The input $u(t)$ is persistently exciting of a sufficiently high order.
    A3: The disturbance $v(t)$ is a stationary stochastic process with rational spectral density. It can thus be uniquely described as

$$v(t) = H(q^{-1})e(t) \tag{8.17}$$

where the matrix filter $H(q^{-1})$ is asymptotically stable and invertible, $H(0) = I$, and $Ee(t)e^{\mathrm{T}}(s) = \Lambda\delta_{t,s}$ with the covariance matrix Λ positive definite. (See, for example, Appendix A6.1.)
    A4: The input $u(t)$ and the disturbance $v(t)$ are independent. (The system operates in open loop.)
    A5: The model (8.1) and the true system (8.6) have the same transfer function

Section 8.2 *Theoretical analysis*

matrix (if and) only if $\theta = \theta_0$. (That is, there exists a unique member in the model set having the same noise-free input–output relation as the true system.)

A6: The IV matrix $Z(t)$ and the disturbance $v(s)$ are uncorrelated for all t and s.

The above assumptions are mild. Note that in order to satisfy A5 a canonical parametrization of the system dynamics may be used. This assumption can be rephrased as: the set $D_T(\mathcal{S}, \mathcal{M})$ (see (6.44)) consists of exactly one point. If instead a pseudo-canonical parametrization (see the discussion following Example 6.3) is used, some computational advantages may be obtained. However, A5 will then fail to be satisfied for some (but only a few) systems (see Complement C6.1, and Söderström and Stoica, 1983, for details). Concerning assumption A6, the matrix $Z(t)$ is most commonly obtained by various operations (filtering, delaying, etc.) on the input $u(t)$. Provided the system operates in open loop, (see A4), assumption A6 will then be satisfied. Assumption A6 can be somewhat relaxed (see Stoica and Söderström, 1983b). However, to keep the analysis simple we will use it in the above form.

Consistency

Introduce the notation

$$q_N = \frac{1}{N} \sum_{t=1}^{N} Z(t) F(q^{-1}) v(t) \qquad (8.18)$$

Then it is readily established from the system description (8.6) and the definitions (8.14b, c) that

$$r_N = R_N \theta_0 + q_N \qquad (8.19)$$

Therefore the estimation error is given by

$$\hat{\theta} - \theta_0 = (R_N^T Q R_N)^{-1} R_N^T Q q_N \qquad (8.20)$$

Under the given assumptions R_N and q_N converge as $N \to \infty$, and

$$\lim_{N \to \infty} R_N = E Z(t) F(q^{-1}) \varphi^T(t) \triangleq R \qquad (8.21a)$$

$$\lim_{N \to \infty} q_N = E Z(t) F(q^{-1}) v(t) \triangleq q \qquad (8.21b)$$

(see Lemma B.2).

The matrix R can be expressed as

$$R = E Z(t) F(q^{-1}) \bar{\varphi}^T(t) \qquad (8.22)$$

where $\bar{\varphi}(t)$ is the noise-free part of $\varphi(t)$. To see this, recall that the elements of $\varphi(t)$ consist of delayed input and output components. If the effect of the disturbance $v(\cdot)$ (which is uncorrelated with $Z(t)$) is subtracted from the output components in $\varphi(t)$, then the noise-free part $\bar{\varphi}(t)$ is obtained. More formally, $\bar{\varphi}(t)$ can be defined as a conditional mean:

$$\bar{\varphi}(t) = E[\varphi(t)|u(t-1), u(t-2), \ldots] \qquad (8.23)$$

As an illustration the following example shows how $\tilde{\phi}(t)$ can be obtained in the SISO case.

Example 8.3 *The noise-free regressor vector*
Consider the SISO system

$$A(q^{-1})y(t) = B(q^{-1})u(t) + v(t) \tag{8.24a}$$

for which

$$\phi(t) = (-y(t-1) \ \ldots \ -y(t-na) \ \ u(t-1) \ \ldots \ u(t-nb))^{\mathrm{T}} \tag{8.24b}$$

Then $\tilde{\phi}(t)$ is given by

$$\tilde{\phi}(t) = (-x(t-1) \ \ldots \ -x(t-na) \ \ u(t-1) \ \ldots \ u(t-nb))^{\mathrm{T}} \tag{8.24c}$$

where $x(t)$, the noise-free part of the output, is given by

$$A(q^{-1})x(t) = B(q^{-1})u(t) \tag{8.24d}$$

∎

Return now to the consistency analysis. From (8.20) and (8.21) it can be seen that the IV estimate (8.13) is consistent (i.e. $\lim_{N\to\infty} \hat{\theta} = \theta_0$) if

---

R has full rank ($= n\theta$)	(8.25a)
$EZ(t)F(q^{-1})v(t) = 0$	(8.25b)

It follows directly from assumption A6 that (8.25b) is satisfied. The condition (8.25a) is more intricate to analyze. It is easy to see that (8.25a) implies that

$$nz \geq n\theta \qquad \text{rank } EZ(t)Z^{\mathrm{T}}(t) \geq n\theta \tag{8.26}$$

(cf. Problem 8.9).

Generic consistency

The condition (8.26) is not sufficient for (8.25a) to hold, but only necessary. It can be shown, though, that when (8.26) is satisfied, then the condition (8.25a) is generically true. Thus, under weak conditions (see (8.26)) the IV estimate is 'generically consistent'. The precise meaning of this is explained in the following.

Definition 8.1
Let Ω be an open set. A statement s, which depends on the elements ϱ of Ω, is *generically true with respect to* Ω if the set

$$M = \{\varrho | \varrho \in \Omega, s \text{ is not true}\}$$

has Lebesgue measure zero in Ω.

∎

The Lebesgue measure is discussed, for example, in Cramér (1946) and Pearson (1974). In loose terms, we require M to have a smaller dimension than Ω. If s is generically true with respect to Ω and $\varrho \in \Omega$ is chosen randomly, then the probability that $\varrho \in M$ is zero, i.e. the probability that s is true is one. (In particular, if s is true for all $\varrho \in \Omega$, it is trivially generically true with respect to Ω.) Thus the statements 's is generically true' and 's is true with probability one' are equivalent.

The idea is now to consider the matrix R as a function of a parameter vector ϱ. A possibility for doing this is illustrated by the following example.

Example 8.4 *A possible parameter vector ϱ*

Consider the IV variant (8.15). Assume that $F(q^{-1})$ is a finite-order (rational) filter specified by the parameters $f_1, \ldots, f_{nf}$. Further, let the input spectral density be specified by the parameters $v_1, \ldots, v_{nv}$. For example, suppose that the input $u(t)$ has a rational spectral density. Then $\{v_i\}$ could be the coefficients of the ARMA representation of $u(t)$. Now take

$$\varrho = (a_1 \ldots a_{na} \ b_1 \ldots b_{nb} \ f_1 \ldots f_{nf} \ c_1 \ldots c_{nc} \ d_1 \ldots d_{nd} \ v_1 \ldots v_{nv})^T$$

The set Ω is the subset of $\mathcal{R}^{na+nb+nf+nc+nd+nv}$ given by the constraints:

$A(z)$ and $C(z)$ have all zeros strictly outside the unit circle
$F(z)$ has all poles strictly outside the unit circle
$A(z)$ and $B(z)$ are coprime
$C(z)$ and $D(z)$ are coprime

Additional and similar constraints may be added on $v_1, \ldots, v_{nv}$, depending on how these parameters are introduced.

Note that all elements of the matrix R will be analytic functions of every element of ϱ. ∎

In the general case, assume that the parametrization of R by ϱ is such that $R(\varrho)$ is analytic in ϱ. Furthermore, assume that there exists a vector $\varrho^* \in \Omega$ which is such that

$$R(\varrho^*) = E\tilde{\phi}(t)\tilde{\phi}^T(t) \tag{8.27}$$

Note that for the example above,

$$\varrho^* = (a_1 \ldots a_{na} \ b_1 \ldots b_{nb} \ 0 \ldots 0 \ a_1 \ldots a_{na} \ b_1 \ldots b_{nb} \ v_1 \ldots v_{nv})^T$$

which gives $F(q^{-1}) \equiv 1$ and $Z(t) = \tilde{\phi}(t)$. The matrix in (8.27) is nonsingular if the model is not overparametrized and the input is persistently exciting (see Complement C6.2). It should be noted that the condition (8.27) is merely a restriction on the structure of the IV matrix $Z(t)$. For some choices of $Z(t)$ (and ϱ) one cannot achieve $Z(t) = \tilde{\phi}(t)$, which implies (8.27) (and essentially, is implied by (8.27)). In such cases (8.27) can be relaxed: it is required only that there exists a vector $\varrho^* \in \Omega$ such that $R(\varrho^*)$ has full rank (equal to $n\theta$.) This is a weak condition indeed.

Now set

$$f(\varrho) = \det[R^T(\varrho)R(\varrho)] \tag{8.28}$$

Here $f(\varrho)$ is analytic in ϱ and $f(\varrho^*) \ne 0$ for some vector ϱ^*. It then follows from the uniqueness theorem for analytic functions that $f(\varrho) \ne 0$ almost everywhere in Ω (see Söderström and Stoica, 1983, 1984, for details). This result means that R has full rank for almost any value of ϱ in Ω. In other words, if the system parameters, the prefilter, the parameters describing the filters used to create the instruments and the input spectral density are chosen at random (in a set Ω of feasible values), then (8.25a) is true with probability one. The counterexamples where R does not have full rank belong to a subset of lower dimension in the parameter space. The condition (8.25a) is therefore not a problem from the practical point of view as far as consistency is concerned. However, if R is nearly rank deficient, inaccurate estimates will be obtained. This follows for example from (8.29), (8.30) below. This phenomenon is illustrated in Example 8.6.

Asymptotic distribution

Next consider the accuracy of the IV estimates. The following result holds for the asymptotic distribution of the parameter estimate $\hat{\theta}$:

$$\sqrt{N}(\hat{\theta} - \theta_0) \xrightarrow{\text{dist}} \mathcal{N}(0, P_{\text{IV}}) \tag{8.29}$$

with the covariance matrix P_{IV} given by

$$P_{\text{IV}} = (R^{\text{T}}QR)^{-1} R^{\text{T}} Q \left\{ E\left[\sum_{i=0}^{\infty} Z(t+i) K_i \right] \right.$$
$$\left. \times \Lambda \left[\sum_{j=0}^{\infty} K_j^{\text{T}} Z^{\text{T}}(t+j) \right] \right\} QR(R^{\text{T}}QR)^{-1} \tag{8.30}$$

Here R is given by (8.22) and $\{K_i\}_{i=0}^{\infty}$ are defined by

$$\sum_{i=0}^{\infty} K_i z^i \equiv F(z) H(z) \tag{8.31}$$

In the scalar case ($ny = 1$) the matrix in (8.30) between curly brackets can be written as

$$E\left[\sum_{i=0}^{\infty} Z(t+i) K_i \right] \Lambda \left[\sum_{j=0}^{\infty} K_j^{\text{T}} Z^{\text{T}}(t+j) \right]$$
$$= \Lambda E[F(q^{-1}) H(q^{-1}) Z(t)][F(q^{-1}) H(q^{-1}) Z(t)]^{\text{T}} \tag{8.32}$$

The proof of these results is given in Appendix A8.1.

Violation of the assumptions

It is of interest to discuss what happens when some of the basic assumptions are not satisfied. Assumption A4 (open loop operation) is not necessary. Chapter 10 contains a discussion of how systems operating in closed loop can be identified using the IV method.

Assume for a moment that A5 is not satisfied, so that the true system is not included in the considered model structure. For such a case it holds that

$$\hat{\theta} \to \theta^* \triangleq \arg \min_{\theta} \|\{EZ(t)F(q^{-1})\phi^T(t)\}\theta - EZ(t)F(q^{-1})y(t)\|_Q^2 \qquad (8.33a)$$

In the general case it is difficult to say anything more specific about the limit θ^* (see, however, Problem 8.12). If $nz = n\theta$, the limiting estimate θ^* can be written as

$$\theta^* = R^{-1}r \qquad (8.33b)$$

where

$$r = EZ(t)F(q^{-1})y(t)$$

In this case, the estimates will still be asymptotically Gaussian distributed,

$$\sqrt{N}(\hat{\theta} - \theta^*) \xrightarrow{\text{dist}} \mathcal{N}(0, P_{\text{IV}})$$
$$P_{\text{IV}} = R^{-1}SR^{-T} \qquad (8.34)$$

where

$$S = \lim_{N \to \infty} \frac{1}{N} E \left[\sum_{t=1}^{N} Z(t)F(q^{-1})\{y(t) - \phi^T(t)\theta^*\} \right]$$
$$\times \left[\sum_{s=1}^{N} Z(s)F(q^{-1})\{y(s) - \phi^T(s)\theta^*\} \right]^T$$

This can be shown in the same way as (8.29), (8.30) are proved in Appendix A8.1. A more explicit expression for the matrix S cannot in general be obtained unless more specific assumptions on the data are introduced. For the case $nz > n\theta$ it is more difficult to derive the asymptotic distribution of the parameter estimates.

Numerical illustrations

The asymptotic distribution (8.29), (8.30) is considered in the following numerical examples. The first example presents a Monte Carlo analysis for verifying the theoretical expression (8.30) of the asymptotic covariance matrix. Example 8.6 illustrates the concept of generic consistency and the fact that a nearly singular R matrix leads to poor accuracy of the parameter estimates.

Example 8.5 *Comparison of sample and asymptotic covariance matrices of IV estimates*

The following two systems were simulated:

$$\mathscr{S}_1: (1 - 0.5q^{-1})y(t) = 1.0u(t - 1) + (1 + 0.5q^{-1})e(t)$$

$$\mathscr{S}_2: (1 - 1.5q^{-1} + 0.7q^{-2})y(t) = (1.0q^{-1} + 0.5q^{-2})u(t) + (1 - 1.0q^{-1} + 0.2q^{-2})e(t)$$

generating for each system 200 realizations, each of length 600. The input $u(t)$ and $e(t)$ were in all cases mutually independent white noise sequences with zero means and unit variances.

The systems were identified in the natural model structure (8.2) taking $na = nb = 1$ for $\mathscr{S}_1$ and $na = nb = 2$ for $\mathscr{S}_2$.

For both systems two IV variants were tried, namely

$$\mathscr{J}_1: Z(t) = (u(t-1) \ldots u(t - na - nb))^T \quad \text{(see (8.15c))}$$

and

$$\mathscr{J}_2: Z(t) = \frac{1}{A_0(q^{-1})}(u(t-1) \ldots u(t - na - nb))^T \quad \text{(see (8.15a, b))}$$

where A_0 is the A polynomial of the true system. From the estimates obtained in each of the 200 independent realizations the sample mean and the sample normalized covariance matrix were evaluated as

$$\bar{\theta} = \frac{1}{m}\sum_{i=1}^{m} \hat{\theta}^i$$

$$\hat{P} = \frac{N}{m}\sum_{i=1}^{m} (\hat{\theta}^i - \bar{\theta})(\hat{\theta}^i - \bar{\theta})^T \tag{8.35}$$

where $\hat{\theta}^i$ denotes the estimate obtained from realization i, $N = 600$ is the number of data points in each realization, and $m = 200$ is the number of realizations. When m and N tend to infinity, $\bar{\theta} \to \theta_0$, $\hat{P} \to P_{IV}$. The deviations from the expected limits for a finite value of m can be evaluated as described in Section B.9 of Appendix B. It follows from the theory developed there that

$$\sqrt{N}(\bar{\theta} - \theta_0) \xrightarrow{dist} \mathcal{N}(0, P_{IV}/m) \quad \text{as } N \to \infty$$

A scalar $\mathcal{N}(0, 1)$ random variable lies within the interval $(-1.96, 1.96)$ with probability 0.95. Hence, by considering the above result for $\bar{\theta} - \theta_0$ componentwise, it is found that, with 95 percent probability,

$$|\bar{\theta}_j - \theta_{0j}| \leq 1.96 \left(\frac{(P_{IV})_{jj}}{mN}\right)^{1/2}$$

The right-hand side is of magnitude 0.01 in the present example.

To evaluate the discrepancy $\hat{P} - P_{IV}$, use is made of the result (B.76) which implies that

$$E[\hat{P}_{jk} - P_{IVjk}]^2 = \frac{1}{m}P_{IVjk}^2 + \frac{1 - 1/m}{m}P_{IVjj}P_{IVkk}$$

Section 8.2   Theoretical analysis

If m is reasonably large application of the central limit theorem implies that $\hat{P}_{jk}$ is asymptotically Gaussian distributed. Then with 95 percent probability,

$$|\hat{P}_{jk} - P_{IVjk}| \leq 1.96 \left(\frac{P_{IVjj} P_{IVkk} + P_{IVjk}^2}{m}\right)^{1/2}$$

For the diagonal elements of $\hat{P}$ this relation can be rewritten as

$$\left|\frac{\hat{P}_{jj} - P_{IVjj}}{P_{IVjj}}\right| \leq \frac{2.77}{\sqrt{m}}$$

In the present example the right-hand side has the value 0.196. This means that the relative error in $\hat{P}_{jj}$ should not be larger than 20 percent, with a probability of 0.95.

The numerical results obtained from the simulations are shown in Table 8.1. They are well in accordance with the theory. This indicates that the asymptotic result (8.29), (8.30) can be applied also for reasonable lengths, say a few hundred points, of the data series. ∎

Example 8.6 *Generic consistency*

For convenience in the analysis the ϱ vector will consist of a single element. Consider the scalar system

$$y(t) = \frac{1.0 q^{-1}}{1 - 2\varrho q^{-1} + \varrho^2 q^{-2}} u(t) + e(t)$$

where ϱ is a parameter and the input signal $u(t)$ obeys

$$u(t) = w(t) - 2\varrho^2 w(t-2) + \varrho^4 w(t-4)$$

TABLE 8.1 Comparison between asymptotic and sample behavior of two IV estimators. The sample behavior shown is estimated from 200 realizations of 600 data pairs each

Distribution parameters: means and normalized covariances	System $\mathscr{S}_1$				System $\mathscr{S}_2$			
	Variant $\mathscr{J}_1$		Variant $\mathscr{J}_2$		Variant $\mathscr{J}_1$		Variant $\mathscr{J}_2$	
	Asympt. expect. values	Sample estim. values	Asympt. expect. values	Sample estim. values	Asympt. expect. values	Sample estim. values	Asympt. expect. values	Sample estim. values
$E\hat{a}_1$	−0.50	−0.50	−0.50	−0.50	−1.50	−1.51	−1.50	−1.50
$E\hat{a}_2$	−	−	−	−	0.70	0.71	0.70	0.70
$E\hat{b}_1$	1.00	1.00	1.00	1.00	1.00	1.00	1.00	1.00
$E\hat{b}_2$	−	−	−	−	0.50	0.49	0.50	0.50
P_{11}	1.25	1.26	1.31	1.37	5.19	6.29	0.25	0.27
P_{12}	−0.50	−0.63	−0.38	−0.56	−7.27	−8.72	−0.22	−0.23
P_{13}	−	−	−	−	−0.24	−0.98	−0.08	0.02
P_{14}	−	−	−	−	6.72	6.74	0.71	0.65
P_{22}	1.25	1.25	1.25	1.23	10.38	12.27	0.20	0.22
P_{23}	−	−	−	−	0.27	1.46	0.06	−0.01
P_{24}	−	−	−	−	−9.13	−9.08	−0.59	−0.54
P_{33}	−	−	−	−	2.04	2.36	2.04	2.15
P_{34}	−	−	−	−	−1.44	−1.97	−1.28	−1.03
P_{44}	−	−	−	−	10.29	8.74	3.21	2.76

Here $e(t)$ and $w(t)$ are mutually independent white noise sequences of unit variance. Let the instrumental variable vector $Z(t)$ be given by (8.15a, b) with

$$na = 2 \quad nb = 1 \quad C(q^{-1}) = 1 + 2\varrho q^{-1} + \varrho^2 q^{-2} \quad D(q^{-1}) = q^{-1}$$

Some tedious but straightforward calculations show that for this example (see Problem 8.5)

$$R = R(\varrho) = \begin{pmatrix} 1 - 4\varrho^2 + \varrho^4 & -2\varrho(1 - \varrho^2) & -4\varrho^3 \\ 2\varrho(1 - \varrho^2) & 1 - 4\varrho^2 + \varrho^4 & \varrho^2(2 - \varrho^4) \\ 4\varrho^3 & \varrho^2(2 - \varrho^4) & 1 + 4\varrho^4 + \varrho^8 \end{pmatrix}$$

$$P_{IV} = P_{IV}(\varrho) = R^{-1}(\varrho)S(\varrho)[R^{-1}(\varrho)]^T$$

where the (3|3) symmetric matrix $S(\varrho)$ is given by

$$S(\varrho)_{11} = S(\varrho)_{22} = 1 + 16\varrho^2 + 36\varrho^4 + 16\varrho^6 + \varrho^8$$

$$S(\varrho)_{12} = -4\varrho - 24\varrho^3 - 24\varrho^5 - 4\varrho^7$$

$$S(\varrho)_{13} = 2\varrho - 4\varrho^3 - 24\varrho^5 - 4\varrho^7 + 2\varrho^9$$

$$S(\varrho)_{23} = \varrho^2 + 16\varrho^4 + 6\varrho^6 - 8\varrho^8 - \varrho^{10}$$

$$S(\varrho)_{33} = 1 + 4\varrho^2 + \varrho^4 + 16\varrho^6 + \varrho^8 + 4\varrho^{10} + \varrho^{12}$$

The determinant of $R(\varrho)$ is given by

$$\det R(\varrho) = (1 - 3\varrho^2 + \varrho^4)(1 - \varrho^2 + 6\varrho^4 - \varrho^6 + \varrho^8)$$

which is zero for

$$\varrho = \tilde{\varrho} \triangleq [(3 - \sqrt{5})/2]^{1/2} \approx 0.6180$$

and nonzero for all other values of ϱ in the interval $(0, 1)$.

For ϱ close to but different from $\tilde{\varrho}$, although the parameter estimates are consistent (since $R(\varrho)$ is nonsingular), they will be quite inaccurate (since the elements of $P_{IV}(\varrho)$ are large).

For numerical illustration some Monte Carlo simulations were performed. For various values of ϱ, 50 different realizations each of length $N = 1000$ were generated. The IV method was applied to each of these realizations and the sample variance calculated for each parameter estimate according to (8.35).

A graphical illustration of the way in which $P_{IV}(\varrho)$ and $\hat{P}_{IV}$ vary with ϱ is given in Figure 8.2.

The plots demonstrate clearly that quite poor accuracy of the parameter estimates can be expected when the matrix $R(\varrho)$ is almost singular. Note also that the results obtained by simulations are quite close to those predicted by the theory. ∎

Optimal IV method

Note that the matrix P_{IV} depends on the choice of $Z(t)$, $F(q^{-1})$ and Q. Interestingly enough, there exists an achievable lower bound on P_{IV}:

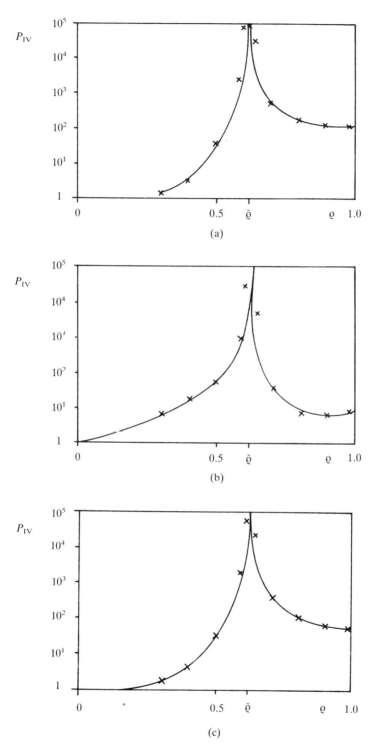

FIGURE 8.2 (a) $P_{IV}(\varrho)_{1,1}$ and $\hat{P}_{IV1,1}$ versus ϱ. (b) $P_{IV}(\varrho)_{2,2}$ and $\hat{P}_{IV2,2}$ versus ϱ. (c) $P_{IV}(\varrho)_{3,3}$ and $\hat{P}_{IV3,3}$ versus ϱ.

$$\boxed{P_{IV} \geqslant P_{IV}^{opt}} \qquad (8.36)$$

($P_{IV} - P_{IV}^{opt}$ is nonnegative definite) where

$$P_{IV}^{opt} = \{E[H^{-1}(q^{-1})\tilde{\phi}^T(t)]^T \Lambda^{-1}[H^{-1}(q^{-1})\tilde{\phi}^T(t)]\}^{-1} \qquad (8.37)$$

$\tilde{\phi}(t)$ being the noise-free part of $\phi(t)$. Moreover, equality in (8.36) is obtained, for example, by taking

$$\boxed{Z(t) = [\Lambda^{-1}H^{-1}(q^{-1})\tilde{\phi}^T(t)]^T \qquad F(q^{-1}) = H^{-1}(q^{-1}) \qquad Q = I} \qquad (8.38)$$

These important results are proved as follows. Introduce the notation

$$w(t) = R^T Q Z(t)$$

$$\alpha(t) = \sum_{i=0}^{\infty} w(t+i) K_i$$

$$\beta(t) = [H^{-1}(q^{-1})\tilde{\phi}^T(t)]^T$$

with $\{K_i\}$ as in (8.31). Then

$$R^T Q R = Ew(t)F(q^{-1})\tilde{\phi}^T(t) = Ew(t) \sum_{i=0}^{\infty} K_i q^{-i} H^{-1}(q^{-1})\tilde{\phi}^T(t)$$

$$= \sum_{i=0}^{\infty} Ew(t+i) K_i H^{-1}(q^{-1})\tilde{\phi}^T(t) = E\alpha(t)\beta^T(t)$$

Thus

$$P_{IV} = [E\alpha(t)\beta^T(t)]^{-1}[E\alpha(t)\Lambda\alpha^T(t)][E\beta(t)\alpha^T(t)]^{-1}$$

From (8.37),

$$P_{IV}^{opt} = [E\beta(t)\Lambda^{-1}\beta^T(t)]^{-1}$$

The inequality (8.36) which, according to the calculations above, can be written as

$$[E\alpha(t)\Lambda\alpha^T(t)] - [E\alpha(t)\beta^T(t)][E\beta(t)\Lambda^{-1}\beta^T(t)]^{-1}[E\beta(t)\alpha^T(t)] \geqslant 0$$

now follows from Lemma A.4 and its remark (by taking $Z_1(t) = \alpha(t)\Lambda^{1/2}$, $Z_2(t) = \beta(t)\Lambda^{-1/2}$). Finally, it is trivial to prove that the choice (8.38) of the instruments and the prefilter gives equality in (8.36).

Approximate implementation of the optimal IV method

The optimal IV method as defined by (8.38) requires knowledge of the undisturbed output (in order to form $\tilde{\phi}(t)$) as well as of the noise covariance Λ and shaping filter

$H(q^{-1})$. In practice a bootstrapping technique must therefore be used to implement the optimal IV estimate. The basic idea of such an algorithm is to combine in an iterative manner the optimal IV method with a procedure for estimation of the noise parameters.

Specifically, introduce the following model structure that extends (8.1):

$$y(t) = \phi^T(t)\theta + H(q^{-1}; \theta, \beta)e(t, \theta, \beta) \tag{8.39}$$

Here, $H(q^{-1}; \theta, \beta)$ is a filter that should describe the correlation properties of the disturbances (cf. (8.17)). It depends on the parameter vector θ as well as on some additional parameters collected in the vector β. An illustration is provided in the following example.

Example 8.7 *Model structures for SISO systems*
For SISO systems two typical model structures are

$$\mathcal{M}_1: A(q^{-1})y(t) = B(q^{-1})u(t) + \frac{C(q^{-1})}{D(q^{-1})} e(t) \tag{8.40}$$

and

$$\mathcal{M}_2: y(t) = \frac{B(q^{-1})}{A(q^{-1})} u(t) + \frac{C(q^{-1})}{D(q^{-1})} e(t) \tag{8.41}$$

where in both cases

$$\begin{aligned} C(q^{-1}) &= 1 + c_1 q^{-1} + \ldots + c_{nc} q^{-nc} \\ D(q^{-1}) &= 1 + d_1 q^{-1} + \ldots + d_{nd} q^{-nd} \end{aligned} \tag{8.42}$$

Note that both these model structures are special cases of the general model set (6.14) introduced in Example 6.2. Introduce

$$\beta = (c_1 \ldots c_{nc} \; d_1 \ldots d_{nd})^T$$

Then for the model structure $\mathcal{M}_1$,

$$H(q^{-1}; \theta, \beta) = \frac{C(q^{-1})}{D(q^{-1})} \tag{8.43a}$$

while $\mathcal{M}_2$ corresponds to

$$H(q^{-1}; \theta, \beta) = \frac{A(q^{-1})C(q^{-1})}{D(q^{-1})} \tag{8.43b}$$

∎

It will generally be assumed that the filter $H(q^{-1}; \theta, \beta)$ satisfies

$$H(q^{-1}; \theta_0, \beta) \equiv H(q^{-1}) \Leftrightarrow \beta = \beta_0 \tag{8.44a}$$

for a unique vector β_0 which will be called the true noise parameter vector. Further, introduce the filter $\bar{H}(q^{-1}; \theta, \beta)$ through

$$H(q^{-1}; \theta, \beta) \equiv A(q^{-1})\bar{H}(q^{-1}; \theta, \beta) \tag{8.44b}$$

The parametrization of $\bar{H}$ will turn out to be of some importance in the following.

Appendix A8.2 provides a comparison of the optimal IV estimate (8.38) and the prediction error estimate for the model structure (8.39). Since the PEM estimate is statistically efficient for Gaussian distributed disturbances, the optimal IV estimate will never have better accuracy. This means that

$$P_{IV}^{opt} \geq P_{PEM} \tag{8.45}$$

When $\bar{H}(q^{-1}; \theta, \beta)$ does not depend on θ, the optimal IVM and PEM will have the same asymptotic distribution (i.e. equality holds in (8.45)), as shown in Appendix A8.2. Note for example that for the structure $\mathcal{M}_2$, (8.41), the filter $\bar{H}(q^{-1}; \theta, \beta)$ does not depend on θ.

An approximate implementation of the optimal IV estimate (8.38) can be carried out as follows.

Step 1. Apply an arbitrary IV method using the model structure (8.1). As a result, a consistent estimate $\hat{\theta}_1$ is obtained.

Step 2. Compute

$$\hat{v}(t) = y(t) - \phi^T(t)\hat{\theta}_1$$

and determine the parameter vector β in the model structure (cf. (8.39))

$$y(t) = \phi^T(t)\hat{\theta}_1 + H(q^{-1}; \hat{\theta}_1, \beta)e(t, \hat{\theta}_1, \beta)$$

using a prediction error method (or any other statistically efficient method). Call the result $\hat{\beta}_2$. Note that when the parametrization $H(q^{-1}; \hat{\theta}_1, \beta) = 1/D(q^{-1})$ is chosen this step becomes a simple least squares estimation.

Step 3. Compute the optimal IV estimate (8.38) using $\hat{\theta}_1$ to form $\bar{\phi}(t)$ and $\hat{\theta}_1, \hat{\beta}_2$ to form $H(q^{-1})$ and Λ. Call the result $\hat{\theta}_3$.

Step 4. Repeat Step 2 with $\hat{\theta}_3$ replacing $\hat{\theta}_1$. The resulting estimate is called $\hat{\beta}_4$.

In practice one can repeat Steps 3 and 4 a number of times until convergence is attained. Theoretically, though, this should not be necessary if the number of data is large (see below).

In Söderström and Stoica (1983) (see also Stoica and Söderström, 1983a, b) a detailed analysis of the above four-step algorithm is provided, as well as a comparison of the accuracy of the estimates of θ and β obtained by the above algorithm, with the accuracy of the estimates provided by a prediction error method applied to the model structure (8.39). See also Appendix A8.2. The main results can be summarized as follows:

- All the parameter estimates $\hat{\theta}_1, \hat{\beta}_2, \hat{\theta}_3$ and $\hat{\beta}_4$ are consistent. They are also asymptotically Gaussian distributed. For $\hat{\theta}_3$ the asymptotic covariance matrix is

$$P_{IV} = P_{IV}^{opt} \tag{8.46}$$

- Assume that the model structure (8.39) is such that $\bar{H}(q^{-1}; \theta, \beta)$ (see (8.44b)) does not depend on θ. (This is the case, for example, for $\mathcal{M}_2$ in (8.41)). Then the algorithm converges in *three* steps in the sense that $\hat{\beta}_4$ has the same (asymptotic) covariance matrix as $\hat{\beta}_2$. Moreover, $(\hat{\theta}_3^T \; \hat{\beta}_2^T)^T$ has the same (asymptotic) covariance matrix as the

Section 8.3   Computational aspects   277

prediction error estimate of $(\theta^T \; \beta^T)^T$. Thus the approximate algorithm is as accurate as a prediction error method for such a model structure.
- Assume that $\bar{H}(q^{-1}; \theta, \beta)$ does depend on θ. Then $\hat{\beta}_4$ will be more accurate than $\hat{\beta}_2$. The estimate $(\hat{\theta}_3^T \; \hat{\beta}_4^T)^T$ will be less accurate than the PEM estimate of $(\theta^T \; \beta^T)^T$.

One may question what is gained by using the optimal IV method (implemented as the above four-step algorithm) compared to a prediction error method. Recall that the optimal IV four-step algorithm relies on a statistically efficient estimation of β in steps 2 and 4. The problem of estimating β is a simple LS problem for the model structure (8.40) or (8.41) with $C(q^{-1}) \equiv 1$. (To get a good fit for this model structure a high order of $D(q^{-1})$ may be needed.) For general model structures nonlinear optimization cannot be avoided. However, this optimization problem has a smaller dimension (namely dim β) than when using a prediction error method for estimating both θ and β (which gives a problem of dimension dim θ + dim β). A PEM will often give a better accuracy than an IV method (at least asymptotically), although the difference in many cases may be reasonably small.

Complements C8.4 and C8.5 discuss some other optimization problems for IV methods, and in Example 10.3 it is shown that the IV methods can be applied to systems operating in closed loop, after some modifications.

8.3 Computational aspects

Multivariable systems

For most model structures considered for multivariable systems the different components of $y(t)$ can be described by *independent* parameter vectors θ_i, $i = 1, \ldots, ny$, so that (8.1) is equivalent to

$$y_i(t) = \varphi_i^T(t)\theta_i + \varepsilon_i(t) \qquad i = 1, \ldots, ny \qquad (8.47)$$

and thus

$$\phi(t) = \begin{pmatrix} \varphi_1(t) & & 0 \\ & \ddots & \\ 0 & & \varphi_{ny}(t) \end{pmatrix} \qquad \theta = \begin{pmatrix} \theta_1 \\ \vdots \\ \theta_{ny} \end{pmatrix} \qquad (8.48)$$

This was indeed the case for the full polynomial form (see (8.5)). Assume that $Z(t)$, $F(q^{-1})$ and Q are constrained so that they also have a diagonal structure, i.e.

$$Z(t) = \text{diag}[z_i(t)] \qquad F(q^{-1}) = \text{diag}[f_i(q^{-1})]$$
$$Q = \text{diag}[Q_i] \qquad (8.49)$$

with dim $z_i(t) \geq$ dim $\varphi_i(t)$, $i = 1, \ldots, ny$. Then the IV estimate (8.13) reduces to

$$\hat{\theta}_i = \arg\min_{\theta_i} \left\| \left[\sum_{t=1}^{N} z_i(t)f_i(q^{-1})\varphi_i^T(t)\right]\theta_i - \left[\sum_{t=1}^{N} z_i(t)f_i(q^{-1})y_i(t)\right] \right\|_{Q_i}^2 \qquad (8.50)$$
$$i = 1, \ldots, ny$$

The form (8.50) has advantages over (8.13) since ny small least squares problems are to be solved instead of one large. The number of operations needed to solve the least squares problem (8.50) for one output is proportional to $(\dim \theta_i)^2 (\dim z_i)$. Hence the reduction in the computational load when solving (8.50) for $i = 1, \ldots, ny$ instead of (8.13) is in the order of ny. A further simplification is possible if none of $\varphi_i(t)$, $z_i(t)$ or $f_i(q^{-1})$ varies with the index i. Then the matrix multiplying θ_i in (8.50) will be independent of i.

Solving overdetermined systems of equations

Consider the estimate $\hat{\theta}$ defined by (8.13). Assume that $nz > n\theta$. Then the following overdetermined linear system of equations must be solved in a least squares sense to obtain $\hat{\theta}$ ($Q^{1/2}$ denotes a square root of Q: $Q^{T/2} Q^{1/2} = Q$)

$$Q^{1/2} \left[\sum_{t=1}^{N} Z(t) F(q^{-1}) \phi^T(t) \right] \theta = Q^{1/2} \left[\sum_{t=1}^{N} Z(t) F(q^{-1}) y(t) \right] \tag{8.51}$$

(there are nz equations and $n\theta$ unknowns). It is straightforward to solve this problem analytically (see (8.14)). However, the solution computed using the so-called normal equation approach (8.14) will be numerically sensitive (cf. Section A.4). This means that numerical errors (such as unavoidable rounding errors made during the computations) will accumulate and have a significant influence on the result. A numerically sound way to solve such systems is the orthogonal triangularization method (see Section A.3 for details).

Nested model structures

The system of equations from which the basic IV estimate (8.12) is found, is composed of covariance elements between $Z(t)$ and $(\phi^T(t) y(t))$. When the model structure is expanded, only a few new covariance elements are used in addition to the old ones. Hence, it is natural that there are simple relations between the estimates corresponding to nested model structures. The following example illustrates this issue.

Example 8.8 *IV estimates for nested structures*
Consider the model structure

$$\mathcal{M}: y(t) = \phi^T(t)\theta + \bar{\varepsilon}(t) \tag{8.52a}$$

Let $\phi^T(t)$ and θ be partitioned as

$$\phi^T(t) = (\phi_1^T(t) \quad \phi_2^T(t)) \qquad \theta = (\theta_1^T \quad \theta_2^T)^T \tag{8.52b}$$

Consider also the smaller model structure

$$\mathcal{M}': y(t) = \phi_1^T(t)\theta_1 + \varepsilon(t) \tag{8.52c}$$

which can be seen as a subset of $\mathcal{M}$. A typical case of such nested structures is when $\mathcal{M}$ is of a higher order than $\mathcal{M}'$. In such a case the elements of $\phi_2(t)$ will have larger time lags than the elements of $\phi_1(t)$.

Let the matrix of instruments $Z(t)$ used for $\mathcal{M}$ be partitioned as

$$Z(t) = \begin{pmatrix} Z_1(t) \\ Z_2(t) \end{pmatrix} \tag{8.52d}$$

where $Z_1(t)$ has the same dimensions as $\phi_1(t)$.

Assume now that the following have been determined

$$R = \frac{1}{N}\sum_{t=1}^{N} Z(t)\phi^T(t) = \frac{1}{N}\sum_{t=1}^{N} \begin{pmatrix} Z_1(t) \\ Z_2(t) \end{pmatrix}(\phi_1^T(t) \quad \phi_2^T(t)) \triangleq \begin{pmatrix} R_{11} & R_{12} \\ R_{21} & R_{22} \end{pmatrix} \tag{8.52e}$$

and the basic IV estimate for the model structure $\mathcal{M}'$:

$$\hat{\theta}'_1 = R_{11}^{-1} \frac{1}{N}\sum_{t=1}^{N} Z_1(t)y(t) \tag{8.52f}$$

Then it is required to find the basic IV estimate for the larger model structure $\mathcal{M}$, i.e.

$$\hat{\theta} = R^{-1} \frac{1}{N}\sum_{t=1}^{N} Z(t)y(t) \tag{8.52g}$$

Using the formula for the inverse of a partitioned matrix (see Lemma A.2),

$$\begin{aligned}
\hat{\theta} &= \begin{pmatrix} \hat{\theta}_1 \\ \hat{\theta}_2 \end{pmatrix} \\
&= \left\{ \begin{pmatrix} R_{11}^{-1} & 0 \\ 0 & 0 \end{pmatrix} + \begin{pmatrix} R_{11}^{-1}R_{12} \\ -I \end{pmatrix}[R_{22} - R_{21}R_{11}^{-1}R_{12}]^{-1}(R_{21}R_{11}^{-1} \quad -I) \right\} \\
&\quad \times \frac{1}{N}\sum_{t=1}^{N} Z(t)y(t) \\
&= \begin{pmatrix} \hat{\theta}'_1 \\ 0 \end{pmatrix} - \begin{pmatrix} R_{11}^{-1}R_{12} \\ -I \end{pmatrix}[R_{22} - R_{21}R_{11}^{-1}R_{12}]^{-1}\left\{ \frac{1}{N}\sum_{t=1}^{N} Z_2(t)\{y(t) - \phi_1^T(t)\hat{\theta}'_1\} \right\}
\end{aligned} \tag{8.52h}$$

These relations can be rewritten as

$$\varepsilon(t) = y(t) - \phi_1^T(t)\hat{\theta}'_1$$

$$r = \frac{1}{N}\sum_{t=1}^{N} Z_2(t)\varepsilon(t) = \frac{1}{N}\sum_{t=1}^{N} Z_2(t)y(t) - R_{21}\hat{\theta}'_1$$

$$\boxed{\begin{aligned}
\hat{\theta}_2 &= [R_{22} - R_{21}R_{11}^{-1}R_{12}]^{-1} r \\
\hat{\theta}_1 &= \hat{\theta}'_1 - R_{11}^{-1}R_{12}\hat{\theta}_2
\end{aligned}} \tag{8.52i}$$

$$R^{-1} = \begin{pmatrix} R_{11}^{-1} & 0 \\ 0 & 0 \end{pmatrix} + \begin{pmatrix} R_{11}^{-1}R_{12} \\ -I \end{pmatrix}[R_{22} - R_{21}R_{11}^{-1}R_{12}]^{-1}(R_{21}R_{11}^{-1} \quad -I)$$

Two expressions for r are given above. The first expression allows some interesting interpretations of the above relations, as shown below. The second expression for r

should be used for implementation since it requires a smaller amount of computation. In particular, the equation errors $\varepsilon(t)$ are then not explicitly needed. If the parameter estimates are sought for only two model structures (and not a longer sequence) then the updating of R^{-1} can be dispensed with.

Note that $\varepsilon(t)$ is the equation error for the smaller model structure $\mathcal{M}'$. If $\mathcal{M}'$ is adequate one can expect $\varepsilon(t)$ to be uncorrelated with the instruments $Z_2(t)$. In such a case r will be close to zero and the estimate for $\mathcal{M}$ will satisfy $\hat{\theta}_1 \approx \hat{\theta}'_1$, $\hat{\theta}_2 \approx 0$.

The scheme above is particularly useful when $\mathcal{M}$ contains one parameter more than $\mathcal{M}'$. Then $\phi_2(t)$ and $Z_2(t)$ become row vectors. Further, $R_{22} - R_{21} R_{11}^{-1} R_{12}$ becomes a scalar, and matrix inversions can thus be avoided completely when determining the IV estimates for $\mathcal{M}$. Note that R_{11}^{-1} is available from the previous stage of estimating the parameters of $\mathcal{M}'$.

Note that for certain model sets which possess a special structure it is possible to simplify the equations (8.52i) considerably. Complements C8.2, C8.3 and C8.6 show how this simplification can be made for AR, ARMA and multivariate regression models. ∎

Remark. Example 8.8 is based on pure algebraic manipulations which hold for a general $Z(t)$ matrix. In particular, setting $Z(t) = \phi(t)$ produces the corresponding relations for the least squares method. ∎

Summary

Instrumental variable methods (IVMs) were introduced in Section 8.1. Extended IVMs include prefiltering of the data and an augmented IV matrix. They are thus considerably more general than the basic IVMs.

The properties of IVMs were analyzed in Section 8.2. The main restriction in the analysis is that the system operates in open loop and that it belongs to the set of models considered. Under such assumptions and a few weak additional ones, it was shown that the parameter estimates are not only consistent but also asymptotically Gaussian distributed. It was further shown how the covariance matrix of the parameter estimates can be optimized by appropriate choices of the prefilter and the IV matrix. It was also discussed how the optimal IV method can be approximately implemented and how it compares to the prediction error method. Computational aspects were presented in Section 8.3.

Problems

Problem 8.1 *An expression for the matrix R*
Prove the expression (8.22) for R.

Problems 281

Problem 8.2 *Linear nonsingular transformations of instruments*
Study the effect of a linear nonsingular transformation of the IV matrix $Z(t)$ on the basic IV estimate (8.12) and the extended IV estimate (8.13). Show that the transformation $Z(t) \to TZ(t)$ will leave the basic IV estimate unchanged while the extended IV estimate will change.

Problem 8.3 *Evaluation of the R matrix for a simple system*
Consider the system

$$y(t) + ay(t-1) = b_1 u(t-1) + b_2 u(t-2) + v(t)$$

whose parameters are estimated with the basic IV method using delayed inputs as instruments

$$Z(t) = (u(t-1) \quad u(t-2) \quad u(t-3))^T$$

(see (8.12), (8.15c)). Consider the matrix

$$R = EZ(t)\phi^T(t)$$

$$\phi(t) = (-y(t-1) \quad u(t-1) \quad u(t-2))^T$$

We are interested in cases where R is nonsingular since then the IV estimate is consistent. Assume in particular that $u(t)$ is white noise of zero mean and unit variance, and independent of $v(s)$ for all t and s.

(a) Evaluate the matrix R.
(b) Examine if (and when) R is singular. Compare with the general assumptions made in the chapter. Will the result hold also for other (persistently exciting) input signals?

Problem 8.4 *IV estimation of the transfer function parameters from noisy input–noisy output systems*
Consider the system shown in the figure, where $u(t)$ and $y(t)$ denote noisy measurements of the input and output, respectively. It is assumed that the input disturbance $w(t)$ is white noise and that $x(t)$, $w(p)$ and $v(s)$ are mutually uncorrelated for all t, p and s. The parameters of the transfer function $B(q^{-1})/A(q^{-1})$ are estimated using an IVM with the following vector of instruments:

$$Z(t) = (u(t - n\tau - 1) \ \ldots \ u(t - n\tau - na - nb))^T$$

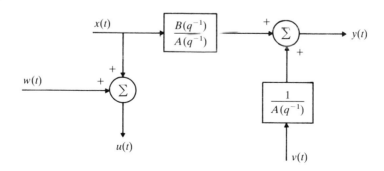

Show how $n\tau$ can be chosen so that the consistency condition (8.25b) is satisfied. Discuss how $n\tau$ should be chosen to satisfy also the condition (8.25a).

Problem 8.5 *Generic consistency*
Prove the expressions for $R(\varrho)$ and $\det R(\varrho)$ in Example 8.6.

Problem 8.6 *Parameter estimates when the system does not belong to the model structure*
Consider the system

$$y(t) = u(t - 1) + u(t - 2) + e(t)$$

where the input $u(t)$ and $e(t)$ are mutually independent white noise sequences of zero means and variances σ^2 and λ^2, respectively. Assume that the system is identified using the model structure

$$y(t) + ay(t - 1) = bu(t - 1) + \varepsilon(t)$$

Derive the asymptotic (for $N \to \infty$) expressions of the LS estimate and the basic IV estimate based on the instruments

$$z(t) = (u(t - 1) \quad u(t - 2))^T$$

of the parameters a and b. Examine also the stability properties of the models so obtained.

Problem 8.7 *Accuracy of a simple IV method applied to a first-order ARMAX system*
Consider the system

$$y(t) + ay(t - 1) = bu(t - 1) + e(t) + ce(t - 1)$$

where $u(t)$ and $e(t)$ are mutually uncorrelated white noise sequences of zero means and variances σ^2 and λ^2, respectively. Assume that a and b are estimated with an IV method using delayed input values as instruments:

$$z(t) = (u(t - 1) \quad u(t - 2))^T$$

No prefiltering is used. Derive the normalized covariance matrix of the parameter estimates.

Problem 8.8 *Accuracy of an optimal IV method applied to a first-order ARMAX system*
(a) Consider the same system as in Problem 8.7. Let a and b be estimated with an optimal IV method. Calculate the normalized covariance matrix of the parameter estimates.
(b) Compare with the corresponding covariance matrix P_{IV} for the basic IV method of Problem 8.7. Show that $P_{IV} > P_{IV}^{opt}$ for $c \neq 0$.
(c) Compare with the corresponding covariance matrix for the prediction error method, P_{PEM}, see Problem 7.16. Show that

$$\tilde{P} \triangleq P_{IV}^{opt} - P_{PEM} \geq 0$$

and that $\tilde{P}$ always is singular (hence positive semidefinite), and equal to zero for $a = c$.

Problem 8.9 *A necessary condition for consistency of IV methods*
Show that (8.26) are necessary conditions for the consistency of IV methods.
  Hint. Consider the solution of

$$x^{\mathrm{T}}[EZ(t)Z^{\mathrm{T}}(t)]x = 0$$

with respect to the nz-dimensional vector x. (The solutions will span a subspace – of what order?) Observe that

$$x^{\mathrm{T}}R = 0$$

and use this observation to make a deduction about the rank of R.

Problem 8.10 *Sufficient conditions for consistency of IV methods, extension of Problem 8.3*
Consider the difference equation model (8.2), (8.3) of a scalar (SISO) system. Assume that the system is described by

$$A^0(q^{-1})y(t) = B^0(q^{-1})u(t) + v(t)$$

where the polynomials A^0 and B^0 have the same structure and degree as A and B in (8.3), and where $v(t)$ and $u(s)$ are uncorrelated for all t and s. Let $u(t)$ be a white noise sequence with zero mean and variance σ_u^2. Finally, assume that the polynomials $A^0(q^{-1})$ and $B^0(q^{-1})$ have no common zeros.
  Show that under the above conditions the IV vector (8.15c) satisfies the consistency conditions (8.25a, b) for any stable prefilter $F(q^{-1})$ with $F(0) \neq 0$.
  Hint. Rewrite the vector $\phi(t)$ in the manner of the calculations made in Complement C5.1.

Problem 8.11 *The accuracy of extended IV methods does not necessarily improve when the number of instruments increases*
Consider an ARMA (1, 1) process

$$y(t) + ay(t-1) = e(t) + ce(t-1) \qquad |a| < 1; \qquad Ee(t)e(s) = \delta_{t,s}$$

Let $\hat{a}$ denote the IV estimate of a obtained from

$$\left\| \left[\sum_{t=1}^{N} Z(t)y(t-1) \right] \hat{a} + \left[\sum_{t=1}^{N} Z(t)y(t) \right] \right\|^2 = \min$$

where N denotes the number of data, and

$$Z(t) = (y(t-2) \ldots y(t-1-n))^{\mathrm{T}} \qquad n \geq 1$$

Verify that the IV estimate $\hat{a}$ is consistent. The asymptotic variance of $\hat{a}$ is given by (8.30)–(8.32):

$$P_n \triangleq [R^{\mathrm{T}}R]^{-1}R^{\mathrm{T}}[EC(q^{-1})Z(t)C(q^{-1})Z^{\mathrm{T}}(t)]R[R^{\mathrm{T}}R]^{-1} \qquad (i)$$

where

$$R = EZ(t)y(t-1)$$
$$C(q^{-1}) = 1 + cq^{-1}$$

In (i) the notation stresses the dependence of P on n (the number of instruments). Evaluate P_1 and P_2. Show that the inequality $P_1 > P_2$ is not necessarily true, as might be expected.

Remark. Complement C8.1 contains a detailed discussion of the type of IV estimates considered in this problem.

Problem 8.12 *A weighting sequence matching property of the IV method*
Consider the linear system

$$y(t) = G(q^{-1})u(t) + v(t)$$

where the input $u(t)$ is *white noise* which is uncorrelated with the stationary disturbance $v(s)$, for all t and s. The transfer function $G(q^{-1})$ is assumed to be stable but otherwise is not restricted in any way. The model considered is (8.2); note that the system does not necessarily belong to the model set. The model parameters are estimated using the basic IV estimate (8.12) with the IV vector given by (8.15c).

Let $\{h_k\}$ denote the system weighting function sequence (i.e. $\Sigma_{k=1}^{\infty} h_k q^{-k} = G(q^{-1})$). Similarly, let $\{\hat{h}_k\}$ denote the weighting sequence of the model provided by the IV method above, for $N \to \infty$. Show that

$$\hat{h}_k = h_k \quad \text{for } k = 1, \ldots, na + nb$$

That is to say, the model weighting sequence exactly matches the first $(na + nb)$ coefficients of the system weighting sequence. Comment on this property.

Bibliographical notes

The IV method was apparently introduced by Reiersøl (1941) and has been popular in the statistical field for quite a long period. It has been applied to and adapted for dynamic systems both in econometrics and in control engineering. In the latter field pioneering work has been carried out by Wong and Polak (1967), Young (1965, 1970), Mayne (1967), Rowe (1970) and Finigan and Rowe (1974). A well written historical background to the use of IV methods in econometrics and control can be found in Young (1976), who also discusses some refinements of the basic IV method.

For a more recent and detailed appraisal of IV methods see the book Söderström and Stoica (1983) and the papers Söderström and Stoica (1981c), Stoica and Söderström (1983a, b). In particular, detailed proofs of the results stated in this chapter as well as many additional references on IV methods can be found in these works. See also Young (1984) for a detailed treatment of IV methods.

The optimal IV method given by (8.38) has been analyzed by Stoica and Söderström (1983a, b). For the approximate implementation see also the schemes presented by Young and Jakeman (1979), Jakeman and Young (1979).

For additional topics (not covered in this chapter) on IV estimation the following papers may be consulted: Söderström, Stoica and Trulsson (1987) (IVMs for closed loop systems), Stoica *et al.* (1985b), Stoica, Söderström and Friedlander (1985) (optimal IV estimation of the AR parameters of an ARMA process), Stoica and Söderström (1981) (analysis of the convergence and accuracy properties of some bootstrapping IV schemes), Stoica and Söderström (1982d) (IVMs for a certain class of nonlinear systems), Benveniste and Fuchs (1985) (extension of the consistency analysis to nonstationary disturbances), and Young (1981) (IV methods applied to continuous time models).

Appendix A8.1
Covariance matrix of IV estimates

A proof is given here of the results (8.29)–(8.32) on the asymptotic distribution of the IV estimates. From (8.6) and (8.14),

$$\sqrt{N}(\hat{\theta} - \theta_0) = (R_N^T Q R_N)^{-1} R_N^T Q \frac{1}{\sqrt{N}} \sum_{t=1}^{N} Z(t) F(q^{-1}) v(t) \qquad (A8.1.1)$$

Now, from assumption A6 and Lemma B.3,

$$\frac{1}{\sqrt{N}} \sum_{t=1}^{N} Z(t) F(q^{-1}) v(t) \xrightarrow{\text{dist}} \mathcal{N}(0, P_0) \text{ as } N \to \infty \qquad (A8.1.2)$$

where

$$P_0 = \lim_{N \to \infty} \frac{1}{N} E \sum_{t=1}^{N} Z(t) F(q^{-1}) v(t) \sum_{s=1}^{N} (F(q^{-1}) v(s))^T Z^T(s)$$

The convergence in distribution (8.29) then follows easily from the convergence of R_N to R and Lemma B.4. It remains to show that P_0 as defined by (A8.1.2) is given by

$$P_0 = \left[\sum_{i=0}^{\infty} Z(t+i) K_i \right] \Lambda \left[\sum_{j=0}^{\infty} K_j^T Z^T(t+j) \right] \qquad (A8.1.3)$$

For convenience, let $K_i = 0$ for $i < 0$. Some straightforward calculations give

$$P_0 = \lim_{N \to \infty} \frac{1}{N} E \sum_{t=1}^{N} Z(t) \sum_{i=0}^{\infty} K_i e(t-i) \sum_{s=1}^{N} \sum_{j=0}^{\infty} e^T(s-j) K_j^T Z^T(s)$$

$$= \lim_{N \to \infty} \frac{1}{N} \sum_{t=1}^{N} \sum_{s=1}^{N} \sum_{i=0}^{\infty} \sum_{j=0}^{\infty} E Z(t) K_i \Lambda \delta_{t-i,s-j} K_j^T Z^T(s)$$

$$= \lim_{N \to \infty} \frac{1}{N} E \sum_{\tau=-N}^{N} \sum_{i=0}^{\infty} (N - |\tau|) Z(t) K_i \Lambda K_{i+\tau}^T Z^T(t+\tau)$$

$$= \sum_{\tau=-\infty}^{\infty} \sum_{i=0}^{\infty} EZ(t+i) K_i \Lambda K_{i+\tau}^{\mathrm{T}} Z^{\mathrm{T}}(t+i+\tau)$$

$$- \lim_{N \to \infty} \frac{1}{N} \sum_{\tau=-N}^{N} |\tau| \sum_{i=0}^{\infty} EZ(t) K_i \Lambda K_{i+\tau}^{\mathrm{T}} Z^{\mathrm{T}}(t+\tau)$$

$$= E\left[\sum_{i=0}^{\infty} Z(t+i) K_i\right] \Lambda \left[\sum_{j=0}^{\infty} K_j^{\mathrm{T}} Z^{\mathrm{T}}(t+j)\right] \tag{A8.1.4}$$

which proves (A8.1.3). The second equality of (A8.1.4) made use of assumption A6. The last equality in (A8.1.4) follows from the assumptions, since

$$\left\| EZ(t) \left(\sum_{i=0}^{\infty} K_i \Lambda K_{i+\tau}^{\mathrm{T}} \right) Z^{\mathrm{T}}(t+\tau) \right\| \leq C\alpha^{|\tau|}$$

for some constants $C > 0$, $0 < \alpha < 1$. Hence

$$\left\| \frac{1}{N} \sum_{\tau=-N}^{N} |\tau| \sum_{i=0}^{\infty} EZ(t) K_i \Lambda K_{i+\tau}^{\mathrm{T}} Z^{\mathrm{T}}(t+\tau) \right\|$$

$$\leq \frac{1}{N} \sum_{\tau=-N}^{N} |\tau| C\alpha^{|\tau|} \leq \frac{2}{N} \sum_{\tau=0}^{\infty} C|\tau|\alpha^{|\tau|} \to 0, \text{ as } N \to \infty$$

Next turn to (8.32). In the scalar case ($ny = 1$), K_i and Λ are scalars, so they commute and

$$P_0 = \Lambda \sum_{i=0}^{\infty} \sum_{j=0}^{\infty} K_i K_j EZ(t+i) Z^{\mathrm{T}}(t+j)$$

$$= \Lambda \sum_{i=0}^{\infty} \sum_{j=0}^{\infty} K_i K_j EZ(t+i-i-j) Z^{\mathrm{T}}(t+j-i-j)$$

$$= \Lambda E\left[\sum_{j=0}^{\infty} K_j Z(t-j)\right]\left[\sum_{i=0}^{\infty} K_i Z^{\mathrm{T}}(t-i)\right] \tag{A8.1.5}$$

$$= \Lambda E[K(q^{-1})Z(t)][K(q^{-1})Z(t)]^{\mathrm{T}}$$

$$= \Lambda E[F(q^{-1})H(q^{-1})Z(t)][F(q^{-1})H(q^{-1})Z(t)]^{\mathrm{T}}$$

which is (8.32).

Appendix A8.2
Comparison of optimal IV and prediction error estimates

This appendix will give a derivation of the covariance matrix P_{PEM} of the prediction error estimates of the parameters θ in the model (8.39), as well as a comparison of the matrices P_{PEM} and $P_{\mathrm{IV}}^{\mathrm{opt}}$. To evaluate the covariance matrix P_{PEM} the gradient of the prediction error must be found. For the model (8.39),

$$\varepsilon(t, \eta) = H^{-1}(q^{-1}; \theta, \beta)[y(t) - \phi^T(t)\theta] \tag{A8.2.1}$$

$$\eta \triangleq (\theta^T \ \beta^T)^T$$

and after some straightforward calculations

$$\left.\frac{\partial \varepsilon(t, \eta)}{\partial \theta_i}\right|_{\eta=\eta_0} = -H^{-1}(q^{-1}) \left.\frac{\partial H(q^{-1}; \eta)}{\partial \theta_i}\right|_{\eta=\eta_0} \varepsilon(t, \eta_0)$$

$$- H^{-1}(q^{-1}) \frac{\partial}{\partial \theta_i} \phi^T(t)\theta \tag{A8.2.2a}$$

$$= -H^{-1}(q^{-1}) \left.\frac{\partial H(q^{-1}; \eta)}{\partial \theta_i}\right|_{\eta=\eta_0} e(t) - H^{-1}(q^{-1})\phi^T(t)e_i$$

$$\left.\frac{\partial \varepsilon(t, \eta)}{\partial \beta_i}\right|_{\eta=\eta_0} = -H^{-1}(q^{-1}) \left.\frac{\partial H(q^{-1}; \eta)}{\partial \beta_i}\right|_{\eta=\eta_0} e(t) \tag{A8.2.2b}$$

where e_i denotes the ith unit vector. (Note that $\varepsilon(t, \eta_0) = e(t)$.) Clearly the gradient w.r.t. θ will consist of two independent parts, one being a filtered input and the other filtered noise. The gradient w.r.t. β is independent of the input. To facilitate the forthcoming calculations, introduce the notation

$$\varepsilon_\theta(t) \triangleq \left.\frac{\partial \varepsilon(t, \eta)}{\partial \theta}\right|_{\eta=\eta_0} = \varepsilon_\theta^u(t) + \varepsilon_\theta^e(t)$$

$$\varepsilon_\beta(t) \triangleq \left.\frac{\partial \varepsilon(t, \eta)}{\partial \beta}\right|_{\eta=\eta_0} \tag{A8.2.3}$$

In (A8.2.3) $\varepsilon_\theta^u(t)$ denotes the part depending on the input. It is readily found from (A8.2.2a) that

$$\varepsilon_\theta^u(t) = -H^{-1}(q^{-1})\tilde{\phi}^T(t) \tag{A8.2.4}$$

Introduce further the notation

$$Q_{\theta\theta} \triangleq E\varepsilon_\theta^T(t)\Lambda^{-1}\varepsilon_\theta(t)$$

$$Q_{\theta\theta}^u \triangleq E(\varepsilon_\theta^u(t))^T\Lambda^{-1}\varepsilon_\theta^u(t)$$

$$Q_{\theta\theta}^e \triangleq E(\varepsilon_\theta^e(t))^T\Lambda^{-1}\varepsilon_\theta^e(t)$$

$$Q_{\theta\beta} \triangleq E\varepsilon_\theta^T(t)\Lambda^{-1}\varepsilon_\beta(t) \tag{A8.2.5}$$

$$Q_{\beta\theta} \triangleq Q_{\theta\beta}^T$$

$$Q_{\beta\beta} \triangleq E\varepsilon_\beta^T(t)\Lambda^{-1}\varepsilon_\beta(t)$$

$$\bar{Q} \triangleq Q_{\theta\theta}^e - Q_{\theta\beta}Q_{\beta\beta}^{-1}Q_{\beta\theta}$$

Then it follows from (8.37) that

$$(P_{IV}^{opt})^{-1} = Q_{\theta\theta}^u \tag{A8.2.6}$$

Further, the matrix

$$\begin{pmatrix} Q^e_{\theta\theta} & Q_{\theta\beta} \\ Q_{\beta\theta} & Q_{\beta\beta} \end{pmatrix} = E \begin{pmatrix} (\varepsilon^e_\theta(t))^T \\ \varepsilon^T_\beta(t) \end{pmatrix} \Lambda^{-1} (\varepsilon^e_\theta(t) \quad \varepsilon_\beta(t))$$

is obviously nonnegative definite, which implies

$$\bar{Q} \geq 0 \tag{A8.2.7}$$

(see Lemma A.3). The covariance matrix of the optimal prediction error estimates of the parameter vector η can be found from (7.72). Using the above notation it can be written as

$$\text{cov} \begin{pmatrix} \hat{\theta} \\ \hat{\beta} \end{pmatrix}_{\text{PEM}} = \begin{pmatrix} (P^{\text{opt}}_{\text{IV}})^{-1} + Q^e_{\theta\theta} & Q_{\theta\beta} \\ Q_{\beta\theta} & Q_{\beta\beta} \end{pmatrix}^{-1} \tag{A8.2.8}$$

In particular, for the $\hat{\theta}$ parameters (see Lemma A.3),

$$\boxed{P_{\text{PEM}} = [(P^{\text{opt}}_{\text{IV}})^{-1} + \bar{Q}]^{-1} \leq P^{\text{opt}}_{\text{IV}}} \tag{A8.2.9}$$

The inequality (A8.2.9) confirms the conjecture that the optimal prediction error estimate is (asymptotically) at least as accurate as an optimal IV estimate. It is also found that these two estimates have the same covariance matrix if and only if

$$\bar{Q} = 0 \tag{A8.2.10}$$

A typical case when (A8.2.10) applies is when

$$\varepsilon^e_\theta(t) = 0 \tag{A8.2.11}$$

since then $Q^e_{\theta\theta} = 0$, $Q_{\theta\beta} = 0$. Note that using the filter $\bar{H}$ defined by (8.44b), the model structure (8.39) can be rewritten as

$$A(q^{-1}; \theta)y(t) = B(q^{-1}; \theta)u(t) + A(q^{-1}; \theta)\bar{H}(q^{-1}; \theta, \beta)e(t) \tag{A8.2.12}$$

Hence the prediction error becomes

$$\varepsilon(t, \theta, \beta) = \bar{H}^{-1}(q^{-1}; \theta, \beta)[y(t) - A^{-1}(q^{-1}; \theta)B(q^{-1}; \theta)u(t)] \tag{A8.2.13}$$

from which it is easily seen that if the filter $\bar{H}(q^{-1}; \theta, \beta)$ *does not depend on* θ then (A8.2.11) is satisfied. Hence in such a case equality holds in (A8.2.9).

Complement C8.1
Yule–Walker equations

This complement will discuss, in two separate subsections, the Yule–Walker equation approach to estimating the AR parameters and the MA parameters of ARMA processes, and the relationship of this approach to the IV estimation techniques.

AR parameters

Consider the following scalar ARMA process:

$$A(q^{-1})y(t) = C(q^{-1})e(t) \tag{C8.1.1}$$

where

$$A(q^{-1}) = 1 + a_1 q^{-1} + \ldots + a_{na} q^{-na}$$
$$C(q^{-1}) = 1 + c_1 q^{-1} + \ldots + c_{nc} q^{-nc}$$
$$Ee(t) = 0 \quad Ee(t)e(s) = \lambda^2 \delta_{t,s}$$

According to the spectral factorization theorem it can be assumed that

$$A(z)C(z) \neq 0 \quad \text{for } |z| \leq 1$$

Let

$$r_k \triangleq Ey(t)y(t-k) \quad k = 0, \pm 1, \pm 2, \ldots$$

Next note that

$$EC(q^{-1})e(t)y(t-k) = 0 \quad \text{for } k > nc \tag{C8.1.2}$$

Thus, multiplying both sides of (C8.1.1) by $y(t-k)$ and taking expectations give

$$r_k + a_1 r_{k-1} + \ldots + a_{na} r_{k-na} = 0 \quad k = nc+1, nc+2, \ldots \tag{C8.1.3}$$

This is a linear system in the AR parameters $\{a_i\}$. The equations (C8.1.3) are often referred to as Yule–Walker equations. Consider the first m equations of (C8.1.3), which can be written compactly as

$$\boxed{Ra = -r} \tag{C8.1.4}$$

where $m \geq na$, and

$$a = (a_1 \ldots a_{na})^T$$
$$r = (r_{nc+1} \ldots r_{nc+m})^T$$
$$R = \begin{pmatrix} r_{nc} & \ldots & r_{nc+1-na} \\ \vdots & & \vdots \\ r_{nc+m-1} & \ldots & r_{nc+m-na} \end{pmatrix}$$

The matrix R has full rank. The proof of this fact can be found in Stoica (1981c) and Stoica, Söderström and Friedlander (1985).

The covariance elements of R and r can be estimated from the observations available, say $\{y(1), \ldots, y(N)\}$. Let $\hat{R}$ and $\hat{r}$ denote consistent estimates of R and r. Then $\hat{a}$ obtained from

$$\hat{R}\hat{a} = -\hat{r} \tag{C8.1.5}$$

is a consistent estimate of a. There is some terminology associated with (C8.1.5):

- For $m = na$, $\hat{a}$ is called a minimal YWE (Yule–Walker estimate).
- For $m > na$, the system is overdetermined and has to be solved in the least squares sense. Then, the estimate $\hat{a}$ satisfying

$$\|\hat{R}\hat{a} + \hat{r}\|^2 = \min$$

is called an overdetermined YWE, and the one which satisfies

$$\|\hat{R}\hat{a} + \hat{r}\|_Q^2 = \min$$

for some positive definite matrix $Q \neq I$ is called a weighted overdetermined YWE.

Next note that a possible form for $\hat{R}$ and $\hat{r}$ in (C8.1.5) is the following:

$$\hat{R} = \frac{1}{N}\sum_{t=1}^{N}\begin{pmatrix} y(t-nc-1) \\ \vdots \\ y(t-nc-m) \end{pmatrix}(y(t-1) \ldots y(t-na))$$

$$\hat{r} = \frac{1}{N}\sum_{t=1}^{N}\begin{pmatrix} y(t-nc-1) \\ \vdots \\ y(t-nc-m) \end{pmatrix}y(t) \tag{C8.1.6}$$

The estimate $\hat{a}$ corresponding to (C8.1.6) can be interpreted as an IV estimate (IVE). Indeed, (C8.1.1) can be rewritten as

$$y(t) = -(y(t-1) \ldots y(t-na))a + C(q^{-1})e(t)$$

and the IV estimation of the vector a using the following IV vector:

$$z(t) = (-y(t-nc-1) \ldots -y(t-nc-m))^T$$

leads precisely to (C8.1.5), (C8.1.6). Other estimators of $\hat{R}$ and $\hat{r}$ are asymptotically (as $N \to \infty$) equivalent to (C8.1.6) and therefore lead to estimates which are asymptotically equal to the IVE (C8.1.5), (C8.1.6). The conclusion is that the asymptotic properties of the YW estimators follow from the theory developed in Section 8.2 for IVMs. A detailed analysis for ARMA processes can be found in Stoica, Söderström and Friedlander (1985).

For pure AR processes it is possible also to obtain a simple formula for λ^2. We have

$$\lambda^2 \triangleq Ee^2(t) = Ee(t)A(q^{-1})y(t) = Ee(t)y(t)$$
$$= EA(q^{-1})y(t)y(t) = r_0 + a_1 r_1 + \ldots + a_{na}r_{na} \tag{C8.1.7}$$

Therefore consistent estimates of the parameters $\{a_i\}$ and $\{\lambda\}$ of an AR process can be obtained replacing $\{r_i\}$ by sample covariances in

$$\begin{pmatrix} r_0 & r_1 & \ldots & r_{na} \\ r_1 & r_0 & & r_{na-1} \\ \vdots & & & \vdots \\ r_{na} & r_{na-1} & \ldots & r_0 \end{pmatrix}\begin{pmatrix} 1 \\ a_1 \\ \vdots \\ a_{na} \end{pmatrix} = \begin{pmatrix} \lambda^2 \\ 0 \\ \vdots \\ 0 \end{pmatrix} \tag{C8.1.8}$$

(cf. (C8.1.4), (C8.1.7)). It is not difficult to see that the estimate of a so obtained is, for large samples, equal to the asymptotically efficient least squares estimate of a. A computationally efficient way of solving the system (C8.1.8) is presented in Complement C8.2, while a similar algorithm for the system (C8.1.4) is given in Complement C8.3.

MA parameters

As shown above, see (C8.1.3), the AR parameters of an ARMA satisfy a set of linear equations, called YW equations. The coefficients of these equations are equal to the covariances of the ARMA process. To obtain a similar set of equations for the MA parameters, let us introduce

$$\gamma_k \triangleq \frac{1}{(2\pi)^2} \int_{-\pi}^{\pi} \frac{1}{\phi(\omega)} e^{i\omega k} d\omega \qquad (C8.1.9)$$

where $\phi(\omega)$ is the spectral density of the ARMA process (C8.1.1). γ_k is called the *inverse* covariance of $y(t)$ at lag k (see e.g. Cleveland, 1972; Chatfield, 1979). It can also be viewed as the ordinary covariance of the ARMA process $x(t)$ given by

$$C(q^{-1})x(t) = A(q^{-1})\varepsilon(t) \qquad E\varepsilon(t)\varepsilon(s) = \frac{1}{\lambda^2} \delta_{t,s} \qquad (C8.1.10)$$

Note the inversion of the MA and AR parts. Indeed, $x(t)$ defined by (C8.1.10) has the spectral density equal to $1/[(2\pi)^2 \phi(\omega)]$ and therefore its covariances are $\{\gamma_k\}$. It is now easy to see that the MA parameters $\{c_i\}$ obey the so-called 'inverse' YW equations:

$$\gamma_k + c_1 \gamma_{k-1} + \ldots + c_{nc} \gamma_{k-nc} = 0 \qquad k \geq na + 1 \qquad (C8.1.11)$$

To use these equations for estimation of $\{c_i\}$ one must first estimate $\{\gamma_k\}$. Methods for doing this can be found in Bhansali (1980). Essentially all these methods use some estimate of $\phi(\omega)$ (either a nonparametric estimate or one based on high-order AR approximation) to approximate $\{\gamma_k\}$ from (C8.1.9).

The use of long AR models to estimate the inverse covariances can be done as follows. First, fit a large-order, say n, AR model to the ARMA sequence $\{y(t)\}$:

$$L(q^{-1})y(t) = w(t) \qquad (C8.1.12)$$

where

$$L(q^{-1}) = \hat{\alpha}_0 + \hat{\alpha}_1 q^{-1} + \ldots + \hat{\alpha}_n q^{-n}$$

and where $w(t)$ are the AR model residuals. The coefficients $\{\hat{\alpha}_i\}$ can be determined using the YW equations (C8.1.5). For large n, the AR model (C8.1.12) will be a good description of the ARMA process (C8.1.1), and therefore $w(t)$ will be close to a white noise of zero mean and unit variance. Since L in (C8.1.12) can be interpreted as an approximation to $A/(C\lambda)$ in the ARMA equation, it follows that the closer the zeros of C are to the unit circle, the larger n has to be for (C8.1.12) to be a good approximation of the ARMA process (C8.1.1).

The discussion above implies that, for sufficiently large n a good approximation of the inverse spectral density function is given by

$$1/\hat{\phi}(\omega) = 2\pi |L(e^{i\omega})|^2$$

which readily gives the following estimates for the inverse covariances:

$$\hat{\gamma}_k = \sum_{i=0}^{n-k} \hat{a}_i \hat{a}_{i+k} \tag{C8.1.13}$$

Inserting (C8.1.13) into the inverse YW equations, one can obtain estimates of the MA parameters of the ARMA process.

It is interesting to note that the above approach to estimation of the MA parameters of an ARMA process is quite related to a method introduced by Durbin in a different way (see Durbin, 1959; Anderson, 1971).

Complement C8.2
The Levinson–Durbin algorithm

Consider the following linear systems of equations with $\{a_{k,i}\}_{i=1}^{k}$ and σ_k as unknowns:

$$\begin{aligned} R_k \theta_k &= -r_k \\ \varrho_0 + \theta_k^T r_k &= \sigma_k \end{aligned} \quad k = 1, \ldots, n \tag{C8.2.1}$$

where

$$R_k = \begin{pmatrix} \varrho_0 & \varrho_1 & \cdots & \varrho_{k-1} \\ \varrho_1 & \varrho_0 & \cdots & \varrho_{k-2} \\ \vdots & \vdots & \ddots & \vdots \\ \varrho_{k-1} & \varrho_{k-2} & \cdots & \varrho_0 \end{pmatrix} \quad r_k = \begin{pmatrix} \varrho_1 \\ \varrho_2 \\ \vdots \\ \varrho_k \end{pmatrix} \tag{C8.2.2}$$

$$\theta_k = (a_{k,1} \ldots a_{k,k})^T$$

and where $\{\varrho_i\}_{i=0}^{k}$ are given. The matrices $\{R_k\}$ are assumed to be nonsingular. Such systems of equations appear, for example, when fitting AR models of orders $k = 1, 2, \ldots, n$ to given (or estimated) covariances $\{\varrho_i\}$ of the data by the YW method (Complement C8.1). In this complement a number of interesting aspects pertaining to (C8.2.1) are discussed. First it is shown that the parameters $\{\theta_k, \sigma_k\}_{k=1}^{m}$ readily determine a Cholesky factorization of R_{m+1}^{-1}. Next a numerically efficient algorithm, the so-called Levinson–Durbin algorithm (LDA), is presented for solving equations (C8.2.1). Finally some properties of the LDA and some of its applications will be given.

Cholesky factorization of R_{m+1}^{-1}

Equations (C8.2.1) can be written in the following compact form:

$$\begin{pmatrix} 1 & a_{m,1} & \cdots & a_{m,m} \\ & \ddots & & \vdots \\ 0 & & 1 & a_{1,1} \\ & & & 1 \end{pmatrix} \begin{pmatrix} \varrho_0 & \varrho_1 & \cdots & \varrho_m \\ \varrho_1 & \varrho_0 & \cdots & \varrho_{m-1} \\ \vdots & \vdots & \ddots & \varrho_1 \\ \varrho_m & \varrho_{m-1} & \cdots & \varrho_0 \end{pmatrix} \begin{pmatrix} 1 & & & \\ a_{m,1} & 1 & & 0 \\ \vdots & & \ddots & 1 \\ a_{m,m} & \cdots & a_{1,1} & 1 \end{pmatrix}$$
$$\underbrace{}_{U_{m+1}^T} \qquad \underbrace{}_{R_{m+1}} \qquad \underbrace{}_{U_{m+1}}$$

$$= \underbrace{\begin{pmatrix} \sigma_m & & & \\ & \ddots & & 0 \\ & & \sigma_1 & \\ 0 & & & \varrho_0 \end{pmatrix}}_{D_{m+1}}$$
(C8.2.3)

To see this, note first that (C8.2.3) clearly holds for $m = 1$ since
$$\begin{pmatrix} 1 & a_{1,1} \\ 0 & 1 \end{pmatrix} \begin{pmatrix} \varrho_0 & \varrho_1 \\ \varrho_1 & \varrho_0 \end{pmatrix} \begin{pmatrix} 1 & 0 \\ a_{1,1} & 1 \end{pmatrix} = \begin{pmatrix} \sigma_1 & 0 \\ 0 & \varrho_0 \end{pmatrix}$$

Next it will be shown that if (C8.2.3) holds for $m + 1 = k$, i.e.
$$U_k^T R_k U_k = D_k$$
then it also holds for $m + 1 = k + 1$,
$$U_{k+1}^T R_{k+1} U_{k+1} = \begin{pmatrix} 1 & \theta_k^T \\ 0 & U_k^T \end{pmatrix} \begin{pmatrix} \varrho_0 & r_k^T \\ r_k & R_k \end{pmatrix} \begin{pmatrix} 1 & 0 \\ \theta_k & U_k \end{pmatrix}$$
$$= \begin{pmatrix} \varrho_0 + \theta_k^T r_k & r_k^T + \theta_k^T R_k \\ U_k^T r_k & U_k^T R_k \end{pmatrix} \begin{pmatrix} 1 & 0 \\ \theta_k & U_k \end{pmatrix}$$
$$= \begin{pmatrix} \sigma_k & 0 \\ U_k^T r_k & U_k^T R_k \end{pmatrix} \begin{pmatrix} 1 & 0 \\ \theta_k & U_k \end{pmatrix} = \begin{pmatrix} \sigma_k & 0 \\ U_k^T (r_k + R_k \theta_k) & U_k^T R_k U_k \end{pmatrix}$$
$$= D_{k+1}$$

which proves the assertion. Thus by induction, (C8.2.3) holds for $m = 1, 2, \ldots, n$.
Now, from (C8.2.3),

$$\boxed{R_{m+1}^{-1} = U_{m+1} D_{m+1}^{-1} U_{m+1}^T} \qquad (C8.2.4)$$

which shows that $U_{m+1} D_{m+1}^{-1/2}$ is the lower triangular Cholesky factor of the inverse covariance matrix R_{m+1}^{-1}.

The Levinson–Durbin algorithm

The following derivation of the Levinson–Durbin algorithm (LDA) relies on the Cholesky factorization above. For other derivations, see Levinson (1947), Durbin

(1960), Söderström and Stoica (1983), Nehorai and Morf (1985), Demeure and Scharf (1987).

Introduce the following convention. If $x = (x_1 \ldots x_n)^T$ then the reversed vector $(x_n\ x_{n-1} \ldots x_1)^T$ is denoted by x^R. It can be readily verified that due to the Toeplitz structure of the matrices R_k, equations (C8.2.1) give

$$\theta_{k+1}^R = -R_{k+1}^{-1} r_{k+1}^R \tag{C8.2.5}$$

Inserting the factorization (C8.2.4) of R_{k+1}^{-1} into (C8.2.5) gives

$$\begin{aligned}
\theta_{k+1}^R &= -\begin{pmatrix} 1 & 0 \\ \theta_k & U_k \end{pmatrix}\begin{pmatrix} \sigma_k^{-1} & 0 \\ 0 & D_k^{-1} \end{pmatrix}\begin{pmatrix} 1 & \theta_k^T \\ 0 & U_k^T \end{pmatrix}\begin{pmatrix} \varrho_{k+1} \\ r_k^R \end{pmatrix} \\
&= -\begin{pmatrix} 1/\sigma_k & \theta_k^T/\sigma_k \\ \theta_k/\sigma_k & R_k^{-1} + \theta_k \theta_k^T/\sigma_k \end{pmatrix}\begin{pmatrix} \varrho_{k+1} \\ r_k^R \end{pmatrix} \\
&= \begin{pmatrix} -(\varrho_{k+1} + \theta_k^T r_k^R)/\sigma_k \\ -\theta_k(\varrho_{k+1} + \theta_k^T r_k^R)/\sigma_k + \theta_k^R \end{pmatrix}
\end{aligned} \tag{C8.2.6}$$

Next note that making use of (C8.2.1), (C8.2.6) one can write

$$\begin{aligned}
\sigma_{k+1} &= \varrho_0 + \theta_{k+1}^T r_{k+1} = \varrho_0 + (\theta_{k+1}^R)^T r_{k+1}^R \\
&= \varrho_0 + (a_{k+1,k+1}\ (\theta_k^R)^T + a_{k+1,k+1}\theta_k^T)\begin{pmatrix} \varrho_{k+1} \\ r_k^R \end{pmatrix} \\
&= \underbrace{\varrho_0 + (\theta_k^R)^T r_k^R}_{\sigma_k} + \underbrace{a_{k+1,k+1}(\varrho_{k+1} + \theta_k^T r_k^R)}_{-\sigma_k a_{k+1,k+1}} \\
&= \sigma_k - a_{k+1,k+1}^2 \sigma_k
\end{aligned} \tag{C8.2.7}$$

It now follows from (C8.2.6) and (C8.2.7) that the following recursion, called the LDA, holds (for $k = 1, \ldots, n - 1$):

$$\boxed{\begin{aligned}
a_{k+1,k+1} &= -(\varrho_{k+1} + a_{k,1}\varrho_k + \ldots + a_{k,k}\varrho_1)/\sigma_k \\
a_{k+1,i} &= a_{k,i} + a_{k+1,k+1}a_{k,k+1-i} \quad i = 1, \ldots, k \\
\sigma_{k+1} &= \sigma_k(1 - a_{k+1,k+1}^2)
\end{aligned}} \tag{C8.2.8}$$

with the initial values

$$a_{1,1} = -\varrho_1/\varrho_0$$

$$\sigma_1 = \varrho_0 + \varrho_1 a_{1,1} = \varrho_0 - \varrho_1^2/\varrho_0 = \varrho_0(1 - a_{1,1}^2)$$

Implementation of the LDA requires $O(k)$ arithmetic operations at iteration k. Thus, determination of $\{\theta_k, \sigma_k\}_{k=1}^n$ requires $O(n^2)$ operations. If the system of equations (C8.2.1) is treated using a standard routine not exploiting the Toeplitz structure, then $O(k^3)$ arithmetic operations are required for determining (θ_k, σ_k), leading to a total of

$O(n^4)$ operations for computation of all $\{\theta_k, \sigma_k\}_{k=1}^n$. Hence for large values of n, the LDA will be much faster than a standard routine for solving linear systems.

Note that the LDA should be stopped if for some k, $a_{k+1,k+1} = 1$. In numerical calculations, however, such a situation is unlikely to occur. For more details on the LDA and its numerical properties see Cybenko (1980), Friedlander (1982).

It is worth noting that if only θ_n and σ_n are required then the LDA can be further simplified. The so-called split Levinson algorithm, which determines θ_n and σ_n, is about twice faster than the LDA; see Delsarte and Genin (1986).

Some properties of the LDA

The following statements are equivalent:

(i) $|a_{k,k}| < 1$ for $k = 1, \ldots, n$; and $\varrho_0 > 0$
(ii) $R_{n+1} > 0$
(iii) $A_n(z) \triangleq 1 + a_{n,1}z + \ldots + a_{n,n}z^n \neq 0$ for $|z| \leq 1$; and $\varrho_0 > 0$

Proof. First consider the equivalence (i) ⇔ (ii). It follows from (C8.2.3) that

$$\det R_{m+1} = \varrho_0 \prod_{k=1}^{m} \sigma_k$$

which implies (since $\det R_1 = \varrho_0$),

$$\det R_{m+1} = \sigma_m \det R_m \quad m = 1, 2 \ldots$$

Thus the matrix R_{n+1} is positive definite if and only if $\varrho_0 > 0$ and $\sigma_k > 0$, $k = 1, \ldots, n$. However the latter condition is equivalent to (i).

Next consider the equivalence (i) ⇔ (iii). We make use of Rouché's theorem to prove this equivalence. This theorem (see Pearson, 1974, p. 253) states that if $f(z)$ and $g(z)$ are two functions which are analytic inside and on a closed contour C and $|g(z)| < |f(z)|$ for $z \in C$, then $f(z)$ and $f(z) + g(z)$ have the same number of zeros inside C. It follows from (C8.2.8) that

$$A_{k+1}(z) = A_k(z) + a_{k+1,k+1}z^{k+1}A_k(z^{-1}) \tag{C8.2.9}$$

where $A_k(z)$ and $A_{k+1}(z)$ are defined similarly to $A_n(z)$. If (i) holds then, for $|z| = 1$,

$$|A_k(z)| = |z^{k+1}A_k(z^{-1})| > |a_{k+1,k+1}z^{k+1}A_k(z^{-1})|$$

Thus, according to Rouché's theorem $A_{k+1}(z)$ and $A_k(z)$ have the same number of zeros inside the unit circle. Since $A_1(z) = 1 + a_{1,1}z$ has no zero inside the unit circle when (i) holds, it follows that $A_k(z)$, $k = 1, \ldots, n$ have all their zeros strictly outside the unit disc.

Next assume that (iii) holds. This clearly implies $|a_{n,n}| < 1$. It also follows that

$$A_{n-1}(z) \neq 0 \quad \text{for } |z| \leq 1 \tag{C8.2.10}$$

For if $A_{n-1}(z)$ has zeros within the unit circle, then Rouché's theorem, (C8.2.9) and $|a_{n,n}| < 1$ would imply that $A_n(z)$ also must have zeros inside the unit circle, which is a contradiction. From (C8.2.10) it follows that $|a_{n-1,n-1}| < 1$, and we can continue in the same manner as above to show that (i) holds. ∎

For more details on the above properties of the LDA and, in particular, their extension to the 'boundary case' in which $|a_{k,k}| = 1$ for some k, see Söderström and Stoica (1983), Stoica and Söderström (1985).

Some applications of the LDA

As discussed above, the LDA is a numerically efficient tool for:

- Computing the UDU^T factorization of the inverse of some symmetric Toeplitz matrix R_{n+1}.
- Solving the linear systems of YW equations (C8.2.1) in AR model fitting.
- Testing whether a given sequence $\{\varrho_k\}_{k=0}^n$ is positive definite or not.

In the following we will show two other applications of the LDA.

Reparametrization for stability

For some optimization problems in system identification, the independent variables are coefficients of certain polynomials (see e.g. Chapter 7). Quite often these coefficients need to be constrained in the course of optimization to guarantee that the corresponding polynomial is stable. More specifically, let

$$A(z) = 1 + a_1 z + \ldots + a_n z^n$$

and

$$\mathcal{D} = \{\{a_i\}_{i=1}^n \mid A(z) \neq 0 \quad \text{for } |z| \leq 1\}$$

It is required that $\{a_i\}$ belong to $\mathcal{D}$. To handle this problem it would be convenient to reparametrize $A(z)$ in terms of some new variables, say $\{\alpha_i\}_{i=1}^n$, which are such that when $\{\alpha_i\}$ span $\mathcal{R}^n$, the corresponding coefficients $\{a_i\}$ of $A(z)$ will span $\mathcal{D}$. Then the optimization could be carried out *without constraint*, with respect to $\{\alpha_i\}$. Such a reparametrization of $A(z)$ can be obtained using the LDA.

Let

$$\phi_k = \frac{1 - e^{\alpha_k}}{1 + e^{\alpha_k}} \quad k = 1, \ldots, n \quad \alpha_k \in \mathcal{R}$$

$$a_{k,k} = \phi_k \quad k = 1, \ldots, n$$

$$a_{k+1,i} = a_{k,i} + a_{k+1,k+1} a_{k,k+1-i} \quad i = 1, \ldots, k \text{ and } k = 1, \ldots, n-1$$

(cf. Jones (1980)). Note that the last equation above is the second one in the LDA (C8.2.8). When $\{\alpha_k\}$ span $\mathcal{R}^n$, the $\{\phi_k\}$ span the interval $(-1, 1)$. Then the coefficients $\{a_{n,i}\}$ of $A_n(z)$ span $\mathcal{D}$, according to the equivalence (i) ⇔ (iii) proved earlier in this complement.

Reparametrization for positive definiteness

Some problems in system identification need optimization with respect to covariance coefficients, say $\varrho_0, \varrho_1, \ldots, \varrho_n$ (see Söderström and Stoica, 1983; Goodwin and Payne, 1973; Stoica and Söderström, 1982b, for some examples). In the course of optimization $\{\varrho_i\}$ need to be constrained to make them belong to the following set:

$$\mathcal{D} = \{\varrho_i | R_{n+1} > 0\}$$

Let

$$\phi_k = \frac{1 - e^{\alpha_k}}{1 + e^{\alpha_k}} \quad k = 1, \ldots, n, \quad \alpha_k \in \mathcal{R}$$

$$a_{k,k} = \phi_k \quad k = 1, \ldots, n$$

By rearranging the LDA it is possible to determine the sequence $\{\varrho_i\}_{i=0}^n$ corresponding to $\{a_{k,k}\}_{k=1}^n$. The rearranged LDA runs as follows. For $k = 1, \ldots, n-1$,

$$\begin{cases} \varrho_{k+1} = -\sigma_k a_{k+1,k+1} - a_{k,1}\varrho_k - \ldots - a_{k,k}\varrho_1 \\ \sigma_{k+1} = \sigma_k(1 - a_{k+1,k+1}^2) \\ a_{k+1,i} = a_{k,i} + a_{k+1,k+1}a_{k,k+1-i} \quad i = 1, \ldots, k \end{cases} \quad (C8.2.11)$$

with initial values (setting the 'scaling parameter' ϱ_0 to $\varrho_0 = 1$)

$$\varrho_1 = -a_{1,1}$$
$$\sigma_1 = 1 - a_{1,1}^2$$

When $\{\alpha_k\}$ span $\mathcal{R}^n$, the $\{\psi_k = a_{k,k}\}$ span the interval $(-1, 1)$ and, therefore, $\{\varrho_i\}$ span $\mathcal{D}$ (cf. the previous equivalence (i) ⇔ (ii)). Thus the optimization may be carried out without constraints with respect to $\{\alpha_i\}$.

Note that the reparametrizations in terms of $\{a_{k,k}\}$ discussed above also suggest simple means for obtaining the nonnegative definite approximation of a given sequence $\{\varrho_0, \ldots, \varrho_n\}$, or the stable approximant of a given $A(z)$ polynomial. For details on these approximation problems we refer to Stoica and Moses (1987). Here we note only that to approach the second approximation problem mentioned above we need a rearrangement of the LDA which makes it possible to get the sequence $\{a_{k,k}\}_{k=1}^n$ from a given sequence of coefficients $\{a_{n,i}\}_{i=1}^n$. For this purpose, observe that (C8.2.9) implies

$$z^{k+1}A_{k+1}(z^{-1}) = z^{k+1}A_k(z^{-1}) + \phi_{k+1}A_k(z) \quad (C8.2.12)$$

Eliminating the polynomial $z^{k+1}A_k(z^{-1})$ from (C8.2.9) and (C8.2.12), (assuming that $|\phi_{k+1}| \neq 1$), gives

$$A_k(z) = [A_{k+1}(z) - \phi_{k+1}z^{k+1}A_{k+1}(z^{-1})]/(1 - \phi_{k+1}^2) \quad k = n-1, \ldots, 1$$

$$(C8.2.13)$$

or, equivalently

$$a_{k,i} = (a_{k+1,i} - a_{k+1,k+1}a_{k+1,k+1-i})/(1 - a_{k+1,k+1}^2) \quad \begin{array}{l} i = 1, \ldots, k \text{ and} \\ k = n-1, \ldots, 1 \end{array}$$

$$(C8.2.14)$$

According to the equivalence (i) ⇔ (iii) proved above, the stability properties of the given polynomial $A_n(z)$ are completely determined by the sequence $\{\phi_k = a_{k,k}\}_{k=1}^n$ provided by (C8.2.14). In fact, the backward recursion (C8.2.14) is one of the possible implementations of *the Schur–Cohn procedure for testing the stability of* $A_n(z)$ (see Vieira and Kailath, 1977; Stoica and Moses, 1987, for more details on this interesting connection between the (backward) LDA and the Schur–Cohn test procedure).

Complement C8.3
A Levinson-type algorithm for solving nonsymmetric Yule–Walker systems of equations

Consider the following linear systems of equations with $\{a_{n,i}\}_{i=1}^n$ as unknowns:

$$R_n \theta_n = -\varrho_n \qquad n = 1, 2, \ldots, M \qquad (C8.3.1)$$

where

$$\theta_n = (a_{n,1} \ \ldots \ a_{n,n})^T$$

$$R_n = \begin{pmatrix} r_m & r_{m-1} & \cdots & r_{m+1-n} \\ r_{m+1} & r_m & & r_{m+2-n} \\ \vdots & & & \vdots \\ r_{m+n-1} & r_{m+n-2} & \cdots & r_m \end{pmatrix} \qquad \varrho_n = \begin{pmatrix} r_{m+1} \\ \vdots \\ r_{m+n} \end{pmatrix}$$

and where m is some fixed integer and $\{r_i\}$ are given reals. Such systems of equations appear for example when determining the AR parameters of ARMA-models of orders $(1, m), (2, m), \ldots, (M, m)$ from the given (or estimated) covariances $\{r_i\}$ of the data by the Yule–Walker method (see Complement C8.1).

By exploiting the Toeplitz structure of the matrix R_n it is possible to solve the systems (C8.3.1) in $3M^2$ operations (see Trench, 1964; Zohar, 1969, 1974) (an operation is defined as one multiplication plus one addition). Note that use of a procedure which does not exploit the special structure of (C8.3.1) would require $O(M^4)$ operations to solve for $\theta_1, \ldots, \theta_M$. An algorithm is presented here which exploits not only the Toeplitz structure of R_n but also the specific form of ϱ_n. This algorithm needs $2M^2$ operations to provide $\theta_1, \ldots, \theta_M$ and is quite simple conceptually. Faster algorithms for solving (C8.3.1), which require in the order of $M(\log M)^2$ operations, have been proposed recently (see Kumar, 1985). However, these algorithms impose some unnecessarily strong conditions on the sequence of $\{r_i\}$. If these conditions (requiring the non-singularity of some matrices (others than R_n) formed from $\{r_i\}$) are not satisfied, then the algorithms cannot be used.

Throughout this complement it is assumed that

$$\det R_n \neq 0 \qquad \text{for } n = 1, \ldots, M \qquad (C8.3.2)$$

This is a minimal requirement if all $\{\theta_n\}_{n=1}^M$ are to be determined. However, if only θ_M is to be determined then it is only necessary to assume that $\det R_M \neq 0$. In such cases condition (C8.3.2) is stronger than necessary (e.g. note that for $r_m = 0$ and $r_{m-1}r_{m+1} \neq 0$, det $R_2 \neq 0$ while det $R_1 = 0$).

As a starting point, note the following nested structures of R_{n+1} and ϱ_{n+1}:

$$R_{n+1} = \begin{pmatrix} R_n & \beta_n \\ \tilde{\varrho}_n^T & r_m \end{pmatrix} = \begin{pmatrix} r_m & \tilde{\beta}_n^T \\ \varrho_n & R_n \end{pmatrix}$$

$$\varrho_{n+1} = \begin{pmatrix} \varrho_n \\ r_{m+n+1} \end{pmatrix}$$

(C8.3.3)

where

$$\beta_n = (r_{m-n} \ldots r_{m-1})^T$$

(C8.3.4)

and where the notation $\tilde{x} = (x_n \ldots x_1)^T$ denotes the vector $x = (x_1 \ldots x_n)^T$ with the elements in reversed order. Note also that

$$\beta_{n+1} = \begin{pmatrix} r_{m-n-1} \\ \beta_n \end{pmatrix}$$

(C8.3.5)

Using Lemma A.2 and (C8.3.3)–(C8.3.5) above,

$$\theta_{n+1} = -R_{n+1}^{-1}\varrho_{n+1}$$

$$= -\left\{ \begin{pmatrix} I \\ 0 \end{pmatrix} R_n^{-1}(I \ 0) + \begin{pmatrix} -R_n^{-1}\beta_n \\ 1 \end{pmatrix}(-\tilde{\varrho}_n^T R_n^{-1} \ 1)/(r_m - \tilde{\varrho}_n^T R_n^{-1}\beta_n) \right\} \begin{pmatrix} \varrho_n \\ r_{m+n+1} \end{pmatrix}$$

$$= \begin{pmatrix} \theta_n \\ 0 \end{pmatrix} - \frac{\alpha_n}{\sigma_n} \begin{pmatrix} \Delta_n \\ 1 \end{pmatrix}$$

where

$$\Delta_n \triangleq -R_n^{-1}\beta_n$$

$$\alpha_n \triangleq r_{m+n+1} - \tilde{\varrho}_n^T R_n^{-1}\varrho_n = r_{m+n+1} + \tilde{\varrho}_n^T \theta_n$$

$$\sigma_n \triangleq r_m - \tilde{\varrho}_n^T R_n^{-1}\beta_n = r_m + \tilde{\varrho}_n^T \Delta_n$$

An order update for α_n does not seem possible. However, order recursions for Δ_n and σ_n can be derived easily. Thus, from Lemma A.2 and (C8.3.3)–(C8.3.5),

$$\Delta_{n+1} = -R_{n+1}^{-1}\beta_{n+1}$$

$$= -\left\{ \begin{pmatrix} 0 \\ I \end{pmatrix} R_n^{-1}(0 \ I) + \begin{pmatrix} 1 \\ -R_n^{-1}\varrho_n \end{pmatrix}(1 \ -\tilde{\beta}_n^T R_n^{-1})/(r_m - \tilde{\beta}_n^T R_n^{-1}\varrho_n) \right\}$$

$$\times \begin{pmatrix} r_{m-n-1} \\ \beta_n \end{pmatrix}$$

$$= \begin{pmatrix} 0 \\ \Delta_n \end{pmatrix} - \frac{\mu_n}{\bar{\sigma}_n} \begin{pmatrix} 1 \\ \theta_n \end{pmatrix}$$

and

$$\sigma_{n+1} = r_m - \tilde{\varrho}_{n+1}^T R_{n+1}^{-1} \beta_{n+1} = r_m + (r_{m+n+1} \quad \tilde{\varrho}_n^T) \left\{ \begin{pmatrix} 0 \\ \Delta_n \end{pmatrix} - \frac{\mu_n}{\bar{\sigma}_n} \begin{pmatrix} 1 \\ \theta_n \end{pmatrix} \right\}$$

$$= r_m + \tilde{\varrho}_n^T \Delta_n - \frac{\mu_n}{\bar{\sigma}_n}(r_{m+n+1} + \tilde{\varrho}_n^T \theta_n) = \sigma_n - \alpha_n \mu_n / \bar{\sigma}_n$$

where

$$\mu_n \triangleq r_{m-n-1} - \tilde{\beta}_n^T R_n^{-1} \beta_n = r_{m-n-1} + \tilde{\beta}_n^T \Delta_n$$

$$\bar{\sigma}_n \triangleq r_m - \tilde{\beta}_n^T R_n^{-1} \varrho_n$$

A simple order update formula for μ_n does not seem to exist.

Next note the following readily verified property. Let x be an $(n|1)$ vector, and set

$$y = R_n x \quad \text{(or } x = R_n^{-1} y\text{)}$$

Then

$$\tilde{y} = R_n^T \tilde{x} \quad \text{(or } \tilde{x} = R_n^{-T} \tilde{y}\text{)}$$

This property implies that

$$\bar{\sigma}_n = r_m - \varrho_n^T R_n^{-T} \tilde{\beta}_n = r_m + \varrho_n^T \tilde{\Delta}_n = r_m + \tilde{\varrho}_n^T \Delta_n = \sigma_n$$

With this observation the derivation of the algorithm is complete. The algorithm is summarized in Table C8.3.1, for easy reference. Initial values for θ, Δ and σ are obtained from the definitions. The algorithm of Table C8.3.1 requires about $4n$ operations per iteration and thus a total of about $2M^2$ operations to compute all $\{\theta_n\}$, $n = 1, \ldots, M$.

Next observe that the algorithm (C8.3.6)–(C8.3.10) should be stopped if $\sigma_n = 0$ is obtained. Thus the algorithm works if and only if

$$\sigma_n \neq 0 \quad \text{for} \quad n = 1, \ldots, M - 1 \quad \text{and} \quad r_m \neq 0 \tag{C8.3.11}$$

TABLE C8.3.1 A Levinson-type algorithm for solving the nonsymmetric Toeplitz systems of equations (C8.3.1)

Initialize:
$\theta_1 = -r_{m+1}/r_m$
$\Delta_1 = -r_{m-1}/r_m$
$\sigma_1 = r_m(1 - \theta_1 \Delta_1)$

For $n = 1$ to $(M - 1)$ compute:

$$\alpha_n = r_{m+n+1} + \tilde{\varrho}_n^T \theta_n \tag{C8.3.6}$$

$$\mu_n = r_{m-n-1} + \tilde{\beta}_n^T \Delta_n \tag{C8.3.7}$$

$$\theta_{n+1} = \begin{pmatrix} \theta_n \\ 0 \end{pmatrix} - \frac{\alpha_n}{\sigma_n} \begin{pmatrix} \Delta_n \\ 1 \end{pmatrix} \tag{C8.3.8}$$

$$\Delta_{n+1} = \begin{pmatrix} 0 \\ \Delta_n \end{pmatrix} - \frac{\mu_n}{\sigma_n} \begin{pmatrix} 1 \\ \theta_n \end{pmatrix} \tag{C8.3.9}$$

$$\sigma_{n+1} = \sigma_n - \alpha_n \mu_n / \sigma_n \tag{C8.3.10}$$

The condition above may be expected to be equivalent to (C8.3.2). This equivalence is proved in the following.

Assume that (C8.3.2) holds. Then the solutions $\{\theta_n\}_{n=1}^{M}$ to (C8.3.1) exist and

$$R_{n+1}\begin{pmatrix} 1 & 0 \\ \theta_n & I \end{pmatrix} = \begin{pmatrix} r_m & \tilde{\beta}_n^T \\ \varrho_n & R_n \end{pmatrix}\begin{pmatrix} 1 & 0 \\ \theta_n & I \end{pmatrix} = \begin{pmatrix} \sigma_n & \tilde{\beta}_n^T \\ 0 & R_n \end{pmatrix} \quad (C8.3.12)$$

which implies that

$$\det R_{n+1} \neq 0 \Rightarrow \{\sigma_n \neq 0 \text{ and } \det R_n \neq 0\} \quad (C8.3.13)$$

From (C8.3.13) it follows easily that (C8.3.2) implies (C8.3.11).

To prove the implication (C8.3.11) $\Rightarrow$ (C8.3.2), observe that when (C8.3.11) holds, θ_1 exists. This implies that (C8.3.12) holds for $n = 1$, which together with $\sigma_1 \neq 0$ implies that $\det R_2 \neq 0$. Thus θ_2 exists; and so on.

Note that under the condition (C8.3.11) (or (C8.1.2)) it also follows from (C8.3.12) that

$$\begin{pmatrix} 1 & \tilde{\Delta}_n^T \\ 0 & I \end{pmatrix} R_{n+1} \begin{pmatrix} 1 & 0 \\ \theta_n & I \end{pmatrix} = \begin{pmatrix} \sigma_n & 0 \\ 0 & R_n \end{pmatrix} \quad (C8.3.14)$$

This equation resembles a factorization equation which holds in the case of the Levinson–Durbin algorithm (see Complement C8.2).

It is interesting to note that for $m = 0$ and $r_{-i} = r_i$, the algorithm above reduces to the LDA presented in Complement C8.2. Indeed, under the previous conditions the matrix R_n is symmetric and $\beta_n = \tilde{\varrho}_n$. Thus $\Delta_n = \theta_n$ and $\mu_n = \alpha_n$, which implies that the equations (C8.3.7) and (C8.3.9) in Table C8.3.1 are redundant and should be eliminated. The remaining equations are precisely those of the LDA.

In the case where $m \neq 0$ or/and $r_{-i} \neq r_i$, the algorithm of Table C8.3.1 is more complex than the LDA, as might be expected. (In fact the algorithm (C8.3.6)–(C8.3.10) resembles more the multivariable version of the LDA (which will be discussed in Complement C8.6) than the scalar LDA.) To see this more clearly, introduce the notation

$$A_n(z) \triangleq 1 + a_{n,1}z + \ldots + a_{n,n}z^n = 1 + (z \ \ldots \ z^n)\theta_n \quad (C8.3.15a)$$

and

$$B_n(z) \triangleq (1 \ z \ \ldots \ z^{n-1})\Delta_n + z^n \quad (C8.3.15b)$$

Then premultiplying (C8.3.8) by $(z \ \ldots \ z^{n+1})$ and (C8.3.9) by $(1 \ z \ \ldots \ z^n)$ and adding respectively 1 and z^{n+1} to both sides of the resulting identities,

$$A_{n+1}(z) = A_n(z) - \tilde{k}_n z B_n(z)$$

$$B_{n+1}(z) = z B_n(z) - \tilde{k}_n A_n(z)$$

or equivalently

$$\begin{pmatrix} A_{n+1}(z) \\ B_{n+1}(z) \end{pmatrix} = \begin{pmatrix} 1 & -\tilde{k}_n \\ -\tilde{k}_n & 1 \end{pmatrix}\begin{pmatrix} A_n(z) \\ zB_n(z) \end{pmatrix} \quad (C8.3.16)$$

where

$$\bar{k}_n = \frac{\alpha_n}{\sigma_n} \qquad \tilde{k}_n = \frac{\mu_n}{\sigma_n}$$

According to the discussion preceeding (C8.3.15), in the symmetric case the recursion (C8.3.16) reduces to

$$\begin{pmatrix} A_{n+1}(z) \\ \tilde{A}_{n+1}(z) \end{pmatrix} = \begin{pmatrix} 1 & -k_n \\ -k_n & 1 \end{pmatrix} \begin{pmatrix} A_n(z) \\ z\tilde{A}_n(z) \end{pmatrix} \qquad \text{(C8.3.17)}$$

where $k_n = \alpha_n/\sigma_n = -a_{n+1,n+1}$ and where $\tilde{A}_n(z)$ denotes the 'reversed' $A_n(z)$ polynomial (compare with (C8.2.9)). Evidently (C8.3.17), which involves one sequence of scalars $\{k_n\}$ and one sequence of polynomials $\{A_n(z)\}$ only, is less complex than (C8.3.16). Note that $\{k_n\}$ in the symmetric case are often called 'reflection coefficients', a term which comes from 'reflection seismology' applications of the LDA. By analogy, $\bar{k}_n$ and $\tilde{k}_n$ in (C8.3.16) may also be called 'reflection coefficients'.

Next, recall that in the symmetric case there exists a simple relation between the sequence of reflection coefficients $\{k_n\}$ and the location of the zeros of $A_n(z)$, namely: $|k_p| < 1$ for $p = 1, \ldots, n \Leftrightarrow A_n(z) \neq 0$ for $|z| \leq 1$ (see Complement C8.2). By analogy it might be thought that a similar result would hold in the nonsymmetric case. However, this does not seem to be true, as is illustrated by the following example.

Example C8.3.1 *Constraints on the nonsymmetric reflection coefficients and location of the zeros of* $A_n(z)$

A straightforward calculation gives

$$A_1(z) = 1 - \bar{k}_0 z$$
$$B_1(z) = -\tilde{k}_0 + z$$

and

$$A_2(z) = 1 + (\bar{k}_1 \tilde{k}_0 - \bar{k}_0)z - \bar{k}_1 z^2$$

where

$$\bar{k}_0 = r_{m+1}/r_m \qquad \tilde{k}_0 = r_{m-1}/r_m$$
$$\bar{k}_1 = (r_{m+2} - r_{m+1}\tilde{k}_0)/[r_m(1 - \bar{k}_0 \tilde{k}_0)]$$

Thus $A_2(z)$ has its roots outside the unit circle if and only if

$$|\bar{k}_1| < 1$$
$$\bar{k}_1(1 - \tilde{k}_0) < 1 - \bar{k}_0 \qquad \text{(C8.3.18)}$$
$$\bar{k}_1(1 + \tilde{k}_0) < 1 + \bar{k}_0$$

Observe that in the symmetric case where $\bar{k}_i = \tilde{k}_i \triangleq k_i$, the inequalities (C8.3.18) reduce to the known condition $|k_0| < 1$, $|k_1| < 1$. In the nonsymmetric case, however, in general $\bar{k}_i \neq \tilde{k}_i$ so that (C8.3.18) does not seem to reduce to a 'simple' condition on $\bar{k}_0$, $\tilde{k}_0$ and $\bar{k}_1$. ∎

Complement C8.4
Min-max optimal IV method

Consider a scalar difference equation of the form (8.1), (8.2) whose parameters are estimated using the IV method (8.13) with $nz = n\theta$ and $Q = I$. The covariance matrix of the parameter estimates so obtained is given by (8.30). Under the conditions introduced above, (8.30) simplifies to

$$P_{IV} = \lambda^2 R^{-1}[EK(q^{-1})z(t)K(q^{-1})z^T(t)]R^{-T}$$

where

$$R = Ez(t)F(q^{-1})\tilde{\varphi}^T(t) \quad K(q^{-1}) = H(q^{-1})F(q^{-1})$$

(see (8.32)).

The covariance matrix P_{IV} depends on the noise shaping filter $H(q^{-1})$. If one tries to obtain optimal instruments $\tilde{z}(t)$ by minimizing some suitable scalar or multivariable function of P_{IV} then in general the optimal $\tilde{z}(t)$ will also depend on $H(q^{-1})$ (see e.g. (8.38)). In some applications this may be an undesirable feature which may prevent the use of the optimal instruments $\tilde{z}(t)$.

A conceptually simple way of overcoming the above difficulty is to formulate the problem of accuracy optimization on a min-max basis. For example,

$$\tilde{z}(t) = \arg \min_{z(t) \in C_z} \max_{H(q^{-1}) \in C_H} P_{IV} \quad (C8.4.1)$$

where

$$C_z = \{z(t) | \det R \neq 0\} \quad (C8.4.2a)$$
$$C_H = \{H(q^{-1}) | E[H(q^{-1})e(t)]^2 \leq \alpha\} \quad (C8.4.2b)$$

for some given positive real α. It is clearly necessary to constrain the variance of the disturbance $H(q^{-1})e(t)$, for example directly as in (C8.4.2b). Otherwise the inner optimization of (C8.4.1) would be meaningless. More exactly, without a constraint on the variance of the disturbance, the 'worst' covariance matrix given by the inner 'optimization' in (C8.4.1) would be of infinite magnitude. It is shown in Stoica and Söderström (1982c) that the min-max approach (C8.4.1), (C8.4.2) to the optimization of the accuracy of IV methods does not lead to a neat solution in the general case.

In the following the problem is reformulated by redefining the class C_H of feasible $H(q^{-1})$ filters. Specifically, consider

$$C_H = \{H(q^{-1}) | |H(e^{i\omega})|^2 \leq |\tilde{H}(e^{i\omega})|^2 \text{ for } \omega \in [-\pi, \pi]\} \quad (C8.4.3)$$

where $\tilde{H}(e^{i\omega})$ is some given suitable function of ω. The min-max problem (C8.4.1), (C8.4.3) can be solved easily. First, however, compare the conditions (C8.4.2b) and (C8.4.3) on $H(q^{-1})$. Let $\tilde{H}(q^{-1})$ be such that

$$\frac{\lambda^2}{2\pi} \int_{-\pi}^{\pi} |\tilde{H}(e^{i\omega})|^2 \, d\omega = \alpha$$

Then (C8.4.3) implies (C8.4.2b) since

$$E[H(q^{-1})e(t)]^2 = \frac{\lambda^2}{2\pi} \int_{-\pi}^{\pi} |H(e^{i\omega})|^2 \, d\omega \leq \frac{\lambda^2}{2\pi} \int_{-\pi}^{\pi} |\tilde{H}(e^{i\omega})|^2 \, d\omega = \alpha$$

The converse is not necessarily true. There are filters $H(q^{-1})$ which satisfy (C8.4.2b) but whose magnitude $|H(e^{i\omega})|$ takes arbitrarily large values at some frequencies, thus violating (C8.4.3). Thus the explicit condition in (C8.4.3) on $H(q^{-1})$ is more restrictive than the implicit condition in (C8.4.2b). This was to be expected since (C8.4.3) assumes a more detailed knowledge of the noise properties than does (C8.4.2b). However, these facts should not be seen as drawbacks of (C8.4.3). In applications, disturbances with arbitrarily large power at some frequencies are unlikely to occur. Furthermore, if there is no *a priori* knowledge available on the noise properties then $\tilde{H}$ in (C8.4.3) can be set to

$$\tilde{H}(e^{i\omega}) = \beta \qquad \omega \in (-\pi, \pi) \tag{C8.4.4}$$

for some arbitrary β. As will be shown, the result of the min-max optimization problem (C8.4.1), (C8.4.3), (C8.4.4) does not depend on β.

Turn now to the min-max optimization problem (C8.4.1), (C8.4.3). Since the matrix

$$E\tilde{H}(q^{-1})F(q^{-1})z(t)\tilde{H}(q^{-1})F(q^{-1})z^T(t) - EK(q^{-1})z(t)K(q^{-1})z^T(t)$$

$$= \int_{-\pi}^{\pi} [|\tilde{H}(e^{i\omega})|^2 - |H(e^{i\omega})|^2]|F(e^{i\omega})|^2 \phi_z(e^{i\omega}) \, d\omega$$

where $\phi_z(e^{i\omega})$ denotes the spectral density matrix of $z(t)$, is nonnegative definite for all $H(q^{-1}) \in C_H$, it follows that

$$\tilde{z}(t) = \arg \min_{z(t) \in C_z} \{R^{-1}[E\tilde{H}(q^{-1})F(q^{-1})z(t)\tilde{H}(q^{-1})F(q^{-1})z^T(t)]R^{-T}\}$$

This problem is of the same type as that treated in Section 8.2. It follows from (8.38) that

$$\boxed{\tilde{z}(t) = \tilde{H}^{-1}(q^{-1})\tilde{\varphi}(t) \qquad \text{and} \qquad F(q^{-1}) = \tilde{H}^{-1}(q^{-1})} \tag{C8.4.5}$$

The closer $\tilde{H}(q^{-1})$ is to the true $H(q^{-1})$, the smaller are the estimation errors associated with the min-max optimal instruments and prefilter above.

Note that in the case of $\tilde{H}(q^{-1})$ given by (C8.4.4)

$$\tilde{z}(t) = \tilde{\varphi}(t) \qquad F(q^{-1}) = 1 \tag{C8.4.6}$$

(since the value of β is clearly immaterial). Thus (C8.4.6) are the min-max optimal instruments and prefilter when *no a priori* knowledge on the noise properties is available (see also Wong and Polak, 1967; Stoica and Söderström, 1982c).

Observe, however, that (C8.4.6) assumes knowledge of the noise-free output (which enters via $\tilde{\varphi}(t)$). One way of overcoming this difficulty is to use a bootstrapping iterative procedure for implementation of the IVE corresponding to (C8.4.6). This procedure should be fed with some initial values of the noise-free output (or equivalently of the system transfer function coefficients). Another *non*-iterative possibility to overcome the aforementioned difficulty associated with the implementation of the min-max optimal IVE corresponding to (C8.4.6) is described in Stoica and Nehorai (1987a).

For the extension of the min-max optimality results above to multivariable systems, see de Gooijer and Stoica (1987).

Complement C8.5
Optimally weighted extended IV method

Consider the covariance matrix P_{IV}, (8.30), in the case of single-output systems ($ny = 1$). According to (8.32) the matrix P_{IV} then takes the following more convenient form:

$$P_{IV} = \lambda^2 (R^T Q R)^{-1} R^T Q S Q R (R^T Q R)^{-1}$$

where

$$S = E K(q^{-1}) z(t) K(q^{-1}) z^T(t)$$

$$K(q^{-1}) = H(q^{-1}) F(q^{-1})$$

$$R = E z(t) F(q^{-1}) \varphi^T(t) = E z(t) F(q^{-1}) \tilde{\varphi}^T(t)$$

In Section 8.2 the following lower bound P_{IV}^{opt} on P_{IV} was derived:

$$P_{IV} \geq \lambda^2 [E H^{-1}(q^{-1}) \tilde{\varphi}(t) H^{-1}(q^{-1}) \tilde{\varphi}^T(t)]^{-1} \triangleq P_{IV}^{opt} \tag{C8.5.1}$$

and it was shown that equality holds for

$$z(t) = H^{-1}(q^{-1}) \tilde{\varphi}(t) \quad (nz = n\theta)$$

$$F(q^{-1}) = H^{-1}(q^{-1}) \tag{C8.5.2}$$

$$Q = I$$

(see (8.36)–(8.38)). The IV estimate corresponding to the choices (C8.5.2) of the IV vector z, the prefilter F and the weighting matrix Q is the following:

$$\hat{\theta} = \left[\sum_{t=1}^{N} H^{-1}(q^{-1}) \tilde{\varphi}(t) H^{-1}(q^{-1}) \varphi^T(t) \right]^{-1}$$

$$\times \left[\sum_{t=1}^{N} H^{-1}(q^{-1}) \tilde{\varphi}(t) H^{-1}(q^{-1}) y(t) \right] \tag{C8.5.3}$$

Here the problem of minimizing the covariance matrix P_{IV} of the estimation errors is approached in a different way. First the following lower bound on P_{IV} is derived with respect to Q:

$$P_{IV} \geq \lambda^2 (R^T S^{-1} R)^{-1} \triangleq \tilde{P}_{nz}$$

with equality for

$$Q = S^{-1}$$

Proof. It can be easily checked that

$$P_{IV} - \tilde{P}_{nz} = \lambda^2 [(R^T Q R)^{-1} R^T Q - (R^T S^{-1} R)^{-1} R^T S^{-1}] S [(R^T Q R)^{-1} R^T Q$$
$$- (R^T S^{-1} R)^{-1} R^T S^{-1}]^T$$

Since $S > 0$, the result follows. ∎

Next it is shown that if the IV vectors of dimension nz and $nz + 1$ are nested so that

$$z_{nz+1}(t) = \begin{pmatrix} z_{nz}(t) \\ x \end{pmatrix} \tag{C8.5.4}$$

where x (as well as ψ, ϱ and σ below) denote some quantity whose exact expression is not important for the present discussion, then

$$\tilde{P}_{nz} \geq \tilde{P}_{nz+1}$$

Proof. The nested structure (C8.5.4) of the IV vectors induces a similar structure on the matrices R and S, say

$$R_{nz+1} = \begin{pmatrix} R_{nz} \\ \psi^T \end{pmatrix}$$

$$S_{nz+1} = \begin{pmatrix} S_{nz} & \varrho \\ \varrho^T & \sigma \end{pmatrix}$$

Thus, using the result of Lemma A.2,

$$\tilde{P}_{nz+1}^{-1} \triangleq (1/\lambda^2)(R_{nz+1}^T S_{nz+1}^{-1} R_{nz+1})$$

$$= (1/\lambda^2)(R_{nz}^T \ \psi) \left\{ \begin{pmatrix} I \\ 0 \end{pmatrix} S_{nz}^{-1}(I \ 0) + \begin{pmatrix} S_{nz}^{-1} \varrho \\ -1 \end{pmatrix} \right.$$

$$\left. \times (\sigma - \varrho^T S_{nz}^{-1} \varrho)^{-1} (\varrho^T S_{nz}^{-1} \ -1) \right\} \begin{pmatrix} R_{nz} \\ \psi^T \end{pmatrix}$$

$$= \tilde{P}_{nz}^{-1} + (1/\lambda^2)[R_{nz}^T S_{nz}^{-1} \varrho - \psi][R_{nz}^T S_{nz}^{-1} \varrho - \psi]^T/(\sigma - \varrho^T S_{nz}^{-1} \varrho)$$

Since S_{nz+1} is positive definite it follows from Lemma A.3 that $\sigma - \varrho^T S_{nz}^{-1} \varrho > 0$. Hence $\tilde{P}_{nz+1}^{-1} \geq \tilde{P}_{nz}^{-1}$, which completes the proof. ∎

Thus, the positive definite matrices $\{\tilde{P}_{nz}\}$ form a monotonically nonincreasing sequence for $nz = n\theta, n\theta + 1, n\theta + 2, \ldots$ In particular, this means that $\tilde{P}_{nz}$ has a limit $\tilde{P}_\infty$ as $nz \to \infty$. In view of (8.36), this limiting matrix must be bounded below by P_{IV}^{opt}:

$$\tilde{P}_{nz} \geq P_{IV}^{opt} \quad \text{for all } nz \tag{C8.5.5}$$

This inequality will be examined more closely, and in particular, conditions will be given under which the limit $\tilde{P}_\infty$ equals the lower bound P_{IV}^{opt}. The vector $\tilde{\varphi}(t)$ can be written (see Complement C5.1) as

$$\tilde{\varphi}(t) = \mathscr{S}(-B, A)\varphi_u(t) \tag{C8.5.6}$$

where $\mathscr{S}(-B, A)$ is the Sylvester matrix associated with the polynomials $-B(z)$ and $A(z)$, and

$$\varphi_u(t) = \frac{1}{A(q^{-1})}(u(t-1) \ldots u(t-na-nb))^T \tag{C8.5.7}$$

The inequality (C8.5.5) can therefore be rewritten as

$$R^T S^{-1} R \leq E H^{-1}(q^{-1})\tilde{\varphi}(t) H^{-1}(q^{-1})\tilde{\varphi}^T(t)$$
$$\Leftrightarrow \mathscr{S}(-B, A)[EF(q^{-1})\varphi_u(t)z^T(t)][EK(q^{-1})z(t)K(q^{-1})z^T(t)]^{-1}$$
$$\times [Ez(t)F(q^{-1})\varphi_u^T(t)]\mathscr{S}^T(-B, A)$$
$$\leq \mathscr{S}(-B, A)[EH^{-1}(q^{-1})\varphi_u(t)H^{-1}(q^{-1})\varphi_u^T(t)]\mathscr{S}^T(-B, A)$$

or equivalently (since the matrix $\mathscr{S}(-B, A)$ is nonsingular; see Lemma A.30),

$$[EF(q^{-1})\varphi_u(t)z^T(t)][EK(q^{-1})z(t)K(q^{-1})z^T(t)]^{-1}[Ez(t)F(q^{-1})\varphi_u^T(t)]$$
$$\leq [EH^{-1}(q^{-1})\varphi_u(t)H^{-1}(q^{-1})\varphi_u^T(t)] \tag{C8.5.8}$$

Assume that the prefilter is chosen as

$$F(q^{-1}) = H^{-1}(q^{-1}) \tag{C8.5.9}$$

and set

$$\bar{\varphi}(t) = H^{-1}(q^{-1})\varphi_u(t)$$

Then (C8.5.8) becomes

$$[E\bar{\varphi}(t)z^T(t)][Ez(t)z^T(t)]^{-1}[Ez(t)\bar{\varphi}^T(t)] \leq E\bar{\varphi}(t)\bar{\varphi}^T(t) \tag{C8.5.10}$$

(Note in passing that this inequality follows easily by generalizing Lemma A.4 to $\dim \bar{\varphi} \neq \dim z$.) Assume further that there is a matrix M of dimension $(na + nb)|nz$ such that

$$\bar{\varphi}(t) = Mz(t) + x(t) \tag{C8.5.11}$$

where

$$E\|x(t)\|^2 \to 0 \quad \text{as } nz \to \infty$$

Set

$$R_z = Ez(t)z^T(t)$$
$$R_{zx} = Ez(t)x^T(t)$$
$$R_x = Ex(t)x^T(t)$$

Under the assumptions introduced above, the difference between the two sides of (C8.5.10) tends to zero as $nz \to \infty$, as shown below.

$$[E\bar{\varphi}(t)\bar{\varphi}^T(t)] - [E\bar{\varphi}(t)z^T(t)][Ez(t)z^T(t)]^{-1}[Ez(t)\bar{\varphi}^T(t)]$$
$$= [MR_zM^T + MR_{zx} + R_{xz}M^T + R_x]$$
$$\quad - [MR_z + R_{xz}]R_z^{-1}[R_zM^T + R_{zx}]$$
$$= R_x - R_{xz}R_z^{-1}R_{zx} \quad\quad\quad\quad (C8.5.12)$$
$$\to 0 \quad \text{as } nz \to \infty$$

Here it has been assumed implicitly that R_z^{-1} remains bounded when nz tends to infinity. This is true under weak conditions (cf. Complement C5.2). The conclusion to draw from (C8.5.12) is that under the assumptions (C8.5.9), (C8.5.11),

$$\lim_{nz \to \infty} \tilde{P}_{nz} = P_{IV}^{opt} \quad\quad\quad\quad (C8.5.13)$$

Consider now the assumption (C8.5.11). According to the previous assumption (C8.5.4) on the nested structure of $z(t)$, it is reasonable to choose the instruments as

$$z(t) = L(q^{-1}) \begin{pmatrix} u(t-1) \\ \vdots \\ u(t-nz) \end{pmatrix} \quad\quad\quad\quad (C8.5.14)$$

where $L(q^{-1})$ is an asymptotically stable and invertible filter. It will be shown that $z(t)$ given by (C8.5.14) satisfies (C8.5.11). Define $M(q^{-1})$ and $\{m_j\}$ by

$$M(q^{-1}) = \sum_{j=0}^{\infty} m_j q^{-j} \triangleq 1/[A(q^{-1})H(q^{-1})L(q^{-1})]$$

and set

$$M = \begin{pmatrix} m_0 & m_1 & \cdots & & m_{nz-1} \\ & m_0 & m_1 & \cdots & m_{nz-2} \\ 0 & & \ddots & & \vdots \\ & & & m_0 & \cdots & m_{nz-na-nb} \end{pmatrix} \quad ((na+nb)|nz) \quad (C8.5.15)$$

Then

$$\bar{\varphi}(t) = Mz(t) + x(t)$$

where the ith element of $x(t)$ is given by

$$x_i(t) = \sum_{k=nz-i+1}^{\infty} m_k L(q^{-1}) u(t - i - k) \quad i = 1, \ldots, n\theta = na + nb$$

Since $M(q^{-1})$ is asymptotically stable by assumption, it follows that m_k converges exponentially to zero as $k \to \infty$. Combining this observation with the expression above for $x(t)$, it is seen that (C8.5.11) holds for $z(t)$ given by (C8.5.14).

In the calculation leading to (C8.5.13) it was an essential assumption that the prefilter was chosen as in (C8.5.9). When $F(q^{-1})$ differs from $H^{-1}(q^{-1})$ there will in general be strict inequality in (C8.5.5), even when nz tends to infinity. As an illustration consider the following simple case:

$$F(q^{-1}) = 1 + fq^{-1} \quad 0 < |f| < 1$$

$$H(q^{-1}) = 1$$

$u(t)$ is white noise of unit variance

$$z(t) = \frac{1}{1 + fq^{-1}} (u(t-1) \ldots u(t-nz))^{\mathrm{T}}$$

$$na = 0 \quad nb = 1; \quad \tilde{\psi}(t) = u(t-1)$$

In this case

$$R = Ez(t)F(q^{-1})u(t-1) = \begin{pmatrix} 1 - f^2 \\ f \\ 0 \\ \vdots \\ 0 \end{pmatrix}$$

$$S = I$$

and for $nz \geq 2$,

$$\tilde{P}_{nz} = \lambda^2 (R^{\mathrm{T}} S^{-1} R)^{-1} = \frac{\lambda^2}{(1-f^2)^2 + f^2} = \frac{\lambda^2}{1 - f^2(1-f^2)} \quad \text{(C8.5.16)}$$

$$> \lambda^2 = P_{\mathrm{IV}}^{\mathrm{opt}}$$

The discussion so far can be summarized as follows. The IV estimate obtained by solving, in the least squares sense, the following overdetermined system of equations

$$S^{-1/2} \left[\sum_{t=1}^{N} z(t) F(q^{-1}) \varphi^{\mathrm{T}}(t) \right] \hat{\theta} = S^{-1/2} \left[\sum_{t=1}^{N} z(t) F(q^{-1}) y(t) \right] \quad \text{(C8.5.17)}$$

has asymptotic covariance matrix equal to $\tilde{P}_{nz}$. Furthermore, the asymptotic accuracy of $\hat{\theta}$ improves with increasing nz, and for a sufficiently large nz the estimate $\hat{\theta}$ and the optimal IV estimate (C8.5.3) are expected to have comparable accuracies. Note that in (C8.5.17) the prefilter F should be taken as $F(q^{-1}) = H^{-1}(q^{-1})$ and the IV vector z should be of the form (C8.5.14).

The estimate $\hat{\theta}$ is called the optimally weighted extended IV estimate. Both (C8.5.3) and (C8.5.17) rely on knowledge of the noise shaping filter $H(q^{-1})$. As in the case of (C8.5.3), replacement of $H(q^{-1})$ in (C8.5.17) by a consistent estimate will not worsen the asymptotic accuracy (for $N \to \infty$) (provided nz does not increase too fast compared to N).

The main difference between the computational burdens associated with (C8.5.3) and (C8.5.17) lies in the following operations.

For (C8.5.3):
- Computation of the optimal IV vector $\{H^{-1}(q^{-1})\tilde{\varphi}(t)\}_{t=1}^{N}$, which needs O($N$) operations. Also, some stability tests on the filters which enter in $\tilde{\varphi}(t)$ are needed to prevent the elements of $\tilde{\varphi}(t)$ from exploding. When those filters are unstable they should be replaced by some stable approximants.

For (C8.5.17):
- Computation of the matrix S from the data, which needs O($nz \times N$) arithmetic operations. Note that for the IV vector $z(t)$ given by (C8.5.14), the matrix S is Toeplitz and symmetric. In particular this means that the inverse square root matrix $S^{-1/2}$ can be computed from S in O(nz^2) operations using an efficient algorithm such as the Levinson–Durbin algorithm described in Complement C8.2.
- Creating and solving an overdetermined system of nz equations with $n\theta$ unknowns. Comparing this with the task of creating and solving a square system of $n\theta$ equations, which is associated with (C8.5.3), there is a difference of about O(nz^2) + O($nz \cdot N$) operations (for $nz \gg na$).

Thus from a computational standpoint the IV estimate (C8.5.17) may be less attractive than (C8.5.3). However, (C8.5.17) does not need a *stable* estimate of the system transfer function $B(q^{-1})/A(q^{-1})$, as does (C8.5.3). Furthermore, in the case of pure time series models where there is no external input signal, the optimal IV estimate (C8.5.3) cannot be applied; while the IV estimate (C8.5.17) can be used after some minor adaptations.

Some results on the optimally weighted extended IV estimate (C8.5.17) in the case of ARMA processes are reported in Stoica, Söderström and Friedlander (1985) and Stoica, Friedlander and Söderström (1985b). For dynamic systems with exogenous inputs there is as yet no experience in using this IV estimator.

Complement C8.6
The Whittle–Wiggins–Robinson algorithm

The Whittle–Wiggins–Robinson algorithm (WWRA) is the extension to the multivariate case of the Levinson–Durbin algorithm discussed in Complement C8.2. To motivate the need for this extension, consider the following problem.

Let $y(t)$, $t = 1, 2, 3, \ldots$, be an ny-dimensional stationary stochastic process. An autoregressive model (also called a linear prediction model) of order n of $y(t)$ will have the form

$$y(t) = -A_{n,1} y(t-1) - \ldots - A_{n,n} y(t-n) + \varepsilon_n(t)$$

where $\{A_{n,i}\}$ are $(ny|ny)$ matrices and $\varepsilon_n(t)$ is the prediction error (of order n) at time instant t. The matrix coefficients $\{A_{n,i}\}$ of the least squares predictor minimize the covariance matrix of the (forward) prediction errors

$$Q_n = E\varepsilon_n(t)\varepsilon_n^T(t)$$

and are given by (C7.3.6):

$$\Theta \triangleq (A_{n,1} \ldots A_{n,n}) = -[Ey(t)\varphi^T(t)][E\varphi(t)\varphi^T(t)]^{-1} \quad (C8.6.1)$$

where $\varphi^T(t) = (y^T(t-1) \ldots y^T(t-n))$. The corresponding minimal value of Q_n is

$$Q_n = E[y(t) + \Theta\varphi(t)][y^T(t) + \varphi^T(t)\Theta^T]$$
$$= Ey(t)y^T(t) + \Theta E\varphi(t)y^T(t) \quad (C8.6.2)$$

Let

$$R_k = Ey(t)y^T(t-k) \quad k = 0, 1, 2, \ldots$$

denote the covariance of $y(t)$ at lag k. Note that $R_{-k} = R_k^T$. The equations (C8.6.1), (C8.6.2) can be written as

$$(A_{n,1} \ldots A_{n,n}) \begin{pmatrix} R_0 & R_1 & \ldots & R_{n-1} \\ R_{-1} & R_0 & & \vdots \\ \vdots & & \ddots & R_1 \\ R_{-(n-1)} & \ldots & \ldots & R_0 \end{pmatrix} = -(R_1 \ldots R_n) \quad (C8.6.3)$$

and

$$Q_n = R_0 + A_{n,1}R_{-1} + \ldots + A_{n,n}R_{-n}$$

which can be rearranged in a single block-Toeplitz matrix equation

$$(I \; A_{n,1} \; \ldots \; A_{n,n}) \underbrace{\begin{pmatrix} R_0 & R_1 & \ldots & R_n \\ R_{-1} & & & \vdots \\ \vdots & & \ddots & R_1 \\ R_{-n} & \ldots & R_{-1} & R_0 \end{pmatrix}}_{\Gamma_n} = (Q_n \; 0 \; \ldots \; 0) \quad (C8.6.4)$$

This Yule–Walker system of equations is the multivariate extension to that treated in Complement C8.2 (see equation (C8.2.1)).

The algorithm

An efficient solution to (C8.6.4) was presented by Whittle (1963) and Wiggins and Robinson (1966). Here the derivation of the WWRA follows Friedlander (1982). Note that there is no need to maintain the assumption that $\{R_i\}$ are theoretical (or sample) covariances. Such an assumption is not required when deriving the WWRA. It was used above only to provide a motivation for a need to solve linear systems of the type

(C8.6.3). In the following it is assumed only that $R_k = R_{-k}^T$ for $k = 0, 1, 2, \ldots$, and that the (symmetric) matrices Γ_{n-1} are nonsingular such that (C8.6.3) has a unique solution for all n of interest.

Efficient computation of the prediction coefficients $\{A_{n,i}\}$ involves some auxiliary quantities. Define $\{B_{n,i}\}$ and S_n by

$$(B_{n,n} \ \ldots \ B_{n,1} \ I)\Gamma_n = (0 \ \ldots \ 0 \ S_n) \tag{C8.6.5}$$

where Γ_n was introduced in (C8.6.4). These quantities can be given the following interpretation. Consider the *backward* prediction model

$$y(t) + B_{n,1}y(t+1) + \ldots + B_{n,n}y(t+n) = \tilde{\varepsilon}(t)$$

Minimizing the covariance matrix of the backward prediction errors

$$S_n = E\tilde{\varepsilon}(t)\tilde{\varepsilon}^T(t)$$

with respect to the matrix coefficients $\{B_{n,i}\}$ will lead precisely to (C8.6.5); note the similarity to (C8.6.4), which gives the optimal *forward* prediction coefficients.

The WWRA computes the solution of (C8.6.4) recursively in n. Assume that the nth-order solution is known. Introduce

$$P_n = R_{n+1} + A_{n,1}R_n + \ldots + A_{n,n}R_1 \tag{C8.6.6}$$

and note that

$$P_n = R_{n+1} + \Theta \begin{pmatrix} R_n \\ \vdots \\ R_1 \end{pmatrix} = R_{n+1} - (R_1 \ \ldots \ R_n)\Gamma_{n-1}^{-1}\begin{pmatrix} R_n \\ \vdots \\ R_1 \end{pmatrix}$$

$$= \left\{ R_{-(n+1)} - (R_{-n} \ \ldots \ R_{-1})\Gamma_{n-1}^{-1}\begin{pmatrix} R_{-1} \\ \vdots \\ R_{-n} \end{pmatrix} \right\}^T$$

$$= \left\{ R_{-(n+1)} + (B_{n,n} \ \ldots \ B_{n,1})\begin{pmatrix} R_{-1} \\ \vdots \\ R_{-n} \end{pmatrix} \right\}^T \tag{C8.6.7}$$

Combining (C8.6.4)–(C8.6.7) gives

$$\begin{pmatrix} I & A_{n,1} & \ldots & A_{n,n} & 0 \\ 0 & B_{n,n} & \ldots & B_{n,1} & I \end{pmatrix}\Gamma_{n+1} = \begin{pmatrix} Q_n & 0 & \ldots & 0 & P_n \\ P_n^T & 0 & \ldots & 0 & S_n \end{pmatrix} \tag{C8.6.8}$$

The $(n+1)$th-order solutions satisfy

$$\begin{pmatrix} I & A_{n+1,1} & \ldots & A_{n+1,n} & A_{n+1,n+1} \\ & B_{n+1,n+1} & \ldots & B_{n+1,1} & I \end{pmatrix}\Gamma_{n+1} = \begin{pmatrix} Q_{n+1} & 0 & \ldots & 0 & 0 \\ 0 & 0 & \ldots & 0 & S_{n+1} \end{pmatrix} \tag{C8.6.9}$$

(see (C8.6.4), (C8.6.5)). Therefore the idea is to modify (C8.6.8) so that zeros occur in the positions occupied by P_n and P_n^T. This can be done by a simple linear transformation of (C8.6.8):

$$\begin{pmatrix} I & -P_n S_n^{-1} \\ -P_n^T Q_n^{-1} & I \end{pmatrix} \begin{pmatrix} I & A_{n,1} & \cdots & A_{n,n} & 0 \\ 0 & B_{n,n} & \cdots & B_{n,1} & I \end{pmatrix} \Gamma_{n+1}$$
$$= \begin{pmatrix} Q_n - P_n S_n^{-1} P_n^T & 0 & \cdots & 0 & 0 \\ 0 & 0 & \cdots & 0 & S_n - P_n^T Q_n^{-1} P_n \end{pmatrix} \quad \text{(C8.6.10)}$$

where Q_n and S_n are assumed to be nonsingular. (Note from (C8.6.13) below that the nonsingularity of Q_n and S_n for $n = 1, 2, \ldots$ is equivalent to the nonsingularity of Γ_n for $n = 1, 2, \ldots$.) Assuming that Γ_n is nonsingular, it follows from (C8.6.9), (C8.6.10) that, for $n = 1, 2, \ldots$,

$$\begin{pmatrix} I & A_{n+1,1} & \cdots & A_{n+1,n} & A_{n+1,n+1} \\ B_{n+1,n+1} & B_{n+1,n} & \cdots & B_{n+1,1} & I \end{pmatrix}$$
$$= \begin{pmatrix} I & -P_n S_n^{-1} \\ -P_n^T Q_n^{-1} & I \end{pmatrix} \begin{pmatrix} I & A_{n,1} & \cdots & A_{n,n} & 0 \\ 0 & B_{n,n} & \cdots & B_{n,1} & I \end{pmatrix} \quad \text{(C8.6.11)}$$
$$Q_{n+1} = Q_n - P_n S_n^{-1} P_n^T$$
$$S_{n+1} = S_n - P_n^T Q_n^{-1} P_n$$

The initial values for the recursion are the following:

$$A_{1,1} = -R_1 R_0^{-1} \qquad B_{1,1} = -R_1^T R_0^{-1}$$
$$Q_1 = R_0 + A_{1,1} R_1^T \qquad S_1 = R_0 + B_{1,1} R_1$$

(or equivalently, $Q_0 = R_0$, $S_0 = R_0$, $P_0 = R_1$). The quantities

$$\bar{K}_{n+1} = P_n S_n^{-1}$$
$$\tilde{K}_{n+1} = Q_n^{-1} P_n$$

are the so-called reflection coefficients (also called partial correlation coefficients).

To write the WWR recursion (C8.6.11) in a more compact form, introduce the following matrix polynomials:

$$A_n(z) = I + A_{n,1} z + \ldots + A_{n,n} z^n$$
$$B_n(z) = B_{n,n} + B_{n,n-1} z + \ldots + B_{n,1} z^{n-1} + I z^n$$

Postmultiplying (C8.6.11) by $(I \ \ Iz \ \ldots \ Iz^{n+1})^T$, gives

$$\begin{pmatrix} A_{n+1}(z) \\ B_{n+1}(z) \end{pmatrix} = \begin{pmatrix} I & -\bar{K}_{n+1} \\ -\tilde{K}_{n+1}^T & I \end{pmatrix} \begin{pmatrix} A_n(z) \\ z B_n(z) \end{pmatrix}$$

These recursions induce a lattice structure, as depicted in Figure C8.6.1, for computing the polynomials $A_n(z)$ and $B_n(z)$.

The upper output of the kth lattice section is the kth-order predictor polynomial $A_k(z)$. Thus, the WWRA provides not only the nth-order solution $A_n(z)$, but also all

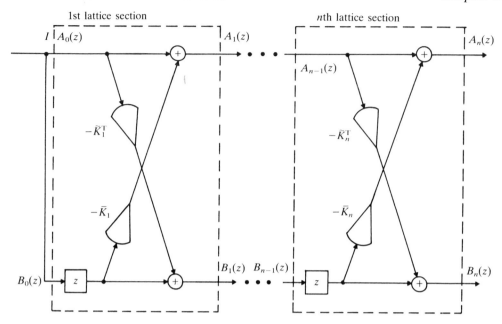

FIGURE C8.6.1 The lattice filter implementation of the WWRA.

the lower-order solutions $A_k(z)$, $k = 0, \ldots, n$, which can be quite useful in some applications.

Cholesky factorization of Γ_n^{-1}

Introduce the following notation:

$$U = \begin{pmatrix} I & A_{n,1} & \cdots & \cdots & A_{n,n} \\ & I & \cdots & \cdots & A_{n-1,n-1} \\ & & \ddots & & \vdots \\ & 0 & & I & A_{1,1} \\ & & & & I \end{pmatrix}$$

Using the defining property of $\{A_{k,i}\}$ it can be verified that

$$U\Gamma_n = \begin{pmatrix} Q_n & & & 0 \\ & \ddots & & \\ & & Q_1 & \\ X & & & R_0 \end{pmatrix} \qquad (C8.6.12)$$

where X denotes elements whose exact expressions are not important for this discussion. From (C8.6.12) and the fact that $U\Gamma_n U^T$ is a symmetric matrix it follows immediately that

$$U\Gamma_n U^{\mathrm{T}} = \begin{pmatrix} Q_n & & & 0 \\ & \ddots & & \\ 0 & & Q_1 & \\ & & & R_0 \end{pmatrix} \qquad \text{(C8.6.13)}$$

The implications of this result are important. It shows that:

$$\{\Gamma_n > 0\} \Leftrightarrow \{R_0 > 0 \text{ and } Q_i > 0 \text{ for } i = 1, \ldots, n\}$$

The implication '$\Rightarrow$' is immediate. Indeed, if $\Gamma_n > 0$ then U exists; hence (C8.6.13) holds, which readily implies that $R_0 > 0$ and $Q_i > 0$, $i = 1, \ldots, n$. To prove the implication '$\Leftarrow$', assume that $R_0 > 0$ and $Q_i > 0$ for $i = 1, \ldots, n$. Since $R_0 > 0$ it follows that $A_{1,1}$ exists; thus the factorization (C8.6.13) holds for $n = 1$. Combining this observation with the fact that $Q_1 > 0$, it follows that $\Gamma_1 > 0$. Thus $A_{2,1}$ and $A_{2,2}$ exist; and so on.

It also follows from (C8.6.13) that if $\Gamma_n > 0$ then the matrix

$$U^{\mathrm{T}} \begin{pmatrix} Q_n^{-1/2} & & & 0 \\ & \ddots & & \\ & & Q_1^{-1/2} & \\ 0 & & & R_0^{-1/2} \end{pmatrix}$$

is a block lower-triangular Cholesky factor of Γ_n^{-1}. Note that since the matrix U is triangular it can be efficiently inverted to produce the block-triangular Cholesky factor of the covariance matrix Γ_n.

Similarly, it can be shown that the matrix coefficients $\{B_{n,i}\}$ provide a block upper-triangular Cholesky factorization of Γ_n^{-1}. To see this, define

$$V = \begin{pmatrix} I & & & & 0 \\ B_{1,1} & I & & & \\ \vdots & \vdots & \ddots & & \\ B_{n,n} & B_{n,n-1} & \cdots & B_{n,1} & I \end{pmatrix}$$

Following the arguments leading to (C8.6.13) it can be readily verified that

$$V\Gamma_n V^{\mathrm{T}} = \begin{pmatrix} R_0 & & & 0 \\ & S_1 & & \\ & & \ddots & \\ 0 & & & S_n \end{pmatrix} \qquad \text{(C8.6.14)}$$

which concludes the proof of the assertion made above. Similarly to the preceding result, it follows from (C8.6.14) that

$$\{\Gamma_n > 0\} \Leftrightarrow \{R_0 > 0 \text{ and } S_i > 0 \text{ for } i = 1, \ldots, n\}$$

Some properties of the WWRA

The following statements are equivalent:

(i) $\Gamma_n > 0$
(ii) $R_0 > 0$ and $Q_i > 0$ for $i = 1, \ldots, n$
(iii) $R_0 > 0$ and $S_i > 0$ for $i = 1, \ldots, n$
(iv) $R_0 > 0$, $Q_n > 0$ and $\det A_n(z) \neq 0$ for $|z| \leq 1$

Proof. The equivalences (i) $\Leftrightarrow$ (ii) and (i) $\Leftrightarrow$ (iii) were proved in the previous subsection.

Consider the equivalence (i) $\Leftrightarrow$ (iv). Introduce the notation

$$A_c = \begin{pmatrix} -A_{n,1}^T & I & & 0 \\ \vdots & & \ddots & \\ & 0 & & I \\ -A_{n,n}^T & 0 & \ldots & 0 \end{pmatrix}$$

Note that A_c is a block companion matrix associated with the polynomial $A_n(z)$. It is well known that $\det z^n A_n(z^{-1})$ is the characteristic polynomial of A_c (see e.g. Kailath, 1980). Hence $\det A_n(z)$ has all its roots outside the unit circle if and only if all the eigenvalues of A_c lie strictly within the unit circle. Let λ denote an arbitrary eigenvalue of A_c and let

$$\mu = \begin{pmatrix} \mu_1 \\ \vdots \\ \mu_n \end{pmatrix} \neq 0$$

be an associated eigenvector, where μ_i are ny-vectors. Thus

$$A_c \mu = \lambda \mu$$

or in a more detailed form,

$$\mu_2 - A_{n,1}^T \mu_1 = \lambda \mu_1$$
$$\ldots$$
$$\mu_n - A_{n,n-1}^T \mu_1 = \lambda \mu_{n-1} \quad \text{(C8.6.15)}$$
$$-A_{n,n}^T \mu_1 = \lambda \mu_n$$

It is easy to see that $\mu \neq 0$ implies $\mu_1 \neq 0$. For if $\mu_1 = 0$ then it follows from (C8.6.15) that $\mu = 0$. The set of equations (C8.6.15) can be written more compactly as

$$\begin{pmatrix} \mu \\ 0 \end{pmatrix} = \begin{pmatrix} I \\ \Theta^T \end{pmatrix} \mu_1 + \lambda \begin{pmatrix} 0 \\ \mu \end{pmatrix} \quad \text{(C8.6.16)}$$

Recall from (C8.6.1) that $\Theta = (A_{n,1} \ldots A_{n,n})$. It follows from (C8.6.16), (C8.6.4) and the Toeplitz structure of Γ_n that

$$\mu^* \Gamma_{n-1} \mu = (\mu^* \quad 0) \Gamma_n \begin{pmatrix} \mu \\ 0 \end{pmatrix}$$

$$= \{\mu_1^*(I \quad \Theta) + \lambda^*(0 \quad \mu^*)\} \Gamma_n \left\{ \begin{pmatrix} I \\ \Theta^T \end{pmatrix} \mu_1 + \lambda \begin{pmatrix} 0 \\ \mu \end{pmatrix} \right\}$$

$$= \mu_1^* Q_n \mu_1 + |\lambda|^2 \mu^* \Gamma_{n-1} \mu$$

(where the asterisk denotes the transpose complex conjugate). Thus,

$$|\lambda|^2 = 1 - \frac{\mu_1^* Q_n \mu_1}{\mu^* \Gamma_{n-1} \mu} \tag{C8.6.17}$$

Since $\Gamma_n > 0$ implies $\Gamma_{n-1} > 0$ and $Q_n > 0$, it follows from (C8.6.17) that $|\lambda| < 1$, and the proof of the implication (i) $\Rightarrow$ (iv) is concluded.

To prove the implication (iv) $\Rightarrow$ (i), consider the following AR model associated with $A_n(z)$:

$$A_n(q^{-1})x(t) = w(t) \tag{C8.6.18}$$

where $Ew(t)w^T(s) = Q_n \delta_{t,s}$. Note that Q_n is a valid covariance matrix by assumption. Since the AR equation above is stable by assumption, $x(t)$ can be written as an infinite moving average in $w(t)$,

$$x(t) = w(t) + C_1 w(t-1) + C_2 w(t-2) + \ldots$$

which implies that

$$\begin{aligned} Ew(t)x^T(t) &= Q_n \\ Ew(t)x^T(t-k) &= 0 \quad \text{for } k \geq 1 \end{aligned} \tag{C8.6.19}$$

Let $\{R_k^x\}$ denote the covariance sequence of $x(t)$,

$$R_k^x = Ex(t)x^T(t-k)$$

With (C8.6.19) in mind, postmultiply (C8.6.18) by $x^T(t-k)$ for $k = 0, 1, 2, \ldots$ and take expectations to get

$$R_0^x + A_{n,1} R_{-1}^x + \ldots + A_{n,n} R_{-n}^x = Q_n \tag{C8.6.20a}$$

$$(A_{n,1} \ldots A_{n,n}) \Gamma_{n-1}^x = -(R_1^x \ldots R_n^x) \tag{C8.6.20b}$$

and

$$R_{n+k}^x = -A_{n,1} R_{n+k-1}^x \ldots -A_{n,n} R_k^x \quad k \geq 1 \tag{C8.6.20c}$$

where Γ_{n-1}^x is defined similarly to Γ_{n-1}. The solution $\{R_0^x \ldots R_n^x\}$ of (C8.6.20a, b), for given $\{A_{n,i}\}$ and Q_n, is unique. To see this, assume that there are two solutions to (C8.6.20a, b). These two solutions lead through (C8.6.20c) to two different infinite covariance sequences associated with the same (AR) process, which is impossible.

Next recall that $\{A_{n,i}\}$ in (C8.6.20a, b) are given by (C8.6.4). With this in mind, it is easy to see that the unique solution of (C8.6.20a, b) is given by

$$R_k^x = R_k \quad k = 0, \ldots, n \tag{C8.6.21}$$

This concludes the proof. ∎

Remark 1. Note that the condition $Q_n > 0$ in (iv) is really necessary. That is to say, it is not implied by the other conditions of (iv). To see this, consider the following first-order bivariate autoregression:

318 *Instrumental variable methods*   Chapter 8

$$A_1(q^{-1})y(t) = \begin{pmatrix} e(t) \\ e(t) \end{pmatrix}$$

where

$$A_1(q^{-1}) = I + A_{1,1}q^{-1} = \begin{pmatrix} 1 + aq^{-1} & aq^{-1} \\ 0 & 1 + aq^{-1} \end{pmatrix}; \quad |a| < 1$$

$$Ee(t)e(s) = \delta_{t,s}$$

Let $R_k = Ey(t)y^T(t-k)$. Some straightforward calculations give

$$R_0 = \frac{1}{(1-a^2)^3}\begin{pmatrix} 1+a^2 & 1-a^2 \\ 1-a^2 & (1-a^2)^2 \end{pmatrix}$$

which is positive definite. Furthermore, by the consistency properties of the Yule–Walker estimates, the solution of (C8.6.3) for $n = 1$ is given by $A_{1,1}$ above. The corresponding $A_1(z)$ polynomial (see above) is stable by construction. Thus, two of the conditions in (iv) (the first and third) are satisfied. However, the matrix Γ_1 is not positive definite (it is only positive semidefinite) as can be seen, for example from the fact that

$$(I \quad A_{1,1})\Gamma_1 \begin{pmatrix} I \\ A_{1,1}^T \end{pmatrix} = E\begin{pmatrix} e(t) \\ e(t) \end{pmatrix}(e(t) \quad e(t)) = \begin{pmatrix} 1 & 1 \\ 1 & 1 \end{pmatrix} = Q_1$$

is singular. Note that for the case under discussion the second condition of (iv) (i.e. $Q_1 > 0$) is not satisfied. ∎

Remark 2. It is worth noting from the above proof of the implication (iv) ⇒ (i), that the following result of independent interest holds:

If $\Gamma_n > 0$ then the autoregression (C8.6.18), where $\{A_{n,i}\}$ and Q_n are given by (C8.6.4), matches the given covariance sequence $\{R_0, \ldots, R_n\}$ exactly ∎

It is now shown that the conditions (ii) and (iii) above are equivalent to requiring that some matrix reflection coefficients (to be defined shortly) have singular values less than one. This is a direct generalization of the result presented in Complement C8.2 for the scalar case.

For $Q_n > 0$ and $S_n > 0$, define

$$K_{n+1} = Q_n^{-1/2} P_n S_n^{-T/2} \tag{C8.6.22}$$

The matrix square root $Q^{1/2}$ of the symmetric positive definite matrix Q is defined by

$$Q^{1/2}(Q^{1/2})^T = Q^{1/2}Q^{T/2} = Q$$

Then

$$Q_{n+1} = Q_n^{1/2}(I - K_{n+1}K_{n+1}^T)Q_n^{T/2} \tag{C8.6.23}$$

$$S_{n+1} = S_n^{1/2}(I - K_{n+1}^T K_{n+1})S_n^{T/2} \tag{C8.6.24}$$

(cf. (C8.6.11)). Let σ_{n+1} denote the maximum singular value of K_{n+1}. Clearly (ii) or (iii) imply

$$\{R_0 > 0 \text{ and } \sigma_i < 1, i = 1, \ldots, n\} \tag{C8.6.25}$$

Using a simple inductive reasoning it can be shown that the converse is also true. If $R_0 = Q_0 = S_0 > 0$ and $\sigma_1 < 1$, then (C8.6.23), (C8.6.24) give $Q_1 > 0$ and $S_1 > 0$, which together with $\sigma_2 < 1$ give $Q_2 > 0$ and $S_2 > 0$, and so on. Thus, the proof that the statements (i), (ii) or (iii) above are equivalent to (C8.6.25) is complete.

The properties of the WWRA derived above are analogous to those presented in Complement C8.2 for the (scalar) Levinson–Durbin algorithm, and therefore they may find applications similar to those described there.

Chapter 9
RECURSIVE IDENTIFICATION METHODS

9.1 Introduction

In recursive (also called on-line) identification methods, the parameter estimates are computed recursively in time. This means that if there is an estimate $\hat{\theta}(t-1)$ based on data up to time $t-1$, then $\hat{\theta}(t)$ is computed by some 'simple modification' of $\hat{\theta}(t-1)$. The counterparts of on-line methods are the so-called off-line or batch methods, in which all the recorded data are used simultaneously to find the parameter estimates. Various off-line methods were studied in Chapters 4, 7 and 8.

Recursive identification methods have the following general features:

- They are a central part of adaptive systems (used, for example, for control or signal processing) where the (control, filtering, etc.) action is based on the most recent model.
- Their requirement on primary memory is quite modest, since not all data are stored.
- They can be easily modified into real-time algorithms, aimed at tracking time-varying parameters.
- They can be the first step in a fault detection algorithm, which is used to find out whether the system has changed significantly.

Most adaptive systems, for example adaptive control systems (see Figure 9.1), are based (explicitly or implicitly) on recursive identification. Then a current estimated model of the process is available at all times. This time-varying model is used to determine the parameters of the (also time-varying) regulator. In this way the regulator will be dependent on the previous behavior of the process (through the information flow: process → model → regulator). If an appropriate principle is used to design the regulator then the regulator should adapt to the changing characteristics of the process.

As stated above, a recursive identification method has a small requirement on primary memory since only a modest amount of information is stored. This amount will not increase with time.

There are many variants of recursive identification algorithms designed for systems with time-varying parameters. For such algorithms the parameter estimate $\hat{\theta}(t)$ will not converge as t tends to infinity, even for time-invariant systems, since the algorithm disregards information from (very) old data points in order to respond to process change.

Fault detection schemes can be used in several ways. One form of application is failure

Section 9.2   *The recursive least squares method*   321

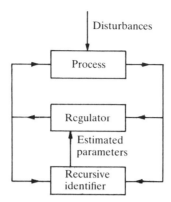

FIGURE 9.1 A general scheme for adaptive control.

diagnosis, where it is desired to find out on-line whether the system malfunctions. Fault detection is also commonly used with real-time identification methods designed for handling abrupt changes of the system. When such a change occurs it should be noticed by the fault detection algorithm. The identification algorithm should then be modified so that previous data points have less effect upon the current parameter estimates.

Many recursive identification methods are derived as approximations of off-line methods. It may therefore happen that the price paid for the approximation is a reduction in accuracy. It should be noted, however, that the user seldom chooses between off-line and on-line methods, but rather between different off-line methods or between different on-line methods.

9.2 The recursive least squares method

To illustrate the derivation of recursive algorithms, there follows a very simple example.

Example 9.1 *Recursive estimation of a constant*
Consider the model

$$y(t) = b + e(t)$$

where $e(t)$ denotes a disturbance of variance λ^2.

In Example 4.4 it was shown that the least squares estimate of $\theta = b$ is the arithmetic mean,

$$\hat{\theta}(t) = \frac{1}{t} \sum_{s=1}^{t} y(s) \qquad (9.1)$$

This expression can be reformulated as a recursive algorithm. Some straightforward calculations give

$$\hat{\theta}(t) = \frac{1}{t}\left[\sum_{s=1}^{t-1} y(s) + y(t)\right] = \frac{1}{t}[(t-1)\hat{\theta}(t-1) + y(t)] \qquad (9.2)$$

$$= \hat{\theta}(t-1) + \frac{1}{t}[y(t) - \hat{\theta}(t-1)]$$

The result is quite appealing. The estimate of θ at time t is equal to the previous estimate (at time $t-1$) plus a correction term. The correction term is proportional to the deviation of the 'predicted' value $\hat{\theta}(t-1)$ from what is actually observed at time t, namely $y(t)$. Moreover, the prediction error is weighted by the factor $1/t$, which means that the magnitude of the changes of the estimate will decrease with time. This is natural since, as t increases, the old information as condensed in the estimate $\hat{\theta}(t-1)$ will become more reliable.

The variance of $\hat{\theta}(t)$ is given, neglecting a factor of λ^2, by

$$P(t) = (\Phi^T\Phi)^{-1} = \frac{1}{t} \qquad \Phi^T = (1 \ldots 1) \qquad (1|t) \qquad (9.3)$$

(see Lemma 4.2). The algorithm (9.2) can be complemented with a recursion for $P(t)$. From (9.3) it follows that

$$P^{-1}(t) = t = P^{-1}(t-1) + 1$$

and hence

$$P(t) = \frac{1}{P^{-1}(t-1) + 1} = \frac{P(t-1)}{1 + P(t-1)} \qquad (9.4)$$

∎

To start the formal discussion of the recursive least squares (RLS) method, consider the scalar system (dim $y = 1$) given by (7.2). Then the parameter estimate is given by

$$\hat{\theta}(t) = \left[\sum_{s=1}^{t} \varphi(s)\varphi^T(s)\right]^{-1}\left[\sum_{s=1}^{t} \varphi(s)y(s)\right] \qquad (9.5)$$

(cf. (4.10) or (7.4)). The argument t has been used to stress the dependence of $\hat{\theta}$ on time. The expression (9.5) can be computed in a recursive fashion. Introduce the notation

$$P(t) = \left[\sum_{s=1}^{t} \varphi(s)\varphi^T(s)\right]^{-1} \qquad (9.6)$$

Since trivially

$$P^{-1}(t) = P^{-1}(t-1) + \varphi(t)\varphi^T(t) \qquad (9.7)$$

it follows that

$$\hat{\theta}(t) = P(t)\left[\sum_{s=1}^{t-1} \varphi(s)y(s) + \varphi(t)y(t)\right]$$

$$= P(t)[P^{-1}(t-1)\hat{\theta}(t-1) + \varphi(t)y(t)]$$

$$= \hat{\theta}(t-1) + P(t)\varphi(t)[y(t) - \varphi^T(t)\hat{\theta}(t-1)]$$

Thus

$$\hat{\theta}(t) = \hat{\theta}(t - 1) + K(t)\varepsilon(t) \tag{9.8a}$$

$$K(t) = P(t)\varphi(t) \tag{9.8b}$$

$$\varepsilon(t) = y(t) - \varphi^T(t)\hat{\theta}(t - 1) \tag{9.8c}$$

Here the term $\varepsilon(t)$ should be interpreted as a prediction error. It is the difference between the measured output $y(t)$ and the one-step-ahead prediction $\hat{y}(t|t - 1; \hat{\theta}(t - 1))$ $= \varphi^T(t)\hat{\theta}(t - 1)$ of $y(t)$ made at time $t - 1$ based on the model corresponding to the estimate $\hat{\theta}(t - 1)$. If $\varepsilon(t)$ is small, the estimate $\hat{\theta}(t - 1)$ is 'good' and should not be modified very much. The vector $K(t)$ in (9.8b) should be interpreted as a weighting or gain factor showing how much the value of $\varepsilon(t)$ will modify the different elements of the parameter vector.

To complete the algorithm, (9.7) must be used to compute $P(t)$ which is needed in (9.8b). However, the use of (9.7) needs a matrix inversion at each time step. This would be a time-consuming procedure. Using the matrix inversion lemma (Lemma A.1), however, (9.7) can be rewritten in a more useful form. Then an updating equation for $P(t)$ is obtained, namely

$$P(t) = P(t - 1) - P(t - 1)\varphi(t)\varphi^T(t)P(t - 1)/[1 + \varphi^T(t)P(t - 1)\varphi(t)] \tag{9.9}$$

Note that in (9.9) there is now a scalar division (a scalar inversion) instead of a matrix inversion. The algorithm consisting of (9.8)–(9.9) can be simplified further. From (9.8b), (9.9),

$$K(t) = P(t - 1)\varphi(t) - P(t - 1)\varphi(t)\varphi^T(t)P(t - 1)\varphi(t)/[1 + \varphi^T(t)P(t - 1)\varphi(t)]$$
$$= P(t - 1)\varphi(t)/[1 + \varphi^T(t)P(t - 1)\varphi(t)] \tag{9.10}$$

This form for $K(t)$ is more convenient to use for implementation than (9.8b). The reason is that the right-hand side of (9.10) must anyway be computed in the updating of $P(t)$ (see (9.9)).

The derivation of the recursive LS (RLS) method is now complete. The RLS algorithm consists of (9.8a), (9.8c), (9.9) and (9.10). The algorithm also needs initial values $\hat{\theta}(0)$ and $P(0)$. For convenience the choice of these quantities is discussed in the next section.

For illustration, consider what the general algorithm (9.8)–(9.10) becomes in the simple case discussed in Example 9.1. The equation (9.9) for $P(t)$ becomes, since $\varphi(t) \equiv 1$,

$$P(t) = P(t - 1) - \frac{P^2(t - 1)}{1 + P(t - 1)} = \frac{P(t - 1)}{1 + P(t - 1)}$$

which is precisely (9.4). Thus

$$P(t) = K(t) = 1/t$$

(cf. (9.8b)). Moreover, (9.8a, c) give

$$\hat{\theta}(t) = \hat{\theta}(t-1) + \frac{1}{t}[y(t) - \hat{\theta}(t-1)]$$

which coincides with (9.2).

9.3 Real-time identification

This section presents some modifications of the recursive LS algorithm which are useful for tracking time-varying parameters. Then we may speak about 'real-time identification'. When the properties of the process may change (slowly) with time, the recursive algorithm should be able to track the time-varying parameters describing such a process.

There are two common approaches to modifying the recursive LS algorithm to a real-time method:

- Use of a forgetting factor.
- Use of a Kalman filter as a parameter estimator.

A third possibility is described in Problem 9.10.

Forgetting factor

The approach in this case is to change the loss function to be minimized. Let the modified loss function be

$$V_t(\theta) = \sum_{s=1}^{t} \lambda^{t-s} \varepsilon^2(s) \tag{9.11}$$

The loss function used earlier had $\lambda = 1$ (see Chapters 4 and 7 and Example 9.1) but now it contains the *forgetting factor* λ, a number somewhat less than 1 (for example 0.99 or 0.95). This means that with increasing t the measurements obtained previously are discounted. The smaller the value of λ, the quicker the information in previous data will be forgotten. One can rederive the RLS method for the modified criterion (9.11) (see problem 9.1). The calculations are straightforward. The result is a special case of a recursive prediction error method presented in Section 9.5. The recursive LS method with a forgetting factor is

$$\begin{aligned}
\hat{\theta}(t) &= \hat{\theta}(t-1) + K(t)\varepsilon(t) \\
\varepsilon(t) &= y(t) - \varphi^T(t)\hat{\theta}(t-1) \\
K(t) &= P(t)\varphi(t) = P(t-1)\varphi(t)/[\lambda + \varphi^T(t)P(t-1)\varphi(t)] \\
P(t) &= \{P(t-1) - P(t-1)\varphi(t)\varphi^T(t)P(t-1)/[\lambda + \varphi^T(t)P(t-1)\varphi(t)]\}/\lambda
\end{aligned} \tag{9.12}$$

Kalman filter as a parameter estimator

Assuming that the parameters are constant, the underlying model

$$y(t) = \varphi^T(t)\theta + e(t) \tag{9.13}$$

can be described as a state space equation

$$x(t + 1) = x(t) \tag{9.14a}$$

$$y(t) = \varphi^T(t)x(t) + e(t) \tag{9.14b}$$

where the 'state vector' $x(t)$ is given by

$$x(t) = (a_1 \ldots a_{na} \ b_1 \ldots b_{nb})^T = \theta \tag{9.15}$$

The optimal state estimate $\hat{x}(t + 1)$ can be computed as a function of the measurements till time t using the Kalman filter (see Section B.7 for a brief discussion). Note that usually the Kalman filter is presented for state space equations whose matrices may be time varying but do not depend on the data. The latter condition fails in the case of (9.14) since $\varphi(t)$ depends on data up to (and inclusive of) the time $(t - 1)$. However, it can be shown that also in such cases the Kalman filter provides the optimal (mean square) estimate of the system state vector (see Ho, 1963, Bohlin, 1970, and Åström and Wittenmark, 1971 for details).

Applying the Kalman filter to the state model (9.14) will give precisely the basic recursive LS algorithm. One way of modifying the algorithm so that time-varying parameters can be tracked better is to change the state equation (9.14a) to

$$x(t + 1) = x(t) + v(t) \qquad Ev(t)v^T(s) = R_1\delta_{t,s} \tag{9.16}$$

This means that the parameter vector is modeled as a random walk or a drift. The covariance matrix R_1 can be used to describe how fast the different components of θ are expected to vary. Applying the Kalman filter to the model (9.16), (9.14b) gives the following recursive algorithm:

$$\boxed{\begin{aligned}
\hat{\theta}(t) &= \hat{\theta}(t - 1) + K(t)\varepsilon(t) \\
\varepsilon(t) &= y(t) - \varphi^T(t)\hat{\theta}(t - 1) \\
K(t) &= P(t)\varphi(t) = P(t - 1)\varphi(t)/[1 + \varphi^T(t)P(t - 1)\varphi(t)] \\
P(t) &= P(t - 1) - P(t - 1)\varphi(t)\varphi^T(t)P(t - 1)/[1 + \varphi^T(t)P(t - 1)\varphi(t)] + R_1
\end{aligned}} \tag{9.17}$$

Observe that for both algorithms (9.12) and (9.17) the basic method has been modified so that $P(t)$ will no longer tend to zero. In this way $K(t)$ also is prevented from decreasing to zero. The parameter estimates will therefore change continually.

In the algorithm (9.12) λ is a design variable to be chosen by the user. The matrix R_1 in the algorithm (9.17) has a role similar to that of λ in (9.12). These design variables should be chosen by a trade-off between alertness and ability to track time variations of the parameters (which requires λ 'small' or R_1 'large') on the one hand, and good convergence properties and small variances of the estimates for time-invariant systems (which requires λ close to 1 or R_1 close to 0) on the other. Note that the algorithm (9.17) offers more flexibility than (9.12) does, since the whole matrix R_1 can be set by the user. It is for example possible to choose it to be a diagonal matrix with different diagonal elements. This choice of R_1 may be convenient for describing different time variations for the different parameters.

Initial values

The Kalman filter interpretation of the RLS algorithm is also useful in another respect. It provides suggestions for the choice of the initial values $\hat{\theta}(0)$ and $P(0)$. These values are necessary to start the algorithm. Since $P(t)$ (times λ^2) is the covariance matrix of $\hat{\theta}(t)$ it is reasonable to take for $\hat{\theta}(0)$ an *a priori* estimate of θ and to let $P(0)$ reflect the confidence in this initial estimate $\hat{\theta}(0)$. If $P(0)$ is small then $K(t)$ will be small for all t and the parameter estimates will therefore not change too much from $\hat{\theta}(0)$. On the other hand, if $P(0)$ is large, the parameter estimates will quickly jump away from $\hat{\theta}(0)$. Without any *a priori* information it is common practice to take

$$\hat{\theta}(0) = 0 \qquad P(0) = \varrho I \qquad (9.18)$$

where ϱ is a 'large' number.

The effect of the initial values $\hat{\theta}(0)$ and $P(0)$ on the estimate $\hat{\theta}(t)$ can be analyzed algebraically. Consider the basic RLS algorithm (with $\lambda = 1$), (9.8)–(9.10). (For the case $\lambda < 1$ see Problem 9.4.) Equation (9.7) gives

$$P^{-1}(t) = P^{-1}(0) + \sum_{s=1}^{t} \varphi(s)\varphi^T(s)$$

Now set

$$x(t) = P^{-1}(t)\hat{\theta}(t)$$

Then

$$\begin{aligned} x(t) &= P^{-1}(t)\hat{\theta}(t-1) + \varphi(t)\varepsilon(t) \\ &= [P^{-1}(t-1) + \varphi(t)\varphi^T(t)]\hat{\theta}(t-1) + \varphi(t)[y(t) - \varphi^T(t)\hat{\theta}(t-1)] \\ &= x(t-1) + \varphi(t)y(t) \\ &= x(0) + \sum_{s=1}^{t} \varphi(s)y(s) \end{aligned}$$

and hence

$$\begin{aligned} \hat{\theta}(t) &= P(t)x(t) \\ &= \left[P^{-1}(0) + \sum_{s=1}^{t} \varphi(s)\varphi^T(s) \right]^{-1} \left[P^{-1}(0)\hat{\theta}(0) + \sum_{s=1}^{t} \varphi(s)y(s) \right] \end{aligned} \qquad (9.19)$$

If $P^{-1}(0)$ is small, (i.e. $P(0)$ is large), then $\hat{\theta}(t)$ is close to the off-line estimate

$$\hat{\theta}_{\text{off}}(t) = \left[\sum_{s=1}^{t} \varphi(s)\varphi^{T}(s)\right]^{-1} \left[\sum_{s=1}^{t} \varphi(s)y(s)\right] \tag{9.20}$$

The expression for $P^{-1}(t)$ can be used to find an appropriate $P(0)$. First choose a time t_0 when $\hat{\theta}(t)$ should be approximately $\hat{\theta}_{\text{off}}(t)$. In practice one may take $t_0 = 10\text{--}25$. Then choose $P(0)$ so that

$$P^{-1}(0) \ll \sum_{s=1}^{t_0} \varphi(s)\varphi^{T}(s) \approx t_0 E\varphi(t)\varphi^{T}(t) \tag{9.21}$$

For example, if the elements of $\varphi(t)$ have minimum variance, say σ^2, then $P(0)$ can be taken as in (9.18) with $\varrho \gg 1/[t_0\sigma^2]$.

The methods discussed in this section are well suited to systems that vary slowly with time. In such cases λ is chosen close to 1 or R_1 as a small nonnegative definite matrix. If the system exhibits some fast parameter changes that seldom occur, some modified methods are necessary. A common idea in many such methods is to use a 'fault detector' which tests for the occurrence of significant parameter changes. If a change is detected the algorithm can be restarted, at least partly. One way of doing this is to decrease the forgetting factor temporarily or to increase R_1 or parts of the $P(t)$ matrix.

9.4 The recursive instrumental variable method

Consider next the basic instrumental variable estimate (8.12). For simplicity, take the case of a scalar system (dim $y = 1$) given by (7.2). The IV estimate can be written as

$$\hat{\theta}(t) = \left[\sum_{s=1}^{t} z(s)\varphi^{T}(s)\right]^{-1} \left[\sum_{s=1}^{t} z(s)y(s)\right] \tag{9.22}$$

Note the algebraic similarity with the least squares estimate (9.5). Going through the derivation of the RLS algorithm, it can be seen that the estimate (9.22) can be computed recursively as

$$\begin{aligned}
\hat{\theta}(t) &= \hat{\theta}(t-1) + K(t)\varepsilon(t) \\
\varepsilon(t) &= y(t) - \varphi^{T}(t)\hat{\theta}(t-1) \\
K(t) &= P(t)z(t) = P(t-1)z(t)/[1 + \varphi^{T}(t)P(t-1)z(t)] \\
P(t) &= P(t-1) - P(t-1)z(t)\varphi^{T}(t)P(t-1)/[1 + \varphi^{T}(t)P(t-1)z(t)]
\end{aligned} \tag{9.23}$$

This is similar to the recursive LS estimate: the only difference is that $\varphi(t)$ has been changed to $z(t)$ while $\varphi^{T}(t)$ is kept the same as before. Note that this is true also for the 'off-line' form (9.22).

Examples of IV methods were given in Chapter 8. One particular IV variant is tied to an adaptive way of generating the instrumental variables. Suppose it is required that

$$z(t) = \tilde{\varphi}(t) = (-x(t-1) \ \ldots \ -x(t-na) \ \ u(t-1) \ \ldots \ u(t-nb))^{\mathrm{T}} \quad (9.24)$$

where $x(t)$ is the noise-free output of the process given by

$$A_0(q^{-1})x(t) = B_0(q^{-1})u(t) \quad (9.25)$$

The signal $x(t)$ is not available by measurements and cannot be computed from (9.25) since $A_0(q^{-1})$ and $B_0(q^{-1})$ are unknown. However, $x(t)$ could be estimated using the parameter vector $\hat{\theta}(t)$ in the following adaptive fashion:

$$\begin{aligned} z(t) &= (-\hat{x}(t-1) \ \ldots \ -\hat{x}(t-na) \ \ u(t-1) \ \ldots \ u(t-nb))^{\mathrm{T}} \\ \hat{x}(t) &= z^{\mathrm{T}}(t)\hat{\theta}(t) \end{aligned} \quad (9.26)$$

The initialization of the recursive instrumental variable method as well as the derivation of real-time variants (for tracking time-varying parameters) are similar to those described for the recursive least squares method.

The recursive algorithm (9.23) applies to the basic IV method. In Complement C9.1 a recursive extended IV algorithm is derived. Note also that in Section 8.3 it was shown that IV estimation for multivariable systems with ny outputs often decouples into ny estimation problems of MISO (multiple input, single output) type. In all such cases the algorithm (9.23) can be applied.

9.5 The recursive prediction error method

This discussion of a recursive prediction error method (RPEM) will include the use of a forgetting factor in the problem formulation, and we will treat the general case of multivariable models of an unspecified form. For convenience the following criterion function will be used:

$$V_t(\theta) = \frac{1}{2} \sum_{s=1}^{t} \lambda^{t-s} \varepsilon^{\mathrm{T}}(s, \theta) Q \varepsilon(s, \theta) \quad (9.27)$$

where Q is a positive definite weighting matrix. (For more general criterion functions that can be used within the PEM, see Section 7.2.) For $\lambda = 1$, (9.27) reduces to the loss function (7.15a) corresponding to the choice $h_1(R) = \mathrm{tr}\ R$, which was discussed previously (see Example 7.1). Use of the normalization factor $1/2$ instead of $1/t$ will be more convenient in what follows. This simple rescaling of the problem will not affect the final result. Note that so far no assumptions on the model structure have been introduced.

The off-line estimate, $\hat{\theta}_t$, which minimizes $V_t(\theta)$ cannot be found analytically (except for the LS case). Instead a numerical optimization must be performed. Therefore it is not possible to derive an exact recursive algorithm (of moderate complexity) for computing $\hat{\theta}_t$. Instead some sort of approximation must be used. The approximations to be made are such that they hold exactly for the LS case. This is a sensible way to proceed since

Section 9.5 The recursive prediction error method 329

the LS estimate, which is a special case of the PEM (with $\varepsilon(t, \theta)$ a linear function of θ), can be computed exactly with a recursive algorithm as shown in Section 9.2.

Assume that $\hat{\theta}(t-1)$ minimizes $V_{t-1}(\theta)$ and that the minimum point of $V_t(\theta)$ is close to $\hat{\theta}(t-1)$. Then it is reasonable to approximate $V_t(\theta)$ by a second-order Taylor series expansion around $\hat{\theta}(t-1)$:

$$V_t(\theta) \approx V_t(\hat{\theta}(t-1)) + V'_t(\hat{\theta}(t-1))(\theta - \hat{\theta}(t-1))$$
$$+ \frac{1}{2}(\theta - \hat{\theta}(t-1))^T V''_t(\hat{\theta}(t-1))(\theta - \hat{\theta}(t-1)) \tag{9.28}$$

The right-hand side of (9.28) is a quadratic function of θ. Minimize this with respect to θ and let the minimum point constitute the new parameter estimate $\hat{\theta}(t)$. Thus

$$\hat{\theta}(t) = \hat{\theta}(t-1) - [V''_t(\hat{\theta}(t-1))]^{-1}[V'_t(\hat{\theta}(t-1))]^T \tag{9.29}$$

which corresponds to one step with the Newton–Raphson algorithm (see (7.82)) initialized in $\hat{\theta}(t-1)$. In order to proceed, recursive relationships for the loss function $V_t(\theta)$ and its derivatives are needed. From (9.27) it is easy to show that

$$V_t(\theta) = \lambda V_{t-1}(\theta) + \frac{1}{2}\varepsilon^T(t, \theta) Q \varepsilon(t, \theta) \tag{9.30a}$$

$$V'_t(\theta) = \lambda V'_{t-1}(\theta) + \varepsilon^T(t, \theta) Q \varepsilon'(t, \theta) \tag{9.30b}$$

$$V''_t(\theta) = \lambda V''_{t-1}(\theta) + [\varepsilon'(t, \theta)]^T Q \varepsilon'(t, \theta) + \varepsilon^T(t, \theta) Q \varepsilon''(t, \theta) \tag{9.30c}$$

The last term in (9.30c) is written in an informal way since $\varepsilon''(t, \theta)$ is a tensor if $\varepsilon(t, \theta)$ is vector-valued. The correct interpretation of the last term in this case is

$$\varepsilon^T(t, \theta) Q \varepsilon''(t, \theta) = \sum_i \sum_j \varepsilon_i(t, \theta) Q_{ij} \varepsilon''_j(t, \theta) \tag{9.30d}$$

where $\varepsilon_i(t, \theta)$ and Q_{ij} are scalars with obvious meanings, and $\varepsilon''_j(t, \theta)$ is the matrix of second-order derivatives of $\varepsilon_j(t, \theta)$.

Now make the following approximations:

$$V'_{t-1}(\hat{\theta}(t-1)) = 0 \tag{9.31}$$

$$V''_{t-1}(\hat{\theta}(t-1)) = V''_{t-1}(\hat{\theta}(t-2)) \tag{9.32}$$

$$\varepsilon^T(t, \theta) Q \varepsilon''(t, \theta) \text{ (in 9.30c) is negligible} \tag{9.33}$$

The motivation for (9.31) is that $\hat{\theta}(t-1)$ is assumed to be the minimum point of $V_{t-1}(\theta)$. The approximation (9.32) means that the second-order derivative $V''_{t-1}(\theta)$ varies slowly with θ. The reason for (9.33) is that $\varepsilon(t, \theta)$ at the true parameter vector will be a white process and hence

$$E\varepsilon^T(t, \theta) Q \varepsilon''(t, \theta) = 0$$

This implies that at least for large t and θ close to the minimum point, one can indeed neglect the influence of the last term in (9.30c) on $V''_t(\theta)$. Neglecting the last term of the Hessian in (9.30c) means that a Gauss–Newton algorithm (cf (7.87)) is used instead of the Newton–Raphson algorithm (9.29) (based on the exact Hessian). Note that all the approximations (9.31)–(9.33) hold exactly in the least squares case.

Using (9.30)–(9.33), the following algorithm is derived from (9.29):

$$\hat{\theta}(t) = \hat{\theta}(t-1) - [V_t''(\hat{\theta}(t-1))]^{-1}[\varepsilon'(t, \hat{\theta}(t-1))]^T \\ \times Q\varepsilon(t, \hat{\theta}(t-1)) \quad (9.34a)$$

$$V_t''(\hat{\theta}(t-1)) = \lambda V_{t-1}''(\hat{\theta}(t-2)) + [\varepsilon'(t, \hat{\theta}(t-1))]^T Q[\varepsilon'(t, \hat{\theta}(t-1))] \quad (9.34b)$$

This algorithm is as it stands not well suited as a recursive algorithm. There are two reasons for this:

- The inverse of V_t'' is needed in (9.34a), while the matrix itself (and not its inverse), is updated in (9.34b).
- For many model structures, calculations of $\varepsilon(t, \hat{\theta}(t-1))$ and its derivative will for every t require a processing of all the data up to time t.

The first problem can be solved by applying the matrix inversion lemma to (9.34b), as described below. The second problem is tied to the model structure used. Note that so far the derivation has not been specialized to any particular model structure. To produce a feasible recursive algorithm, some additional approximations must be introduced. Let

$$\varepsilon(t) \approx \varepsilon(t, \hat{\theta}(t-1)) \\ \psi(t) \approx -[\varepsilon'(t, \hat{\theta}(t-1))]^T \quad (9.35)$$

denote approximations which can be evaluated on-line. The actual way of implementing these approximations will depend on the model structure. Example 9.2 will illustrate how to construct $\varepsilon(t)$ and $\psi(t)$ for a scalar ARMAX model. Note that in (9.35) the quantities have the following dimensions:

$\varepsilon(t)$ is an $(ny|1)$ vector
$\psi(t)$ is an $(n\theta|ny)$ matrix
($ny = \dim y$, $n\theta = \dim \theta$)

In particular, for scalar systems, $(ny = 1)$, $\varepsilon(t)$ becomes a scalar and $\psi(t)$ an $n\theta$-dimensional vector.

Introduce the notation

$$P(t) = [V_t''(\hat{\theta}(t-1))]^{-1} \quad (9.36)$$

Then from (9.34b) and (9.35),

$$P^{-1}(t) = \lambda P^{-1}(t-1) + \psi(t)Q\psi^T(t) \quad (9.37)$$

which can be rewritten by using the matrix inversion lemma (Lemma A.1), to give

$$P(t) = \{P(t-1) - P(t-1)\psi(t)[\lambda Q^{-1} \\ + \psi^T(t)P(t-1)\psi(t)]^{-1}\psi^T(t)P(t-1)\}/\lambda \quad (9.38)$$

Note that at each time step it is now required to invert a matrix of dimension $(ny|ny)$ instead of an $(n\theta|n\theta)$ matrix, as earlier. Normally the number of parameters $(n\theta)$ is much larger than the number of outputs (ny). Therefore the use of the equation (9.38) leads indeed to an improvement of the algorithm (9.34).

A general recursive prediction error algorithm has now been derived, and is summarized as follows:

$$\hat{\theta}(t) = \hat{\theta}(t-1) + K(t)\varepsilon(t) \quad (9.39a)$$

$$K(t) = P(t)\psi(t)Q \quad (9.39b)$$

$$P(t) = \{P(t-1) - P(t-1)\psi(t)[\lambda Q^{-1} + \psi^T(t)P(t-1)\psi(t)]^{-1}\psi^T(t)P(t-1)\}/\lambda \quad (9.39c)$$

Similarly to the case of the scalar RLS previously discussed, the gain $K(t)$ in (9.39b) can be rewritten in a more convenient computational form. From (9.38),

$$K(t) = P(t-1)\psi(t)Q/\lambda - P(t-1)\psi(t)[\lambda Q^{-1} + \psi^T(t)P(t-1)\psi(t)]^{-1}$$
$$\times \psi^T(t)P(t-1)\psi(t)Q/\lambda$$
$$= P(t-1)\psi(t)/\lambda[\lambda Q^{-1} + \psi^T(t)P(t-1)\psi(t)]^{-1}$$
$$\times \{[\lambda Q^{-1} + \psi^T(t)P(t-1)\psi(t)]Q - \psi^T(t)P(t-1)\psi(t)Q\}$$
$$= P(t-1)\psi(t)Q[\lambda I + \psi^T(t)P(t-1)\psi(t)Q]^{-1}$$

that is,

$$K(t) = P(t-1)\psi(t)[\lambda Q^{-1} + \psi^T(t)P(t-1)\psi(t)]^{-1} \quad (9.40)$$

The algorithm (9.39) is applicable to a variety of model structures. The model structure will influence the way in which the quantities $\varepsilon(t)$ and $\psi(t)$ in the algorithm are computed from the data and the previously computed parameter estimates. The following example illustrates the computation of $\varepsilon(t)$ and $\psi(t)$.

Example 9.2 *RPEM for an ARMAX model*
Consider the model structure

$$A(q^{-1})y(t) = B(q^{-1})u(t) + C(q^{-1})e(t)$$

where $ny = nu = 1$, and

$$A(q^{-1}) = 1 + a_1 q^{-1} + \ldots + a_n q^{-n}$$
$$B(q^{-1}) = b_1 q^{-1} + \ldots + b_n q^{-n} \quad (9.41)$$
$$C(q^{-1}) = 1 + c_1 q^{-1} + \ldots + c_n q^{-n}$$

332 *Recursive identification methods* Chapter 9

This model structure was considered in Examples 6.1, 7.3 and 7.8, where the following relations were derived:

$$\varepsilon(t, \theta) = \frac{A(q^{-1})y(t) - B(q^{-1})u(t)}{C(q^{-1})} \tag{9.42a}$$

$$\varepsilon'(t, \theta) = -(-y^F(t-1, \theta) \ldots -y^F(t-n, \theta) \tag{9.42b}$$
$$u^F(t-1, \theta) \ldots u^F(t-n, \theta) \quad \varepsilon^F(t-1, \theta) \ldots \varepsilon^F(t-n, \theta))$$

where

$$y^F(t, \theta) = \frac{1}{C(q^{-1})} y(t) \tag{9.42c}$$

$$u^F(t, \theta) = \frac{1}{C(q^{-1})} u(t) \tag{9.42d}$$

$$\varepsilon^F(t, \theta) = \frac{1}{C(q^{-1})} \varepsilon(t, \theta) \tag{9.42e}$$

It is clear that to compute $\varepsilon(t, \theta)$ and $\varepsilon'(t, \theta)$ for any value of θ, it is necessary to process all data up to time t. A reasonable approximate way to compute these quantities recursively is to use time-varying filtering as follows:

$$\varepsilon(t) = y(t) + \hat{a}_1(t-1)y(t-1) + \ldots + \hat{a}_n(t-1)y(t-n)$$
$$- \hat{b}_1(t-1)u(t-1) - \ldots - \hat{b}_n(t-1)u(t-n) \tag{9.43a}$$
$$- \hat{c}_1(t-1)\varepsilon(t-1) - \ldots - \hat{c}_n(t-1)\varepsilon(t-n)$$

$$\psi(t) = (-y^F(t-1) \ldots -y^F(t-n) \quad u^F(t-1) \ldots u^F(t-n) \quad \varepsilon^F(t-1) \ldots \varepsilon^F(t-n))^T \tag{9.43b}$$

$$y^F(t) = y(t) - \hat{c}_1(t)y^F(t-1) - \ldots - \hat{c}_n(t)y^F(t-n) \tag{9.43c}$$

$$u^F(t) = u(t) - \hat{c}_1(t)u^F(t-1) - \ldots - \hat{c}_n(t)u^F(t-n) \tag{9.43d}$$

$$\varepsilon^F(t) = \varepsilon(t) - \hat{c}_1(t)\varepsilon^F(t-1) - \ldots - \hat{c}_n(t)\varepsilon^F(t-n) \tag{9.43e}$$

The idea behind (9.43) is simply to formulate (9.42a, c–e) as difference equations. Then these equations are iterated only once using previously calculated values for initialization. Note that 'exact' computation of $\varepsilon(s, \cdot)$ and $\varepsilon'(s, \cdot)$ would require iteration of (9.43) with $\hat{\theta}(t-1)$ *fixed*, from $t = 1$ to $t = s$.

The above equations can be modified slightly. When updating $\hat{\theta}(t)$ in (9.39a) it is necessary to compute $\varepsilon(t)$. Then, of course, one can only use the parameter estimate $\hat{\theta}(t-1)$ as in (9.43a). However, $\varepsilon(t)$ is also needed in (9.43a, b, e). In these relations it would be possible to use $\hat{\theta}(t)$ for computing $\varepsilon(t)$ in which case more recent information could be utilized. If the prediction errors computed from $\hat{\theta}(t)$ are denoted by $\bar{\varepsilon}(t)$, then

$$\varepsilon(t) = y(t) + \hat{a}_1(t-1)y(t-1) + \ldots + \hat{a}_n(t-1)y(t-n)$$
$$- \hat{b}_1(t-1)u(t-1) - \ldots - \hat{b}_n(t-1)u(t-n) \tag{9.44a}$$
$$- \hat{c}_1(t-1)\bar{\varepsilon}(t-1) - \ldots - \hat{c}_n(t-1)\bar{\varepsilon}(t-n)$$

Section 9.5   *The recursive prediction error method*   333

$$\bar{\varepsilon}(t) = y(t) + \hat{a}_1(t)y(t-1) + \ldots + \hat{a}_n(t)y(t-n)$$
$$- \hat{b}_1(t)u(t-1) - \ldots - \hat{b}_n(t)u(t-n) \qquad (9.44b)$$
$$- \hat{c}_1(t)\bar{\varepsilon}(t-1) - \ldots - \hat{c}_n(t)\bar{\varepsilon}(t-n)$$

$$\psi(t) = (-y^F(t-1) \ldots -y^F(t-n) \quad u^F(t-1) \ldots u^F(t-n) \quad \bar{\varepsilon}^F(t-1) \ldots \bar{\varepsilon}^F(t-n))^T$$
$$(9.44c)$$

$$\bar{\varepsilon}^F(t) = \bar{\varepsilon}(t) - \hat{c}_1(t)\bar{\varepsilon}^F(t-1) \ldots - \hat{c}_n(t)\bar{\varepsilon}^F(t-n) \qquad (9.44d)$$

The equations (9.43c, d) remain the same.

It turns out that, in practice, the RPEM algorithm corresponding to (9.44) often has superior properties (for example, a faster rate of convergence) than the algorithm using (9.43). ∎

There is a popular recursive identification method called pseudolinear regression (PLR) (also known as the extended least squares (ELS) method or the approximate ML method) that can be viewed as a simplified version of RPEM. This is explained in the following example, which considers ARMAX models.

Example 9.3 *PLR for an ARMAX model*
Consider again the ARMAX model

$$A(q^{-1})y(t) = B(q^{-1})u(t) + C(q^{-1})e(t) \qquad (9.45)$$

which can be written as a (pseudo-) linear regression

$$y(t) = \varphi^T(t)\theta + e(t) \qquad (9.46a)$$

$$\varphi(t) = (-y(t-1) \ldots -y(t-n) \quad u(t-1) \ldots u(t-n) \quad e(t-1) \ldots e(t-n))^T$$
$$(9.46b)$$

$$\theta = (a_1 \ldots a_n \quad b_1 \ldots b_n \quad c_1 \ldots c_n)^T \qquad (9.46c)$$

One could try to apply the LS method to this model. The problem is, of course, that the noise terms $e(t-1), \ldots, e(t-n)$ in $\varphi(t)$ are not known. However, they can be replaced by the estimated prediction errors. This approach gives the algorithm

$$\boxed{\begin{aligned}
\hat{\theta}(t) &= \hat{\theta}(t-1) + K(t)\varepsilon(t) \\
\varepsilon(t) &= y(t) - \varphi^T(t)\hat{\theta}(t-1) \\
K(t) &= P(t)\varphi(t) = P(t-1)\varphi(t)/[1 + \varphi^T(t)P(t-1)\varphi(t)] \\
P(t) &= P(t-1) - P(t-1)\varphi(t)\varphi^T(t)P(t-1)/[1 + \varphi^T(t)P(t-1)\varphi(t)] \\
\varphi(t) &= (-y(t-1) \ldots -y(t-n) \quad u(t-1) \ldots u(t-n) \quad \varepsilon(t-1) \ldots \varepsilon(t-n))^T
\end{aligned}} \qquad (9.47)$$

This algorithm can be seen as an approximation of the RPEM algorithm (9.39), (9.43). The 'only' difference between (9.47) and RPEM is that the filtering in (9.42c–e) is neglected. This simplification is not too important for the amount of computations involved in updating the variables $\hat{\theta}(t)$ and $P(t)$. It can, however, have a considerable influence on the behavior of the algorithm, as will be shown later in this chapter (cf. Examples 9.4 and 9.5). The possibility of using the $\bar{\varepsilon}(t)$ in $\varphi(t)$ as in (9.44) still remains. ∎

9.6 Theoretical analysis

To summarize the development so far, the following general algorithm has been derived:

$$\hat{\theta}(t) = \hat{\theta}(t-1) + K(t)\varepsilon(t) \tag{9.48a}$$

$$K(t) = P(t)Z(t)Q \tag{9.48b}$$
$$= P(t-1)Z(t)[\lambda Q^{-1} + \psi^T(t)P(t-1)Z(t)]^{-1}$$

$$P(t) = [P(t-1) - P(t-1)Z(t)[\lambda Q^{-1} \tag{9.48c}$$
$$+ \psi^T(t)P(t-1)Z(t)]^{-1}\psi^T(t)P(t-1)]/\lambda$$

The equations are to be used in the order (c), (b), (a). For all methods discussed earlier, except the instrumental variable method, $Z(t) = \psi(t)$.

The algorithm (9.48) must be complemented with a part that is dependent on the model structure where it is specified how to compute $\varepsilon(t)$ and $\psi(t)$.

The algorithm above is a set of nonlinear stochastic difference equations which depend on the data $u(1), y(1), u(2), y(2), \ldots$ It is therefore difficult to analyze it except for the case of the LS and IV methods, where the algorithm is just an exact reformulation of the corresponding off-line estimate. To gain some insights into the properties of the algorithm (9.48), consider some simulation examples.

Example 9.4 *Comparison of some recursive algorithms*
The following system was simulated:

$$y(t) = \frac{1.0q^{-1}}{1 - 0.9q^{-1}} u(t) + e(t)$$

where $u(t)$ was a square wave with amplitude 1 and period 10. The white noise sequence $e(t)$ had zero mean and variance 1. The system was identified using the following estimation methods: recursive least squares (RLS), recursive instrumental variables (RIV), PLR and RPEM. For RLS and RIV the model structure was

$$y(t) + ay(t-1) = bu(t-1) + e(t)$$

$$\theta = (a\ b)^T$$

For PLR and RPEM the model structure was

$$y(t) + ay(t-1) = bu(t-1) + e(t) + ce(t-1)$$

$$\theta = (a \ b \ c)^T$$

In all cases initial values were taken to be $\hat{\theta}(0) = 0$, $P(0) = 10I$. In the RIV algorithm the instruments were chosen as $Z(t) = (u(t-1) \ u(t-2))^T$. The results obtained from 300 samples are shown in Figure 9.2.

The results illustrate some of the general properties of the four methods:

- RLS does not give consistent parameter estimates for systems with correlated equation errors. This is so since RLS is equivalent to an off-line LS algorithm which according to the analysis in Section 7.1 should not give consistency. For low-order systems the deviations of the estimates from the true values are often smaller than in this example. For high-order systems the deviations are often more substantial.
- In contrast to RLS, the RIV algorithm gives consistent parameter estimates. Again, this follows from previous analysis (Chapter 8), since RIV is equivalent to an off-line IV method.
- Both PLR and RPEM often give consistent parameter estimates of a, b and c. The estimates $\hat{a}$ and $\hat{b}$ converge more quickly than $\hat{c}$. The behavior of PLR in the transient phase may be better than that of RPEM. In particular, the PLR estimates of c may converge faster than for RPEM. ∎

Example 9.5 *Effect of the initial values*
The following system was simulated:

$$y(t) - 0.9y(t-1) = 1.0u(t-1) + e(t)$$

where $u(t)$ was a square wave with amplitude 1 and period 10. The white noise sequence $e(t)$ had zero mean and variance 1. The system was identified using RLS and a first-order model

$$y(t) + ay(t-1) = bu(t-1) + e(t)$$

The algorithm was initialized with

$$\hat{\theta}(0) = \begin{pmatrix} \hat{a}(0) \\ \hat{b}(0) \end{pmatrix} = 0 \qquad P(0) = \varrho I$$

Various values of ϱ were tried. The results obtained from 300 samples are shown in Figure 9.3.

It can be seen from the figure that large and moderate values of ϱ (i.e. $\varrho = 10$ and $\varrho = 1$) lead to similar results. In both cases little confidence is given to $\hat{\theta}(0)$, and $\{\hat{\theta}(t)\}$ departs quickly from this value. On the other hand, a small value of ϱ (such as 0.1 or 0.01) gives a slower convergence. The reason is that a small ϱ implies a small $K(t)$ for $t \geq 0$ and thus the algorithm can only make small corrections at each time step. ∎

The behavior of RLS exhibited in Example 9.5 is essentially valid for all methods. The choice of ϱ necessary to produce reasonable performance (including satisfactory

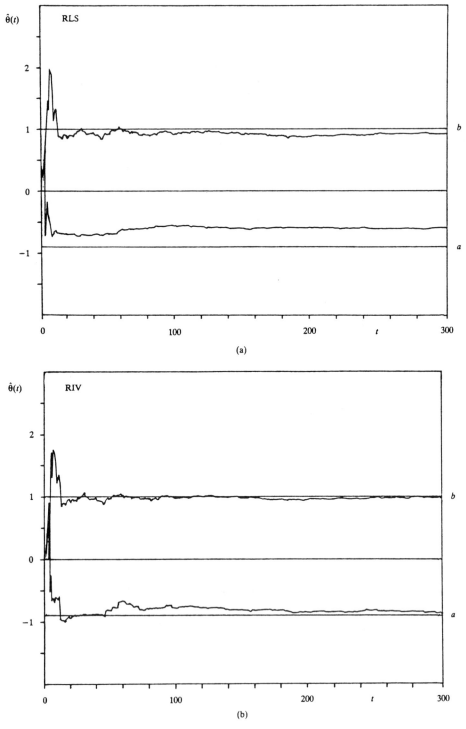

FIGURE 9.2 Parameter estimates and true values for: (a) RLS, (b) RIV, (c) PLR (d) RPEM, Example 9.4.

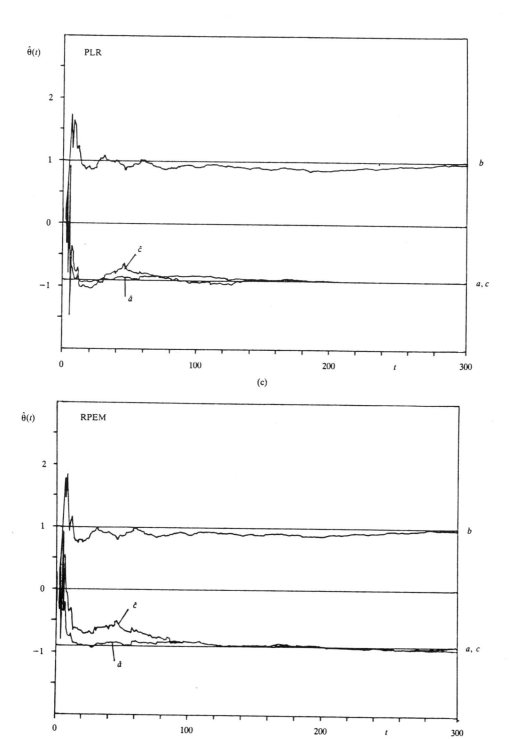

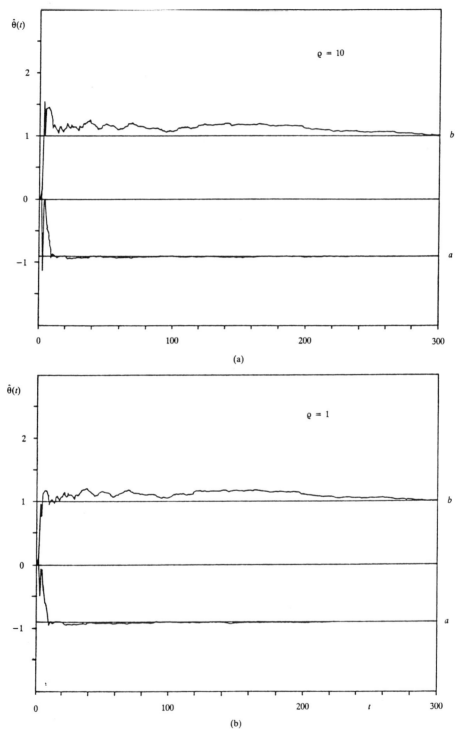

FIGURE 9.3 Parameter estimates and true values for: (a) $\varrho = 10$, (b) $\varrho = 1$, (c) $\varrho = 0.1$, (d) $\varrho = 0.01$, Example 9.5.

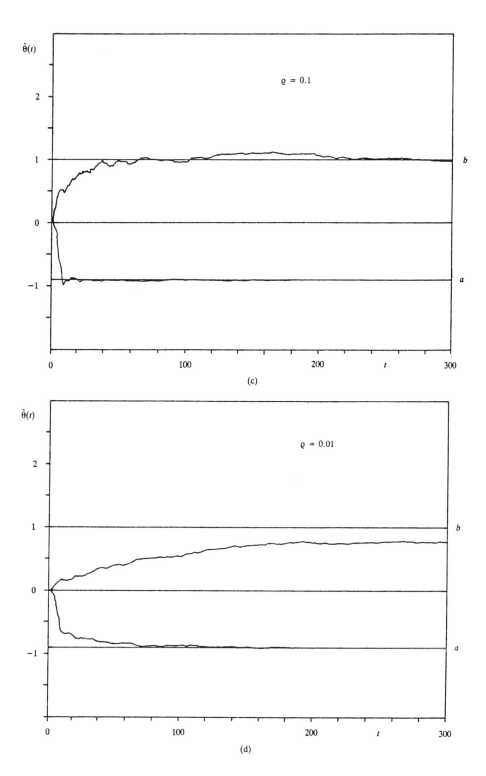

convergence rate) has been discussed previously (see (9.18)–(9.21)). See also Problem 9.4 for a further analysis.

Example 9.6 *Effect of the forgetting factor*
The following ARMA process was simulated:

$$y(t) - 0.9y(t-1) = e(t) + 0.9e(t-1)$$

The white noise sequence $e(t)$ had zero mean and variance 1. The system was identified using RPEM and a first-order ARMA model

$$y(t) + ay(t-1) = e(t) + ce(t-1)$$

The algorithm was initialized with $\hat{\theta}(0) = 0$, $P(0) = 100I$.

Different values of the forgetting factor λ were tried. The results are shown in Figure 9.4. It is clear from the figure that a decrease in the forgetting factor leads to two effects:

- The correction steps taken by the algorithm increase and the parameter estimates approach the true values more rapidly. This is particularly so for the c parameter.
- The algorithm becomes more sensitive to noise. When $\lambda < 1$ the parameter estimates do not converge, but oscillate around the true values. As λ decreases, the oscillations become larger.

It is worth remarking that the $\hat{c}$ estimates in many cases converge quicker than in this example. ∎

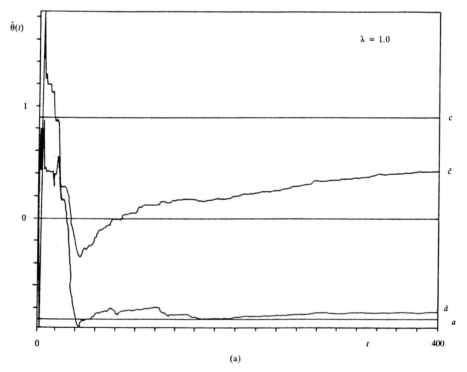

FIGURE 9.4 Parameter estimates and true values: (a) $\lambda = 1.0$, (b) $\lambda = 0.99$, (c) $\lambda = 0.95$, Example 9.6.

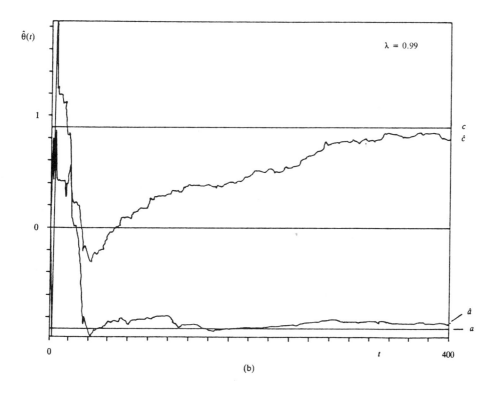

(b)

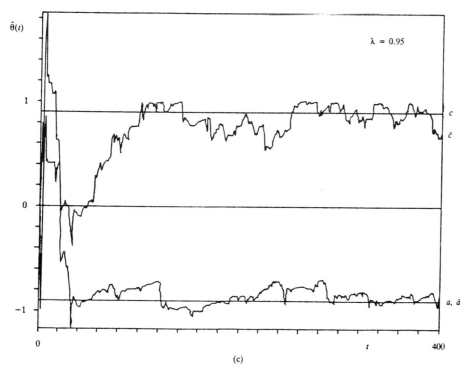

(c)

The features exhibited in Example 9.6 are valid for other methods and model structures. These features suggest the use of a time-varying forgetting factor $\lambda(t)$. More exactly, $\lambda(t)$ should be small (that is slightly less than 1) for small t to make the transient phase short. After some time $\lambda(t)$ should get close to 1 (in fact $\lambda(t)$ should at least tend to 1 as t tends to infinity) to enable convergence and to decrease the oscillations around the true values. More details on the use of variable forgetting factors are presented in Section 9.7.

Simulation no doubt gives useful insight. However, it is also clear that it does not permit generally valid conclusions to be drawn. Therefore it is only a complement to theory. The scope of a theoretical analysis would in particular be to study whether the parameter estimates $\hat{\theta}(t)$ *converge* as t tends to infinity; if so, to what *limit*; and also possibly to establish the *limiting distribution*.

Recall that for the recursive LS and IV algorithms the same parameter estimates are obtained asymptotically, as in the off-line case, provided the forgetting factor is 1 (or tends to 1 exponentially). Thus *the (asymptotic) properties of the RLS and RIV estimates follow from the analysis developed in Chapters 7 and 8.*

Next consider the algorithm (9.48) in the general case. Suppose that the forgetting factor λ is time varying and set $\lambda = \lambda(t)$. Assume that $\lambda(t)$ tends to 1 as t increases. The algorithm (9.48) with λ replaced by $\lambda(t)$ corresponds to minimization of the criterion

$$V_t(\theta) = \sum_{s=1}^{t} \left[\prod_{k=s+1}^{t} \lambda(k) \right] \varepsilon^T(s, \theta) Q \varepsilon(s, \theta)$$

with the convention $\prod_{k=t+1}^{t} \lambda(k) = 1$ (see Problem 9.9).

Introduce the 'step length' or 'gain sequence' $\gamma(t)$ through

$$\gamma(t) = \frac{\gamma(t-1)}{\lambda(t) + \gamma(t-1)} \quad (9.49)$$

$$\gamma(1) = 1$$

The reason for using the name 'step length' will become apparent later (see (9.51)). Roughly speaking, computing $\hat{\theta}(t)$ when $\hat{\theta}(t-1)$ is known, corresponds to taking one Gauss–Newton step, and $\gamma(t)$ controls the length of this step.

Note that $\lambda(t) \equiv 1$ gives $\gamma(t) = 1/t$. It is also possible to show that if $\lambda(t) \to 1$, then $t\gamma(t) \to 1$.

Next define the matrix $\tilde{R}(t)$ by

$$\tilde{R}(t) = \gamma(t) P^{-1}(t) \quad (9.50)$$

The matrix $P(t)$ will usually be decreasing. If $\lambda(t) \equiv 1$ it will behave as $1/t$ for large t. The matrix $\tilde{R}(t)$ will under weak assumptions have a nonzero finite limit as $t \to \infty$.

Now (9.48a) can be rewritten as

$$\hat{\theta}(t) = \hat{\theta}(t-1) + \gamma(t) \tilde{R}^{-1}(t) Z(t) Q \varepsilon(t) \quad (9.51a)$$

Moreover, if (9.48c) is rewritten in a form similar to (9.37), this gives

$$P^{-1}(t) = \lambda(t) P^{-1}(t-1) + Z(t) Q \psi^T(t)$$

With the substitutions (9.49) and (9.50) this becomes

$$\tilde{R}(t) = \gamma(t)\lambda(t)\frac{1}{\gamma(t-1)}\tilde{R}(t-1) + \gamma(t)Z(t)Q\psi^T(t) \quad (9.51b)$$
$$= \tilde{R}(t-1) + \gamma(t)[Z(t)Q\psi^T(t) - \tilde{R}(t-1)]$$

Note that both equations (9.51a), (9.51b) have the form

$$x(t) = x(t-1) + \gamma(t)\zeta(t) \quad (9.52a)$$

where $\zeta(t)$ is a correction term. By iterating (9.52a) it can be shown that

$$x(t) = x(0) + \sum_{k=1}^{t}\gamma(k)\zeta(k) \quad (9.52b)$$

Note that $\sum_{k=1}^{t}\gamma(k)$ diverges. (Since $\gamma(k) = 1/k$ asymptotically, $\sum_{k=1}^{t}\gamma(k) \approx \log t$.) Hence *if* (9.52) converges we must have

$$E\zeta(t) = 0 \quad (9.53)$$

which is to be evaluated for a constant argument θ corresponding to the convergence point. This kind of analysis is sometimes called the *principle of averaging*. To apply this idea to the algorithm (9.51), first introduce the vector $f(\theta)$ and matrix $G(\theta)$ by

$$f(\theta) = EZ(t)Q\varepsilon(t, \theta)$$
$$G(\theta) = EZ(t)Q\psi^T(t, \theta) \quad (9.54)$$

The evaluation of (9.54) is for a constant (time-invariant) parameter vector θ.

Associated differential equations

Assume now that $\hat{\theta}(t)$ in (9.51a) converges to θ^* and $\tilde{R}(t)$ in (9.51b) to R^*. Applying the relation (9.53) to the algorithm (9.51) then gives

$$(R^*)^{-1}f(\theta^*) = 0$$
$$G(\theta^*) - R^* = 0 \quad (9.55a)$$

In particular, it is found that the possible limit points of (9.51a) must satisfy

$$f(\theta^*) = 0 \quad (9.55b)$$

To proceed, consider the following set of ordinary differential equations (ODEs)

$$\frac{d}{d\tau}\theta_\tau = R_\tau^{-1} f(\theta_\tau)$$
$$\frac{d}{d\tau} R_\tau = G(\theta_\tau) - R_\tau \qquad (9.56)$$

Assume that (9.56) is solved numerically by a Euler method and computed for $\tau = t_1, t_2, \ldots$ Then

$$\theta_{t_2} \approx \theta_{t_1} + (t_2 - t_1) R_{t_1}^{-1} f(\theta_{t_1})$$
$$R_{t_2} \approx R_{t_1} + (t_2 - t_1)[G(\theta_{t_1}) - R_{t_1}] \qquad (9.57)$$

Note the similarity between (9.57) and the algorithm (9.51). This similarity suggests that the solutions to the deterministic ODE will be close to the paths corresponding to the algorithm (9.48) if the ODE is used with the time scale

$$\tau = \sum_{k=1}^{t} \gamma(k) \approx \log t \qquad (9.58)$$

The above analysis is quite heuristic. The link between the algorithm (9.51) and the ODE (9.56) can be established formally, but the analysis will be quite technical. For detailed derivations of the following results, see Ljung (1977a, b), Ljung and Söderström (1983).

1. The possible limit points of $\{\hat{\theta}(t)\}$ as t tends to infinity are the stable stationary points of the differential equation (9.56). The estimates $\hat{\theta}(t)$ converge *locally* to stable stationary points. (If $\hat{\theta}(t_0)$ is close to a stable stationary point θ^* and the gains $\{K(t)\}$, $t \geq t_0$ are small enough, then $\hat{\theta}(t)$ converges to θ^*.)
2. The trajectories of the solutions to the differential equations are expected paths of the algorithm (9.48).
3. Assume that there is a positive function $V(\theta, R)$ such that along the solutions of the differential equation (9.56) we have $\frac{d}{d\tau} V(\theta_\tau, R_\tau) \leq 0$. Then as $t \to \infty$, the estimates $\hat{\theta}(t)$, $\tilde{R}(t)$ either tend to the set

$$D_c = \left\{ \theta, R \,\bigg|\, \frac{d}{d\tau} V(\theta_\tau, R_\tau) = 0 \right\}$$

or to the boundary of the set $\mathcal{D}$ given by (6.2). (It is assumed that the updating of $\hat{\theta}(t)$ in (9.48a) is modified when necessary to guarantee that $\hat{\theta}(t) \in \mathcal{D}$, i.e. the model corresponding to $\hat{\theta}(t)$ gives a stable predictor; this may be done, for instance, by reducing the step length according to a certain rule, whenever $\hat{\theta}(t) \notin \mathcal{D}$.)

Next examine the (locally) stable stationary points of (9.56), which constitute the possible limits of $\hat{\theta}(t)$. Let θ^*, $R^* = G(\theta^*)$ be a stationary point of the ODE (9.56). Set $r^* = \text{vec}(R^*)$, $r_\tau = \text{vec}(R_\tau)$, cf. Definition A.10. If the ODE is linearized around (θ^*, R^*), then (recall (9.55))

Section 9.6   Theoretical analysis

$$\frac{d}{d\tau}\begin{pmatrix} \theta_\tau - \theta^* \\ r_\tau - r^* \end{pmatrix} = \begin{pmatrix} (R^*)^{-1}\dfrac{\partial f(\theta)}{\partial \theta}\bigg|_{\theta=\theta^*} & 0 \\ X & -I \end{pmatrix}\begin{pmatrix} \theta_\tau - \theta^* \\ r_\tau - r^* \end{pmatrix} \quad (9.59)$$

In (9.59) the block marked X is apparently of no importance for the local stability properties. The stationary point θ^*, R^* will be stable if and only if the matrix

$$L(\theta^*) = [G(\theta^*)]^{-1}\frac{\partial f(\theta)}{\partial \theta}\bigg|_{\theta=\theta^*} \quad (9.60)$$

has all eigenvalues in the left half-plane.

The above theoretical results will be used to analyze two typical recursive algorithms, namely a recursive prediction error method applied to a general model structure and a pseudolinear regression applied to an ARMAX model. In both cases it is assumed that the system is included in the model structure. It then holds that $\varepsilon(t, \theta_0) = e(t)$ is a white noise process, where θ_0 is the true parameter vector. This readily implies that $f(\theta_0) = 0$ and hence that $(\theta_0, G(\theta_0))$ is a stationary point of the differential equations (9.56). The local and global stability properties remain to be investigated. These properties will indeed depend on the identification method and the model structure used, as shown in the following two examples.

Example 9.7 *Convergence analysis of the RPEM*
Consider the possible stationary points of (9.56). It is easy to derive that

$$\begin{aligned} 0 = f(\theta) &= E\psi(t, \theta)Q\varepsilon(t, \theta) \\ &= -\frac{1}{2}\left[\frac{\partial}{\partial \theta}E\varepsilon^T(t, \theta)Q\varepsilon(t, \theta)\right]^T \end{aligned} \quad (9.61)$$

Hence only the stationary points of the asymptotic loss function

$$V_\infty(\theta) = E\varepsilon^T(t, \theta)Q\varepsilon(t, \theta) \quad (9.62)$$

can be possible convergence points of $\hat{\theta}(t)$. If $V_\infty(\theta)$ has a unique minimum at $\theta = \theta_0$ (so that no false local minima exist – this can be shown for some special cases such as ARMA processes; cf. Section 12.8), then convergence can take place to the true parameters only. Further, for the true parameter vector

$$\begin{aligned} \frac{\partial f(\theta)}{\partial \theta}\bigg|_{\theta=\theta_0} &= \frac{\partial}{\partial \theta}[E\psi(t, \theta)Q\varepsilon(t, \theta)]\bigg|_{\theta=\theta_0} \\ &= E\frac{\partial \psi}{\partial \theta}(t, \theta)\bigg|_{\theta=\theta_0}Q\varepsilon(t, \theta_0) + E\psi(t, \theta_0)Q\frac{\partial \varepsilon(t, \theta)}{\partial \theta}\bigg|_{\theta=\theta_0} \\ &= -E\psi(t, \theta_0)Q\psi^T(t, \theta_0) = -G(\theta_0) \end{aligned} \quad (9.63a)$$

Thus

$$L(\theta_0) = -I \quad (9.63b)$$

which evidently has all eigenvalues in the left half-plane. Strictly speaking, in the multivariable case ($ny > 1$), $\psi(t, \theta)$ is a matrix (of dimension ($n\theta | ny$)) and thus $\partial \psi/\partial \theta$ a tensor. The calculations made in (9.63) can, however, be justified since $\partial \psi/\partial \theta$ and $\varepsilon(t, \theta_0)$ are certainly uncorrelated. The exact interpretation of the derivative $\partial \psi/\partial \theta$ is similar to that of (9.30d).

As shown above, $(\theta_0, G(\theta_0))$ is a locally stable solution to (9.56). To investigate global stability, use result 3 quoted above and take $V_\infty(\theta)$ as a candidate function for $V(\theta, R)$. Clearly $V_\infty(\theta)$ is positive by construction. Furthermore, from (9.61),

$$\frac{d}{d\tau} V_\infty(\theta_\tau) = \frac{\partial}{\partial \theta_\tau} V_\infty(\theta_\tau) \frac{d}{d\tau} \theta_\tau$$
$$= -2f^T(\theta_\tau) R_\tau^{-1} f(\theta_\tau) \leq 0$$
(9.64)

Hence the RPEM will *converge globally* to the set

$$\begin{aligned} D_c &= \{\theta, R | f(\theta) = 0\} \\ &= \left\{\theta, R \left| \frac{\partial}{\partial \theta} V_\infty(\theta) = 0, R = G(\theta) \right.\right\} \end{aligned}$$
(9.65)

This set consists of the stationary points of the criterion $V_\infty(\theta)$. If θ_0 is a unique stationary point, it follows that the RPEM gives consistent parameter estimates under weak conditions.

It can also be shown that the RPEM parameter estimates are asymptotically Gaussian distributed with the same distribution as that of the off-line estimates (see (7.59), (7.66), (7.72)). Note the word 'asymptotic' here. The theory cannot provide the value of t for which it is applicable. The result as such does not mean that off- and on-line PEM estimates are equally accurate, at least not for a finite number of data points. ∎

Example 9.8 *Convergence analysis of PLR for an ARMAX model*

In order to analyze the convergence properties of PLR for an ARMAX model, consider

$$V(\theta, R) = (\theta - \theta_0)^T R (\theta - \theta_0)$$
(9.66)

as a tentative (Lyapunov) function for application of result 3 above. The true system, which corresponds to θ_0, will be denoted by

$$A_0(q^{-1}) y(t) = B_0(q^{-1}) u(t) + C_0(q^{-1}) e(t)$$
(9.67)

By straightforward differentiation of $V(\theta, R)$ (along the paths of (9.56))

$$\begin{aligned} \frac{d}{d\tau} V(\theta_\tau, R_\tau) &= f^T(\theta_\tau)(\theta_\tau - \theta_0) + (\theta_\tau - \theta_0)^T (G(\theta_\tau) - R_\tau)(\theta_\tau - \theta_0) \\ &\quad + (\theta_\tau - \theta_0)^T f(\theta_\tau) \end{aligned}$$
(9.68)

It will be useful to derive a convenient formula for $f(\theta_\tau)$. Let $\varphi_0(t)$ denote the vector $\varphi(t)$ corresponding to the true system (see (9.46b)). Then

$$f(\theta_\tau) = E\varphi(t)\varepsilon(t) = E\varphi(t)\{\varepsilon(t) - e(t)\} \tag{9.69}$$

This follows since $\varphi(t)$ depends only on the data up to time $t - 1$. It is hence uncorrelated with $e(t)$. It now follows from (9.46), (9.47) that

$$\begin{aligned}\varepsilon(t) - e(t) &= \{y(t) - \varphi^T(t)\theta_\tau\} - \{y(t) - \varphi_0^T(t)\theta_0\} \\ &= -\varphi^T(t)(\theta_\tau - \theta_0) - \{\varphi(t) - \varphi_0(t)\}^T\theta_0 \\ &= -\varphi^T(t)(\theta_\tau - \theta_0) - \{C_0(q^{-1}) - 1\}\{\varepsilon(t) - e(t)\}\end{aligned}$$

and hence

$$\varepsilon(t) - e(t) = \frac{-1}{C_0(q^{-1})} \varphi^T(t)(\theta_\tau - \theta_0) \tag{9.70}$$

Define

$$\tilde{G}(\theta_\tau) = E\varphi(t)\frac{1}{C_0(q^{-1})}\varphi^T(t) \tag{9.71}$$

Inserting (9.70) into (9.69) gives

$$f(\theta_\tau) = -\tilde{G}(\theta_\tau)(\theta_\tau - \theta_0)$$

Then get from (9.68),

$$\frac{d}{d\tau} V(\theta_\tau, R_\tau) = -(\theta_\tau - \theta_0)^T[\tilde{G}^T(\theta_\tau) - G(\theta_\tau) + R_\tau + \tilde{G}(\theta_\tau)](\theta_\tau - \theta_0) \tag{9.72}$$

Thus a sufficient convergence condition is obtained when the matrix

$$H \triangleq \tilde{G}^T(\theta_\tau) + \tilde{G}(\theta_\tau) - G(\theta_\tau) + R_\tau \tag{9.73}$$

is positive definite. Let x be an arbitrary $n\theta$ vector, set $z(t) = \varphi^T(t)x$ and let $\phi_z(\omega)$ be the spectral density of $z(t)$. Then

$$\begin{aligned}x^T H x &= 2Ez(t)\frac{1}{C_0(q^{-1})}z(t) - Ez^2(t) + x^T R_\tau x \\ &\geq 2E\left[z(t)\left\{\frac{1}{C_0(q^{-1})} - \frac{1}{2}\right\}z(t)\right] \\ &= 2\int_{-\pi}^{\pi} \phi_z(\omega)\left\{\frac{1}{C_0(e^{i\omega})} - \frac{1}{2}\right\}d\omega \\ &= 2\int_{-\pi}^{\pi} \phi_z(\omega)\operatorname{Re}\left\{\frac{1}{C_0(e^{i\omega})} - \frac{1}{2}\right\}d\omega\end{aligned} \tag{9.74}$$

So if

$$\boxed{\operatorname{Re}\left\{\frac{1}{C_0(e^{i\omega})} - \frac{1}{2}\right\} > 0 \quad \text{all } \omega} \tag{9.75}$$

then it can be concluded that $\hat{\theta}(t)$ converges to θ_0 globally. The condition (9.75) is often expressed in words as '$1/C_0 - 1/2$ is a strictly positive real filter'. It is not satisfied for all possible polynomials $C_0(q^{-1})$. See, for example, Problem 9.12.

Consider also the matrix $L(\theta_0)$ in (9.60), which determines the local convergence properties. In a similar way to the derivation of (9.63a),

$$\left.\frac{\partial}{\partial\theta}f(\theta)\right|_{\theta=\theta_0} = \left.\frac{\partial}{\partial\theta}E[\varepsilon(t)\varphi(t)]\right|_{\theta=\theta_0}$$

$$= \left.E\left[\varphi(t)\frac{\partial\varepsilon(t)}{\partial\theta}\right]\right|_{\theta=\theta_0} + \left.E\left[e(t)\frac{\partial\varphi(t)}{\partial\theta}\right]\right|_{\theta=\theta_0}$$

$$= -E\varphi(t)\psi^T(t)|_{\theta=\theta_0}$$

where (cf. (9.42), (9.46))

$$\psi(t) = \frac{1}{C(q^{-1})}\varphi(t)$$

Hence for PLR applied to ARMAX models the matrix $L(\theta_0)$ is given by

$$L(\theta_0) = -[E\varphi_0(t)\varphi_0^T(t)]^{-1}\left[E\varphi_0(t)\frac{1}{C_0(q^{-1})}\varphi_0^T(t)\right] \qquad (9.76)$$

For certain specific cases the eigenvalues of $L(\theta_0)$ can be determined explicitly and hence conditions for local convergence established. A pure ARMA process is such a case. Then the eigenvalues of $L(\theta_0)$ are

-1 with multiplicity n

$-\dfrac{1}{C_0(\alpha_k)} \qquad k = 1, \ldots, n$

where $\{\alpha_k\}$ are the zeros of the polynomial $z^n A_0(z^{-1})$.

Note in particular that the local convergence condition imposed by the locations of the eigenvalues in the left half-plane is weaker than the positive real condition (9.75). ∎

9.7 Practical aspects

The preceding sections have presented some theoretical tools for the analysis of recursive identification methods. The analysis is complemented here with a discussion of some practical aspects concerning:

- Search direction.
- Choice of forgetting factor.
- Numerical implementation.

Search direction

The general recursive estimation method (9.48) may be too complicated to use for some applications. One way to reduce its complexity is to use a simpler gain vector than $K(t)$ in (9.48). In turn this will modify the search direction, i.e. the updating direction in (9.48a). The following simplified algorithm, which is often referred to as 'stochastic approximation' (although 'stochastic gradient method' would be a more appropriate name) can be used:

$$\hat{\theta}(t) = \hat{\theta}(t-1) + K(t)\varepsilon(t)$$
$$K(t) = Z(t)Q/r(t) \tag{9.77}$$
$$r(t) = \lambda r(t-1) + \text{tr}\{Z(t)Q\psi^T(t)\}$$

In the algorithm (9.77) the scalar $r(t)$ corresponds in some sense to $\text{tr } P^{-1}(t)$. In fact from (9.48c)

$$P^{-1}(t) = \lambda P^{-1}(t-1) + Z(t)Q\psi^T(t)$$

If $r(t)$ is introduced as $\text{tr} P^{-1}(t)$ it follows easily that it satisfies the recursion above. The amount of computation for updating the variable $r(t)$ is less than the amount of computation for updating of $P(t)$ in (9.48). The price is a slower convergence of the parameter estimates.

The LMS (least mean square) estimate (see Widrow and Stearns, 1985) is popular in many signal processing applications. It corresponds to the algorithm (9.77) for linear regressions, although it is often implemented with a constant $r(t)$, thus reducing the computational load further.

Forgetting factor

The choice of the forgetting factor λ in the algorithm is often very important. Theoretically one must have $\lambda = 1$ to get convergence. On the other hand, if $\lambda < 1$ the algorithm becomes more sensitive and the parameter estimates can change quickly. For that reason it is often an advantage to allow the forgetting factor to vary with time. Therefore substitute λ everywhere in (9.48) by $\lambda(t)$. A typical choice is to let $\lambda(t)$ tend exponentially to 1. This can be written as

$$\lambda(t) = 1 - \lambda_0^t[1 - \lambda(0)]$$

which is easily implemented as a recursion

$$\lambda(t) = \lambda_0 \lambda(t-1) + (1 - \lambda_0) \tag{9.78}$$

Typical values for λ_0 and $\lambda(0)$ are $\lambda_0 = 0.99$ and $\lambda(0) = 0.95$. Using (9.78) one can improve the transient behavior of the algorithm (the behavior for small or medium values of t).

Numerical implementation

The updating (9.39c) for $P(t)$ can lead to numerical problems. Rounding errors may accumulate and make the computed $P(t)$ indefinite, even though $P(t)$ theoretically is always positive definite. When $P(t)$ becomes indefinite the parameter estimates tend to diverge (see, for example, Problem 9.3). A way to overcome this difficulty is to use a square root algorithm. Define $S(t)$ through

$$P(t) = S(t)S^T(t) \tag{9.79}$$

and update $S(t)$ instead of $P(t)$. Then the matrix $P(t)$ as given by (9.79) will automatically be positive definite. Consider for simplicity the equation (9.39c) for the scalar output case with $Q = 1$ and $\lambda = 1$, i.e.

$$P(t) = P(t-1) - P(t-1)\psi(t)\psi^T(t)P(t-1)/[1 + \psi^T(t)P(t-1)\psi(t)] \tag{9.80}$$

The updating of $S(t)$ then consists of the following equations:

$$
\begin{aligned}
f(t) &= S^T(t-1)\psi(t) \\
\beta(t) &= 1 + f^T(t)f(t) \\
\alpha(t) &= 1/[\beta(t) + \sqrt{\beta(t)}] \\
L(t) &= S(t-1)f(t) \\
S(t) &= S(t-1) - \alpha(t)L(t)f^T(t)
\end{aligned}
\tag{9.81}
$$

The vector $L(t)$ is related to the gain $K(t)$ by

$$K(t) = L(t)/\beta(t) \tag{9.82}$$

In practice it is not advisable to compute $K(t)$ as in (9.82). Instead the updating of the parameter estimates can be done using

$$\hat{\theta}(t) = \hat{\theta}(t-1) + L(t)[\varepsilon(t)/\beta(t)] \tag{9.83}$$

Proceeding in this way requires one division only, instead of $n\theta$ divisions as in (9.82).

In many practical cases the use of (9.80) will not result in any numerical difficulties. However, a square root algorithm or a similar implementation is recommended (see Ljung and Söderström, 1983, for more details), since the potential numerical difficulties will then be avoided.

In Complements C9.2 and C9.3 some lattice filter implementations of RLS for AR models and for multivariate linear regressions are presented, respectively. The lattice algorithms are fast in the sense that they require less computation per time step than does the algorithm (9.81), (9.83). The difference in computational load can be significant when the dimension of the parameter vector θ is large.

Summary

In this chapter a number of recursive identification algorithms have been derived. These algorithms have very similar algebraic structure. They have small requirements in terms of computer time and memory.

In Section 9.2 the recursive least squares method (for both static linear regressions and dynamic systems) was considered; in Section 9.4 the recursive instrumental variable method was presented; while the recursive prediction error method was derived in Section 9.5.

The recursive algorithms can easily be modified to track time-varying parameters, as was shown in Section 9.3.

Although the recursive identification algorithms consist of rather complicated nonlinear transformations of the data, it is possible to analyze their asymptotic properties. The theory is quite elaborate. Some examples of how to use specific theoretical tools for analysis were presented in Section 9.6. Finally, some practical aspects were discussed in Section 9.7.

Problems

Problem 9.1 *Derivation of the real-time RLS algorithm*
Show that the set of equations (9.12) recursively computes the minimizing vector of the weighted LS criterion (9.11).

Problem 9.2 *Influence of forgetting factor on consistency properties of parameter estimates*
Consider the static-gain system

$$y(t) = bu(t) + e(t) \quad t = 1, 2, 3 \ldots$$

where

$$Ee(t) = 0 \quad Ee(t)e(s) = \delta_{t,s}$$

and $u(t)$ is a persistently exciting nonrandom signal. The unknown parameter b is estimated by

$$\hat{b} = \arg\min_{b} \sum_{t=1}^{N} \lambda^{N-t}[y(t) - bu(t)]^2 \tag{i}$$

where N denotes the number of data points, and the forgetting factor λ satisfies $0 < \lambda \leq 1$. Determine $\text{var}(\hat{b}) = E(\hat{b} - b)^2$. Show that for $\lambda = 1$,

$$\text{var}(\hat{b}) \to 0 \quad \text{as } N \to \infty \tag{ii}$$

(i.e. that $\hat{b}$ is a consistent estimate, in the mean square sense). Also show that for $\lambda < 1$, there are signals $u(t)$ for which (ii) does not hold.

Hint. For showing the inconsistency of $\hat{b}$ in the case $\lambda < 1$, consider $u(t) = $ constant.

Remark. For more details on the topic of this problem, see Zarrop (1983) and Stoica and Nehorai (1988).

Problem 9.3 *Effects of $P(t)$ becoming indefinite*
As an illustration of the effects of $P(t)$ becoming indefinite, consider the following simple example:

System: $y(t) = \theta_0 + e(t)$

$$Ee(t)e(s) = \lambda^2 \delta_{t,s}$$

Model: $y(t) = \theta + \varepsilon(t)$

Let the system be identified using RLS with initial values $\hat{\theta}(0) = \hat{\theta}_0$ and $P(0) = P_0$.

(a) Derive an expression for the mean square error

$$V(t) = E[\hat{\theta}(t) - \theta_0]^2$$

(b) Assume that $P_0 < 0$ (which may be caused by rounding errors from processing previous data points). Then show that $V(t)$ will be increasing when t increases to (at least) $t = -1/P_0$. (If P_0 is small and negative, $V(t)$ will hence be increasing for a long period!)

Problem 9.4 *Convergence properties and dependence on initial conditions of the RLS estimate*
Consider the model

$$y(t) = \varphi^T(t)\theta + \varepsilon(t)$$

Let the off-line weighted LS estimate of θ based on $y(1), y(2), \ldots, y(t), \varphi(1), \ldots, \varphi(t)$ be denoted by $\hat{\theta}_t$.

$$\hat{\theta}_t = \left[\sum_{s=1}^{t} \lambda^{t-s}\varphi(s)\varphi^T(s)\right]^{-1} \left[\sum_{s=1}^{t} \lambda^{t-s}\varphi(s)y(s)\right]$$

Consider also the RLS algorithm (9.12) which provides recursive (on-line) estimates $\hat{\theta}(t)$ of θ.

(a) Derive difference equations for $P^{-1}(t)$ and $P^{-1}(t)\hat{\theta}(t)$. Solve these equations to find how $\hat{\theta}(t)$ depends on the initial values $\hat{\theta}(0)$, $P(0)$ and on the forgetting factor λ.
   Hint. Generalize the calculations in (9.18)–(9.20) to the case of $\lambda \leq 1$.
(b) Let $P(0) = \varrho I$. Prove that for every t for which $\hat{\theta}_t$ exists,

$$\lim_{\varrho \to \infty} \hat{\theta}(t) = \hat{\theta}_t$$

(c) Suppose that $\hat{\theta}_t$ is bounded and that $\lambda^t P(t) \to 0$, as $t \to \infty$. Prove that

$$\lim_{t \to \infty} [\hat{\theta}(t) - \hat{\theta}_t] = 0$$

Problem 9.5 *Updating the square root of $P(t)$*
Verify the square root algorithm (9.81).

Problem 9.6 *On the condition for global convergence of PLR*
A sufficient condition for global convergence of PLR is (9.75): $\text{Re}\{1/C(e^{i\omega})\} - 1/2 > 0$ for $\omega \in (-\pi, \pi)$.

(a) Show that (9.75) is equivalent to

$$|C(e^{i\omega}) - 1| < 1 \tag{i}$$

Comment on the fact that (i) is a sufficient condition for global convergence of PLR, in view of the interpretation of PLR as an 'approximate' RPEM (see Section 9.5).
(b) Determine the set (for $n = 2$)

$$D_g = \left\{ c_1, c_2 \middle| \operatorname{Re}\left[\frac{1}{1 + c_1 e^{i\omega} + c_2 e^{2i\omega}}\right] - \frac{1}{2} > 0 \quad \text{for } \omega \in (-\pi, \pi) \right\}$$

Compare with the local convergence set D_l of Problem 9.12.

Problem 9.7 *Updating the prediction error loss function*
Consider the recursive prediction error algorithm (9.39). Show that the minimum value of the loss function can, with the stated approximations, be updated according to the following recursion:

$$V_t(\hat{\theta}(t)) = \lambda V_{t-1}(\hat{\theta}(t-1)) + \frac{\lambda}{2} \varepsilon^T(t)[\lambda Q^{-1} + \psi^T(t)P(t-1)\psi(t)]^{-1}\varepsilon(t)$$

Problem 9.8 *Analysis of a simple RLS algorithm*
Let $\{y(t)\}$, $t = 1, 2, \ldots$ be a sequence of independent random variables with means $\theta(t) = Ey(t)$ and unit variances. Therefore

$$y(t) = \theta(t) + e(t)$$

where $Ee(t)e(s) = \delta_{t,s}$. Consider the following recursive algorithm for estimating $\theta(t)$:

$$\hat{\theta}(t+1) = \hat{\theta}(t) + \gamma(t)[y(t+1) - \hat{\theta}(t)] \qquad \hat{\theta}(0) = 0 \qquad \text{(i)}$$

where either

$$\gamma(t) = \frac{1}{t+1} \qquad \text{(ii)}$$

or

$$\gamma(t) = \gamma \in (0, 1) \qquad \text{(iii)}$$

Show that $\hat{\theta}(t)$ given by (i), (ii) minimizes the LS criterion

$$\sum_{k=1}^{t} \lambda^{t-k}[y(k) - \theta]^2 \qquad \text{(iv)}$$

with $\lambda = 1$; while $\hat{\theta}(t)$ given by (i), (iii) is asymptotically equal to the minimizer of (iv) with $\lambda = 1 - \gamma$. Then, assuming that $\theta(t) = \theta = $ constant, use the results of Problem 9.2 to establish the convergence properties of (i), (ii) and (i), (iii). Finally, assuming that $\theta(t)$ is slowly time-varying, discuss the advantage of using (iii) in such a case over the choice of $\gamma(t)$ in (ii).

Problem 9.9 *An alternative form for the RPEM*
Consider the loss function

$$V_t(\theta) = \frac{1}{2}\sum_{s=1}^{t}\left[\prod_{k=s+1}^{t}\lambda(k)\right]\varepsilon^T(s, \theta)Q\varepsilon(s, \theta)$$

with the convention $\Pi_{t+1}^t \lambda(k) = 1$ (cf. the expression before (9.49)). Show that the RPEM approximately minimizing $V_t(\theta)$ is given by (9.48) with $Z(t) = \psi(t)$ and λ replaced by $\lambda(t)$:

$$\hat{\theta}(t) = \hat{\theta}(t-1) + K(t)\varepsilon(t)$$

$$K(t) = P(t)\psi(t)Q$$
$$= P(t-1)\psi(t)[\lambda(t)Q^{-1} + \psi^T(t)P(t-1)\psi(t)]^{-1}$$

$$P(t) = [P(t-1) - K(t)\psi^T(t)P(t-1)]/\lambda(t)$$

Problem 9.10 *An RLS algorithm with a sliding window*
Consider the parameter estimate

$$\hat{\theta}(t) = \arg\min_{\theta} \sum_{s=t-m+1}^{t} \varepsilon^2(s, \theta)$$

$$\varepsilon(s, \theta) = y(s) - \varphi^T(s)\theta$$

The number m of prediction errors used in the criterion remains constant. Show that $\hat{\theta}(t)$ can be computed recursively as

$$\hat{\theta}(t) = \hat{\theta}(t-1) + K_1(t)\varepsilon(t, \hat{\theta}(t-1)) - K_2(t)\varepsilon(t-m, \hat{\theta}(t-1))$$

$$(K_1(t) \quad K_2(t)) = P(t-1)(\varphi(t) \quad \varphi(t-m))$$
$$\times \left\{ I + \begin{pmatrix} \varphi^T(t) \\ -\varphi^T(t-m) \end{pmatrix} P(t-1)(\varphi(t) \quad \varphi(t-m)) \right\}^{-1}$$

$$P(t) = P(t-1) - (K_1(t) \quad K_2(t)) \begin{pmatrix} \varphi^T(t) \\ -\varphi^T(t-m) \end{pmatrix} P(t-1)$$

Remark. The identification algorithm above can be used for real-time applications, since by its construction it has a finite memory. See Young (1984) for an alternative treatment of this problem.

Problem 9.11 *On local convergence of the RPEM*
Consider a single output system identified with the RPEM. The model structure is not necessarily rich enough to include the true system. Show that any local minimum point θ^* of the asymptotic loss function

$$V_\infty(\theta) = E\varepsilon^2(t, \theta)$$

is a possible local convergence point (i.e. a point for which (9.55b) holds and $L(\theta^*)$ has all eigenvalues in the left half-plane). Also show that other types of stationary points of $V_\infty(\theta)$ cannot be local convergence points to the RPEM.

Hint. First show the following result. Let $A > 0$ and B be two symmetric matrices. Then AB has all eigenvalues in the right half-plane if and only if $B > 0$.

Problem 9.12 *On the condition for local convergence of PLR*

A necessary and sufficient condition for local convergence of the PLR algorithm is that the matrix $-L(\theta_0)$ (see (9.76)) has all its eigenvalues located in the right half-plane. Consider the matrix

$$M = A^{-1}\left[E\varphi(t)\,\frac{1}{C(q^{-1})}\,\varphi^T(t)\right]$$

where A is a positive definite matrix, $\varphi(t)$ is some stationary full-rank random vector and

$$C(q^{-1}) = 1 + c_1 q^{-1} + \ldots + c_n q^{-n}$$

Clearly the class of matrices M includes $-L(\theta_0)$ as a special case.

(a) Show that if

$$\operatorname{Re} C(e^{i\omega}) > 0 \quad \text{for } \omega \in (-\pi, \pi)$$

then the eigenvalues of M belong to the right half-plane.

Hint. Since $A > 0$ by assumption, there exists a nonsingular matrix B such that $A^{-1} = BB^T$. The matrix M has the same eigenvalues as

$$\tilde{M} = B^{-1}MB = B^T\left[E\varphi(t)\,\frac{1}{C(q^{-1})}\,\varphi^T(t)\right]B$$

To evaluate the location of the eigenvalues of $\tilde{M}$, first establish, by a calculation similar to (9.74), that $\tilde{M}$ has the following property:

$$h^T\tilde{M}h > 0 \tag{i}$$

for any real vector $h \neq 0$.

(b) Determine the set (for $n = 2$)

$$D_l = \{c_1, c_2 | \operatorname{Re}(1 + c_1 e^{i\omega} + c_2 e^{2i\omega}) > 0 \quad \text{for } \omega \in (-\pi, \pi)\}$$

Compare with the stability set (derived in Problem 6.1)

$$\{c_1, c_2 | 1 + c_1 z + c_2 z^2 \neq 0 \quad \text{for } |z| \leq 1\}$$

Problem 9.13 *Local convergence of the PLR algorithm for a first-order ARMA process*

Consider a first-order ARMA process

$$y(t) + ay(t-1) = e(t) + ce(t-1)$$

and assume that a and c are estimated using the PLR algorithm (9.47).

Evaluate the matrix $L(\theta_0)$ in (9.76), and its eigenvalues. Compare with the result on the eigenvalues of $L(\theta_0)$ given at the end of Section 9.6.

Problem 9.14 *On the recursion for updating $P(t)$*

Show that equation (9.9) for updating the matrix $P(t)$ is equivalent to

$$P(t) = [I - K(t)\varphi^T(t)]P(t-1)[I - K(t)\varphi^T(t)]^T + K(t)K^T(t) \tag{i}$$

where $K(t)$ is defined in (9.12). A clear advantage of (i) over (9.9) is that use of (i) preserves the nonnegative definiteness of $P(t)$, whereas due to numerical errors which affect the computations, $P(t)$ given by (9.9) may be indefinite. The recursion (i) appears to have an additional advantage over (9.9), as now explained. Let ΔK denote a small perturbation of $K(t)$ in (9.9) or (i). Let ΔP denote the resulting perturbation in $P(t)$. Show that

$$\Delta P = O(\|\Delta K\|) \quad \text{for (9.9)}$$

and

$$\Delta P = O(\|\Delta K\|^2) \quad \text{for (i)}$$

Problem 9.15 *One-step-ahead optimal input design for RLS identification*
Consider the difference equation model (9.13),

$$y(t) = \varphi^T(t)\theta + e(t)$$

where

$$\theta = (b_1 \ \ldots \ b_{nb} \ \ a_1 \ \ldots \ a_{na})^T$$

$$\varphi(t) = (u(t-1) \ \ldots \ u(t-nb) \ -y(t-1) \ \ldots \ -y(t-na))^T$$

Assume that the RLS algorithm (9.12) is used to estimate the unknown parameters θ. Find $\hat{u}(t)$ at each time instant t such that

$$\hat{u}(t) = \arg \min_{u(t);\ |u(t)| \leq 1} \det P(t+1)$$

Since for t sufficiently large, $P(t+1)$ is proportional to the covariance matrix of the estimation errors, it makes sense to design the input signal as above.

Problem 9.16 *Illustration of the convergence rate for stochastic approximation algorithms*
Consider the following noisy static-gain system:

$$y(t) = \theta u(t) + \varepsilon(t)$$

where θ is the unknown gain to be estimated, $u(t)$ is constant

$$u(t) = \alpha$$

and the measurement disturbance $\varepsilon(t)$ is zero-mean white noise

$$E\varepsilon(t) = 0 \quad E\varepsilon(t)\varepsilon(s) = \lambda^2 \delta_{t,s}$$

Let $\alpha > 0$ (the case $\alpha < 0$ can be treated similarly). The unknown parameter θ is recursively estimated using the following stochastic approximation (SA) algorithm:

$$\hat{\theta}(t) = \hat{\theta}(t-1) + \frac{1}{t}[y(t) - \alpha\hat{\theta}(t-1)] \tag{i}$$

where $\hat{\theta}(t)$ denotes the estimated parameter at time t.
Let

$$\sigma_N = E[\hat{\theta}(N) - \theta]^2$$

Show that asymptotically (for $N \to \infty$) the following results hold:

(a) For $\alpha > 1/2$: $\sigma_N = \lambda^2/(N(2\alpha - 1))$ (thus the estimation error $[\hat{\theta}(N) - \theta]$ is asymptotically of the order $1/\sqrt{N}$).
(b) For $\alpha = 1/2$: $\sigma_N = \lambda^2(\log N)/N$ (thus $\hat{\theta}(N) - \theta$ is of the order of $\sqrt{\log N}/\sqrt{N}$).
(c) For $\alpha < 1/2$: the estimation error $[\hat{\theta}(N) - \theta]$ is of the order $1/N^\alpha$.

Compare the convergence rates of the SA algorithm above to the convergence rate of the recursive least squares (LS) estimator which corresponds to (i) with $1/t$ replaced by $1/(\alpha t)$.

Hint. For large N it can be shown that

$$\sum_{k=1}^{N} k^\mu \approx \begin{cases} N^{(\mu+1)}/(\mu + 1) & \text{for } \mu > -1 \\ \log N & \text{for } \mu = -1 \\ \text{constant} & \text{for } \mu < -1 \end{cases} \quad \text{(ii)}$$

(see, for example, Zygmund, 1968).

Remark. By rescaling the problem as

$$\bar{u}(t) = 1 \qquad \bar{\theta} = \alpha\theta$$

the SA algorithm of (i) becomes

$$\bar{\theta}(t) = \bar{\theta}(t - 1) + \frac{\alpha}{t}[y(t) - \bar{\theta}(t - 1)] \quad \text{(iii)}$$

Here the parameter α which drastically influences the accuracy of the estimates, appears as a factor in the gain sequence. Note that for the RLS algorithm α/t is replaced by $1/t$ in (iii). ∎

Bibliographical notes

Recursive identification is a rich field with many references. Some early survey papers are Saridis (1974), Isermann *et al.* (1974), Söderström *et al.* (1978), Dugard and Landau (1980). The paper by Söderström *et al.* (1978) also contains an analysis based on the ODE approach which was introduced by Ljung (1977a, b). For a thorough treatment of recursive identification in all its aspects the reader is referred to the book by Ljung and Söderström (1983). There is also a vast literature on adaptive systems for control and signal processing, for which recursive identification is a main topic. The reader is referred to Åström and Wittenmark (1989), Åström *et al.* (1977), Landau (1979), Egardt (1979), Goodwin and Sin (1984), Macchi (1984), Widrow and Stearns (1985), Honig and Messerschmitt (1984), Chen (1985), Haykin (1986), Alexander (1986), Treichler *et al.* (1987) and Goodwin, Ramadge and Caines (1980). Åström (1983b, 1987) and Seborg *et al.* (1986) have written excellent tutorial survey papers on adaptive control, while Basseville and Benveniste (1986) survey the area of fault detection in connection with recursive identification.

(*Section 9.3*). Some further aspects on real-time identification are given by Bohlin (1976), Benveniste and Ruget (1982), Benveniste (1984, 1987), and Benveniste *et al.* (1987). Identification of systems subject to large parameter changes is discussed for example by Hägglund (1984), Andersson (1983), Fortescue *et al.* (1981).

(*Section 9.4*). The adaptive IV method given by (9.26) was proposed by Wong and Polak (1967) and Young (1965, 1970). See also Young (1984) for a comprehensive discussion. The extended IV estimate, (8.13), can also be rewritten as a recursive algorithm; see Friedlander (1984) and Complement C9.1.

(*Section 9.5*). The recursive prediction error algorithm (9.39) was first derived for scalar ARMAX models in Söderström (1973), based on an idea by Åström. For similar and independent work, see Furht (1973) and Gertler and Bányász (1974). The RPEM has been extended to a much more general setting, for example, by Ljung (1981). Early proposals of the PLR algorithm (9.47) were made by Panuska (1968) and Young (1968), while Solo (1979) coined the name PLR and also gave a detailed analysis.

(*Section 9.6*). The ODE approach to analysis is based on Ljung (1977a, 1977b); see also Ljung (1984). The result on the location of the eigenvalues of $L(\theta_0)$ is given by Holst (1977), Stoica, Holst and Söderström (1982). The analysis of PLR in Example 9.6 follows Ljung (1977a). Alternative approaches for analysis, based on martingale theory and stochastic Lyapunov functions, have been given by Solo (1979), Moore and Ledwich (1980), Kushner and Kumar (1982), Chen and Guo (1986); see also Goodwin and Sin (1984). The asymptotic distribution of the parameter estimates has been derived for RPEM by Ljung (1980); see also Ljung and Söderström (1983). Solo (1980) and Benveniste and Ruget (1982) give some distribution results for PLR. A detailed comparison of the convergence and accuracy properties of PLR and its iterative off-line version is presented by Stoica, Söderström, Ahlén and Solbrand (1984, 1985).

(*Section 9.7*). The square root algorithm (9.81) is due to Potter (1963). Peterka (1975) also discusses square root algorithms for recursive identification schemes. For some other sophisticated and efficient ways (in particular the so-called U-D factorization) of implementing the update of factorized $P(t)$ like (9.81), see Bierman (1977), Ljung and Söderström (1983). For more details about lattice algorithms see Friedlander (1982), Samson (1982), Ljung (1983), Cybenko (1984), Haykin (1986), Ljung and Söderström (1983), Honig and Messerschmitt (1984), Benveniste and Chaure (1981), Lee *et al.* (1982), Porat *et al.* (1982), Ölçer *et al.* (1986), and Karlsson and Hayes (1986, 1987). Numerical properties and effects of unavoidable round-off errors for recursive identification algorithms are described in Ljung and Ljung (1985) and Cioffi (1987).

Complement C9.1
The recursive extended instrumental variable method

Consider the extended IV estimate (8.13). Denote by $\hat{\theta}(t)$ the estimate based on t data points. For simplicity of notation assume that $F(q^{-1}) \equiv 1$, $Q = I$ and that the system is scalar. Then

$$\hat{\theta}(t) = P(t)R^{\mathrm{T}}(t)r(t)$$

where

$$r(t) = \sum_{s=1}^{t} z(s)y(s)$$

$$R(t) = \sum_{s=1}^{t} z(s)\varphi^T(s)$$

$$P(t) = [R^T(t)R(t)]^{-1}$$

and where dim z > dim φ = dim θ. A recursive algorithm for computing $\hat{\theta}(t)$ is presented in the following.

Firstly,

$$\hat{\theta}(t) = \hat{\theta}(t-1) + P(t)R^T(t)[r(t) - R(t)\hat{\theta}(t-1)]$$

Next note that

$$R^T(t)[r(t) - R(t)\hat{\theta}(t-1)]$$
$$= [R^T(t-1) + \varphi(t)z^T(t)]\{r(t-1) + z(t)y(t)$$
$$\quad - [R(t-1) + z(t)\varphi^T(t)]\hat{\theta}(t-1)\}$$
$$= R^T(t-1)[r(t-1) - R(t-1)\hat{\theta}(t-1)]$$
$$\quad + R^T(t-1)z(t)[y(t) - \varphi^T(t)\hat{\theta}(t-1)]$$
$$\quad + \varphi(t)z^T(t)[r(t-1) - R(t-1)\hat{\theta}(t-1)$$
$$\quad + z(t)\{y(t) - \varphi^T(t)\hat{\theta}(t-1)\}] \quad (C9.1.1)$$

The first term in (C9.1.1) is equal to zero by the definition of $\hat{\theta}(t-1)$. The remaining terms can be written more compactly as

$$(\varphi(t) \quad R^T(t-1)z(t) + \varphi(t)z^T(t)z(t))\begin{pmatrix} z^T(t)\{r(t-1) - R(t-1)\hat{\theta}(t-1)\} \\ y(t) - \varphi^T(t)\hat{\theta}(t-1) \end{pmatrix}$$

$$= (R^T(t-1)z(t) \quad \varphi(t))\begin{pmatrix} 0 & 1 \\ 1 & z^T(t)z(t) \end{pmatrix}\left\{\begin{pmatrix} z^T(t)r(t-1) \\ y(t) \end{pmatrix} - \begin{pmatrix} w^T(t) \\ \varphi^T(t) \end{pmatrix}\hat{\theta}(t-1)\right\}$$

$$= \phi(t)\Lambda^{-1}(t)[v(t) - \phi^T(t)\hat{\theta}(t-1)]$$

where

$$w(t) = R^T(t-1)z(t) \qquad (n\theta|1)$$
$$\phi(t) = (w(t) \quad \varphi(t)) \qquad (n\theta|2)$$
$$\Lambda^{-1}(t) = \begin{pmatrix} 0 & 1 \\ 1 & z^T(t)z(t) \end{pmatrix} \qquad (2|2)$$

and

$$v(t) = \begin{pmatrix} z^T(t)r(t-1) \\ y(t) \end{pmatrix} \qquad (2|1)$$

Turn now to the recursive computation of $P(t)$. Note that

$$P^{-1}(t) = [R^{\mathrm{T}}(t-1) + \varphi(t)z^{\mathrm{T}}(t)][R(t-1) + z(t)\varphi^{\mathrm{T}}(t)]$$
$$= P^{-1}(t-1) + \varphi(t)w^{\mathrm{T}}(t) + w(t)\varphi^{\mathrm{T}}(t) + \varphi(t)z^{\mathrm{T}}(t)z(t)\varphi^{\mathrm{T}}(t)$$
$$= P^{-1}(t-1) + (w(t) \quad \varphi(t))\begin{pmatrix} 0 & 1 \\ 1 & z^{\mathrm{T}}(t)z(t) \end{pmatrix}\begin{pmatrix} w^{\mathrm{T}}(t) \\ \varphi^{\mathrm{T}}(t) \end{pmatrix}$$
$$= P^{-1}(t-1) + \phi(t)\Lambda^{-1}(t)\phi^{\mathrm{T}}(t)$$

Thus, from the matrix inversion Lemma A.1,

$$P(t) = P(t-1) - P(t-1)\phi(t)[\Lambda(t) + \phi^{\mathrm{T}}(t)P(t-1)\phi(t)]^{-1}\phi^{\mathrm{T}}(t)P(t-1)$$

The above equation implies that

$$P(t)\phi(t) = P(t-1)\phi(t)[\Lambda(t) + \phi^{\mathrm{T}}(t)P(t-1)\phi(t)]^{-1}\Lambda(t)$$

Combining the above equations provides a complete set of recursions for computing the extended IV estimate $\hat{\theta}$. The extended RIV algorithm is summarized in the following.

$$\hat{\theta}(t) = \hat{\theta}(t-1) + K(t)[v(t) - \phi^{\mathrm{T}}(t)\hat{\theta}(t-1)] \quad (n\theta|1)$$
$$K(t) = P(t-1)\phi(t)[\Lambda(t) + \phi^{\mathrm{T}}(t)P(t-1)\phi(t)]^{-1} \quad (n\theta|2)$$
$$\phi(t) = (w(t) \quad \varphi(t)) \quad (n\theta|2)$$
$$w(t) = R^{\mathrm{T}}(t-1)z(t) \quad (n\theta|1)$$
$$\Lambda(t) = \begin{pmatrix} -z^{\mathrm{T}}(t)z(t) & 1 \\ 1 & 0 \end{pmatrix} \quad (2|2)$$
$$v(t) = \begin{pmatrix} z^{\mathrm{T}}(t)r(t-1) \\ y(t) \end{pmatrix} \quad (2|1)$$
$$R(t) = R(t-1) + z(t)\varphi^{\mathrm{T}}(t) \quad (nz|n\theta)$$
$$r(t) = r(t-1) + z(t)y(t) \quad (nz|1)$$
$$P(t) = P(t-1) - K(t)\phi^{\mathrm{T}}(t)P(t-1) \quad (n\theta|n\theta)$$

A simple initialization procedure which does not require extra computations is given by

$$\hat{\theta}(0) = 0 \quad P(0) = \varrho I \quad \varrho = \text{some large positive number}$$
$$r(0) = 0 \quad R(0) = 0$$

The extended RIV algorithm above is more complex numerically than the basic RIV recursion (9.23). However, in some applications, especially in the signal processing field, the accuracy of $\hat{\theta}(t)$ increases with increasing dim $z = nz$. In such applications use of the extended RIV is well worth the effort. See Friedlander (1984) for a more detailed discussion of the extended RIV algorithm (called overdetermined RIV there) and for a description of some applications.

Complement C9.2
Fast least squares lattice algorithm for AR modeling

Let $y(t)$ be a stationary process. Consider the following nth-order autoregressive model of $y(t)$ (also called a linear (forward) prediction model):

$$y(t) = a_{n,1} y(t-1) + \ldots + a_{n,n} y(t-n) + \varepsilon_n(t) \tag{C9.2.1}$$

Here $\{a_{n,i}\}$ denote the coefficients and $\varepsilon_n(t)$ the residual (or the prediction error) of the nth-order model. Writing (C9.2.1) for $t = 1, \ldots, N$ gives

$$\underbrace{\begin{pmatrix} y(1) \\ y(2) \\ \vdots \\ y(N) \end{pmatrix}}_{y_N} = \underbrace{\begin{pmatrix} 0 & \cdots & \cdots & 0 \\ y(1) & & & \vdots \\ \vdots & & & 0 \\ & & y(1) & \\ \vdots & & & \vdots \\ y(N-1) & \cdots & & y(N-n) \end{pmatrix}}_{\Phi_n(N)} \underbrace{\begin{pmatrix} a_{n,1} \\ \vdots \\ \vdots \\ a_{n,n} \end{pmatrix}}_{\theta_n} + \begin{pmatrix} \varepsilon_n(1) \\ \vdots \\ \vdots \\ \varepsilon_n(N) \end{pmatrix} \tag{C9.2.2}$$

For convenience, assume zero initial conditions. For large N this will have only a negligible effect on the results that follow. The vector θ_n in (C9.2.2) is determined by the least squares (LS) method. Thus θ_n is given by

$$\theta_n(N) = [\Phi_n^T(N) \Phi_n(N)]^{-1} \Phi_n^T(N) y_N \tag{C9.2.3}$$

(cf. (4.7)), where the dependence of θ_n on N is shown explicitly. From (C9.2.2),

$$\varepsilon_n(N) = \mu_N^T [y_N - \Phi_n(N) \theta_n(N)] \tag{C9.2.4}$$

where μ_N is the Nth unit vector:

$$\mu_N = (0 \ \ldots \ 0 \ 1)^T$$

With

$$P_n(N) = I - \Phi_n(N) [\Phi_n^T(N) \Phi_n(N)]^{-1} \Phi_n^T(N) \tag{C9.2.5}$$

equation (C9.2.4) can be written more compactly as

$$\varepsilon_n(N) = \mu_N^T P_n(N) y_N \tag{C9.2.6}$$

The problem is to determine $\varepsilon_n(N)$ and $\theta_n(N)$ for $n = 1, \ldots, M$ (say) and $N = 1, 2, \ldots$ Use of a standard on-line LS algorithm for this purpose requires $O(M^3)$ operations (multiplications and additions) per time step. Any algorithm which provides the aforementioned quantities in less than $O(M^3)$ operations is called a fast algorithm. In the following a fast lattice LS algorithm is presented.

The problem of computing $\varepsilon_n(N)$ for $n = 1, \ldots, M$ and $N = 1, 2, \ldots$ is considered first. Determination of $\varepsilon_n(N)$ is important for prediction applications and for other applications of AR models (e.g. in geophysics) for which $\theta_n(N)$ is not necessarily needed. As will be shown, it is possible to determine $\varepsilon_n(N)$ for $n = 1, \ldots, M$ and given

N, in O(M) operations. In other applications, however, such as system identification and spectral estimation, one needs to determine $\theta_n(N)$ rather than $\varepsilon_n(N)$. For large N a good *approximation* to $\theta_n(N)$ may be obtained from the quantities needed to compute $\varepsilon_n(N)$ (see below). The exact computation of $\theta_n(N)$ is also possible and an algorithm for doing this will be presented which requires $O(M^2)$ operations per time step. The derivation of this exact algorithm relies on results obtained when discussing the problem of updating $\varepsilon_n(N)$ in time and order.

Updating the prediction errors

The matrix $P_n(N)$ plays an important role in the definition of $\varepsilon_n(N)$. The analysis begins by studying various possible updates of this matrix. In the following calculations repeated use is made of Lemma A.2. To simplify the notation various indexes will be omitted when this does not lead to any confusion. Later it will be shown that the recursive formulas for updating $P_n(N)$ with respect to n and N can be used as a basis for deriving a fast algorithm.

Order update

Let

$$\psi_n(N) = (\underbrace{0 \ldots 0}_{n+1} \ y(1) \ \ldots \ y(N-n-1))^T$$

Then

$$\Phi_{n+1}(N) = (\Phi_n(N) \quad \psi_n(N))$$

which implies that

$$P_{n+1} = I - (\Phi_n \quad \psi_n) \left\{ \begin{pmatrix} \Phi_n^T \\ \psi_n^T \end{pmatrix} (\Phi_n \quad \psi_n) \right\}^{-1} \begin{pmatrix} \Phi_n^T \\ \psi_n^T \end{pmatrix}$$

$$= I - (\Phi_n \quad \psi_n) \left\{ \begin{pmatrix} I \\ 0 \end{pmatrix} (\Phi_n^T \Phi_n)^{-1} (I \quad 0) + \begin{pmatrix} -(\Phi_n^T \Phi_n)^{-1} \Phi_n^T \psi_n \\ 1 \end{pmatrix} \right.$$

$$\left. \times (-\psi_n^T \Phi_n (\Phi_n^T \Phi_n)^{-1} \quad 1)/[\psi_n^T \psi_n - \psi_n^T \Phi_n (\Phi_n^T \Phi_n)^{-1} \Phi_n^T \psi_n] \right\} \begin{pmatrix} \Phi_n^T \\ \psi_n^T \end{pmatrix}$$

$$= P_n - P_n \psi_n \psi_n^T P_n / (\psi_n^T P_n \psi_n)$$

(C9.2.7)

With

$$\sigma_n(N) \triangleq \psi_n^T(N) P_n(N) \psi_n(N) \qquad (C9.2.8)$$

this gives

$$P_{n+1}(N) = P_n(N) - P_n(N)\psi_n(N)\psi_n^T(N)P_n(N)/\sigma_n(N) \qquad (C9.2.9)$$

Time update

The time update formula is the following:

$$P_n(N) - P_n(N)\mu_N\mu_N^T P_n(N)/\gamma_n(N) = \begin{pmatrix} P_n(N-1) & 0 \\ 0 & 0 \end{pmatrix} \quad \text{(C9.2.10)}$$

where

$$\gamma_n(N) = \mu_N^T P_n(N)\mu_N \quad \text{(C9.2.11)}$$

The result (C9.2.10) will be useful in the following even though it is not a true time update (it does not give a method for computing $P_n(N)$ from $P_n(N-1)$).

To prove (C9.2.10) note from (C9.2.7)–(C9.2.9) that the left-hand side of (C9.2.10) is equal to

$$\Omega \triangleq I - (\Phi(N) \quad \mu_N)\left\{\begin{pmatrix} \Phi^T(N) \\ \mu_N^T \end{pmatrix}(\Phi(N) \quad \mu_N)\right\}^{-1}\begin{pmatrix} \Phi^T(N) \\ \mu_N^T \end{pmatrix}$$

(where the index n has been omitted). Let φ^T denote the last row of $\Phi(N)$. Note that

$$\Phi^T(N)\mu_N = \varphi$$

and

$$\Phi^T(N)\Phi(N) = \Phi^T(N-1)\Phi(N-1) + \varphi\varphi^T$$

Thus

$$\Omega = I - (\Phi(N) \quad \mu_N)\left\{\begin{pmatrix} 0 \\ 1 \end{pmatrix}(0 \quad 1)\right.$$

$$\left. + \begin{pmatrix} I \\ -\varphi^T \end{pmatrix}[\Phi^T(N)\Phi(N) - \varphi\varphi^T]^{-1}(I \quad -\varphi)\right\}\begin{pmatrix} \Phi^T(N) \\ \mu_N^T \end{pmatrix}$$

$$= I - \mu_N\mu_N^T - [\Phi(N) - \mu_N\varphi^T][\Phi^T(N-1)\Phi(N-1)]^{-1}[\Phi(N) - \mu_N\varphi^T]^T$$

$$= I - \begin{pmatrix} 0 & 0 \\ 0 & 1 \end{pmatrix} - \begin{pmatrix} \Phi(N-1) \\ 0 \end{pmatrix}[\Phi^T(N-1)\Phi(N-1)]^{-1}(\Phi^T(N-1) \quad 0)$$

$$= \begin{pmatrix} P(N-1) & 0 \\ 0 & 0 \end{pmatrix}$$

which concludes the proof of (C9.2.10).

Time and order update

The matrix $\Phi_{n+1}(N+1)$ can be partitioned as

$$\Phi_{n+1}(N+1) = \begin{pmatrix} 0 \\ \hline y_N \mid \Phi_n(N) \end{pmatrix}$$

Thus

$$P_{n+1}(N+1) = I - \begin{pmatrix} 0 \\ \hline y_N \mid \Phi_n(N) \end{pmatrix} \left\{ \begin{pmatrix} 0 \mid y_N^T \\ \hline \mid \Phi_n^T(N) \end{pmatrix} \begin{pmatrix} 0 \\ \hline y_N \mid \Phi_n(N) \end{pmatrix} \right\}^{-1} \begin{pmatrix} 0 \mid y_N^T \\ \hline \mid \Phi_n^T(N) \end{pmatrix}$$

$$= I - \begin{pmatrix} 0 \\ \hline y_N \mid \Phi_n(N) \end{pmatrix} \left\{ \begin{pmatrix} 0 \\ I \end{pmatrix} [\Phi_n^T(N)\Phi_n(N)]^{-1}(0 \quad I) \right.$$

$$+ \begin{pmatrix} 1 \\ -\theta_n(N) \end{pmatrix}(1 \quad -\theta_n^T(N))/(y_N^T y_N - y_N^T \Phi_n(N)$$

$$\times [\Phi_n^T(N)\Phi_n(N)]^{-1}\Phi_n^T(N)y_N) \Bigg\} \begin{pmatrix} 0 \mid y_N^T \\ \hline \mid \Phi_n^T(N) \end{pmatrix}$$

$$= I - \begin{pmatrix} 0 \\ \Phi_n(N) \end{pmatrix} [\Phi_n^T(N)\Phi_n(N)]^{-1}(0 \quad \Phi_n^T(N))$$

$$- \begin{pmatrix} 0 \\ P_n(N)y_N \end{pmatrix}(0 \quad y_N^T P_n(N))/y_N^T P_n(N)y_N \qquad (C9.2.12)$$

With

$$\varrho_n(N) \triangleq y_N^T P_n(N) y_N \qquad (C9.2.13)$$

this becomes

$$P_{n+1}(N+1) = \begin{pmatrix} 1 & 0 \\ 0 & P_n(N) - P_n(N)y_N y_N^T P_n(N)/\varrho_n(N) \end{pmatrix} \qquad (C9.2.14)$$

It is important to note that the order update and time- and order update formulas, (C9.2.9) and (C9.2.14), hold also for $n = 0$ provided that $P_0(N) = I$. With this convention it is easy to see that the time update formula (C9.2.10) also holds for $n = 0$.

The update formulas derived above are needed to generate the desired recursions for the prediction errors. It should already be quite clear that in order to get a complete set of recursions it will be necessary to introduce several additional quantities. For easy reference, Table C9.2.1 collects together the definitions of all the auxiliary variables needed to update ε. Some of these variables have interesting interpretations. However, to keep the discussion reasonably brief, these variables are considered here as auxiliary quantities (refer to Friedlander, 1982, for a discussion of their 'meanings').

TABLE C9.2.1 The scalar variables used in the fast prediction error algorithm

$$\varepsilon_n(N) = \mu_N^T P_n(N) y_N$$
$$\sigma_n(N) = \psi_n^T(N) P_n(N) \psi_n(N)$$
$$\gamma_n(N) = \mu_N^T P_n(N) \mu_N$$
$$\varrho_n(N) = y_N^T P_n(N) y_N$$
$$\beta_n(N) = \mu_N^T P_n(N) \psi_n(N)$$
$$\delta_n(N) = \psi_n^T(N) P_n(N) y_N$$

Note that some of the auxiliary variables can be updated in several different ways. Thus, the recursions for updating ε can be organized in a variety of ways to provide a complete algorithm. The following paragraphs present all these alternative ways and indicate which seem preferable. While all these possible implementations of the update equations for ε are mathematically equivalent, their numerical properties and computational burdens may be different. This aspect, however, is not considered here.

Equations will be derived for updating the quantities introduced in Table C9.2.1. For each variable in the table all of the possible updating equations for $P_n(N)$ will be presented. It is implicitly understood that if an equation for updating $P_n(N)$ is not considered, then this means that it cannot be used.

From (C9.2.6) and (C9.2.9),

$$\varepsilon_{n+1}(N) = \varepsilon_n(N) - \beta_n(N)\delta_n(N)/\sigma_n(N) \tag{C9.2.15}$$

where β, δ and σ are as defined in Table C9.2.1.

Next consider the update of β. Observing that

$$\psi_{n+1}(N + 1) = \begin{pmatrix} 0 \\ \psi_n(N) \end{pmatrix} \tag{C9.2.16}$$

and

$$\mu_{N+1} = \begin{pmatrix} 0 \\ \mu_N \end{pmatrix} \tag{C9.2.17}$$

it follows from (C9.2.14) that

$$\beta_{n+1}(N + 1) = \beta_n(N) - \varepsilon_n(N)\delta_n(N)/\varrho_n(N) \tag{C9.2.18}$$

Next consider the problem of updating δ. As

$$\psi_n(N) = \begin{pmatrix} \psi_n(N - 1) \\ y(N - n - 1) \end{pmatrix} \tag{C9.2.19}$$

and

$$y_N = \begin{pmatrix} y_{N-1} \\ y(N) \end{pmatrix} \tag{C9.2.20}$$

then from (C9.2.10),

$$\delta_n(N) = \delta_n(N - 1) + \beta_n(N)\varepsilon_n(N)/\gamma_n(N) \tag{C9.2.21}$$

Next we discuss the update of ϱ. Either (C9.2.9) or (C9.2.10) can be used: use of (C9.2.9) gives

$$\varrho_{n+1}(N) = \varrho_n(N) - \delta_n^2(N)/\sigma_n(N) \tag{C9.2.22a}$$

while use of (C9.2.10) and (C9.2.20) leads to

$$\varrho_n(N) = \varrho_n(N - 1) + \varepsilon_n^2(N)/\gamma_n(N) \tag{C9.2.22b}$$

We recommend the use of (C9.2.22a). The reason may be explained as follows. The recursion (C9.2.22a) holds for $n = 0$ provided

$$\varrho_0(N) = y_N^T y_N \qquad \delta_0(N) = \psi_0^T(N) y_N \quad \text{and} \quad \sigma_0(N) = \psi_0^T(N)\psi_0(N)$$

Note that

$$\varrho_0(N) = \varrho_0(N-1) + y^2(N)$$

and similarly for $\delta_0(N)$ and $\sigma_0(N)$. Thus (C9.2.22a) can be initialized simply and exactly. This is not true for (C9.2.22b). The larger n is, the more computations are needed to determine exact initial values for (C9.2.22b). Equation (C9.2.22b) can be initialized by setting $\varrho_n(0) = 0$ for $n = 1, \ldots, M$. However, this initialization is somewhat arbitrary and may lead to long transients.

Next consider σ. Either (C9.2.10) or (C9.2.14) can be used. From (C9.2.10) and (C9.2.19),

$$\sigma_n(N) = \sigma_n(N-1) + \beta_n^2(N)/\gamma_n(N) \tag{C9.2.23a}$$

while (C9.2.14) and (C9.2.16) give

$$\sigma_{n+1}(N+1) = \sigma_n(N) - \delta_n^2(N)/\varrho_n(N) \tag{C9.2.23b}$$

For reasons similar to those stated above when discussing the choice between (C9.2.22a) and (C9.2.22b), we tend to prefer (C9.2.23b) to (C9.2.23a).

Finally consider the update equations for γ. From (C9.2.9),

$$\gamma_{n+1}(N) = \gamma_n(N) - \beta_n^2(N)/\sigma_n(N) \tag{C9.2.24a}$$

and from (C9.2.14) using (C9.2.17),

$$\gamma_{n+1}(N+1) = \gamma_n(N) - \varepsilon_n^2(N)/\varrho_n(N) \tag{C9.2.24b}$$

Both (C9.2.24a) and (C9.2.24b) are convenient to use but we prefer (C9.2.24a) which seems to require somewhat simpler programming.

A complete set of recursions has been derived for updating $\varepsilon_n(N)$ and the supporting variables of Table C9.2.1. Table C9.2.2 summarizes the least squares prediction algorithm derived above, in a form that should be useful for reference and coding. The initial values for the variables in Table C9.2.2 follow from their definitions (recalling that $P_0(N) = I$).

Next note the following facts. Introduce the notation

$$\begin{aligned}\bar{k}_n(N) &= \delta_n(N)/\sigma_n(N) \\ \tilde{k}_n(N) &= \delta_n(N)/\varrho_n(N)\end{aligned} \tag{C9.2.25}$$

Then the updating equations for ε and β can be written as

$$\begin{aligned}\varepsilon_{n+1}(N) &= \varepsilon_n(N) - \tilde{k}_n(N)\beta_n(N) \\ \beta_{n+1}(N) &= \beta_n(N-1) - \bar{k}_n(N-1)\varepsilon_n(N-1)\end{aligned} \tag{C9.2.26}$$

These equations define a lattice filter for computation of the prediction errors, which is depicted in Figure C9.2.1. Note that $\bar{k}$ and $\tilde{k}$ are so-called reflection coefficients.

The lattice filter of Figure C9.2.1 acts on data measured at time instant N. Its parameters change with N. Observe that the first n sections of the lattice filter give the

Complement C9.2

TABLE C9.2.2 The least squares lattice prediction algorithm

Given: M

Set: $\varrho_0(1) = y^2(1) \neq 0$; $\delta_0(1) = \beta_0(1) = \sigma_0(1) = 0$; $\varepsilon_0(1) = y(1)$; $\delta_n(1) = 0$ for $n = 1, \ldots, M$ *

Perform the following calculation, for every $N = 2, 3, \ldots$:

Initialize:

$\varepsilon_0(N) = y(N)$

$\beta_0(N) = y(N - 1)$

$\varrho_0(N) = \varrho_0(N - 1) + y^2(N)$

$\sigma_0(N) = \varrho_0(N - 1)$

$\delta_0(N) = \delta_0(N - 1) + y(N - 1)y(N)$

$\gamma_0(N) = 1$

For $n = 0$ to $\min[M - 1; N - 2]$, compute:

$\beta_{n+1}(N) = \beta_n(N - 1) - \varepsilon_n(N - 1)\delta_n(N - 1)/\varrho_n(N - 1)$

$\sigma_{n+1}(N) = \sigma_n(N - 1) - \delta_n^2(N - 1)/\varrho_n(N - 1)$

$\varepsilon_{n+1}(N) = \varepsilon_n(N) - \beta_n(N)\delta_n(N)/\sigma_n(N)$

$\varrho_{n+1}(N) = \varrho_n(N) - \delta_n^2(N)/\sigma_n(N)$

$\gamma_{n+1}(N) = \gamma_n(N) - \beta_n^2(N)/\sigma_n(N)$

$\delta_{n+1}(N) = \delta_{n+1}(N - 1) + \beta_{n+1}(N)\varepsilon_{n+1}(N)/\gamma_{n+1}(N)$

* This initialization is approximate.

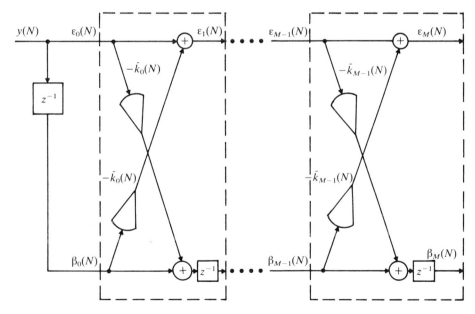

FIGURE C9.2.1 The lattice filter implementation of the fast LS prediction algorithm for AR models.

nth-order prediction errors $\varepsilon_n(N)$. Due to this nested modular structure the filter of Figure C9.2.1 is called a lattice or ladder filter.

The lattice LS algorithm introduced above requires only $\sim 10\,M$ operations per time step. However, it does not compute the LS parameter estimates. As will be shown in the next subsection, computation of $\theta_n(N)$ for $n = 1, \ldots, M$ and given N needs $O(M^2)$ operations. This may still be a substantial saving compared to the standard on-line LS method, which requires $O(M^3)$ operations per time step.

Note that for sufficiently large N a good approximation to $\theta_n(N)$ can be obtained in the following way. From (C9.2.1) it follows that

$$\varepsilon_n(N) = A_n^N(q^{-1})y(N)$$

where q^{-1} denotes the unit time delay, and

$$A_n^N(z) = 1 - a_{n,1}^N z - \ldots - a_{n,n}^N z^n$$

Next introduce

$$\begin{pmatrix} b_{n,1}^N \\ \vdots \\ b_{n,n}^N \end{pmatrix} = [\Phi_n^T(N)\Phi_n(N)]^{-1}\Phi_n^T(N)\psi_n(N)$$

and

$$B_n^N(z) = -b_{n,1}^N - b_{n,2}^N z - \ldots - b_{n,n}^N z^{n-1} + z^n$$

Then

$$\beta_n(N) = y(N - n - 1) - (y(N-1) \ldots y(N-n)) \begin{pmatrix} b_{n,1}^N \\ \vdots \\ b_{n,n}^N \end{pmatrix}$$

$$= B_n^N(q^{-1})y(N-1)$$

The above equation is the backward prediction model of $y(\cdot)$ and $\{\beta_n(N)\}$ are the so-called backward prediction errors.

Inserting the above expressions for ε and β in the lattice recursions (C9.2.26) gives

$$[A_{n+1}^N(q^{-1}) - A_n^N(q^{-1}) + \tilde{k}_n(N)q^{-1}B_n^N(q^{-1})]y(N) = 0$$
$$[B_{n+1}^N(q^{-1}) - q^{-1}B_n^{N-1}(q^{-1}) + \tilde{k}_n(N-1)A_n^{N-1}(q^{-1})]y(N-1) = 0$$
(C9.2.27)

Note that since the 'filters' acting on $y(\cdot)$ in (C9.2.27) are time-varying it cannot be concluded that they are equal to zero despite the fact that their outputs are identically zero. However, for large N the lattice filter parameters will approximately converge and then the following identities will approximately hold (the index N is omitted):

$$\boxed{\begin{aligned} A_{n+1}(z) &= A_n(z) - \tilde{k}_n z B_n(z) \\ B_{n+1}(z) &= z B_n(z) - \tilde{k}_n A_n(z) \end{aligned} \qquad (A_0(z) = B_0(z) = 1) \qquad \text{(C9.2.28)}}$$

The recursions (C9.2.28) may be compared to those encountered in Complement C8.3 (see equations (C8.3.16) and (C8.3.17)). They can be used to determine $\{\theta_n\}$ from a knowledge of the reflection coefficients $\{\bar{k}_n\}$ and $\{\tilde{k}_n\}$. This computation requires $O(M^2)$ operations. However, it may be done only when necessary. For instance, in system identification and spectral estimation applications, estimates of the model parameters may be needed at a (much) slower rate than the recording of data.

Updating the parameter estimates

Define
$$R_n(N) = [\Phi_n^T(N)\Phi_n(N)]^{-1}\Phi_n^T(N) \qquad (C9.2.29)$$

Since $R_n(N)$ plays a central role in the definition of $\theta_n(N)$ it is necessary to discuss various possible updates of this matrix. To simplify the calculations, use will be made of the results for updating $P_n(N)$ derived in the previous subsection. Note that from (C9.2.5) and (C9.2.29) it follows that
$$P_n(N) = I - \Phi_n(N)R_n(N)$$

Order update

From (C9.2.9) it follows that
$$\Phi_{n+1}R_{n+1} = \Phi_n R_n + [\psi_n - \Phi_n R_n \psi_n]\psi_n^T P_n/\sigma_n \qquad (C9.2.30)$$

Next note that
$$R_{n+1}\Phi_{n+1} = R_{n+1}(\Phi_n \quad \psi_n) = I \qquad (C9.2.31)$$

Thus, premultiplying (C9.2.30) by R_{n+1},

$$R_{n+1}(N) = \begin{pmatrix} R_n(N) \\ 0 \end{pmatrix} + \begin{pmatrix} -R_n(N)\psi_n(N) \\ 1 \end{pmatrix} \psi_n^T(N) P_n(N)/\sigma_n(N) \qquad (C9.2.32)$$

Time update

It follows from (C9.2.10) that
$$\Phi(N)R(N) + [\mu_N - \Phi(N)R(N)\mu_N]\mu_N^T P(N)/\gamma(N) = \begin{pmatrix} \Phi(N-1)R(N-1) & 0 \\ 0 & 1 \end{pmatrix}$$

Premultiplying this identity by the $((N-1)|N)$ matrix $(I \quad 0)$, the following equation is obtained:
$$\Phi(N-1)R(N) - \Phi(N-1)R(N)\mu_N\mu_N^T P(N)/\gamma(N) = (\Phi(N-1)R(N-1) \quad 0)$$

which after premultiplication by $R(N-1)$ gives

370 *Recursive identification methods* Chapter 9

$$R_n(N) - R_n(N)\mu_N\mu_N^T P_n(N)/\gamma_n(N) = (R_n(N-1) \quad 0) \qquad (C9.2.33)$$

Time and order update

It follows from the calculation (C9.2.12) that

$$R_{n+1}(N+1) = \binom{0}{I}[\Phi_n^T(N)\Phi_n(N)]^{-1}(0 \quad \Phi_n^T(N))$$

$$+ \binom{1}{-R_n(N)y_N}(0 \quad y_N^T P_n(N))/\varrho_n(N)$$

which gives

$$R_{n+1}(N+1) = \begin{pmatrix} 0 & 0 \\ 0 & R_n(N) \end{pmatrix} + \binom{1}{-R_n(N)y_N}(0 \quad y_N^T P_n(N))/\varrho_n(N) \qquad (C9.2.34)$$

The formulas derived above can be used to update $\theta_n(N)$ and any supporting quantities. As will be shown, the update of $\theta_n(N)$ may be based on the recursions derived earlier for updating $\varepsilon_n(N)$. It will be attempted to keep the number of additional recursions needed for updating $\theta_n(N)$ to a minimum.

Note that either (C9.2.32) or (C9.2.33) can be used to update $\theta_n(N)$. First consider the use of (C9.2.32). It follows from (C9.2.32) that

$$\theta_{n+1}(N) = \binom{\theta_n(N)}{0} + \binom{-b_n(N)}{1}\delta_n(N)/\sigma_n(N) \qquad (C9.2.35)$$

where δ and σ are as defined previously, and

$$b_n(N) \triangleq R_n(N)\psi_n(N)$$

is the vector of the coefficients of the backward prediction model (see the discussion at the end of the previous subsection).

To update $b_n(N)$, either (C9.2.33) or (C9.2.34) can be used. From (C9.2.33) and (C9.2.19),

$$b_n(N) = b_n(N-1) + c_n(N)\beta_n(N)/\gamma_n(N) \qquad (C9.2.36a)$$

where

$$c_n(N) \triangleq R_n(N)\mu_N$$

The update of c is discussed a little later. Using equation (C9.2.34) and (C9.2.16) gives

$$b_{n+1}(N+1) = \binom{0}{b_n(N)} + \binom{1}{-\theta_n(N)}\delta_n(N)/\varrho_n(N) \qquad (C9.2.36b)$$

Next consider the use of (C9.2.33) to update $\theta_n(N)$. Using this equation and (C9.2.20),

$$\theta_n(N) = \theta_n(N-1) + c_n(N)\varepsilon_n(N)/\gamma_n(N) \quad \text{(C9.2.37)}$$

Concerning the update of c, either (C9.2.32) or (C9.2.34) can be used. Use of (C9.2.32) results in

$$c_{n+1}(N) = \begin{pmatrix} c_n(N) \\ 0 \end{pmatrix} + \begin{pmatrix} -b_n(N) \\ 1 \end{pmatrix} \beta_n(N)/\sigma_n(N) \quad \text{(C9.2.38a)}$$

while (C9.2.34) gives

$$c_{n+1}(N+1) = \begin{pmatrix} 0 \\ c_n(N) \end{pmatrix} + \begin{pmatrix} 1 \\ -\theta_n(N) \end{pmatrix} \varepsilon_n(N)/\varrho_n(N) \quad \text{(C9.2.38b)}$$

The next step is to discuss the selection of a complete set of recursions from those above, for updating the LS parameter estimates. As discussed earlier, one should try to avoid selection of the time update equations since these need to be initialized rather arbitrarily. Furthermore, in the present case one may wish to avoid time or time and order recursions since parameter estimates may not be needed at every sampling point. With these facts in mind, begin by defining the following quantities:

$$b_n^*(N) \triangleq b_n(N) - c_n(N)\beta_n(N)/\gamma_n(N) \quad \text{(C9.2.39)}$$

$$\theta_n^*(N) \triangleq \theta_n(N) - c_n(N)\varepsilon_n(N)/\gamma_n(N) \quad \text{(C9.2.40)}$$

Note from (C9.2.36a) and (C9.2.37) that

$b_n^*(N) = b_n(N-1)$

$\theta_n^*(N) = \theta_n(N-1)$

With these definitions (C9.2.36b) can be written as

$$b_{n+1}(N) = \begin{pmatrix} 0 \\ b_n^*(N) \end{pmatrix} + \begin{pmatrix} 1 \\ -\theta_n^*(N) \end{pmatrix} \delta_n(N-1)/\varrho_n(N-1) \quad \text{(C9.2.41)}$$

The following equations can be used to update the LS parameter estimates: (C9.2.35), (C9.2.38a), (C9.2.39), (C9.2.40) and (C9.2.41). Note that these equations have the following important feature: they do not contain any time update and thus may be used to compute $\theta_n(N)$ when desired. The auxiliary scalar quantities δ, σ, β, γ, ε, and ϱ which appear in the equations above are available from the LS lattice prediction algorithm. The initial values $\theta_1(N)$, $b_1(N)$ and $c_1(N)$ for the algorithm proposed are given by

$$\theta_1(N) = \frac{\sum_{t=2}^{N} y(t-1)y(t)}{\sum_{t=1}^{N-1} y^2(t)} = \frac{\delta_0(N)}{\sigma_0(N)} \tag{C9.2.42a}$$

$$b_1(N) = \frac{\sum_{t=2}^{N-1} y(t-1)y(t)}{\sum_{t=1}^{N-1} y^2(t)} = \frac{\delta_0(N-1)}{\sigma_0(N)} \tag{C9.2.42b}$$

and

$$c_1(N) = \frac{y(N-1)}{\sum_{t=1}^{N-1} y^2(t)} = \frac{\beta_0(N)}{\sigma_0(N)} \tag{C9.2.42c}$$

where δ_0, σ_0 and β_0 are also available from the LS lattice prediction algorithm.

Note that the fast lattice LS parameter estimation algorithm introduced above requires $\sim 2.5\,M^2$ operations at each time sampling point for which it is applied. If the LS parameter estimates are desired at *all* time sampling points, then somewhat simpler algorithms requiring $\sim M^2$ operations per time step may be used. For example, one may use the update formulas (C9.2.35), (C9.2.36b) to compute $\theta_n(N)$ for $n = 1, \ldots, M$ and $N = 1, 2, 3, \ldots$ The initial values for (C9.2.35) and (C9.2.36b) are given by (C9.2.42a, b). It might seem that (C9.2.36b) needs to be initialized arbitrarily by setting $b_n(0) = 0$, $n = 1, \ldots, M$. However, this is not so since (C9.2.36b) is iterated for n from 1 to $\min(M-1, N-2)$. This iteration process is illustrated in Table C9.2.3.

Finally consider the following simple alternative to the fast LS *parameter estimation* algorithms described above. Update the covariances

$$\hat{r}_k^N = \frac{1}{N}\sum_{t=1}^{N-k} y(t)y(t+k) = \hat{r}_k^{N-1} + \frac{1}{N}[y(N-k)y(N) - \hat{r}_k^{N-1}]$$

as each new data point is collected. Then solve the Yule–Walker system of equations associated with (C9.2.1) (see equation (C8.2.1) in Complement C8.2), when desired, using the Levinson–Durbin algorithm (Complement C8.2), to obtain estimates of

TABLE C9.2.3 Illustration of the iteration of (C9.2.36b) for $M = 4$

N	2	3	4	5	6	
$\{b_n(N)\}$	$b_1(2)$	$b_1(3)$	$b_1(4)$	$b_1(5)$	$b_1(6)$	} equation (C9.2.42b)
		$b_2(3)$	$b_2(4)$	$b_2(5)$	$b_2(6)$	
			$b_3(4)$	$b_3(5)$	$b_3(6)$	} equation (C9.2.36b)
				$b_4(5)$	$b_4(6)$	

$\theta_1, \ldots, \theta_M$. This requires $\sim M^2$ operations. For large N the estimates obtained in this way will be close to those provided by the fast LS algorithms described previously.

Complement C9.3
Fast least squares lattice algorithm for multivariate regression models

Consider the following multivariate nth-order regression model:

$$x(t) = C_{n,1} y(t-1) + \ldots + C_{n,n} y(t-n) + e_n(t) \qquad t = 1, 2, \ldots \qquad (C9.3.1)$$

where x is an $(nx|1)$ vector, y is an $(ny|1)$ vector, the $(nx|ny)$ matrices $C_{n,i}$ denote the coefficients, and the nx-vector e_n denotes the prediction errors (or the residuals) of the model. As it stands equation (C9.3.1) may be viewed as a multivariate (truncated) weighting function model. Several other models of interest in system identification can be obtained as particular cases of (C9.3.1). For $x(t) = y(t)$, equation (C9.3.1) becomes a multivariate autoregression. A little calculation shows that a difference equation model can also be written in the form (C9.3.1). To see this, consider the following multivariable full polynomial form difference equation: this was discussed in Example 6.3 (see (6.22)–(6.25)), and its use in identification was described in Complement C7.3. This model structure is given by

$$x(t) + A_1 x(t-1) + \ldots + A_n x(t-n) = B_1 u(t-1) + \ldots$$
$$+ B_n u(t-n) + e_n(t) \qquad (C9.3.2)$$

where x, u and e_n are vectors, and A_i and B_i are matrices of appropriate dimensions. With

$$y(t) = \begin{pmatrix} x(t) \\ u(t) \end{pmatrix}$$

and

$$C_{n,i} = (-A_i \quad B_i) \qquad i = 1, \ldots, n$$

(C9.3.2) can be rewritten as (C9.3.1).

Writing (C9.3.1) for $t = 1, \ldots, N$ and assuming zero initial conditions for convenience,

$$\underbrace{(x(1) \ldots x(N))}_{\substack{x_N^T \\ (nx|N)}} = \underbrace{(C_{n,1} \ldots C_{n,n})}_{\substack{\theta_n^T \\ (nx|ny \cdot n)}} \underbrace{\begin{pmatrix} 0 & y(1) & \ldots & y(N-1) \\ \vdots & & \ddots & \vdots \\ 0 & \ldots & 0 & y(1) & \ldots & y(N-n) \end{pmatrix}}_{\substack{\Phi_n^T(N) \\ (ny \cdot n|N)}} \qquad (C9.3.3)$$

$$+ (e_n(1) \ldots e_n(N))$$

The matrix θ_n of unknown parameters is determined as the minimizer of the following exponentially weighted least squares (LS) criterion:

$$\sum_{t=1}^{N} \lambda^{N-t} e_n^T(t) e_n(t) \tag{C9.3.4}$$

where $\lambda \in (0, 1)$ is a forgetting factor (cf. (9.11)). It follows from (4.7) and some simple calculation that θ_n is given by

$$\theta_n(N) = [\Phi_n^T(N) \Lambda_N \Phi_n(N)]^{-1} \Phi_n^T(N) \Lambda_N x_N \tag{C9.3.5}$$

where the dependence of θ_n on N is shown explicitly, and where

$$\Lambda_N = \begin{pmatrix} \lambda^{N-1} & & & 0 \\ & \ddots & & \\ & & \lambda & \\ 0 & & & 1 \end{pmatrix} \tag{C9.3.6}$$

Introduce the following notation:

$$\mu_N = (0 \ldots 0 \ 1)^T \quad \text{(the } N\text{th unit vector)}$$

$$\Lambda_N^{1/2} = \text{the square root of } \Lambda_N$$

$$\tilde{\Phi}_n(N) = \Lambda_N^{1/2} \Phi_n(N) \tag{C9.3.7}$$

$$\tilde{x}_N = \Lambda_N^{1/2} x_N$$

$$P_n(N) = I - \tilde{\Phi}_n(N) [\tilde{\Phi}_n^T(N) \tilde{\Phi}_n(N)]^{-1} \tilde{\Phi}_n^T(N)$$

Then, from (C9.3.3), (C9.3.5) and (C9.3.7),

$$\begin{aligned} e_n(N) &= [x_N^T - \theta_n^T(N) \Phi_n^T(N)] \mu_N \\ &= x_N^T \{I - \Lambda_N \Phi_n(N) [\Phi_n^T(N) \Lambda_N \Phi_n(N)]^{-1} \Phi_n^T(N)\} \mu_N \\ &= x_N^T \Lambda_N^{1/2} P_n(N) \Lambda_N^{-1/2} \mu_N \\ &= \tilde{x}_N^T P_n(N) \mu_N \end{aligned}$$

The problem considered in this complement is the determination of $e_n(N)$ and $\theta_n(N)$ for $N = 1, 2, \ldots$ and $n = 1$ to M (some maximum order). This problem is a significant generalization of the scalar problem treated in Complement C9.2. Both the model and the LS criterion considered here are more general. Note that by an appropriate choice of λ in (C9.3.4), it is possible to discount exponentially the old measurements, which are given smaller weights than more recent measurements.

In the following a solution is provided to the problem of computing $e_n(N)$ for $n = 1, \ldots, M$ and $N = 1, 2, \ldots$. As will be shown, a good approximation (for large N) of the LS parameter estimates $\theta_n(N)$ may be readily obtained from the parameters of the lattice filter which computes $e_n(N)$. Exact computation of $\theta_n(N)$ is also possible and an exact lattice LS parameter estimation algorithm may be derived as in the scalar case (see, e.g., Friedlander, 1982; Porat et al., 1982). However, in the interest of brevity we concentrate on the lattice prediction filter and leave the derivation of the exact lattice parameter estimation algorithm as an exercise for the reader.

Since the matrix $P_n(N)$ plays a key role in the definition of $e_n(N)$ we begin by studying various possible updates of this matrix. Note that $P_n(N)$ has the same structure as in the scalar case. Thus, the calculations leading to update formulas for $P_n(N)$ are quite similar to those made in Complement C9.2. Some of the details will therefore be omitted.

Order update

Let

$$\psi_n(N) = (\underbrace{0 \ldots 0}_{n+1 \text{ columns}} \; y(1) \; \ldots \; y(N - n - 1))^{\mathrm{T}}$$

$$\tilde{\psi}_n(N) = \Lambda_N^{1/2} \psi_n(N)$$

Then

$$\tilde{\Phi}_{n+1}^{\mathrm{T}}(N) = \begin{pmatrix} \tilde{\Phi}_n^{\mathrm{T}}(N) \\ \tilde{\psi}_n^{\mathrm{T}}(N) \end{pmatrix}$$

which implies that

$$P_{n+1}(N) = P_n(N) - P_n(N)\tilde{\psi}_n(N)\sigma_n^{-1}(N)\tilde{\psi}_n^{\mathrm{T}}(N)P_n(N) \qquad (C9.3.8)$$

where

$$\sigma_n(N) = \tilde{\psi}_n^{\mathrm{T}}(N) P_n(N) \tilde{\psi}_n(N)$$

Time update

Let φ denote the last column of $\tilde{\Phi}_n^{\mathrm{T}}(N)$. Then

$$\tilde{\Phi}_n^{\mathrm{T}}(N)\mu_N = \varphi$$

$$\tilde{\Phi}_n^{\mathrm{T}}(N) - \varphi\mu_N^{\mathrm{T}} = (\lambda^{1/2}\tilde{\Phi}_n^{\mathrm{T}}(N-1) \; \; 0)$$

$$\tilde{\Phi}_n^{\mathrm{T}}(N)\tilde{\Phi}_n(N) = \lambda \tilde{\Phi}_n^{\mathrm{T}}(N-1)\tilde{\Phi}_n(N-1) + \varphi\varphi^{\mathrm{T}}$$

With these equations in mind, the following time update formula is obtained in exactly the same way as for the scalar case (see Complement C9.2 for details):

$$P_n(N) = \begin{pmatrix} P_n(N-1) & 0 \\ 0 & 0 \end{pmatrix} + P_n(N)\mu_N\mu_N^{\mathrm{T}}P_n(N)/\gamma_n(N) \qquad (C9.3.9)$$

where

$$\gamma_n(N) = \mu_N^{\mathrm{T}} P_n(N) \mu_N$$

Time and order update

We have

$$\Phi_{n+1}^{\mathrm{T}}(N+1) = \begin{pmatrix} 0 & y_N^{\mathrm{T}} \\ \hline & \Phi_n^{\mathrm{T}}(N) \end{pmatrix}$$

where
$$y_N^T = (y(1) \ \ldots \ y(N))$$

Thus
$$\tilde{\Phi}_{n+1}^T(N+1) = \begin{pmatrix} 0 & y_N^T \\ \hline 0 & \Phi_n^T(N) \end{pmatrix} \begin{pmatrix} \lambda^{N/2} & 0 \\ \hline 0 & \Lambda_N^{1/2} \end{pmatrix} = \begin{pmatrix} 0 & \bar{y}_N^T \\ \hline 0 & \tilde{\Phi}_n^T(N) \end{pmatrix}$$

where
$$\bar{y}_N = \Lambda_N^{1/2} y_N$$

The nested structure of $\tilde{\Phi}_{n+1}(N+1)$ above leads exactly as in the scalar case to the following update formula:

$$P_{n+1}(N+1) = \begin{pmatrix} 1 & 0 \\ 0 & P_n(N) - P_n(N)\bar{y}_N \varrho_n^{-1}(N) \bar{y}_N^T P_n(N) \end{pmatrix} \quad \text{(C9.3.10)}$$

where
$$\varrho_n(N) = \bar{y}_N^T P_n(N) \bar{y}_N$$

Note that the update formulas introduced above also hold for $n = 0$ provided $P_0(N) = I$.

The update formulas for $P_n(N)$ derived previously can be used to update $e_n(N)$ and any supporting quantities. Table C9.3.1 summarizes the definitions of all the variables needed in the algorithm for updating $e_n(N)$.

Several variables in Table C9.3.1 may be updated in order or in time as well as in time and order simultaneously. Thus, as in the scalar case treated in Complement C9.2, the recursive-in-time-and-order calculation of $e_n(N)$ may be organized in many different ways. The following paragraphs present one particular implementation of the recursive algorithm for computing the prediction errors $\{e_n(N)\}$. (See Complement C9.2 for details of other possible implementations.)

TABLE C9.3.1 Definitions of the variables used in the fast lattice LS prediction algorithm

$e_n(N) = \bar{x}_N^T P_n(N) \mu_N$	$(nx\|1)$
$\sigma_n(N) = \bar{\psi}_n^T(N) P_n(N) \bar{\psi}_n(N)$	$(ny\|ny)$
$\gamma_n(N) = \mu_N^T P_n(N) \mu_N$	$(1\|1)$
$\varrho_n(N) = \bar{y}_N^T P_n(N) \bar{y}_N$	$(ny\|ny)$
$\beta_n(N) = \bar{\psi}_n^T(N) P_n(N) \mu_N$	$(ny\|1)$
$\delta_n(N) = \bar{\psi}_n^T(N) P_n(N) \bar{y}_N$	$(ny\|ny)$
$d_n(N) = \bar{x}_N^T P_n(N) \bar{\psi}_n(N)$	$(nx\|ny)$
$\varepsilon_n(N) = \bar{y}_N^T P_n(N) \mu_N$	$(ny\|1)$

First note the following facts:

$$\tilde{\psi}_{n+1}^T(N+1) = (0 \quad \tilde{\psi}_n^T(N))$$
$$\mu_{N+1}^T = (0 \quad \mu_N^T)$$
$$\tilde{\psi}_n^T(N) = (\lambda^{1/2}\tilde{\psi}_n^T(N-1) \quad y(N-n-1))$$
$$\tilde{y}_N^T = (\lambda^{1/2}\tilde{y}_{N-1}^T \quad y(N))$$
$$\tilde{x}_N^T = (\lambda^{1/2}\tilde{x}_{N-1}^T \quad x(N))$$

With the above identities in mind, a straightforward application of the update formulas for $P_n(N)$ produces the following recursions:

$$e_{n+1}(N) = e_n(N) - d_n(N)\sigma_n^{-1}(N)\beta_n(N)$$
$$\beta_{n+1}(N+1) = \beta_n(N) - \delta_n(N)\varrho_n^{-1}(N)\varepsilon_n(N)$$
$$\delta_n(N) = \lambda\delta_n(N-1) + \beta_n(N)\varepsilon_n^T(N)/\gamma_n(N)$$
$$\varrho_{n+1}(N) = \varrho_n(N) - \delta_n^T(N)\sigma_n^{-1}(N)\delta_n(N)$$
$$\sigma_{n+1}(N+1) = \sigma_n(N) - \delta_n(N)\varrho_n^{-1}(N)\delta_n^T(N)$$
$$\gamma_{n+1}(N) = \gamma_n(N) - \beta_n^T(N)\sigma_n^{-1}(N)\beta_n(N)$$
$$d_n(N) = \lambda d_n(N-1) + e_n(N)\beta_n^T(N)/\gamma_n(N)$$
$$\varepsilon_{n+1}(N) = \varepsilon_n(N) - \delta_n^T(N)\sigma_n^{-1}(N)\beta_n(N)$$

The initial values for the recursions above follow from the definitions of the involved quantities and the convention that $P_0(N) = I$:

$$e_0(N) = x(N)$$
$$\beta_0(N) = y(N-1)$$
$$\varrho_0(N) = \lambda\varrho_0(N-1) + y(N)y^T(N); \quad \varrho_0(0) = 0$$
$$\sigma_0(N) = \varrho_0(N-1)$$
$$\gamma_0(N) = 1$$
$$\varepsilon_0(N) = y(N)$$
$$\delta_0(N) = \lambda\delta_0(N-1) + y(N-1)y^T(N)$$
$$d_0(N) = \lambda d_0(N-1) + x(N)y^T(N-1)$$
$$\delta_n(0) = 0, \quad n = 0, \ldots, M$$
$$d_n(0) = 0, \quad n = 0, \ldots, M$$

Next introduce the so-called matrix reflection coefficients,

$$K_{n+1}(N) = d_n(N)\sigma_n^{-1}(N)$$
$$\tilde{K}_{n+1}(N) = \delta_n(N)\varrho_n^{-1}(N)$$
$$\bar{K}_{n+1}(N) = \delta_n^T(N)\sigma_n^{-1}(N)$$

Then the recursions for e, β, and ε can be written as

$$e_{n+1}(N) = e_n(N) - K_{n+1}(N)\beta_n(N)$$
$$\beta_{n+1}(N+1) = \beta_n(N) - \tilde{K}_{n+1}(N)\varepsilon_n(N) \quad \text{(C9.3.11)}$$
$$\varepsilon_{n+1}(N) = \varepsilon_n(N) - \bar{K}_{n+1}(N)\beta_n(N)$$

These equations define a lattice filter for computation of the prediction errors, as depicted in Figure C9.3.1. Note that, as compared to the (scalar) case analyzed in Complement C9.2, the lattice structure of Figure C9.3.1 has one more line which involves the parameters e and K.

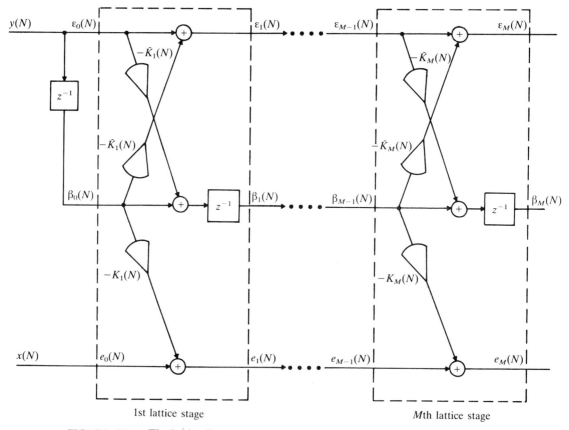

FIGURE C9.3.1 The lattice filter implementation of the fast LS prediction algorithm for multivariable regression models.

Next it will be shown that a good approximation (for large N) of the LS parameter estimates $\{\theta_n(N)\}$ can be obtained from the parameters of the fast lattice predictor introduced above. To see how this can be done, introduce the following notation:

$$C_n^N(q^{-1}) = C_{n,1}^N q^{-1} + \ldots + C_{n,n}^N q^{-n}$$

$$(B_{n,1}^N \ldots B_{n,n}^N) = -\psi_n^T(N)\Lambda_N \Phi_n(N)[\Phi_n^T(N)\Lambda_N \Phi_n(N)]^{-1}$$

$$B_n^N(q^{-1}) = B_{n,1}^N + \ldots + B_{n,n}^N q^{-(n-1)} + q^{-n}I$$

$$(A_{n,1}^N \ldots A_{n,n}^N) = -y_N^T \Lambda_N \Phi_n(N)[\Phi_n^T(N)\Lambda_N \Phi_n(N)]^{-1}$$

$$A_n^N(q^{-1}) = I + A_{n,1}^N q^{-1} + \ldots + A_{n,n}^N q^{-n}$$

Recall from (C9.3.5) that

$$(C_{n,1}^N \ldots C_{n,n}^N) = x_N^T \Lambda_N \Phi_n(N)[\Phi_n^T(N)\Lambda_N \Phi_n(N)]^{-1}$$

Using these definitions and the definitions of e, β and ε, one may write

$$e_n(N) - x(N) - C_n^N(q^{-1})y(N)$$

$$\beta_n(N) = y(N - n - 1) + (B_{n,1}^N \ldots B_{n,n}^N)\begin{pmatrix} y(N-1) \\ \vdots \\ y(N-n) \end{pmatrix} = B_n^N(q^{-1})y(N-1)$$

and

$$\varepsilon_n(N) = y(N) + (A_{n,1}^N \ldots A_{n,n}^N)\begin{pmatrix} y(N-1) \\ \vdots \\ y(N-n) \end{pmatrix} = A_n^N(q^{-1})y(N)$$

Substituting the above expressions for e, β and ε into the lattice recursions (C9.3.11),

$$[C_{n+1}^N(q^{-1}) - C_n^N(q^{-1}) - K_{n+1}(N)q^{-1}B_n^N(q^{-1})]y(N) = 0$$
$$[B_{n+1}^{N+1}(q^{-1}) - q^{-1}B_n^N(q^{-1}) + \tilde{K}_{n+1}(N)A_n^N(q^{-1})]y(N) = 0 \qquad \text{(C9.3.12)}$$
$$[A_{n+1}^N(q^{-1}) - A_n^N(q^{-1}) + \bar{K}_{n+1}(N)q^{-1}B_n^N(q^{-1})]y(N) = 0$$

As the number of data points increases, the predictor parameters will approximately converge. Then, it can be concluded from (C9.3.12) that the following identities will approximately hold (the superscript N is omitted):

$$\boxed{\begin{aligned} C_{n+1}(z) &= C_n(z) + K_{n+1}zB_n(z) \\ B_{n+1}(z) &= zB_n(z) - \tilde{K}_{n+1}A_n(z) \\ A_{n+1}(z) &= A_n(z) - \bar{K}_{n+1}zB_n(z) \end{aligned}} \qquad \text{(C9.3.13)}$$

These recursions initialized by $A_0(z) = B_0(z) = I$, $C_0(z) = 0$, and iterated for $n = 0$ to $M - 1$, can be used to determine $\{\theta_n\}_{n=1}^M$ from the reflection coefficients $\{K_n, \tilde{K}_n$ and

380 *Recursive identification methods* *Chapter 9*

$\bar{K}_n\}$ which are provided by the lattice prediction algorithm. Note that this computation of $\{\theta_n(N)\}$ may be performed when desired (i.e. for some specified values of N). Note also that the hardware implementation of (C9.3.13) may be done using a lattice filter similar to that of Figure C9.3.1.

Chapter 10
IDENTIFICATION OF SYSTEMS OPERATING IN CLOSED LOOP

10.1 Introduction

The introductory examples of Chapter 2 demonstrated that the result of an identification can be very poor for certain experimental conditions. The specification of the experimental condition $\mathscr{X}$ includes such things as the choice of prefiltering, sampling interval, and input generation. This chapter looks at the effect of the feedback from output to input on the results of an identification experiment.

Many systems work under feedback control. This is typical, for example, in the process industry for the production of paper, cement, glass, etc. There are many other systems in non-technical areas where feedback mechanisms normally are in force on the systems, for example in many biological and economical systems. For technical systems the open loop system may be unstable or so poorly damped that no identification experiment can be performed in open loop. Safety and production restrictions can also be strong reasons for not allowing experiments in open loop.

It is very important to know if and how the open loop system can be identified when it must operate under feedback control during the experiment. It will be shown that the feedback can cause difficulties but also that these can be circumvented. Sometimes there are certain practical restrictions on identification experiments that must be met. These can include bounds on the input and output variances. In such situations it can even be an advantage to use a feedback control during the experiment.

Two different topics are considered in this chapter:

- The identifiability properties for systems operating in closed loop will be investigated. Consideration will then be given to the conditions under which it is possible to identify a system under feedback control. Explicit results will be presented on how to proceed when identification is possible. This topic is dealt with in Sections 10.2–10.5.
- The second topic concerns accuracy aspects, which will be dealt with mainly by means of examples. Then an experimental condition that gives optimal accuracy of the identified model is presented. It will turn out that the best experimental condition often includes a feedback control. This is shown in Section 10.6.

10.2 Identifiability considerations

The situation to be discussed in this and the following sections is depicted in Figure 10.1.
The open loop system is assumed to be given by

$$y(t) = G_s(q^{-1})u(t) + H_s(q^{-1})e(t) \qquad Ee(t)e^T(\bar{t}) = \Lambda_s \delta_{t,\bar{t}} \qquad (10.1)$$

where $e(t)$ is white noise. The input $u(t)$ is determined through feedback as

$$u(t) = -F(q^{-1})y(t) + L(q^{-1})v(t) \qquad (10.2a)$$

In (10.2a) the signal $v(t)$ can be a reference value, a setpoint or noise entering the regulator. $F(q^{-1})$ and $L(q^{-1})$ are matrix filters of compatible dimensions (the notations dim $y(t)$ = dim $e(t)$ = ny, dim $u(t)$ = nu, dim $v(t)$ = nv are used in what follows). Most often the goal of identification of the system above is the determination of the filters $G_s(q^{-1})$ and $H_s(q^{-1})$. Sometimes one may also wish to determine the filter $F(q^{-1})$ of the feedback path.

A number of cases will be considered, including:

- The feedback $F(q^{-1})$ may or may not be known.
- The external signal $v(t)$ may or may not be measurable.

Later (see (10.17)), the feedback (10.2a) will be extended to a shift between r different time-invariant regulators:

$$u(t) = -F_i(q^{-1})y(t) + L_i(q^{-1})v(t) \qquad (10.2b)$$

The reason for considering such an extension is that this special form of time-varying regulator will be shown to give identifiability under weak conditions.

For the system (10.1) with the feedback (10.2a) the closed loop system can be shown to be

$$\begin{aligned} y(t) &= [I + G_s(q^{-1})F(q^{-1})]^{-1}[G_s(q^{-1})L(q^{-1})v(t) + H_s(q^{-1})e(t)] \\ u(t) &= [L(q^{-1}) - F(q^{-1})\{I + G_s(q^{-1})F(q^{-1})\}^{-1}G_s(q^{-1})L(q^{-1})]v(t) \\ &\quad - F(q^{-1})\{I + G_s(q^{-1})F(q^{-1})\}^{-1}H_s(q^{-1})e(t) \end{aligned} \qquad (10.3)$$

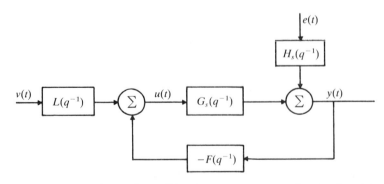

FIGURE 10.1 Experimental condition for a system operating in closed loop.

Section 10.2   *Identifiability considerations*   383

The following general assumptions are made:

- The open loop system is strictly proper (i.e. it does not contain any direct term). This means that $G_s(0) = 0$. This assumption is weak and is introduced to avoid algebraic loops in the closed loop system. (An algebraic loop occurs if neither $G_s(q^{-1})$ nor $F(q^{-1})$ contains a delay. Then $y(t)$ depends on $u(t)$, which in turn depends on $y(t)$. To avoid such a situation it is assumed that the system has a delay so that $y(t)$ depends only on *past* input values.)
- The subsystems from v and e to y of the closed loop system are asymptotically stable and have no unstable hidden modes. This implies that the filters $L(q^{-1})$, $H_s(q^{-1})$ and $[I + G_s(q^{-1})F(q^{-1})]^{-1}$ are asymptotically stable.
- The external signal $v(t)$ is stationary and persistently exciting of a sufficient order. What 'sufficient' means will depend on the system. Note that it is not required that the signal $w(t) \triangleq L(q^{-1})v(t)$ is persistently exciting. For example, if $v(t)$ is scalar and the filter $L(q^{-1})$ is chosen to have zeros on the unit circle that exactly match the frequencies for which $\phi_v(\omega)$ is nonzero, then $w(t)$ will be persistently exciting of a lower order than $v(t)$ (see Section 5.4). It is convenient in the analysis which follows to assume that $v(t)$ is persistently exciting, and to allow $L(q^{-1})$ to have an arbitrary form.
- The external signal $v(t)$ and the disturbance $e(s)$ are independent for all t and s.

Spectral analysis

The following two general examples illustrate that a straightforward use of spectral analysis will not give identifiability.

Example 10.1 *Application of spectral analysis*
Consider a SISO system ($nu = nv = ny = 1$) as in Figure 10.1. Assume that $L(q^{-1}) \equiv 1$. For convenience, introduce the signal

$$z(t) = F(q^{-1})H_s(q^{-1})e(t) \tag{10.4a}$$

From (10.3) the following descriptions of the input and output signals are obtained:

$$y(t) = \frac{1}{1 + G_s(q^{-1})F(q^{-1})} \left[G_s(q^{-1})v(t) + \frac{1}{F(q^{-1})}z(t) \right] \tag{10.4b}$$

$$u(t) = \frac{1}{1 + G_s(q^{-1})F(q^{-1})}[v(t) - z(t)] \tag{10.4c}$$

Hence

$$\phi_u(\omega) = \frac{1}{|1 + G_s(e^{i\omega})F(e^{i\omega})|^2}[\phi_v(\omega) + \phi_z(\omega)] \tag{10.4d}$$

$$\phi_{yu}(\omega) = \frac{1}{|1 + G_s(e^{i\omega})F(e^{i\omega})|^2}\left[G_s(e^{-i\omega})\phi_v(\omega) - \frac{1}{F(e^{-i\omega})}\phi_z(\omega)\right] \tag{10.4e}$$

Assuming that the spectral densities $\phi_u(\omega)$ and $\phi_{yu}(\omega)$ can be estimated exactly, which might be true at least asymptotically as the number of data points tends to infinity, it is found that the spectral analysis estimate of $G_s(e^{-i\omega})$ is given by

$$\hat{G}(e^{-i\omega}) = \frac{\phi_{yu}(\omega)}{\phi_u(\omega)} = \frac{G_s(e^{-i\omega})\phi_v(\omega) - \phi_z(\omega)/F(e^{-i\omega})}{\phi_v(\omega) + \phi_z(\omega)} \quad (10.4f)$$

(cf. (3.33)). If there are no disturbances then $z(t) \equiv 0$, $\phi_z(\omega) = 0$ and (10.4f) simplifies to

$$\hat{G}(e^{-i\omega}) = G_s(e^{-i\omega}) \quad (10.4g)$$

i.e. it is possible to identify the true system dynamics.

However, in the other extreme case when there is no external input ($v(t) \equiv 0$), $\phi_v(\omega) = 0$ and (10.4f) becomes

$$\hat{G}(e^{-i\omega}) = -\frac{1}{F(e^{-i\omega})} \quad (10.4h)$$

Here the result is the negative inverse of the feedback. In the general case it follows from (10.4f) that

$$\hat{G}(e^{-i\omega}) - G_s(e^{-i\omega}) = -\frac{1 + G_s(e^{-i\omega})F(e^{-i\omega})}{F(e^{-i\omega})} \frac{\phi_z(\omega)}{\phi_v(\omega) + \phi_z(\omega)} \quad (10.4i)$$

which shows how the spectral densities $\phi_v(\omega)$ and $\phi_z(\omega)$ influence the deviation of $\hat{G}(e^{-i\omega})$ from the true value $G_s(e^{-i\omega})$.

For the special case when $F(q^{-1}) \equiv 0$, equation (10.4i) cannot be used as it stands. However, from (10.4a),

$$\phi_z(\omega) = |F(e^{i\omega})|^2 |H_s(e^{i\omega})|^2 \phi_e(\omega) \quad (10.4j)$$

and (10.4i) gives for this special case

$$\hat{G}(e^{-i\omega}) - G_s(e^{-i\omega}) = -[1 + G_s(e^{-i\omega}) \times 0]F(e^{i\omega}) \frac{|H_s(e^{i\omega})|^2 \phi_e(\omega)}{\phi_v(\omega)} \bigg|_{F \equiv 0}$$

$$= 0 \quad (10.4k)$$

This means that for open loop operation ($F(q^{-1}) \equiv 0$) the system is identifiable (cf. Section 3.5).

To summarize, the spectral analysis applied in the usual way gives biased estimates if there is a feedback acting on the system. ■

The next example extends the examination of spectral analysis to the multivariable case.

Example 10.2 *Application of spectral analysis in the multivariable case*
Assume first that $ny = nu$. Then $F(q^{-1})$ is a square matrix. Further, let $v(t) \equiv 0$. Since $u(t)$ is determined as

$$u(t) = -F(q^{-1})y(t) \quad (10.5)$$

the spectral densities satisfy

$$\phi_u(\omega) = F(e^{-i\omega})\phi_y(\omega)F^T(e^{i\omega})$$

$$\phi_{yu}(\omega) = -\phi_y(\omega)F^T(e^{i\omega}) \quad (10.6)$$

Section 10.2 *Identifiability considerations* 385

When using spectral analysis for estimation of $G_s(q^{-1})$ one determines

$$\hat{G}(e^{-i\omega}) = \hat{\phi}_{yu}(\omega)\hat{\phi}_u^{-1}(\omega) \tag{10.7}$$

(see (3.33)). Assuming that the spectral density estimates $\hat{\phi}_{yu}$ and $\hat{\phi}_u$ are exact, (10.6), (10.7) give

$$\begin{aligned}\hat{G}(e^{-i\omega}) &= [-\phi_y(\omega)F^T(e^{i\omega})][F(e^{-i\omega})\phi_y(\omega)F^T(e^{i\omega})]^{-1} \\ &= -F^{-1}(e^{-i\omega})\end{aligned} \tag{10.8}$$

This result is a generalization of (10.4h).

Consider next the more general case where ny may differ from nu, and $v(t)$ can be nonzero. In such a case the estimate of G will no longer be given by (10.8), but it will still differ from the true system. To see this, make the following evaluation. Set

$$\begin{aligned}e_y(t) &= H_s(q^{-1})e(t) \\ e_u(t) &= -F(q^{-1})\{I + G_s(q^{-1})F(q^{-1})\}^{-1}H_s(q^{-1})e(t)\end{aligned} \tag{10.9a}$$

These are the parts of $y(t)$ and $u(t)$, respectively, that depend on the disturbances $e(\cdot)$, see (10.3). Since

$$y(t) - G_s(q^{-1})u(t) = H_s(q^{-1})e(t) = e_y(t)$$

one gets

$$\phi_{yu}(\omega) - G_s(e^{-i\omega})\phi_u(\omega) = \phi_{e_y,u}(\omega) = \phi_{e_y,e_u}(\omega) \tag{10.9b}$$

This expression must be zero if the estimate (10.7) is to be consistent. It is easily seen that the expression is zero if $F(q^{-1}) \equiv 0$ (i.e. no feedback), or expressed in other terms, if the disturbance $H_s(q^{-1})e(t)$ and the input $u(t)$ are uncorrelated. In the general case of $F(q^{-1}) \neq 0$, however, (10.9b) will be different from zero. This shows that the spectral analysis fails to give identifiability of the open loop system transfer function.

Sometimes, however, a *modified spectral analysis* can be used. Assume that the external input signal $v(t)$ is measurable. A simple calculation gives, (cf. (10.1))

$$\begin{aligned}\phi_{yv}(\omega) &= G_s(e^{-i\omega})\phi_{uv}(\omega) + H_s(e^{-i\omega})\phi_{ev}(\omega) \\ &= G_s(e^{-i\omega})\phi_{uv}(\omega)\end{aligned} \tag{10.10}$$

If $v(t)$ has the same dimension as $u(t)$ (i.e. $nv = nu$) $G_s(q^{-1})$ can therefore be estimated by

$$\boxed{\hat{G}(e^{-i\omega}) = \hat{\phi}_{yv}(\omega)\hat{\phi}_{uv}^{-1}(\omega)} \tag{10.11}$$

This estimate of $G_s(e^{-i\omega})$ will work, unlike (10.7). It can be seen as a natural extension of the 'open loop formula' (10.7) to the case of systems operating under feedback. ∎

It is easy to understand the reason for the difficulties encountered when applying spectral analysis as in Examples 10.1 and 10.2. The model (10.8) provided by the method, that is

$$y(t) = -F^{-1}(q^{-1})u(t)$$

gives a valid description of the relation between the signals $u(t)$ and $y(t)$. This relation corresponds to the *inverse* of the feedback law $y(t) \to u(t)$ (see (10.5)). Note that:

(i) The feedback path is noise-free, while the relation of interest $u(t) \to y(t)$ corresponding to the direct path is corrupted by noise.
(ii) Within spectral analysis, the noise part of the output is not modeled.
(iii) The nonparametric model used by the spectral analysis method by its very definition has no structural restrictions. Hence it cannot eliminate certain true but uninteresting relationships between $u(t)$ and $y(t)$ (such as the inverse feedback law model (10.8)).

The motivation for the results obtained by the spectral analysis identification method lies in the facts noted above.

The situation should be different if a parametric model is used. As will be shown later in this chapter, the system will be identifiable under weak conditions if a parametric identification method is used. Of the methods considered in the book, the prediction error method shows most promise since the construction of IV methods normally is based on the assumption of open loop experiments. (See, however, the next subsection.)

Instrumental variable methods

The IV methods can be extended to closed loop systems with a *measurable external input*. A similar extension of the spectral analysis was presented in Example 10.2. This extension of IV methods is illustrated in the following general example.

Example 10.3 *IV method for closed loop systems*
Consider the scalar system

$$A(q^{-1})y(t) = B(q^{-1})u(t) + w(t) \tag{10.12}$$

where the input $u(t)$ is determined through feedback,

$$R(q^{-1})u(t) = -S(q^{-1})y(t) + T(q^{-1})v(t) \tag{10.13}$$

in which $v(t)$ is a *measurable* external input that is independent of the disturbance $w(s)$ for all t and s. In (10.12), (10.13) $A(q^{-1})$, $B(q^{-1})$, $R(q^{-1})$, etc., are polynomials in q^{-1}. Comparing with the general description (10.1), (10.2a), it is seen that in this case

$$G_s(q^{-1}) = \frac{B(q^{-1})}{A(q^{-1})}$$

$$F(q^{-1}) = \frac{S(q^{-1})}{R(q^{-1})}$$

$$L(q^{-1}) = \frac{T(q^{-1})}{R(q^{-1})}$$

One can apply IV estimators to (10.12) as in Chapter 8. *The IV vector $Z(t)$ will now consist of filtered and delayed values of the external input $v(t)$.* Then the second consistency condition (8.25b) ($EZ(t)w(t) = 0$) is automatically satisfied. Fulfilment of the first consistency condition, rank$[EZ(t)\tilde{\phi}^T(t)] = \dim \theta$, will depend on the system, the experimental condition and the instruments used. As in the open loop case, this condition will be generically satisfied under weak conditions. The 'noise-free' vector $\tilde{\phi}(t)$ is in this case the part of $\phi(t)$ that depends on the external input. To be more exact, note that from (10.12) and (10.13) the following description for the closed loop system is obtained:

$$[A(q^{-1})R(q^{-1}) + B(q^{-1})S(q^{-1})]y(t) = B(q^{-1})T(q^{-1})v(t) + R(q^{-1})w(t)$$
$$[A(q^{-1})R(q^{-1}) + B(q^{-1})S(q^{-1})]u(t) = A(q^{-1})T(q^{-1})v(t) - S(q^{-1})w(t) \quad (10.14)$$

which implies that

$$\phi(t) = (-y(t-1) \ldots -y(t-na) \quad u(t-1) \ldots u(t-nb))^T \quad (10.15a)$$

$$\tilde{\phi}(t) = (-\tilde{y}(t-1) \ldots -\tilde{y}(t-na) \quad \tilde{u}(t-1) \ldots \tilde{u}(t-nb))^T \quad (10.15b)$$

$$\tilde{u}(t) = \frac{A(q^{-1})T(q^{-1})}{A(q^{-1})R(q^{-1}) + B(q^{-1})S(q^{-1})} v(t) \quad (10.15c)$$

$$\tilde{y}(t) = \frac{B(q^{-1})T(q^{-1})}{A(q^{-1})R(q^{-1}) + B(q^{-1})S(q^{-1})} v(t) = \frac{B(q^{-1})}{A(q^{-1})} \tilde{u}(t) \quad (10.15d)$$

It is also possible to extend the analysis of optimal IV methods developed in Chapter 8 to this situation. With $\tilde{\phi}(t)$ defined as in (10.15b) the results given in Chapter 8 extend to closed loop operation. Note that in order to apply the optimal IV it is necessary to know not only the true system but also the regulator, since R, S, and T appear in (10.15b–d).

∎

Use of a prediction error method

In the following it will generally be assumed that a prediction error method (PEM) is used. In most cases it is not necessary to assume that the external input is measurable, which makes a PEM more attractive than an IV method. A further advantage is that a PEM gives statistically efficient estimates under mild conditions. The disadvantage from a practical point of view is that a PEM is more computationally demanding than an IV method.

As a simple illustration of the usefulness of the PEM approach to closed loop system identification, recall Example 2.7, where a simple PEM was applied to a first-order system. It was shown that the appropriate requirement on the experimental condition in order to guarantee identifiability is

$$E \begin{pmatrix} -y(t) \\ u(t) \end{pmatrix} (-y(t) \quad u(t)) > 0 \quad (10.16)$$

This condition is violated if and only if $u(t)$ is not persistently exciting of order 1 or if $u(t)$ is determined as a static (i.e. zero order) linear output feedback, $u(t) = -ky(t)$. It is easy to see that for a proportional regulator the matrix in (10.16) will be singular (i.e. positive semidefinite). Conversely, to make the matrix singular the signals $y(t)$ and $u(t)$ must be linearly dependent, which implies that $u(t)$ is proportional to $y(t)$. For a linear higher-order or a nonlinear feedback the condition (10.16) will be satisfied and the PEM will give consistent estimates of the open loop system parameters.

In the following a generalization will be made of the experimental condition (10.2a). As already stated a certain form of time-varying regulator will be allowed. Assume that during the experiment, r different constant regulators are used such that

$$u(t) = -F_i(q^{-1})y(t) + L_i(q^{-1})v(t) \qquad i = 1, \ldots, r \qquad (10.17)$$

during a proportion γ_i of the total experiment time. Then

$$\gamma_i \geq 0 \qquad i = 1, \ldots, r$$

$$\sum_{i=1}^{r} \gamma_i = 1 \qquad (10.18)$$

For example, if one regulator is used for 30 percent of the total experiment time and a second one for the remaining 70 percent, then $\gamma_1 = 0.3$, $\gamma_2 = 0.7$.

When dealing with parametric methods, the following model structure will be used (cf. (6.1)):

$$y(t) = G(q^{-1}; \theta)u(t) + H(q^{-1}; \theta)e(t) \qquad Ee(t)e^T(s) = \Lambda(\theta)\delta_{t,s} \qquad (10.19a)$$

Assume that $G(q^{-1}; \theta)$ is strictly proper, i.e. that $G(0; \theta) = 0$ for all θ. Also, assume that $H(0; \theta) = I$ for all θ. Equation (10.19a) will be abbreviated as

$$y(t) = \hat{G}u(t) + \hat{H}e(t) \qquad Ee(t)e^T(t) = \hat{\Lambda} \qquad (10.19b)$$

For convenience the subscript s will be omitted in the true system description in the following calculations. Note that the model parametrization in (10.19a) will not be specified. This means that we will deal with system identifiability (SI) (cf. Section 6.4). We will thus be satisfied if the identification method used is such that

$$\hat{G} \equiv G \qquad \text{(i.e. } G(q^{-1}; \theta) \equiv G_s(q^{-1})\text{)}, \qquad \hat{H} \equiv H \qquad (10.20)$$

It is then a matter of the parametrization of the model only, and not of the experimental condition, whether there is a *unique* parameter vector θ which satisfies (10.20). In Chapter 6 (see (6.44)) the set $D_T(\mathscr{S}, \mathscr{M})$ was introduced to describe the parameter vectors satisfying (10.20). If $D_T(\mathscr{S}, \mathscr{M})$ consists of exactly one point then the system is parameter identifiable (PI) (cf. Section 6.4).

In the following sections three different approaches to identifying a system working in closed loop will be analyzed:

- *Direct identification*. The existence of possible feedback is neglected and the recorded data are treated as if the system were operating in open loop.
- *Indirect identification*. It is assumed that the external setpoint $v(t)$ is measurable and that the feedback law is known. First the closed loop system is identified regarding $v(t)$

as the input. Then the open loop system is determined from the known regulator and the identified closed loop system.
- *Joint input–output identification.* The recorded data $u(t)$ and $y(t)$ are regarded as outputs of a multivariable system driven by white noise, i.e. as a multivariable $(ny + nu)$-dimensional time series. This multivariable system is identified using the original parameters as unknowns.

For all the approaches above it will be assumed that a PEM is used for parameter estimation.

10.3 Direct identification

With this approach the recorded values of $\{u(t)\}$ and $\{y(t)\}$ are used in the estimation scheme for finding θ in (10.19a) as if no feedback were present. This is of course an attractive approach if it works, since one does not have to bother about the possible presence of feedback.

To analyze the identifiability properties, first note that the closed loop system corresponding to the ith feedback regulator (10.17) is described by the following equations (cf. (10.3)):

$$y_i(t) = (I + GF_i)^{-1}(GL_i v(t) + He(t))$$
$$u_i(t) = [I - F_i(I + GF_i)^{-1}G]L_i v(t) - F_i(I + GF_i)^{-1}He(t)$$
(10.21)

The corresponding prediction error of the direct path model is given by

$$\varepsilon_i(t) = \hat{H}^{-1}[y_i(t) - \hat{G}u_i(t)]$$
(10.22)

(see (10.19b)). Let I_i denote the time interval(s) when the feedback (10.17) is used. I_i may consist of a union of disjoint time intervals. The length of I_i is $\gamma_i N$. The asymptotic loss function associated with the PEM (see Chapter 7) is given by

$$V = h(R_\infty(\theta))$$

$$R_\infty(\theta) = \lim_{N \to \infty} \frac{1}{N} \sum_{t=1}^{N} E\varepsilon(t)\varepsilon^T(t)$$

where h is a scalar increasing function such as tr or det. Note that if $\varepsilon(t)$ is a stationary process then R_∞ reduces to $R_\infty = E\varepsilon(t)\varepsilon^T(t)$, which was the case dealt with in Chapter 7. In the present case $\varepsilon(t)$ is nonstationary. However, for $t \in I_i$, $\varepsilon(t) = \varepsilon_i(t)$ which is stationary. Thus

$$R_\infty = \lim_{N \to \infty} \frac{1}{N} \sum_{t=1}^{N} E\varepsilon(t)\varepsilon^T(t)$$

$$= \lim_{N \to \infty} \frac{1}{N} \sum_{i=1}^{r} \sum_{t \in I_i} E\varepsilon_i(t)\varepsilon_i^T(t)$$

$$= \lim_{N\to\infty} \frac{1}{N} \sum_{i=1}^{r} \gamma_i N \, E\varepsilon_i(t)\varepsilon_i^T(t)$$

$$= \sum_{i=1}^{r} \gamma_i E\varepsilon_i(t)\varepsilon_i^T(t) \tag{10.23}$$

Recall that as N tends to infinity, the prediction error estimates tend to a global minimum point of the asymptotic loss function $h(R_\infty)$. Thus, to investigate the identifiability (in particular, the consistency) properties of the PEM, one studies the global minima of $h(R_\infty)$.

To see if the direct approach makes sense, consider a simple example.

Example 10.4 *A first-order model*

Let the system be given by

$$y(t) + ay(t-1) = bu(t-1) + e(t) \qquad Ee^2(t) = \lambda^2$$

$e(t)$ being white noise, and the model structure by

$$y(t) + \hat{a}y(t-1) = \hat{b}u(t-1) + \varepsilon(t)$$

(cf. Example 2.8). The input is assumed to be determined from a time-varying proportional regulator,

$$u(t) = \begin{cases} -f_1 y(t) & \text{for a proportion } \gamma_1 \text{ of the total experiment} \\ -f_2 y(t) & \text{for a proportion } \gamma_2 \text{ of the total experiment} \end{cases}$$

Then

$$y_i(t) + (a + bf_i)y_i(t-1) = e(t)$$

$$\varepsilon_i(t) = y_i(t) + (\hat{a} + \hat{b}f_i)y_i(t-1)$$

which gives

$$E\varepsilon_i^2(t) = \lambda^2 \frac{1 + (\hat{a} + \hat{b}f_i)^2 - 2(\hat{a} + \hat{b}f_i)(a + bf_i)}{1 - (a + bf_i)^2}$$

$$= \lambda^2 \left[1 + \frac{(\hat{a} + \hat{b}f_i - a - bf_i)^2}{1 - (a + bf_i)^2} \right]$$

The loss function V becomes

$$V(\hat{a}, \hat{b}) = R_\infty = \lambda^2 + \gamma_1 \lambda^2 \frac{(\hat{a} + \hat{b}f_1 - a - bf_1)^2}{1 - (a + bf_1)^2}$$

$$+ \gamma_2 \lambda^2 \frac{(\hat{a} + \hat{b}f_2 - a - bf_2)^2}{1 - (a + bf_2)^2} \tag{10.24a}$$

It is easy to see that

$$V(\hat{a}, \hat{b}) \geq \lambda^2 = V(a, b) \tag{10.24b}$$

As shown in Section 7.5, the limits of the parameter estimates are the minimum points of the loss function $V(\hat{a}, \hat{b})$. It can be seen from (10.24) that $\hat{a} = a$, $\hat{b} = b$ is certainly a minimum point. To examine whether $\hat{a} = a$, $\hat{b} = b$ is a *unique* minimum, it is necessary to solve the equation $V(\hat{a}, \hat{b}) = \lambda^2$ with respect to $\hat{a}$ and $\hat{b}$. Then from (10.24a),

$$\hat{a} + \hat{b}f_1 - a - bf_1 = 0$$
$$\hat{a} + \hat{b}f_2 - a - bf_2 = 0 \tag{10.24c}$$

or

$$\begin{pmatrix} 1 & f_1 \\ 1 & f_2 \end{pmatrix} \begin{pmatrix} \hat{a} \\ \hat{b} \end{pmatrix} = \begin{pmatrix} a + bf_1 \\ a + bf_2 \end{pmatrix} \tag{10.24d}$$

which has a unique solution if and only if $f_1 \neq f_2$. The use of two different constant regulators is therefore sufficient in this case to give parameter identifiability.

The above result can also be derived in the following way. Let

$$V_i = \gamma_i E\varepsilon_i^2(t) \qquad i = 1, 2 \tag{10.25a}$$

Then

$$V_i = \gamma_i \lambda^2 \left[1 + \frac{(\hat{a} + \hat{b}f_i - a - bf_i)^2}{1 - (a + bf_i)^2} \right] \tag{10.25b}$$

and

$$V = V_1 + V_2 \tag{10.25c}$$

Now V_1 does not have a unique minimum point. It is minimized for all points on the line (in the $\hat{a}$, $\hat{b}$ space) given by

$$\hat{a} + \hat{b}f_1 = a + bf_1 \tag{10.25d}$$

The true parameters $\hat{a} = a$, $\hat{b} = b$ are a point on this line. Similarly, the minimum points of V_2 are situated on the line

$$\hat{a} + \hat{b}f_2 = a + bf_2 \tag{10.25e}$$

The intersection of these two lines will thus give the minimum point of the total loss function V and is the solution to (10.24d). The condition $f_1 \neq f_2$ is necessary to get an intersection. ∎

Return now to the general case. It is not difficult to see that (using $G(0) = \hat{G}(0) = 0$, $H(0) = \hat{H}(0) = I$)

$y_i(t) = e(t)$ + term independent of $e(t)$

$u_i(t)$ is independent of $e(s)$ for $t < s$

$\varepsilon_i(t) = e(t)$ + term independent of $e(t)$

Using these results a lower bound on the matrix R_∞ (10.23) can be derived:

$$R_\infty = \sum_{i=1}^{r} \gamma_i E\varepsilon_i(t)\varepsilon_i^T(t)$$

$$\geq \sum_{i=1}^{r} \gamma_i Ee(t)e^T(t) = Ee(t)e^T(t) = \Lambda$$

If equality can be achieved then this will characterize the (global) minimum points of any suitable scalar function $V = h(R_\infty)$.

To have equality it is required that

$$\varepsilon_i(t) = e(t) \qquad i = 1, \ldots, r \tag{10.26}$$

From (10.21) and (10.22) it follows that

$$\begin{aligned}\varepsilon_i(t) &= \hat{H}^{-1}(I + GF_i)^{-1}(GL_i v(t) + He(t)) \\ &\quad - \hat{H}^{-1}\hat{G}[I - F_i(I + GF_i)^{-1}G]L_i v(t) + \hat{H}^{-1}\hat{G}F_i(I + GF_i)^{-1}He(t) \\ &= [\hat{H}^{-1}(I + GF_i)^{-1}GL_i - \hat{H}^{-1}\hat{G}\{I - F_i(I + GF_i)^{-1}G\}L_i]v(t) \\ &\quad + [\hat{H}^{-1}(I + GF_i)^{-1}H + \hat{H}^{-1}\hat{G}F_i(I + GF_i)^{-1}H]e(t)\end{aligned}$$

Since the disturbances $e(t)$ are assumed to be independent of $v(s)$ for all t and s, it follows that (10.26) holds if and only if

$$\hat{H}^{-1}[(I + GF_i)^{-1}G$$
$$- \hat{G}\{I - F_i(I + GF_i)^{-1}G\}]L_i v(t) \equiv 0 \tag{10.27}$$
$$\hat{H}^{-1}(I + \hat{G}F_i)(I + GF_i)^{-1}H \equiv I \qquad i = 1, \ldots, r \tag{10.28}$$

These two conditions describe the identifiability properties of the direct approach. Note that (10.27), (10.28) are satisfied by the 'desired solution' $\hat{G} = G$, $\hat{H} = H$. The crucial problem is, of course, whether other solutions exist or not. The special cases examined in the following two examples will give useful insights into the general identifiability properties.

Example 10.5 *No external input*
Assume that $v(t) \equiv 0$ and $r = 1$. Then (10.27) carries no information and (10.28) gives

$$\hat{H}^{-1}(I + \hat{G}F_1)(I + GF_1)^{-1}H \equiv I \tag{10.29}$$

However, this relation is not sufficient to conclude (10.20), i.e. to get identifiability. It depends on the parametrization and the regulator whether or not

$$\hat{G} \equiv G \qquad \hat{H} \equiv H$$

are obtained as a unique solution to (10.29). Compare also Example 2.7 where the identity (10.29) is solved explicitly with respect to the parameter vector θ.

The identity (10.29) can be rewritten as

$$\hat{H}^{-1}(I + \hat{G}F_1) \equiv H^{-1}(I + GF_1)$$

If the regulator F_1 has low order, then the transfer function $\hat{H}^{-1}(I + \hat{G}F_1)$ may also be of low order and the identity above may then give too few equations to determine the parameter vector uniquely. Recall the discussion around (10.16) where it was also seen, in a different context, that a low-order regulator can create identifiability problems.

On the other hand, if F_1 has a sufficiently high order, then the identity above will lead to sufficient equations to determine θ uniquely. Needless to say, it is not known *a priori* what exactly a 'sufficiently high' order means.

Complement C10.1 analyzes in detail the identifiability properties of an ARMAX system operating under feedback with no external input. ∎

Example 10.6 *External input*
Assume that $r = 1$ and $L_1(q^{-1})v(t)$ is persistently exciting (pe) of a sufficiently high order. Set

$$M = (I + GF_1)^{-1}G - \hat{G}\{I - F_1(I + GF_1)^{-1}G\}$$

Then (10.27) becomes

$$\hat{H}^{-1}ML_1v(t) = 0$$

Since $L_1v(t)$ is pe it follows from Property 4 of pe signals (Section 5.4) that $\hat{H}^{-1}M \equiv 0$, which implies that $M \equiv 0$. Now use the fact that

$$(I + GF_1)^{-1}G \equiv G(I + F_1G)^{-1}$$

Then the identity $M \equiv 0$ can be rewritten as

$$G(I + F_1G)^{-1} - \hat{G}\{I - F_1G(I + F_1G)^{-1}\} \equiv 0$$

or

$$[G - \hat{G}\{(I + F_1G) - F_1G\}](I + F_1G)^{-1} \equiv 0$$

which implies

$$\hat{G} \equiv G$$

Then (10.28) gives

$$\hat{H}^{-1}H \equiv I$$

or

$$\hat{H} \equiv H$$

For this case (where use is made of an external persistently exciting signal) identifiability is obtained despite the presence of feedback. ∎

Returning to the general case, it is necessary to analyze (10.27), (10.28) for $i = 1, \ldots, r$. It can be assumed that $v(t)$ is persistently exciting of an arbitrary order. The reason is that $L_i(q^{-1})$ can be used to describe any possible nonpersistent excitation property of the signal $L_i(q^{-1})v(t)$ injected in the loop. Then $v(t)$ in (10.27) can be omitted (cf. Property 4 of Section 5.4). Using again the relations

$$(I + GF_i)^{-1}G = G(I + F_iG)^{-1}$$
$$I - F_i(I + GF_i)^{-1}G = I - F_iG(I + F_iG)^{-1} = (I + F_iG)^{-1}$$

(10.27) can be rewritten as

$$(\hat{H}^{-1} - H^{-1})G(I + F_iG)^{-1}L_i - (\hat{H}^{-1}\hat{G} - H^{-1}G)(I + F_iG)^{-1}L_i \equiv 0 \quad (10.30)$$

Similarly, (10.28) can be written as

$$(\hat{H}^{-1} - H^{-1})(I + GF_i)^{-1}H + (\hat{H}^{-1}\hat{G} - H^{-1}G)F_i(I + GF_i)^{-1}H \equiv 0$$

or

$$(\hat{H}^{-1} - H^{-1}) + (\hat{H}^{-1}\hat{G} - H^{-1}G)F_i \equiv 0 \quad (10.31)$$

Next write (10.30), (10.31) as

$$\begin{aligned}
0 &\equiv (\hat{H}^{-1} - H^{-1} \quad \hat{H}^{-1}\hat{G} - H^{-1}G)\begin{pmatrix} I & G(I + F_iG)^{-1}L_i \\ F_i & -(I + F_iG)^{-1}L_i \end{pmatrix} \\
&\equiv (\hat{H}^{-1} - H^{-1} \quad \hat{H}^{-1}\hat{G} - H^{-1}G)\begin{pmatrix} I & 0 \\ F_i & L_i \end{pmatrix}\begin{pmatrix} I & G(I + F_iG)^{-1}L_i \\ 0 & -I \end{pmatrix}
\end{aligned} \quad (10.32)$$

The last matrix appearing in (10.32) is clearly nonsingular. Thus considering simultaneously all the r parts of the experiment,

$$0 \equiv (\hat{H}^{-1} - H^{-1} \quad \hat{H}^{-1}\hat{G} - H^{-1}G)\begin{pmatrix} I & \dots & I & 0 & \dots & 0 \\ F_1 & \dots & F_r & L_1 & \dots & L_r \end{pmatrix} \quad (10.33)$$

Hence if the matrix

$$\mathscr{F} \triangleq \begin{pmatrix} I & \dots & I & 0 & \dots & 0 \\ F_1 & \dots & F_r & L_1 & \dots & L_r \end{pmatrix} \quad (10.34)$$

has full (row) rank $= nu + ny$ almost everywhere, then

$$(\hat{H}^{-1} - H^{-1} \quad \hat{H}^{-1}\hat{G} - H^{-1}G) \equiv (0 \quad 0) \quad \text{almost everywhere}$$

which trivially gives the desired relation (10.20). This rank condition is fairly mild. Recall that the block matrices in $\mathscr{F}$ have the following dimensions:

I has dimension $(ny|ny)$
F_i has dimension $(nu|ny)$
0 has dimension $(ny|nv)$
L_i has dimension $(nu|nv)$

Since F_i and L_i in $\mathscr{F}$ are filters in the operator q^{-1}, the rank condition above should be explained. Regard for the moment q^{-1} as a complex variable. Then we require that rank $\mathscr{F} = nu + ny$ holds for almost every q^{-1} (as already stated immediately following (10.34)).

Since the $\mathcal{F}$ matrix is of dimension $((ny + nu)|(r \cdot ny + r \cdot nv))$, the following is a necessary condition for the rank requirement to hold:

$$ny + nu \le r(ny + nv) \tag{10.35}$$

This gives a lower bound on the number r of different regulators which can guarantee identifiability. If in particular $nu = nv$, then $r \ge 1$. The case $nu = nv, r = 1$ was examined in Example 10.6. Another special case is when $nv = 0$ (no additional external input). Then $r \ge 1 + nu/ny$. If in particular $ny = nu$ (as in Example 10.4), then $r \ge 2$, and it is thus necessary and in fact also sufficient (see (10.34)) to use two different proportional regulators.

The identifiability results for the direct approach can be summarized as follows:

- Identifiability cannot be guaranteed if the input is determined through a noise-free linear low-order feedback from the output.
- Identifiability can be obtained by using a high-order noise-free linear feedback. The order required will, however, depend on the order of the true (and unknown) system. See Complement C10.1 for more details.
- Simple ways to achieve identifiability are: (1) to use an additional (external) input such as a time-varying setpoint, and/or (2) to use a regulator that shifts between different settings during the identification experiment. The necessary number of settings depends only on the dimensions of the input, the output and the external input.

10.4 Indirect identification

For the indirect identification approach, it must be assumed that $v(t)$ is measurable. This approach consists of two steps:

Step 1. Identify the closed loop system using $v(t)$ as input and $y(t)$ as output.
Step 2. Determine the open loop system parameters from the closed loop model obtained in step 1, using the knowledge of the feedback.

According to (10.21) the closed loop system is given by

$$y_i(t) = \bar{G}_i(q^{-1})v(t) + \bar{H}_i(q^{-1})e(t) \tag{10.36}$$

where

$$\begin{aligned} \bar{G}_i(q^{-1}) &\triangleq (I + GF_i)^{-1} GL_i \\ \bar{H}_i(q^{-1}) &\triangleq (I + GF_i)^{-1} H \end{aligned} \quad i = 1, \ldots, r \tag{10.37}$$

The parameters of the system (10.36) can be estimated using a PEM (or any other parametric method giving consistent estimates) as in the open loop case, since $v(t)$ is assumed to be persistently exciting and independent of $e(t)$.

If $\bar{G}_i$ and $\bar{H}_i$ are parametrized by the original parameters of G and H (L_i and F_i in the expressions of $\bar{G}_i$ and $\bar{H}_i$ are known), then the second step above is not needed since the parameters of interest are estimated directly. In the general case, however, when unconstrained 'standard' parametrizations are used for $\bar{G}_i$ and $\bar{H}_i$, step 2 becomes necessary. It is not difficult to see that both these ways to parametrize $\bar{G}_i$ and $\bar{H}_i$ lead to estimation methods with identical identifiability properties. In the following, for conciseness, it is assumed that the two-step procedure above is used. Therefore assume that, as a result of step 1, estimates $\hat{\bar{G}}_i$ and $\hat{\bar{H}}_i$ of $\bar{G}_i$ and $\bar{H}_i$ are known. In step 2 the equations

$$(I + \hat{G}F_i)^{-1}\hat{G}L_i \equiv \hat{\bar{G}}_i$$
$$(I + \hat{G}F_i)^{-1}\hat{H} \equiv \hat{\bar{H}}_i \qquad i = 1, \ldots, r \tag{10.38}$$

must be solved with respect to the vector θ which parametrizes $\hat{G}$ and $\hat{H}$.

Assume that the estimates of step 1 are exact, so that $\hat{\bar{G}}_i = \bar{G}_i$, $\hat{\bar{H}}_i = \bar{H}_i$. In view of (10.37) the second part of (10.38) becomes

$$\hat{H}^{-1}(I + \hat{G}F_i) \equiv \bar{H}_i^{-1} \equiv H^{-1}(I + GF_i) \tag{10.39}$$

which obviously is equivalent to (10.28). Consider next the first part of (10.38). This equation can also be written

$$\hat{G}L_i \equiv (I + \hat{G}F_i)\bar{G}_i \equiv (I + \hat{G}F_i)G(I + F_iG)^{-1}L_i$$

or

$$[\hat{G}\{I - F_iG(I + F_iG)^{-1}\} - G(I + F_iG)^{-1}]L_i \equiv 0 \tag{10.40}$$

which is equivalent to (10.27) for persistently exciting $v(t)$. Thus *the same identifiability relations*, namely (10.27), (10.28), are obtained *as for the direct identification*. In particular, the analysis carried out for the direct approach, leading to the rank requirement on $\mathcal{F}$ in (10.34) is still perfectly valid. This equivalence between the direct and the indirect approaches must be interpreted appropriately, however. The identifiability properties are the same, but this does not mean that both methods give the same result in the finite sample case or that they are equally easy to apply. An advantage of the direct approach is that only one step is needed. For the indirect approach it is not obvious how the 'identities' (10.38) should be solved. In the finite sample case there may very well be no exact solution with respect to θ. It is then an open question in what sense these 'identities' should be 'solved'. Accuracy can easily be lost if the identities are not solved in the 'best' way. (See Problem 10.2 for a simple illustration.) To avoid this complication of the two-step indirect approach $\bar{G}_i$ and $\bar{H}_i$ can be parametrized directly in terms of the original parameters of G and H. Then solving the equations (10.38) is no longer necessary, as explained above. Note, however, that the use of constrained parametrizations for $\bar{G}_i$ and $\bar{H}_i$ complicates the PEM algorithm to some extent. Furthermore, there is a general drawback of the indirect identification method that remains. This drawback is associated with the need to know the regulators and to be able to measure $v(t)$, which, moreover, should be a persistently exciting signal.

10.5 Joint input–output identification

The third approach is the joint input–output identification. Regard

$$z(t) \triangleq \begin{pmatrix} y(t) \\ u(t) \end{pmatrix} \tag{10.41}$$

as an $(ny + nu)$-dimensional time series, and apply standard PEM techniques for estimating the parameters in an appropriately structured model of $z(t)$. To find the identifiability properties one must first specify the representation of $z(t)$ that is used. As long as one is dealing with prediction error methods for the estimation it is most natural to use the 'innovation model' see (6.1), (6.2) and Example 6.5. This model/representation is unique and is given by

$$\begin{aligned} z(t) &= \mathcal{H}(q^{-1}; \theta) \tilde{z}(t) \\ E\tilde{z}(t)\tilde{z}^T(s) &= \Lambda_{\tilde{z}}(\theta) \delta_{t,s} \end{aligned} \tag{10.42}$$

where

$$\begin{aligned} &\mathcal{H}^{-1}(q^{-1}; \theta) \text{ is asymptotically stable} \\ &\mathcal{H}(0; \theta) = I \end{aligned} \tag{10.43}$$

Since the closed loop system is asymptotically stable it must be assumed that $\mathcal{H}(q^{-1}; \theta)$ is also asymptotically stable.

The way in which the filter $\mathcal{H}(q^{-1}; \theta)$ and the covariance matrix $\Lambda_{\tilde{z}}(\theta)$ depend on the parameter vector θ will be determined by the original parametrization of the model. Using the notational convention of (10.19), it follows from the uniqueness of (10.42) as a model for $z(t)$ and from the consistency of the PEM applied to (10.42), that the relations determining the identifiability properties are

$$\boxed{\hat{\mathcal{H}} \equiv \mathcal{H} \qquad \hat{\Lambda}_{\tilde{z}} = \Lambda_{\tilde{z}}} \tag{10.44}$$

Similarly to the indirect approach, the joint input–output method could be applied in two ways:

- Parametrize the innovation model (10.42) using the original parameters. Determine these parameters by applying a prediction error method to the model structure (10.42).
- The second alternative is a two-step procedure. In the first step the innovation model (10.42) is estimated using an arbitrary parametrization. This gives estimates, say $\bar{\mathcal{H}}(q^{-1})$ and $\bar{\Lambda}_{\tilde{z}}$. Then in the second step the identities

$$\bar{\mathcal{H}}(q^{-1}) \equiv \mathcal{H}(q^{-1}; \theta) \qquad \bar{\Lambda}_{\tilde{z}} = \Lambda_{\tilde{z}}(\theta) \tag{10.45}$$

are solved with respect to the parameter vector θ.

The identifiability properties of both these methods are given by (10.44). For practical use the first way seems more attractive since the second step (consisting of solving (10.45), which can be quite complicated) is avoided.

For illustration of the identity (10.44) and of the implied identifiability properties, consider the following simple example.

Example 10.7 *Joint input–output identification of a first-order system*
Consider the system

$$y(t) + ay(t-1) = bu(t-1) + e(t) + ce(t-1) \qquad (10.46a)$$

$$|c| < 1, \; e(t) \text{ white noise}, \; Ee^2(t) = \lambda_e^2$$

operating under the feedback

$$u(t) = -fy(t) + v(t) \qquad (10.46b)$$

where $v(t)$ is white noise $(Ev^2(t) = \lambda_v^2)$ and is independent of $e(s)$ for all t and s. Assume that the closed loop system is asymptotically stable, i.e.

$$\alpha = a + bf \qquad (10.46c)$$

has modulus less than 1.

It follows that the closed loop system is described by

$$(1 + \alpha q^{-1})y(t) = bq^{-1}v(t) + (1 + cq^{-1})e(t)$$

and hence

$$z(t) = \frac{1}{1 + \alpha q^{-1}} \begin{pmatrix} 1 + cq^{-1} & bq^{-1} \\ -f(1 + cq^{-1}) & 1 + aq^{-1} \end{pmatrix} \begin{pmatrix} e(t) \\ v(t) \end{pmatrix} \qquad (10.47a)$$

This is 'almost' an innovation form. The inverse of the (2|2) matrix filter in (10.47a) has a denominator given by $(1 + cq^{-1})(1 + \alpha q^{-1})$ and is hence asymptotically stable. However, the leading (zeroth-order) term of the filter differs from I. So to obtain the innovation form the description (10.47a) must be modified. This can be done in the following way:

$$z(t) = \frac{1}{1 + \alpha q^{-1}} \begin{pmatrix} 1 + cq^{-1} & bq^{-1} \\ -f(1 + cq^{-1}) & 1 + aq^{-1} \end{pmatrix} \begin{pmatrix} 1 & 0 \\ f & 1 \end{pmatrix} \begin{pmatrix} 1 & 0 \\ -f & 1 \end{pmatrix} \begin{pmatrix} e(t) \\ v(t) \end{pmatrix}$$

$$= \frac{1}{1 + \alpha q^{-1}} \begin{pmatrix} 1 + (c + bf)q^{-1} & bq^{-1} \\ f(a - c)q^{-1} & 1 + aq^{-1} \end{pmatrix} \begin{pmatrix} e(t) \\ -fe(t) + v(t) \end{pmatrix} \qquad (10.47b)$$

from which

$$\mathcal{H}(q^{-1}; \theta) = \frac{1}{1 + \alpha q^{-1}} \begin{pmatrix} 1 + (c + bf)q^{-1} & bq^{-1} \\ f(a - c)q^{-1} & 1 + aq^{-1} \end{pmatrix}$$

$$\Lambda(\theta) = E \begin{pmatrix} e(t) \\ -fe(t) + v(t) \end{pmatrix} (e(t) \; -fe(t) + v(t)) = \begin{pmatrix} \lambda_e^2 & -f\lambda_e^2 \\ -f\lambda_e^2 & f^2\lambda_e^2 + \lambda_v^2 \end{pmatrix} \qquad (10.48)$$

Now assume that a PEM is used to fit a first-order ARMA model to $z(t)$. Note that in doing so any unique parametrization of the first-order ARMA model may be used; in other words, there is no need to bother about the parametrization (10.48) at this stage. Next, using the estimates $\hat{\mathcal{H}}$ and $\hat{\Lambda}$ provided by the previous step, determine the unknown parameters a, b, etc., of the original system, by solving the identities (10.44). At this stage the parametrization (10.48) is, of course, important. The quantities a, b, c and λ_e^2 are regarded as unknowns, as are f and λ_v^2 (which characterize the feedback

(10.46b)). The identifiability relations (10.44) become

$$\frac{1}{1+(\hat{a}+\hat{b}\hat{f})q^{-1}}\begin{pmatrix}1+(\hat{c}+\hat{b}\hat{f})q^{-1} & \hat{b}q^{-1} \\ \hat{f}(\hat{a}-\hat{c})q^{-1} & 1+\hat{a}q^{-1}\end{pmatrix}$$

$$\equiv \frac{1}{1+(a+bf)q^{-1}}\begin{pmatrix}1+(c+bf)q^{-1} & bq^{-1} \\ f(a-c)q^{-1} & 1+aq^{-1}\end{pmatrix} \quad (10.49)$$

$$\begin{pmatrix}\hat{\lambda}_e^2 & -\hat{f}\hat{\lambda}_e^2 \\ -\hat{f}\hat{\lambda}_e^2 & \hat{f}^2\hat{\lambda}_e^2+\hat{\lambda}_v^2\end{pmatrix} = \begin{pmatrix}\lambda_e^2 & -f\lambda_e^2 \\ -f\lambda_e^2 & f^2\lambda_e^2+\lambda_v^2\end{pmatrix}$$

From (10.49) it follows immediately that

$$\hat{a}=a \quad \hat{b}=b \quad \hat{c}=c \quad \hat{f}=f \quad \hat{\lambda}_e^2=\lambda_e^2 \quad \hat{\lambda}_v^2=\lambda_v^2 \quad (10.50)$$

Thus both the system and the regulator parameters can be identified consistently. ∎

As a general case consider the following feedback for the system (10.1):

$$u(t) = -F(q^{-1})y(t) + v(t)$$
$$v(t) = K(q^{-1})w(t) \quad Ew(t)w^T(s) = \Lambda_w \delta_{t,s} \quad (10.51)$$

Assume that:

- The closed loop system is asymptotically stable.
- $K(q^{-1})$ and $K^{-1}(q^{-1})$ are asymptotically stable, $K(0)=I$.
- $w(t)$ and $e(s)$ are independent for all t and s.
- $H^{-1}(q^{-1})G(q^{-1})$ and $K^{-1}(q^{-1})F(q^{-1})$ are asymptotically stable.

The model to be considered is

$$y(t) = G(q^{-1};\theta)u(t) + H(q^{-1};\theta)e(t) \quad Ee(t)e^T(t) = \Lambda_e(\theta)$$
$$u(t) = -F(q^{-1};\theta)y(t) + K(q^{-1};\theta)w(t) \quad Ew(t)w^T(t) = \Lambda_w(\theta) \quad (10.52)$$

One thus allows both the regulator $F(q^{-1})$ and the filter $K(q^{-1})$ to be (at least partly) unknown. Note that since the parametrization will not be specified, the situations when $F(q^{-1})$ and/or $K(q^{-1})$ are known can be regarded as special cases of the general model (10.52).

In Appendix A10.1 it is shown that for the above case *one can identify both the system, the regulator and the external input shaping filter*. More specifically, it is shown that the identifiability relations (10.44) imply

$$\hat{G} \equiv G \quad \hat{H} \equiv H \quad \hat{F} \equiv F \quad \hat{K} \equiv K$$
$$\hat{\Lambda}_e = \Lambda_e \quad \hat{\Lambda}_w = \Lambda_w \quad (10.53)$$

The joint input–output approach gives identifiability under essentially the same conditions as the direct and indirect identification approaches. It is computationally more demanding than the direct approach, but on the other hand it can simultaneously identify the open loop system, the regulator and the spectral characteristics of the external signal $v(t)$.

To get a more specific relation to the direct approach, consider again the case given by (10.52). Assume that some parameters, say θ_1, were used to model G and H, while others, say θ_2, were used to model F and K. If a direct approach is used for estimation of θ_1, the matrix

$$R_N(\theta_1) = \frac{1}{N} \sum_{t=1}^{N} \varepsilon_1(t, \theta_1) \varepsilon_1^T(t, \theta_1) \tag{10.54}$$

is to be minimized in a suitable sense as described in Chapter 7. The prediction error in (10.54) is given by

$$\varepsilon_1(t, \theta_1) = H^{-1}(q^{-1}; \theta_1)[y(t) - G(q^{-1}; \theta_1)u(t)] \tag{10.55}$$

For the joint input–output method some appropriate scalar function of the matrix

$$S_N(\theta_1, \theta_2) = \begin{pmatrix} \frac{1}{N} \sum_{t=1}^{N} \varepsilon_1(t, \theta_1) \varepsilon_1^T(t, \theta_1) & \frac{1}{N} \sum_{t=1}^{N} \varepsilon_1(t, \theta_1) \varepsilon_2^T(t, \theta_1, \theta_2) \\ \frac{1}{N} \sum_{t=1}^{N} \varepsilon_2(t, \theta_1, \theta_2) \varepsilon_1^T(t, \theta_1) & \frac{1}{N} \sum_{t=1}^{N} \varepsilon_2(t, \theta_1, \theta_2) \varepsilon_2^T(t, \theta_1, \theta_2) \end{pmatrix} \tag{10.56}$$

is to be minimized, where the prediction error is given by (see (10.42) and Appendix A10.1)

$$\begin{pmatrix} \varepsilon_1(t, \theta_1) \\ \varepsilon_2(t, \theta_1, \theta_2) \end{pmatrix} = \begin{pmatrix} H^{-1}(q^{-1}; \theta_1)[y(t) - G(q^{-1}; \theta_1)u(t)] \\ -F(0)\varepsilon_1(t, \theta_1) + K^{-1}(q^{-1}; \theta_2)[u(t) + F(q^{-1}; \theta_2)y(t)] \end{pmatrix} \tag{10.57}$$

Let

$$\bar{R}_N(\theta_2) = \frac{1}{N} \sum_{t=1}^{N} \tilde{\varepsilon}_2(t, \theta_2) \tilde{\varepsilon}_2^T(t, \theta_2) \tag{10.58}$$

$$\tilde{\varepsilon}_2(t, \theta_2) = K^{-1}(q^{-1}; \theta_2)[u(t) + F(q^{-1}; \theta_2)y(t)] \tag{10.59}$$

Assume that the regulator has a time delay such that $F(0) = 0$. Then

$$S_N(\theta_1, \theta_2) = \frac{1}{N} \sum_{t=1}^{N} \begin{pmatrix} \varepsilon_1(t, \theta_1) \\ \tilde{\varepsilon}_2(t, \theta_2) \end{pmatrix} (\varepsilon_1^T(t, \theta_1) \quad \tilde{\varepsilon}_2^T(t, \theta_2)) \tag{10.60}$$

Choose the criterion functions, for some weighting matrices Q_1 and Q_2, as

$$V_{\text{dir}}(\theta_1) = \text{tr } Q_1 R_N(\theta_1) \tag{10.61}$$

$$V_{\text{joint}}(\theta_1, \theta_2) = \text{tr}\left\{ \begin{pmatrix} Q_1 & 0 \\ 0 & Q_2 \end{pmatrix} S_N(\theta_1, \theta_2) \right\}$$

$$= \text{tr}[Q_1 R_N(\theta_1) + Q_2 \bar{R}_N(\theta_2)] \tag{10.62}$$

$$= V_{\text{dir}}(\theta_1) + \text{tr}[Q_2 \bar{R}_N(\theta_2)]$$

Then the corresponding estimates $\hat{\theta}_1$ (which minimize the two loss functions) are identical. Note that (10.62) holds for finite $\vec{N}$ and for all parameter vectors. Thus, in this case the direct and the joint input–output approaches are equivalent in the finite sample case, and not only asymptotically (for $N \to \infty$).

10.6 Accuracy aspects

The previous sections looked at ways in which feedback can influence the identifiability properties. This section considers, by means of examples, how feedback during the identification experiment can influence the accuracy.

One might believe that it is always an advantage to use open loop experiments. A reason could be that with well-tuned feedback the signals in the loop will not have extensive variations. Since then less information may be present in the signals, a model with reduced accuracy may result. However, such a comparison is not fair. It is more appropriate to compare the accuracy, measured in some sense, under the constraint that the input or output variance is bounded. Then it will turn out that an optimal experimental condition (i.e. one that gives optimal accuracy) will often include feedback.

Although the following example considers a first-order system, the calculations are quite extensive. It is in fact difficult to generalize the explicit results to higher-order systems. (See, however, Problem 10.10.)

Example 10.8 *Accuracy for a first-order system*
Let the system be given by

$$y(t) + ay(t-1) = bu(t-1) + e(t) \qquad |a| < 1 \tag{10.63}$$

where $e(t)$ is white noise of zero mean and variance λ^2.

For this system the output variance for any type of input, including inputs determined by a causal regulator, will satisfy

$$\begin{aligned} Ey^2(t) &= E[-ay(t-1) + bu(t-1) + e(t)]^2 \\ &= E[-ay(t-1) + bu(t-1)]^2 + E[e(t)]^2 \\ &\geq E[e(t)]^2 = \lambda^2 \end{aligned} \tag{10.64}$$

Moreover, equality in (10.64) is obtained precisely for the so-called minimum variance regulator,

$$u(t) = \frac{a}{b} y(t) \tag{10.65}$$

Assume that the parameters a and b of (10.63) are estimated using a PEM applied in a model structure of the same form as (10.63). The PEM will then, of course, be the simple LS method. Take the determinant of the normalized covariance matrix of the parameter estimates as a scalar measure of the accuracy. The accuracy criterion is thus given by

$$V = \det\left\{\lambda^2 \begin{pmatrix} r_y(0) & -r_{yu}(0) \\ -r_{yu}(0) & r_u(0) \end{pmatrix}^{-1}\right\} \tag{10.66}$$

$$= \frac{\lambda^4}{r_y(0)r_u(0) - r_{yu}^2(0)}$$

with $r_y(0) = Ey^2(t)$, $r_u(0) = Eu^2(t)$, $r_{yu}(0) = Ey(t)u(t)$ (see (7.68)). The problem is to find the experimental conditions that minimize this criterion. Clearly V can be made arbitrarily small if the output or input variance can be chosen sufficiently large. Therefore it is necessary to introduce a constraint. Consider a constrained output variance. In view of (10.64), impose the following constraint:

$$r_y(0) \leq \lambda^2(1 + \delta) \tag{10.67}$$

where $\delta > 0$ is a given value.

First the optimum of V under the constraint (10.67) will be determined. Later it will be shown how the optimal experimental condition can be realized. This means that the optimization problem will be solved first without specifying explicitly the experimental condition. This will give a characterization of any optimal experimental condition in terms of $r_y(0)$, $r_{yu}(0)$ and $r_u(0)$. The second step is to see for some specific experimental conditions if (and how) this characterization can be met.

For convenience introduce the variables r and z through

$$\begin{aligned} r_y(0) &= \lambda^2(1 + b^2 r) \\ Ey(t)y(t+1) &= \lambda^2 bz \end{aligned} \tag{10.68}$$

The constraint (10.67) then becomes

$$0 \leq r \leq \delta/b^2 \tag{10.69}$$

Moreover, using the system representation (10.63),

$$r_{yu}(0) = Ey(t)u(t) = \frac{1}{b} Ey(t)[y(t+1) + ay(t) - e(t+1)]$$

$$= \frac{1}{b}[\lambda^2 bz + a\lambda^2(1 + b^2 r)] = \lambda^2\left(z + \frac{a}{b} + abr\right)$$

$$r_u(0) = \frac{1}{b^2} E[y(t+1) + ay(t) - e(t+1)]^2$$

$$= \frac{1}{b^2}[(1 + a^2)\lambda^2(1 + b^2 r) + \lambda^2 + 2a\lambda^2 bz - 2\lambda^2]$$

$$= \lambda^2[a^2 + a^2 b^2 r + 2abz + b^2 r]/b^2$$

One can now describe the criterion V, (10.66), using r and z as free variables. The result is

Section 10.6 Accuracy aspects

$$V = \frac{1}{(1 + b^2r)(a^2 + a^2b^2r + 2abz + b^2r)/b^2 - (z + a/b + abr)^2}$$
$$= \frac{1}{b^2r^2 + r - z^2} \qquad (10.70)$$

Consider now minimization of V, (10.70), under the constraint (10.69). It is obvious that the optimum is obtained exactly when

$$r = \delta/b^2 \qquad z = 0 \qquad (10.71a)$$

The corresponding optimal value is

$$V = \frac{b^2}{\delta(1 + \delta)} \qquad (10.71b)$$

So far the optimization problem has been discussed in general terms. It now remains to see how the optimal condition (10.71) can be realized. In view of (10.68), equation (10.71a) can also be written as

$$Ey^2(t) = \lambda^2(1 + \delta) \qquad Ey(t)y(t + 1) = 0 \qquad (10.72)$$

Consider first *open loop operation*. Then one must have

$$\frac{\lambda^2}{1 - a^2} \le \lambda^2(1 + \delta)$$

i.e.

$$\delta \ge \frac{a^2}{1 - a^2} \qquad (10.73)$$

since otherwise the constraint (10.67) cannot be met at all. Introduce now

$$w(t) = \frac{1}{1 + aq^{-1}} u(t)$$

Then

$$y(t) = bw(t - 1) + \frac{1}{1 + aq^{-1}} e(t)$$

In view of (10.72) the following conditions must be met:

$$b^2 Ew^2(t) = \lambda^2(1 + \delta) - \frac{\lambda^2}{1 - a^2} = \frac{\lambda^2}{1 - a^2}(\delta - \delta a^2 - a^2)$$
$$b^2 Ew(t)w(t - 1) = -\frac{-a\lambda^2}{1 - a^2} \qquad (10.74)$$

For this solution to be realizable,

$$|Ew(t)w(t - 1)| \le Ew^2(t)$$

must hold, which gives

$$|a| \leq \delta(1 - a^2) - a^2$$

or

$$\delta \geq \frac{|a|}{1 - |a|}$$

To summarize, for open loop operation there are three possibilities:

- $\delta < a^2/(1 - a^2)$. Then the constraint (10.67) cannot be met by any choice of the input (cf. (10.73)).
- $a^2/(1 - a^2) \leq \delta < |a|/(1 - |a|)$. Then the overall optimal value (10.71b) of the criterion cannot be achieved.
- $|a|/(1 - |a|) \leq \delta$. Then (10.74) characterizes precisely the conditions that the optimal input must satisfy. The optimal input can then be realized in different ways. One possibility is to take $w(t)$ as an AR(1) process

$$w(t) - \frac{a}{\delta - \delta a^2 - a^2} w(t-1) = \tilde{e}(t) \qquad E\tilde{e}^2(t) = \frac{\lambda^2}{b^2} \frac{\delta^2 - a^2(1+\delta)^2}{\delta - a^2(1+\delta)}$$

but several other possibilities exist; see, e.g., Problem 10.6.

Next consider an experimental condition of the form

$$u(t) = \frac{a}{b} y(t) + v(t) \tag{10.75}$$

which is a *minimum variance regulator* plus an additive white noise $v(t)$ of variance λ_v^2, which is independent of $e(s)$ for all t and s. The closed loop system then becomes

$$y(t) = bv(t-1) + e(t)$$

which satisfies the optimality condition (10.72) if

$$b^2 \lambda_v^2 + \lambda^2 = \lambda^2(1 + \delta) \tag{10.76}$$

which gives $\lambda_v^2 = \lambda^2 \delta / b^2$. Note that by an appropriate choice of λ_v^2 one can achieve the optimal accuracy (10.71b) for any value of δ. This is in contrast to the case of open loop operation.

As a third possibility consider the case of *shifting between two proportional regulators*

$$u(t) = -k_i y(t) \qquad \text{to be used in proportion } \gamma_i, \ i = 1, 2 \tag{10.77}$$

The closed loop system becomes

$$y_i(t) + (a + bk_i) y_i(t-1) = e(t)$$

Since the data are not stationary in this case the covariances which occur in the definition of V, (10.66), should be interpreted with care. Analogously to (10.23),

$$r_y(0) = \gamma_1 E y_1^2(t) + \gamma_2 E y_2^2(t), \text{ etc.}$$

Since r and z depend linearly on the covariance elements, (10.71a) gives

$$r = \gamma_1 r_1 + \gamma_2 r_2 = \delta/b^2 \qquad z = \gamma_1 z_1 + \gamma_2 z_2 = 0 \tag{10.78}$$

where r_i and z_i are the quantities defined by (10.68) for $y_i(t)$, $i = 1, 2$.

The constraint (10.67) must hold for both the regulators (10.77). This gives

$$r_i \leq \delta/b^2 \qquad i = 1, 2 \tag{10.79}$$

Now (10.78), (10.79) give $r_1 = r_2 = \delta/b^2$, which leads to

$$\lambda^2(1 + \delta) = Ey_i^2(t) = \frac{\lambda^2}{1 - (a + bk_i)^2}$$

or

$$k_{1,2} = \left\{ \pm\left(\frac{\delta}{1+\delta}\right)^{1/2} - a \right\} \bigg/ b \tag{10.80}$$

The second requirement in (10.78) gives

$$\gamma_1 \frac{(a + bk_1)}{1 - (a + bk_1)^2} \lambda^2 + \gamma_2 \frac{(a + bk_2)}{1 - (a + bk_2)^2} \lambda^2 = 0$$

which, in view of (10.80), reduces to $\gamma_1 - \gamma_2 = 0$, or

$$\gamma_1 = \gamma_2 = 0.5 \tag{10.81}$$

Thus this alternative consists of using two different proportional regulators, each to be used during 50 percent of the total experiment time, and each giving an output variance

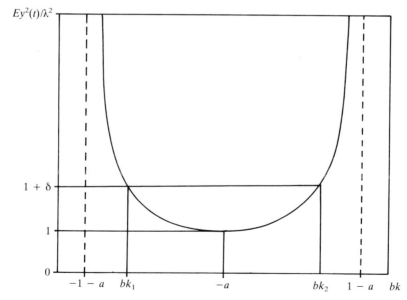

FIGURE 10.2 The output variance versus the regulator gain k, Example 10.8.

equal to the bound $\lambda^2(1 + \delta)$. Note again, that by using feedback in this way one can achieve the optimal accuracy (10.71b) for all values of δ.

The optimal experimental condition (10.80), (10.81) is illustrated in Figure 10.2, where the output variance is plotted as a function of the regulator gain for a proportional feedback $u(t) = -ky(t)$.

The minimum variance regulator $k = -a/b$ gives the minimal value λ^2 of this variance. The curve happens to be symmetric around $k = -a/b$. The two values of k that give $Ey^2(t) = \lambda^2(1 + \delta)$ are precisely those which constitute the optimal gains given by (10.80). ■

The above example shows that the experimental conditions involving feedback can really be beneficial for the accuracy of the estimated model. In practice the power of the input or the output signal must be constrained in some way. In order to get maximally informative experiments it may then be necessary to generate the input by feedback. Note that the example illustrates that the optimal experimental condition is not unique, and thus that the optimal accuracy can be obtained in a number of different ways. Open loop experiments can sometimes be used to achieve optimal accuracy, but not always. In contrast to this, several types of closed loop experiments are shown to give optimal accuracy.

Summary

Section 10.2 showed that for an experimental condition including feedback it is not possible to use (in a straightforward way) nonparametric identification methods such as spectral analysis. Instead a parametric method for which the corresponding model has an inherent structure must be used. Three parametric approaches based on the PEM to identify systems which operate in closed loop were described:

- Direct identification (using the input–output data as in the open loop case, therefore neglecting the possible existence of feedback; Section 10.3).
- Indirect identification (first identifying the closed loop system, next determining the open loop part assuming that the feedback is known; Section 10.4).
- Joint input–output identification (regarding $u(t)$ and $y(t)$ as a multivariable $(nu + ny)$-dimensional time series and using an appropriately structured model for it; with this approach both the open loop process and the regulator can be identified; Section 10.5).

These approaches will all give identifiability under weak and essentially the same conditions. From a computational point of view, the direct approach will in most cases be the simplest one.

It was shown in Section 10.6 that the use of feedback during the experiment can be beneficial if the accuracy of the estimates is to be optimized under constrained variance of the output signal.

Problems

Problem 10.1 *The estimates of parameters of 'nearly unidentifiable' systems have poor accuracy*

From a practical point of view it is not sufficient to require identifiability, as illustrated in this problem.

Consider the first-order system

$$y(t) + ay(t-1) = bu(t-1) + e(t)$$

where $e(t)$ is white noise of zero mean and variance λ^2. Assume that the input is given by

$$u(t) = -fy(t) + v(t)$$

where $v(t)$ is an external signal that is independent of $e(s)$ for all t and s. Then the system is identifiable if $v(t)$ is pe of order 1.

(a) Let $v(t)$ be white noise of zero mean and variance σ^2. Assume that the parameters a and b are estimated using the LS method (the direct approach). Show that the variances of these parameter estimates $\hat{a}$ and $\hat{b}$ will tend to infinity when σ^2 tends to zero. (Hence a small positive value of σ^2 will give identifiability but a poor accuracy.)

(b) The closed loop system will have a pole in $-\alpha$, with $\alpha = a + bf$. Assume that the model with parameters $\hat{a}$ and $\hat{b}$ is used for pole assignment in $-\alpha$. The regulator gain will then be $\hat{f} = (\alpha - \hat{a})/\hat{b}$. Determine the variance of $\hat{f}$ and examine what happens when σ^2 tends to zero.

Hint.

$$\hat{f} - f = \frac{\alpha - \hat{a}}{\hat{b}} - \frac{\alpha - a}{b} \approx \frac{b(\alpha - \hat{a}) - \hat{b}(\alpha - a)}{b^2}$$

Hence $\mathrm{var}(\hat{f}) = (-1/b \quad -(\alpha - a)/b^2) \mathrm{cov}\begin{pmatrix}\hat{a}\\\hat{b}\end{pmatrix}\begin{pmatrix}-1/b\\-(\alpha - a)/b^2\end{pmatrix}$

Problem 10.2 *On the use and accuracy of indirect identification*

Consider the simple system

$$y(t) = bu(t-1) + e(t)$$

with $e(t)$ white noise of zero mean and variance λ^2. The input is determined through the feedback

$$u(t) = -ky(t) + v(t)$$

where k is assumed to be known and $v(t)$ is white noise of zero mean and variance σ^2. The noises $v(t)$ and $e(s)$ are independent for all t and s.

(a) Assume that direct identification (with the LS method) is used to estimate b in the model structure

$$y(t) = \hat{b}u(t-1) + \varepsilon(t)$$

What is the asymptotic normalized variance of $\hat{b}$?

(b) Assume that indirect identification is used and that in the first step the parameters $\bar{a}$ and $\bar{b}$ of the model of the closed loop system

$$y(t) + \bar{a}y(t-1) = \bar{b}v(t-1) + \varepsilon(t)$$

are estimated by the LS method. Show how $\bar{a}$ and $\bar{b}$ depend on b. Then motivate the following estimates of b (assuming k is known):

$$\hat{b}_1 = \bar{b}$$

$$\hat{b}_2 = \frac{k\bar{a} + \bar{b}}{1 + k^2}$$

$$\hat{b}_3 = \frac{(k \;\; 1) Q \begin{pmatrix} \bar{a} \\ \bar{b} \end{pmatrix}}{(k \;\; 1) Q \begin{pmatrix} k \\ 1 \end{pmatrix}} \quad \text{with } Q \text{ a positive definite matrix}$$

(c) Evaluate the normalized asymptotic variances of the estimates $\hat{b}_1$, $\hat{b}_2$ and $\hat{b}_3$. For $\hat{b}_3$ choose

$$Q = \left[\text{cov} \begin{pmatrix} \bar{a} \\ \bar{b} \end{pmatrix} \right]^{-1}$$

Compare with the variance corresponding to the direct identification.

Remark. The estimate $\hat{b}_3$ with Q as in (c) can be viewed as an indirect PEM as described in Problem 7.14.

Problem 10.3 *Persistent excitation of the external signal*
Consider the system

$$y(t) + ay(t-1) = bu(t-1) + e(t) + ce(t-1)$$

controlled by a minimum variance regulator with an additional external signal:

$$u(t) = -\frac{c - a}{b} y(t) + v(t)$$

Assume that $v(s)$ and the white noise $e(t)$ are independent.

The system parameters are estimated using the direct PEM applied to a first-order ARMAX model structure. Examine the identifiability condition in the following three cases:

(a) $v(t) \equiv 0$
(b) $v(t) \equiv m$ (a nonzero constant)
(c) $v(t) = \alpha \sin \omega t$ $(0 < \omega < \pi, \alpha > 0)$

Problem 10.4 *On the use of the output error method for systems operating under feedback*
The output error method (OEM) (mentioned briefly in Section 7.4 and discussed in some detail in Complement C7.5) cannot generally be used for identification of systems operating in closed loop. This is so since the properties of the method depend critically on the assumption that the input and the additive disturbance on the output are uncorrelated. This assumption fails to hold, in general, for closed loop systems. To illustrate the difficulties in using OEM for systems operating under feedback, consider the following simple closed loop configuration:

direct path: $\quad y(t) = gu(t-1) + v(t) \quad\quad |g| < 1$

feedback path: $u(t) = -y(t) + r(t)$

The assumption $|g| < 1$ guarantees the stability of the closed loop system. For simplicity, consider the following special (but not peculiar) disturbance $v(t)$ and reference signal $r(t)$:

$$v(t) = (1 + gq^{-1})e(t)$$
$$r(t) = (1 + gq^{-1})\varepsilon(t)$$

where

$$Ee(t)e(s) = \sigma_e^2 \delta_{t,s}$$
$$E\varepsilon(t)\varepsilon(s) = \sigma_\varepsilon^2 \delta_{t,s}$$
$$Ee(t)\varepsilon(s) = 0 \text{ for all } t, s$$

The output error estimate $\hat{g}$ of g determined from input–output data $\{u(t), y(t)\}$ is asymptotically (i.e. for an infinite number of data points) given as the minimum point

$$E[y(t) - \hat{g}u(t-1)]^2$$

Determine $\hat{g}$ defined above and show that in general $\hat{g} \neq g$.

Problem 10.5 *Identifiability properties of the PEM applied to ARMAX systems operating under a minimum variance control feedback*

(a) Consider the system

$$A_0(q^{-1})y(t) = B_0(q^{-1})u(t) + C_0(q^{-1})e(t)$$

where

$$A_0(q^{-1}) = 1 + a_{01}q^{-1} + \ldots + a_{0n}q^{-n}$$
$$B_0(q^{-1}) = b_{01}q^{-1} + \ldots + b_{0n}q^{-n} \quad\quad b_{01} \neq 0$$
$$C_0(q^{-1}) = 1 + c_{01}q^{-1} + \ldots + c_{0n}q^{-n}$$

are assumed to have no common factor. Assume that $z^n B_0(z^{-1})$ has all zeros inside the unit circle and that $z = 0$ is the only common zero to $B_0(z)$ and $A_0(z) - C_0(z)$. Assume further that the system is controlled with a minimum variance regulator. This regulator is given by

$$u(t) = -\frac{C_0(q^{-1}) - A_0(q^{-1})}{B_0(q^{-1})} y(t)$$

Suppose that the system is identified using a direct PEM in the model structure (an nth-order ARMAX model)

$$\mathcal{M}: A(q^{-1})y(t) = B(q^{-1})u(t) + C(q^{-1})e(t)$$

Show that the system is not identifiable. Moreover, show that if b_1 and $C(q^{-1})$ are fixed to arbitrary values then *biased* estimates are obtained of $A(q^{-1})$ and the remaining parameters of $B(q^{-1})$. Show that the minimum variance regulator for the biased model coincides with the minimum variance regulator for the true system. (Treat the asymptotic case $N \to \infty$.)

(b) Consider the system

$$y(t) + a_0 y(t-1) = b_0 u(t-1) + e(t) + c_0 e(t-1)$$

Assume that it is identified using the recursive LS method in the model structure

$$\mathcal{M}: y(t) + ay(t-1) = \beta u(t-1) + e(t)$$

where β has a fixed value (not necessarily $= b_0$).

The input is determined by a minimum variance regulator in an adaptive fashion using the current model

$$u(t) = \frac{\hat{a}(t)}{\beta} y(t)$$

Investigate the possible limits of the estimate $\hat{a}(t)$. What are the corresponding values of the output variance? Compare with the minimal variance value, which can be obtained when the true system is known.

Hint. The results of Complement C10.1 can be used to answer the questions on identifiability.

Problem 10.6 *Another optimal open loop solution to the input design problem of Example 10.8*
Consider the optimization problem defined in Example 10.8. Suppose that

$$\delta \geq \frac{|a|}{1 - |a|}$$

is satisfied so that an open loop solution can be found. Show that an optimal input can be generated as a sinusoid

$$u(t) = A \sin \omega t$$

Determine the amplitude A and the frequency ω.

Problem 10.7 *On parametrizing the information matrix in Example 10.8*
Consider the system (10.63) and the variables r and z introduced by (10.68).

(a) Show that for any experimental condition
$$r \geq 0 \quad z^2 \leq r(1 + b^2 r)$$

(b) Use the result of (a) to show that any values of r and z can be obtained by using a shift between two proportional regulators.

Problem 10.8 *Optimal accuracy with bounded input variance*
Consider the system (10.63) with $|a| < 1$ and the criterion V, (10.66). Assume that V is to be minimized under the constraint

$$Eu^2(t) \leq \frac{\lambda^2}{b^2} \delta \quad (\delta > 0)$$

(a) Use the variables r and z, introduced in (10.68), to solve the problem. Show that optimal accuracy is achieved for

$$r = \frac{\delta(1 + a^2) + a^2(1 - a^2)}{b^2(1 - a^2)^2}$$

$$z = -\frac{a(2\delta + 1 - a^2)}{b(1 - a^2)^2}$$

(b) Show that the optimal accuracy can always be achieved with open loop operation.

Problem 10.9 *Optimal input design for weighting coefficient estimation*
The covariance matrix of the least squares estimate of $\{h(k)\}$ in the weighting function equation model

$$y(t) = \sum_{k=1}^{M} h(k)u(t - k) + e(t) \quad t = 1, \ldots, N \quad Ee(t)e(s) = \lambda^2 \delta_{t,s}$$

is asymptotically (for $N \to \infty$) given by

$$P = \lambda^2 \left\{ E \begin{pmatrix} u(t-1) \\ \vdots \\ u(t-M) \end{pmatrix} (u(t-1) \ldots u(t-M)) \right\}^{-1}$$

Introduce the following class of input signals:

$$\mathcal{C} = \{u(t) | Eu^2(t) \leq 1\}$$

Show that both $\det P$ and $[P]_{ii}$ as well as $\lambda_{\max}(P)$ achieve their minimum values over $\mathcal{C}$ if $u(t)$ is white noise of zero mean and unit variance.
  Hint. Apply Lemma A.35.

Problem 10.10 *Maximal accuracy estimation with output power constraint may require closed loop experiments*
To illustrate the claim in the title consider the following all-pole system:

$$y(t) + a_1 y(t-1) + \ldots + a_n y(t-n) = bu(t-1) + e(t) \tag{i}$$

412 *Systems operating in closed loop*   Chapter 10

where $e(t)$ is zero-mean white noise with variance λ^2. The parameters $\{a_i\}_{i=1}^n$ and b are estimated using the least squares method. For a large number of data points the estimation errors $\{\sqrt{N}(\hat{a}_i - a_i)\}_{i=1}^n$, $\sqrt{N}(\hat{b} - b)$ are Gaussian distributed with zero mean and covariance matrix P given in Example 7.6. The optimal input design problem considered here is to minimize det P under output power constraint

$$\hat{u}(t) = \arg \min_{u(t);\, Ey^2(t) \leq \lambda^2(1+\delta)} \det P \quad (\delta \geq 0)$$

Show that a possible solution is to generate $\hat{u}(t)$ by a minimum variance feedback control together with a white noise setpoint perturbation of variance $\lambda^2 \delta / b^2$, i.e.

$$\hat{u}(t) = [a_1 y(t) + a_2 y(t-1) + \ldots + a_n y(t-n+1)]/b + w(t)$$
$$Ew(t)w(s) = (\lambda^2 \delta / b^2)\delta_{t,s} \qquad Ew(t)e(s) = 0 \text{ for all } t, s \tag{ii}$$

Hint. Use Lemma A.5 to evaluate det P and Lemma A.35 to solve the optimization problem.

Remark. Note that this problem generalizes Example 10.8.

Bibliographical notes

The chapter is primarily based on Gustavsson *et al.* (1977, 1981). Some further results have been given by Anderson and Gevers (1982). See also Gevers and Anderson (1981, 1982), Sin and Goodwin (1980), Ng *et al.* (1977) and references in the above works.

The use of IV methods for systems operating in closed loop is described and analyzed by Söderström, Stoica and Trulsson (1987).

Sometimes it can be of interest to design an open loop optimal input. Fedorov (1972) and Pazman (1986) are good general sources for design of optimal experiments. For more specific treatments of such problems see Mehra (1974, 1976, 1981), Goodwin and Payne (1977), Zarrop (1979a, b), Stoica and Söderström (1982b). For a further discussion of experiment design, see also Goodwin (1987), and Gevers and Ljung (1986).

For some practical aspects on the choice of the experimental condition, see, for example, Isermann (1980).

Appendix A10.1
Analysis of the joint input–output identification

In this appendix the result (10.53) is proved under the assumptions introduced in Section 10.5.

The identifiability properties are given by (10.44). To evaluate them one first has to find the innovation representation (10.42). The calculations for achieving this will resemble those made in Example 10.7. Set

$$F_0 = F(0) \qquad \hat{F}_0 = F(0, \theta)$$

(the direct terms of the regulator and of its model).

For the closed loop system (10.1), (10.51) one can write (cf. (10.21))

$$z(t) = \begin{pmatrix} y(t) \\ u(t) \end{pmatrix} = \begin{pmatrix} (I + GF)^{-1}(GKw(t) + He(t)) \\ \{I - F(I + GF)^{-1}G\}Kw(t) - F(I + GF)^{-1}He(t) \end{pmatrix}$$

$$= \begin{pmatrix} (I + GF)^{-1}He(t) + G(I + FG)^{-1}Kw(t) \\ -F(I + GF)^{-1}He(t) + (I + FG)^{-1}Kw(t) \end{pmatrix}$$

$$= \begin{pmatrix} (I + GF)^{-1}H & G(I + FG)^{-1}K \\ -F(I + GF)^{-1}H & (I + FG)^{-1}K \end{pmatrix} \begin{pmatrix} I & 0 \\ F_0 & I \end{pmatrix} \begin{pmatrix} I & 0 \\ -F_0 & I \end{pmatrix} \begin{pmatrix} e(t) \\ w(t) \end{pmatrix}$$

$$= \begin{pmatrix} (I + GF)^{-1}(H + GKF_0) & G(I + FG)^{-1}K \\ (I + FG)^{-1}(-FH + KF_0) & (I + FG)^{-1}K \end{pmatrix} \begin{pmatrix} e(t) \\ -F_0e(t) + w(t) \end{pmatrix} \quad \text{(A10.1.1)}$$

Next it will be shown that (A10.1.1) is the innovation representation.

Let $\mathcal{H}$ denote the filter appearing in (A10.1.1) and set

$$\tilde{z}(t) = \begin{pmatrix} e(t) \\ -F_0e(t) + w(t) \end{pmatrix}$$

Since the closed loop system is asymptotically stable, then so also is the filter $\mathcal{H}$. Using the variant (A.4) of Lemma A.2,

$$\mathcal{H}^{-1} = \begin{pmatrix} 0 & 0 \\ 0 & K^{-1}(I + FG) \end{pmatrix} + \begin{pmatrix} I \\ -K^{-1}(I + FG)(I + FG)^{-1}(-FH + KF_0) \end{pmatrix}$$

$$\times \{(I + GF)^{-1}(H + GKF_0) - G(I + FG)^{-1}KK^{-1}(I + FG)(I + FG)^{-1}$$

$$\times (-FH + KF_0)\}^{-1}(I \quad -G(I + FG)^{-1}KK^{-1}(I + FG))$$

$$= \begin{pmatrix} 0 & 0 \\ 0 & K^{-1}(I + FG) \end{pmatrix} + \begin{pmatrix} I \\ K^{-1}(FH - KF_0) \end{pmatrix} \{(I + GF)^{-1}$$

$$\times (H + GKF_0 + GFH - GKF_0)\}^{-1}(I \quad -G)$$

$$= \begin{pmatrix} H^{-1} & -H^{-1}G \\ K^{-1}(FH - KF_0)H^{-1} & K^{-1}(I + FG) - K^{-1}(FH - KF_0)H^{-1}G \end{pmatrix}$$

$$= \begin{pmatrix} H^{-1} & -H^{-1}G \\ K^{-1}F - F_0H^{-1} & K^{-1} + F_0H^{-1}G \end{pmatrix}$$

$$= \begin{pmatrix} I & 0 \\ -F_0 & I \end{pmatrix} \begin{pmatrix} H^{-1} & -H^{-1}G \\ K^{-1}F & K^{-1} \end{pmatrix}$$

Thus the filter $\mathcal{H}^{-1}$ is asymptotically stable since by assumption H^{-1}, $H^{-1}G$, K^{-1} and $K^{-1}F$ are asymptotically stable.

Since $G(0) = 0$, $H(0) = I$, $K(0) = I$, $F(0) = F_0$, it is easy to verify that $\mathcal{H}(0) = I$. It is also trivial to see that $\tilde{z}(t)$ is a sequence of zero mean and uncorrelated vectors (a white noise process). Its covariance matrix is equal to

$$E\tilde{z}(t)\tilde{z}^T(t) = \begin{pmatrix} \Lambda_e & -\Lambda_e F_0^T \\ -F_0 \Lambda_e & F_0 \Lambda_e F_0^T + \Lambda_w \end{pmatrix} \tag{A10.1.2}$$

Thus it is proven that (A10.1.1) is the innovation representation. Using (A10.1.1), (A10.1.2) and the simplified notation $\hat{G} = G(q^{-1}; \theta)$, $\hat{H} = H(q^{-1}; \theta)$, $\hat{F} = F(q^{-1}; \theta)$, $\hat{K} = K(q^{-1}; \theta)$, $\hat{\Lambda}_w = \Lambda_w(\theta)$, $\hat{\Lambda}_e = \Lambda_e(\theta)$, the identifiability relations (10.44) give

$$\hat{\Lambda}_e = \Lambda_e \tag{A10.1.3a}$$

$$\hat{F}_0 \hat{\Lambda}_e = F_0 \Lambda_e \tag{A10.1.3b}$$

$$\hat{F}_0 \hat{\Lambda}_e \hat{F}_0^T + \hat{\Lambda}_w = F_0 \Lambda_e F_0^T + \Lambda_w \tag{A10.1.3c}$$

$$(I + \hat{G}\hat{F})^{-1}(\hat{H} + \hat{G}\hat{K}\hat{F}_0) \equiv (I + GF)^{-1}(H + GKF_0) \tag{A10.1.3d}$$

$$\hat{G}(I + \hat{F}\hat{G})^{-1}\hat{K} \equiv G(I + FG)^{-1}K \tag{A10.1.3e}$$

$$(I + \hat{F}\hat{G})^{-1}(-\hat{F}\hat{H} + \hat{K}\hat{F}_0) \equiv (I + FG)^{-1}(-FH + KF_0) \tag{A10.1.3f}$$

$$(I + \hat{F}\hat{G})^{-1}\hat{K} \equiv (I + FG)^{-1}K \tag{A10.1.3g}$$

It follows from (A10.1.3a–c) that

$$\hat{\Lambda}_e = \Lambda_e \quad \hat{\Lambda}_w = \Lambda_w \quad \hat{F}_0 = F_0 \tag{A10.1.4}$$

Now, equation (A10.1.3e) gives

$$(I + \hat{G}\hat{F})^{-1}\hat{G}\hat{K} = (I + GF)^{-1}GK$$

or

$$\begin{aligned}(\hat{G}\hat{K} - GK) &\equiv [(I + \hat{G}\hat{F})(I + GF)^{-1} - I]GK \\ &\equiv [(I + \hat{G}\hat{F}) - (I + GF)](I + GF)^{-1}GK \\ &\equiv (\hat{G}\hat{F} - GF)(I + GF)^{-1}GK \end{aligned} \tag{A10.1.5a}$$

Similarly, (A10.1.3g) gives

$$\begin{aligned}\hat{K} - K &\equiv [(I + \hat{F}\hat{G})(I + FG)^{-1} - I]K \\ &\equiv [(I + \hat{F}\hat{G}) - (I + FG)](I + FG)^{-1}K \\ &\equiv (\hat{F}\hat{G} - FG)(I + FG)^{-1}K \end{aligned} \tag{A10.1.5b}$$

Then (A10.1.3d) can be rewritten, using (A10.1.4) and (A10.1.5a), as

$$\begin{aligned}(\hat{H} - H) &\equiv -\hat{G}\hat{K}F_0 - H + (I + \hat{G}\hat{F})(I + GF)^{-1}(H + GKF_0) \\ &\equiv -(\hat{G}\hat{K} - GK)F_0 + [(I + \hat{G}\hat{F})(I + GF)^{-1} - I] \\ &\quad \times (H + GKF_0) \\ &\equiv -(\hat{G}\hat{F} - GF)(I + GF)^{-1}GKF_0 \\ &\quad + (\hat{G}\hat{F} - GF)(I + GF)^{-1}(H + GKF_0) \\ &\equiv (\hat{G}\hat{F} - GF)(I + GF)^{-1}H \end{aligned} \tag{A10.1.5c}$$

Equation (A10.1.3f) gives (recall from (A10.1.4) that $\hat{F}_0 = F_0$)

$$-(\hat{F}\hat{H} - FH) + (\hat{K} - K)F_0 = FH - KF_0 + (I + \hat{F}\hat{G})(I + FG)^{-1}$$
$$\times (-FH + KF_0)$$
$$= [(I + \hat{F}\hat{G})(I + FG)^{-1} - I] \quad \text{(A10.1.5d)}$$
$$\times (-FH + KF_0)$$
$$= (\hat{F}\hat{G} - FG)(I + FG)^{-1}(-FH + KF_0)$$

Using

$$\hat{G}\hat{K} - GK = \hat{G}(\hat{K} - K) + (\hat{G} - G)K$$

we get from (A10.1.5a, b)

$$\hat{G}(\hat{F}\hat{G} - FG)(I + FG)^{-1}K + (\hat{G} - G)K \equiv (\hat{G}\hat{F} - GF)G(I + FG)^{-1}K$$

or, in rewritten form

$$[\hat{G}(\hat{F}\hat{G} - FG) + (\hat{G} - G)(I + FG) - (\hat{G}\hat{F} - GF)G](I + FG)^{-1}K \equiv 0$$

or

$$(I + \hat{G}\hat{F})(\hat{G} - G)(I + FG)^{-1}K \equiv 0$$

which implies

$$\hat{G} = G \quad \text{(A10.1.6a)}$$

Similarly, using

$$\hat{F}\hat{H} - FH = \hat{F}(\hat{H} - H) + (\hat{F} - F)H$$

we get from (A10.1.5b, c, d)

$$-\hat{F}(\hat{G}\hat{F} - GF)(I + GF)^{-1}H - (\hat{F} - F)H + (\hat{F}\hat{G} - FG)(I + FG)^{-1}KF_0$$
$$- (\hat{F}\hat{G} - FG)(I + FG)^{-1}(-FH + KF_0) \equiv 0$$

or, in rewritten form

$$[-\hat{F}(\hat{G}\hat{F} - GF) - (\hat{F} - F)(I + GF) + (\hat{F}\hat{G} - FG)F](I + GF)^{-1}H \equiv 0$$

or

$$(I + \hat{F}\hat{G})(\hat{F} - F)(I + GF)^{-1}H \equiv 0$$

which implies

$$\hat{F} \equiv F \quad \text{(A10.1.6b)}$$

Then (A10.1.5b, c), (A10.1.6a, b) give easily

$$\hat{H} \equiv H \quad \hat{K} \equiv K \quad \text{(A10.1.6c)}$$

The conclusion is that the system is identifiable using the joint input–output approach. The identities (A10.1.4) and (A10.1.6) are precisely the stated identifiability result (10.53).

Complement C10.1
Identifiability properties of the PEM applied to ARMAX systems operating under general linear feedback

Consider the ARMAX system

$$A(q^{-1})y(t) = q^{-k}B(q^{-1})u(t) + C(q^{-1})e(t) \qquad (C10.1.1)$$

where

$$\begin{aligned}
A(q^{-1}) &= 1 + a_1 q^{-1} + \ldots + a_{na} q^{-na} \\
B(q^{-1}) &= b_0 + b_1 q^{-1} + \ldots + b_{nb} q^{-nb} \\
C(q^{-1}) &= 1 + c_1 q^{-1} + \ldots + c_{nc} q^{-nc}
\end{aligned} \qquad (C10.1.2)$$

The integer k, $k \geq 1$, accounts for the time delay of the system. Assume that the system is controlled by the linear feedback

$$R(q^{-1})u(t) = -S(q^{-1})y(t) \qquad (C10.1.3)$$

where

$$\begin{aligned}
R(q^{-1}) &= 1 + r_1 q^{-1} + \ldots + r_{nr} q^{-nr} \\
S(q^{-1}) &= s_0 + s_1 q^{-1} + \ldots + s_{ns} q^{-ns}
\end{aligned} \qquad (C10.1.4)$$

Note that the feedback (C10.1.3) does not include any external signal. From the viewpoint of identifiability this is a disadvantage, as shown in Section 10.3. The closed loop system is readily found from (C10.1.1) and (C10.1.3). It is described by

$$[A(q^{-1})R(q^{-1}) + q^{-k}B(q^{-1})S(q^{-1})]y(t) = R(q^{-1})C(q^{-1})e(t) \qquad (C10.1.5)$$

The open loop system (C10.1.1) is assumed to be identified by using a direct PEM approach applied to an ARMAX model structure.

We make the following assumptions:

A1. There is no common factor to A, B and C; C has all its zeros outside the unit circle.
A2. There is no common factor to R and S.
A3. The closed loop system (C10.1.5) is asymptotically stable.
A4. The integers na, nb, nc and k are known.

Assumptions A1–A3 are all fairly weak. Assumption A4 is more restrictive from a practical point of view. However, the use of A4 makes the details in the analysis much easier, and the general type of conclusions will hold also if the polynomial degrees are overestimated. For a detailed analysis of the case when A4 is relaxed, see Söderström et al. (1975).

Write the estimated model in the form

$$\hat{A}(q^{-1})y(t) = q^{-k}\hat{B}(q^{-1})u(t) + \hat{C}(q^{-1})\varepsilon(t) \qquad (C10.1.6)$$

where $\varepsilon(t)$ is the model innovation. The estimated polynomials $\hat{A}$, $\hat{B}$ and $\hat{C}$ are (asymptotically, when the number of data points tends to infinity) given by the identifiability identity (10.26):

$$\varepsilon(t) = e(t) \tag{C10.1.7}$$

It follows easily from (C10.1.3) and (C10.1.6) that

$$\varepsilon(t) = \frac{\hat{A}(q^{-1})y(t) - q^{-k}\hat{B}(q^{-1})u(t)}{\hat{C}(q^{-1})}$$
$$= \frac{\hat{A}(q^{-1})R(q^{-1}) + q^{-k}\hat{B}(q^{-1})S(q^{-1})}{R(q^{-1})\hat{C}(q^{-1})} y(t) \tag{C10.1.8}$$

Comparing (C10.1.5) and (C10.1.8), it can be seen that (C10.1.7) is equivalent to

$$\frac{\hat{A}R + q^{-k}\hat{B}S}{\hat{C}} \equiv \frac{AR + q^{-k}BS}{C} \tag{C10.1.9}$$

where we have dropped the polynomial arguments for convenience. This is the identity that must be analyzed in order to find the identifiability properties. Consider first two examples.

Example C10.1.1 *Identifiability properties of an nth-order system with an mth-order regulator*
Assume that $na = nb = nc = n$, $nr = ns = m$. Assume further that C and $AR + q^{-k}BS$ are coprime. The latter assumption implies that the numerators and the denominators of (C10.1.9) must satisfy

$$\hat{C} \equiv C$$
$$\hat{A}R + q^{-k}\hat{B}S \equiv AR + q^{-k}BS \tag{C10.1.10a}$$

The latter identity gives

$$(\hat{A} - A)R \equiv -q^{-k}(\hat{B} - B)S \tag{C10.1.10b}$$

This identity can easily be rewritten as a system of linear equations by equating the different powers of q^{-1}. There are $2n + 1$ unknowns (the coefficients of $\hat{A}$ and $\hat{B}$) and $k + n + m$ equations. Hence, in order to have a unique solution

$$2n + 1 \leq k + n + m$$

which is equivalent to

$$\boxed{k + m \geq n + 1} \tag{C10.1.10c}$$

Hence (C10.1.10c) is a necessary identifiability condition. Next it is shown that it is also a sufficient condition for identifiability. Assume that it is satisfied. Since R and $q^{-k}S$

are coprime (Assumption A2) it follows from (C10.1.10b) that

$q^{-k}S$ is a factor of $\hat{A} - A$

R is a factor of $\hat{B} - B$ (C10.1.10d)

In view of the degree condition (C10.1.10c) this is impossible. Thus (C10.1.10b) implies $\hat{A} - A \equiv 0$, $\hat{B} - B \equiv 0$, which proves sufficiency.

To summarize,

$$\hat{A} \equiv A \qquad \hat{B} \equiv B \qquad \hat{C} \equiv C$$

if and only if the degree condition (C10.1.10c) holds. Thus the system is identifiable if and only if the delay and/or the regulator order are large enough. ■

Example C10.1.2 *Identifiability properties of a minimum-phase system under minimum variance control*

Assume that the system is minimum phase ($B(z)$ has all zeros outside the unit circle) and is controlled with the minimum variance regulator (Åström, 1970). Let the polynomials F and G be defined by

$$C \equiv AF + q^{-k}G \qquad (C10.1.11a)$$

Here deg $F = k - 1$ and deg $G = \max(nc - k, na - 1)$. Then the minimum variance regulator is given by

$$R \equiv BF \qquad S \equiv G$$

For this type of regulator there is cancellation in the right-hand side of the identity (C10.1.9), which reduces to

$$\frac{\hat{A}BF + q^{-k}\hat{B}G}{\hat{C}} \equiv B$$

or equivalently

$$B(\hat{A}F - \hat{C}) \equiv -q^{-k}\hat{B}G \qquad (C10.1.11b)$$

Since in this case $AR + q^{-k}BS = BC$, a basic assumption of the analysis in Example C10.1.1 is violated here. The polynomial identity (C10.1.11b) can be written as a system of linear equations with $na + nb + nc + 1$ unknowns. The number of equations is $\max(nb + na + k - 1, nb + nc, k + nb + \deg G) = \max(na + nb + k - 1, nb + nc, nb + nc, na + k + nb - 1) = \max(nb + nc, na + nb + k - 1)$. A necessary condition for identifiability is then readily established as

$$na + nb + nc + 1 \leq \max(nb + nc, na + nb + k - 1)$$

which gives

$$\boxed{k \geq nc + 2} \qquad (C10.1.11c)$$

Assume now that G, as defined by (C10.1.11a), and B are coprime. (This assumption is most likely to be satisfied.) Since B is a factor of the left-hand side of (C10.1.11b) it must also appear in the right-hand side. This is only possible by taking $\hat{B} \equiv B$. Next,

$$\hat{C} \equiv \hat{A}F + q^{-k}G$$

and from (C10.1.11a),

$$(\hat{C} - C) \equiv (\hat{A} - A)F \tag{C10.1.11d}$$

If (C10.1.11c) is satisfied then F cannot be a factor of $\hat{C} - C$ and it follows immediately that $\hat{A} \equiv A$ and $\hat{C} \equiv C$.

To summarize, (C10.1.11c) is a necessary and sufficient condition for identifiability Note that, similarly to Example C10.1.1, the system will be identifiable if the delay is large enough. Also note that, due to the pole-zero cancellation in the closed loop system caused by the regulator (C10.1.11a), one must now require a larger delay than before to guarantee identifiability. ∎

Now consider the general case. Consider again the identity (C10.1.9). Assume that C and $AR + q^{-k}BS$ have exactly nh common zeros (with $nh \geq 0$). Define a polynomial H which has precisely these zeros. This means that

$$\begin{aligned} C &\equiv C_0 H \\ AR + q^{-k}BS &\equiv P_0 H \end{aligned} \tag{C10.1.12}$$

with C_0 and P_0 being coprime.

Then it follows from (C10.1.9) that

$$C_0(\hat{A}R + q^{-k}\hat{B}S) \equiv P_0 \hat{C} \tag{C10.1.13}$$

Since C_0 and P_0 are coprime, this implies that

$$\begin{aligned} \hat{C} &\equiv C_0 M \\ \hat{A}R + q^{-k}\hat{B}S &\equiv P_0 M \end{aligned} \tag{C10.1.14}$$

for some polynomial M. The degree of M cannot exceed nh. Some straightforward calculations give

$$H(\hat{A}R + q^{-k}\hat{B}S) \equiv HP_0 M \equiv M(AR + q^{-k}BS)$$

which implies

$$R(H\hat{A} - MA) \equiv q^{-k}S(-H\hat{B} + MB) \tag{C10.1.15}$$

First we derive a *necessary* condition by requiring that in (C10.1.13) the number of unknowns is at most equal to the number of equations (i.e. powers of q^{-1}). This principle gives

$$na + nb + nc + 1 \leq \max(\deg P_0 + nc, nc - nh + \deg(\hat{A}R + q^{-k}\hat{B}S))$$
$$= nc - nh + \max(na + nr, k + nb + ns)$$

which is easily rewritten as

$$nh \leq \max(nr - nb, k + ns \quad na) - 1 \qquad (C10.1.16)$$

Next we show that this condition is also *sufficient* for identifiability. We would first like to conclude from (C10.1.15) that

$$H\hat{A} - MA \equiv 0 \qquad H\hat{B} - MB \equiv 0 \qquad (C10.1.17)$$

Since R and $q^{-k}S$ are coprime by assumption, (C10.1.17) follows from (C10.1.15) provided that

$$\deg(H\hat{A} - MA) < \deg q^{-k}S \quad \text{or} \quad \deg(H\hat{B} - MB) < \deg R$$

which can be rewritten as

$$na + nh < k + ns \quad \text{or} \quad nh + nb < nr$$

which is equivalent to (C10.1.16). Hence (C10.1.16) and (C10.1.15) imply (C10.1.17). Also, from (C10.1.12) and (C10.1.14),

$$H\hat{C} \equiv HC_0M \equiv CM \qquad (C10.1.18)$$

Since M by assumption A1 is the *only* common factor to AM, BM and CM, it follows from (C10.1.17), (C10.1.18) that $M \equiv H$ and hence

$$\hat{A} \equiv A \qquad \hat{B} \equiv B \qquad \hat{C} \equiv C$$

This completes the proof that (C10.1.16) is not only a necessary but also a sufficient identifiability condition.

Note that, as before, identifiability is secured only if the delay and/or the order of the regulator are large enough.

In the general case when the model polynomials have degrees $\hat{n}a \geq na$, $\hat{n}b \geq nb$, $\hat{n}c \geq nc$ and delay $\hat{k} \leq k$ the situation can never improve (see Söderström *et al.* (1975) for details). The reason is easy to understand. When the model degrees are increased, the number of unknowns will automatically increase. However, the number of 'independent' equations that can be generated from the identity (C10.1.9) may or may not increase. It will certainly not increase faster than the number of unknowns.

For illustration, consider the main result (C10.1.16) for the two special cases discussed previously in the examples.

In Example C10.1.1, (C10.1.16) becomes

$$0 \leq \max(m - n, k + m - n) - 1 = k + m - n - 1$$

which is the previous result (C10.1.10c).

In Example C10.1.2, (C10.1.16) becomes

$$nc \leq \max(nb + k - 1 - nb, k + ns - na) - 1$$

with

$$ns = \max(nc - k, na - 1)$$

This condition can be rewritten as

$$nc \leq \max(k-1, k-na+nc-k, k-na+na-1) - 1 = \max(k-1, nc-na) - 1$$

or equivalently

$$nc \leq k-2$$

This condition is the same as the previously obtained (C10.1.11c).

Chapter 11

MODEL VALIDATION AND MODEL STRUCTURE DETERMINATION

11.1 Introduction

In system identification both the determination of model structure and model validation are important aspects. An overparametrized model structure can lead to unnecessarily complicated computations for finding the parameter estimates and for using the estimated model. An underparametrized model may be very inaccurate. The purpose of this chapter is to present some basic methods that can be used to find an appropriate model structure.

The choice of model structure in practice is influenced greatly by the intended use of the model. A stabilizing regulator can often be based on a crude low-order model, whereas more complex and detailed models are necessary if the model is aimed at giving physical insight into the process.

In practice one often performs identification for an increasing set of model orders (or more generally, structural indices). Then one must know when the model order is appropriate, i.e. when to stop. Needless to say, any real-life data set cannot be modeled exactly by a linear finite-order model. Nevertheless such models often give good approximations of the true dynamics. However, the methods for finding the 'correct' model order are based on the statistical assumption that the data come from a true system within the model class considered.

When searching for the 'correct' model order one can raise different questions, which are discussed in the following sections.

- Is a given model flexible enough? (Section 11.2)
- Is a given model too complex? (Section 11.3)
- Which model structure of two or more candidates should be chosen? (Section 11.5)

Note that such questions are relevant also to model reduction.

It should be remarked here that some simulation examples involving the model validation phase are given in Chapter 12 (see Sections 12.3, 12.4 and 12.10).

11.2 Is a model flexible enough?

The question that is asked in this section can also be phrased as: 'Is the model structure large enough to cover the true system?' There are basically two ways to approach this question:

- Use of plots and common sense.
- Use of statistical tests on the prediction errors $\varepsilon(t, \hat{\theta}_N)$.

Concerning the first approach, it is often useful to plot the measured data and the model output. The model output $y_m(t)$ is defined as the output of the model excited by the measured input and with no disturbances added. This means that (see (6.1)),

$$y_m(t) = G(q^{-1}; \hat{\theta}_N)u(t) \tag{11.1}$$

For a good model, $y_m(t)$ should resemble the measured output. The deviations of $y_m(t)$ from $y(t)$ are due both to modeling errors and to the disturbances. It is therefore important to realize that if the data are noisy then $y_m(t)$ should differ from $y(t)$. It should only describe the part of the output that is due to the input signal.

There are several statistical tests on the prediction errors $\varepsilon(t, \hat{\theta}_N)$. The prediction errors evaluated at the parameter estimate $\hat{\theta}_N$ are often called the *residuals*. To simplify the notation, in this chapter the residuals will frequently be written just as $\varepsilon(t)$.

The methods for model structure determination based on tests of the residuals are tied to model structures and identification methods where the disturbances are explicitly modeled. This means that they are not suitable if, for example, an instrumental variable method is used for parameter estimation, but they are well adapted to validate models obtained by a prediction error method.

For the following model assumptions (or, in statistical terms, the null hypotheses H_0) several tests can be constructed:

A1: $\varepsilon(t)$ is a zero mean white noise
A2: $\varepsilon(t)$ has a symmetric distribution
A3: $\varepsilon(t)$ is independent of past inputs ($E\varepsilon(t)u(s) = 0$, $t > s$).
A4: $\varepsilon(t)$ is independent of all inputs ($E\varepsilon(t)u(s) = 0$, all t and s) (applicable if the process operates in open loop).

Some typical tests will be described in Examples 11.1–11.3. The tests are formulated for single output systems ($ny = 1$). For multivariable systems the tests have to be generalized. This can be done in a fairly straightforward manner, but the discussion will be confined to the scalar case to avoid cumbersome notation. The statistical properties of the tests will be analyzed under the null hypothesis that the model assumptions actually are satisfied. Thus *all the distribution results presented below hold under a certain null hypothesis*.

Example 11.1 *An autocorrelation test*
This test is based on assumption A1. If $\varepsilon(t)$ is white noise then its covariance function is zero except at $\tau = 0$:

$$r_\varepsilon(\tau) = 0 \quad \tau \neq 0 \tag{11.2}$$

First construct estimates of the covariance function as

$$\hat{r}_\varepsilon(\tau) = \frac{1}{N}\sum_{t=1}^{N-\tau} \varepsilon(t+\tau)\varepsilon(t) \quad (\tau \geq 0) \tag{11.3}$$

(cf. (3.25)). Under assumption A1, it follows from Lemma B.2 that

$$\begin{aligned}\hat{r}_\varepsilon(\tau) &\to 0 \quad \tau \neq 0 \\ \hat{r}_\varepsilon(0) &\to \lambda^2 = E\varepsilon^2(t)\end{aligned} \quad \text{as } N \to \infty \tag{11.4}$$

To get a normalized test quantity, consider

$$x_\tau = \frac{\hat{r}_\varepsilon(\tau)}{\hat{r}_\varepsilon(0)} \tag{11.5}$$

According to (11.4) one can expect x_τ to be small for $\tau \neq 0$ and large N provided $\varepsilon(t)$ is white noise. However, what does 'small' mean? To answer that question a more detailed analysis is necessary. The analysis will be based on assumption A1. Define

$$r = \frac{1}{N}\sum_{t=1}^{N}\begin{pmatrix}\varepsilon(t-1)\\ \vdots \\ \varepsilon(t-m)\end{pmatrix}\varepsilon(t) = \begin{pmatrix}\hat{r}_\varepsilon(1)\\ \vdots \\ \hat{r}_\varepsilon(m)\end{pmatrix} \tag{11.6}$$

where, for convenience, the inferior limit of the sums was set to 1 (for large N this will have a negligible effect). Lemma B.3 shows that r is asymptotically Gaussian distributed,

$$\sqrt{N}\, r \xrightarrow{\text{dist}} \mathcal{N}(0, P) \tag{11.7}$$

where the covariance matrix is

$$P = \lim_{N \to \infty} ENrr^T$$

The (i, j) element of P $(i, j = 1, \ldots, m)$ can be evaluated as

$$\begin{aligned}P_{i,j} &= \lim_{N \to \infty} \frac{1}{N}\sum_{t=1}^{N}\sum_{s=1}^{N} E\varepsilon(t-i)\varepsilon(t)\varepsilon(s)\varepsilon(s-j) \\ &= \lim_{N \to \infty} \Bigg\{\frac{1}{N}\sum_{t=1}^{N}\sum_{s=1}^{t-1} E\varepsilon(t-i)\varepsilon(t)\varepsilon(s)\varepsilon(s-j) \\ &\quad + \frac{1}{N}\sum_{t=1}^{N}\sum_{s=t+1}^{N} E\varepsilon(t-i)\varepsilon(t)\varepsilon(s)\varepsilon(s-j) \\ &\quad + \frac{1}{N}\sum_{t=1}^{N} E\varepsilon(t-i)\varepsilon^2(t)\varepsilon(t-j)\Bigg\} \\ &= \lim_{N \to \infty}\Bigg\{0 + 0 + \frac{1}{N}\sum_{t=1}^{N}\lambda^2 E\varepsilon(t-i)\varepsilon(t-j)\Bigg\} = \lambda^4 \delta_{i,j}\end{aligned}$$

Hence

$$P = \lambda^4 I \tag{11.8}$$

The result (11.7) implies that

$$Nr^T P^{-1} r = Nr^T r / r_\varepsilon^2(0) \xrightarrow{\text{dist}} \chi^2(m)$$

(see Lemma B.12). Hence (cf. Lemma B.4)

$$\frac{N}{\hat{r}_\varepsilon^2(0)} \sum_{i=1}^{m} \hat{r}_\varepsilon^2(i) = Nr^T r / \hat{r}_\varepsilon^2(0) \xrightarrow{\text{dist}} \chi^2(m) \tag{11.9}$$

From (11.7), (11.8) it also follows for $\tau > 0$ that

$$\sqrt{N} r_\tau = \sqrt{N} \frac{\hat{r}_\varepsilon(\tau)}{\hat{r}_\varepsilon(0)} \xrightarrow{\text{dist}} \mathcal{N}(0, 1) \tag{11.10}$$

∎

How to apply the test

It should be stressed once more that the distribution of the statistics presented above holds under the null hypothesis H_0 (asserting that $\varepsilon(t)$ is white). The typical way of using the test statistics for model validation may be described as follows. Consider the test quantity $Nr^T r / \hat{r}_\varepsilon^2(0)$ (see (11.9)). Let x denote a random variable which is χ^2 distributed with m degrees of freedom. Define $\chi_\alpha^2(m)$ by

$$\alpha = P(x > \chi_\alpha^2(m)) \tag{11.11}$$

for some given α which typically is chosen between 0.01 and 0.1. Then, if

$Nr^T r / \hat{r}_\varepsilon^2(0) > \chi_\alpha^2(m)$ reject H_0 (and thus invalidate the model)
$Nr^T r / \hat{r}_\varepsilon^2(0) \leq \chi_\alpha^2(m)$ accept H_0 (and thus validate the model)

Evidently the risk of rejecting H_0 when H_0 holds (which is called the first type of risk) is equal to α. The risk of accepting H_0 when it is not true depends on how much the properties of the tested model differ from H_0. This second type of risk cannot, in general, be determined for the statistics introduced previously, unless one restricts considerably the class of alternative hypotheses against which H_0 is tested. Thus, in applications the value of α (or equivalently, the value of the test threshold) should be chosen by considering only the first type of risk. When doing so it should, of course, be kept in mind that when α decreases the first type of risk decreases, *but* the second type of risk increases. A frequently used value of α is $\alpha = 0.05$. For some numerical values of $\chi_\alpha^2(m)$, see Table B.1 in Appendix B.

The test can also be applied to the quantity x_τ, (11.5), for a single τ value. Since $\sqrt{N}x_\tau \xrightarrow{\text{dist}} \mathcal{N}(0, 1)$ we have (for large N) $x_\tau \sim \mathcal{N}(0, 1/N)$ and hence $P(|x_\tau| \leq 1.96/\sqrt{N}) = 0.95$. The null hypothesis (that $r_\varepsilon(\tau) = 0$) can therefore be accepted if $|x_\tau| \leq 1.96/\sqrt{N}$ (with an unknown risk), and otherwise rejected (with a risk of 0.05).

Example 11.2 *A cross-correlation test*
This test is based on assumptions A3 and A4. For its evaluation assumption A1 will also be needed. If the residuals and the input are independent,

$$r_{\varepsilon u}(\tau) = E\varepsilon(t + \tau)u(t) = 0 \tag{11.12}$$

There are good reasons for considering both positive and negative values of τ when testing the cross-correlation between input and residuals. If the model is not an accurate representation of the system, it is to be expected that $r_{\varepsilon u}(\tau)$ for $\tau \geq 0$ is far from zero. For $\tau < 0$, $r_{\varepsilon u}(\tau)$ may or may not be zero depending on the autocorrelation of $u(t)$; for example, if $u(t)$ is white then $r_{\varepsilon u}(\tau) = 0$ for $\tau < 0$ even if the model is inaccurate. Next, assume that the model is an exact description of the system. Then $r_{\varepsilon u}(\tau) = 0$ for $\tau \geq 0$. However, it should be the case that $r_{\varepsilon u}(\tau) \neq 0$ for $\tau < 0$, if there is feedback acting on the process during the identification experiment. Hence one can also use the test for checking the existence of possible feedback (see Section 12.10 for a further discussion of this point.)

As a normalized test quantity take

$$x_\tau = \frac{\hat{r}_{\varepsilon u}(\tau)}{[\hat{r}_\varepsilon(0)\hat{r}_u(0)]^{1/2}} \tag{11.13}$$

where

$$\hat{r}_{\varepsilon u}(\tau) = \frac{1}{N} \sum_{t=1-\min(0,\tau)}^{N-\max(\tau,0)} \varepsilon(t + \tau)u(t) \tag{11.14}$$

One can expect x_τ to be 'small' when N is large and assumptions A3 and A4 are satisfied. To examine what 'small' means, one must perform an analysis similar to that of Example 11.1. For that purpose introduce

$$\hat{R}_u = \frac{1}{N} \sum_{t=m+1}^{N} \begin{pmatrix} u(t-1) \\ \vdots \\ u(t-m) \end{pmatrix} (u(t-1) \ \ldots \ u(t-m)) \tag{11.15}$$

$$r = (\hat{r}_{\varepsilon u}(\bar{\tau} + 1) \ \ldots \ \hat{r}_{\varepsilon u}(\bar{\tau} + m))^\text{T}$$

where $\bar{\tau}$ is a given integer. Note that (assuming $u(t) = 0$ for $t \leq 0$)

$$r = \frac{1}{N} \sum_{t=1}^{N} \begin{pmatrix} u(t - \bar{\tau} - 1) \\ \vdots \\ u(t - \bar{\tau} - m) \end{pmatrix} \varepsilon(t)$$

Assume that $\varepsilon(t)$ is white noise (Assumption A1). In a similar way to the analysis in Example 11.1, one obtains

$$\sqrt{N}r \xrightarrow{\text{dist}} \mathcal{N}(0, P) \tag{11.16}$$

with

$$P = \lim_{N\to\infty} E\, Nrr^T = \lambda^2 E \begin{pmatrix} u(t-1) \\ \vdots \\ u(t-m) \end{pmatrix} (u(t-1) \;\cdots\; u(t-m)) \tag{11.17}$$

Thus from Lemma B.12

$$Nr^T P^{-1} r \xrightarrow{\text{dist}} \chi^2(m)$$

and also

$$Nr^T[\hat{r}_\varepsilon(0)\hat{R}_u]^{-1} r \xrightarrow{\text{dist}} \chi^2(m) \tag{11.18}$$

Concerning x_τ, note from (11.16), (11.17) and $\sqrt{N}\hat{r}_{\varepsilon u}(\tau) \xrightarrow{\text{dist}} \mathcal{N}(0, r_\varepsilon(0)r_u(0))$ which implies $\sqrt{N}x_\tau \xrightarrow{\text{dist}} \mathcal{N}(0, 1)$. ∎

Remark 1. The residuals $\varepsilon(t, \hat{\theta}_N)$ differ slightly from the 'true' white noise sequence $\varepsilon(t, \theta_0)$. It turns out that this deviation, even if it is small and in fact vanishes as $N \to \infty$, invalidates the expressions (11.8) and (11.17) for the P matrices. As shown in Appendix A11.1, if $\varepsilon(t, \hat{\theta}_N)$ are considered, with $\hat{\theta}_N$ being the PEM estimate of θ_0, then the true (asymptotic) covariance matrices are smaller than (11.8), (11.17). This means that if the tests are used as described above the risk of rejecting H_0 when it holds is smaller than expected but the risk of accepting H_0 when it is not true is larger. Expressed in another way, if, in the tests above, $\varepsilon(t, \hat{\theta}_N)$ is used rather than $\varepsilon(t, \theta_0)$, then the correct threshold value, corresponding to a first type of risk equal to α, is smaller than $\chi^2_\alpha(m)$. This fact is not a serious practical problem, especially if the number of estimated parameters $n\theta$ is much less than m (see Appendix A11.1 and in particular (A11.1.18) for calculations supporting this claim). Since the exact tests based on the expressions of the (asymptotic) covariance matrices P derived in Appendix A11.1 are more complicated than (11.9) and (11.18), in practice the tests are often performed as outlined in Examples 11.1 and 11.2. The tests will also be used in that way in some subsequent examples in this and the next chapter. ∎

Remark 2. The number m in the above correlation tests could be chosen from 5 up to $N/4$. Sometimes the integer $\bar{\tau}$ in the cross-correlation test must be chosen with care. The reason is that for some methods, for example the least squares method, $\hat{r}_{\varepsilon u}(\tau)$ is constrained by construction to be zero for some values of τ. (For instance for the 'least squares' model (6.12), $\hat{r}_{\varepsilon u}(\tau) = 0$ for $\tau = 1, \ldots, nb$ by the definition of the LS estimate; cf. Problem 11.1. These values of τ must not be considered in the r-vector.) ∎

Remark 3. It is often illustrative to plot the correlation functions $\hat{r}_\varepsilon(\tau)/\hat{r}_\varepsilon(0)$ and $\hat{r}_{\varepsilon u}(\tau)/[\hat{r}_\varepsilon(0)\hat{r}_u(0)]^{1/2}$ since they will show in a readily perceived way how well the hypothesis H_0 is satisfied. Some examples where this is done are given below and in Chapter 12. ∎

Next consider another simple test of the null hypothesis.

Example 11.3 *Testing changes of sign*
This test is based on assumptions A1 and A2. Let $\bar{x}_N$ be the number of changes of sign in the sequence $\varepsilon(1), \varepsilon(2), \ldots, \varepsilon(N)$. One would intuitively expect that (under assumptions A1 and A2) the residual sequence will, on average, change sign every second time step. Hence one expects that $\bar{x}_N \approx N/2$. However, a more precise statistical analysis is required to see how well this holds. For that purpose introduce the random variables $\{\delta_i\}_{i=1}^{N-1}$ by

$$\delta_i = \begin{cases} 1 & \text{if } \varepsilon(i)\varepsilon(i+1) < 0 \quad \text{(i.e. a change of sign from time } i \text{ to } i+1) \\ 0 & \text{if } \varepsilon(i)\varepsilon(i+1) > 0 \quad \text{(i.e. no change of sign from time } i \text{ to } i+1) \end{cases}$$

Then

$$\bar{x}_N = \sum_{i=1}^{N-1} \delta_i$$

Now observe that:

- $\delta_i = 0$ with probability 0.5 and $\delta_i = 1$ with probability 0.5. (This follows from assumptions A1 and A2.)
- δ_i and δ_j are identically distributed and independent random variables. (The changes of sign are independent events since $\varepsilon(t)$ is white noise.)

Hence one can apply the central limit theorem to $\bar{x}_N$. Thus $\bar{x}_N$ is asymptotically Gaussian distributed,

$$\bar{x}_N \xrightarrow{\text{dist}} \mathcal{N}(m, P) \tag{11.19}$$

with

$$m = E\bar{x}_N = (N-1)E\delta_i = (N-1)/2 \approx N/2$$

$$P = E(\bar{x}_N - m)^2 = E\bar{x}_N^2 - m^2 = E\sum_{i=1}^{N-1}\sum_{j=1}^{N-1}\delta_i\delta_j - (N-1)^2/4$$

$$= [(N-1)^2 - (N-1)](E\delta_i)^2 + (N-1)E(\delta_i^2) - (N-1)^2/4$$

$$= (N-1)[E(\delta_i^2) - (E\delta_i)^2] = (N-1)/4$$

$$\approx N/4$$

According to this analysis (for large N)

$$\frac{\bar{x}_N - N/2}{\sqrt{N/2}} \xrightarrow{\text{dist}} \mathcal{N}(0, 1) \tag{11.20}$$

Hence a 95 percent confidence interval for $\bar{x}_N$ is given by

Section 11.2   Is a model flexible enough?   429

$$\left|\frac{\bar{x}_N - N/2}{\sqrt{N/2}}\right| \leq 1.96$$

Equivalently stated, the inequalities

$$\frac{N}{2} - \frac{1.96}{2}\sqrt{N} \leq \bar{x}_N \leq \frac{N}{2} + \frac{1.96}{2}\sqrt{N} \qquad (11.21)$$

hold with 95 percent probability.   ■

The following example illustrates the use of the above tests.

Example 11.4 *Use of the statistical tests on residuals*
The program package IDPAC (see Wieslander, 1976) was used for generation of data and performing the tests. The number m is automatically set in IDPAC to $m = 10$. Similarly the time lag used for the cross-correlation tests is set automatically. Using a random number generator 300 data points of a sequence $\{\varepsilon(t)\}$ were generated, where $\varepsilon(t)$ is Gaussian white noise of zero mean and unit variance. A test was made of the hypothesis H_0: $\{\varepsilon(t)\}$ is white noise. The estimated autocorrelation function of $\varepsilon(t)$ is shown in Figure 11.1a.

The figure shows the normalized covariance function $\hat{r}_\varepsilon(\tau)/\hat{r}_\varepsilon(0)$ versus τ and a 95% confidence interval for $x_\tau \triangleq \hat{r}_\varepsilon(\tau)/\hat{r}_\varepsilon(0)$. Since $x_\tau \sim \mathcal{N}(0, 1/N)$ the lines in the diagram are drawn at $x = \pm 1.96/\sqrt{N}$. It can be seen from the figure that x_τ lies in this interval. One can hence expect that $\varepsilon(t)$ is a white process.

Next consider the numerical values of the test quantities. The number of changes of sign is 149. This is the variable $\bar{x}_N$ in (11.21). Since the interval defined by (11.21) is (132,166) it can be seen that the number of changed signs falls well inside this interval. Hence using this test quantity one should indeed accept that $\varepsilon(t)$ can be white noise.

The test quantity (11.9) is computed for $m = 10$. Its numerical value was 11.24. This variable is, under the null hypothesis H_0, approximately $\chi^2(10)$ distributed. According to Table B.1 for $\alpha = 0.05$ the threshold value is 18.3. The null hypothesis is therefore accepted.

Next a signal $u(t)$ was generated by linear filtering of $\varepsilon(t)$ as

$$u(t) - 0.8u(t - 1) = \varepsilon(t - 1)$$

This signal is an autoregressive process with a low-frequency content. The tests for whiteness were applied to this signal. The correlation function is shown in Figure 11.1b. The plot indicates clearly that the signal $u(t)$ is *not* white noise.

The numerical values of the test quantities are as follows. The number of changes of sign was 53, while the interval defined by (11.21) still is (132, 166). Hence based on $\bar{x}_N$ one can clearly reject the hypothesis that $u(t)$ is white noise. The numerical value of test quantity (11.9) was 529.8. This is too large to come from a $\chi^2(10)$ distribution. Hence one finds once more that $u(t)$ is not white noise.

Finally the cross-correlation between $\varepsilon(t)$ and $u(t)$ was tested. $\varepsilon(t)$ was treated as residual and $u(t)$ as input. The cross-correlation function is depicted in Figure 11.1c.

From the generation of $u(t)$ it is known that $u(t)$ and $\varepsilon(s)$ are uncorrelated for $t \leq s$ and correlated for $t > s$. The figure shows the normalized correlation function $\hat{r}_{\varepsilon u}(\tau)/[\hat{r}_u(0)\hat{r}_\varepsilon(0)]^{1/2}$ versus τ. As expected, it is close to zero for $\tau \geq 0$.

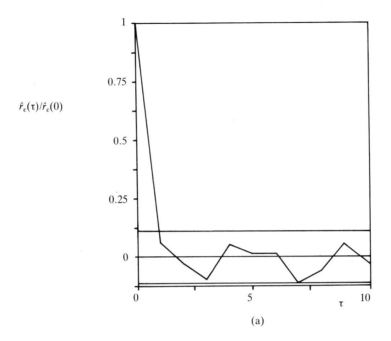

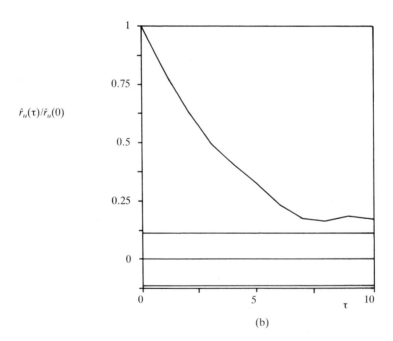

FIGURE 11.1 (a) Normalized covariance function of $\varepsilon(t)$. (b) Normalized covariance function of $u(t)$. (c) Normalized cross-covariance function, Example 11.4.

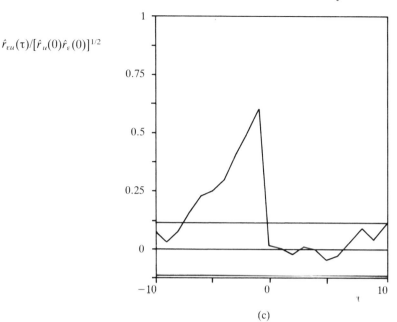

(c)

FIGURE 11.1 continued

The test quantity (11.18) is applied twice. First, it is used for $\bar{\tau} = 1$ and $m = 10$. The numerical value obtained is 20.2. Since the test quantity is approximately $\chi^2(10)$ distributed, one finds from Table B.1 that the threshold value is 18.3 for $\alpha = 0.05$ and 20.5 for $\alpha = 0.025$. One can accept the null hypothesis ($u(t)$ and $\varepsilon(s)$ uncorrelated for $t \leq s$) with a significance level of 0.025.

Second, the test is applied for $\bar{\tau} = -11$ and $m = 10$. The numerical value obtained was 296.7. Under the null hypothesis ($u(t)$ and $\varepsilon(s)$ uncorrelated) the test variable is ,approximately $\chi^2(10)$ distributed. Table B.1 gives the threshold value 25.2 for $\alpha = 0.005$. The null hypothesis is therefore rejected. ∎

It should be noted that the role of the statistical tests should not be exaggerated. A drastic illustration of the fact that statistical tests should not replace the study of plots of measured data and model output and the use of common sense is given in the following example.

Example 11.5 *Statistical tests and common sense*
The data were measured at a distillation column, $u(t)$ being the reflux ratio and $y(t)$ the top product composition. The following model structure was fitted to the data:

$$A(q^{-1})y(t) = B(q^{-1})u(t) + \frac{1}{D(q^{-1})} e(t)$$

with $na = nb = nc = 2$. A prediction error method was used to estimate the model parameters. It was implemented using a generalized least squares algorithm (cf.

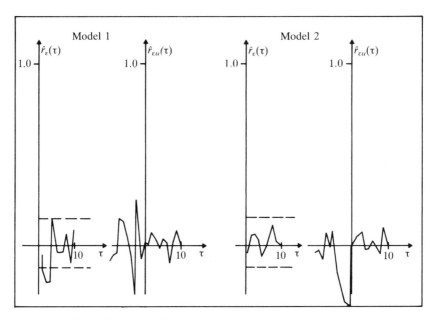

FIGURE 11.2 Normalized covariance functions. The dashed lines give the 95 percent significance interval.

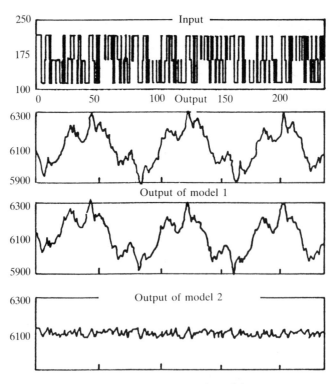

FIGURE 11.3 Input, output and model outputs.

Complement C7.4). Depending on the initial conditions $\hat{\theta}^{(0)}$ for the numerical optimization, two different models were obtained. The residual correlation functions corresponding to these two models are shown in Figure 11.2. From this figure both models seem quite reasonable, since they both give estimated autocorrelation functions $\hat{r}_\varepsilon(\tau)/\hat{r}_\varepsilon(0)$ that almost completely lie inside the 95 percent significance interval.

However, a plot of the measured signals and the model outputs (Figure 11.3) reveals that model 1 should be selected. For model 2 the characteristic oscillating pattern in the output, which is due to the periodic input used (the period of $u(t)$ is about 80), is modeled as a disturbance. A closer analysis of the loss function and its local minima corresponding to the two models is given by Söderström (1974).

This example shows that one should not rely on statistical tests only, but should also use graphical plots and common sense as complements. Note, however, that this example is extreme. In most practical cases the tests on the correlation functions give a very good indication of the relevance of the model. ∎

11.3 Is a model too complex?

This section describes some methods for testing model structures, based on properties of overparametrized models. Such models often give rise to pole-zero cancellations and singular information matrices. The exact conditions for this, however, are dependent on the type of model structure used. The following example illustrates the basic ideas of this approach to structure testing.

Example 11.6 *Effects of overparametrizing an ARMAX system*
Consider an ARMAX system

$$A_0(q^{-1})y(t) = B_0(q^{-1})u(t) + C_0(q^{-1})e(t) \tag{11.22}$$

where $A_0(q^{-1})$, $B_0(q^{-1})$ and $C_0(q^{-1})$ are assumed to be coprime. The system is further assumed to be identified in the model structure

$$A(q^{-1})y(t) = B(q^{-1})u(t) + C(q^{-1})\varepsilon(t) \tag{11.23}$$

where

$$na \geq na^0 \quad nb \geq nb^0 \quad nc \geq nc^0 \tag{11.24}$$

Then it is easily seen that all models satisfying

$$\begin{aligned}
A(q^{-1}) &\equiv A_0(q^{-1})M(q^{-1}) \\
B(q^{-1}) &\equiv B_0(q^{-1})M(q^{-1}) \\
C(q^{-1}) &\equiv C_0(q^{-1})M(q^{-1}) \\
M(q^{-1}) &= 1 + m_1 q^{-1} + \ldots + m_{nm} q^{-nm} \\
nm &= \min(na - na^0, nb - nb^0, nc - nc^0)
\end{aligned} \tag{11.25}$$

$(M(q^{-1}) = 1$ if $nm = 0)$,

434 *Validation and structure determination*   *Chapter 11*

give a correct description of the system, i.e. the corresponding values of θ lie in the set $D_T(\mathcal{S}, \mathcal{M})$ (see Example 6.7). Note that $m_1, \ldots, m_{nm}$ are arbitrary and that strict inequalities (11.24) are crucial for $nm \geq 1$ to hold.

Assume that a prediction error method is used for the parameter estimation. The effect of the overparametrization will be analyzed in two ways:

- Determination of the global minimum points of the asymptotic loss function.
- The singularity properties or, more generally, the null space of the information matrix.

To simplify the analysis the following assumptions are made:

- The input signal is persistently exciting of order $\max(na + nb_0, na_0 + nb)$. This type of assumption was introduced in Chapter 7 (see assumption A2 in Section 7.5) when analyzing the identifiability properties of PEMs.
- The system operates in open loop. It was noted in Chapter 10 that a low-order feedback will prevent identifiability. The assumption of open loop operation seems only partially necessary for the result that follows to hold, but it will make the analysis considerably simpler.

The two assumptions above are sufficient to guarantee that identifiability is not lost due to the experimental condition. The point of interest here is to see that parameter identifiability (see Section 6.4) cannot hold true due to overparametrization of the model.

First evaluate the global minimum points. The asymptotic loss function is given by

$$
\begin{aligned}
V(\theta) = E\varepsilon^2(t) &= E\left[\frac{1}{C(q^{-1})}\{A(q^{-1})y(t) - B(q^{-1})u(t)\}\right]^2 \\
&= E\left[\frac{A(q^{-1})}{C(q^{-1})}\left\{\frac{B_0(q^{-1})}{A_0(q^{-1})}u(t) + \frac{C_0(q^{-1})}{A_0(q^{-1})}e(t)\right\} - \frac{B(q^{-1})}{C(q^{-1})}u(t)\right]^2 \\
&= E[e(t)]^2 + E\left[\frac{A(q^{-1})B_0(q^{-1}) - A_0(q^{-1})B(q^{-1})}{A_0(q^{-1})C(q^{-1})}u(t)\right]^2 \\
&\quad + E\left[\frac{A(q^{-1})C_0(q^{-1}) - A_0(q^{-1})C(q^{-1})}{A_0(q^{-1})C(q^{-1})}e(t)\right]^2
\end{aligned}
\tag{11.26}
$$

The last equality follows since $e(t)$ is white noise and therefore is independent of $e(s)$, $s \leq t - 1$. Thus the global minimum points must satisfy

$$
\begin{aligned}
\frac{[A(q^{-1})B_0(q^{-1}) - A_0(q^{-1})B(q^{-1})]}{A_0(q^{-1})C(q^{-1})} u(t) &= 0 \quad \text{(w.p. 1)} \\
\frac{[A(q^{-1})C_0(q^{-1}) - A_0(q^{-1})C(q^{-1})]}{A_0(q^{-1})C(q^{-1})} e(t) &= 0 \quad \text{(w.p. 1)}
\end{aligned}
\tag{11.27}
$$

The true parameter values for which $A(q^{-1}) \equiv A_0(q^{-1})$, $B(q^{-1}) \equiv B_0(q^{-1})$, $C(q^{-1}) \equiv C_0(q^{-1})$, clearly form a possible solution to (11.27).

Since $e(t)$ is persistently exciting of any order (see Property 2 of Section 5.4), and $u(t)$ is persistently exciting of order $p = \max(na + nb_0, na_0 + nb)$, it follows from Property 4 of Section 5.4 that

Section 11.3   Is a model too complex?   435

$$\frac{1}{A_0(q^{-1})C(q^{-1})} u(t) \quad \text{and} \quad \frac{1}{A_0(q^{-1})C(q^{-1})} e(t)$$

are persistently exciting of order p and of any finite order, respectively. From Property 5 of Section 5.4 and (11.27) it follows that

$$A(q^{-1})B_0(q^{-1}) - A_0(q^{-1})B(q^{-1}) \equiv 0 \tag{11.28a}$$

$$A(q^{-1})C_0(q^{-1}) - A_0(q^{-1})C(q^{-1}) \equiv 0 \tag{11.28b}$$

However, the identities (11.28) are precisely the same as (6.48). Therefore the solution is given by (11.25). Thus (11.25) describes *all* the global minimum points of the asymptotic loss function.

Next consider the information matrix. At this point assume the noise to be Gaussian distributed. Since the information matrix, except for a scale factor, is equal to the Hessian (the matrix of second-order derivatives) of the asymptotic loss function one can expect its singularity properties to be closely related to the properties of the set of global minima. In particular, this matrix must be singular when there are infinitely many non-isolated global minimum points.

The information matrix for the system (11.22) is given by (see (B.29) and (7.79)–(7.81)):

$$J = \frac{N}{\lambda^2} E\psi(t)\psi^T(t) \tag{11.29}$$

where

$$\psi^T(t) = \frac{1}{C_0(q^{-1})} \tag{11.30}$$
$$\times (y(t-1) \ldots y(t-na) \quad -u(t-1) \ldots -u(t-nb) \quad -e(t-1) \ldots -e(t-nc))$$

Let

$$x = (\alpha_1 \ldots \alpha_{na} \quad \beta_1 \ldots \beta_{nb} \quad \gamma_1 \ldots \gamma_{nc})^T$$

be an arbitrary element of the null space of J. Set

$$\alpha(q^{-1}) = \sum_{i=1}^{na} \alpha_i q^{-i} \quad \beta(q^{-1}) = \sum_{i=1}^{nb} \beta_i q^{-i} \quad \gamma(q^{-1}) = \sum_{i=1}^{nc} \gamma_i q^{-i}$$

The null space of J is investigated by solving

$$Jx = 0 \tag{11.31}$$

This equation can be equivalently rewritten as

$$x^T J x = 0 \Leftrightarrow E[\psi^T(t)x]^2 = 0$$
$$\Leftrightarrow \psi^T(t)x = 0 \quad (\text{w.p. } 1) \tag{11.32}$$
$$\Leftrightarrow \alpha(q^{-1})y(t) - \beta(q^{-1})u(t) - \gamma(q^{-1})e(t) = 0$$

It can be seen from (11.32) that if the system is controlled with a noise-free (low-order) linear feedback $\tilde{\beta}(q^{-1})u(t) = \tilde{\alpha}(q^{-1})y(t)$ with $n\tilde{\alpha} < n\alpha$ and $n\tilde{\beta} < n\beta$ then J is singular.

Indeed, in such a case (11.32) will have (at least) one nontrivial solution $\alpha(q^{-1}) = q^{-1}\tilde{\alpha}(q^{-1})$, $\beta(q^{-1}) = q^{-1}\tilde{\beta}(q^{-1})$ and $\gamma(q^{-1}) = 0$. Since J is proportional to the inverse of the covariance matrix of the parameter estimates, its singularity implies that the system is not identifiable. This is another derivation of the result of Section 10.3 that a system controlled by a low-order noise-free linear feedback is not identifiable.

Now (11.22) and (11.32) give, under the general assumptions introduced above,

$$\left[\alpha(q^{-1})\frac{B_0(q^{-1})}{A_0(q^{-1})} - \beta(q^{-1})\right]u(t) = 0 \quad \text{(w.p. 1)}$$

$$\left[\alpha(q^{-1})\frac{C_0(q^{-1})}{A_0(q^{-1})} - \gamma(q^{-1})\right]e(t) = 0 \quad \text{(w.p. 1)}$$

from which one can conclude as before ($u(t)$, $e(t)$ being pe) that

$$\begin{aligned} \alpha(q^{-1})B_0(q^{-1}) - A_0(q^{-1})\beta(q^{-1}) &\equiv 0 \\ \alpha(q^{-1})C_0(q^{-1}) - A_0(q^{-1})\gamma(q^{-1}) &\equiv 0 \end{aligned} \quad (11.33)$$

These identities are similar to (11.28). To see this more clearly, consider again (11.28) and make the following substitutions:

$$A(q^{-1}) = A_0(q^{-1}) + \alpha(q^{-1})$$
$$B(q^{-1}) = B_0(q^{-1}) + \beta(q^{-1})$$
$$C(q^{-1}) = C_0(q^{-1}) + \gamma(q^{-1})$$

Then (11.28) is easily transformed into (11.33). Thus the properties of the null space of J can be deduced from (11.25).

Note especially that if $nm = 0$ in (11.25) then $x = 0$ is the only solution to (11.31). This means that the information matrix is nonsingular. However, if $nm \geq 1$ then the general solution to (11.31) is given by

$$\alpha(q^{-1}) \equiv A_0(q^{-1})L(q^{-1})$$
$$\beta(q^{-1}) \equiv B_0(q^{-1})L(q^{-1})$$
$$\gamma(q^{-1}) \equiv C_0(q^{-1})L(q^{-1})$$

where

$$L(q^{-1}) = l_1 q^{-1} + \ldots + l_{nm} q^{-nm} = M(q^{-1}) - 1$$

is arbitrary. Thus the solutions to (11.31) lie in an nm-dimensional subspace of $\mathcal{R}^{na+nb+nc}$.

To summarize this example, note that:

- If $nm = 0$ then no pole-zero cancellation occurs in the model. There is a unique global minimum point of $V(\theta)$, which is given by the true parameters. The information matrix is nonsingular.
- If $nm > 0$ then pole-zero cancellations appear. The function $V(\theta)$ has a set of nonisolated global minima. The information matrix is singular.

Section 11.3   Is a model too complex?   437

Note that when $nc = nc^0 = 0$ ('the least squares case') there will be no pole-zero cancellations or singular information matrices. In this case $nm = 0$ due to $nc = nc^0$ (see also Problem 11.2).   ∎

An approach to model structure determination

The results described above can be used for model structure selection as follows. Estimate the parameters in an increasing set of model structures. For each model make a test for pole-zero cancellations and/or singularity of the information matrix. When cancellation occurs or the information matrix has become singular, then the model structure has been chosen too large and the preceding one should be selected. An approach for a systematic pole-zero cancellation test has been presented by Söderström (1975c).

Modification for instrumental variable methods

As described above, this approach is tied to the use of prediction error methods. However, it can also be used in a modified form for instrumental variable methods. Then instead of the information matrix, an instrumental product moment matrix is to be used. Such a matrix can be written as (cf. (8.21))

$$R = EZ(t)F(q^{-1})\phi^T(t) \tag{11.34}$$

This matrix is singular if the model order is chosen to be greater than the order of the true system. The matrix R will otherwise be nonsingular (see Chapter 8). In practice, R is not available and one has to work with an estimate

$$\hat{R} = \frac{1}{N} \sum_{t=1}^{N} Z(t)F(q^{-1})\phi^T(t)$$

In most cases it is too cumbersome to test in a well-defined statistical way whether $\hat{R}$ is (nearly) singular (for some simple cases such as scalar ARMA models, however, this is feasible; see for example Stoica, 1981c and Fuchs, 1987). Instead one can for example compute det $\hat{R}$ for increasing model orders. When det $\hat{R}$ drops from a large to a small value, the model order corresponding to the large value of det $\hat{R}$ is chosen. An alternative is to use cond$(\hat{R}) = \|\hat{R}\| \|\hat{R}^{-1}\|$ and look for a significant increase in this quantity. When the order is overestimated cond$(\hat{R})$ can be expected to increase drastically.

Note that the 'instruments' $Z(t)$ and the filter $F(q^{-1})$ can be chosen in many ways. It is not necessary to make the same choice when computing the IV parameter estimates $\hat{\theta}$ and when searching for the best model order, but using the same instruments and prefilter is highly convenient from a computational viewpoint. One possibility that has shown some popularity is to take $F(q^{-1}) \equiv I$ and to let $Z(t)$ be an estimate of $\bar{\phi}(t)$, the noise-free part of $\phi(t)$. Then R in (11.34) would be the covariance matrix of $\bar{\phi}(t)$. Note that an estimate of $\bar{\phi}(t)$ requires an estimated parameter vector $\hat{\theta}$ for its evaluation.

Numerical problems

The main drawback of the methods described in this section lies in the difficulty of designing good statistical test criteria and in the numerical problems associated with evaluation of the rank of a matrix. The use of a singular value decomposition seems to be the numerically soundest way for the test of nonsingularity or more generally for determining the rank of a given matrix. (See Section A.2 for a detailed discussion of singular value decomposition.) Due to their drawbacks these methods cannot be recommended as a first choice for SISO systems. For multivariable systems, however, there are often fewer alternatives and rank tests have been advocated in many papers.

11.4 The parsimony principle

The parsimony principle is a useful rule when determining an appropriate model order. This principle says that out of two or more competing models which all explain the data well, the model with the smallest number of independent parameters should be chosen. Such a choice will be shown to give the best accuracy. This rule is quite in line with common sense: 'Do not use extra parameters for describing a dynamic phenomenon if they are not needed.' The parsimony principle is discussed, for example, by Box and Jenkins (1976). A theoretical justification is given in Complement C11.1. Here its use is illustrated in a simple case which will be useful in the subsequent discussion in this chapter.

Consider a single output system. The general case of multi-output systems is covered in Problem 11.10. Assume that $\hat{\theta}_N$ denotes a parameter vector estimated from past data using the PEM. Assume further that the system belongs to the model structure considered. This means that there exists a true (not necessarily unique) parameter vector θ_0.

A scalar measure will be used to assess the goodness of the model associated with $\hat{\theta}_N$. Such a measure should be a function of $\hat{\theta}_N$. It will be denoted by $W(\hat{\theta}_N)$ or W_N, where the dependence on the number of data is emphasized. The assessment measure $W(\theta)$ must be a smooth function of θ and be minimized by the true parameter vector θ_0:

$$W(\theta) \geq W(\theta_0) \quad \text{all } \theta \tag{11.35}$$

(see Figure 11.4).

When the estimate $\hat{\theta}_N$ deviates a little from θ_0, the criterion W will increase somewhat above its minimum value $W(\theta_0)$. This increase, $W(\hat{\theta}_N) - W(\theta_0)$, will be taken as a scalar measure of the performance of the model.

The assessment criterion can be chosen in several ways, including the following:

- The variance of the one-step prediction errors, when the model is applied to future data. This possibility will be considered later in this section.
- The variance of the multi-step prediction error (cf. Problem 11.5).
- The deviation of the estimated transfer function from that of the true system. Such a deviation can be expressed in the frequency domain.

Section 11.4   The parsimony principle   439

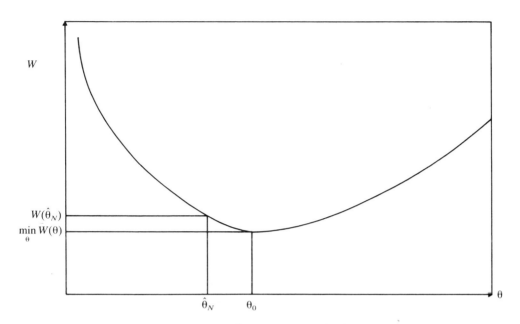

FIGURE 11.4 An assessment criterion.

- Assume that an optimal regulator is based on the identified model and applied to the (true) system. If the model were perfect, the closed loop system would perform optimally. Due to the deviation of the model from the true system the performance of the closed loop system will deteriorate somewhat. One can take the performance of the closed loop system as an assessment criterion. (See Problem 11.4 for an illustration.)

In what follows a specific choice of the assessment criterion is discussed. Let W_N be the prediction error variance when the model corresponding to $\hat{\theta}_N$ is used to predict *future* data. This means that

$$W_N = E\varepsilon^2(t, \hat{\theta}_N) \tag{11.36}$$

In (11.36) and also in (11.37) below, the expectation is conditional with respect to past data in the sense that

$$W_N = E[\varepsilon^2(t, \hat{\theta}_N)|\hat{\theta}_N]$$

If the estimate $\hat{\theta}_N$ were exact, i.e. $\hat{\theta}_N = \theta_0$, then the prediction errors $\{\varepsilon(t, \hat{\theta}_N)\}$ would be white noise, $\{e(t)\}$, and have minimum variance Λ. Consider now how much the prediction error variance W_N is increased due to the deviation of $\hat{\theta}_N$ from the true value θ_0. A Taylor series expansion of $\varepsilon(t, \hat{\theta}_N)$ around θ_0 gives

$$W_N \approx E\left[\varepsilon(t, \theta_0) + \frac{\partial \varepsilon(t, \theta)}{\partial \theta}\bigg|_{\theta=\theta_0}(\hat{\theta}_N - \theta_0)\right]^2$$
$$\triangleq E[e(t) - \psi^T(t, \theta_0)(\hat{\theta}_N - \theta_0)]^2$$

$$= \Lambda + E\psi^T(t, \theta_0)(\hat\theta_N - \theta_0)(\hat\theta_N - \theta_0)^T\psi(t, \theta_0)$$
$$= \Lambda + (\hat\theta_N - \theta_0)^T[E\psi(t, \theta_0)\psi^T(t, \theta_0)](\hat\theta_N - \theta_0) \tag{11.37}$$

The second term in (11.37) shows the increase over the minimal value Λ due to the deviation of the estimate $\hat\theta_N$ from the true value θ_0. Taking expectation of W_N with respect to the past data, which are implicit in $\hat\theta_N$, gives

$$EW_N \approx \Lambda + \text{tr}[E(\hat\theta_N - \theta_0)(\hat\theta_N - \theta_0)^T][E\psi(t, \theta_0)\psi^T(t, \theta_0)] \tag{11.38}$$

According to (7.59) the estimate $\hat\theta_N$ is asymptotically Gaussian distributed,

$$\sqrt{N}(\hat\theta_N - \theta_0) \xrightarrow{\text{dist}} \mathcal{N}(0, \Lambda[E\psi_p(t, \theta_0)\psi_p^T(t, \theta_0)]^{-1}) \tag{11.39}$$

Note that $\psi(t, \theta_0)$ in (11.37), (11.38) is to be evaluated for the future (fictitious) data, while $\psi_p(t, \theta_0)$ in (11.39) refers to the past data, from which the estimate $\hat\theta_N$ was found. If the *same experimental condition* is assumed for the past and the future data, then the second order properties of $\psi_p(t, \theta_0)$ and $\psi(t, \theta_0)$ are identical. For such a case (11.38), (11.39) give

$$\boxed{EW_N \approx \Lambda(1 + p/N)} \tag{11.40}$$

where $p = \dim \theta$.

This expression is remarkable in its simplicity. Note that its derivation has not been tailored to any special model structure. It says that the expected prediction error variance increases with a relative amount of p/N. Thus, there is a penalty in using models with unnecessarily many parameters. This can be seen as a formal statement of the parsimony principle.

In Complement C11.1 it is shown that the parsimony principle holds true under much more general conditions. The expected value of a fairly general assessment criterion $W(\hat\theta_N)$ will, under very mild conditions, increase when the model structure is expanded beyond the true structure.

11.5 Comparison of model structures

This section describes methods that can be used to compare two or more model structures. For such comparisons a discriminating criterion is needed. The discussion will be confined to the case when a prediction error method (PEM) is used for estimating the parameters. Recall that for a PEM the parameter estimate $\hat\theta_N$ is obtained as the minimizing element of a loss function $V_N(\theta)$ (see (7.17)).

When the model structure is expanded so that more parameters are included in the parameter vector, the minimal value of $V_N(\theta)$ naturally decreases since new degrees of freedom have been added to the optimization problem, or, in other words, the set over which optimization is done has been enlarged. The comparison of model structures can be interpreted as a test for a *significant* decrease in the minimal values of the loss function

Section 11.5   *Comparison of model structures*   441

associated with the (nested) model structures in question. This is conceptually the same problem as that discussed in Section 4.4 for linear regressions.

The *F*-test

Let $\mathcal{M}_1$ and $\mathcal{M}_2$ be two model structures, such that $\mathcal{M}_1 \subset \mathcal{M}_2$ ($\mathcal{M}_1$ is a subset of $\mathcal{M}_2$; for example, $\mathcal{M}_1$ corresponds to a lower-order model than $\mathcal{M}_2$). In such a case they are called hierarchical model structures. Further let V_N^i denote the minimum of $V_N(\theta)$ in the structure $\mathcal{M}_i$ ($i = 1, 2$) and let $\mathcal{M}_i$ have p_i parameters. As for the static case treated in Chapter 4, one may try

$$x = N \frac{V_N^1 - V_N^2}{V_N^2} \tag{11.41}$$

as a test quantity for comparing the model structures $\mathcal{M}_1$ and $\mathcal{M}_2$. If x is 'large' then one concludes that the decrease in the loss function from V_N^1 to V_N^2 is significant and hence that the model structure $\mathcal{M}_2$ is significantly better than $\mathcal{M}_1$. On the other hand, when x is 'small', the conclusion is that $\mathcal{M}_1$ and $\mathcal{M}_2$ are almost equivalent and according to the parsimony principle the smaller model structure $\mathcal{M}_1$ should be chosen as the more appropriate one.

The discussion above leads to a qualitative procedure for discriminating between $\mathcal{M}_1$ and $\mathcal{M}_2$. To get a quantitative test procedure it is necessary to be more exact about what is meant by saying that x is 'large' or 'small'. This is done in the following.

First consider the case when $\mathcal{S} \notin \mathcal{M}_1$, i.e. $\mathcal{M}_1$ is not large enough to include the true system. Then the decrease $V_N^1 - V_N^2$ in the criterion function will be $O(1)$ (that is, it does not go to zero as $N \to \infty$) and therefore the test quantity x, (11.41), will be of magnitude N.

Next assume that $\mathcal{M}_1$ is large enough to include the true system, i.e.

$$\mathcal{S} \in \mathcal{M}_1 \subset \mathcal{M}_2 \tag{11.42}$$

Then it is possible to prove (see Appendix A11.2 for details) that

$$x = N \frac{V_N^1 - V_N^2}{V_N^2} \xrightarrow{\text{dist}} \chi^2(p_2 - p_1) \tag{11.43}$$

The result (11.43) can be used to conceive a simple test for model structure selection. At a significance level α (typical values in practice could range from 0.01 to 0.1) the smaller model structure $\mathcal{M}_1$ is selected over $\mathcal{M}_2$ if

$$x \leq \chi_\alpha^2(p_2 - p_1) \tag{11.44}$$

where $\chi_\alpha^2(p_2 - p_1)$ is defined by (11.11). Otherwise $\mathcal{M}_2$ is selected.

In view of (11.41), the inequality (11.44) can be rewritten as

$$V_N^1 \leq V_N^2[1 + (1/N)\chi_\alpha^2(p_2 - p_1)] \tag{11.45}$$

The test (11.44) is closely related to the F-test which was developed in some detail in Section 4.4 for the static linear regression case. It was shown there that

$$x \frac{N - p_2}{N(p_2 - p_1)}$$

is exactly $F(p_2 - p_1, N - p_2)$ distributed. This implies that x is asymptotically (for $N \to \infty$) $\chi^2(p_2 - p_1)$ distributed as in (11.43) (see Section B.2 of Appendix B). Although the result on F-distribution refers to static linear regressions, the test procedure is often called 'F-test' also for the 'dynamic' case where the χ^2 distribution is used.

Criteria with complexity terms

Another approach to model structure selection consists of using a criterion for assessment of the model structures under study. Such a criterion may for example be obtained by penalizing in some way the decrease of the loss function $V_N(\hat{\theta}_N)$ with increasing model sets. *The model structure giving the smallest value of this criterion is selected.* A general form of this type of criterion is the following:

$$W_N = V_N(\hat{\theta}_N)[1 + \beta(N, p)] \tag{11.46}$$

where $\beta(N, p)$ is a function of N and the number of parameters p in the model, which should increase with p (in order to penalize too complex (overparametrized) model structures in view of the parsimony principle) but should tend to zero as $N \to \infty$ (to guarantee that the penalizing term in (11.46) will not obscure the decrease of the loss function $V_N(\hat{\theta}_N)$ with increasing underparametrized model structures). A typical choice is $\beta(N, p) = 2p/N$.

An alternative is to use the criterion

$$W_N = N \log V_N(\hat{\theta}_N) + \gamma(N, p) \tag{11.47}$$

where the additional term $\gamma(N, p)$ should penalize high-order models. The choice $\gamma(N, p) = 2p$ will give the widely used Akaike's information criterion (AIC) (a justification for this choice is presented in Example 11.7 below):

$$\text{AIC} = N \log V_N(\hat{\theta}_N) + 2p \tag{11.48}$$

It is not difficult to see that the criteria (11.46) and (11.47) are asymptotically equivalent provided $\gamma(N, p) = N\beta(N, p)$. Indeed, for large N

$$\log\{V_N(\hat{\theta}_N)[1 + \beta(N, p)]\} = \log V_N(\hat{\theta}_N) + \log[1 + \beta(N, p)]$$

$$\approx \frac{1}{N}[N \log V_N(\hat{\theta}_N) + N\beta(N, p)]$$

which shows that the logarithm of (11.46) asymptotically is proportional to (11.47). Since the logarithm function is monotonically increasing, it follows that for large N the two criteria will be minimized by the same model structures.

Several proposals for the terms $\gamma(N, p)$ in (11.47) have appeared in the literature. As mentioned above, the choice $\gamma(N, p) = 2p$ corresponds to Akaike's information criterion. Other choices, such as $\gamma(N, p) = p \log N$ and $\gamma(N, p) = 2pc \log(\log N)$, with $c \geq 1$ being a constant, allow $\gamma(N, p)$ to grow slowly with N. As will be shown later in this section, this is a way of obtaining consistent order estimates (cf. Example 11.10).

Example 11.7 *The FPE criterion*
Consider a single output system. The assessment criterion W_N is required to express the expected prediction error variance when the prediction of future data is based on the model determined from past data. This means that

$$W_N = E\varepsilon^2(t, \hat{\theta}_N) \tag{11.49}$$

where the expectation is with respect to both future *and* past data. Since W_N in (11.49) is not directly measurable, some approximations have to be made. It was shown in Section 11.4 that (for a flexible enough structure)

$$W_N \approx \Lambda(1 + p/N) \tag{11.50}$$

Now, Λ in (11.50) is unknown and must be replaced by some known quantity. Recall that the loss function used in the estimation is

$$V_N(\theta) = \frac{1}{N} \sum_{t=1}^{N} \varepsilon^2(t, \theta) \tag{11.51}$$

Hence the variance Λ could possibly be substituted by $V_N(\hat{\theta}_N)$. However, $V_N(\hat{\theta}_N)$ deviates from Λ due to the fit of $\hat{\theta}_N$ to the given realization. More exactly, due to the fitting of the model $\mathcal{M}(\hat{\theta}_N)$ to the realization at hand, $V_N(\hat{\theta}_N)$ will be a biased estimate of Λ. To investigate the asymptotic bias $V_N(\hat{\theta}_N) - \Lambda$, note that

$$V_N(\hat{\theta}_N) = V_\infty(\theta_0) + [V_N(\theta_0) - V_\infty(\theta_0)] + [V_N(\hat{\theta}_N) - V_N(\theta_0)]$$

$$\approx \Lambda + \left[\frac{1}{N}\sum_{t=1}^{N} e^2(t) - \Lambda\right] + \left[-V'_N(\hat{\theta}_N)(\theta_0 - \hat{\theta}_N)\right.$$

$$\left. - \frac{1}{2}(\theta_0 - \hat{\theta}_N)^T V''_N(\hat{\theta}_N)(\theta_0 - \hat{\theta}_N)\right]$$

$$\approx \Lambda + \left[\frac{1}{N}\sum_{t=1}^{N} e^2(t) - \Lambda\right] - \frac{1}{2}\left[(\theta_0 - \hat{\theta}_N)^T V''_\infty(\theta_0)(\theta_0 - \hat{\theta}_N)\right] \tag{11.52}$$

Here the first term is a constant. The expected value of the second term is zero, and the expected value of the third term can easily be evaluated. Recall from (7.59) that

$$\sqrt{N}(\hat{\theta}_N - \theta_0) \xrightarrow{\text{dist}} \mathcal{N}(0, 2\Lambda[V''_\infty(\theta_0)]^{-1})$$

Hence

$$-E\frac{1}{2}(\theta_0 - \hat{\theta}_N)^T V''_\infty(\theta_0)(\theta_0 - \hat{\theta}_N)$$

$$= -\frac{1}{2} E \operatorname{tr}\{(\theta_0 - \hat{\theta}_N)(\theta_0 - \hat{\theta}_N)^T V''_\infty(\theta_0)\}$$

$$= -\frac{1}{2} \operatorname{tr}\{2\Lambda [V''_\infty(\theta_0)]^{-1} V''_\infty(\theta_0)/N\}$$

$$= -\Lambda\, p/N$$

It follows from the above calculations that an asymptotically unbiased estimate of Λ is given by

$$\hat{\Lambda} = V_N(\hat{\theta}_N)/(1 - p/N) \tag{11.53}$$

The above estimate is identical to the unbiased estimate of Λ derived in the linear regression case (see (4.13)). Note that in contrast to the present case, for linear regressions the unbiasedness property holds for all N (not only for $N \to \infty$).

Combining (11.50) and (11.53) now gives a realizable criterion, namely

$$W_N = V_N(\hat{\theta}_N) \frac{1 + p/N}{1 - p/N} \triangleq \text{FPE} \tag{11.54}$$

This is known as the final prediction error (FPE) criterion (Davisson, 1965, Akaike, 1969). Note that for large N (11.54) can be approximated by

$$W_N = V_N(\hat{\theta}_N)\left[1 + \frac{2p/N}{1 - p/N}\right] \approx V_N(\hat{\theta}_N)\left[1 + \frac{2p}{N}\right]$$

which is of the form (11.46) with $\beta(N, p) = 2p/N$. The corresponding criterion (11.47) is

$$W_N = N \log V_N(\hat{\theta}_N) + 2p$$

which, as already stated, is the so-called Akaike information criterion (AIC).

Finally, there is a more subtle aspect pertaining to (11.54). As derived, the FPE criterion (11.54) holds for the model structures which are flexible enough to include the true system. In a model structure selection application one will also have to consider underparametrized model structures. For such structures (11.54) loses its interpretation as an estimate of the 'final prediction error'. Nevertheless, (11.54) can still be used to assess the difference between the prediction ability of various underparametrized model structures. This is so since this difference is $O(1)$ and therefore can be consistently estimated by using (11.54) (or even $V_N(\hat{\theta}_N)$). ∎

Equivalence of the F-test and the criteria with complexity terms

The F-test, or rather the χ^2 test, (11.41), will now be compared to the model structure determination procedure based on minimization of the criteria W_N given by (11.46) and (11.47).

Section 11.5 *Comparison of model structures* 445

When the test quantity x is used, the smaller model structure, $\mathcal{M}_1$, is selected if

$$V_N^1 \leq V_N^2[1 + (1/N)\chi_\alpha^2(p_2 - p_1)] \tag{11.55}$$

(see (11.45)).

Assume that one of the criteria W_N, (11.46), (11.47), is used to select one of the *two* model structures $\mathcal{M}_1$ and $\mathcal{M}_2$ with $\mathcal{M}_1 \subset \mathcal{M}_2$. If the criterion (11.46) is used, $\mathcal{M}_1$ is selected if

$$V_N^1[1 + \beta(N, p_1)] \leq V_N^2[1 + \beta(N, p_2)]$$

i.e. if

$$V_N^1 \leq V_N^2 \frac{1 + \beta(N, p_2)}{1 + \beta(N, p_1)} \tag{11.56}$$

Comparing (11.55) and (11.56), it is seen that the criterion (11.46) used as above can be interpreted as a χ^2 test with a prespecified significance level. In fact it is easy to show that

$$\boxed{\chi_\alpha^2(p_2 - p_1) = N \frac{\beta(N, p_2) - \beta(N, p_1)}{1 + \beta(N, p_1)}} \tag{11.57}$$

Similarly, using the criterion (11.47) $\mathcal{M}_1$ is selected if

$$N \log V_N^1 + \gamma(N, p_1) \leq N \log V_N^2 + \gamma(N, p_2)$$

i.e. if

$$V_N^1 \leq V_N^2 \exp[\{\gamma(N, p_2) - \gamma(N, p_1)\}/N] \tag{11.58}$$

By comparing (11.55) and (11.58) it can be seen that for the problem formulation considered, the χ^2 test and the approach based on (11.47) are equivalent provided

$$\boxed{\chi_\alpha^2(p_2 - p_1) = N\,(\exp[\{\gamma(N, p_2) - \gamma(N, p_1)\}/N] - 1)} \tag{11.59}$$

The above calculations show that certain significance levels can be associated with the FPE and AIC criteria when these are used to select between two model structures. The details are developed in the following example.

Example 11.8 *Significance levels for FPE and AIC*

Consider first the FPE criterion. From (11.46) and (11.54),

$$\beta(N, p) = 2p/(N - p)$$

Hence from (11.57),

$$\chi_\alpha^2(p_2 - p_1) = N \frac{2p_2/(N - p_2) - 2p_1/(N - p_1)}{1 + 2p_1/(N - p_1)}$$

$$= N \frac{2N(p_2 - p_1)}{(N + p_1)(N - p_2)} \approx 2(p_2 - p_1) \tag{11.60}$$

where the approximate equality holds for large values of N.

If in particular $p_2 - p_1 = 1$ then for large values of N, $\chi_\alpha^2(1) = 2$, which gives $\alpha = 0.157$. This means that the risk of choosing the larger structure $\mathcal{M}_2$ when $\mathcal{M}_1$ (with $p_2 - p_1 = 1$) is more appropriate, will asymptotically be 15.7 percent.

The AIC criterion can be analyzed in a similar fashion. In this case

$$\gamma(N, p) = 2p$$

and (11.59) gives

$$\chi_\alpha^2(p_2 - p_1) = N[e^{2(p_2-p_1)/N} - 1] \approx 2(p_2 - p_1) \tag{11.61}$$

where again the approximation holds for large values of N. When $p_2 = p_1 + 1$ the risk of overfitting is 15.7 percent, exactly as in the case of the FPE (this was expected since as shown previously, AIC and FPE are asymptotically equivalent). Note that the above risk of overfitting does not vanish when $N \to \infty$, which means that neither the AIC nor the FPE estimates of the order are consistent. ∎

As a further comparison of the model order determination methods introduced above, consider the following example based on Monte Carlo simulations.

Example 11.9 *Numerical comparisons of the F-test, FPE and AIC*

The following first-order AR process was simulated, in which $e(t)$ is white noise of zero mean and unit variance:

$$y(t) - 0.9y(t-1) = e(t) \tag{11.62}$$

generating 100 different long realizations ($N = 500$) as well as 100 short ones ($N = 25$). First- and second-order AR models, denoted below by $\mathcal{M}_1$ and $\mathcal{M}_2$, were fitted to the data using the least squares method. The criteria FPE and AIC as well as the F-test quantity were computed for each realization. Using the preceding analysis it will be shown that the following quantities are theoretically all asymptotically $\chi^2(1)$ distributed:

$$f \triangleq \frac{V_N^1 - V_N^2}{V_N^2}(N-2) \tag{11.63}$$

$$N[\text{FPE}(\mathcal{M}_1) - \text{FPE}(\mathcal{M}_2)] + 2 \tag{11.64}$$

$$\text{AIC}(\mathcal{M}_1) - \text{AIC}(\mathcal{M}_2) + 2 \tag{11.65}$$

That $f \xrightarrow{\text{dist}} \chi^2(1)$ follows from (11.43).

To derive the asymptotic distribution of (11.64) note that $1 = \Lambda \approx V_N^1 \approx V_N^2$ and use (11.43) to get

Section 11.5 — Comparison of model structures

$$N[\text{FPE}(\mathcal{M}_1) - \text{FPE}(\mathcal{M}_2)] + 2$$

$$= N\left[V_N^1\left(1 + \frac{2p_1}{N - p_1}\right) - V_N^2\left(1 + \frac{2p_2}{N - p_2}\right)\right]/\Lambda + 2$$

$$= N\frac{V_N^1 - V_N^2}{\Lambda} + 2N\left[V_N^1\frac{p_1}{N - p_1} - V_N^2\frac{p_2}{N - p_2}\right]/\Lambda + 2$$

$$\approx f - 2[p_2 - p_1] + 2 = f \xrightarrow{\text{dist}} \chi^2(1)$$

To derive the asymptotic distribution of (11.65), use (11.43) once more and proceed as follows.

$$\text{AIC}(\mathcal{M}_1) - \text{AIC}(\mathcal{M}_2) + 2 = N \log V_N^1 + 2p_1 - N \log V_N^2 - 2p_2 + 2$$

$$= N \log \frac{V_N^1}{V_N^2} = N \log(1 + f/(N-2)) \approx f \xrightarrow{\text{dist}} \chi^2(1)$$

The experimental distributions (the histograms) of the quantities (11.63)–(11.65), obtained from the 100 realizations considered, are illustrated in Figure 11.5.

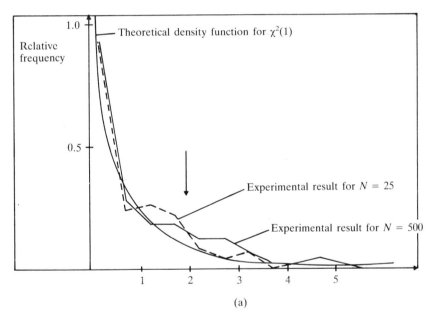

(a)

FIGURE 11.5 (a) Normalized curves for AIC $(\mathcal{M}_1)$ − AIC $(\mathcal{M}_2)$ + 2 (the AIC will select first-order models for the realizations to the left of the arrow). (b) Normalized curves for $N[\text{FPE}(\mathcal{M}_1) - \text{FPE}(\mathcal{M}_2)] + 2$ (the FPE criterion will select first-order models for the realizations to the left of the arrow). (c) Normalized curves for f (the F-test with a 5 percent significance level will select first-order models for the realizations to the left of the arrow).

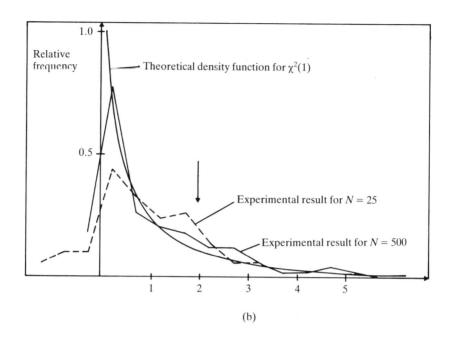

(b)

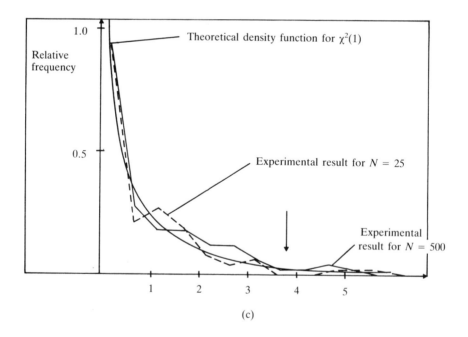

(c)

FIGURE 11.5 continued

Section 11.5
Comparison of model structures

It can be seen from Figure 11.5 that the experimental distributions, except for that of the FPE for the case $N = 25$, show very good agreement with the theoretical $\chi^2(1)$ distribution. Note also that the quite short realizations with $N = 25$ gave a result close to the asymptotically valid theoretical results.

It was evaluated explicitly how many times a second (or higher) order model has been chosen. The numerical results, given in Table 11.1, are quite congruent with the theory.

TABLE 11.1 Numerical evaluation of the risk of overfitting when using AIC, FPE and the F-test for Example 11.9

Number of samples	Criterion	Proportion of realizations giving a second-order model	
		Theoretical	Experimental
$N = 500$	AIC	0.16	0.20
	FPE	0.16	0.20
	F-test	0.05	0.04
$N = 25$	AIC	(0.16)	0.18
	FPE	(0.16)	0.23
	F test	(0.05)	0.09

■

Consistency analysis

In the previous subsection it was found that the FPE and AIC do not give consistent estimates of the model order. There is a nonzero risk, even for a large number of data points, of choosing too high a model order. However, this should not be seen as too serious a drawback. Both the FPE and AIC enjoy certain properties in the practically important case where the system does not belong to the class of model structures considered (see Shibata, 1980; Stoica, Eykhoff *et al.*, 1986). More exactly, both AIC and FPE determine good prediction models no matter whether or not the system belongs to the model set.

In the following it will be shown that it is possible to get consistent estimates of the model order by using criteria of the type (11.46) or (11.47). (An order selection rule is said to be consistent if the probability of selecting a wrong order tends to zero as the number of data points tends to infinity.)

It was found previously that the selection between two model structures with the criteria (11.46) and (11.47) is equivalent to the examination of the inequality (11.55) for $\chi_\alpha^2(p_2 - p_1)$ given by (11.57) and (11.59), respectively. The analysis that follows is based on this simple but essential observation.

Consider first the risk of overfitting of the selection rules (11.46) or (11.47). Therefore assume $\mathscr{S} \in \mathscr{M}_1 \subset \mathscr{M}_2$. The probability of choosing the too large model structure $\mathscr{M}_2$ will then be

$$P(V_N^1 > V_N^2(1 + \chi_\alpha^2(p_2 - p_1)/N)) = P(x > \chi_\alpha^2(p_2 - p_1)) = \alpha$$

where x is the F-test quantity defined in (11.43). To eliminate the risk of overfitting α must tend to zero as N tends to infinity, or equivalently

$$\chi_\alpha^2(p_2 - p_1) \to \infty \quad \text{as } N \to \infty \tag{11.66}$$

Next consider the risk of underfitting. So assume that $\mathscr{S} \notin \mathscr{M}_1 \subset \mathscr{M}_2$. The probability of choosing the too small model structure $\mathscr{M}_1$ will be

$$P(V_N^1 < V_N^2(1 + \chi_\alpha^2(p_2 - p_1)/N))$$

$$= P\left(N \frac{V_N^1 - V_N^2}{V_N^2} < \chi_\alpha^2(p_2 - p_1)\right)$$

In this case, when the system does not belong to the model set, the difference $V_N^1 - V_N^2$ should be significant. More exactly, $V_N^1 - V_N^2 = O(1)$, i.e. $V_N^1 - V_N^2$ does not tend to zero as N tends to infinity; otherwise the two models $\mathscr{M}_1$ and $\mathscr{M}_2$ would be (asymptotically) equivalent which would be a contradiction to the working hypothesis. This implies that $N(V_N^1 - V_N^2)/V_N^2$ is of order N and thus $\chi_\alpha^2(p_2 - p_1)$ must be small compared to N, i.e.

$$\chi_\alpha^2(p_2 - p_1)/N \to 0 \quad \text{as } N \to \infty \tag{11.67}$$

The following example demonstrates how the requirements (11.66), (11.67) for a consistent order determination can be satisfied by appropriate choices of $\beta(N, p)$ or $\gamma(N, p)$.

Example 11.10 *Consistent model structure determination*
Consider first the criterion (11.46). Choose the term $\beta(N, p)$ as

$$\beta(N, p) = \frac{2p}{N} f(N) \tag{11.68a}$$

If $f(N) = 1$ this gives approximately the FPE criterion. Now suppose that $f(N)$ is chosen such that

$$f(N) \to \infty \quad \text{and} \quad f(N)/N \to 0 \quad \text{as } N \to \infty \tag{11.68b}$$

There are many ways to satisfy this condition. The functions $f_1(N) = \sqrt{N}$ and $f_2(N) = \log N$ are two possibilities.

Inserting (11.68a) into (11.57), for large N

$$\chi_\alpha^2(p_2 - p_1) \approx N2(p_2 - p_1)f(N)/N = 2(p_2 - p_1)f(N) \tag{11.69}$$

It is then found from (11.68b) that the consistency conditions (11.66), (11.67) are satisfied.

Next consider the criterion (11.47). Let the term $\gamma(N, p)$ be given by

$$\gamma(N, p) = 2pg(N) \tag{11.70a}$$

The choice $g(N) = 1$ would give the AIC. Now choose $g(N)$ such that

$$g(N) \to \infty \quad \text{and} \quad g(N)/N \to 0 \quad \text{as } N \to \infty \tag{11.70b}$$

From (11.59) it is found that for large N

$$\chi_\alpha^2(p_2 - p_1) \approx N[\gamma(N, p_2) - \gamma(N, p_1)]/N = 2(p_2 - p_1)g(N) \tag{11.71}$$

It is then easily seen from (11.70b) that the consistency conditions (11.66), (11.67) are satisfied. ∎

Summary

Model structure determination and model validation are often very important steps in system identification. For the determination of an appropriate model structure, it is recommended to use a combination of statistical tests and plots of relevant signals. This has been demonstrated in the chapter, where also the following tests have been described:

- Tests of whiteness of the prediction errors and of uncorrelatedness of the prediction errors and the input (assuming that the prediction errors are available after the parameter estimation phase.)
- Tests for detecting a too complex model structure, for example by means of pole-zero cancellations and/or singular information matrices. (The application of such tests is to some extent dependent on the chosen class of model structures.)
- Tests on the values of the loss functions corresponding to different model structures. (These tests require the use of a prediction error method.) The χ^2-test (sometimes also called – improperly, in the dynamic case – the F-test) is of this type. It can be used to test whether a decrease of the loss function corresponding to an increase of the model structure is significant or not. The other tests of this type which were discussed utilize structure-dependent terms to penalize the decrease of the loss function with increasing model structure. These tests were shown to be closely related to the χ^2-test. By properly selecting some user parameters in these tests, it is possible to obtain consistent structure selection rules.

Problems

Problem 11.1 *On the use of the cross-correlation test for the least squares method*
Consider the ARX model

$$A(q^{-1})y(t) = B(q^{-1})u(t) + v(t)$$

where

$$A(q^{-1}) = 1 + a_1 q^{-1} + \ldots + a_{na} q^{-na}$$

$$B(q^{-1}) = b_1 q^{-1} + \ldots + b_{nb} q^{-nb}$$

Assume that the least squares method is used to estimate the parameters $a_1, \ldots, a_{na}, b_1, \ldots, b_{nb}$. Show that

$$\hat{r}_{\varepsilon u}(\tau) = 0 \quad \tau = 1, \ldots, nb$$

where $\varepsilon(t)$ are the model residuals.

Hint. Use the normal equations to show that $\Sigma \varphi(t)\varepsilon(t) = 0$.

Remark. Note that the problem illustrates Remarks 1 and 2 after (11.18). In particular it is obvious that $\hat{r}_{\varepsilon u}(\tau)$ and $\hat{r}_{vu}(\tau)$ have different distributions.

Problem 11.2 *Identifiability results for ARX models*
Consider the system

$$A_0(q^{-1})y(t) = B_0(q^{-1})u(t) + e(t)$$

A_0, B_0 coprime

$\deg A_0 = na_0$, $\deg B_0 = nb_0$

$e(t)$ white noise

identified by the least squares method in the model structure

$$A(q^{-1})y(t) = B(q^{-1})u(t) + \varepsilon(t)$$

$\deg A = na \geq na_0$, $\deg B = nb \geq nb_0$

Assume that the system operates in open loop and that the input signal is persistently exciting of order nb. Prove that the following results hold:

(a) The asymptotic loss function, $E\varepsilon^2(t)$, has a unique minimum and the corresponding polynomials A and B are coprime.
(b) The information matrix is nonsingular.

Compare with the corresponding properties of ARMAX models, see Example 11.6.

Problem 11.3 *Variance of the prediction error when future and past experimental conditions differ*
Consider the system

$$y(t) + ay(t-1) = bu(t-1) + e(t)$$

Assume that the system is identified by the least squares method using a white noise input $u(t)$ of zero mean and variance σ^2. Consider the one-step prediction error variance as an assessment criterion. Evaluate the expected value of this assessment criterion when $u(t)$ is white noise of zero mean and variance $\alpha^2 \sigma^2$ (α being a scalar). Compare with the result (11.40).

Remark. Note that in this case $\psi(t, \theta_0) \neq \psi_p(t, \theta_0)$ in (11.38), (11.39).

Problem 11.4 *An assessment criterion for closed loop operation*
Consider a system identified in the model structure

$$y(t) + ay(t-1) = bu(t-1) + e(t) + ce(t-1)$$

The minimum variance regulator based on the first-order ARMAX model above is

$$u(t) = \frac{a-c}{b} y(t) \qquad \text{(i)}$$

Assume that the true (open loop) system is given by

$$y(t) + a_0 y(t-1) = b_0 u(t-1) + e(t) + c_0 e(t-1)$$
$$Ee^2(t) = \lambda^2 \quad e(t) \text{ white noise} \qquad \text{(ii)}$$

Consider the closed loop system described by (i), (ii) and assume it is stable. Evaluate the variance of the output. Show that it satisfies the property (11.35) of assessment criteria W_N. Also show that the minimum of W_N in this case is reached for many models.

Problem 11.5 *Variance of the multi-step prediction error as an assessment criterion*

(a) Consider an autoregressive process

$$A_0(q^{-1}) y(t) = e(t)$$

and assume that an identified model

$$A(q^{-1}) y(t) = \varepsilon(t)$$

of the process has been obtained with the least squares method. Let $\hat{y}(t+k|t)$ denote the mean square optimal k-step predictor based on the model. Show that when this predictor is applied to the (true) process, the prediction error variance is

$$W_N = E[F_0(q^{-1}) e(t)]^2 + E\left[\frac{G_0(q^{-1}) - G(q^{-1})}{A_0(q^{-1})} e(t)\right]^2$$

where F_0, G_0 and G are defined through

$$1 \equiv A_0(q^{-1}) F_0(q^{-1}) + q^{-k} G_0(q^{-1})$$
$$1 \equiv A(q^{-1}) F(q^{-1}) + q^{-k} G(q^{-1})$$
$$\deg F_0 = \deg F = k - 1$$

(Cf. Complement C7.2.)

(b) Let $k = 2$ and assume that the true process is of first order. Show that for a large number of data points (N), for a first-order model

$$EW_N^{(1)} = \lambda^2 (1 + a_0^2) + \frac{\lambda^2}{N} (4 a_0^2)$$

and for a second-order model

$$EW_N^{(2)} = \lambda^2 (1 + a_0^2) + \frac{\lambda^2}{N} (1 + 4 a_0^2)$$

In these two expressions E denotes expectation with respect to the parameter estimates. Note that $W_N^{(1)} < W_N^{(2)}$ on average, which is a form of the parsimony principle.

Problem 11.6 *Misspecification of the model structure and prediction ability*
Consider a first-order moving average process

$$y(t) = e(t) + ce(t-1) \quad Ee^2(t) = \lambda^2 \quad e \text{ white noise}$$

Assume that this process is identified by the least squares method as an autoregressive process

$$y(t) + a_1 y(t-1) + \ldots + a_n y(t-n) = \varepsilon(t)$$

Consider the asymptotic case ($N \to \infty$).

(a) Find the prediction error variance $E\varepsilon^2(t)$ for the model when $n = 1, 2, 3$. Compare with the optimal prediction error variance based on the true system. Generalize this comparison to an arbitrary value of n.
  Hint. The variance corresponding to the model is given by

$$E\varepsilon^2(t) = \min_{a_1,\ldots,a_n} E[y(t) + a_1 y(t-1) + \ldots + a_n y(t-n)]^2$$

(b) By what percentage is the prediction error variance deteriorated for $n = 1, 2, 3$ (as compared to the optimal value) in the following two cases?

  Case I: $c = 0.5$

  Case II: $c = 1.0$

Problem 11.7 *An illustration of the parsimony principle*
(a) Consider the following AR(1) process

$$\mathscr{S}: y(t) + a_0 y(t-1) = e(t) \quad |a_0| < 1$$

where $e(t)$ is a zero-mean white noise with variance λ^2. Also, consider the following two models of $\mathscr{S}$:

  $\mathscr{M}_1$ (an AR(1)): $y(t) + ay(t-1) = \varepsilon(t)$

  $\mathscr{M}_2$ (an AR(2)): $y(t) + a_1 y(t-1) + a_2 y(t-2) = \varepsilon(t)$

Let $\hat{a}$ and $\hat{a}_1$ denote the LS estimates of a in $\mathscr{M}_1$ and a_1 in $\mathscr{M}_2$. Determine the asymptotic variances of the estimation errors $\sqrt{N}(\hat{a} - a_0)$ and $\sqrt{N}(\hat{a}_1 - a_0)$, and show that $\text{var}(\hat{a}) < \text{var}(\hat{a}_1)$ (i.e. the 'parsimony principle' applies).
(b) Generalize (a) to an autoregressive process of order n, $\mathscr{M}_1$ and $\mathscr{M}_2$ being autoregressions of orders n_1 and n_2, with $n_2 > n_1 \geq n$.
(c) Show that the result of (b) can be obtained as a special case of the parsimony principle introduced in Complement C11.1.

Problem 11.8 *The parsimony principle does not necessarily apply to nonhierarchical model structures*
Consider the following scalar system:

$\mathcal{S}: y(t) = bu(t-1) + e(t)$

where $e(t)$ and $u(t)$ are mutually independent zero-mean white noise sequences. The variance of $e(t)$ is λ^2. Consider also the two model structures

$$\mathcal{M}_1: y(t) + \hat{a}y(t-1) = \hat{b}u(t-1) + \varepsilon_{\mathcal{M}_1}(t)$$

$$\mathcal{M}_2: y(t) = \hat{b}_1 u(t-1) + \hat{b}_2 u(t-2) + \hat{b}_3 u(t-3) + \varepsilon_{\mathcal{M}_2}(t)$$

The estimates of the parameters of $\mathcal{M}_1$ and $\mathcal{M}_2$ are obtained by using the least squares method (which is identical to the PEM in this case).

(a) Let $Eu^2(t) = \sigma^2$. Determine the asymptotic covariance matrices of the estimation errors

$$\delta_1 = \sqrt{N}/\lambda \begin{pmatrix} \hat{a} \\ \hat{b} - b \end{pmatrix} \quad \text{for } \mathcal{M}_1$$

$$\delta_2 = \sqrt{N}/\lambda \begin{pmatrix} \hat{b}_1 - b \\ \hat{b}_2 \\ \hat{b}_3 \end{pmatrix} \quad \text{for } \mathcal{M}_2$$

where N denotes the number of data points.

(b) Let the adequacy of a model structure be expressed by its ability to predict the system's output one step ahead, when $Eu^2(t) = s^2 \neq \sigma^2$ (recall that σ^2 was the variance of $u(t)$ during the estimation stage):

$$A_{\mathcal{M}} = E\{E[\varepsilon_{\mathcal{M}}(t, \hat{\theta}_{\mathcal{M}})]^2\}$$

The inner expectation is with respect to $e(t)$, and the outer one with respect to $\hat{\theta}_{\mathcal{M}}$. Determine asymptotically (for $N \to \infty$) valid approximations for $A_{\mathcal{M}_1}$ and $A_{\mathcal{M}_2}$. Show that the inequality $A_{\mathcal{M}_1} \leq A_{\mathcal{M}_2}$ does not necessarily hold. Does this fact invalidate the parsimony principle introduced in Complement C11.1?

Problem 11.9 *On testing cross-correlations between residuals and input*
It was shown in Example 11.2 that

$$\sqrt{N} x_\tau \xrightarrow{\text{dist}} \mathcal{N}(0, 1)$$

where

$$x_\tau = \frac{\hat{r}_{\varepsilon u}(\tau)}{[\hat{r}_\varepsilon(0)\hat{r}_u(0)]^{1/2}}$$

Hence, for every τ it holds asymptotically with 95 percent probability that $|x_\tau| \leq 1.96/\sqrt{N}$. By analogy with (11.9), it may be tempting to define and use the following test quantity:

$$y \triangleq N \sum_{k=1}^{m} x_{\tau+k}^2 = \frac{N}{\hat{r}_\varepsilon(0)\hat{r}_u(0)} \sum_{k=1}^{m} \hat{r}_{\varepsilon u}^2(\tau + k)$$

instead of (11.18). Compare the test quantities y above and (11.18). Evaluate their means and variances.

Hint. First prove the following result using Lemma B.9. Let $x \sim \mathcal{N}(0, P)$ and set $z = x^T Q x$. Then $Ez = \text{tr } PQ$, $\text{var}(z) = 2 \text{ tr } PQPQ$. Note that z will in general *not* be χ^2 distributed.

Problem 11.10 *Extension of the prediction error formula (11.40) to the multivariable case*
Let $\{\varepsilon(t, \theta_0)\}$ denote the white prediction errors of a multivariable system with (true) parameters θ_0. An estimate $\hat{\theta}$ of θ_0 is obtained by using the optimal PEM (see (7.76) and the subsequent discussion) as the minimizer of $\det\{\sum_{t=1}^{N} \varepsilon(t, \theta) \varepsilon^T(t, \theta)\}$, where N denotes the number of data points. To assess the prediction performance of the model introduce

$$W_N(\hat{\theta}) = \det\{E\varepsilon(t, \hat{\theta}) \varepsilon^T(t, \hat{\theta})\}$$

where the expectation is with respect to the future data used for assessment. Show that the expectation of $W_N(\hat{\theta})$ with respect to the past data used to determine $\hat{\theta}$, is asymptotically (for $N \to \infty$) given by

$$EW_N(\hat{\theta}) = (\det \Lambda)(1 + p/N)$$

(here Λ is the covariance matrix of $\varepsilon(t, \theta_0)$, and $p = \dim \theta$) *provided* the past experimental conditions used to determine $\hat{\theta}$ and the future ones used to assess the model have the same probability characteristics.

Bibliographical notes

Some general comparisons between different methods for model validation and model order determination have been presented by van den Boom and van den Enden (1974), Unbehauen and Göhring (1974), Söderström (1977), Anděl *et al.* (1981), Jategaonkar *et al.* (1982), Freeman (1985), Leontaritis and Billings (1987). Short tutorial papers on the topic have been written by Bohlin (1987) and Söderström (1987). A rather extensive commented list of references may be found in Stoica, Eykhoff *et al.* (1986). Lehmann (1986) is a general text on statistical tests.

(*Section 11.2*) Tests on the correlation functions have been analyzed in a more general setting by Bohlin (1971, 1978). Example 11.5 is taken from Söderström (1974). The statistic (11.9) (the so-called 'portmanteau' statistic) was first analyzed by Box and Pierce (1970). An extension to nonlinear models has been given by Billings and Woon (1986).

(*Section 11.3*) The conditions for singularity of the information matrix have been examined in Söderström (1975a), Stoica and Söderström (1982e). Tests based on the instrumental product matrix are described by Wellstead (1978), Wellstead and Rojas (1982), Young *et al.* (1980), Söderström and Stoica (1983), Stoica (1981b, 1981c, 1983), Fuchs (1987).

(*Section 11.4*) For additional results on the parsimony principle see Stoica *et al.* (1985a), Stoica and Söderström (1982f), and Complement C11.1.

(*Section 11.5*) The FPE and AIC criteria were proposed by Davisson (1965) and Akaike (1969, 1971). See also Akaike (1981) and Butash and Davisson (1986) for further discussions. For a statistical analysis of the AIC test, see Shibata (1976), Kashyap (1980).

The choice $\gamma(N, p) = p \log N$ has appeared in Schwarz (1978), Rissanen (1978, 1979, 1982), Kashyap (1982), while Hannan and Quinn (1979), Hannan (1980, 1981), Hannan and Rissanen (1982) have suggested $\gamma(N, p) = 2\, cp \log(\log N)$. Another more pragmatic extension of the AIC has been proposed and analyzed in Bhansali and Downham (1977).

Appendix A11.1
Analysis of tests on covariance functions

This appendix analyzes the statistical properties of the covariance function tests introduced in Section 11.2. To get the problem in a more general setting consider random vectors of the form

$$r = \frac{1}{\sqrt{N}} \sum_{t=1}^{N} z(t, \hat{\theta}) \varepsilon(t, \hat{\theta}) \tag{A11.1.1}$$

where $z(t, \hat{\theta})$ is an $(m|1)$ vector that might depend on the estimated parameter vector $\hat{\theta}$ of dimension $n\theta$. The prediction errors evaluated at $\hat{\theta}$, i.e. the residuals, are denoted $\varepsilon(t, \hat{\theta})$. For the autocovariance test the vector $z(t, \hat{\theta})$ consists of delayed residuals

$$z(t, \hat{\theta}) = \begin{pmatrix} \varepsilon(t - 1, \hat{\theta}) \\ \vdots \\ \varepsilon(t - m, \hat{\theta}) \end{pmatrix} \tag{A11.1.2a}$$

while for the cross-covariance test $z(t, \hat{\theta})$ consists of shifted input values

$$z(t, \hat{\theta}) = \begin{pmatrix} u(t - t_1 - 1) \\ \vdots \\ u(t - t_1 - m) \end{pmatrix} \tag{A11.1.2b}$$

Note that other choices of z and ε besides those above are possible. The elements of $z(t, \hat{\theta})$ may for example contain delayed and filtered residuals or outputs. Moustakides and Benveniste (1986), Basseville and Benveniste (1986) have used a test quantity of the form (A11.1.1) to detect changes in the system dynamics from a nominal model by using delayed outputs in $z(t)$.

In Section 11.2 the asymptotic properties of r were examined under the assumption that $\hat{\theta} = \theta_0$. As will be shown in this appendix, a small (asymptotically vanishing) deviation of the estimate $\hat{\theta}$ from the true value θ_0 will in fact change the asymptotic properties of the vector r.

In the analysis the following mild assumptions will be made.

A1: The parameter estimate $\hat{\theta}$ is obtained by the prediction error method. Then asymptotically for large N (cf. (7.58))

$$\hat{\theta} - \theta_0 \approx R_\psi^{-1} \frac{1}{N} \sum_{t=1}^{N} \psi(t) e(t) \tag{A11.1.3}$$

where

$$R_\psi = E\psi(t)\psi^T(t) \qquad \psi(t) = -\left(\frac{\partial \varepsilon(t, \theta)}{\partial \theta}\bigg|_{\theta=\theta_0}\right)^T \qquad (A11.1.4)$$

Note that $\psi(s)$ is independent of $\varepsilon(t, \theta_0) = e(t)$ for all $s \leq t$.

A2: The vector $z(s, \theta)$ is independent of $e(t)$ for all $s \leq t$. This is a weak assumption that certainly holds true for the specific covariance tests mentioned before.

Next form a Taylor series expansion of (A11.1.1) around θ_0. Since (A11.1.3) implies that $\hat{\theta} - \theta_0$ is of the order $1/\sqrt{N}$, it will be sufficient to retain terms that are constant or linear in $\hat{\theta} - \theta_0$. Then

$$r = \frac{1}{\sqrt{N}} \sum_{t=1}^{N} \left[z(t, \theta_0) + \frac{\partial z(t, \theta)}{\partial \theta}\bigg|_{\theta=\theta_0} (\hat{\theta} - \theta_0) + \ldots \right][e(t) - \psi^T(t)(\hat{\theta} - \theta_0) + \ldots]$$

$$\approx \frac{1}{\sqrt{N}} \sum_{t=1}^{N} [z(t, \theta_0) e(t)] - [Ez(t, \theta_0)\psi^T(t)]\sqrt{N}(\hat{\theta} - \theta_0) \qquad (A11.1.5)$$

$$\approx (I - R_{z\psi} R_\psi^{-1}) \frac{1}{\sqrt{N}} \sum_{t=1}^{N} \binom{z(t, \theta_0)}{\psi(t)} e(t)$$

where

$$R_{z\psi} = Ez(t, \theta_0)\psi^T(t) \qquad (A11.1.6)$$

It follows from Lemma B.3 that r is asymptotically Gaussian distributed,

$$r \xrightarrow{\text{dist}} \mathcal{N}(0, P_r) \qquad (A11.1.7)$$

where

$$P_r = \lambda^2(I \quad -R_{z\psi}R_\psi^{-1}) \begin{pmatrix} R_z & R_{z\psi} \\ R_{\psi z} & R_\psi \end{pmatrix} \begin{pmatrix} I \\ -R_\psi^{-1}R_{\psi z} \end{pmatrix}$$

$$= \lambda^2(R_z - R_{z\psi}R_\psi^{-1}R_{\psi z}) \qquad (A11.1.8)$$

and $R_z = Ez(t, \theta_0)z^T(t, \theta_0)$, $R_{\psi z} = R_{z\psi}^T$ and $\lambda^2 = Ee^2(t)$. Note that P_r differs from $\lambda^2 R_z$, which is the result that would be obtained by neglecting the deviation of $\hat{\theta}$ from θ_0 (cf. Section 11.2 for detailed derivations). A result of the form (A11.1.8) for pure time series models has also been obtained by McLeod (1978).

Next examine the test statistic

$$x = r^T(\lambda^2 R_z)^{-1} r \qquad (A11.1.9)$$

which was introduced and analyzed in Section 11.2 under the idealized assumption $\hat{\theta} = \theta_0$. In practice the test quantity (A11.1.9) is often used (with λ^2 and R_z replaced by estimates) instead of the 'exact' χ^2 test quantity $r^T P_r^{-1} r$, which is more difficult to evaluate. It would then be valuable to know how to set the threshold for the test based on (A11.1.9). To answer that question, introduce x_α through

$$P(x \leq x_\alpha) = 1 - \alpha \qquad (A11.1.10)$$

In the idealized case where $P_r = \lambda^2 R_z$ one would apparently have $x_\alpha = \chi_\alpha^2(m)$. For (A11.1.7), (A11.1.8) this holds true only if $R_{z\psi} = 0$. In general this condition is not satisfied with the exception of the autocovariance test for an output error model:

model: $y(t) = G(q^{-1}; \theta)u(t) + \varepsilon(t, \theta)$

z-vector: delayed residuals as in (A11.1.2a)

In the following an upper and a lower bound are derived on the threshold x_α introduced in (A11.1.10). To do so assume that $P_r > 0$. This weak assumption implies that (cf. Lemmas A.3 and A.4) $R_z > 0$ (note that if R_z is singular then the test statistic (A11.1.9) cannot be used). The condition $P_r > 0$ may sometimes be violated if the system is overparametrized, as illustrated in the following example.

Example A11.1.1 *A singular P_r matrix due to overparametrization*
Consider a first-order autoregression

$$y(t) - ay(t-1) = e(t) \quad |a| < 1, \ e(t) \text{ white noise} \quad Ee^2(t) = 1 \quad (A11.1.11)$$

identified as a second-order autoregressive process. Let the $z(t)$ vector consist of delayed prediction errors, (cf. (A11.1.2a)). Then

$$R_z = I$$

$$R_{\psi z} = \begin{pmatrix} 1 & a & \ldots & a^{m-1} \\ 0 & 1 & \ldots & a^{m-2} \end{pmatrix}$$

$$R_\psi = \frac{1}{1-a^2} \begin{pmatrix} 1 & a \\ a & 1 \end{pmatrix}$$

Thus the first row of P_r becomes

$$(1 \ 0 \ \ldots \ 0) - (1 \ 0) \frac{1-a^2}{1-a^2} \begin{pmatrix} 1 & -a \\ -a & 1 \end{pmatrix} \begin{pmatrix} 1 & a & \ldots & a^{m-1} \\ 0 & 1 & \ldots & a^{m-2} \end{pmatrix}$$

$$= (1 \ 0 \ \ldots \ 0) - (1 \ 0) \begin{pmatrix} 1 & 0 & 0 & \ldots & 0 \\ -a & 1-a^2 & a(1-a^2) & \ldots & a^{m-2}(1-a^2) \end{pmatrix} = 0$$

Hence the matrix P_r is singular. A more direct way to realize this is to note that the elements of the vector

$$(z^T(t) \ \psi^T(t)) = (e(t-1) \ \ldots \ e(t-m) \ y(t-1) \ y(t-2))$$

are linearly dependent, due to (A11.1.11). ∎

Next introduce a normalized r vector as

$$\bar{r} = P_r^{-1/2} r \quad (A11.1.12)$$

where $P_r^{1/2}$ is a symmetric square root of P_r. Then

$$\bar{r} \xrightarrow{\text{dist}} \mathcal{N}(0, I) \quad (A11.1.13)$$

and

$$x = \bar{r}^T J \bar{r} \qquad (A11.1.14a)$$

where

$$J = P_r^{1/2}(\lambda^2 R_z)^{-1} P_r^{1/2} \qquad (A11.1.14b)$$

Since an orthogonal transformation of $\bar{r}$ does not change the result (A11.1.13) it is no restriction to assume that J is diagonal. Thus it is of interest to examine the eigenvalues of J. By construction of J they are all real and strictly positive. Assume that $m > n\theta$. The eigenvalues of J, denoted $s(J)$, are the solutions to

$$0 = \det[sI - J] = \det[sI - P_r^{1/2} J P_r^{-1/2}] = \det[sI - P_r(\lambda^2 R_z)^{-1}]$$
$$= \det[sI - I + R_{z\psi} R_\psi^{-1} R_{\psi z} R_z^{-1}]$$
$$= (s-1)^m \det\left[I + \frac{1}{s-1} R_{z\psi} R_\psi^{-1} R_{\psi z} R_z^{-1}\right]$$
$$= (s-1)^{m-n\theta} \det[(s-1)I + R_\psi^{-1} R_{\psi z} R_z^{-1} R_{z\psi}] \qquad (A11.1.15)$$

Here use has been made of the fact that J and TJT^{-1} for some nonsingular matrix T, have the same eigenvalues, and also Corollary 1 of Lemma A.5. From (A11.1.15) it can be concluded that J has $m - n\theta$ eigenvalues at $s = 1$. The remaining $n\theta$ eigenvalues satisfy

$$s(J) = 1 - s[R_\psi^{-1} R_{\psi z} R_z^{-1} R_{z\psi}]$$
$$= 1 - s[R_\psi^{-1/2} R_{\psi z} R_z^{-1} R_{z\psi} R_\psi^{-1/2}] \leq 1 \qquad (A11.1.16)$$

They are hence located in the interval $(0, 1]$. As a consequence,

$$x \leq \bar{r}^T I_m \bar{r} \xrightarrow{\text{dist}} \chi^2(m)$$
$$x \geq \bar{r}^T \begin{pmatrix} I_{m-n\theta} & 0 \\ 0 & 0 \end{pmatrix} \bar{r} \xrightarrow{\text{dist}} \chi^2(m - n\theta) \qquad (A11.1.17)$$

(cf. Lemma B.13). Equation (A11.1.17) implies in turn

$$\boxed{\chi_\alpha^2(m - n\theta) \leq x_\alpha \leq \chi_\alpha^2(m)} \qquad (A.11.1.18)$$

Remark. If $R_{z\psi}$ has rank smaller than $n\theta$ and this is known, then a tighter lower bound on x_α can be derived. When rank $R_{z\psi} = q < n\theta$, the matrix $R_\psi^{-1/2} R_{\psi z} R_z^{-1} R_{z\psi} R_\psi^{-1/2}$ which appears in (A11.1.16) will be of rank q and will hence have $n\theta - q$ eigenvalues in the origin. Hence one can substitute $m - n\theta$ by $m -$ rank $R_{z\psi}$ in (A11.1.17), (A11.1.18). ∎

Roughly speaking, the system parameters will determine where in the interval (A11.1.18) x_α will be situated. There is one important case where more can be said. Consider a pure *linear time series model* (no exogenous input) and let $z(t)$ consist of delayed prediction errors (hence the *autocorrelation test* is examined). Split the vector $\psi(t)$ into two parts, one that depends on $z(t)$ and a second that depends on older prediction errors. The two parts will therefore be uncorrelated. Thus

$$\psi(t) = Bz(t) + \tilde{\psi}(t)$$

$$R_z = \lambda^2 I$$

$$R_{z\psi} = \lambda^2 B^T$$

$$R_\psi = \lambda^2 BB^T + \lambda^2 \Delta$$

where B is an $(n\theta|m)$ constant matrix and $\|\Delta\|$ tends to zero exponentially as m tends to infinity, since the elements of $\tilde{\psi}$ do so. From (A11.1.16),

$$\begin{aligned}s(J) &= 1 - s[(BB^T + \Delta)^{-1}BB^T] \\ &= s[(BB^T + \Delta)^{-1}\Delta]\end{aligned} \quad (A11.1.19)$$

which tends to zero exponentially as $m \to \infty$. Hence in this case

$$x_\alpha \approx \chi_\alpha^2(m - n\theta) \quad \text{for } m \text{ large} \quad (A11.1.20)$$

Such a result has been derived by Box and Pierce (1970) for ARIMA models using a different approach. (See Section 12.3 for a discussion of such models). The text based on x_α is in such cases often called the 'portmanteau test' in the statistical literature.

Another case where (A11.1.20) holds is the *cross-correlation test* for an *output error model*. Then $z(t, \theta)$ is given by (A11.1.2b) with $t_1 = 0$. In this case it is also possible to write

$$\psi(t) = Bz(t) + \tilde{\psi}(t)$$

where B is an $(n\theta|m)$ constant matrix and $\|E\tilde{\psi}(t)\tilde{\psi}^T(t)\| \to 0$ as m tends to infinity. Some straightforward calculation then shows that

$$\begin{aligned}R_\psi^{-1}R_{\psi z}R_z^{-1}R_{z\psi} = I - R_\psi^{-1}[&E\tilde{\psi}(t)\tilde{\psi}^T(t) \\ &- \{E\tilde{\psi}(t)z^T(t)\}\{Ez(t)z^T(t)\}^{-1}\{Ez(t)\tilde{\psi}^T(t)\}]\end{aligned}$$

Hence $s(J)$ in (A11.1.16) tends to zero as m tends to infinity. Therefore (A11.1.20) holds also in this case.

Appendix A11.2
Asymptotic distribution of the relative decrease in the criterion function

This appendix is devoted to proving the result (11.43).

Let θ_0^i denote the true parameter vector in model structure $\mathcal{M}_i$ and let $\hat{\theta}_N^i$ denote the estimate, $i = 1, 2$. For both model structures θ_0^i describes exactly the true system and thus $\varepsilon(t, \theta_0^i)$ is the white noise process of the system. Hence

$$V_N^1(\theta_0^1) = V_N^2(\theta_0^2) \quad (A11.2.1)$$

where $V_N^i(\cdot)$ denotes the loss function associated with $\mathcal{M}_i$.

Next consider the following Taylor series expansion:

$$V_N^i(\theta_0^i) \approx V_N^i(\hat{\theta}_N^i) + V_N'^i(\hat{\theta}_N^i)(\theta_0^i - \hat{\theta}_N^i) + \frac{1}{2}(\theta_0^i - \hat{\theta}_N^i)^T V_N''^i(\hat{\theta}_N^i)(\theta_0^i - \hat{\theta}_N^i)$$

$$= V_N^i + \frac{1}{2}(\theta_0^i - \hat{\theta}_N^i)^T V_N''^i(\hat{\theta}_N^i)(\theta_0^i - \hat{\theta}_N^i) \quad (A11.2.2)$$

$$\approx V_N^i + \frac{1}{2}(\theta_0^i - \hat{\theta}_N^i)^T V_\infty''^i(\theta_0^i)(\theta_0^i - \hat{\theta}_N^i)$$

Here use has been made of the fact that $V_N'^i(\hat{\theta}_N^i) = 0$ since $\hat{\theta}_N^i$ minimizes $V_N^i(\theta^i)$. Note also that the substitution above of $V_N''^i(\hat{\theta}_N^i)$ by $V_\infty''^i(\theta_0^i)$ has only a higher-order effect. Now recall from (7.58) that

$$\hat{\theta}_N - \theta_0 \approx - [V_\infty''(\theta_0)]^{-1}[V_N'(\theta_0)]^T \quad (A11.2.3)$$

Using (A11.2.1)–(A11.2.3),

$$V_N^1 - V_N^2 \approx \frac{1}{2}(\theta_0^2 - \hat{\theta}_N^2)^T V_\infty''^2(\theta_0^2)(\theta_0^2 - \hat{\theta}_N^2)$$

$$- \frac{1}{2}(\theta_0^1 - \hat{\theta}_N^1)^T V_\infty''^1(\theta_0^1)(\theta_0^1 - \hat{\theta}_N^1)$$

$$\approx \frac{1}{2} V_N'^2(\theta_0^2)[V_\infty''^2(\theta_0^2)]^{-1}[V_N'^2(\theta_0^2)]^T$$

$$- \frac{1}{2} V_N'^1(\theta_0^1)[V_\infty''^1(\theta_0^1)]^{-1}[V_N'^1(\theta_0^1)]^T \quad (A11.2.4)$$

To continue it is necessary to exploit some relation between $V_N^1(\theta^1)$ and $V_N^2(\theta^2)$. Recall that $\mathcal{M}_1$ is a subset of $\mathcal{M}_2$. One can therefore describe the parameter vectors in $\mathcal{M}_1$ by appropriately constraining the parameter vectors in $\mathcal{M}_2$. For this purpose introduce the function $g: \mathcal{M}_1 \to \mathcal{M}_2$ (here $\mathcal{M}_i$ are viewed as vector sets) which is such that the model $\mathcal{M}_2$ with parameters $\theta^2 = g(\theta^1)$ is identical to $\mathcal{M}_1$ with parameters θ^1. Define the derivative

$$S(\theta^1) = \frac{\partial g(\theta^1)}{\partial \theta^1} \quad (A11.2.5)$$

which is a $p_2 | p_1$ matrix. We can now write

$$\theta_0^2 = g(\theta_0^1)$$
$$V_N^1(\theta^1) = V_N^2(g(\theta^1))$$
$$V_N'^1(\theta^1) = V_N'^2(g(\theta^1))S(\theta^1)$$
$$V_\infty''^1(\theta^1) = S^T(\theta^1)V_\infty''^2(g(\theta^1))S(\theta^1) + V_\infty'^2(g(\theta^1))\frac{\partial S(\theta^1)}{\partial \theta^1}$$

Note that the second term in the last equation above is written in an informal way since $\partial S(\theta^1)/\partial \theta^1$ is a tensor; this term is not important for what follows since it vanishes when evaluated in θ_0^1. (See (C11.1.11) for a formal expression of this term.)

Evaluation for the true parameter vector gives

$$V_N'^1(\theta_0^1) = V_N'^2(\theta_0^2)S(\theta_0^1)$$
$$V_\infty''^1(\theta_0^1) = S^T(\theta_0^1)V_\infty''^2(\theta_0^2)S(\theta_0^1) \quad \text{(A11.2.6)}$$

Using the notations

$$P = [V_\infty''^2(\theta_0^2)]^{-1}$$
$$V_N' = V_N'^2(\theta_0^2)$$
$$S = S(\theta_0^1)$$

we get from (A11.2.4), (A11.2.6)

$$V_N^1 - V_N^2 \approx \frac{1}{2} V_N' P(V_N')^T - \frac{1}{2} V_N' S(S^T P^{-1} S)^{-1} S^T (V_N')^T$$
$$= \frac{1}{2} V_N' [P - S(S^T P^{-1} S)^{-1} S^T](V_N')^T \quad \text{(A11.2.7)}$$

Now we recall the result, cf. (7.58)–(7.60)

$$\sqrt{N}(V_N')^T \xrightarrow{\text{dist}} \mathcal{N}(0, 2\lambda^2 P^{-1}) \quad \text{(A11.2.8)}$$

Set

$$z = \frac{1}{\sqrt{(2\lambda)}} P^{1/2} \sqrt{N}(V_N')^T \quad \text{(A11.2.9)}$$

where $P^{1/2}$ is a matrix square root of P (i.e. $P \triangleq P^{T/2}P^{1/2}$). Then from (A11.2.8), (A11.2.9)

$$z \xrightarrow{\text{dist}} \mathcal{N}(0, I) \quad \text{(A11.2.10)}$$

and from (A11.2.7) and (A11.2.9)

$$V_N^1 - V_N^2 \approx \frac{1}{2} \frac{2\lambda^2}{N} z^T P^{-T/2}[P - S(S^T P^{-1} S)^{-1} S^T] P^{-1/2} z$$
$$= \frac{\lambda^2}{N} z^T [I - A(A^T A)^{-1} A^T] z \quad \text{(A11.2.11)}$$

where

$$A = P^{-T/2} S$$

The matrix $[I - A(A^T A)^{-1} A^T]$ appearing in (A11.2.11) is idempotent, see Appendix A (Example A.2). Furthermore, we have

$$\text{rank}[I - A(A^T A)^{-1} A^T] = \text{tr}[I - A(A^T A)^{-1} A^T]$$
$$= \text{tr } I - \text{tr}(A^T A)(A^T A)^{-1} = p_2 - p_1 \quad \text{(A11.2.12)}$$

It then follows from Lemma B.13 that

$$\frac{N}{\lambda^2}(V_N^1 - V_N^2) \sim N \frac{V_N^1 - V_N^2}{V_N^2} = x \xrightarrow{\text{dist}} \chi^2(p_2 - p_1) \tag{A11.2.13}$$

which is exactly the stated result (11.43).

Complement C11.1
A general form of the parsimony principle

The parsimony principle was used for quite a long time by workers dealing with empirical model building. The principle says, roughly, that of two identifiable model structures that fit certain data, the simpler one (that is, the structure containing the smaller number of parameters) will on average give better accuracy (Box and Jenkins, 1976). Here we show that indeed the parsimony principle holds if the two model structures under consideration are *hierarchical* (i.e. one structure can be obtained by constraining the other structure in some way), and if the parameter estimation method used is *the PEM*. As shown in Stoica and Söderström (1982f) the parsimony principle does not necessarily hold if the two assumptions introduced above (hierarchical model structures, and use of the PEM) are relaxed; also, see Problem 11.8.

Let $\{\varepsilon_\mathcal{M}(t, \theta_\mathcal{M})\}$ denote the prediction errors of the model $\mathcal{M}(\theta_\mathcal{M})$. Here $\theta_\mathcal{M}$ denotes the finite-dimensional unknown parameter vector of the model. It is assumed that $\theta_\mathcal{M}$ belongs to an admissible set of parameters, denoted by $D_\mathcal{M}$. The parameters $\theta_\mathcal{M}$ are estimated using the PEM

$$\hat{\theta}_\mathcal{M} = \arg \min_{\theta \in D_\mathcal{M}} \det \left[\sum_{t=1}^{N} \varepsilon_\mathcal{M}(t, \theta) \varepsilon_\mathcal{M}^T(t, \theta) \right] \tag{C11.1.1}$$

where N denotes the number of data points. Introduce the following assumption:

A: There exists a unique parameter vector in D, say $\theta_\mathcal{M}^0$, such that $\varepsilon_\mathcal{M}(t, \theta_\mathcal{M}) = e(t)$ where $e(t)$ is zero-mean white noise with covariance matrix $\Lambda > 0$ (this assumption essentially means that the true system belongs to $\mathcal{M}$).

Then it follows from (7.59), (7.75), (7.76) that the asymptotic covariance matrix of the estimation error $\sqrt{N}(\hat{\theta}_\mathcal{M} - \theta_\mathcal{M}^0)$ is given by

$$P_\mathcal{M} = \left\{ E \left[\frac{\partial \varepsilon_\mathcal{M}(t, \theta)}{\partial \theta} \right]^T_{\theta=\theta_\mathcal{M}^0} \Lambda^{-1} \frac{\partial \varepsilon_\mathcal{M}(t, \theta)}{\partial \theta} \bigg|_{\theta=\theta_\mathcal{M}^0} \right\}^{-1} \tag{C11.1.2}$$

Next we discuss how to assess the accuracy of a model structure $\mathcal{M}$. This aspect is essential since it is required to compare model structures which may contain different numbers of parameters. A fairly general measure of accuracy can be introduced in the following way. Let $W_\mathcal{M}(\theta)$ be an assessment criterion as introduced in (11.35). It is hence a scalar-valued function of $\theta \in D_\mathcal{M}$, which is such that

$$\inf_{\theta \in D_\mathcal{M}} W_\mathcal{M}(\theta) = W_\mathcal{M}(\theta^0_\mathcal{M}) \tag{C11.1.3}$$

Any norm of the prediction error covariance matrix is an example of such a function $W_\mathcal{M}$, but there are many other possible choices (see e.g. Goodwin and Payne, 1973; Söderström et al., 1974). $W_\mathcal{M}(\hat{\theta}_\mathcal{M})$ can be used to express the accuracy of the model $\mathcal{M}(\hat{\theta}_\mathcal{M})$. Clearly $W_\mathcal{M}(\hat{\theta}_\mathcal{M})$ depends on the realization from which $\hat{\theta}_\mathcal{M}$ was determined. To obtain an accuracy measure which is independent of the realization, introduce

$$EW_\mathcal{M}(\hat{\theta}_\mathcal{M}) \tag{C11.1.4}$$

where the expectation is with respect to $\hat{\theta}_\mathcal{M}$. In the general case this measure cannot be evaluated exactly. However, an asymptotically valid approximation of (C11.1.4) can readily be obtained (cf. (11.37)). For N sufficiently large, a simple Taylor series expansion around $\theta^0_\mathcal{M}$ gives

$$W_\mathcal{M}(\hat{\theta}_\mathcal{M}) = W_\mathcal{M}(\theta^0_\mathcal{M}) + W'_\mathcal{M}(\theta^0_\mathcal{M})(\hat{\theta}_\mathcal{M} - \theta^0_\mathcal{M}) \\ + \frac{1}{2}(\hat{\theta}_\mathcal{M} - \theta^0_\mathcal{M})^T W''_\mathcal{M}(\theta^0_\mathcal{M})(\hat{\theta}_\mathcal{M} - \theta^0_\mathcal{M}) + O(\|\hat{\theta}_\mathcal{M} - \theta^0_\mathcal{M}\|^3) \tag{C11.1.5}$$

The second term in (C11.1.5) is equal to zero, since according to (C11.1.3) $W'_\mathcal{M}(\theta^0_\mathcal{M}) = 0$. Define

$$F_\mathcal{M} \triangleq \lim_{N \to \infty} NE[W_\mathcal{M}(\hat{\theta}_\mathcal{M}) - W_\mathcal{M}(\theta^0_\mathcal{M})]$$

According to the calculations above,

$$\boxed{F_\mathcal{M} = \frac{1}{2} \operatorname{tr} W''_\mathcal{M}(\theta^0_\mathcal{M}) P_\mathcal{M}} \tag{C11.1.6}$$

where $P_\mathcal{M}$ is given by (C11.1.2).

Now, let $\mathcal{M}_1$ and $\mathcal{M}_2$ be two model structures, which satisfy assumption A. Furthermore, let $\mathcal{M}_1, \mathcal{M}_2$ be such that there exists a differentiable function

$$g(\theta): D_{\mathcal{M}_1} \to D_{\mathcal{M}_2}$$

for which it holds that

$$\varepsilon_{\mathcal{M}_2}(t, g(\theta)) \equiv \varepsilon_{\mathcal{M}_1}(t, \theta) \quad \text{for all } \theta \in D_{\mathcal{M}_1} \tag{C11.1.7}$$

(Note that the function $g(\theta)$ introduced above was also used in Appendix A11.2.) That is to say, any model in the set $\mathcal{M}_1$ belongs also to the set $\mathcal{M}_2$, which can be written as

$$\mathcal{M}_1 \subset \mathcal{M}_2 \tag{C11.1.8}$$

The model structures $\mathcal{M}_1$ and $\mathcal{M}_2$ satisfying (C11.1.8) will be called hierarchical or nested. Clearly for hierarchical model structures the term $W_\mathcal{M}(\theta^0_\mathcal{M})$ in (C11.1.6) takes the same value. Hence, the criterion $F_\mathcal{M}$ can be used for comparing $\mathcal{M}_1$ and $\mathcal{M}_2$.

We can now state the main result.

Lemma C11.1.1
Consider the hierarchical model structures $\mathcal{M}_1$ and $\mathcal{M}_2$ (with $\mathcal{M}_1 \subset \mathcal{M}_2$) the parameters of which are estimated by the PEM (C11.1.1). Then $\mathcal{M}_1$ leads on the average to more accurate results than $\mathcal{M}_2$ does, in the sense that

$$F_{\mathcal{M}_1} \leq F_{\mathcal{M}_2} \tag{C11.1.9}$$

Proof. Introduce the following Jacobian matrix:

$$S \triangleq \left[\frac{\partial g(\theta)}{\partial \theta}\right]\bigg|_{\theta = \theta^0_{\mathcal{M}_1}}$$

(cf. (A11.2.5)). From (C11.1.7),

$$\frac{\partial \varepsilon_{\mathcal{M}_1}(t, \theta)}{\partial \theta}\bigg|_{\theta = \theta^0_{\mathcal{M}_1}} = \frac{\partial \varepsilon_{\mathcal{M}_2}(t, g)}{\partial g}\bigg|_{g(\theta) = g(\theta^0_{\mathcal{M}_1})} \times S \tag{C11.1.10}$$

Similarly, from

$$W_{\mathcal{M}_1}(\theta) = W_{\mathcal{M}_2}(g(\theta)) \quad \text{all } \theta \in D_{\mathcal{M}_1}$$

straightforward differentiations give

$$\frac{\partial W_{\mathcal{M}_1}(\theta)}{\partial \theta} = \frac{\partial W_{\mathcal{M}_2}(g)}{\partial g} \frac{\partial g(\theta)}{\partial \theta}$$

and

$$\frac{\partial^2 W_{\mathcal{M}_1}(\theta)}{\partial \theta^2} = \left[\frac{\partial g(\theta)}{\partial \theta}\right]^T \frac{\partial^2 W_{\mathcal{M}_2}(g)}{\partial g^2} \frac{\partial g(\theta)}{\partial \theta} + \left[\frac{\partial W_{\mathcal{M}_2}(g)}{\partial g} \otimes I\right]\left[\frac{\partial}{\partial \theta} \text{vec}\left\{\frac{\partial g(\theta)}{\partial \theta}\right\}\right] \tag{C11.1.11}$$

Since $\theta^0_{\mathcal{M}_2} = g(\theta^0_{\mathcal{M}_1})$ and since $W_{\mathcal{M}_2}$ achieves its minimum value at $\theta^0_{\mathcal{M}_2}$, it follows that $\partial W_{\mathcal{M}_2}(g)/\partial g|_{g = g(\theta^0_{\mathcal{M}_1})} = 0$. Thus, from (C11.1.11),

$$W''_{\mathcal{M}_1}(\theta^0_{\mathcal{M}_1}) = S^T W''_{\mathcal{M}_2}(\theta^0_{\mathcal{M}_2}) S \tag{C11.1.12}$$

It follows from (C11.1.2), (C11.1.6), (C11.1.10) and (C11.1.12) that the following inequality must be proved:

$$\text{tr}[(S^T H S)(S^T P^{-1} S)^{-1}] \leq \text{tr}[HP] \tag{C11.1.13}$$

where

$$H \triangleq W''_{\mathcal{M}_2}(\theta^0_{\mathcal{M}_2})$$

$$P \triangleq P_{\mathcal{M}_2}$$

are positive definite matrices. Define

$$\delta \triangleq \text{tr}[HP] - \text{tr}[(S^THS)(S^TP^{-1}S)^{-1}]$$
$$= \text{tr}\{H[P - S(S^TP^{-1}S)^{-1}S^T]\} \tag{C11.1.14}$$

Note the similarity to expression (A11.2.7) in Appendix A11.2.

Let $P = G^TG$, where the matrix G is nonsingular since $P > 0$. Using this factorization of P one can write

$$\Omega \triangleq P - S(S^TP^{-1}S)^{-1}S^T = G^T\{I - (G^{-T}S)(S^TG^{-1}G^{-T}S)^{-1}(S^TG^{-1})\}G$$
$$= G^T[I - A(A^TA)^{-1}A^T]G$$

where $A = G^{-T}S$. Since the matrix $I - A(A^TA)^{-1}A^T$ is idempotent, it follows that Ω is nonnegative definite (see Section A.5 of Appendix A). Thus, Ω can be factored as $\Omega = \Gamma^T\Gamma$. From this observation and (C11.1.14) the result follows immediately:

$$\delta = \text{tr } H\Omega = \text{tr } H\Gamma^T\Gamma = \text{tr } \Gamma H \Gamma^T \geq 0 \qquad \blacksquare$$

The parsimony result established above is quite strong. The class of models to which it applies is fairly general. The class of accuracy criteria considered is also fairly general. The experimental conditions which are implicit in P_u (the experimental condition during the estimation stage) and in W'''_u (the experimental condition used for assessment of the model structure) were minimally constrained. Furthermore, these two experimental conditions may be different. That is to say, it is allowed to estimate the parameters using some experimental conditions, and then to assess the model obtained on another set of experimental conditions.

Chapter 12
SOME PRACTICAL ASPECTS

12.1 Introduction

The purpose of this chapter is to give some guidelines for performing identification in practice. The chapter is organized around the following issues:

- Design of the experimental condition $\mathscr{X}$.
- Precomputations.
- Determination of the model structure $\mathscr{M}$.
- Time delays.
- Initial conditions of the model.
- Choice of the identification method $\mathscr{I}$.
- Local minima of the loss function.
- Robustness of the parameter estimates.
- Model validation.
- Software.

Note that for obvious reasons the user has to choose $\mathscr{X}$ before the experiment. After the experiment and the data acquisition have been done the user can still try different model structures and identification methods.

12.2 Design of the experimental condition $\mathscr{X}$

The design of $\mathscr{X}$ involves several factors. The most important are:

- Choice of input signal.
- Choice of sampling interval.

Concerning the choice of input signal there are several aspects to consider. Certain identification methods require a special type of input. This is typically the case for several of the nonparametric methods discussed in Chapter 3. For frequency analysis the input must be a sinusoid, for transient analysis a step or an impulse, and for correlation analysis a white noise or a pseudorandom sequence.

For other types of method it is only required that the input signal is persistently exciting (pe). To identify the parameters of an nth-order model it is typically necessary that $u(t)$ is pe of order $2n$, but the exact order can vary from method to method. Recall that a signal which is pe of order n has a spectral density that is nonzero in at least n

points (see Property 3 in Section 5.4). If a signal is chosen which has strictly positive spectral density for all frequencies, this will guarantee persistent excitation of a sufficient order.

The discussion so far has concerned the *form* of the input. It is also of relevance to discuss the choice of its *amplitude*. When choosing the amplitude, the user must bear in mind the following points:

- There may be constraints on how much variation of the signals (for example, $u(t)$ and $y(t)$) is allowed during the experiment. For safety or economic reasons it may not be possible to introduce too large fluctuations in the process.
- There are also other constraints on the input amplitude. Suppose one wishes to estimate the parameters of a linear model. In practice most processes are nonlinear and a linear model can hence only be an approximation. A linearization of the nonlinear dynamics will be valid only in some region. To estimate the parameters of the linearized model too large an amplitude of the input should not be chosen. On the other hand, it could be of great value to make a second experiment with a larger amplitude to test the linearity of the process, i.e. to investigate over which region the linearized model can be regarded as appropriate.
- On the other hand, there are reasons for using a large input amplitude. One can expect that the accuracy of the estimates will improve when the input amplitude is increased. This is natural since by increasing $u(t)$, the signal-to-noise ratio will increase and the disturbances will play a smaller role.

A further important comment on the choice of input must be made. In practice the user can hardly assume that the true process is linear and of finite order. Identification must therefore be considered as a method of model approximation. As shown in Chapter 2, the resulting model will then depend on the experimental condition. A simple but general rule can therefore be formulated: choose the experimental condition that resembles the condition under which the model will be used in the future. Expressed more formally: *let the input have its major energy in the frequency band which is of interest for the intended application of the model.* This general rule is illustrated in the following simple example.

Example 12.1 *Influence of the input signal on model performance*
Consider the system

$$y(t) + ay(t-1) = bu(t-1) + e(t) \tag{12.1}$$

where $e(t)$ is white noise with variance λ^2. Also consider the following two experimental conditions:

$\mathscr{X}_1$: $u(t)$ is white noise of variance σ^2
$\mathscr{X}_2$: $u(t)$ is a step function with amplitude σ

The first input has an even frequency content, while the second is of extremely low frequency. Note that both inputs have the same power.

The asymptotic covariance matrix of the LS estimates of the parameters a and b is given by, (cf. (7.66))

470 Some practical aspects   Chapter 12

$$\text{cov}\begin{pmatrix} \hat{a} \\ \hat{b} \end{pmatrix} = \frac{\lambda^2}{N} \begin{pmatrix} Ey^2(t) & -Ey(t)u(t) \\ -Ey(t)u(t) & Eu^2(t) \end{pmatrix}^{-1} \triangleq P_\theta \qquad (12.2)$$

Consider first the (noise-free) step response of the system. The continuous time counterpart of (12.1) is (omitting the noise $e(t)$)

$$\dot{y} + \alpha y = \beta u \qquad (12.3)$$

where

$$\alpha = -\frac{1}{h}\log(-a)$$

$$\beta = -\frac{b}{(1+a)h}\log(-a) \qquad (12.4)$$

and where h is the sampling interval. The step response is easily deduced from (12.3). It is given by

$$y(t) = \frac{\beta}{\alpha}[1 - e^{-\alpha t}] \qquad (12.5)$$

Next examine the variance of the step response (12.5), due to the deviation $(\hat{a} - a \quad \hat{b} - b)^T$. Denoting the step responses associated with the system and the model by $y(t, \theta)$ and $y(t, \hat{\theta})$, respectively,

$$y(t, \hat{\theta}) \approx y(t, \theta) + \left[\frac{\partial}{\partial \theta} y(t, \theta)\right](\hat{\theta} - \theta)$$

and

$$\text{var}[y(t, \hat{\theta})] \approx \left[\frac{\partial}{\partial \theta} y(t, \theta)\right] E\left[(\hat{\theta} - \theta)(\hat{\theta} - \theta)^T\right]\left[\frac{\partial}{\partial \theta} y(t, \theta)\right]^T$$

$$= \left[\frac{\partial}{\partial \theta} y(t, \theta)\right] P_\theta \left[\frac{\partial}{\partial \theta} y(t, \theta)\right]^T \qquad (12.6)$$

The covariance matrix P_θ can be evaluated as in (7.68) (see also Examples 2.3 and 2.4). Figure 12.1 shows the results of a numerical evaluation of the step response and its accuracy given by (12.6) for the two inputs $\mathscr{X}_1$ and $\mathscr{X}_2$. It is apparent from the figure that $\mathscr{X}_2$ gives the best accuracy for large t. The static gain is more accurately estimated for this input. On the other hand, $\mathscr{X}_1$ gives a much more accurate step response for small and medium-sized t.

Next consider the frequency response. Here

$$\frac{\hat{b}e^{i\omega}}{1 + \hat{a}e^{i\omega}} - \frac{be^{i\omega}}{1 + ae^{i\omega}} \approx \frac{e^{i\omega}}{(1 + ae^{i\omega})^2}[\hat{b}(1 + ae^{i\omega}) - b(1 + \hat{a}e^{i\omega})]$$

$$= \frac{e^{i\omega}}{(1 + ae^{i\omega})^2}(-be^{i\omega} \quad 1 + ae^{i\omega})\begin{pmatrix} \hat{a} - a \\ \hat{b} - b \end{pmatrix}$$

and a criterion for evaluation of the model accuracy can be taken as

Section 12.2   Design of the experimental condition $\mathscr{X}$ 471

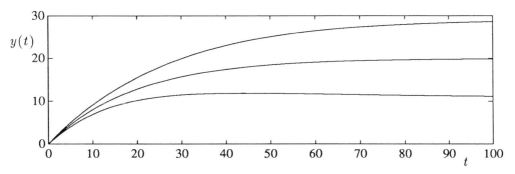

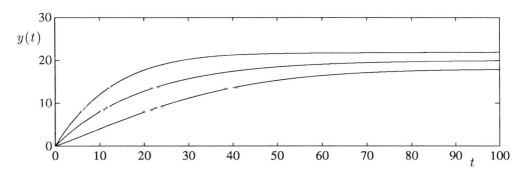

FIGURE 12.1 Step responses of the models obtained using white noise ($\mathscr{X}_1$, upper part) and a step function ($\mathscr{X}_2$, lower part) as input signals. For each part the curves shown are the exact step responses and the response ± one standard deviation. The parameter values are $\sigma^2 = Eu^2(t) = 1$, $\lambda^2 = Ee^2(t) = 1$, $a = -0.95$, $b = 1$, $N = 100$.

$$Q(\omega) \triangleq E \left| \frac{\hat{b}e^{i\omega}}{1 + \hat{a}e^{i\omega}} - \frac{be^{i\omega}}{1 + ae^{i\omega}} \right|^2$$
$$\approx \frac{1}{|1 + ae^{i\omega}|^4} (-be^{i\omega} \quad 1 + ae^{i\omega}) P_\theta \begin{pmatrix} -be^{-i\omega} \\ 1 + ae^{-i\omega} \end{pmatrix} \tag{12.7}$$

This criterion is plotted as a function of ω in Figure 12.2 for the two inputs $\mathscr{X}_1$ and $\mathscr{X}_2$. Note that the low-frequency input $\mathscr{X}_2$ gives a good accuracy of the very low-frequency model behavior ($Q(\omega)$ is small for $\omega \approx 0$), while the medium and high-frequency behavior of the model (as expressed by $Q(\omega)$) is best for $\mathscr{X}_1$. ∎

Several attempts have been made in the literature to provide a satisfactory solution to the problem of choosing optimal experimental conditions. It was seen in Chapter 10 that it can sometimes pay to use a closed loop experiment if the accuracy must be optimized under constrained output variance. If it is required that the process operates in open loop then the optimal input can often be synthesized as a sum of sinusoids. The number of sinusoids is equal to or greater than the order of the model. The optimal choice of the

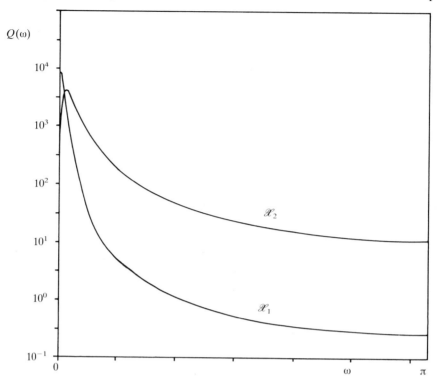

FIGURE 12.2 The functions $Q(\omega)$ for $u(t)$ white noise ($\mathscr{X}_1$) and step ($\mathscr{X}_2$). The parameter values are $\sigma^2 = Eu^2(t) = 1$, $\lambda^2 = Ee^2(t) = 1$, $a = -0.95$, $b = 1$, $N = 100$. Note the logarithmic scale.

amplitudes and frequencies of the sinusoidal input signal is not, however, an easy matter. This optimal choice will depend on the unknown parameters of the system to be identified. The topic of optimal input design will not be pursued here; refer to Chapter 10 (which contains several simple results on this problem) and its bibliographical notes.

Next consider the choice of sampling interval. Several issues should then be taken into account:

- Prefiltering of data is often necessary to avoid aliasing (folding the spectral density). Analog filters should be used prior to the sampling. The bandwidth of the filter should be somewhat smaller than half the sampling frequency. The filter should for low and medium-sized frequencies have a constant gain and a phase close to zero in order not to distort the signal unnecessarily. For high frequencies the gain should drop quickly. For a filter designed in this way, high-frequency disturbances in the data are filtered out. This will decrease the aliasing efect and may also increase the signal-to-noise ratio. Output signals should always be considered for prefiltering. In case the input signal is not in a sampled form (held constant over the sampling intervals), it may be useful to prefilter it as well. Otherwise the high-frequency variations of the input can cause a deterioration of the identification result (see Problem 12.15).
- Assume that the total time interval for an identification experiment is fixed. Then it

Section 12.2   Design of the experimental condition $\mathscr{X}$ 473

may be useful to sample the record at a high sampling rate, since then more measurements from the system are collected.
- Assume that the total number of data points is fixed. The sampling interval must then be chosen by a trade-off. If it is very large, the data will contain very little information about the high-frequency dynamics. If it is very small, the disturbances may have a relatively large influence. Furthermore, in this last case, the sampled data may contain little information on the low-frequency dynamics of the system.
- As a rough rule of thumb one can say that the sampling interval should be taken as 10% of the settling time of a step response. It may often be much worse to select the sampling interval too large than too small (see Problem 12.14 for a simple illustration).
- Very short sampling intervals will often give practical problems: all poles cluster around the point $z = 1$ in the complex plane and the model determination becomes very sensitive. A system with a pole excess of two or more becomes nonminimum phase when sampled (very) fast (see Åström et al., 1984). This will cause special problems when designing regulators.

Having collected the data, the user often has to perform some precomputations. It is for example advisable to perform some filtering to reduce the effect of noise.

As already stated, high-frequency noise can cause trouble if it is not filtered out before sampling the signals (due to the aliasing effect of sampling). The remedy is to use analog lowpass filters before the signals are sampled. The filters should be designed so that the high-frequency content of the signals above half the sampling frequency is well damped and the low-frequency content (the interesting part) is not very much affected.

As a precomputation it is common practice to filter the recorded data with a discrete-time filter. The case when both the input and the output are filtered with the same filter can be given some interesting interpretation. Consider a SISO model structure

$$y(t) = G(q^{-1}; \theta)u(t) + H(q^{-1}; \theta)e(t) \tag{12.8}$$

and assume that the data are filtered so that

$$y^F(t) = F(q^{-1})y(t) \qquad u^F(t) = F(q^{-1})u(t) \tag{12.9}$$

are available. Using the filtered data in the model structure (12.8) will give the prediction errors

$$\begin{aligned} \varepsilon_2(t) &= H^{-1}(q^{-1}; \theta)[y^F(t) - G(q^{-1}; \theta)u^F(t)] \\ &= F(q^{-1})[H^{-1}(q^{-1}; \theta)\{y(t) - G(q^{-1}; \theta)u(t)\}] \end{aligned} \tag{12.10}$$

Let $\varepsilon_1(t)$ denote the prediction error for the model (12.8). Then

$$\varepsilon_2(t) = F(q^{-1})\varepsilon_1(t) \tag{12.11}$$

Let the unknown parameters θ of the model be determined by the PEM. Then, for a large number of data points, θ is determined by minimizing

$$\int_{-\pi}^{\pi} \phi_{\varepsilon_1}(\omega; \theta) d\omega \tag{12.12a}$$

in the case of unfiltered data, and

$$\int_{-\pi}^{\pi} |F(e^{i\omega})|^2 \phi_{\varepsilon_1}(\omega; \theta) d\omega \qquad (12.12b)$$

in the filtered data case. Here $\phi_{\varepsilon_1}(\omega; \theta)$ denotes the spectral density of $\varepsilon_1(t)$. Thus the flexibility obtained by the introduction of $F(q^{-1})$ can be used to choose the frequency bands in which the model should be a good approximation of the system. This is quite useful in practice where the system is always more complex than the model, and its 'global' approximation is never possible.

12.3 Treating nonzero means and drifts in disturbances

Low-frequency noise (drift) and nonzero mean data can cause problems unless proper action is taken. The effect of nonzero mean values on the result of identification is illustrated by a simple example.

Example 12.2 *Effect of nonzero means*
The system

$$\begin{aligned} x(t) - 0.8x(t-1) &= 1.0u(t-1) + e(t) + 0.7e(t-1) \\ y(t) &= x(t) + m \end{aligned} \qquad (12.13)$$

was simulated, generating $u(t)$ as a PRBS of amplitude ± 1 and $e(t)$ as white noise of zero mean and unit variance. The number of data points was $N = 1000$. The number m accounts for the mean value of the output. The prediction error method was applied for two cases $m = 0$ and $m = 10$. The results are shown in Table 12.1 and Figure 12.3. (The plots show the first 100 data points.)

TABLE 12.1a Parameter estimates for Example 12.2, $m = 0$

Model order	Parameter	True value	Estimated value
1	a	-0.8	-0.784
	b	1.0	0.991
	c	0.7	0.703

TABLE 12.1b Parameter estimates for Example 12.2, $m = 10$

Model order	Parameter	True value	Estimated value
1	a	-0.8	-0.980
	b	1.0	1.076
	c	0.7	0.618
2	a_1	-1.8	-1.788
	a_2	0.8	0.788
	b_1	1.0	1.002
	b_2	-1.0	-0.988
	c_1	-0.3	-0.315
	c_2	-0.7	-0.622

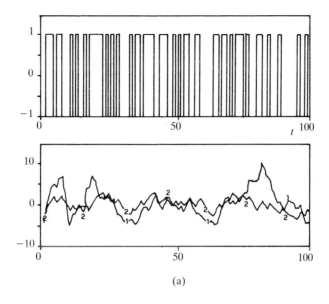

(a)

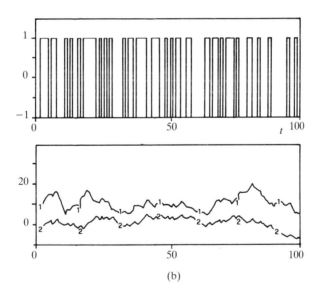

(b)

FIGURE 12.3 Input (upper part), output (1, lower part) and model output (2, lower part) for Example 12.2. (a) $m = 0$. (b) $m = 10$, model order = 1. (c) $m = 10$, model order = 2. (d) Prediction errors for $m = 0$ (upper), $m = 10$, first-order model (middle), and $m = 10$, second-order model (lower).

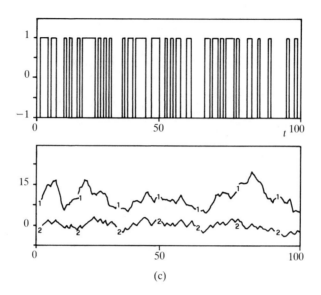

(c)

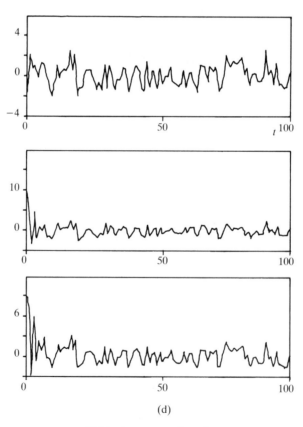

(d)

FIGURE 12.3 continued

Section 12.3 Treating nonzero means and drifts in disturbances

It is clear from the numerical results that good estimates are obtained for a zero mean output ($m = 0$), whereas $m \neq 0$ gives rise to a substantial bias. Note also the large spike in the residuals for $t = 1$, when $m = 10$.

The true system is of first order. The second-order model obtained for $m = 10$ has an interesting property. The estimated polynomials (with coefficients given in Table 12.1b) can be written as

$$A(q^{-1}) = (1 - q^{-1})(1 - 0.788q^{-1}) + 0.000q^{-2}$$
$$B(q^{-1}) = (1 - q^{-1})1.002 + 0.015q^{-2} \quad (12.14)$$
$$C(q^{-1}) = (1 - q^{-1})(1 + 0.685q^{-1}) + 0.063q^{-2}$$

The small second-order terms can be ignored and the following model results:

$$(1 - 0.788q^{-1})(1 - q^{-1})y(t) = 1.002(1 - q^{-1})u(t - 1)$$
$$+ (1 + 0.685q^{-1})(1 - q^{-1})e(t) \quad (12.15)$$

A theoretical justification can be given for the above result. The system (12.13) with $m = 10$ can be written as

$$y(t) - 0.8y(t - 1) = 1.0u(t - 1) + 2 + e(t) + 0.7e(t - 1)$$

Multiplying with $(1 - q^{-1})$ will eliminate the constant term (i.e. 2) and gives

$$(1 - 1.8q^{-1} + 0.8q^{-2})y(t) = (1.0q^{-1} - 1.0q^{-2})u(t)$$
$$+ (1 - 0.3q^{-1} - 0.7q^{-2})e(t) \quad (12.16)$$

The 'true values' given in Table 12.1b were in fact found in this way.

The model (12.14) (or (12.15)) is a good approximation of the above second-order description of the system. Nevertheless, as shown in Figure 12.3c, the model output is far from the system output. This seemingly paradoxical behavior may be explained as follows. The second-order model gives no indication on the mean value m of $y(t)$. When its output was computed the initial values were set to zero. However, the 'correct' initial values are equal to the (unknown) mean m. Since the model has a pole very close to 1, the wrong initialization is not forgotten, which makes its output look like Figure 12.3c (which is quite similar to Figure 12.3b). Note that the spikes which occur at $t = 1$ in the middle and lower parts of Figure 12.3d can also be explained by the wrong initialization with zero of the LS predictor.

By introducing the difference or delta operator

$$\Delta = 1 - q^{-1} \quad (12.17a)$$

the model (12.16) can be written as

$$(1 - 0.8q^{-1})\Delta y(t) = 1.0\Delta u(t - 1) + (1 + 0.7q^{-1})\Delta e(t) \quad (12.17b)$$

The implications of writing the model in this way, as compared to the original representation (12.13) are the following:

- The model describes the relation between differenced data rather than between the original input and output.

- The constant level m disappears. Instead the initial value (at $t = 0$) of (12.17b) due to the integrated mode, $\Delta = 1 - q^{-1}$ will determine the level of the output $y(t)$ for $t > 0$.

Remark 1. In what follows an integrated process, denoted by

$$y(t) = \frac{1}{\Delta} u(t) \qquad (12.18a)$$

will mean

$$y(t) - y(t - 1) = u(t) \qquad t \geq 1 \qquad (12.18b)$$

As a consequence

$$y(t) = y(0) + \sum_{k=1}^{t} u(k) \qquad (12.18c)$$

Note the importance of starting the 'integration' at time $t = 0$ (or any other fixed time). A convention such as

$$\frac{1}{\Delta} u(t) = \sum_{k=-\infty}^{t} u(k)$$

would not be suitable, since even with $u(t)$ a stationary process, $y(t)$ will not have finite variance for any finite t. ∎

Remark 2. Consider a general ARMAX model for differenced data. It can be written as

$$A(q^{-1})\Delta y(t) = B(q^{-1})\Delta u(t) + C(q^{-1})\Delta e(t) \qquad (12.19a)$$

To model a possible drift disturbance on the output, add a second white noise term, $v(t)$, to the right-hand side of (12.19a). Then

$$A(q^{-1})\Delta y(t) = B(q^{-1})\Delta u(t) + C(q^{-1})\Delta e(t) + v(t) \qquad (12.19b)$$

Note that the part of the output corresponding to $v(t)$ is $1/[A(q^{-1})\Delta]v(t)$, which is a nonstationary process. The variance of this term will increase approximately linearly with time, assuming the integration starts at $t = 0$. When $A(q^{-1}) = 1$, such a process is often called a random walk (cf. Problem 12.13).

Next one can apply spectral factorization to the two noise terms in (12.19b). Note that both terms, i.e. $C(q^{-1})\Delta e(t)$ and $v(t)$, are stationary processes. Hence the spectral density of their sum can be obtained using a single noise source, say $\bar{C}(q^{-1})w(t)$ where $\bar{C}(q^{-1})$ is a polynomial with all zeros outside the unit circle and is given by

$$\bar{C}(z)\bar{C}(z^{-1})\lambda_w^2 = C(z)C(z^{-1})(1 - z)(1 - z^{-1})\lambda_e^2 + \lambda_v^2 \qquad (12.19c)$$

(cf. Appendix A6.1). (λ_w^2 is the variance of the white noise $w(t)$.)

Hence the following model is obtained, which is equivalent to (12.19b):

$$A(q^{-1})\Delta y(t) = B(q^{-1})\Delta u(t) + \bar{C}(q^{-1})w(t) \qquad (12.19d)$$

This model may also be written as

$$A(q^{-1})y(t) = B(q^{-1})u(t) + \frac{\bar{C}(q^{-1})}{1 - q^{-1}} w(t) \tag{12.19e}$$

Such models are often called CARIMA (controlled autoregressive integrated moving average) models or ARIMAX. The disturbance term, $\bar{C}(q^{-1})/[A(q^{-1})(1 - q^{-1})]w(t)$, is called an ARIMA process. Its variance increases without bounds as t increases. Such models are often used in econometric applications, where the time series frequently are nonstationary. They constitute a natural way to describe the effect of drifts and other nonstationary disturbances, within a stochastic framework. Note that even if they are nonstationary due to the integrator in the dynamics, the associated predictors are asymptotically stable since $\bar{C}(z)$ has all zeros strictly outside the unit circle. ∎

Two approaches for treating nonzero means

There are two different approaches that can be used for identifying systems with nonzero mean data:

- Estimate the mean values explicitly. With such an approach a deterministic model is used to describe the means.
- Use models for the differenced data. Equivalently, use an ARIMA model to describe the disturbance part in the output. With this approach no estimates of the mean values are provided.

The following paragraphs describe these two approaches in some further detail. The first approach should generally be preferred. Problem 12.4 asks for some comparisons of these approaches.

Estimation of mean values

The mean values (or their effect on the output) can be estimated in two ways.

- Fit a polynomial trend

$$y^*(t) = \alpha_0 + \alpha_1 t + \ldots + \alpha_r t^r \tag{12.20}$$

to the output by linear regression techniques and similarly for the input. Then compute new ('detrended') data by

$$\bar{y}(t) = y(t) - y^*(t)$$
$$\bar{u}(t) = u(t) - u^*(t)$$

and thereafter apply an identification method to $\bar{y}(t)$, $\bar{u}(t)$. If the degree r in (12.20) is chosen as zero, this procedure means simply that the arithmetic mean values

$$y^* = \frac{1}{N} \sum_{t=1}^{N} y(t)$$

$$u^* = \frac{1}{N} \sum_{t=1}^{N} u(t)$$

are computed and subtracted from the original data. (See Problem 12.5 for an illustration of this approach.) Note that (12.20) for $r > 0$ can also model some drift in the data, not only a constant mean.
- Estimate the means during the parameter estimation phase. To be able to do so a model structure must be used that contains an explicit parameter for describing the effect of some possibly nonzero mean values on $y(t)$. The model will have the form

$$y(t) = G(q^{-1}; \theta)u(t) + H(q^{-1}; \theta)e(t) + m(\theta) \tag{12.21}$$

where $m(\theta)$ is one (or ny in the multivariable case) of the elements in the parameter vector θ. This alternative is illustrated in Problems 12.3–12.5. The model (12.21) can be extended by replacing the constant $m(\theta)$ by a polynomial in t, whose coefficients depend on θ.

Handling nonzero means with stochastic models

In this approach a model structure is used where the noise filter has a pole in $z = 1$. This means that the model structure is

$$y(t) = G(q^{-1}; \theta)u(t) + H(q^{-1}; \theta)e(t) \tag{12.22a}$$

where $H(q^{-1}; \theta)$ has a pole in $q^{-1} = 1$. This is equivalent to saying that the output disturbance is modeled as an ARIMA model, which means that

$$H(q^{-1}; \theta) = \frac{\bar{H}(q^{-1}; \theta)}{1 - q^{-1}} \tag{12.22b}$$

for some filter $\bar{H}(q^{-1}; \theta)$. The model can therefore be written as

$$\Delta y(t) = G(q^{-1}; \theta)\Delta u(t) + \bar{H}(q^{-1}; \theta)e(t) \tag{12.22c}$$

(cf. (12.19d, e)). Therefore, one can compute the differenced data and then use the model structure (12.22c) which does not contain a parameter $m(\theta)$.

Assume that the true system satisfies

$$y(t) = G_s(q^{-1})u(t) + H_s(q^{-1})e(t) + m \tag{12.23}$$

where m accounts for the nonzero mean of the output. The prediction error becomes

$$\begin{aligned}\varepsilon(t, \theta) &= \bar{H}^{-1}(q^{-1}; \theta)[\Delta y(t) - G(q^{-1}; \theta)\Delta u(t)] \\ &= \bar{H}^{-1}(q^{-1}; \theta)[G_s(q^{-1}) - G(q^{-1}; \theta)]\Delta u(t) \\ &\quad + \bar{H}^{-1}(q^{-1}; \theta)H_s(q^{-1})\Delta e(t)\end{aligned} \tag{12.24}$$

It is clearly seen that m does not influence the prediction error $\varepsilon(t, \theta)$. However, $\varepsilon(t, \theta) = e(t)$ cannot hold unless $\bar{H}(q^{-1}; \theta) = H_s(q^{-1})\Delta$. This means that the noise filter must have a zero in $z = 1$. Such a case should be avoided, if at all possible, since the predictor will then be unstable and therefore significantly dependent on unknown initial values. (The parameter vector θ no longer belongs to the set $\mathcal{D}$, (6.2).)

Note that in Example 12.2 by increasing the model order both $G(q^{-1}; \theta)$ and $H(q^{-1}; \theta)$ got approximately a pole and a zero in $z = 1$. Even if it is possible to get rid of nonzero mean values in this way it cannot be recommended since the additional poles and zeros in $z = 1$ will cause problems.

Handling drifting disturbances

The situation of drifting disturbances can be viewed as an extension of the nonzero mean data case. Assume that the system can be described by

$$y(t) = G_s(q^{-1})u(t) + \frac{\bar{H}_s(q^{-1})}{1 - q^{-1}} e(t) \qquad \bar{H}_s(1) \neq 0 \qquad (12.25)$$

Here the disturbance is a nonstationary process (in fact an ARIMA process). The variance of $y(t)$ will increase without bound as t increases.

Consider first the situation where the system (12.25) is identified using an ARMAX model

$$A(q^{-1})y(t) = B(q^{-1})u(t) + C(q^{-1})e(t) \qquad (12.26a)$$

To get a correct model it is required that

$$\frac{C(q^{-1})}{A(q^{-1})} = \frac{\bar{H}_s(q^{-1})}{1 - q^{-1}} \qquad \frac{B(q^{-1})}{A(q^{-1})} = G_s(q^{-1}) \qquad (12.26b)$$

which implies that $\Delta = 1 - q^{-1}$ must be a factor of both $A(q^{-1})$ and $B(q^{-1})$. This should be avoided for several reasons. First, the model is not parsimonious, since it will contain unnecessarily many free parameters, and will hence give a degraded accuracy. Second, and perhaps more important, such models are often not feasible to use. For example, in pole placement design of regulators, an approximately common zero at $z = 1$ for the A- and B-polynomials will make the design problem very ill-conditioned.

A more appealing approach is to identify the system (12.25) using a model structure of the form (12.22). The prediction error is then given by

$$\begin{aligned}\varepsilon(t, \theta) &= \bar{H}^{-1}(q^{-1}; \theta)[\Delta y(t) - G(q^{-1}; \theta)\Delta u(t)] \\ &= \bar{H}^{-1}(q^{-1}; \theta)[G_s(q^{-1}) - G(q^{-1}; \theta)]\Delta u(t) \\ &\quad + \bar{H}^{-1}(q^{-1}; \theta)\Delta \bar{H}_s(q^{-1})\frac{1}{\Delta}e(t)\end{aligned} \qquad (12.27)$$

where it is assumed that $\bar{H}^{-1}(q^{-1}; \theta)$ is asymptotically stable. Here one would like to cancel the factor Δ that appears both in numerator and denominator. This must be done with some care since a pole at $z = 1$ (which is on the stability boundary) can cause a nondecaying 'transient' (which in this case would be a constant). Note however that $(1/\Delta)e(t)$ can be written as (cf. (12.18))

$$\frac{1}{\Delta}e(t) = x_0 + \sum_{i=1}^{t} e(i) \qquad (12.28a)$$

where x_0 is an initial value. Thus

$$\Delta\left(\frac{1}{\Delta}e(t)\right) = \Delta\left[x_0 + \sum_{i=1}^{t} e(i)\right] = e(t) \qquad (12.28b)$$

This calculation justifies the cancellation of the factor Δ in (12.27). Hence

$$\varepsilon(t, \theta) = \bar{H}^{-1}(q^{-1}; \theta)[G_s(q^{-1}) - G(q^{-1}; \theta)]\Delta u(t) \\ + \bar{H}^{-1}(q^{-1}; \theta)\bar{H}_s(q^{-1})e(t) \qquad (12.29)$$

is a stationary process and the PEM loss function is well behaved.

12.4 Determination of the model structure $\mathcal{M}$

Chapter 11 presented some methods for determining an appropriate model structure. Here some of these methods are illustrated by means of simulation examples. Doing so we will verify the theoretical results and show how different techniques can be used to find a reasonable model structure.

Example 12.3 *Model structure determination,* $\mathcal{S} \in \mathcal{M}$
The following second-order system was simulated:

$$y(t) - 1.5y(t-1) + 0.7y(t-2) = 1.0u(t-1) + 0.5u(t-2) + e(t) - 1.0e(t-1) \\ + 0.2e(t-2)$$

where $u(t)$ is a PRBS of amplitude ± 1 and $e(t)$ a zero mean white noise process with unit variance.

The number of data points was taken as $N = 1000$. The system was identified using a prediction error method within an ARMAX model structure

$$A(q^{-1})y(t) = B(q^{-1})u(t) + C(q^{-1})e(t) \qquad (12.30)$$

The polynomials in (12.30) are all assumed to have degree n. This degree was varied from 1 to 3. The results obtained are summarized in Tables 12.2–12.4 and Figure 12.4.

It is obvious from Tables 12.2–12.4 and the various plots that a second-order model should be chosen for these data. The test quantity for the F-test comparing the second- and the third-order models is 4.6. The 5% significance level gives a threshold of 7.81. Hence a second-order model should be chosen. Observe from Table 12.3 that for $n = 3$ the polynomials $\hat{A}$, $\hat{B}$ and $\hat{C}$ have approximately a common zero ($z \approx -0.6$). Note from Table 12.2 and Figure 12.4a, third part, that the estimated second-order model gives a very good description of the true system. ∎

Example 12.4 *Model structure determination,* $\mathcal{S} \notin \mathcal{M}$
This example examines what can happen when the model structure does not contain the true system. The following system was simulated:

$$y(t) - 0.8y(t-1) = 1.0u(t-1) + e(t) + 0.7e(t-1)$$

TABLE 12.2 Parameter estimates for Example 12.3

Model order	Loss	Parameter	True value	Estimated value
1	1713	a		-0.822
		b		0.613
		c		0.389
2	526.6	a_1	-1.5	-1.502
		a_2	0.7	0.703
		b_1	1.0	0.938
		b_2	0.5	0.550
		c_1	-1.0	-0.975
		c_2	0.2	0.159
3	524.3	a_1		-0.834
		a_2		-0.291
		a_3		0.462
		b_1		0.942
		b_2		1.151
		b_3		0.420
		c_1		-0.308
		c_2		-0.455
		c_3		0.069

TABLE 12.3 Zeros of the true and the estimated polynomials, Example 12.3

Polynomial	True system	Estimated models		
		$n = 1$	$n = 2$	$n = 3$
A	$0.750 \pm i0.371$	0.822	$0.751 \pm i0.373$	$0.748 \pm i0.373$
				-0.662
B	-0.500	–	-0.586	$-0.611 \pm i0.270$
C	0.276	-0.389	0.207	-0.615
	0.724		0.768	0.144
				0.779

TABLE 12.4 Test quantities for Example 12.3. The distributions for the correlation tests refer to the approximate analysis of Examples 11.1 and 11.2

Model order	Test quantity	Numerical value	Distributions (under null hypothesis)	95% [99%] confidence level
$n = 1$	changes of sign in $\{\varepsilon(t)\}$	459	$\mathcal{N}(500,250)$	(468,530) [(459,540)]
	test on $r_\varepsilon(\tau) = 0$ $\tau = 1, \ldots, 10$	212.4	$\chi^2(10)$	18.3 [23.2]
	test on $r_{\varepsilon u}(\tau) = 0$ $\tau = 2, \ldots, 11$	461.4	$\chi^2(10)$	18.3 [23.2]
	test on $r_{\varepsilon u}(\tau) = 0$ $\tau = -9, \ldots, 0$	3.0	$\chi^2(10)$	18.3 [23.2]
$n = 2$	changes of sign in $\{\varepsilon(t)\}$	514	$\mathcal{N}(500,250)$	(468,530) [(459,540)]
	test on $r_\varepsilon(\tau) = 0$ $\tau = 1, \ldots, 10$	12.2	$\chi^2(10)$	18.3 [23.2]
	test on $r_{\varepsilon u}(\tau) = 0$ $\tau = 3, \ldots, 12$	4.8	$\chi^2(10)$	18.3 [23.2]
	test on $r_{\varepsilon u}(\tau) = 0$ $\tau = -9, \ldots, 0$	4.7	$\chi^2(10)$	18.3 [23.2]

484   *Some practical aspects*   Chapter 12

where $u(t)$ is a **PRBS** of amplitude ± 1 and $e(t)$ white noise of zero mean and unit variance. The number of data points was $N = 1000$. The system was identified using the least squares method in the model structure

$$A(q^{-1})y(t) = B(q^{-1})u(t) + e(t)$$

with

$$A(q^{-1}) = 1 + a_1 q^{-1} + \ldots + a_n q^{-n}$$
$$B(q^{-1}) = b_1 q^{-1} + \ldots + b_n q^{-n}$$

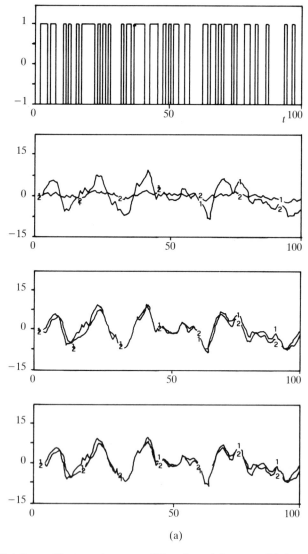

(a)

FIGURE 12.4(a) Input (first part), output (1) and model output (2) for $n = 1$ (second part), $n = 2$ (third part) and $n = 3$ (fourth part), Example 12.3. The first 100 data points are plotted.

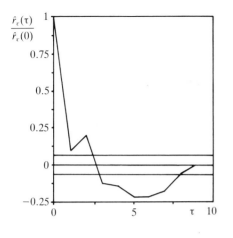
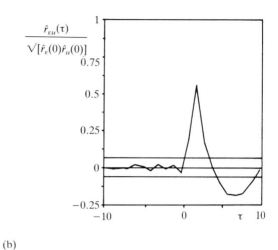

(b)

FIGURE 12.4(b) Normalized covariance functions for the first-order model, Example 12.3. Lefthand curve: the normalized covariance function $\hat{r}_\varepsilon(\tau)$ of the residuals $\varepsilon(t)$. Right-hand curve: the normalized cross-covariance function $\hat{r}_{\varepsilon u}(\tau)$ between $\varepsilon(t)$ and $u(t)$.

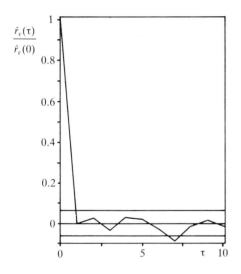
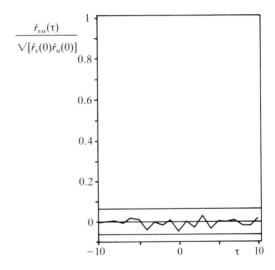

(c)

FIGURE 12.4(c) Same as (b) but for the second-order model.

FIGURE 12.4 continued

The model order was varied from $n = 1$ to $n = 5$. The results obtained are summarized in Tables 12.5, 12.6 and Figures 12.5, 12.6.

When searching for the most appropriate model order the result will depend on how the model fit is assessed.

First compare the deterministic parts of the models (i.e. $[\hat{B}(q^{-1})/\hat{A}(q^{-1})]u(t)$). The model outputs do not differ very much when the model order is increased from 1 to 5. Indeed, Figure 12.5 illustrates that the estimated transfer functions $\hat{B}(q^{-1})/\hat{A}(q^{-1})$, for all n, have similar pole-zero configurations to that of the true system. Hence, as long as only the deterministic part of the model is of importance, it is sufficient to choose a first-order model. However, note from Table 12.5 that the obtained model is slightly biased. Hence, the estimated first-order model may not be sufficiently accurate if it is used for other types of input signals.

If one also considers the stochastic part of the model (i.e. $[1/\hat{A}(q^{-1})]e(t)$) then the situation changes. This part of the model must be considered when evaluating the prediction ability of the model. The test quantities given in Table 12.6 are all based on the stochastic part of the model. Most of them indicate that a fourth-order model would be adequate.

TABLE 12.5 Parameter estimates for Example 12.4

Model order	Loss	Parameter	True value	Estimated value
1	786	a	−0.8	−0.868
		b	1.0	0.917
2	603	a_1		−1.306
		a_2		0.454
		b_1		0.925
		b_2		−0.541
3	551	a_1		−1.446
		a_2		0.836
		a_3		−0.266
		b_1		0.935
		b_2		−0.673
		b_3		0.358
4	533.1	a_1		−1.499
		a_2		0.992
		a_3		−0.530
		a_4		0.169
		b_1		0.938
		b_2		−0.729
		b_3		0.469
		b_4		−0.207
5	530.0	a_1		−1.513
		a_2		1.035
		a_3		−0.606
		a_4		0.281
		a_5		−0.067
		b_1		0.938
		b_2		−0.743
		b_3		0.501
		b_4		−0.258
		b_5		0.100

Section 12.5   Time delays   487

TABLE 12.6 Test quantities for selecting the model order, Example 12.4. For the *F*-test the threshold corresponding to 95% significance level is 5.99. For the tests on the covariance functions the threshold is 18.3 for a significance level of 95% and 23.2 for a significance level of 99%, using the approximate distributions developed in Examples 11.1 and 11.2

Model order	Loss	F-test	Tests of prediction errors		
			No. of sign changes	$r_\varepsilon(\tau)$	$r_{\varepsilon u}(\tau)$
Degrees of freedom		2		10	10
95% confidence interval			(467,531)		
1	787		356	257	15.3
		305			
2	603		438	108	13.1
		94			
3	551		482	50.2	4.1
		34			
4	533.1		494	22.1	4.9
		5.8			
5	530.0		502	20.0	3.3

To illustrate how well the different models describe the stochastic part of the system, the noise spectral densities are plotted in Figure 12.7.

It can be seen from Figure 12.7 that the fourth-order model gives a closer fit than the first-order model to the true spectral density of the stochastic part of the system. ∎

12.5 Time delays

No consideration has yet been given to time delays. Many industrial and other processes have time delays that are not negligible. It could be very important to describe the delay correctly in the model.

First note that the general model structure which has been used, i.e.

$$y(t) = G(q^{-1}; \theta)u(t) + H(q^{-1}; \theta)e(t) \tag{12.31}$$

can in fact cover cases with a time delay. To see this write

$$G(q^{-1}; \theta) = \sum_{i=1}^{\infty} g_i(\theta) q^{-i} \tag{12.32a}$$

where the sequence $\{g_i(\theta)\}$ is the discrete time impulse response. If there should be a delay of k ($k \geq 1$) sampling intervals in the model, it is sufficient to require that the parametrization satisfies

$$g_i(\theta) \equiv 0 \quad i = 1, 2, \ldots, (k-1) \tag{12.32b}$$

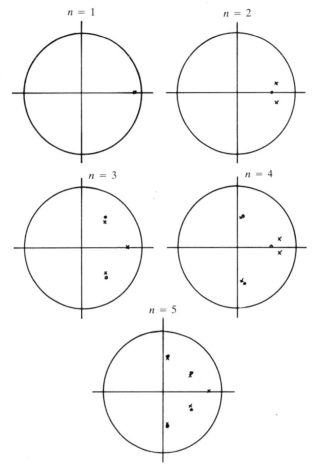

FIGURE 12.5 Pole-zero configurations of the different models, $n = 1, \ldots, 5$, Example 12.4. ($\times$ = pole; $\bigcirc$ = zero)

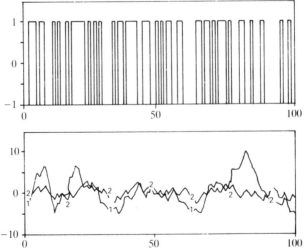

FIGURE 12.6 Input (upper part), output (1, lower part) and model output (2, lower part) for model order $n = 4$, Example 12.4.

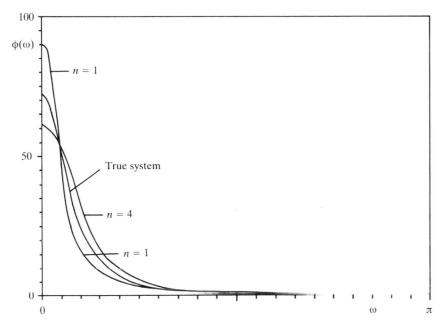

FIGURE 12.7 Noise spectral density $|1 + 0.7e^{i\omega}|^2/(2\pi|1 - 0.8e^{i\omega}|^2)$ of the true system and the estimated spectra $\hat{\lambda}^2/(2\pi|\hat{A}(e^{i\omega})|^2)$ of the first- and fourth-order models.

With this observation it can be concluded that the general theory that has been developed (concerning consistency and asymptotic distribution of the parameter estimates, identifiability under closed loop operation, etc.) will hold also for models with time delays.

If the constraint (12.32b) is not imposed on the model parametrization, the theory still applies since $\mathscr{S} \in \mathscr{M}$ under weak conditions. The critical point is to not consider a parametrized model having a larger time delay than the true one. In such a case $\mathscr{S}$ no longer belongs to $\mathscr{M}$ and the theory based on this assumption collapses.

The following example illustrates how to cope with time delays from a practical point of view.

Example 12.5 *Treating time delays for ARMAX models*
Assume that estimates are required of the parameters of the following model:

$$A(q^{-1})y(t) = q^{-(k-1)}B(q^{-1})u(t) + C(q^{-1})e(t) \tag{12.33}$$

where

$$A(q^{-1}) = 1 + a_1 q^{-1} + \ldots + a_n q^{-n}$$
$$B(q^{-1}) = b_1 q^{-1} + \ldots + b_n q^{-n}$$
$$C(q^{-1}) = 1 + c_1 q^{-1} + \ldots + c_n q^{-n}$$

k being an integer, $k \geq 1$. By writing the model as

$$A(q^{-1})y(t) = B(q^{-1})\{q^{-(k-1)}u(t)\} + C(q^{-1})e(t)$$

it is fairly obvious how to proceed with the parameter estimation:

1. First 'compute' the delayed input according to
$$\bar{u}(t) = q^{-(k-1)}u(t) = u(t - k + 1) \quad t = k, k + 1, k + 2, \ldots, N$$
2. Next estimate the parameters of the ARMAX model applying a standard method (i.e. one designed to work with unit time delay) using $\bar{u}(t)$ as input signal instead of $u(t)$.

Note that one could alternatively shift the output sequence as follows:

1'. Compute $\bar{y}(t) = y(t + k - 1) \quad t = 1, 2 \ldots, N - k + 1$
2'. Estimate the ARMAX model parameters using $\bar{y}(t)$ as output instead of $y(t)$.

Such a procedure is clearly not limited to ARMAX models. It may be repeated for various values of k. The determination of k can be made using the same methods as those used for determination of the model order. Various methods for order estimation were examined above; see also Chapter 11 for a detailed discussion. ∎

It should be noted that some care should be exercised when scanning the delay k and the model order n simultaneously. Indeed, variation of both k and n may lead to non-nested model structures which cannot be compared by using the procedures described in Section 11.5 (such as the F-test and the AIC). To give an example of such a case, note that the model structures (12.33) corresponding to $\{k = 1, n = 1\}$ and to $\{k = 2, n = 2\}$ are not nested. In practice it is preferable to compare the model structures corresponding to a fixed value of n and to varying k. Note that such structures are nested, the one corresponding to a given k being included in those corresponding to *smaller* values of k.

12.6 Initial conditions

When applying a prediction error method to the general model structure
$$y(t) = G(q^{-1}; \theta)u(t) + H(q^{-1}; \theta)e(t) \tag{12.34}$$
it is necessary to compute the prediction errors
$$\varepsilon(t, \theta) = H^{-1}(q^{-1}; \theta)[y(t) - G(q^{-1}; \theta)u(t)] \tag{12.35}$$
and possibly also their gradient with respect to θ. In order to compute $\varepsilon(1, \theta), \ldots, \varepsilon(N, \theta)$ from the measured data $y(1), u(1), \ldots, y(N), u(N)$ using (12.35) some initial conditions for this difference equation are needed. One can then proceed in at least two ways:

- Set the initial values to zero. If *a priori* information is available then other more appropriate values can be used.
- Include the unknown initial values in the parameter vector.

A further, but often more complicated, possibility is to find the exact ML estimate (see

Section 12.6　　　　　　　　　　　　　　　　　　　　　　　　*Initial conditions*　491

Complement C7.7). The following example illustrates the two possibilities above for an ARMAX model.

Example 12.6 *Initial conditions for ARMAX models*
Consider the computation of the prediction errors for an ARMAX model. They are given by the difference equation

$$C(q^{-1})\varepsilon(t, \theta) = A(q^{-1})y(t) - B(q^{-1})u(t) \qquad (12.36)$$

For convenience let all the polynomials $A(q^{-1})$, $B(q^{-1})$ and $C(q^{-1})$ have degree n. The computation of $\varepsilon(t, \theta)$ can then be done as follows:

$$\varepsilon(t, \theta) = y(t) - \varphi^T(t)\theta \qquad (12.37)$$

where

$$\varphi^T(t) = (-y(t-1) \ldots -y(t-n) \quad u(t-1) \ldots u(t-n) \quad \varepsilon(t-1, \theta) \ldots \varepsilon(t-n, \theta))$$

$$\theta = (a_1 \ldots a_n \quad b_1 \ldots b_n \quad c_1 \ldots c_n)^T$$

When proceeding in this way, $\varphi(1)$ is needed to compute $\varepsilon(1, \theta)$ but this vector contains unknown elements. The first possibility would be to set

$$y(t) = 0$$
$$u(t) = 0 \quad \text{for } t \leq 0$$
$$\varepsilon(t, \theta) = 0$$

The second possibility is to include the unknown values needed in the parameter vector, which would give

$$\theta' = \begin{pmatrix} \theta \\ \bar{\theta} \end{pmatrix} \qquad (12.38a)$$

$$\bar{\theta} = (y(0) \ldots y(-n+1) \ u(0) \ldots u(-n+1) \ \varepsilon(0, \theta) \ldots \varepsilon(-n+1, \theta))^T \qquad (12.38b)$$

This makes the new parameter vector θ' of dimension $6n$. This vector is estimated by minimizing the usual loss function (sum of squared prediction errors).

The second possibility above is conceptually straightforward but leads to unnecessary computational complications. Since the computation of the prediction errors is needed as an intermediate step during an optimization with respect to θ (or θ') it is of paramount importance not to increase the dimension of the parameter vector more than necessary. Now regard (12.36) as a dynamic system with $u(t)$ and $y(t)$ as inputs and $\varepsilon(t, \theta)$ as output. This system is clearly of order n. Therefore it should be sufficient to include only n initial values in the parameter vector, which makes a significant reduction as compared to the $3n$ additional values entered in $\bar{\theta}$ (12.38b).

One way to achieve the aforementioned reduction would be to extend θ with the first n prediction errors in the criterion. Consider therefore the modified criterion

$$\bar{V}(\theta') = \sum_{t=n+1}^{N} \varepsilon^2(t, \theta') \qquad (12.39)$$

where

$$\theta' = \begin{pmatrix} \theta \\ \varepsilon(1, \theta) \\ \vdots \\ \varepsilon(n, \theta) \end{pmatrix} \quad (12.40)$$

Clearly, the dimension of θ' in (12.40) is $4n$.

An alternative way to find n appropriate initial values for the computation of $\varepsilon(t, \theta)$, $t \geq 1$, is to write (12.36) in state space form. A convenient possibility is to use the so-called controller form

$$x(t+1) = \begin{pmatrix} -c_1 & 1 & 0 \\ \vdots & & \ddots & \\ & & & 1 \\ -c_n & 0 & 0 \end{pmatrix} x(t) - \begin{pmatrix} b_1 \\ \vdots \\ b_n \end{pmatrix} u(t) + \begin{pmatrix} a_1 - c_1 \\ \vdots \\ a_n - c_n \end{pmatrix} y(t) \quad (12.41)$$

$$\varepsilon(t, \theta) = (1 \ 0 \ \ldots \ 0) x(t) + y(t)$$

A new parameter vector can be taken as

$$\theta' = \begin{pmatrix} \theta \\ x(1) \end{pmatrix} \quad (12.42)$$

which is of dimension $4n$. It is clearly seen from (12.41) how to compute $\varepsilon(1, \theta), \ldots, \varepsilon(N, \theta)$ from the data $y(1), u(1), \ldots, y(N), u(N)$ for any value of θ'. When implementing (12.41) one should, of course, make use of the structure of the companion matrix in order to reduce the computational effort. There is no need to perform multiplication with elements that are known *a priori* to be zero. When using (12.41), (12.42) the parameter estimates are determined by minimizing the criterion

$$V(\theta') = \sum_{t=1}^{N} \varepsilon^2(t, \theta') \quad (12.43)$$

In contrast to (12.39) the first n prediction errors $\varepsilon(1, \theta'), \ldots, \varepsilon(n, \theta')$ are now included in the criterion. ∎

Finally, it should be noted that the initial conditions cannot be consistently estimated using a PEM or another estimation method (see Åström and Bohlin, 1965). See also Problem 12.11. This is in fact quite expected since the consistency is an asymptotic property, and the initial conditions are immaterial asymptotically for stable systems. However, in practice one always processes a finite (sometimes rather small) number of samples, and inclusion of the initial conditions in the unknown parameter vector may improve substantially the performance of the estimated model.

12.7 Choice of the identification method $\mathcal{I}$

For this choice it is difficult to give more than general comments. It is normally obvious from the purpose of identification whether the process should be identified off-line or on-line. On-line identification is needed, for example, if the parameter estimation is a part of an adaptive system or if the purpose is to track (slowly) time-varying parameters. In contrast, off-line identification is used as a batch method which processes all the recorded data 'simultaneously'.

Note that an interesting off-line approach is to apply an on-line algorithm repeatedly to the data. The 'state' of the on-line recursion (for example the vector $\hat{\theta}(t)$ and the matrix $P(t)$; see Chapter 9) obtained after processing the sample of data is employed as 'initial state' for the next pass through the data. However, processing of the available data does not give any information on the initial values for either the prediction errors or their gradient with respect to the parameter vector θ. These variables have to be re-initialized rather arbitrarily at the beginning of each new pass through the data, which may produce some undesirable fluctuations in the parameter estimates. Nevertheless, use of an on-line algorithm for off-line identification in the manner outlined above may lead to some computational time saving as compared to a similar off-line algorithm.

When choosing an identification method, the purpose of the identification is important, since it may specify both what type of model is needed and what accuracy is desired. If the purpose is to tune a PI regulator, a fairly crude model can be sufficient, while high accuracy may be needed if the purpose is to describe the process in detail, to verify theoretical models and the like.

The following methods are ordered in terms of improved accuracy and increased computational complexity:

- Transient analysis.
- Frequency analysis.
- The least squares method.
- The instrumental variable method.
- The prediction error method.

Note that here the choice of model structure is implicitly involved, since the methods above are tied to certain types of model structure.

In summary, the user must make a trade-off between required accuracy and computational effort when choosing the identification method. This must be done with the purpose of the identification in mind. In practice other factors will influence the choice, such as previous experience with various methods, available software, etc.

12.8 Local minima

When a prediction error method is used the parameter estimate is determined as the global minimum point of the loss function. Since a numerical search algorithm must be used there is a potential risk that the algorithm is stuck at a local minimum. Example 11.5 demonstrated how this can happen in practice. Note that this type of problem is linked to the use of prediction error methods. It does not apply to instrumental variable methods.

It is hard to analyze the possible existence of local minima theoretically. Only a few general results are available. They all hold asymptotically, i.e. the number of data, N, is assumed to be large. It is also assumed that the true system is included in the model structure. The following results are known:

- For a scalar ARMA process

$$A(q^{-1})y(t) = C(q^{-1})e(t)$$

all minima are global. There is a unique minimum if the model order is correctly chosen. This result is proved in Complement C7.6.

- For a multivariable MA process

$$y(t) = C(q^{-1})e(t)$$

there is a unique minimum point.

- For the single output system

$$A(q^{-1})y(t) = B(q^{-1})u(t) + \frac{1}{C(q^{-1})}e(t)$$

there is a unique minimum if the signal-to-noise ratio is (very) high, and several local minima if this ratio is (very) low.

- For the SISO output error model

$$y(t) = \frac{B(q^{-1})}{F(q^{-1})}u(t) + e(t)$$

all minima are global if the input signal is white noise. This result is proved in Complement C7.5.

In all the above cases the global minimum of the loss function corresponds to the true system.

If the minimization algorithm is stuck at a local minimum a bad model may be obtained, as was illustrated in Example 11.5. There it was also illustrated how the misfit of such a model can be detected. If it is found that a model is providing a poor description of the data one normally should try a larger model structure. However, if there is reason to believe that the model corresponds to a nonglobal minimum, one can try to make a new optimization of the loss function using another starting point for the numerical search routine.

The practical experience of how often certain PEM algorithms for various model structures are likely to be stuck at local minima is rather limited. It may be said, however, that with standard optimization methods (for example, a variable metric method or a Gauss–Newton method; see (7.87)) the global minimum is usually found for an ARMAX model, while convergence to a false local minimum occasionally may occur for the output error model

$$y(t) = \frac{B(q^{-1})}{F(q^{-1})} u(t) + e(t)$$

When the loss function is expected to be severely multimodal (as, for example, is the PEM loss function associated with the sinusoidal parameter estimation, Stoica, Moses *et al.*, 1986), an alternative to accurate initialization is using a special global optimization algorithm such as the simulated annealing (see, for example, Sharman and Durrani, 1988). For this type of algorithms, the probability of finding the global minimum is close to one even in complicated multimodal environments.

12.9 Robustness

When recording experimental data occasional large measurement errors may occur. Such errors can be caused by disturbances, conversion failures, etc. The corresponding abnormal data points are called outliers. If no specific action is taken, the outliers will influence the estimated model considerably. The outliers tend to appear as spikes in the sequence of prediction errors $\{\varepsilon(t, \theta)\}$, and will hence give large contributions to the loss function. Their effect is illustrated in the following two examples.

Example 12.7 *Effect of outliers on an ARMAX model*
Consider an ARMAX process

$$A(q^{-1})y(t) = B(q^{-1})u(t) + C(q^{-1})e(t) \tag{12.44a}$$

where $e(t)$ is white noise of zero mean and variance λ^2. Assume that there are occasionally some errors on the output, so that one is effectively measuring not $y(t)$ but

$$z(t) = y(t) + v(t) \tag{12.44b}$$

The measurement noise $v(t)$ has the following properties:

- $v(t)$ and $v(s)$ are independent if $t \neq s$.
- $v(t) = 0$ with a large probability.
- $Ev(t) = 0$ and $Ev^2(t) = \sigma^2 < \infty$.

These assumptions imply that $v(t)$ is white noise. From (12.44a, b),

$$A(q^{-1})z(t) = B(q^{-1})u(t) + C(q^{-1})e(t) + A(q^{-1})v(t) \tag{12.44c}$$

Applying spectral factorization gives the following equivalent ARMAX model:

$$A(q^{-1})z(t) = B(q^{-1})u(t) + \bar{C}(q^{-1})w(t) \tag{12.44d}$$

where $w(t)$ is white noise of variance λ_w^2, and $\bar{C}(q^{-1})$ and λ_w^2 are given by

$$\bar{C}(z)\bar{C}(z^{-1})\lambda_w^2 \equiv C(z)C(z^{-1})\lambda^2 + A(z)A(z^{-1})\sigma^2 \tag{12.44e}$$

If the system (12.44a, b) is identified using an ARMAX model then (asymptotically, for an *infinite* number of data points) this will give the model (12.44d). Note that $A(q^{-1})$ and

496   Some practical aspects   *Chapter 12*

$B(q^{-1})$ remain unchanged. Further, from (12.44e), when $\sigma^2 \to 0$ (the effect of outliers tends to zero) then $\bar{C}(z) \approx C(z)$, i.e. the true noise description is found. Similarly, when $\sigma^2 \to \infty$ (the outliers dominate the disturbances in (12.44a)), then $\bar{C}(z) \approx A(z)$ and the filter $H(q^{-1}) = C(q^{-1})/A(q^{-1}) \approx 1$ (as intuitively expected).   ∎

Example 12.8 *Effect of outliers on an ARX model*

The effect of outliers on an ARX model is more complex to analyze. The reason is that an ARX system with outliers can no longer be described exactly within the class of ARX models. Consider for illustration the system

$$y(t) + ay(t-1) = bu(t-1) + e(t)$$
$$z(t) = y(t) + v(t) \tag{12.45a}$$

with $u(t)$, $e(t)$ and $v(t)$ mutually uncorrelated white noise sequences of zero means and variances σ_u^2, λ_e^2, λ_v^2 respectively. Let the system be identified with the LS method in the model structure

$$z(t) + \hat{a}z(t-1) = \hat{b}u(t-1) + \varepsilon(t) \tag{12.45b}$$

The asymptotic values of the parameter estimates are given by

$$\begin{pmatrix} \hat{a} \\ \hat{b} \end{pmatrix} = \begin{pmatrix} Ez^2(t) & -Ez(t)u(t) \\ -Ez(t)u(t) & Eu^2(t) \end{pmatrix}^{-1} \begin{pmatrix} -Ez(t+1)z(t) \\ Ez(t+1)u(t) \end{pmatrix}$$

$$= \begin{pmatrix} Ez^2(t) & 0 \\ 0 & \sigma_u^2 \end{pmatrix}^{-1} \begin{pmatrix} -Ez(t+1)z(t) \\ b\sigma_u^2 \end{pmatrix}$$

Thus it is found (for the specific input assumed) that $\hat{b} = b$ and

$$\hat{a} = \frac{aEy^2(t)}{Ey^2(t) + \lambda_v^2} \triangleq \alpha a \tag{12.45c}$$

The scalar α satisfies $0 \le \alpha \le 1$. Specifically it varies monotonically with λ_v^2, tends to 0 when λ_v^2 tends to infinity, and to 1 when λ_v^2 tends to zero. Next examine how the noise filter $H(q^{-1})$ differs from 1. It will be shown that in the present case, for all frequencies

$$\left| \frac{1}{1 + \hat{a}e^{i\omega}} - 1 \right| \le \left| \frac{1}{1 + ae^{i\omega}} - 1 \right| \tag{12.45d}$$

This means that in the presence of outliers the estimated noise filter $1/\hat{A}(q^{-1})$ is closer to 1 than the true noise filter $1/A(q^{-1})$. The inequality (12.45d) is proved by the following equivalences:

$$(12.45\text{d}) \Leftrightarrow |\hat{a}| \, |1 + ae^{i\omega}| \le |a| \, |1 + \hat{a}e^{i\omega}| \Leftrightarrow$$

$$\alpha^2 |1 + ae^{i\omega}|^2 \le |1 + a\alpha e^{i\omega}|^2 \Leftrightarrow \alpha^2(1 + a^2 + 2a\cos\omega) \le 1 + a^2\alpha^2 + 2a\alpha\cos\omega \Leftrightarrow$$

$$0 \le 1 - \alpha^2 + 2a\alpha(1 - \alpha)\cos\omega \Leftrightarrow 0 \le 1 + \alpha + 2a\alpha\cos\omega \Leftrightarrow$$

$$0 \le (1 - \alpha) + 2\alpha(1 + a\cos\omega)$$

The last inequality is obviously true.   ∎

There are several ways to cope with outliers in the data; for example:

- Test of outliers and adjustment of erroneous data.
- Use of a modified loss function.

In the first approach a model is fitted to the data without any special action. Then the residuals $\varepsilon(t, \hat{\theta})$ of the obtained model are plotted. Possible spikes in the sequence $\{\varepsilon(t, \hat{\theta})\}$ are detected. If $|\varepsilon(t, \hat{\theta})|$ is abnormally large then the corresponding output $y(t)$ is modified. A simple modification could be to take

$$y(t) := 0.5[y(t-1) + y(t+1)]$$

Another possibility is to set $y(t)$ to the predicted value:

$$y(t) := \hat{y}(t|t-1, \hat{\theta})$$

The data string obtained as above is used to get an improved model by making a new parameter estimation.

For explanation of the second approach (the possibility of using a modified loss function) consider single output systems. Then the usual criterion

$$V(\theta) = \sum_{t=1}^{N} \varepsilon^2(t, \theta) \qquad (12.46)$$

can be shown to be (asymptotically) optimal if and only if the disturbances are Gaussian distributed (cf. below).

Under weak assumptions the ML estimate is optimal. More exactly, it is asymptotically statistically efficient. This estimate maximizes the log-likelihood function

$$\log L(\theta) = \log p(y(1) \ldots y(N)|\theta)$$

Using the model assumption that $\{\varepsilon(t, \theta)\}$ is a white noise sequence with probability density function $f(\varepsilon(t))$, it can be shown by applying Bayes' rule that

$$\log L(\theta) = \sum_{t=1}^{N} \log f(\varepsilon(t)) \qquad (12.47)$$

Hence the optimal choice of loss function is

$$V(\theta) = -\sum_{t=1}^{N} \log f(\varepsilon(t)) \qquad (12.48)$$

When the data are Gaussian distributed,

$$f(\varepsilon) = \frac{1}{\sqrt{(2\pi)}\lambda} e^{-\varepsilon^2/2\lambda^2}$$

and hence

$$-\log f(\varepsilon(t, \theta)) = \varepsilon^2(t, \theta) + \theta\text{-independent term}$$

This means that in the Gaussian case, the optimal criterion (12.48) reduces to (12.46). If there are outliers in the data, then $f(\varepsilon)$ is likely to decrease more slowly with $|\varepsilon|$ than in the Gaussian case. This means that for the optimal choice, (12.48), of the loss function the large values of $|\varepsilon(t, \theta)|$ are less penalized than in (12.46). In other words, in (12.48) the large prediction errors have less influence on the loss function than they have in (12.46). There are many *ad hoc* ways to achieve this qualitatively by modifying (12.46) (note that the function $f(\varepsilon)$ in (12.48) is normally unavailable). For example, one can take

$$V(\theta) = \sum_{t=1}^{N} l(\varepsilon(t, \theta)) \qquad (12.49a)$$

with

$$l(\varepsilon) = \frac{\varepsilon^2}{\alpha^2 + \varepsilon^2} \qquad (12.49b)$$

or

$$l(\varepsilon) = \begin{cases} \varepsilon^2 & \text{if } \varepsilon^2 \leq \alpha^2 \\ \alpha^2 & \text{if } \varepsilon^2 > \alpha^2 \end{cases} \qquad (12.49c)$$

Note that for both choices (12.49b) and (12.49c) there is a parameter α to be chosen by the user. This parameter should be set to a value given by an expected amplitude of the prediction errors.

Sometimes the user can have useful *a priori* information for choosing the parameter α. If this is not the case it has to be estimated in some way.

One possibility is to perform a first off-line identification using the loss function (12.46). An examination of the obtained residuals $\{\varepsilon(t, \hat{\theta})\}$, including plots as well as computation of the variance, may give useful information on how to choose α. Then a second off-line estimation has to be carried out based on a modified criterion using the determined value of α.

Another alternative is to choose α in an adaptive manner using on-line identification. To see how this can be done the following example considers a linear regression model for ease of illustration.

Example 12.9 *On-line robustification*
Consider the model

$$y(t) = \varphi^T(t)\theta + \varepsilon(t) \qquad (12.50a)$$

whose unknown parameters θ are to be estimated by minimizing the weighted LS criterion

$$V_t(\theta) = \sum_{s=1}^{t} \beta(s)\varepsilon^2(s, \theta) \qquad (12.50b)$$

The choice of the weights $\beta(s)$ is discussed later. Paralleling the calculations in Section 9.2, it can be shown that the minimizer $\hat{\theta}(t)$ of (12.50b) can be computed recursively in t using the following algorithm:

$$\hat{\theta}(t) = \hat{\theta}(t - 1) + K(t)\varepsilon(t)$$
$$\varepsilon(t) = y(t) - \varphi^T(t)\hat{\theta}(t - 1)$$

$$K(t) = P(t-1)\varphi(t)/[1/\beta(t) + \varphi^T(t)P(t-1)\varphi(t)] \quad (12.50c)$$

$$P(t) = P(t-1) - K(t)\varphi^T(t)P(t-1)$$

The minimal value $V_t(\hat{\theta}(t))$ can also be determined on-line as

$$V_t(\hat{\theta}(t)) = V_{t-1}(\hat{\theta}(t-1)) + \frac{\varepsilon^2(t)}{1/\beta(t) + \varphi^T(t)P(t-1)\varphi(t)} \quad (12.50d)$$

Note that

$$\hat{\lambda}_t = [V_t(\hat{\theta}(t))/t]^{1/2} \quad (12.50e)$$

provides an estimate of the standard deviation of the (weighted) prediction errors.

The following method can now be suggested for reducing the influence of the outliers on the LS estimator (cf (12.49a,c)): use (12.50c,d) with

$$\beta(t) = \begin{cases} 1 & \text{if } |\varepsilon(t)| \leq \gamma\hat{\lambda}_t \\ \dfrac{\gamma^2\hat{\lambda}_t^2}{\varepsilon^2(t)} & \text{if } |\varepsilon(t)| > \gamma\hat{\lambda}_t \end{cases} \quad (12.50f)$$

The value of γ above should be chosen by a trade-off between robustification and estimation accuracy. For a small value of γ, the estimator (12.50c) is quite robust to outliers but its accuracy may be poor, and vice versa for a large γ. Choosing γ in the range 2 to 4 may be suitable.

It should be noted that since $\beta(t)$ was allowed to depend on $\varepsilon(t)$, the estimator (12.50c) no longer provides the exact minimizer of the criterion (12.50b). However, this is not a serious drawback from a practical point of view. ∎

12.10 Model verification

Model verification is concerned with determining whether an obtained model is adequate or not. This question was also discussed in Section 12.4 and in Chapter 11.

The verification should be seen in the light of the intended purpose of the model. Therefore the ultimate verification can only be performed by using the model in practice and checking the results. However, there are a number of ways which can be used to test if the model is likely to describe the system in an adequate way, before using the model effectively. See Chapter 11.

It is also of importance to check the *a priori* assumptions. This can include the following checks:

- *Test of linearity*. If possible the experiment should be repeated with another amplitude (or variance) of the input signal in order to verify for what operating range a linear model is adequate. If for example a transient analysis is applied, the user should try both a positive and a negative step (or impulse, if applicable). Haber (1985) describes a number of tests for examining possible nonlinearities. A simple time domain test runs as follows. Let y_0 be the stationary output for zero input signal. Let $y_1(t)$ be the output response for the input $u_1(t)$, and $y_2(t)$ the output for the input $u_2(t) = \gamma u_1(t)$ (γ being a nonzero scalar). Then form the ratio

$$\delta(t) = \frac{y_2(t) - y_0}{y_1(t) - y_0} \tag{12.51a}$$

and take

$$\eta = \max_t \frac{|\delta(t) - \gamma|}{|\gamma|} \tag{12.51b}$$

as an index for the degree of nonlinearity. For a (noise-free) linear system $\eta = 0$, whereas for a nonlinear system $\eta > 0$.

- *Test of time invariance.* A convenient way of testing time invariance of a system is to use data from two different experiments. (This may be achieved by dividing the recorded data into two parts.) The parameter estimates are determined in the usual way from the first data set. Then the model output is computed for the second set of data using the parameter estimates obtained from the first data set. If the process is time invariant the model should 'explain' the process data equally well for both data sets. This procedure is sometimes called *cross-checking*. Note that it is used not only for checking time invariance: cross-checking may be used for the more general purpose of determining the structure of a model. When used for this purpose it is known as 'cross-validation'. The basic idea can be stated as follows: determine the models (having the structures under study) that fit the first data set, and of these select the one which best fits the second data set. The FPE and AIC criteria may be interpreted as cross-validation procedures for assessing a given model structure (as was shown in Section 11.5; see also Stoica, Eykhoff *et al.*, 1986).

- *Test for the existence of feedback.* In Chapter 11 it was shown how to test the hypothesis

$$r_{\varepsilon u}(\tau) = E\varepsilon(t)u(t - \tau) = 0 \tag{12.52}$$

Such a test can be used to detect possible feedback.

Assume that $u(t)$ is determined by (causal) feedback from the output $y(t)$ and that the residual $\varepsilon(t)$ is a good estimate of the white noise which drives the disturbances. Then the input $u(t)$ at time t will in general be dependent on past residuals but independent of future values of the residuals. This means that

$$r_{\varepsilon u}(\tau) = 0 \quad \tau > 0$$

and in general

$$r_{\varepsilon u}(\tau) \neq 0 \quad \tau \leq 0$$

if there is feedback.

Testing for or detection of feedback can also be done by the following method due to Caines and Chan (1975). Apply the joint input–output method described in Section 10.5 to identify the system. Use the innovation model (10.42), (10.43) normalized in the following way:

$$z(t) = \begin{pmatrix} y(t) \\ u(t) \end{pmatrix} = \mathsf{H}(q^{-1}; \theta)\tilde{z}(t)$$

$$E\tilde{z}(t)\tilde{z}^T(s) = \Lambda\delta_{t,s} \quad \Lambda \text{ block diagonal} \tag{12.53}$$

$$\mathsf{H}^{-1}(q^{-1}; \theta) \text{ is asymptotically stable}$$

$$H(0, \theta) = \begin{pmatrix} I & 0 \\ H_0 & I \end{pmatrix} \quad \text{for some matrix } H_0$$

These requirements give a *unique* innovation representation. The system is then feedback-free if and only if the 21-block of $H(q^{-1}; \theta)$ is zero. This can be demonstrated as follows. From the calculations in (A10.1.1),

$$H(q^{-1}; \theta) = \begin{pmatrix} (I + GF)^{-1}H & G(I + FG)^{-1}K \\ -F(I + GF)^{-1}H & (I + FG)^{-1}K \end{pmatrix}$$

$$\Lambda = \begin{pmatrix} \Lambda_e & 0 \\ 0 & \Lambda_w \end{pmatrix} \tag{12.54}$$

Clearly

$$(H)_{21} = 0 \Leftrightarrow F = 0$$

To apply this approach estimate the spectral density of $z(t)$ by some parametric method and then form the innovation filter $H(q^{-1}; \hat{\theta})$ based on the parametrization used. By using hypothesis testing as described in Chapter 11 it can then be tested whether $(H)_{21}$ is zero.

12.11 Software aspects

When performing identification it is very important to have good software. It is convenient to use an interactive program package. This form of software is in most respects superior to a set of subroutines, for which the user has to provide a main program. There are several good packages developed throughout the world; see Jamshidi and Herget (1985) for a recent survey. Several examples in Chapters 2, 3, 11 and 12 of this book have been obtained using the package IDPAC, developed at Department of Automatic Control, Lund Institute of Technology, Sweden.

The main characteristics of a good program package for system identification include the following:

- It should be possible to run in a command-driven way. For the inexperienced user it is useful if a menu is also available. Such a menu should not impose constraints on what alternative the user can choose. The user interface (that is the syntax of the commands) is of great importance. In recent years several popular packages for control applications have emerged. These packages are often constructed as extensions of MATLAB, thus keeping MATLAB's dialog form for the user. See Ljung (1986), and Jamshidi and Herget (1985) for examples of such packages.
- The package should have flexible commands for plotting and graphical representation.
- The package should have a number of commands for data handling. Such commands include loading and reading data from mass storage, removal of trends, prefiltering of data, copying data files, plotting data, picking out data from a file, adjusting single erroneous data, etc.

502 *Some practical aspects* Chapter 12

- The package should have some commands for performing nonparametric identification (for example spectral analysis and correlation analysis).
- The package should have some commands for performing parametric identification (for example the predicton error method and the instrumental variable method) linked to various model structures.
- The package should have some commands for model validation (for example some tests on residual correlations).
- The package should have some commands for handling of models. This includes simulation as well as transforming a model from one representation to another (for example transformation between transfer function, state space form, and frequency function).

12.12 Concluding remarks

We have now come to the end of the main text. We have developed a theory for identification of linear systems and presented several important and useful results. We have also shown how a number of practical issues can be tackled with the help of the theory. We believe that system identification is still an art to some extent. The experience and skill of the user are important for getting good results. It has been our aim to set a theoretical foundation which, together with a good software package, could be most useful for carrying out system identification in practice.

There are several issues in system identification that are not completely clarified, and areas where more experience should be sought. These include the following:

- Application of system identification techniques to signal processing, fault detection and pattern recognition.
- Identification (or tracking) of time-varying systems, which was touched on in Section 9.3.
- Efficient use of *a priori* information. This is seldom used when black box models like (6.14) are employed.
- Identification of nonlinear and distributed parameter systems.
- Simplified, robust and more efficient numerical algorithms for parameter and structure estimation.

We have in this text tried to share with the readers much of our own knowledge and experience of how to apply system identification to practical problems. We would welcome and expect that the reader will broaden this experience both by using system identification in his or her own field of interest and by new research efforts.

Problems

Problem 12.1 *Step response of a simple sampled-data system*
Determine a formula for the step response of (12.1) (omitting $e(t)$). Compare the result with (12.5) at the sampling points.

Problem 12.2 *Optimal loss function*
Prove equation (12.47).

Problem 12.3 *Least squares estimation with nonzero mean data*
Consider the following system:

$$\tilde{y}(t) + a_0 \tilde{y}(t-1) = b_0 \tilde{u}(t-1) + e(t)$$
$$y(t) = \tilde{y}(t) + m_y$$
$$u(t) = \tilde{u}(t) + m_u$$

where $\tilde{u}(t)$ and $e(t)$ are mutually uncorrelated white noise sequences with zero means and variances σ^2 and λ^2, respectively.

The system is to be identified with the least squares method.

(a) Determine the asymptotic LS estimates of a and b in the model structure

$$\mathcal{M}_1: y(t) + ay(t-1) = bu(t-1) + \varepsilon(t)$$

Show that consistent estimates are obtained if and only if m_y and m_u satisfy a certain condition. Give an interpretation of this condition.

(b) Suppose that the LS method is applied to the model structure

$$\mathcal{M}_2: y(t) + ay(t-1) = bu(t-1) + c + \varepsilon(t)$$

Determine the asymptotic values of the estimates $\hat{a}$, $\hat{b}$ and $\hat{c}$. Discuss the practical consequences of the result.

Problem 12.4 *Comparison of approaches for treating nonzero mean data*
Consider the system

$$\tilde{y}(t) + a\tilde{y}(t-1) = bu(t-1) + e(t)$$
$$y(t) = \tilde{y}(t) + m$$

where $u(t)$ and $e(t)$ are mutually independent white noise sequences of zero means and variances σ^2 and λ^2, respectively.

(a) Assume that the system is identified with the least squares method using the model structure

$$y(t) + ay(t-1) = bu(t-1) + \mu + e(t)$$

Find the asymptotic covariance matrix of the parameter estimates.

(b) Assume that the estimated output mean

$$\bar{y} = \frac{1}{N} \sum_{t=1}^{N} y(t)$$

is computed and subtracted from the output. Then the least squares method is applied to the model structure

$$(y(t) - \bar{y}) + a(y(t-1) - \bar{y}) = bu(t-1) + e(t)$$

Find the asymptotic covariance matrix of the parameter estimates.

Hint. First note that $\bar{y} - m = O(1/\sqrt{N})$ as $N \to \infty$.

(c) Assume that the least squares method is used in the model structure

$$\Delta y(t) + a\Delta y(t-1) = b\Delta u(t-1) + \varepsilon(t)$$

Derive the asymptotic parameter estimates and show that they are biased.

(d) Assume that an instrumental variable method is applied to the model structure

$$\Delta y(t) + a\Delta y(t-1) = b\Delta u(t-1) + v(t)$$

using the instruments

$$z(t) = (u(t-1) \quad u(t-2))^T$$

Show that the parameter estimates are consistent and find their asymptotic covariance matrix. (Note the correlation function of the noise $v(t)$.)

(e) Compare the covariance matrices found in parts (a), (b) and (d).

Problem 12.5 *Linear regression with nonzero mean data*
Consider the 'regression' model

$$y(t) = \varphi^T(t)\theta + \alpha + \varepsilon(t) \qquad t = 1, 2, \ldots \qquad (i)$$

where $y(t)$ and $\varphi(t)$ are given for $t = 1, 2, \ldots, N$, and where the unknown parameters θ and α are to be estimated. The scalar-valued parameter α was introduced in (i) to cover the possible case of nonzero mean residuals (see, for example Problem 12.3). Let

$$\bar{y} = \frac{1}{N}\sum_{t=1}^{N} y(t) \qquad \bar{\varphi} = \frac{1}{N}\sum_{t=1}^{N} \varphi(t)$$

$$\tilde{y}(t) = y(t) - \bar{y} \qquad \tilde{\varphi}(t) = \varphi(t) - \bar{\varphi}$$

$$R = \frac{1}{N}\sum_{t=1}^{N} \tilde{\varphi}(t)\tilde{\varphi}^T(t) \qquad r = \frac{1}{N}\sum_{t=1}^{N} \tilde{\varphi}(t)\tilde{y}(t)$$

and

$$\hat{\theta} = R^{-1}r \qquad \hat{\alpha} = \bar{y} - \bar{\varphi}^T\hat{\theta} \qquad (ii)$$

Show that the LS estimates of θ and α in (i) are given by $\hat{\theta}$ and $\hat{\alpha}$ above. Comment on this result.

Problem 12.6 *Neglecting transients*
Consider the autoregression:

$$(1 - aq^{-1})y(t) = e(t)$$

where $|a| < 1$ and $\{e(t)\}$ is white noise of zero mean and variance λ^2 and the two following representations of $y(t)$:

$$y^{(1)}(t) = \sum_{i=0}^{\infty} a^i e(t-i)$$

and

$$x(t+1) = ax(t) + e(t+1) \quad t \geq 0$$
$$y^{(2)}(t) = x(t)$$

$x(0)$ being a random variable of finite variance and which is uncorrelated with $e(t)$.

Show that the difference $y^{(1)}(t) - y^{(2)}(t)$ has a variance that decays exponentially to zero as t tends to infinity.

Problem 12.7 *Accuracy of PEM and hypothesis testing for an ARMA(1, 1) process*
Consider the ARMA model

$$y(t) + ay(t-1) = e(t) + ce(t-1)$$

where a and c have been estimated with a PEM applied to a time series $y(t)$. Assume that the data is an ARMA(1, 1) process

$$y(t) + a_0 y(t-1) = e(t) + c_0 e(t-1)$$
$$|a_0| < 1, \ |c_0| < 1, \ Ee(t)e(s) = \lambda^2 \delta_{t,s}$$

(a) What is the (asymptotic) variance of $\hat{a} - \hat{c}$? Use the answer to derive a test for the hypothesis $a_0 = c_0$.
(b) Suggest some alternative ways to test the hypothesis $a_0 = c_0$.
    Hint. Observe that for $a_0 = c_0$, $y(t)$ is a white process.
(c) Suppose $a_0 = -0.707$, $c_0 = 0$. How many data points are needed to make the standard deviation of $\hat{a}$ equal to 0.01?

Problem 12.8 *A weighted recursive least squares method*
Prove the results (12.50c, d).

Problem 12.9 *The controller form state space realization of an ARMAX system*
Verify that (12.41) is a state space realization of the ARMAX equation (12.36).
    Hint. Use the following readily verified identity:

$$(1 \ 0 \ \ldots \ 0)(qI - \mathscr{C})^{-1} = (q^{-1} \ \ldots \ q^{-n})/C(q^{-1})$$

where $\mathscr{C}$ is the companion transition matrix occurring in (12.41).

Problem 12.10 *Gradient of the loss function with respect to initial values*
Consider the approach (12.41), (12.42) for computing the prediction errors. Derive the derivative of the loss function

$$V(\theta') = \sum_{t=1}^{N} \varepsilon^2(t, \theta')$$

with respect to the initial value $x(1)$.

Problem 12.11 *Estimates of initial values are not consistent*
Consider a first-order MA process

$$y(t) = e(t) + ce(t-1) \qquad |c| < 1, \; e(t) \text{ white noise}$$

Assume that c is known and the prediction errors are determined as in (12.41):

$$x(t+1) = -cx(t) - cy(t)$$
$$\varepsilon(t) = x(t) + y(t) \qquad t = 1, 2, \ldots, N$$

Let the initial value $x(1)$ be determined as the minimizing element of the loss function

$$V = \sum_{t=1}^{N} \varepsilon^2(t)$$

(a) Determine the estimate $\hat{x}(1)$ as a function of the data $\{y(t)\}_{t=1}^{N}$ and c.
(b) $\hat{x}(1)$ is an estimate of $e(1) - y(1) = -ce(0)$. Evaluate the mean square error

$$W = E[\hat{x}(1) + ce(0)]^2$$

and show that W does not tend to zero as N tends to infinity. Hence $\hat{x}(1)$ is not a consistent estimate of the initial value.

Problem 12.12 *Choice of the input signal for accurate estimation of static gain*
Consider the system

$$y(t) + a_0 y(t-1) = b_0 u(t-1) + e(t) \qquad Ee(t)e(s) = \lambda^2 \delta_{t,s}$$

identified with the LS method in the model structure

$$\mathcal{M}: y(t) + \hat{a} y(t-1) = \hat{b} u(t-1) + \varepsilon(t)$$

The static gain $S = b_0/(1 + a_0)$ can be estimated as

$$\hat{S} = \frac{\hat{b}}{1 + \hat{a}}$$

Compute the variance of $\hat{S}$ for the following two experimental conditions:

$\mathcal{X}_1$: $u(t)$ zero mean white noise of variance σ^2
$\mathcal{X}_2$: $u(t)$ a step of size σ

(Note that in both cases $Eu^2(t) = \sigma^2$.) Which case will give the smallest variance? Evaluate the variances numerically for

$$a_0 = -0.9, \; b_0 = 1, \; \lambda = 1, \; \sigma = 1$$

Hint. The variance of $\hat{S}$ can be (approximately) evaluated, observing that

$$\hat{S} - S = \frac{\hat{b}}{1 + \hat{a}} - \frac{b_0}{1 + a_0} = \frac{\hat{b}(1 + a_0) - (1 + \hat{a})b_0}{(1 + \hat{a})(1 + a_0)}$$

$$= \frac{-b_0(\hat{a} - a_0) + (\hat{b} - b_0)(1 + a_0)}{(1 + \hat{a})(1 + a_0)} \approx \frac{-(\hat{a} - a_0)b_0 + (\hat{b} - b_0)(1 + a_0)}{(1 + a_0)^2}$$

which expresses $\hat{S} - S$ as a *linear* combination of $\hat{\theta} - \theta_0$. Then the variance of $\hat{S}$ can be easily found from the covariance matrix of $\hat{\theta}$.

Remark. This problem is related to Example 12.1.

Problem 12.13 *Variance of an integrated process*
(a) Consider the process $y(t)$ defined by

$$y(t) = \frac{1}{1 - q^{-1}} e(t)$$

$$y(0) = 0$$

where $e(t)$ is white noise with zero mean and unit variance. Show that

$$\text{var}(y(t)) = t$$

(b) Consider the process $y(t)$ defined by

$$y(t) = \frac{1}{(1 - q^{-1})(1 + aq^{-1})} e(t) \qquad |a| < 1$$

$$y(0) = y(-1) = 0$$

where $e(t)$ is white noise with zero mean and unit variance. Show that the variance of $y(t)$ for large t is given by

$$\text{var}(y(t)) = \frac{t}{(1 + a)^2}$$

Hint. First show that $y(t) = \sum_{k=0}^{t-1} h_k e(t - k)$ where

$$h_k = \frac{1 - (-a)^{k+1}}{1 + a}$$

Problem 12.14 *An example of optimal choice of the sampling interval*
Consider the following simple stochastic differential equation:

$$dx = -\alpha x dt + dw \qquad (\alpha > 0) \qquad \text{(i)}$$

which is a special case of (6.37), (6.38). In (i), w is a Wiener process with incremental variance $Edw^2 = rdt$, and α is the unknown parameter to be estimated. Let x be observed at equidistant sampling points. The discrete-time observations of x satisfy the following difference equation (see (6.39)):

$$x(t + 1) = ax(t) + e(t) \qquad t = 1, 2, \ldots \qquad \text{(ii)}$$

where

$$a = e^{-\alpha h}$$

$$Ee(t)e(s) = \sigma^2 \delta_{t,s}$$

$$\sigma^2 = r \int_0^h e^{-2\alpha s} ds = \frac{r}{2\alpha}[1 - e^{-2\alpha h}]$$

508 Some practical aspects   Chapter 12

and where h denotes the sampling interval. Note that to simplify the notation of (ii) h is taken as the time unit.

The parameter a of (ii) is estimated by the LS method. From the estimate $\hat{a}$ of a one can estimate α as

$$\hat{\alpha} = -\frac{1}{h}\log(\hat{a})$$

Determine the (asymptotic) variance of $\hat{\alpha}$ and discuss its dependence on h in the following two cases:

(a) N observations of $x(t)$ are processed ($N \gg 0$).
(b) The observations in a given interval T are processed ($T/h \gg 0$).

Show that in case (a) there exists an optimal choice of h, say h_0, and that var($\hat{\alpha}$) increases rapidly for $h < h_0$ and (much) more slowly for $h > h_0$. In case (b) show that var($\hat{\alpha}$) increases monotonically with increasing h. Give an intuitive explanation of these facts.

Remark. See Åström (1969) for a further discussion of the choice of sampling interval.

Problem 12.15 *Effects on parameter estimates of input variations during the sampling interval*

Consider a first-order continuous time system given by

$$\frac{dy}{dt} + \alpha y = \beta u \qquad (\alpha > 0) \tag{i}$$

Assume that the input is a continuous time stationary stochastic process given by

$$\frac{du}{dt} + \gamma u = e \qquad (\gamma > 0) \tag{ii}$$

where $e(t)$ is white noise with covariance function $\delta(\tau)$.

(a) Sample the system (i) as if the input were constant over the sampling intervals. Show that the sampled system has the form

$$y(t) + ay(t - h) = bu(t - h) \qquad t = h, 2h, \ldots$$

where h is the sampling interval.

(b) Assume that the least squares method is used to identify the system (i) with the input (ii) in the model structure

$$y(t) + ay(t - h) = bu(t - h) + \varepsilon(t)$$

Derive the asymptotic estimates of a and b and compare with the parameters of the model in (a). How will the parameter γ influence the result?

Hint. The system (i), (ii) can be written as

$$\dot{x} = Fx + v$$

where

$$x(t) = \begin{pmatrix} y(t) \\ u(t) \end{pmatrix} \quad F = \begin{pmatrix} -\alpha & \beta \\ 0 & -\gamma \end{pmatrix} \quad v = \begin{pmatrix} 0 \\ e \end{pmatrix}$$

Since $v(t)$ is white noise, the covariance function of the state vector for $\tau \geq 0$ is given by

$$Ex(t + \tau)x^T(t) = e^{F\tau}P$$

where $P = Ex(t)x^T(t)$ see, for example, Åström (1970).

Bibliographical notes

Some papers dealing in general terms with practical aspects of system identification are Åström (1980), Isermann (1980) and Bohlin (1987). Both the role of prefiltering the data prior to parameter estimation and the approximation induced by using a model structure not covering the true system can be analyzed in the frequency domain; see Ljung (1985a), Wahlberg and Ljung (1986), Wahlberg (1987).

(*Section 12.2*) Optimal experimental conditions have been studied in depth by Mehra (1974, 1976), Goodwin and Payne (1977), Zarrop (1979a, b). Some aspects on the choice of sampling interval have been discussed by Sinha and Puthenpura (1985).

(*Section 12.3*) For the use of ARIMA models, especially in econometric applications, see Box and Jenkins (1976), Granger and Newbold (1977).

(*Section 12.5*) Multivariable systems may often have different time delays in the different channels from input to output. Vector difference equations which can accommodate this case of different time delays are discussed, for example, by Janssen (1987a, 1988).

(*Section 12.7*) The possibility of performing a number of passes through the data with an on-line algorithm as an off-line method has been described by Ljung (1982), Ljung and Söderström (1983), Solbrand et al. (1985). It has frequently been used for instrumental variable identification; see e.g. Young (1968, 1976).

(*Section 12.8*) Proofs of the results on local minima have appeared in Åström and Söderström (1974) (see also Complement C7.6), ARMA processes; Stoica and Söderström (1982g), multivariable MA processes; Söderström (1974), the 'GLS' structure; Söderström and Stoica (1982) (see also Complement C7.5), output error models. Stoica and Söderstrom (1984) contains some further results of this type for k-step prediction models of ARMA processes.

(*Section 12.9*) Robust methods that are less sensitive to outliers have been treated for linear regressions by Huber (1973), for off-line methods by Ljung (1978), and for on-line algorithms by Polyak and Tsypkin (1980), Ljung and Soderström (1983) and Tsypkin (1987).

(*Section 12.10*) The problem of testing time invariance can be viewed as a form of fault detection (test for a change of the system dynamics). See Willsky (1976) and Basseville and Benveniste (1986) for some surveys of this field. The problem of testing for the existence of feedback has been treated by, e.g., Bohlin (1971), Caines and Chan (1975, 1976).

(*Section 12.11*) The program package IDPAC is described in Wieslander (1976). Some further aspects and other packages have been discussed by Åström (1983a), van den Boom and Bollen (1984), Furuta *et al.* (1981), Schmid and Unbehauen (1979), Tyssø (1982), Young (1982).

Appendix A
SOME MATRIX RESULTS

This appendix presents several matrix results which are used at various places in the book.

A.1 Partitioned matrices

The first result is the matrix inversion lemma, which naturally fits into the context of partitioned matrices (cf the remark to Lemma A.2 below)

Lemma A.1
Provided the inverses below exist,

$$[A + BCD]^{-1} = A^{-1} - A^{-1}B[C^{-1} + DA^{-1}B]^{-1}DA^{-1} \qquad (A.1)$$

Proof. The statement is verified by direct multiplication.

$$[A + BCD]\{A^{-1} - A^{-1}B[C^{-1} + DA^{-1}B]^{-1}DA^{-1}\}$$
$$= I + BCDA^{-1} - [B + BCDA^{-1}B][C^{-1} + DA^{-1}B]^{-1}DA^{-1}$$
$$= I + BCDA^{-1} - BC[C^{-1} + DA^{-1}B][C^{-1} + DA^{-1}B]^{-1}DA^{-1}$$
$$= I$$

which proves (A.1). ∎

For some remarks on the historical background of this result, see Kailath (1980).
The matrix inversion lemma is closely related to the following result on the inverse of partitioned matrices.

Lemma A.2
Consider the matrix

$$S = \begin{pmatrix} A & B \\ D & C \end{pmatrix} \qquad (A.2)$$

where A and C are square matrices. Assume that A and $C - DA^{-1}B$ are nonsingular. Then

$$S^{-1} = \begin{pmatrix} A^{-1} & 0 \\ 0 & 0 \end{pmatrix} + \begin{pmatrix} -A^{-1}B \\ I \end{pmatrix}(C - DA^{-1}B)^{-1}(-DA^{-1} \quad I) \tag{A.3}$$

Proof. Direct multiplication gives

$$\begin{pmatrix} A & B \\ D & C \end{pmatrix} \left\{ \begin{pmatrix} A^{-1} & 0 \\ 0 & 0 \end{pmatrix} + \begin{pmatrix} -A^{-1}B \\ I \end{pmatrix}(C - DA^{-1}B)^{-1}(-DA^{-1} \quad I) \right\}$$

$$= \begin{pmatrix} I & 0 \\ DA^{-1} & 0 \end{pmatrix} + \begin{pmatrix} 0 \\ C - DA^{-1}B \end{pmatrix}(C - DA^{-1}B)^{-1}(-DA^{-1} \quad I)$$

$$= \begin{pmatrix} I & 0 \\ DA^{-1} & 0 \end{pmatrix} + \begin{pmatrix} 0 \\ I \end{pmatrix}(-DA^{-1} \quad I) = \begin{pmatrix} I & 0 \\ 0 & I \end{pmatrix}$$

which proves (A.3). ∎

Remark. In a similar fashion one can prove that

$$S^{-1} = \begin{pmatrix} 0 & 0 \\ 0 & C^{-1} \end{pmatrix} + \begin{pmatrix} I \\ -C^{-1}D \end{pmatrix}(A - BC^{-1}D)^{-1}(I \quad -BC^{-1}) \tag{A.4}$$

Comparing the upper left blocks of (A.4) and (A.3),

$$(A - BC^{-1}D)^{-1} = A^{-1} + A^{-1}B(C - DA^{-1}B)^{-1}DA^{-1} \tag{A.5}$$

which is just a reformulation of (A.1). ∎

The next result concerns the rank and positive (semi) definiteness of certain symmetric matrices.

Lemma A.3
Consider the symmetric matrix

$$S = \begin{pmatrix} A & B \\ B^T & C \end{pmatrix} \tag{A.6}$$

where A (($n|n$)-dimensional) and C (($m|m$)-dimensional) are positive definite.

(i) The following properties are equivalent:

> S is positive definite (positive semidefinite).
> $A - BC^{-1}B^T$ is positive definite (positive semidefinite).
> $C - B^T A^{-1} B$ is positive definite (positive semidefinite).

(ii) Assume that S is nonnegative definite. Then

$$\text{rank } S = m + \text{rank}(A - BC^{-1}B^T) = n + \text{rank}(C - B^T A^{-1} B) \tag{A.7}$$

Proof. To prove part (i) consider the quadratic form

$$(x_1^T \quad x_2^T) S \begin{pmatrix} x_1 \\ x_2 \end{pmatrix}$$

where x_1 is $(n|1)$ and x_2 is $(m|1)$. If S is positive definite this form is strictly positive for all $(x_1^T \; x_2^T) \neq 0$. If S is positive semidefinite the quadratic form is positive (≥ 0) for all $(x_1^T \; x_2^T)$. Set

$$y = x_2 + C^{-1}B^T x_1$$

and note that

$$\begin{aligned}(x_1^T \; x_2^T) S \begin{pmatrix} x_1 \\ x_2 \end{pmatrix} &= x_1^T A x_1 + x_1^T B x_2 + x_2^T B^T x_1 + x_2^T C x_2 \\ &= x_1^T A x_1 + x_1^T B (y - C^{-1}B^T x_1) + (y - C^{-1}B^T x_1)^T B^T x_1 \\ &\quad + (y - C^{-1}B^T x_1)^T C (y - C^{-1}B^T x_1) \\ &= x_1^T (A - BC^{-1}B^T) x_1 + y^T C y \end{aligned}$$

Since C is positive definite by assumption and since for any x_1 one can choose x_2 such that $y = 0$, it follows easily that S is positive definite (positive semidefinite) if and only if $(A - BC^{-1}B^T)$ has the same property. In a similar fashion it can be shown that S is positive definite if and only if $C - B^T A^{-1} B$ is positive definite (set $z = x_1 + A^{-1}Bx_2$, etc.).

To prove part (ii) consider the equation

$$(x_1^T \; x_2^T) S \begin{pmatrix} x_1 \\ x_2 \end{pmatrix} = 0$$

The solutions span a linear subspace of dimension $n + m -$ rank S. Following the above calculations, first make a nonsingular transformation from $(x_1^T \; x_2^T)^T$ to $(x_1^T \; y^T)^T$. In the $(x_1^T \; y^T)^T$ space the solutions must satisfy

$$x_1^T (A - BC^{-1}B^T) x_1 = 0, \quad y = 0$$

Thus the solutions to the original equation span a subspace of dimension $n -$ rank $(A - BC^{-1}B^T)$. Equating the two expressions for the dimension of the subspace containing the solution gives

$$n + m - \text{rank } S = n - \text{rank}(A - BC^{-1}B^T)$$

which trivially reduces to the first equality in (A.7). The second equality is proved in a similar fashion. ∎

Corollary. Let A and B be two positive definite matrices and assume

$$A \geq B$$

Then

$$A^{-1} \leq B^{-1}$$

Proof. Set

$$S = \begin{pmatrix} A & I \\ I & B^{-1} \end{pmatrix}$$

According to the lemma

$$S \geq 0 \Leftrightarrow A - (B^{-1})^{-1} \geq 0 \Leftrightarrow B^{-1} - A^{-1} \geq 0$$

which proves the result. ∎

The following lemma deals with relations between some covariance matrices. It will be proved using partitioned matrices.

Lemma A.4
Let $z_1(t)$ and $z_2(t)$ be two matrix-valued stationary stochastic processes having the same dimension. Assume that the matrices $Ez_i(t)z_j^T(t)$, $i, j = 1, 2$, are nonsingular. Then

$$[Ez_2(t)z_1^T(t)]^{-1}[Ez_2(t)z_2^T(t)][Ez_1(t)z_2^T(t)]^{-1} \geq [Ez_1(t)z_1^T(t)]^{-1} \tag{A.8}$$

where equality holds if and only if

$$z_1(t) = Mz_2(t) \text{ in mean square} \tag{A.9}$$

M being a constant and nonsingular matrix.

Proof. Clearly,

$$E\begin{pmatrix} z_1(t) \\ z_2(t) \end{pmatrix}(z_1^T(t) \quad z_2^T(t)) \triangleq \begin{pmatrix} Z_{11} & Z_{12} \\ Z_{12}^T & Z_{22} \end{pmatrix} \geq 0$$

which gives (see Lemma A.3)

$$Z_{11} - Z_{12}Z_{22}^{-1}Z_{12}^T \geq 0$$

and therefore (see Corollary of Lemma A.3),

$$(Z_{12}^T)^{-1}Z_{22}Z_{12}^{-1} \geq Z_{11}^{-1}$$

since the inverses are assumed to exist. Thus (A.8) is proved.

It is trivial to see that (A.9) implies equality in (A.8). Consider now the converse situation, i.e. assume that equality is valid in (A.8). Then

$$Z_{11} - Z_{12}Z_{22}^{-1}Z_{12}^T = 0$$

Set $M = Z_{12}Z_{22}^{-1}$ and note that

$$E[z_1(t) - Mz_2(t)][z_1(t) - Mz_2(t)]^T$$
$$= Z_{11} - MZ_{12}^T - Z_{12}M^T + MZ_{22}M^T$$
$$= Z_{12}Z_{22}^{-1}Z_{12}^T - Z_{12}Z_{22}^{-1}Z_{12}^T - Z_{12}Z_{22}^{-1}Z_{12}^T + Z_{12}Z_{22}^{-1}Z_{12}^T = 0$$

which proves (A.9). ∎

The following result concerns determinants of partitioned matrices.

Lemma A.5
Consider the matrix

$$P = \begin{pmatrix} A & B \\ C & D \end{pmatrix}$$

where A and D are square matrices. Assuming the inverses occurring below exist,

$$\det P = \det A \, \det(D - CA^{-1}B)$$
$$= \det D \, \det(A - BD^{-1}C) \tag{A.10}$$

Proof. A proof will be given of the first equality only, since the second one can be shown similarly. Assume that A is nonsingular. Then

$$\det P = \det \left\{ P \begin{pmatrix} I & -A^{-1}B \\ 0 & I \end{pmatrix} \right\}$$

$$= \det \begin{pmatrix} A & 0 \\ C & D - CA^{-1}B \end{pmatrix} = \det A \, \det(D - CA^{-1}B) \qquad \blacksquare$$

Corollary 1. Let A be an $(n|m)$-dimensional and B an $(m|n)$-dimensional matrix. Then

$$\det(I_n + AB) = \det(I_m + BA) \tag{A.11}$$

Proof. Set

$$S = \begin{pmatrix} I_n & A \\ -B & I_m \end{pmatrix}$$

Then from the lemma

$$\det S = \det(I_m + BA) = \det(I_n + AB)$$

which proves (A.11). $\qquad\blacksquare$

Corollary 2. Let P be a symmetric $(n|n)$ matrix and let P_i denote its upper left $(i|i)$ block. Then the following properties are equivalent:

(i) P is positive definite
(ii) $\det P_i > 0 \quad i = 1, \ldots, n$

Proof. First consider the implication (i) $\Rightarrow$ (ii). Clearly $P > 0 \Rightarrow x^T P x > 0$ all $x \neq 0$. By specializing to x-vectors where all but the first i components are zero it follows that $P_i > 0$. This implies that all eigenvalues, say $\lambda_j(P_i)_{j=1}^i$, of P_i are strictly positive. Hence

$$\det P_i = \prod_{j=1}^{i} \{\lambda_j(P_i)\} > 0 \quad i = 1, \ldots, n$$

Next consider the implication (ii) $\Rightarrow$ (i). Clearly $P_1 = \det P_1 > 0$. The result will follow if it can be shown that

$$P_k > 0, \det P_{k+1} > 0 \Rightarrow P_{k+1} > 0$$

However, the matrix P_{k+1} has the following structure:

$$P_{k+1} = \begin{pmatrix} P_k & b \\ b^T & c \end{pmatrix}$$

Thus from the lemma

$$0 < \det P_{k+1} = \det P_k \det(c - b^T P_k^{-1} b)$$
$$= (\det P_k)(c - b^T P_k^{-1} b)$$

Hence $c - b^T P_k^{-1} b > 0$. Then it follows from Lemma A.3 that $P_{k+1} > 0$, which completes the proof. ∎

The final lemma in this section is a result on Cholesky factors of banded block matrices.

Lemma A.6
Let P be a symmetric, banded, positive definite matrix

$$P = \begin{pmatrix} R_1 & S_1 & & & & \\ S_1^T & R_2 & S_2 & & 0 & \\ & S_2^T & R_3 & & & \\ & & & \ddots & & \\ & 0 & & & & S_{m-1} \\ & & & & S_{m-1}^T & R_m \end{pmatrix} \quad (A.12)$$

where all S_i are lower triangular matrices. Then there is a banded Cholesky factor, i.e. a matrix L of the form

$$L = \begin{pmatrix} L_1 & & & & 0 \\ Q_1 & L_2 & & & \\ & Q_2 & L_3 & & \\ & 0 & & \ddots & \\ & & & Q_{m-1} & L_m \end{pmatrix} \quad (A.13)$$

with all L_i lower triangular and nonsingular and all Q_i upper triangular matrices, and where L obeys

$$P = LL^T \quad (A.14)$$

Proof. First block-diagonalize P as follows. Let $\Delta_1 = R_1 > 0$ and define

$$M_1 = \begin{pmatrix} I & & & 0 \\ -S_1^T \Delta_1^{-1} & I & & \\ & & \ddots & \\ 0 & & & I \end{pmatrix}$$

Then some easy calculations show

$$M_1 P M_1^T = \begin{pmatrix} \Delta_1 & 0 & \cdots & & & 0 \\ 0 & \Delta_2 & S_2 & & & \\ \vdots & S_2^T & R_3 & & & \\ & & & \ddots & & S_{m-1} \\ 0 & & & & S_{m-1}^T & R_m \end{pmatrix}$$

where

$$\Delta_2 = R_2 - S_1^T \Delta_1^{-1} S_1$$

Since $P > 0$ it holds that $\Delta_2 > 0$. Proceeding in this way for $k = 1, \ldots, m-1$ with

$$M_k = \begin{pmatrix} I & & & & & 0 \\ & \ddots & & & & \\ & & I & & & \\ & & -S_k^T \Delta_k^{-1} & I & & \\ 0 & & & & \ddots & \\ & & & & & I \end{pmatrix} \quad \text{(block row } k+1\text{)}$$

$$\Delta_{k+1} = R_{k+1} - S_k^T \Delta_k^{-1} S_k$$

one gets

$$M_{m-1} M_{m-2} \cdots M_1 P M_1^T \cdots M_{m-1}^T = \begin{pmatrix} \Delta_1 & & 0 \\ & \ddots & \\ 0 & & \Delta_m \end{pmatrix}$$

By construction all the matrices $\{\Delta_k\}$ are hence symmetric and positive definite.

Next the banded Cholesky factor will be constructed. First standard Cholesky factorization of $\{\Delta_k\}$ is performed:

$$\Delta_k = L_k L_k^T \qquad k = 1, \ldots, m$$

where $\{L_k\}$ are lower triangle matrices. Clearly all the $\{L_k\}$ matrices are nonsingular by construction. Then take L as in (A.13) with

$$Q_k = S_k^T L_k^{-T} \qquad k = 1, \ldots, m-1$$

Since L_k, S_k and L_k^{-1} are lower triangular, it follows that Q_k are upper triangular. The matrix L so constructed is therefore a banded, lower triangular matrix. It remains to verify that L is a Cholesky factor, i.e. $LL^T = P$. Evaluating the blocks of LL^T it is easily found that

$$(LL^T)_{11} = L_1 L_1^T = \Delta_1 = P_{11}$$
$$(LL^T)_{ii} = Q_{i-1} Q_{i-1}^T + L_i L_i^T$$
$$= S_{i-1}^T L_{i-1}^{-T} L_{i-1}^{-1} S_{i-1} + L_i L_i^T$$
$$= S_{i-1}^T \Delta_{i-1}^{-1} S_{i-1} + \Delta_i = R_i = P_{ii} \qquad i = 2, \ldots, m$$

$$(LL^T)_{i+1,i} = Q_i L_i^T = S_i^T L_i^{-T} L_i^T = S_i^T = P_{i+1,i} \qquad i = 1, \ldots, m-1$$
$$(LL^T)_{i+j,i} = 0 = P_{i+j,i} \qquad j > 1 \qquad i = 1, \ldots, m-j$$

Since LL^T is symmetric it follows that $LL^T = P$. It can be noted that the matrix L can be written as

$$L = M_1^{-1} M_2^{-1} \ldots M_{m-1}^{-1} \begin{pmatrix} L_1 & & 0 \\ & \ddots & \\ 0 & & L_m \end{pmatrix}$$

∎

A.2 The least squares solution to linear equations, pseudoinverses and the singular value decomposition

This section gives some results related to the least squares solutions of linear systems of equations. The following notation will be used throughout this section:

- A is an $(n|m)$-dimensional matrix. It is associated with a linear transformation $\mathscr{R}^m \to \mathscr{R}^n$.
- $\mathscr{N}(A)$ is the nullspace of A, $\mathscr{N}(A) = \{x | Ax = 0\}$.
- $\mathscr{R}(A)$ is the range of A, $\mathscr{R}(A) = \{b | Ax = b, x \in \mathscr{R}^m\}$.
- $\mathscr{N}(A)^\perp$ is the orthogonal complement to $\mathscr{N}(A)$, i.e. $\mathscr{N}(A)^\perp = \{x | x^T x_0 = 0 \text{ for all } x_0 \in \mathscr{N}(A)\}$. The orthogonal complement $\mathscr{R}(A)^\perp$ is defined in an analogous way.

The Euclidean spaces $\mathscr{R}^m$ and $\mathscr{R}^n$ can be described as direct sums

$$\begin{aligned} \mathscr{R}^m &= \mathscr{N}(A) \oplus \mathscr{N}(A)^\perp \\ \mathscr{R}^n &= \mathscr{R}(A) \oplus \mathscr{R}(A)^\perp \end{aligned} \qquad (A.15)$$

As a consequence of the decomposition (A.15) the following conventions will be used:

- The (arbitrary) vector b in $\mathscr{R}^n$ is uniquely decomposed as $b = b_1 + b_2$, where $b_1 \in \mathscr{R}(A)$, $b_2 \in \mathscr{R}(A)^\perp$
- The (arbitrary) vector x in $\mathscr{R}^m$ is uniquely decomposed as $x = x_1 + x_2$, where $x_1 \in \mathscr{N}(A)$, $x_2 \in \mathscr{N}(A)^\perp$

As an illustration let $e_1, \ldots, e_k$ be a basis for $\mathscr{R}(A)$ and $e_{k+1}, \ldots, e_n$ a basis for $\mathscr{R}(A)^\perp$. Then $e_1, \ldots, e_n$ is a basis for $\mathscr{R}^n$. Expressing an arbitrary vector b in this basis will give a unique component, called b_1, in $\mathscr{R}(A)$ while $b - b_1 \in \mathscr{R}(A)^\perp$. The above conventions are illustrated in the following example.

Example A.1 *Decomposition of vectors*
Let

$$A = \begin{pmatrix} 1 & 1 \\ 2 & 2 \end{pmatrix} \qquad x = \begin{pmatrix} 1 \\ 0 \end{pmatrix} \qquad b = \begin{pmatrix} 1 \\ 1 \end{pmatrix}$$

In this case the matrix A is singular and has rank 1. Its nullspace, $\mathcal{N}(A)$, is spanned by the vector $(1 \ -1)^T$ while the orthogonal complement is spanned by $(1 \ 1)^T$. Examine now how the above vector x can be written as a linear combination of these two vectors:

$$\begin{pmatrix} 1 \\ 0 \end{pmatrix} = \alpha_1 \begin{pmatrix} 1 \\ -1 \end{pmatrix} + \alpha_2 \begin{pmatrix} 1 \\ 1 \end{pmatrix}$$

which gives $\alpha_1 = \alpha_2 = 0.5$. Hence

$$x_1 = \alpha_1 \begin{pmatrix} 1 \\ -1 \end{pmatrix} = \begin{pmatrix} 0.5 \\ -0.5 \end{pmatrix} \quad x_2 = \alpha_2 \begin{pmatrix} 1 \\ 1 \end{pmatrix} = \begin{pmatrix} 0.5 \\ 0.5 \end{pmatrix}$$

In a similar way it is found that the range $\mathcal{R}(A)$ is spanned by the vector $(1 \ 2)^T$ and its orthogonal complement, $\mathcal{R}(A)^\perp$, by the vector $(2 \ -1)^T$. To decompose the given vector b, examine

$$\begin{pmatrix} 1 \\ 1 \end{pmatrix} = \beta_1 \begin{pmatrix} 1 \\ 2 \end{pmatrix} + \beta_2 \begin{pmatrix} 2 \\ -1 \end{pmatrix}$$

which gives $\beta_1 = 0.6$, $\beta_2 = 0.2$. Hence

$$b_1 = \beta_1 \begin{pmatrix} 1 \\ 2 \end{pmatrix} = \begin{pmatrix} 0.6 \\ 1.2 \end{pmatrix} \quad b_2 = \beta_2 \begin{pmatrix} 2 \\ -1 \end{pmatrix} = \begin{pmatrix} 0.4 \\ -0.2 \end{pmatrix}$$ ∎

Some useful results can now be stated.

Lemma A.7
The orthogonal complements satisfy

$$\mathcal{N}(A)^\perp = \mathcal{R}(A^T)$$
$$\mathcal{R}(A)^\perp = \mathcal{N}(A^T) \tag{A.16}$$

Proof. Since $(\mathcal{N}^\perp)^\perp = \mathcal{N}$ for an arbitrary space $\mathcal{N}$, it is sufficient to prove the last relation. This follows from the following series of equivalences:

$$x \in \mathcal{R}(A)^\perp \Leftrightarrow x^T z = 0, \quad \forall z \in \mathcal{R}(A)$$
$$\Leftrightarrow x^T z = 0, \quad z = Ay, \forall y \in \mathcal{R}^m$$
$$\Leftrightarrow x^T A y = 0, \quad \forall y \in \mathcal{R}^m$$
$$\Leftrightarrow y^T A^T x = 0, \quad \forall y \in \mathcal{R}^m$$
$$\Leftrightarrow A^T x = 0 \Leftrightarrow x \in \mathcal{N}(A^T)$$ ∎

Lemma A.8
The restriction of the linear transformation $A \colon \mathcal{R}^m \to \mathcal{R}^n$ to $\mathcal{R}(A^T) \to \mathcal{R}(A)$ is unique and has an inverse.

Proof. Let b_1 be an arbitrary element of $\mathcal{R}(A)$. Then $b_1 = Ax$ some x. Decompose x as $x = x_1 + x_2$. Since $x_1 \in \mathcal{N}(A)$ it follows that there is an $x_2 \in \mathcal{R}(A^T)$ such that $b_1 = Ax_2$. Assume further that there are two vectors x_2' and x_2'' such that $b_1 = Ax_2' = Ax_2''$. However, this gives $A(x'_2 - x''_2) = b_1 - b_1 = 0$ and hence $x'_2 - x''_2 \in \mathcal{N}(A)$. But as $x'_2, x''_2 \in \mathcal{N}(A)^\perp$ we conclude that $x'_2 - x''_2 = 0$, i.e. $x'_2 = x''_2$. Hence there is a 1–1 correspondence between $\mathcal{R}(A)$ and $\mathcal{R}(A^T)$, which proves the lemma. ∎

Remark. It follows that the spaces $\mathcal{R}(A)$ and $\mathcal{R}(A^T)$ have the same dimension. This dimension is equal to the rank of the matrix A. In particular,

$$\text{rank } A = n \Leftrightarrow \mathcal{N}(A^T) = \{0\}$$

$$\text{rank } A = m \Leftrightarrow \mathcal{N}(A) = \{0\}$$
∎

Now consider the system of equations

$$Ax = b \tag{A.17}$$

Lemma A.9
Consider the system (A.17).

(i) It has a solution for every $b \Leftrightarrow \text{rank } A = n$
(ii) Solutions are unique $\Leftrightarrow \text{rank } A = m$

Proof. Using the notational setup (A.17) can be written as

$$A(x_1 + x_2) = Ax_2 = b_1 + b_2$$

Here Ax_2 and b_1 belong to $\mathcal{R}(A)$ while b_2 belongs to $\mathcal{N}(A^T) = \mathcal{R}(A)^\perp$. Clearly solutions exist if and only if $b_2 = 0$. Hence it is required that $b \in \mathcal{R}(A)$. Since b is arbitrary in $\mathcal{R}^n$, for a solution to exist we must have $\mathcal{R}(A) = \mathcal{R}^n$, i.e. rank $A = n$ (see remark to Lemma A.8).

According to Lemma A.8, the solution is unique if and only if $x_1 = 0$. This means precisely that $\mathcal{N}(A) = \{0\}$, which is equivalent to rank $A = m$. (See remark to Lemma A.8.) ∎

The pseudoinverse (also called the Moore–Penrose generalized inverse) of A is defined as follows.

Definition A.1
The pseudoinverse $A^\dagger$ of A is a linear transformation $A^\dagger: \mathcal{R}^n \to \mathcal{R}^m$ such that

(i) $x \in \mathcal{R}(A^T) \Rightarrow A^\dagger A x = x$
(ii) $x \in \mathcal{N}(A^T) \Rightarrow A^\dagger x = 0$
∎

Remark. $A^\dagger$ is uniquely defined by the above relations. This follows since it is defined for all $x \in \mathcal{R}^n$ (see (A.15)). Also the first relation makes perfect sense according to Lemma A.8. Note that from the definition it follows that

$$\mathcal{N}(A^\dagger) = \mathcal{N}(A^T) \qquad \mathcal{R}(A^\dagger) = \mathcal{R}(A^T)$$

The first relation in Definition A.1 can be seen as a generalization of

Section A.2 — Least squares solution

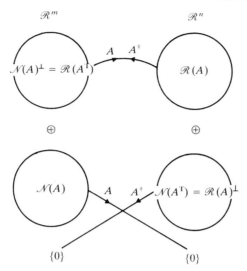

FIGURE A.1 Illustration of the relations $\mathcal{N}(A)^\perp = \mathcal{R}(A^T)$, $\mathcal{N}(A^T) = \mathcal{R}(A)^\perp$ and of the properties of the pseudoinverse $A^\dagger$.

$$A^{-1}Ax = x \qquad \text{all } x \in \mathcal{R}^n$$

which can be used as a definition of A^{-1} for nonsingular square matrices. In fact, if A is square and nonsingular the above definition of $A^\dagger$ easily gives $A^\dagger = A^{-1}$, so the pseudoinverse then becomes the usual inverse. ∎

The description (A.15) of $\mathcal{R}^m$ and $\mathcal{R}^n$ as direct sums and the properties of the pseudoinverse are illustrated in Figure A.1.

Now consider linear systems of equations. Since they may not have any exact solution a least squares solution will be considered. Introduce for this purpose the notation

$$V(x) = \|Ax - b\|^2 = (Ax - b)^T(Ax - b) \tag{A.18}$$

$$\bar{x} = A^\dagger b \tag{A.19}$$

Lemma A.10

(i) $V(x)$ is minimized by $x = \bar{x}$
(ii) If x is another minimum point of $V(x)$ then $\|x\| > \|\bar{x}\|$

Proof. Straightforward calculations give, using the general notation,

$$\begin{aligned}
V(x) - V(\bar{x}) &= \|A(x_1 + x_2) - (b_1 + b_2)\|^2 - \|A\bar{x} - (b_1 + b_2)\|^2 \\
&= \|(Ax_2 - b_1) - b_2\|^2 - \|(A\bar{x} - b_1) - b_2\|^2 \\
&= [\|Ax_2 - b_1\|^2 + \|b_2\|^2] - [\|A\bar{x} - b_1\|^2 + \|b_2\|^2] \\
&= \|Ax_2 - b_1\|^2 - \|AA^\dagger(b_1 + b_2) - b_1\|^2 \\
&= \|Ax_2 - b_1\|^2 \geq 0
\end{aligned}$$

Hence $\bar{x}$ must be a minimum point of $V(x)$. It also follows that if $x = x_1 + x_2$ is another minimum point then $Ax_2 = b_1$. Since $b_1 = A\bar{x}$, $A(x_2 - \bar{x}) = 0$. Now $x_2 - \bar{x} \in \mathcal{N}(A)^\perp$ and it follows that $x_2 = \bar{x}$ and hence $\|x\|^2 = \|x_1 + x_2\|^2 = \|x_1\|^2 + \|\bar{x}\|^2 \geq \|\bar{x}\|^2$ with equality only for $x = x$. ∎

Lemma A.11

(i) Suppose rank $A = m$. Then
$$A^\dagger = (A^T A)^{-1} A^T \qquad (A.20)$$

(ii) Suppose rank $A = n$. Then
$$A^\dagger = A^T (A A^T)^{-1} \qquad (A.21)$$

Proof. It will be shown that (A.20), (A.21) satisfy Definition A.1. For part (i),
$$A^\dagger A = (A^T A)^{-1} A^T A = I$$
which proves (i) of the definition. If $x \in \mathcal{N}(A^T)$ then $A^T x = 0$ and hence
$$A^\dagger x = (A^T A)^{-1} A^T x = 0$$
which completes the proof of (A.20). For part (ii), note that an arbitrary element x of $\mathcal{R}(A^T)$ can be written as $x = A^T y$. Hence
$$A^\dagger A x = A^T (A A^T)^{-1} A A^T y = A^T y = x$$
Since in this case $\mathcal{N}(A^T) = \{0\}$, (A.21) follows. ∎

It is now possible to interpret certain matrices (see below) as orthogonal projectors.

Lemma A.12

(i) $A^\dagger A$ is the orthogonal projector of $\mathcal{R}^m$ onto $\mathcal{R}(A^T)$.
(ii) $I - A^\dagger A$ is the orthogonal projector of $\mathcal{R}^m$ onto $\mathcal{N}(A)$.
(iii) $A A^\dagger$ is the orthogonal projector of $\mathcal{R}^n$ onto $\mathcal{R}(A)$.
(iv) $I - A A^\dagger$ is the orthogonal projector of $\mathcal{R}^n$ onto $\mathcal{N}(A^T)$.

Proof. To prove part (i) let $x = x_1 + x_2 \in \mathcal{R}^m$. Then
$$A^\dagger A x = A^\dagger A x_2 = x_2$$
Part (ii) is then trivial. To prove part (iii) let $b = b_1 + b_2 \in \mathcal{R}^n$. Then
$$A A^\dagger b = A A^\dagger b_1 = b_1$$
and finally part (iv) is trivial once part (iii) is established. ∎

The following lemma presents the singular value decomposition.

Lemma A.13

Let A be an $(n|m)$ matrix. Then there are orthogonal matrices U (of dimension $(n|n)$) and V (of dimension $(m|m)$) such that

Section A.2 *Least squares solution* 523

$$A = U\Sigma V^T \tag{A.22}$$

where Σ is $(n|m)$ dimensional and has the structure

$$\Sigma = \begin{pmatrix} D & 0 \\ 0 & 0 \end{pmatrix} \tag{A.23a}$$

$$D = \begin{pmatrix} \sigma_1 & & & 0 \\ & \sigma_2 & & \\ & & \ddots & \\ 0 & & & \sigma_r \end{pmatrix} \tag{A.23b}$$

$$\sigma_1 \geqslant \sigma_2 \geqslant \ldots \geqslant \sigma_r > 0 \tag{A.23c}$$

and where $r \leqslant \min(n, m)$.

Proof. The matrix $A^T A$ is clearly nonnegative definite. Hence all its eigenvalues are positive or zero. Denote them by $\sigma_1^2, \sigma_2^2, \ldots, \sigma_m^2$ where $\sigma_1 \geqslant \sigma_2 \geqslant \ldots \sigma_r > 0 = \sigma_{r+1} = \ldots = \sigma_m$. Let $v_1, v_2, \ldots, v_m$ be a set of orthonormal eigenvectors corresponding to these eigenvalues and set

$$V_1 = (v_1 \ v_2 \ldots v_r) \quad V_2 = (v_{r+1} \ldots v_m) \quad V = (V_1 \ V_2)$$

and Σ and D as in (A.23). Then

$$A^T A v_i = \sigma_i^2 v_i \quad i = 1, \ldots, r$$
$$A^T A v_i = 0 \quad i = r+1, \ldots, m$$

In matrix form this can be expressed as

$$A^T A V_1 = V_1 D^2 \tag{A.24a}$$
$$A^T A V_2 = 0 \tag{A.24b}$$

The last equation implies

$$0 = V_2^T A^T A V_2 = (AV_2)^T (AV_2)$$

which gives

$$A V_2 = 0 \tag{A.24c}$$

Now introduce the matrix

$$U_1 = A V_1 D^{-1} \tag{A.25a}$$

of dimension $(n|r)$. This matrix satisfies

$$U_1^T U_1 = D^{-1} V_1^T A^T A V_1 D^{-1} = D^{-1} V_1^T V_1 D^2 D^{-1} = I$$

according to (A.24a). Hence the columns of U_1 are orthonormal vectors. Now introduce U_2 such that U is an $(n|n)$ orthogonal matrix

$$U = (U_1 \ U_2)$$

Then

$$U_1^T U_2 = 0 \qquad U_2^T U_2 = I \tag{A.25b}$$

Next consider

$$\begin{aligned} U^T A V &= \begin{pmatrix} U_1^T \\ U_2^T \end{pmatrix} A (V_1 \quad V_2) \\ &= \begin{pmatrix} D^{-1} V_1^T A^T A V_1 & D^{-1} V_1^T A^T A V_2 \\ U_2^T A V_1 & U_2^T A V_2 \end{pmatrix} \\ &= \begin{pmatrix} D^{-1} V_1^T V_1 D^2 & D^{-1} V_1^T 0 \\ U_2^T U_1 D & 0 \end{pmatrix} = \begin{pmatrix} D & 0 \\ 0 & 0 \end{pmatrix} \\ &= \Sigma \end{aligned}$$

using (A.25a), (A.24a), (A.24c), (A.25b), (A.23a). Premultiplying this relation by U and postmultiplying it by V^T finally gives (A.22). ∎

Definition A.2
The factorization (A.22) is called the *singular value decomposition* of the matrix A. The elements $\sigma_1, \ldots, \sigma_r$ and the possibly additional $\min(m, n) - r$ zero diagonal elements of Σ are called the *singular values* of A. ∎

Remark. It follows by the construction that the singular value decomposition (A.22) can be written as

$$A = U \Sigma V^T = (U_1 \quad U_2) \begin{pmatrix} D & 0 \\ 0 & 0 \end{pmatrix} \begin{pmatrix} V_1^T \\ V_2^T \end{pmatrix} = U_1 D V_1^T \tag{A.26}$$

The matrices on the right-hand side have the following dimensions: U_1 is $(n|r)$, D is $(r|r)$ (and nonsingular), V_1^T is $(r|m)$. ∎

Using the singular value decomposition one can easily find the pseudoinverse. This is done in the following lemma.

Lemma A.14
Let A be an $(n|m)$ matrix with a singular value decomposition given by (A.22)–(A.23). Then the pseudoinverse is given by

$$A^\dagger = V \Sigma^\dagger U^T \tag{A.27a}$$

where

$$\Sigma^\dagger = \begin{pmatrix} D^{-1} & 0 \\ 0 & 0 \end{pmatrix} \tag{A.27b}$$

Proof. The proof is based on the characterization of $A^\dagger b$ as the uniquely given shortest vector that minimizes $V(x) = \|Ax - b\|^2$ (cf. Lemma A.10). Now set $y = V^T x$, $c = U^T b$. Since multiplication with an orthogonal matrix does not change the norm of a vector,

$$V(x) = \|U^T(Ax - b)\|^2 = \|U^T U \Sigma V^T x - c\|^2$$

$$= \|\Sigma y - c\|^2 = \sum_{i=1}^{r} \{\sigma_i y_i - c_i\}^2 + \sum_{i=r+1}^{n} c_i^2$$

Minimization can now be carried out with respect to y instead of x. It gives

$y_i = c_i/\sigma_i \quad i = 1, \ldots, r$

y_i arbitrary, $i = r + 1, \ldots, m$

Since $\|y\| = \|x\|$, the *shortest vector*, say $\bar{y}$, must be characterized by

$y_i = 0 \quad i = r + 1, \ldots, m$

which means

$\bar{y} = \Sigma^\dagger c$

using (A.27b). The corresponding vector x is then easily found to be

$\bar{x} = V\bar{y} = V\Sigma^\dagger c = V\Sigma^\dagger U^T b$

Since also $\bar{x} = A^\dagger b$ for any b, (A.27a) follows. ∎

Remark. Using the singular value decomposition (SVD) one can give an interpretation of the spaces $\mathcal{N}(A)$, $\mathcal{R}(A^T)$, etc. Change the basis in $\mathcal{R}^n$ by rotation using U and similarly rotate the basis in $\mathcal{R}^m$ using V^T. The transformation A is thereafter given by the diagonal matrix Σ. The number r of positive singular values is equal to the rank of A. In the transformed $\mathcal{R}^m$ space, the first r components correspond to the subspace $\mathcal{R}(A^T)$ while the last $m - r$ refer to the nullspace $\mathcal{N}(A)$. Similarly in the transformed space $\mathcal{R}^n$, the first r components describe $\mathcal{R}(A)$ while the last $n - r$ are due to $\mathcal{N}(A^T)$. ∎

An alternative characterization of the pseudoinverse is provided by the following lemma.

Lemma A.15
Let A be a given $(n|m)$ matrix. Consider the following equations, where X is an $(m|n)$ matrix:

$$AXA = A$$
$$XAX = X$$
$$(AX)^T = AX \qquad\qquad (A.28)$$
$$(XA)^T = XA$$

The equations (A.28) have a unique solution given by $X = A^\dagger$.

Proof. First transform the equations (A.28) by using the SVD of A, $A = U\Sigma V^T$, and setting $Y = V^T X U$. Then (A.28) is readily found to be equivalent to

$$\Sigma Y \Sigma = \Sigma$$
$$Y \Sigma Y = Y$$
$$Y^T \Sigma^T = \Sigma Y \quad \text{(A.29)}$$
$$\Sigma^T Y^T = Y \Sigma$$

Next partition Σ as

$$\Sigma = \begin{pmatrix} D & 0 \\ 0 & 0 \end{pmatrix}$$

where D is square and nonsingular (cf. (A.23a)). If A is square and nonsingular all the 0 blocks will disappear. If A is rectangular and of full rank two of the 0 blocks will disappear. Now partition Y as:

$$Y = \begin{pmatrix} Y_{11} & Y_{12} \\ Y_{21} & Y_{22} \end{pmatrix}$$

Using these partitioned forms it is easy to deduce that (A.29) becomes

$$\begin{pmatrix} DY_{11}D & 0 \\ 0 & 0 \end{pmatrix} = \begin{pmatrix} D & 0 \\ 0 & 0 \end{pmatrix} \quad \text{(A.30a)}$$

$$\begin{pmatrix} Y_{11}DY_{11} & Y_{11}DY_{12} \\ Y_{21}DY_{11} & Y_{21}DY_{12} \end{pmatrix} = \begin{pmatrix} Y_{11} & Y_{12} \\ Y_{21} & Y_{22} \end{pmatrix} \quad \text{(A.30b)}$$

$$\begin{pmatrix} Y_{11}^T D & 0 \\ Y_{12}^T D & 0 \end{pmatrix} = \begin{pmatrix} DY_{11} & DY_{12} \\ 0 & 0 \end{pmatrix} \quad \text{(A.30c)}$$

$$\begin{pmatrix} DY_{11}^T & DY_{21}^T \\ 0 & 0 \end{pmatrix} = \begin{pmatrix} Y_{11}D & 0 \\ Y_{21}D & 0 \end{pmatrix} \quad \text{(A.30d)}$$

First the 11 block of (A.30a) gives $Y_{11} = D^{-1}$. The 12 block of (A.30c) gives $Y_{12} = 0$. The 21 block of (A.30d) gives $Y_{21} = 0$. Finally, the 22 block of (A.30b) gives $Y_{22} = 0$. Hence

$$Y = \begin{pmatrix} D^{-1} & 0 \\ 0 & 0 \end{pmatrix} = \Sigma^\dagger$$

Therefore (A.28) has a unique solution which is given by

$$X = VYU^T = V\Sigma^\dagger U^T = A^\dagger$$

(cf. (A.27a)). ∎

Remark 1. The result can also be proved by the use of the geometric properties of the pseudoinverse as given in Definition A.1, Figure A.1 and Lemma A.12. ∎

Remark 2. The relations (A.28) are sometimes used as an alternative definition of the pseudoinverse. ∎

A.3 The QR method

The results in this section concern the QR method which was introduced in Chapter 4. The required orthogonal matrix Q will be constructed as a product of 'elementary' orthogonal matrices. Such matrices are introduced in Definitions A.3 and A.4 below.

Definition A.3
Let w be a vector of norm 1, so that $w^T w = 1$. Then the matrix Q given by

$$Q = I - 2ww^T \tag{A.31}$$

is called a Householder transformation. ∎

Lemma A.16
A Householder matrix is symmetric and orthogonal. The transformation $x \to Qx$ means geometrically a reflection with respect to the plane perpendicular to w.

Proof Q is trivially symmetric since

$$Q^T = (I - 2ww^T)^T = I - 2ww^T = Q$$

It is orthogonal since

$$QQ^T = Q^2 = (I - 2ww^T)(I - 2ww^T) = I - 4ww^T + 4ww^T ww^T = I$$

To prove the reflection property note that

$$Qx = x - 2w(w^T x)$$

A geometric illustration is given in Figure A.2.

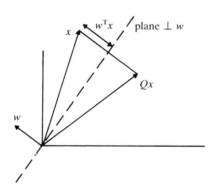

FIGURE A.2 Illustration of the reflection property (Qx is the reflection of x with respect to the plane perpendicular to w). ∎

A second type of 'elementary' orthogonal matrix is provided by the following definition.

Definition A.4
A Given's rotation matrix is defined as

$$Q = \begin{pmatrix} 1 & & & & & & & & & \\ & \ddots & & & & & & 0 & & \\ & & c & \cdots & s & & & & & \\ & & & 1 & & & & & & \\ & & \vdots & & \ddots & & & & & \\ & & & & & 1 & & & & \\ & & -s & & & c & & & & \\ & & & & & & 1 & & & \\ & & 0 & & & & & \ddots & & \\ & & & & & & & & 1 & \\ & & i & & j & & & & & \end{pmatrix} \quad \text{(A.32)}$$

where $c^2 + s^2 = 1$. ∎

Lemma A.17
A Given's rotation matrix is orthogonal.

Proof. Straightforward calculations give

$$QQ^T = \begin{pmatrix} 1 & & & & & \\ & \ddots & & & & \\ & & c & \cdots & s & 0 \\ & & & 1 & & \\ & & \ddots & & & \\ & & -s & \cdots & c & \\ & & & & 1 & \\ & & 0 & & & \ddots \\ & & & & & 1 \end{pmatrix} \begin{pmatrix} 1 & & & & & \\ & \ddots & & & & 0 \\ & & c & \cdots & -s & \\ & & & 1 & & \\ & & \ddots & & & \\ & & s & \cdots & c & \\ & & & & 1 & \\ & & 0 & & & \ddots \\ & & & & & 1 \end{pmatrix}$$

$$= \begin{pmatrix} 1 & & & & & & \\ & \ddots & & & & & \\ & & 1 & & & 0 & \\ & & & c^2+s^2 & & & \\ & & & & 1 & & \\ & & & & & \ddots & \\ & & & & & & 1 \\ & & & & & & & c^2+s^2 \\ & & 0 & & & & & & 1 \\ & & & & & & & & & \ddots \\ & & & & & & & & & & 1 \end{pmatrix} = I$$

∎

Section A.3   The QR method

The next lemmas show how the elementary orthogonal matrices can be chosen so that a transformed vector gets a simple form.

Lemma A.18
Let x be an arbitrary vector. Then there is a Householder transformation such that

$$Qx = \lambda \begin{pmatrix} 1 \\ 0 \\ \vdots \\ 0 \end{pmatrix} \quad (A.33)$$

Here, $\lambda = \|x\|$.

Proof. If $Q = I - 2ww^T$ is a Householder transformation satisfying (A.33) then this relation is equivalent to

$$x = Q\lambda \begin{pmatrix} 1 \\ 0 \\ \vdots \\ 0 \end{pmatrix} = \lambda \begin{pmatrix} 1 - 2w_1^2 \\ -2w_1 w_2 \\ \vdots \\ -2w_1 w_n \end{pmatrix}$$

Therefore

$$w_1 = \left(\frac{1}{2}\left(1 - \frac{x_1}{\lambda}\right)\right)^{1/2}$$

$$w_i = -\frac{x_i}{2w_1 \lambda} \quad i = 2, \ldots, n$$

Further, λ is given by

$$x^T x = x^T Q^T Q x = \lambda^2 (1 \ 0 \ \ldots \ 0) \begin{pmatrix} 1 \\ 0 \\ \vdots \\ 0 \end{pmatrix} = \lambda^2$$

so $\lambda = \|x\|$. In particular, w_1 as given above is well defined. ∎

Lemma A.19
Let x be an arbitrary vector of dimension n. Then there exists an orthogonal matrix of the form

$$Q = Q_{n-1} Q_{n-2} \cdots Q_1 \quad (A.34)$$

with $\{Q_i\}$ being Given's rotation matrices such that

$$Qx = \lambda \begin{pmatrix} 1 \\ 0 \\ \vdots \\ 0 \end{pmatrix} \quad (A.35)$$

Here, $\lambda = \|x\|$.

Proof. First assume that $x_1 \neq 0$. Then for Q_1 choose $i = 1, j = n$ and take

$$c = \frac{x_1}{[x_1^2 + x_n^2]^{1/2}} \quad s = \frac{x_n}{[x_1^2 + x_n^2]^{1/2}}$$

which gives

$$x^{(1)} = \begin{pmatrix} x_1^{(1)} \\ \cdot \\ \cdot \\ \cdot \\ x_n^{(1)} \end{pmatrix} \triangleq Q_1 x = \begin{pmatrix} c & & & s \\ & 1 & & \\ & & \ddots & \\ & & & 1 & \\ -s & & & & c \end{pmatrix} \begin{pmatrix} x_1 \\ \cdot \\ \cdot \\ \cdot \\ x_n \end{pmatrix} = \begin{pmatrix} (x_1^2 + x_n^2)^{1/2} \\ x_2 \\ \vdots \\ x_{n-1} \\ 0 \end{pmatrix}$$

For Q_2 take $i = 1, j = n - 1$ and

$$c = \frac{x_1^{(1)}}{[x_1^{(1)2} + x_{n-1}^{(1)2}]^{1/2}} \quad s = \frac{x_{n-1}^{(1)}}{[x_1^{(1)2} + x_{n-1}^{(1)2}]^{1/2}}$$

Then

$$x^{(2)} \triangleq Q_2 x^{(1)} = Q_2 Q_1 x = \begin{pmatrix} c & & s & & \\ & 1 & & & \\ & & \ddots & & \\ & & & 1 & \\ -s & & c & & \\ & & & & 1 \end{pmatrix} x^{(1)} = \begin{pmatrix} [x_1^{(1)2} + x_{n-1}^{(1)2}]^{1/2} \\ x_2 \\ \vdots \\ x_{n-2} \\ 0 \\ 0 \end{pmatrix}$$

Proceeding in this way for Q_k, $i = 1, j = n + 1 - k$ and

$$c = \frac{x_1^{(k-1)}}{[x_1^{(k-1)2} + x_j^{(k-1)2}]^{1/2}} \quad s = \frac{x_j^{(k-1)}}{[x_1^{(k-1)2} + x_j^{(k-1)2}]^{1/2}}$$

giving

$$x^{(k)} = \begin{pmatrix} [x_1^2 + x_{n-k+1}^2 + \ldots + x_n^2]^{1/2} \\ x_2 \\ \vdots \\ x_{n-k} \\ 0 \\ \vdots \\ 0 \end{pmatrix}$$

Section A.3

For $k = n - 1$,

$$x^{(n-1)} = \begin{pmatrix} \|x\| \\ 0 \\ \vdots \\ 0 \end{pmatrix}$$

The stated result now follows since the product of orthogonal matrices is an orthogonal matrix.

Assume next that $x_1 = 0$, but $x \neq 0$. (Should $x = 0$, $\{Q_i\}$ can be chosen arbitrarily.) Say $x_k \neq 0$. Then first permute elements 1 and k of x by taking $i = 1, j = k$, and using the previous rules to obtain c and s. When thereafter successively zeroing the elements in the vector x, note that there is already a zero in the kth element. This means that x can be transformed to the form (A.35) in $(n - 1)$ steps. (This case illustrates the principle of using pivot elements.) ∎

The two previous lemmas will now be extended to the matrix case, effectively describing the QR method.

Lemma A.20
Let A be an $(n|m)$ matrix, $n \geq m$. Then there exists an orthogonal matrix Q such that

$$QA = \begin{pmatrix} R \\ 0 \end{pmatrix} \tag{A.36}$$

where R is an $(m|m)$-dimensional upper triangular matrix and 0 is an $((n-m)|m)$-dimensional zero matrix.

Proof. The proof will proceed by constructing Q as a product

$$Q = Q_m Q_{m-1} \cdots Q_1$$

Let

$$A = (a_1 \ \ldots \ a_m)$$

First choose Q_1 (using for example Lemma A.18 or Lemma A.19) such that

$$Q_1 a_1 = \begin{pmatrix} a_{11}^{(1)} \\ 0 \\ \vdots \\ 0 \end{pmatrix}$$

which gives

$$Q_1 A \triangleq (a_1^{(1)} \ \ldots \ a_m^{(1)})$$

Then choose Q_2 so that it leaves the first component unchanged but else transform $a_2^{(1)}$ as

$$Q_2 a_2^{(1)} = \begin{pmatrix} a_{21}^{(1)} \\ a_{22}^{(2)} \\ 0 \\ \vdots \\ 0 \end{pmatrix} = Q_2 Q_1 a_2$$

Proceeding in this way,

$$QA = \begin{pmatrix} a_{11}^{(1)} & a_{21}^{(1)} & \cdots & a_{m1}^{(1)} \\ & a_{22}^{(2)} & \cdots & a_{m2}^{(2)} \\ & & \ddots & \vdots \\ 0 & & & a_{mm}^{(m)} \\ & & & 0 \end{pmatrix}$$

which completes the proof. ∎

A.4 Matrix norms and numerical accuracy

In this section the notation $\|x\|$ will be used for the Euclidean vector norm, i.e.

$$\|x\| = [x_1^2 + \ldots + x_n^2]^{1/2} \tag{A.37}$$

The norm of a matrix is defined as follows.

Definition A.5
Let A be an $(n|m)$ matrix. Its norm is given by

$$\|A\| \triangleq \sup_{x \neq 0} \frac{\|Ax\|}{\|x\|} \tag{A.38}$$

∎

Note that it follows directly from this definition that

$$\|Ax\| \leq \|A\| \, \|x\| \quad \text{all } x \tag{A.39}$$

The next lemma establishes that the introduced norm satisfies the usual properties ((i)–(iii) below) of a norm.

Lemma A.21
Let A, B and C be matrices of dimensions $(n|m)$, $(n|m)$, $(m|p)$ respectively and let λ be a scalar. Then

(i) $\|A\| \geq 0$ with equality only for $A = 0$ \hfill (A.40a)

(ii) $\|\lambda A\| = |\lambda| \, \|A\|$ \hfill (A.40b)

Section A.4   *Matrix norms and numerical accuracy*   533

(iii) $\|A + B\| \leq \|A\| + \|B\|$ (A.40c)

(iv) $\|AC\| \leq \|A\|\,\|C\|$ (A.40d)

Proof. The statements (A.40a) and (A.40b) are immediate. Further,

$$\|A + B\| = \sup_{x \neq 0} \frac{\|Ax + Bx\|}{\|x\|} \triangleq \frac{\|Ax^* + Bx^*\|}{\|x^*\|} \leq \frac{\|Ax^*\|}{\|x^*\|} + \frac{\|Bx^*\|}{\|x^*\|}$$

$$\leq \sup_{x \neq 0} \frac{\|Ax\|}{\|x\|} + \sup_{x \neq 0} \frac{\|Bx\|}{\|x\|} = \|A\| + \|B\|$$

which proves (A.40c). Similarly,

$$\|AC\| = \sup_{x \neq 0} \frac{\|ACx\|}{\|x\|} \triangleq \frac{\|ACx^*\|}{\|x^*\|} = \frac{\|ACx^*\|}{\|Cx^*\|} \frac{\|Cx^*\|}{\|x^*\|}$$

$$\leq \|A\|\,\|C\|$$

which proves (A.40d). ∎

Lemma A.22
Let A be an $(n|n)$ orthogonal matrix. Then $\|A\| = 1$.

Proof. Simple calculations give

$$\|x\|^2 = x^T x = x^T A^T A x = \|Ax\|^2$$

Hence, $\|Ax\| = \|x\|$, all x, which gives $\|A\| = 1$ by Definition A.5. ∎

The following lemma shows the relationship between the singular values of a matrix and its norm as previously defined. Some related results will also be presented. In what follows A will be an $(n|m)$-dimensional matrix with singular values

$$\sigma_1 \geq \sigma_2 \geq \ldots \geq \sigma_r > 0 = \sigma_{r+1} = \ldots = \sigma_{\min(m,n)}$$

Lemma A.23
The matrix norms of A and $A^\dagger$ are given by

$$\|A\| = \sigma_1 \tag{A.41a}$$

$$\|A^\dagger\| = 1/\sigma_r \tag{A.41b}$$

Proof. Let the SVD of A be $A = U\Sigma V^T$. Set $y = V^T x$. By Definition A.5,

$$\|A\| = \sup_{x \neq 0} \frac{\|Ax\|}{\|x\|} = \sup_{x \neq 0} \frac{\|U\Sigma V^T x\|}{\|x\|}$$

$$= \sup_{y \neq 0} \frac{\|\Sigma y\|}{\|Vy\|} = \sup_{y \neq 0} \frac{\|\Sigma y\|}{\|y\|}$$

$$= \sup_{y \neq 0} \left[\sum_{i=1}^{r} \sigma_i^2 y_i^2 \bigg/ \sum_{i=1}^{m} y_i^2 \right]^{1/2} \leq \sup_{y \neq 0} \left[\sum_{i=1}^{r} \sigma_1^2 y_i^2 \bigg/ \sum_{i=1}^{m} y_i^2 \right]^{1/2} \leq \sigma_1$$

Note that the equalities hold if $y_1 = 1$ and $y_i = 0$ for $i = 2, \ldots, m$. The norm of A is thus equal to the largest singular value of A. Since $A^\dagger = V\Sigma^\dagger U^T$, the pseudoinverse $A^\dagger$ has singular values

$$1/\sigma_1, \ldots, 1/\sigma_r, 0, \ldots, 0$$

The largest of these is $1/\sigma_r$ which then must be the norm of $A^\dagger$. ∎

Lemma A.24
Let A be an $(n|m)$ matrix of rank m and let $x = Ab$. Then the following bounds apply:

$$\|x\| \leq \sigma_1 \|b\| \tag{A.42a}$$

$$\|x\| \geq \sigma_m \|b\| \tag{A.42b}$$

Proof. The first inequality, (A.42a), is immediate from (A.41a) and (A.39). To verify (A.42b) consider the following calculation with $c = V^T b$:

$$\|x\| = \|Ab\| = \|U\Sigma V^T b\| = \|\Sigma(V^T b)\| = \|\Sigma c\|$$

$$= \left[\sum_{i=1}^{m} \sigma_i^2 c_i^2\right]^{1/2} \geq \sigma_m \left[\sum_{i=1}^{m} c_i^2\right]^{1/2} = \sigma_m \|c\| = \sigma_m \|b\|$$

∎

Definition A.6
The condition number of a matrix A is given by

$$C(A) = \|A\| \, \|A^\dagger\| \tag{A.43}$$

∎

Remark. Referring to Lemma A.23, it follows that

$$C(A) = \sigma_1/\sigma_r \tag{A.44}$$

∎

The next topic concerns the effect of rounding errors on the solution of linear systems of equations. Consider a linear system of equations

$$Ax = b \tag{A.45}$$

where A is an $(n|n)$ nonsingular matrix. When solving (A.45) on a finite word length machine, rounding errors are inherently affecting the computations. In such a case one cannot expect to find the exact solution x of (A.45). All that can be hoped for is to get the exact solution of a (fictitious) system of equations obtained by slightly perturbing A and b in (A.45). Denote the perturbed system of equations by

$$(A + \delta A)(x + \delta x) = (b + \delta b) \tag{A.46}$$

The unknown in (A.46) is denoted by $x + \delta x$ to stress that the solution changes. It is of course desirable that small deviations δA and δb only give a small deviation δx in the solution. This is dealt with in the following lemma.

Lemma A.25
Consider the systems (A.45) and (A.46). Assume that

$$C(A)\|\delta A\|/\|A\| < 1 \tag{A.47}$$

Then

$$\frac{\|\delta x\|}{\|x\|} \leq \frac{C(A)}{1 - C(A)\|\delta A\|/\|A\|} \left\{ \frac{\|\delta A\|}{\|A\|} + \frac{\|\delta b\|}{\|b\|} \right\} \tag{A.48}$$

Proof. Subtraction of (A.45) from (A.46) gives

$$A\delta x = \delta b - \delta A(x + \delta x)$$
$$\delta x = A^{-1}\delta b - A^{-1}\delta A(x + \delta x)$$

Using Lemma A.21 it is easy to show that

$$\|\delta x\| \leq \|A^{-1}\|\{\|\delta b\| + \|\delta A\|\|x + \delta x\|\}$$
$$\leq \|A^{-1}\|\{\|\delta b\| + \|\delta A\|[\|x\| + \|\delta x\|]\}$$

and hence

$$(1 - \|A^{-1}\|\|\delta A\|)\|\delta x\| \leq \|A^{-1}\|\{\|\delta b\| + \|\delta A\|\|x\|\} \tag{A.49a}$$

Note that by Definition A.6 ($A^\dagger = A^{-1}$ now)

$$1 - \|A^{-1}\|\|\delta A\| = 1 - C(A)\|\delta A\|/\|A\|$$

Since $x = A^{-1}b$ it also follows from (A.42b) that

$$\|x\| \geq (1/\sigma_1)\|b\| = \|b\|/\|A\| \tag{A.49b}$$

Combining (A.49a) and (A.49b),

$$(1 - \|A^{-1}\|\|\delta A\|) \frac{\|\delta x\|}{\|x\|} \leq \|A^{-1}\|\{\|\delta b\|/\|x\| + \|\delta A\|\}$$
$$\leq \|A\|\|A^{-1}\| \left\{ \frac{\|\delta b\|}{\|b\|} + \frac{\|\delta A\|}{\|A\|} \right\}$$

which easily can be reformulated as (A.48). ∎

In loose terms, one can say that the relative errors in A and b can be magnified by a factor of $C(A)$ in the solution to (A.46). Note that the condition number is given by the *problem*, while the errors $\|\delta A\|$ and $\|\delta b\|$ depend on the *algorithm* used to solve the linear system of equations. For numerically stable algorithms which include partial pivoting the relative errors of A and b are typically of the form

$$\frac{\|\delta A\|}{\|A\|} = \eta f(n)$$

where η is the machine precision and $f(n)$ is a function that grows moderately with the order n. In particular, the product $\eta C(A)$ gives an indication of how many significant digits can be expected in the solution. For example, if $\eta = 10^{-7}$ and $C(A) = 10$ then $\eta C(A) = 10^{-6}$, and 5 to 6 significant digits can be expected; but if $C(A)$ is equal to 10^4 then $\eta C(A) = 10^{-3}$ and only 2 or 3 significant digits can be expected.

The following result gives a basis for a preliminary discussion of how to cope with the more general problem of a least squares solution to an inconsistent system of equations of the form (A.45).

Lemma A.26

Let A be a matrix with condition number $C(A)$ and Q an orthogonal matrix. Then

$$C(A^T A) = [C(A)]^2 \qquad (A.50a)$$

$$C(QA) = C(A) \qquad (A.50b)$$

Proof. Let A have the singular value decomposition

$$A = U\Sigma V^T$$

Then

$$A^T A = V\Sigma^T U^T U \Sigma V^T = V(\Sigma^T \Sigma) V^T$$

The singular values of $A^T A$ are thus $\sigma_1^2, \sigma_2^2, \ldots$. The result (A.50a) now follows from (A.44). Since QU is an orthogonal matrix the singular value decomposition of QA is given by

$$QA = (QU)\Sigma V^T$$

Hence, QA and A have the same singular values and (A.50b) follows immediately. ∎

Lemma A.26 can be used to illustrate a key rule in determining the LS solution of an overdetermined system of linear equations. This rule states that instead of forming and solving the corresponding so-called normal equations, it is much better from the point of view of numerical accuracy to use a QR method directly on the original system. For simplicity of illustration, consider the system of equations (A.45), where A is square and nonsingular,

$$Ax = b \qquad (A.51)$$

Using a standard method for solving (A.51) the errors are magnified by a factor $C(A)$ (see (A.48)). If instead the normal equations

$$A^T A x = A^T b \qquad (A.52)$$

are solved then the errors are magnified by a factor $[C(A)]^2$ (cf. (A.50a)). There is thus a considerable loss in accuracy when numerically solving (A.52) instead of (A.51). On the other hand if the QR method is applied then instead of (A.51) the equation

$$QAx = Qb$$

is solved, with Q orthogonal. In view of (A.50b) the error is then magnified only by a factor of $C(A)$.

In the discussion so far, A in (A.51) has been square and nonsingular. The general case of a rectangular and possibly rank-deficient matrix is more complicated. Recall that Lemma A.25 is applicable only for square nonsingular matrices. The following result applies to a more general case.

Lemma A.27
Let A be an $(n|m)$-dimensional matrix of rank m. Also let the perturbed matrix $A + \delta A$ have rank m. Consider

$$x = A^\dagger b \tag{A.53a}$$

$$x + \delta x = (A + \delta A)^\dagger (b + \delta b) \tag{A.53b}$$

Then

$$\frac{\|\delta x\|}{\|x\|} \le \frac{C(A)}{1 - C(A)\|\delta A\|/\|A\|} \left\{ \left[1 + C(A) \frac{\|b - Ax\|}{\|A\|\|x\|} \right] \frac{\|\delta A\|}{\|A\|} \right. \\ \left. + \frac{\|b\|}{\|A\|\|x\|} \frac{\|\delta b\|}{\|b\|} \right\} \tag{A.54}$$

Proof. See Lawson and Hanson (1974). ∎

Remark. Note that for A square (and nonsingular), $\|b - Ax\| = 0$ and $\|A\|\|x\| \ge \|b\|$. Hence (A.54) can in this case be simplified to (A.48). ∎

A.5 Idempotent matrices

This section is devoted to the so-called idempotent matrices. Among others, these matrices are useful when analyzing linear regressions.

Definition A.7
A square matrix P is said to be *idempotent* if $P^2 = P$. ∎

Idempotent matrices have some interesting properties.

Lemma A.28
The following statements hold for an arbitrary idempotent matrix P:

(i) All eigenvalues are either zero or one
(ii) Rank $P = \text{tr } P$ \hfill (A.55)
(iii) If P is symmetric then there exists an orthogonal matrix U such that

$$P = U^T \begin{pmatrix} I_r & 0 \\ 0 & 0 \end{pmatrix} U \qquad (A.56)$$

where I_r is the identity matrix of order $r = \text{rank } P$.

(iv) If P is symmetric and of order n then it is the orthogonal projector of $\mathscr{R}^n$ onto the range $\mathscr{R}(P)$.

Proof. Let λ be an arbitrary eigenvalue of P and denote the corresponding eigenvector by e. Then $Pe = \lambda e$. Using $P = P^2$ gives

$$\lambda e = Pe = P^2 e = P(\lambda e) = \lambda Pe = \lambda^2 e$$

Since $e \neq 0$ it follows that $\lambda(\lambda - 1) = 0$, which proves part (i). Let P have eigenvalues $\lambda_1, \ldots, \lambda_n$. The rank of P is equal to the number of nonzero eigenvalues. However,

$$\text{tr } P = \sum_{1}^{n} \lambda_i$$

is, in view of part (i), also equal to this number. This proves part (ii). Part (iii) then follows easily since one can always diagonalize a symmetric matrix. To prove part (iv) let x be an arbitrary vector in $\mathscr{R}^n$. It can then be uniquely decomposed as $x = x_1 + x_2$ where x_1 lies in the nullspace of P, $\mathscr{N}(P)$, and x_2 lies in the range of P, $\mathscr{R}(P)$ (cf. Lemma A.7). The components x_1 and x_2 are orthogonal ($x_1^T x_2 = 0$). Thus $Px_1 = 0$ and $x_2 = Pz$ for some vector z. Then $Px = Px_1 + Px_2 = Px_2 = P^2 z = Pz = x_2$, which shows that P is the orthogonal projector of $\mathscr{R}^n$ onto $\mathscr{R}(P)$. ∎

Example A.2 *An orthogonal projector*
Let F be an $(n|r)$-dimensional matrix, $n \geq r$, of full rank r. Consider

$$P = I_n - F(F^T F)^{-1} F^T \qquad (A.57)$$

Then

$$\begin{aligned} P^2 &= [I_n - F(F^T F)^{-1} F^T][I_n - F(F^T F)^{-1} F^T] \\ &= I_n - 2F(F^T F)^{-1} F^T + F(F^T F)^{-1} F^T F(F^T F)^{-1} F^T \\ &= I_n - F(F^T F)^{-1} F^T = P \end{aligned}$$

Thus P is idempotent and symmetric. Further,

$$\begin{aligned} \text{tr } P &= \text{tr } I_n - \text{tr } F(F^T F)^{-1} F^T \\ &= n - \text{tr}(F^T F)(F^T F)^{-1} \\ &= n - \text{tr } I_r = n - r \end{aligned}$$

This shows that P has rank $n - r$. P can in fact be interpreted as the orthogonal projector of $\mathscr{R}^n$ on the null space $\mathscr{N}(F^T)$. This can be shown as follows. Let $x \in \mathscr{N}(F^T)$. Then $Px = x - F(F^T F)^{-1} F^T x = x$. Also if $x \in \mathscr{R}(F)$, (which is the orthogonal complement to $\mathscr{N}(F^T)$; see Lemma A.7), say $x = Fz$, then $Px = x - F(F^T F)^{-1} F^T F z = x - Fz = 0$. ∎

Section A.5   Idempotent matrices

The following lemma utilizes properties of idempotent matrices. It is used in connection with structure determination for linear regression models.

Lemma A.29
Consider a matrix F of dimension $(n|r)$, $n \geq r$, of full rank r. Let

$$F = \begin{matrix}(F_1 & F_2) & n \\ m & r-m & \end{matrix} \tag{A.58}$$

where F_1 is $(n|m)$ and has full rank m. Further, let

$$\begin{aligned} P_1 &= I_n - F_1(F_1^T F_1)^{-1} F_1^T \\ P_2 &= I_n - F(F^T F)^{-1} F^T \end{aligned} \tag{A.59}$$

Then there exists an orthogonal matrix U of dimension $(n|n)$ such that

$$P_2 = U \begin{pmatrix} I_{n-r} & 0 \\ 0 & 0_r \end{pmatrix} U^T \tag{A.60}$$

$$P_1 - P_2 = U \begin{pmatrix} 0_{n-r} & 0 & 0 \\ 0 & I_{r-m} & 0 \\ 0 & 0 & 0_m \end{pmatrix} U^T \tag{A.61}$$

Proof. The diagonalization of P_2 in (A.60) follows from Lemma A.28 part (iii). Let

$$U = \begin{matrix}(U_1 & U_2) & n \\ n-r & r & \end{matrix}$$

The matrices U_1 and U_2 are not unique. Let V_1 and V_2 denote two matrices whose columns are the orthonormal eigenvectors of P_2 associated with $\lambda = 1$ and, respectively, $\lambda = 0$. Then set $U_1 = V_1$ and $U_2 = V_2 W$ where W is some arbitrary orthogonal matrix. Clearly (A.60) holds for this choice. The matrix W will be selected later in the proof. First note that

$$\begin{aligned} P_2(I - P_1) &= [I - F(F^T F)^{-1} F^T] F_1 (F_1^T F_1)^{-1} F_1^T \\ &= F_1(F_1^T F_1)^{-1} F_1^T - (F_1 \ F_2) \begin{pmatrix} F_1^T F_1 & F_1^T F_2 \\ F_2^T F_1 & F_2^T F_2 \end{pmatrix}^{-1} \begin{pmatrix} F_1^T F_1 \\ F_2^T F_1 \end{pmatrix} (F_1^T F_1)^{-1} F_1^T \\ &= F_1(F_1^T F_1)^{-1} F_1^T - (F_1 \ F_2) \begin{pmatrix} I \\ 0 \end{pmatrix} (F_1^T F_1)^{-1} F_1^T = 0 \end{aligned}$$

This implies in particular that

$$P_2(P_1 - P_2) = P_2 P_1 - P_2^2 = P_2 P_1 - P_2 = -P_2(I - P_1) = 0$$

and

$$P_1 P_2 = (P_2 P_1)^T = P_2^T = P_2$$
$$(P_1 - P_2)^2 = P_1^2 - P_1 P_2 - P_2 P_1 + P_2^2 = P_1 - P_1 P_2 - P_2 P_1 + P_2 = P_1 - P_2$$

The latter calculation shows that $P_1 - P_2$ is idempotent. Then by Lemma A.28

$$\mathrm{rank}(P_1 - P_2) = \mathrm{tr}(P_1 - P_2) = \mathrm{tr}\, P_1 - \mathrm{tr}\, P_2 = (n - m) - (n - r) = r - m$$

Now set

$$P_1 - P_2 = U \begin{pmatrix} A & B \\ B^\mathrm{T} & C \end{pmatrix} U^\mathrm{T} \tag{A.62}$$

where the matrices A, B, C have the dimensions

A: $((n-r)|(n-r))$, B: $((n-r)|r)$ and C: $(r|r)$.

Since $P_2(P_1 - P_2) = 0$ it follows from (A.60), (A.62) that $A = 0$, $B = 0$. Then since $P_1 - P_2$ is idempotent it follows from (A.62) that $C^2 = C$. C is thus an idempotent matrix. Now rank $C = \mathrm{rank}(P_1 - P_2) = r - m$. According to Lemma A.28 one can choose an orthogonal matrix V of dimension $(r|r)$ such that

$$V^\mathrm{T} C V = J \triangleq \begin{pmatrix} I_{r-m} & 0 \\ 0 & 0_m \end{pmatrix} \begin{matrix} \} \, r-m \\ \} \, m \end{matrix}$$
$$\quad\quad\quad r-m \quad m$$

Thus, cf. (A.62),

$$P_1 - P_2 = U \begin{pmatrix} I & 0 \\ 0 & V \end{pmatrix} \begin{pmatrix} 0 & 0 \\ 0 & J \end{pmatrix} \begin{pmatrix} I & 0 \\ 0 & V^\mathrm{T} \end{pmatrix} U^\mathrm{T} = U \begin{pmatrix} 0 & 0 \\ 0 & J \end{pmatrix} U^\mathrm{T}$$

where in the last equality W in U_2 has been redefined as WV. This proves (A.61). ∎

A.6 Sylvester matrices

In this section Sylvester matrices are defined and a result is given on their rank.

Definition A.8
Consider the polynomials

$$A(z) = a_0 z^{na} + a_1 z^{na-1} + \ldots + a_{na}$$
$$B(z) = b_0 z^{nb} + b_1 z^{nb-1} + \ldots + b_{nb}$$

Then the Sylvester matrix $\mathscr{S}(A, B)$ of dimension $((na + nb)|(na + nb))$ is defined as

$$\mathscr{S}(A, B) = \begin{pmatrix} a_0 & a_1 & \cdots & a_{na} & & & 0 \\ & \ddots & & & \ddots & & \\ 0 & & a_0 & a_1 & \cdots & a_{na} \\ b_0 & b_1 & \cdots & b_{nb} & & & 0 \\ & \ddots & & & \ddots & & \\ 0 & & b_0 & b_1 & \cdots & b_{nb} \end{pmatrix} \begin{matrix} \} \\ \} \, nb \text{ rows} \\ \\ \} \\ \} \, na \text{ rows} \end{matrix} \tag{A.63}$$

∎

Lemma A.30
The rank of the Sylvester matrix $\mathscr{S}(A, B)$ is given by

$$\text{rank } \mathscr{S}(A, B) = na + nb - n \qquad (A.64)$$

where n is the number of common zeros to $A(z)$ and $B(z)$.

Proof. Consider the equation

$$x^T \mathscr{S}(A, B) = 0 \qquad (A.65)$$

where

$$x^T = (\tilde{b}_1 \ \ldots \ \tilde{b}_{nb} \ \ \tilde{a}_1 \ \ldots \ \tilde{a}_{na})$$

Let

$$\tilde{A}(z) = \tilde{a}_1 z^{na-1} + \ldots + \tilde{a}_{na} \qquad \tilde{B}(z) = \tilde{b}_1 z^{nb-1} + \ldots + \tilde{b}_{nb}$$

Then some simple calculations show that (A.65) is equivalent to

$$\tilde{B}(z)A(z) + \tilde{A}(z)B(z) \equiv 0 \qquad (A.66)$$

Due to the assumption we can write

$$A(z) \equiv A_0(z)L(z) \qquad B(z) \equiv B_0(z)L(z)$$
$$L(z) \equiv l_0 z^n + l_1 z^{n-1} + \ldots + l_n$$

where $A_0(z)$, $B_0(z)$ are coprime and $\deg A_0 = na - n$, $\deg B_0 = nb - n$.
Thus (A.66) is equivalent to

$$\tilde{B}(z)A_0(z) \equiv -\tilde{A}(z)B_0(z)$$

Since both sides must have the same zeros, it follows that the general solution can be written as

$$\tilde{A}(z) \equiv A_0(z)M(z)$$
$$\tilde{B}(z) \equiv -B_0(z)M(z)$$

where

$$M(z) = m_1 z^{n-1} + \ldots + m_n$$

has arbitrary coefficients. This means that x lies in an n-dimensional subspace. However, this subspace is $\mathscr{N}(\mathscr{S}^T(A, B))$ (cf. (A.65)), and thus its dimension must be equal to $na + nb - \text{rank } \mathscr{S}(A, B)$ (cf. Section A.2). This proves (A.64). ∎

Corollary 1. If $A(z)$ and $B(z)$ are coprime then $\mathscr{S}(A, B)$ is nonsingular.

Proof. When $n = 0$, rank $\mathscr{S}(A, B) = na + nb$, which is equivalent to $\mathscr{S}(A, B)$ being nonsingular. ∎

Corollary 2. Consider polynomials $A(z)$ and $B(z)$ as in Definition A.8, and the following matrix of dimension $((na + nb + 2k)|(na + nb + k))$, where $k \geq 0$:

$$\bar{\mathcal{F}}(A, B) = \begin{pmatrix} a_0 & a_1 & \ldots & a_{na} & & 0 \\ & \ddots & & & \ddots & \\ 0 & & a_0 & a_1 & \ldots & a_{na} \\ b_0 & b_1 & \ldots & b_{nb} & & 0 \\ & \ddots & & & \ddots & \\ 0 & & b_0 & b_1 & \ldots & b_{nb} \end{pmatrix} \begin{matrix} \left.\vphantom{\begin{matrix}a\\a\\a\end{matrix}}\right\} nb + k \text{ rows} \\ \\ \left.\vphantom{\begin{matrix}a\\a\\a\end{matrix}}\right\} na + k \text{ rows} \end{matrix} \qquad (A.67)$$

Then

$$\text{rank } \bar{\mathcal{F}}(A, B) = na + nb + k - n \qquad (A.68)$$

where n is the number of common zeros to $A(z)$ and $B(z)$.

Proof. Introduce the polynomials $\tilde{A}(z)$ of degree $na + k$ and $\tilde{B}(z)$ of degree $nb + k$ by

$$\tilde{A}(z) = z^k A(z) \qquad \tilde{B}(z) = z^k B(z)$$

Clearly $\tilde{A}(z)$ and $\tilde{B}(z)$ have precisely $n + k$ common zeros (k of them are ocated at $z = 0$). Next note that

$$\mathcal{S}(\tilde{A}, \tilde{B}) = (\bar{\mathcal{F}}(A, B) \quad 0)$$

where $\mathcal{S}(\tilde{A}, \tilde{B})$ has dimension $((na + nb + 2k)|(na + nb + 2k))$ and 0 is a null matrix of dimension $(na + nb + 2k)|k$. Thus, from the lemma we get

$$\text{rank } \bar{\mathcal{F}}(A, B) = \text{rank } \mathcal{S}(\tilde{A}, \tilde{B}) = (na + nb + 2k) - (k + n)$$
$$= na + nb + k - n$$

which proves (A.68). ∎

A.7 Kronecker products

This section presents the definition and some properties of Kronecker products.

Definition A.9
Let A be an $(m|n)$ matrix and B an $(\bar{m}|\bar{n})$ matrix. Then the Kronecker product $A \otimes B$ is an $(m\bar{m}|n\bar{n})$ matrix defined in block form by

$$A \otimes B = \begin{pmatrix} a_{11}B & \ldots & a_{1n}B \\ \vdots & & \\ a_{m1}B & \ldots & a_{mn}B \end{pmatrix} \qquad (A.69)$$

∎

Lemma A.31
Assume that A, B, C and D are matrices of compatible dimensions. Then

$$(A \otimes B)(C \otimes D) = AC \otimes BD \tag{A.70}$$

Proof. The *ij* block of the left-hand side is given by

$$(a_{i1}B \ \ldots \ a_{in}B)\begin{pmatrix} c_{1j}D \\ \vdots \\ c_{nj}D \end{pmatrix} = \left(\sum_{k=1}^{n} a_{ik}c_{kj}\right)BD = (AC)_{ij}BD$$

which clearly also is the *ij* block of the right-hand side in (A.70). ∎

Lemma A.32
Let A and B be nonsingular matrices. Then

$$(A \otimes B)^{-1} = A^{-1} \otimes B^{-1} \tag{A.71}$$

Proof. Lemma A.31 gives directly

$$(A \otimes B)(A^{-1} \otimes B^{-1}) = AA^{-1} \otimes BB^{-1} = I \otimes I = I$$

which proves (A.71). ∎

Lemma A.33
Let A and B be two matrices. Then

$$(A \otimes B)^{\mathrm{T}} = A^{\mathrm{T}} \otimes B^{\mathrm{T}} \tag{A.72}$$

Proof. By Definition A.9,

$$(A \otimes B)^{\mathrm{T}} = \begin{pmatrix} a_{11}B & \ldots & a_{1n}B \\ \vdots & & \\ a_{m1}B & \ldots & a_{mn}B \end{pmatrix}^{\mathrm{T}} = \begin{pmatrix} (a_{11}B)^{\mathrm{T}} & \ldots & (a_{m1}B)^{\mathrm{T}} \\ \vdots & & \\ (a_{1n}B)^{\mathrm{T}} & \ldots & (a_{mn}B)^{\mathrm{T}} \end{pmatrix}$$

$$= \begin{pmatrix} a_{11}B^{\mathrm{T}} & \ldots & a_{m1}B^{\mathrm{T}} \\ \vdots & & \\ a_{1n}B^{\mathrm{T}} & \ldots & a_{mn}B^{\mathrm{T}} \end{pmatrix} = A^{\mathrm{T}} \otimes B^{\mathrm{T}}$$

∎

The following definition introduces the notation vec(A) for the vector obtained by stacking the columns of matrix A.

Definition A.10
Let A be an $(m|n)$ matrix, and let a_i denote its *i*th column.

$$A = (a_1 \ \ldots \ a_n)$$

Then the $(mn|1)$ column vector $\text{vec}(A)$ is defined as

$$\text{vec}(A) = \begin{pmatrix} a_1 \\ a_2 \\ \vdots \\ a_n \end{pmatrix}$$

∎

Lemma A.34
Let A, B and C be matrices of compatible dimensions. Then

$$\text{vec}(ABC) = (C^T \otimes A)\,\text{vec}(B) \tag{A.73}$$

Proof. Let b_i denote the ith column of B, c_i the ith column of C, and c_{ij} the i,j element of C. Then

$$\text{vec}(ABC) = \text{vec}(ABc_1 \quad ABc_2 \quad \ldots \quad ABc_n)$$

$$= \begin{pmatrix} ABc_1 \\ \vdots \\ ABc_n \end{pmatrix}$$

Further,

$$ABc_j = A(b_1 \ \ldots \ b_m)\begin{pmatrix} c_{1j} \\ \vdots \\ c_{mj} \end{pmatrix} = \sum_{i=1}^{m} c_{ij}Ab_i$$

$$= (c_{1j}A \ \ldots \ c_{mj}A)\begin{pmatrix} b_1 \\ \vdots \\ b_m \end{pmatrix}$$

$$= [c_j^T \otimes A]\,\text{vec}(B)$$

which proves (A.73). ∎

A.8 An optimization result for positive definite matrices

The following result is useful in certain optimization problems which occur, for example, when designing optimal experimental conditions.

Lemma A.35
Let S be an $(n|n)$ positive definite matrix with unit diagonal elements ($S_{ii} = 1$, $i = 1, \ldots, n$). Then

(i) $\det S^{-1}$;

Section A.8   An optimization result

(ii) $[S^{-1}]_{ii}$, $i = 1, \ldots, n$
(iii) $\lambda_{\max}(S^{-1})$

all achieve their minimum values if and only if $S = I$.

Proof. (Cf. Goodwin and Payne, 1977)

(i) Since the function $\log(\cdot)$ is monotonically increasing, the result is proved if it can be shown that

$$\log \det S = \log\left(\prod_{i=1}^{n} \lambda_i\right) = \sum_{i=1}^{n} \log \lambda_i$$

where $\{\lambda_i\}$ are the eigenvalues of S, achieves its maximum value for $S = I$. Now, for all positive λ,

$$\log \lambda \leq \lambda - 1$$

with equality if and only if $\lambda = 1$. Thus

$$\log \det S \leq \operatorname{tr} S - n = 0$$

with equality if and only if $\lambda_i = 1$, $i = 1, \ldots, n$, which proves the first assertion.

(ii) Partition S as

$$S = \begin{pmatrix} 1 & \psi^T \\ \psi & \bar{S} \end{pmatrix}$$

Then (cf. Lemma A.2)

$$[S^{-1}]_{11} = (1 - \psi^T \bar{S}^{-1} \psi)^{-1}$$

Thus

$$[S^{-1}]_{11} \geq 1$$

with equality if and only if $\psi = 0$. By repeating the above reasoning for $\bar{S}$, etc., one can prove that

$$[S^{-1}]_{ii} \geq 1, \quad i = 1, \ldots, n$$

with equality if and only if $S = I$.

(iii) It follows from (ii) that

$$\operatorname{tr} S^{-1} \geq n \text{ with equality if and only if } S = I$$

But

$$\lambda_{\max}(S^{-1}) \geq \frac{\operatorname{tr} S^{-1}}{n}$$

Thus

$$\lambda_{\max}(S^{-1}) \geq 1$$

with equality if and only if $S = I$. ∎

Bibliographical notes

Most of the material in this appendix is standard. See for example Pearson (1974), Strang (1976), Graybill (1983) or Golub and van Loan (1983) for a further discussion of matrices. The method of orthogonal triangularization (the QR method) is discussed in Stewart (1973). Its application to identification problems has been discussed, for example, by Peterka and Šmuk (1969), and Strejc (1980).

Appendix B
SOME RESULTS FROM PROBABILITY THEORY AND STATISTICS

B.1 Convergence of stochastic variables

In this section some convergence concepts for stochastic variables are presented.

Definition B.1
Let $\{x_n\}$ be an indexed sequence of stochastic variables. Let x^* be a stochastic variable. Then

- $x_n \to x^*$ (as $n \to \infty$) *with probability one* (w.p. 1) if $P(x_n \to x^*, n \to \infty) = 1$.
- $x_n \to x^*$ *in mean square* if $E[x_n - x^*]^2 \to 0$, $n \to \infty$.
- $x_n \to x^*$ *in probability* if for every $\varepsilon > 0$, $P(|x_n - x^*| > \varepsilon) \to 0$ as $n \to \infty$.
- $x_n \to x^*$ *in distribution* if $f_{x_n}(x) \to f_{x^*}(x)$ where $f_{x_n}(x)$, $f_{x^*}(x)$ denote the probability density functions of x_n and x^*, respectively. With some abuse of language it is often stated that $x_n \to f_{x^*}(x)$ in distribution. ∎

Remark. If $f_{x^*}(x)$ is the Gaussian distribution $\mathcal{N}(m, P)$ and $x_n \to x^*$ in distribution it is said that x_n is asymptotically Gaussian distributed, denoted by

$$x_n \xrightarrow{\text{dist}} \mathcal{N}(m, P)$$
∎

The following connections exist between these convergence concepts:

$x_n \to x^*$ w.p. 1
$\searrow$
$\qquad\qquad x_n \to x^*$ in probability $\to x_n \to x^*$ in distribution
$\nearrow$
$x_n \to x^*$ in mean square

There follow some ergodicity results. The following definition will be used.

Definition B.2
Let $x(t)$ be a stationary stochastic process. It is said to be *ergodic* with respect to its first- and second-order moments if

$$\frac{1}{N} \sum_{t=1}^{N} x(t) \to Ex(t)$$

$$\frac{1}{N} \sum_{t=1}^{N} x(t + \tau)x(t) \to Ex(t + \tau)x(t)$$

(B.1)

with probability one as $N \to \infty$. ∎

Ergodicity results are often used when analyzing identification methods. The reason is that many identification methods can be phrased in terms of the sample means and covariances of the data. When the number of data points tends to infinity it would then be very desirable for the analysis if one could substitute the limits by expectations as in (B.1). It will be shown that this is possible under weak assumptions. The following result is useful as a tool for establishing some ergodicity results.

Lemma B.1
Assume that $x(t)$ is a discrete time stationary process with finite variance. If the covariance function $r_x(\tau) \to 0$ as $|\tau| \to \infty$ then

$$\frac{1}{N} \sum_{t=1}^{N} x(t) \to Ex(t) \quad (N \to \infty)$$

(B.2)

with probability one and in mean square.

Proof. See, for example, Gnedenko (1963). ∎

The following lemma provides the main result on ergodicity.

Lemma B.2
Let the stationary stochastic processes $z_1(t)$ and $z_2(t)$ be given by

$$z_1(t) = G(q^{-1})e_1(t)$$
$$z_2(t) = H(q^{-1})e_2(t)$$

(B.3)

where

$$G(q^{-1}) = \sum_{i=0}^{\infty} g_i q^{-i} \quad \sum_{i=0}^{\infty} g_i^2 < \infty$$

$$H(q^{-1}) = \sum_{i=0}^{\infty} h_i q^{-i} \quad \sum_{i=0}^{\infty} h_i^2 < \infty$$

(B.4)

and

$$e(t) = \begin{pmatrix} e_1(t) \\ e_2(t) \end{pmatrix}$$

Section B.1 *Convergence of stochastic variables* 549

is zero mean white noise with covariance matrix

$$Ee(t)e^{\mathrm{T}}(t) = \begin{pmatrix} \sigma_1^2 & \varrho\sigma_1\sigma_2 \\ \varrho\sigma_1\sigma_2 & \sigma_2^2 \end{pmatrix} \quad (B.5)$$

and finite fourth moment. Then

$$\frac{1}{N}\sum_{t=1}^{N} z_1(t)z_2(t) \to Ez_1(t)z_2(t) = \varrho\sigma_1\sigma_2 \sum_{i=0}^{\infty} g_i h_i \quad (B.6)$$

as N tends to infinity, both with probability one and in mean square.

Proof. Define

$$x(t) = z_1(t)z_2(t)$$

which will be a stationary stochastic process with mean value

$$Ex(t) = Ez_1(t)z_2(t) = \sum_{i=0}^{\infty}\sum_{j=0}^{\infty} Eg_i e_1(t-i)h_j e_2(t-j) = \varrho\sigma_1\sigma_2 \sum_{i=0}^{\infty} g_i h_i$$

The convergence of the sum follows from Cauchy–Schwartz lemma since

$$\left|\sum_{i=0}^{\infty} g_i h_i\right|^2 \leq \sum_{i=0}^{\infty} g_i^2 \sum_{i=0}^{\infty} h_i^2 < \infty$$

The idea now is, of course, to apply Lemma B.1. For this purpose one must find the covariance function $r_x(\tau)$ of $x(t)$ and verify that it tends to zero as τ tends to infinity. Now

$$r_x(\tau) = E[x(t+\tau) - Ex(t)][x(t) - Ex(t)]$$
$$= E[x(t+\tau)x(t)] - [Ex(t)]^2$$
$$= E\sum_{i=0}^{\infty}\sum_{j=0}^{\infty}\sum_{k=0}^{\infty}\sum_{l=0}^{\infty} g_i h_j g_k h_l e_1(t+\tau-i)e_2(t+\tau-j)e_1(t-k)e_2(t-l)$$
$$- \left[\varrho\sigma_1\sigma_2 \sum_{i=0}^{\infty} g_i h_i\right]^2$$

However, since $e(t)$ is *white* noise,

$$Ee_1(t+\tau-i)e_2(t+\tau-j)e_1(t-k)e_2(t-l)$$
$$= \varrho^2\sigma_1^2\sigma_2^2\delta_{i,j}\delta_{k,l} + \sigma_1^2\sigma_2^2\delta_{i,k+\tau}\delta_{j,l+\tau}$$
$$+ \varrho^2\sigma_1^2\sigma_2^2\delta_{i,l+\tau}\delta_{j,k+\tau} + [\mu - (2\varrho^2 + 1)\sigma_1^2\sigma_2^2]\delta_{i,j}\delta_{k,l}\delta_{i,k+\tau}$$

where $\mu = Ee_1^2(t)e_2^2(t)$. Using these relations,

$$r_x(\tau) = \sigma_1^2\sigma_2^2\left[\sum_{k=0}^{\infty} g_k g_{k+\tau}\right]\left[\sum_{l=0}^{\infty} h_l h_{l+\tau}\right]$$

$$+ \varrho^2\sigma_1^2\sigma_2^2\left[\sum_{k=0}^{\infty} g_k h_{k+\tau}\right]\left[\sum_{l=0}^{\infty} h_l g_{l+\tau}\right]$$

$$+ [\mu - (2\varrho^2 + 1)\sigma_1^2\sigma_2^2]\left[\sum_{i=0}^{\infty} g_i h_i g_{i+\tau} h_{i+\tau}\right]$$

However, all the sums will tend to zero as $\tau \to \infty$. This can be seen by invoking the Cauchy–Schwartz lemma again. Thus

$$\left|\sum_{k=0}^{\infty} g_k g_{k+\tau}\right|^2 \leq \sum_{k=0}^{\infty} g_k^2 \sum_{j=0}^{\infty} g_{j+\tau}^2 = \sum_{k=0}^{\infty} g_k^2 \sum_{j=\tau}^{\infty} g_j^2 \to 0, \ \tau \to \infty$$

which proves that $\sum_{k=0}^{\infty} g_k g_{k+\tau} \to 0$ as τ tends to infinity. In similar ways it can be proved that the remaining sums also tend to zero.

It has thus been established that $r_x(\tau) \to 0$, $|\tau| \to \infty$. Then the result (B.6) follows from Lemma B.1. ∎

The next result is a variant of the central limit theorem due to Ljung (1977c).

Lemma B.3
Consider

$$X_N = \frac{1}{\sqrt{N}} \sum_{t=1}^{N} z(t) \tag{B.7}$$

where $z(t)$ is a (vector-valued) zero mean stationary process given by

$$z(t) = \phi(t)v(t) \tag{B.8}$$

In (B.8), $\phi(t)$ is a matrix and $v(t)$ a vector. The entries of $\phi(t)$ and $v(t)$ are stationary, possibly correlated, ARMA processes with zero means and underlying white noise sequences with finite fourth-order moments. The elements of $\phi(t)$ may also contain a bounded deterministic term.

Assume that the limit

$$P = \lim_{N \to \infty} E X_N X_N^T \tag{B.9}$$

exists and is nonsingular. Then X_N is asymptotically Gaussian distributed,

$$X_N \xrightarrow{dist} \mathcal{N}(0, P) \tag{B.10}$$

Proof. See Ljung (1977c). ∎

The following result on convergence in distribution will often be useful as a complement to the above lemma.

Lemma B.4

Let $\{x_n\}$ be a sequence of random variables that converges in distribution to $F(x)$. Let $\{A_n\}$ be a sequence of random square matrices that converges in probability to a nonsingular matrix A, and $\{b_n\}$ a sequence of random vectors that converges in probability to b. Define

$$y_n = A_n x_n + b_n \tag{B.11}$$

Then y_n converges in distribution to $F(A^{-1}(y - b))$.

Proof. The lemma is a trivial extension to the multivariable case of the scalar result, given for example by Chung (1968, p. 85) and Cramér (1946, p. 245). ∎

The lemma can be specialized to the case when $F(x)$ corresponds to a Gaussian distribution, as follows.

Corollary. Assume that x_n is asymptotically Gaussian distributed $\mathcal{N}(0, P)$. Then y_n as given by (B.11) converges in distribution to $\mathcal{N}(b, APA^T)$.

Proof. Let $m = \dim x_n = \dim y_n$. The limiting distribution function of y_n is given by (cf. Lemma B.4 and Definition B.3 in the next section)

$$G(y) = \int_{-\infty}^{x} \frac{1}{(2\pi)^{m/2}(\det(P))^{1/2}} \exp\left[-\frac{1}{2}x'^T P^{-1} x'\right] dx' \bigg|_{x = A^{-1}(y-b)}$$

$$= \frac{1}{\det A} \int_{-\infty}^{y} \frac{1}{(2\pi)^{m/2}(\det(P))^{1/2}}$$

$$\times \exp\left[-\frac{1}{2}\{[A^{-1}(y' - b)]^T P^{-1} A^{-1}(y' - b)\}\right] dy'$$

$$= \int_{-\infty}^{y} \frac{1}{(2\pi)^{m/2}[\det(APA^T)]^{1/2}} \exp\left[-\frac{1}{2}(y' - b)^T (APA^T)^{-1}(y' - b)\right] dy'$$

Thus $G(y)$ is the distribution function of $\mathcal{N}(b, APA^T)$. ∎

In (B.11) the new sequence y_n is an affine transformation of x_n. For *rational* functions there is another result that concerns convergence in probability. It is given in the following and is often referred to as Slutsky's lemma.

Lemma B.5

Let $\{x_n\}$ be a sequence of random vectors that converges in probability to a constant x. Let $f(\cdot)$ be a rational function and suppose that $f(x)$ is finite. Then $f(x_n)$ converges in probability to $f(x)$.

Proof. See Cramér (1946, p. 254). ∎

B.2 The Gaussian and some related distributions

Definition B.3
The n-dimensional stochastic variable x is said to be Gaussian (or equivalently normal) distributed, denoted $x \sim \mathcal{N}(m, P)$, where m is a vector and P a positive definite matrix, if its probability density function is

$$f(x) = \frac{1}{(2\pi)^{n/2}(\det P)^{1/2}} \exp\left[-\frac{1}{2}(x-m)^T P^{-1}(x-m)\right] \tag{B.12}$$

∎

The following lemma shows that the parameters m and P are the mean and the covariance matrix, respectively.

Lemma B.6
Assume that $x \sim \mathcal{N}(m, P)$. Then

$$Ex = m \qquad E(x-m)(x-m)^T = P \tag{B.13}$$

Proof. In the integrals below the variable substitution $y = P^{-1/2}(x-m)$ is made.

$$Ex = \int_{\mathcal{R}^n} xf(x)dx = \frac{1}{(2\pi)^{n/2}(\det P)^{1/2}} \int_{\mathcal{R}^n} [m + P^{1/2}y] \exp\left[-\frac{1}{2}y^T y\right](\det P)^{1/2}dy$$

$$= m\frac{1}{(2\pi)^{n/2}} \int_{\mathcal{R}^n} \exp\left[-\frac{1}{2}y^T y\right]dy + \frac{1}{(2\pi)^{n/2}} P^{1/2} \int_{\mathcal{R}^n} y \exp\left[-\frac{1}{2}y^T y\right]dy = m$$

The second integral above is zero by symmetry. (This can be seen by making the variable substitution $y \to -y$). Also

$$E(x-m)(x-m)^T = \int_{\mathcal{R}^n} (x-m)(x-m)^T f(x)dx$$

$$= \int_{\mathcal{R}^n} P^{1/2}yy^T P^{1/2} \frac{1}{(2\pi)^{n/2}} \exp\left[-\frac{1}{2}y^T y\right]dy \triangleq P^{1/2}QP^{1/2}$$

For $k \neq l$

$$Q_{kl} = \int_{\mathcal{R}^n} y_k y_l \frac{1}{(2\pi)^{n/2}} \exp\left[-\frac{1}{2}y^T y\right]dy$$

$$= \int_{\mathcal{R}} y_k \frac{1}{(2\pi)^{1/2}} \exp\left[-\frac{1}{2}y_k^2\right]dy_k \int_{\mathcal{R}} y_l \frac{1}{(2\pi)^{1/2}} \exp\left[-\frac{1}{2}y_l^2\right]dy_l$$

$$\times \left[\prod_{\substack{j=1 \\ j \neq k, l}}^{n} \int_{\mathcal{R}} \frac{1}{(2\pi)^{1/2}} \exp\left[-\frac{1}{2}y_j^2\right]dy_j\right] = 0$$

Section B.2 Gaussian and related distributions 553

while the diagonal elements of Q are given by

$$Q_{kk} = \int_{\mathcal{R}^n} y_k^2 \frac{1}{(2\pi)^{n/2}} \exp\left[-\frac{1}{2} y^T y\right] dy$$

$$= \int_{\mathcal{R}} y_k^2 \frac{1}{(2\pi)^{1/2}} \exp\left[-\frac{1}{2} y_k^2\right] dy_k \left[\prod_{\substack{j=1 \\ j \neq k}}^{n} \int_{\mathcal{R}} \frac{1}{(2\pi)^{1/2}} \exp\left[-\frac{1}{2} y_j^2\right] dy_j\right]$$

The first integral on the right-hand side gives the variance of a $\mathcal{N}(0, 1)$ distributed random variable, while the other integrands are precisely the pdf of a $\mathcal{N}(0, 1)$ distributed random variable. Hence $Q_{kk} = 1$, which implies $Q = I$, and the stated result (B.13) follows. ∎

Remark. An alternative, and shorter proof, is given after Lemma B.10. ∎

To prove that an affine transformation of Gaussian variables gives a Gaussian variable, it is convenient to use moment generating functions, which are defined in the following.

Definition B.4
If x is a (vector-valued) stochastic variable its moment generating function is defined by

$$\varphi(z) = E[e^{z^T x}] = \int_{\mathcal{R}^n} e^{z^T x} f(x) \, dx \qquad (B.14)$$

∎

Remark 1. The moment generating function evaluated at $z = i\omega$ is the Fourier transform of the probability density function. From uniqueness results on Fourier transforms it follows that every probability density function has a unique moment generating function. Thus, if two random variables have the same moment generating function, they will also have the same probability density function. ∎

Remark 2. The name 'moment generating function' is easy to explain for scalar processes. The expected value $E[x^k]$ is called the kth-order moment. By series expansion, (B.14) gives

$$\varphi(z) = E\left[\sum_{k=0}^{\infty} \frac{z^k}{k!} x^k\right] = \sum_{k=0}^{\infty} \frac{z^k}{k!} E[x^k] \qquad (B.15)$$

Thus

$$\left.\frac{\partial^k \varphi(z)}{\partial z^k}\right|_{z=0} = E[x^k].$$

∎

Remark 3. More general moments can conveniently be found by series expansion of $\varphi(z)$, similarly to (B.15). For example, assume $Ex_1 x_2^2$ is sought. Set $z = (z_1 \ z_2)^T$. Then

$$\varphi(z) = E\left[\sum_{k=0}^{\infty} \frac{1}{k!}(z^T x)^k\right]$$

$$= E\left[1 + (z_1 x_1 + z_2 x_2) + \frac{1}{2!}(z_1 x_1 + z_2 x_2)^2 + \frac{1}{3!}(z_1 x_1 + z_2 x_2)^3 + \cdots\right]$$

The term of current interest in the series expansion is $E\left[\frac{1}{3!}3(z_1 x_1)(z_2 x_2)^2\right]$ $= \frac{1}{2} z_1 z_2^2 E(x_1 x_2^2)$. Hence the moment $E x_1 x_2^2$ can be found as twice the coefficient of $z_1 z_2^2$ in the series expansion of $\varphi(z)$. As a further example, consider $E x_1 x_2 x_3 x_4$ and set $z = (z_1 \ z_2 \ z_3 \ z_4)^T$. Then

$$\varphi(z) = E\left[\cdots + \frac{1}{4!}(z_1 x_1 + z_2 x_2 + z_3 x_3 + z_4 x_4)^4 + \cdots\right]$$

$$= \cdots + z_1 z_2 z_3 z_4 E x_1 x_2 x_3 x_4 + \cdots$$

Hence the moment $E x_1 x_2 x_3 x_4$ can be calculated as the coefficient of $z_1 z_2 z_3 z_4$ in the series expansion of $\varphi(z)$. ∎

The following lemma gives the moment generating function for a Gaussian random variable.

Lemma B.7
Assume that $x \sim \mathcal{N}(m, P)$. Then its moment generating function is

$$\varphi(z) = \exp\left[z^T m + \frac{1}{2} z^T P z\right] \tag{B.16}$$

Proof. By direct calculation, using (B.12),

$$\varphi(z) = \int_{\mathcal{R}^n} \exp[z^T x] f(x) \, dx$$

$$= \int_{\mathcal{R}^n} \frac{1}{(2\pi)^{n/2} (\det P)^{1/2}} \exp\left[-\frac{1}{2}(x - m - Pz)^T P^{-1}(x - m - Pz)\right] dx$$

$$\times \exp\left[z^T m + \frac{1}{2} z^T P z\right]$$

$$= \exp\left[z^T m + \frac{1}{2} z^T P z\right]$$

since the integrand is the probability density function of $\mathcal{N}(m + Pz, P)$. ∎

Lemma B.8
Assume that $x \sim \mathcal{N}(m, P)$ and set $y = Ax + b$ for constant A and b of appropriate dimensions. Then $y \sim \mathcal{N}(Am + b, APA^T)$.

Section B.2   *Gaussian and related distributions*

Proof. This lemma is conveniently proved by using moment generating functions (mgfs). Let $\varphi_x(z_1)$ and $\varphi_y(z_2)$ be the mgfs associated with x and y, respectively. Then

$$\varphi_y(z) = E[\exp[z^T(Ax + b)]] = \exp[z^T b] E[\exp[(A^T z)^T x]]$$
$$= \exp[z^T b] \varphi_x(A^T z)$$
$$= \exp[z^T b] \exp\left[z^T A m + \frac{1}{2} z^T A P A^T z\right]$$
$$= \exp\left[z^T(A m + b) + \frac{1}{2} z^T (A P A^T) z\right]$$

By the unique correspondence between mgfs and probability density functions, it follows from Lemma B.7 that $y \sim \mathcal{N}(Am + b, APA^T)$. ∎

Lemma B.9
Let the scalar random variables x_1, x_2, x_3, x_4 be jointly Gaussian with zero mean values. Then

$$Ex_1 x_2 x_3 x_4 = Ex_1 x_2 Ex_3 x_4 + Ex_1 x_3 Ex_2 x_4 + Ex_1 x_4 Ex_2 x_3 \qquad \text{(B.17a)}$$

Proof. Let $\mathscr{C}[f(z)]$ denote the coefficient of $z_1 z_2 z_3 z_4$ in the series expansion of some function $f(z)$. According to Remark 3 of Definition B.4, the moment $Ex_1 x_2 x_3 x_4$ is $\mathscr{C}[\varphi(z)]$. Set $P_{ij} = Ex_i x_j$. Then from Lemma B.7

$$Ex_1 x_2 x_3 x_4 = \mathscr{C}[\varphi(z)] = \mathscr{C}\left[\frac{1}{2!}\left(\frac{1}{2} z^T P z\right)^2\right]$$
$$= \frac{1}{8} \mathscr{C}[z^T P z z^T P z] = P_{12} P_{34} + P_{13} P_{24} + P_{14} P_{23} \qquad ∎$$

Corollary. Let x_1, x_2, x_3, x_4 be jointly Gaussian with mean values $m_i = Ex_i$. Then

$$Ex_1 x_2 x_3 x_4 = Ex_1 x_2 Ex_3 x_4 + Ex_1 x_3 Ex_2 x_4 + Ex_1 x_4 Ex_2 x_3 - 2 m_1 m_2 m_3 m_4 \qquad \text{(B.17b)}$$

Proof. Set $y_i = x_i - m_i$ and $P_{ij} = Ey_i y_j$. Since $Ey_i = 0$, it follows that $Ey_i y_j y_k = 0$. Hence

$$Ex_1 x_2 x_3 x_4 = E(y_1 + m_1)(y_2 + m_2)(y_3 + m_3)(y_4 + m_4)$$
$$= Ey_1 y_2 y_3 y_4 + E[y_1 m_2 m_3 m_4 + y_2 m_1 m_3 m_4 + y_3 m_1 m_2 m_4 + y_4 m_1 m_2 m_3]$$
$$+ E[y_1 y_2 m_3 m_4 + y_1 y_3 m_2 m_4 + y_1 y_4 m_2 m_3 + y_2 y_3 m_1 m_4 + y_2 y_4 m_1 m_3$$
$$+ y_3 y_4 m_1 m_2] + E[y_1 y_2 y_3 m_4 + y_1 y_2 y_4 m_3 + y_1 y_3 y_4 m_2 + y_2 y_3 y_4 m_1]$$
$$+ m_1 m_2 m_3 m_4$$
$$= (P_{12} P_{34} + P_{13} P_{24} + P_{14} P_{23}) + (P_{12} m_3 m_4 + P_{13} m_2 m_4 + P_{14} m_2 m_3$$
$$+ P_{23} m_1 m_4 + P_{24} m_1 m_3 + P_{34} m_1 m_2) + m_1 m_2 m_3 m_4$$

$$= (P_{12} + m_1 m_2)(P_{34} + m_3 m_4) + (P_{13} + m_1 m_3)(P_{24} + m_2 m_4)$$
$$+ (P_{14} + m_1 m_4)(P_{23} + m_2 m_3) - 2m_1 m_2 m_3 m_4$$
$$= Ex_1 x_2 Ex_3 x_4 + Ex_1 x_3 Ex_2 x_4 + Ex_1 x_4 Ex_2 x_3 - 2m_1 m_2 m_3 m_4 \qquad \blacksquare$$

Remark. An extension of Lemma B.9 and its corollary to the case of complex matrix-valued Gaussian random variables $\{x_i\}_{i=1}^4$ has been presented by Janssen and Stoica (1987). $\blacksquare$

Lemma B.10
Uncorrelated Gaussian random variables are independent.

Proof. Let x be a vector of uncorrelated Gaussian random variables. Then the covariance matrix P of x is a diagonal matrix. The probability density function of x will be (cf. (B.12))

$$f(x) = \frac{1}{(2\pi)^{n/2} \left[\prod_{i=1}^n P_{ii}\right]^{1/2}} \exp\left[-\frac{1}{2} \sum_{i=1}^n \frac{(x_i - m_i)^2}{P_{ii}}\right]$$

$$= \prod_{i=1}^n \frac{1}{(2\pi P_{ii})^{1/2}} \exp\left[-\frac{1}{2} \frac{(x_i - m_i)^2}{P_{ii}}\right] = \prod_{i=1}^n f_i(x_i)$$

where $f_i(x_i)$ is the probability density function for the variable x_i. This proves the lemma. $\blacksquare$

Remark 1. Note that a different proof of the above lemma can be found in Section B.6 (see Corollary to Lemma B.17). $\blacksquare$

Remark 2. An alternative proof of Lemma B.6 can now be made as follows. Set $y = P^{-1/2}(x - m)$. By Lemma B.8, $y \sim \mathcal{N}(0, I)$. Hence by Lemma B.10, $y_i \sim \mathcal{N}(0, 1)$ and different y_i:s are independent. Since $Ey_i = 0$, $\text{var}(y_i) = 1$, it holds that $Ey = 0$, $\text{cov}(y) = I$. Finally since $x = m + P^{1/2}y$ it follows that $Ex = m$, $\text{cov}(x) = P$. $\blacksquare$

The next definition introduces the χ^2 distribution.

Definition B.5
Let $\{y_i\}_{i=1}^n$ be uncorrelated $\mathcal{N}(0, 1)$ variables. Then $x = \sum_{i=1}^n y_i^2$ is said to be χ^2 distributed with n degrees of freedom, written as $x \sim \chi^2(n)$. $\blacksquare$

Lemma B.11
Let $x \sim \chi^2(n)$. Then its probability density function is

$$f(x) = k e^{-x/2} x^{(n/2 - 1)} \qquad (B.18)$$

where k is a normalizing constant.

Proof. Set $x = y^T y$ where $y \sim \mathcal{N}(0, I)$ is of dimension n. Then $f(x)$ can be calculated as follows

$$f(x) = \frac{d}{dx} P(y^T y \leq x) = \frac{d}{dx} \int_{y^T y \leq x} \frac{1}{(2\pi)^{n/2}} \exp\left[-\frac{1}{2} y^T y\right] dy$$

$$= \frac{d}{dx} \int_{r \leq \sqrt{x}} \frac{1}{(2\pi)^{n/2}} \exp\left[-\frac{1}{2} r^2\right] r^{n-1} dr S_n$$

where S_n is the area of the unit sphere in $\mathcal{R}^n$. Continuing the calculations,

$$f(x) = \frac{S_n}{(2\pi)^{n/2}} \frac{d}{dx} \int_0^{\sqrt{x}} \exp\left[-\frac{1}{2} r^2\right] r^{n-1} dr$$

$$= \frac{S_n}{(2\pi)^{n/2}} \frac{1}{2\sqrt{x}} \exp\left[-\frac{1}{2} x\right] x^{(n-1)/2}$$

This proves (B.18). ∎

Lemma B.12
Let $x \sim \mathcal{N}(m, P)$ be of dimension n. Then

$$(x - m)^T P^{-1} (x - m) \sim \chi^2(n) \tag{B.19}$$

Proof. Set $z = P^{-1/2}(x - m)$. According to Lemma B.8, $z \sim \mathcal{N}(0, I)$. Furthermore, $(x - m)^T P^{-1}(x - m) = z^T z$ and (B.19) follows from Definition B.5. ∎

Lemma B.13
Let x be an n-dimensional Gaussian vector, $x \sim \mathcal{N}(0, I)$ and let P be an $(n|n)$-dimensional idempotent matrix of rank r. Then $x^T P x$ is $\chi^2(r)$ distributed.

Proof. By Lemma A.28 one can write

$$P = U^T \Lambda U$$

with U orthogonal and

$$\Lambda = \begin{pmatrix} I_r & 0 \\ 0 & 0 \end{pmatrix}$$

Now set $y = Ux$. Clearly $y \sim \mathcal{N}(0, I)$. Furthermore,

$$x^T P x = x^T U^T \Lambda U x = y^T \Lambda y = \sum_{i=1}^r y_i^2$$

and the lemma follows from the definition of the χ^2 distribution. ∎

The next definition introduces the F distribution.

Definition B.6
Let $x_1 \sim \chi^2(n_1)$, $x_2 \sim \chi^2(n_2)$ be independent. Then $(x_1/n_1)/(x_2/n_2)$ is said to be F distributed with n_1 and n_2 degrees of freedom, written as

$$\frac{x_1}{n_1} \frac{n_2}{x_2} \sim F(n_1, n_2)$$ ∎

Lemma B.14
Let $y \sim F(n_1, n_2)$. Then

$$n_1 y \xrightarrow{\text{dist}} \chi^2(n_1) \quad \text{as } n_2 \to \infty \tag{B.20}$$

Proof. Set

$$y = \frac{x_1}{n_1} \frac{n_2}{x_2}$$

where x_1, x_2, n_1 and n_2 are as in Definition B.6. According to Lemma B.2, $x_2/n_2 \to 1$ as $n_2 \to \infty$, with probability one. Hence, according to Lemma B.4, in the limit, $n_1 y \xrightarrow{\text{dist}} x_1 \sim \chi^2(n_1)$. ∎

Remark. The consequence of (B.20) is that for large values of n_2, the $F(n_1, n_2)$ distribution can be approximated with the $\chi^2(n_1)$ distribution, normalized by n_1. ∎

This section ends with some numerical values pertaining to the χ^2 distribution. Define $\chi^2_\alpha(n)$ by

$$P(x > \chi^2_\alpha(n)) = \alpha \quad \text{where } x \sim \chi^2(n)$$

This quantity is useful when testing statistical hypotheses (see Chapters 4, 11 and 12). Table B.1 gives some numerical values of $\chi^2_\alpha(n)$ as a function of α and n.

As a rule of thumb one may approximately take

$$\chi^2_{0.05}(n) \approx n + 1.645\sqrt{(2n)}$$

It is not difficult to justify this approximation. According to the central limit theorem, the χ^2 distributed random variable $x = \sum_{i=1}^n y_i^2$ (see Definition B.5) is approximately (for large enough n) Gaussian distributed with mean

$$Ex = \sum_{i=1}^n Ey_i^2 = n$$

and variance

$$E(x - n)^2 = E \sum_{i=1}^n \sum_{j=1}^n y_i^2 y_j^2 - n^2$$

$$= \sum_{\substack{i=1 \\ i \neq j}}^n \sum_{j=1}^n (Ey_i^2)(Ey_j^2) + \sum_{i=1}^n Ey_i^4 - n^2 = n(n-1) + 3n - n^2 = 2n$$

TABLE B.1 Numerical values of $\chi_\alpha^2(n)$

α	0.05	0.025	0.01	0.005
n				
1	3.84	5.02	6.63	7.88
2	5.99	7.38	9.21	10.6
3	7.81	9.35	11.3	12.8
4	9.49	11.1	13.3	14.9
5	11.1	12.8	15.1	16.7
6	12.6	14.4	16.8	18.5
7	14.1	16.0	18.5	20.3
8	15.5	17.5	20.1	22.0
9	16.9	19.0	21.7	23.6
10	18.3	20.5	23.2	25.2
11	19.7	21.9	24.7	26.8
12	21.0	23.3	26.2	28.3
13	22.4	24.7	27.7	29.8
14	23.7	26.1	29.1	31.3
15	25.0	27.5	30.6	32.8
16	26.3	28.8	32.0	34.3
17	27.6	30.2	33.4	35.7
18	28.9	31.5	34.8	37.2
19	30.1	32.9	36.2	38.6
20	31.4	34.2	37.6	40.0

Thus for sufficiently large n

$$\chi^2(n) \sim \mathcal{N}(n, 2n)$$

from which the above approximation of $\chi_{0.05}^2(n)$ readily follows. Approximation formulas for $\chi_\alpha^2(n)$ for other values of α can be obtained similarly.

B.3 Maximum a posteriori and maximum likelihood parameter estimates

Let x denote the vector of observations of a stochastic variable and let $p(x, \theta)$ denote the probability density function (pdf) of x. The form of $p(x, \theta)$ is assumed to be known. The vector θ of unknown parameters, which completely describes the pdf, is to be estimated.

The maximum *a posteriori* (MAP) approach to θ-estimation treats θ as a random vector while the sample of observations x is considered to be given.

Definition B.7
The MAP estimate of θ is

$$\hat{\theta}_{MAP} = \arg \max_\theta p(\theta|x) \tag{B.21}$$

The conditional pdf $p(\theta|x)$ is called the *a posteriori* pdf of θ. ∎

From Bayes rule,

$$p(\theta|x) = \frac{p(x|\theta)p(\theta)}{p(x)} \tag{B.22}$$

The conditional pdf $p(x|\theta)$, with x fixed, is called the likelihood function. It can be interpreted as giving a measure of the plausibility of the data under different parameters. To evaluate $p(\theta|x)$ we also need the 'prior pdf' of the parameters, $p(\theta)$. While $p(x|\theta)$ can be derived relatively easily for a variety of situations, the choice of $p(\theta)$ is a controversial topic of the MAP approach to estimation (see e.g. Peterka, 1981, for a discussion). Finally, the pdf $p(x)$ which also occurs in (B.22) may be evaluated as a marginal distribution

$$p(x) = \int_{\mathcal{R}^{\dim\theta}} p(x, \theta)d\theta \qquad p(x, \theta) = p(x|\theta)p(\theta)$$

Note that while $p(x)$ is needed to evaluate $p(\theta|x)$, it is not necessary for determining $\hat{\theta}_{MAP}$ (since only the numerator of (B.22) depends on θ).

The maximum likelihood (ML) approach to θ-estimation is conceptually different from the MAP approach. Now x is treated as a random variable and θ as unknown but fixed parameters.

Definition B.8
The ML estimate of θ is

$$\hat{\theta}_{ML} = \arg\max_{\theta} p(x|\theta) \tag{B.23}$$

∎

Thus within the ML approach the value of θ is chosen which makes the data most plausible as measured by the likelihood function. Note that in cases where the prior pdf has a small influence on the *a posteriori* pdf, one may expect that $\hat{\theta}_{ML}$ is close to $\hat{\theta}_{MAP}$. In fact, for noninformative priors, i.e. for $p(\theta) = $ constant, it follows from (B.22) that $\hat{\theta}_{ML} = \hat{\theta}_{MAP}$.

B.4 The Cramér–Rao lower bound

There is a famous lower bound on the covariance matrix of unbiased estimates. This is known as the Cramér–Rao lower bound.

Lemma B.15
Let x be a stochastic vector-valued variable, the distribution of which depends on an unknown vector θ. Let $L(x, \theta)$ denote the likelihood function, and let $\hat{\theta} = \hat{\theta}(x)$ be an arbitrary unbiased estimate of θ determined from x. Then

$$\text{cov}(\hat{\theta}) \geq \left[E\left(\frac{\partial \log L}{\partial \theta}\right)^T \frac{\partial \log L}{\partial \theta}\right]^{-1} = -\left[E \frac{\partial^2 \log L}{\partial \theta^2}\right]^{-1} \tag{B.24}$$

Proof. As $L(x, \theta)$ is a probability density function,

$$1 = \int L(x, \theta) dx \tag{B.25}$$

(integration over $\mathscr{R}^n$, $n = \dim x$). The assumption on unbiasedness can be written as

$$\theta = \int \hat{\theta}(x) L(x, \theta) dx \tag{B.26}$$

Differentiation of (B.25) and (B.26) with respect to θ gives

$$0 = \int \frac{\partial L}{\partial \theta}(x, \theta) dx = \int \frac{\partial \log L(x, \theta)}{\partial \theta} L(x, \theta) dx = E \frac{\partial \log L(x, \theta)}{\partial \theta} \tag{B.27}$$

$$I = \int \hat{\theta}(x) \frac{\partial L}{\partial \theta}(x, \theta) dx = \int \hat{\theta}(x) \frac{\partial \log L(x, \theta)}{\partial \theta} L(x, \theta) dx = E \hat{\theta}(x) \frac{\partial \log L(x, \theta)}{\partial \theta}$$

It follows in particular from these two expressions that

$$I = E[\hat{\theta}(x) - \theta] \frac{\partial \log L(x, \theta)}{\partial \theta} \tag{B.28}$$

Now the matrix

$$E \left(\begin{array}{c} \hat{\theta}(x) - \theta \\ \left(\frac{\partial \log L}{\partial \theta} \right)^T \end{array} \right) \left((\hat{\theta}(x) - \theta)^T \quad \frac{\partial \log L}{\partial \theta} \right)$$

is by construction nonnegative definite. This implies that (see Lemma A.4)

$$E[\hat{\theta}(x) - \theta][\hat{\theta}(x) - \theta]^T - \left\{ E[\hat{\theta}(x) - \theta] \frac{\partial \log L}{\partial \theta} \right\}$$
$$\times \left\{ E \left(\frac{\partial \log L}{\partial \theta} \right)^T \left(\frac{\partial \log L}{\partial \theta} \right) \right\}^{-1} \left\{ E \left(\frac{\partial \log L}{\partial \theta} \right)^T [\hat{\theta}(x) - \theta]^T \right\} \geq 0$$

which can be rewritten using (B.28) as

$$E[\hat{\theta}(x) - \theta][\hat{\theta}(x) - \theta]^T \geq \left[E \left(\frac{\partial \log L}{\partial \theta} \right)^T \frac{\partial \log L}{\partial \theta} \right]^{-1}$$

which is precisely the inequality in (B.24).

It remains to show the equality in (B.24). Differentiation of (B.27) gives

$$0 = \int \frac{\partial^2 \log L(x, \theta)}{\partial \theta^2} L(x, \theta) dx + \int \left(\frac{\partial \log L(x, \theta)}{\partial \theta} \right)^T \left(\frac{\partial \log L(x, \theta)}{\partial \theta} \right) L(x, \theta) dx$$

or

$$-E \frac{\partial^2 \log L}{\partial \theta^2} = E \left(\frac{\partial \log L}{\partial \theta} \right)^T \left(\frac{\partial \log L}{\partial \theta} \right)$$

which proves the equality in (B.24). ∎

Remark 1. The matrix

$$J = E\left[\left(\frac{\partial \log L}{\partial \theta}\right)^{\mathrm{T}}\left(\frac{\partial \log L}{\partial \theta}\right)\right] \tag{B.29}$$

is called the *(Fisher) information matrix*. The inequality (B.24) can be written

$$\mathrm{cov}(\hat{\theta}) \geq J^{-1}$$

The right-hand side of (B.24) is called the *Cramér–Rao lower bound*. ∎

Remark 2. If the estimate $\hat{\theta}$ is biased, a similar result applies. Assume that

$$E\hat{\theta} = \gamma(\theta) \tag{B.30}$$

Then

$$\mathrm{cov}(\hat{\theta}) \geq \left[\frac{\partial \gamma(\theta)}{\partial \theta}\right] J^{-1} \left[\frac{\partial \gamma(\theta)}{\partial \theta}\right]^{\mathrm{T}} \tag{B.31}$$

The equality in (B.24) is still applicable. This result is easily obtained by generalizing the proof of Lemma B.15. When doing this it is necessary to substitute (B.26) by

$$\gamma(\theta) = \int \hat{\theta}(x) L(x, \theta) dx$$

and (B.28) by

$$\frac{\partial \gamma(\theta)}{\partial \theta} = E[\hat{\theta}(x) - \theta] \frac{\partial \log L(x, \theta)}{\partial \theta}$$

∎

Remark 3. An estimate for which equality holds in (B.24) is said to be *(statistically) efficient*. ∎

For illustration the following examples apply the lemma to some linear regression problems.

Example B.1 *Cramér–Rao lower bound for a linear regression with uncorrelated residuals*
Consider the linear regression equation given by

$$Y = \Phi\theta + e \tag{B.32}$$

where e is Gaussian distributed with zero mean and covariance matrix $\lambda^2 I$. Let N be the dimension of Y. Assume that θ and λ^2 are unknown. The Cramér–Rao lower bound can be calculated for any unbiased estimates of these quantities.

In this case

Section B.4 *The Cramér–Rao lower bound*

$$L(Y, \theta, \lambda^2) = \frac{1}{(2\pi)^{N/2}(\det \lambda^2 I_N)^{1/2}} \exp\left[-\frac{1}{2}(Y - \Phi\theta)^T(\lambda^2 I_N)^{-1}(Y - \Phi\theta)\right]$$

where I_N is the identity matrix of order N. Hence

$$\log L(Y, \theta, \lambda^2) = -\frac{1}{2\lambda^2}(Y - \Phi\theta)^T(Y - \Phi\theta) - \frac{N}{2}\log 2\pi - \frac{N}{2}\log \lambda^2$$

and the following expressions are obtained for the derivatives:

$$\frac{\partial}{\partial \theta}\log L(Y, \theta, \lambda^2) = \frac{1}{\lambda^2}(Y - \Phi\theta)^T\Phi$$

$$\frac{\partial}{\partial \lambda^2}\log L(Y, \theta, \lambda^2) = \frac{1}{2\lambda^4}(Y - \Phi\theta)^T(Y - \Phi\theta) - \frac{N}{2\lambda^2}$$

$$\frac{\partial^2}{\partial \theta^2}\log L(Y, \theta, \lambda^2) = -\frac{1}{\lambda^2}\Phi^T\Phi$$

$$\frac{\partial}{\partial \theta}\frac{\partial}{\partial \lambda^2}\log L(Y, \theta, \lambda^2) = \frac{1}{\lambda^4}(Y - \Phi\theta)^T\Phi$$

$$\frac{\partial^2}{\partial (\lambda^2)^2}\log L(Y, \theta, \lambda^2) = -\frac{1}{\lambda^6}(Y - \Phi\theta)^T(Y - \Phi\theta) + \frac{N}{2\lambda^4}$$

Thus

$$J = E\begin{pmatrix} -\dfrac{\partial^2}{\partial \theta^2}\log L(Y, \theta, \lambda^2) & -\dfrac{\partial}{\partial \lambda^2}\dfrac{\partial}{\partial \theta}\log L(Y, \theta, \lambda^2) \\ -\dfrac{\partial}{\partial \theta}\dfrac{\partial}{\partial \lambda^2}\log L(Y, \theta, \lambda^2) & -\dfrac{\partial^2}{\partial (\lambda^2)^2}\log L(Y, \theta, \lambda^2) \end{pmatrix}$$

$$= \begin{pmatrix} \dfrac{1}{\lambda^2}\Phi^T\Phi & 0 \\ 0 & \dfrac{N}{2\lambda^4} \end{pmatrix}$$

Therefore for all unbiased estimates,

$$\text{cov}(\hat{\theta}) \geq \lambda^2(\Phi^T\Phi)^{-1}$$

$$\text{var}(\hat{\lambda}^2) \geq \frac{2\lambda^4}{N}$$

It can now be seen that the least squares estimate (4.7) is efficient. In fact, note from the above expression for $\partial \log L/\partial \theta$ that under the stated assumptions the LS estimate of θ coincides with the ML estimate.

Consider the unbiased estimate s^2 of λ^2 as introduced in Lemma 4.2,

$$s^2 = \frac{e^T P e}{N - n} \qquad P = I - \Phi(\Phi^T\Phi)^{-1}\Phi^T$$

where $n = \dim \theta$. The variance of s^2 can be calculated as follows.
$$\text{var}(s^2) = E[s^2 - \lambda^2]^2 = E[(s^2)^2] - \lambda^4$$
Noting that e is Gaussian and P symmetric and idempotent,
$$(N - n)^2 E(s^2)^2 = E \sum_{i=1}^{N} \sum_{j=1}^{N} \sum_{k=1}^{N} \sum_{l=1}^{N} e_i P_{ij} e_j e_k P_{kl} e_l$$
$$= \sum_i \sum_j \sum_k \sum_l P_{ij} P_{kl} [\delta_{ij}\delta_{kl} + \delta_{ik}\delta_{jl} + \delta_{il}\delta_{jk}]\lambda^4$$
$$= \left[\sum_i P_{ii} \sum_k P_{kk} + 2 \sum_i \sum_k P_{ik} P_{ki} \right]\lambda^4$$
$$= [(\text{tr } P)^2 + 2 \text{ tr } P]\lambda^4$$

Since by Example A.2 $\text{tr } P = N - n$, it follows that
$$\text{var}(s^2) = \frac{(N - n)^2 + 2(N - n)}{(N - n)^2}\lambda^4 - \lambda^4 = \frac{2\lambda^4}{N - n}$$

This is slightly larger than the Cramér–Rao lower bound. The ratio is $[\lambda^4/(N - n)]/[\lambda^4/N] = N/(N - n)$, which becomes close to one for large N. Thus the estimate s^2 is *asymptotically efficient*. ∎

Example B.2 *Cramér–Rao lower bound for a linear regression with correlated residuals*
Consider now the more general case of a linear regression equation with correlated errors,
$$Y = \Phi\theta + e \quad e \sim \mathcal{N}(0, R) \text{ with } R \text{ known} \tag{B.33}$$
The aim is to find the Cramér–Rao lower bound on the covariance matrix of any unbiased estimator of θ. Simple calculations give
$$L(Y, \theta) = \frac{1}{(2\pi)^{N/2}(\det R)^{1/2}} \exp\left[-\frac{1}{2}(Y - \Phi\theta)^T R^{-1}(Y - \Phi\theta)\right]$$
$$\log L(Y, \theta) = -\frac{1}{2}(Y - \Phi\theta)^T R^{-1}(Y - \Phi\theta) - \frac{N}{2}\log 2\pi - \frac{1}{2}\log(\det R)$$
$$\frac{\partial}{\partial \theta}\log L(Y, \theta) = (Y - \Phi\theta)^T R^{-1}\Phi$$
$$\frac{\partial^2}{\partial \theta^2}\log L(Y, \theta) = -\Phi^T R^{-1}\Phi$$

Then (B.24) implies that for any unbiased estimate $\hat\theta$
$$\text{cov}(\hat\theta) \geq (\Phi^T R^{-1}\Phi)^{-1}$$

Thus the estimate given by (4.17), (4.18) is not only BLUE but also the best *nonlinear* unbiased estimate. The price paid for this much stronger result is that it is necessary to assume that the disturbance e is Gaussian distributed. It is interesting to note from the expression above for $\partial \log L(Y, \theta)/\partial\theta$ that under the stated assumptions the ML estimate of θ is identical to the Markov estimate (4.17), (4.18). ∎

B.5 Minimum variance estimation

Let x and y be two random vectors which are jointly distributed. The minimum (conditional) variance estimate (MVE) of x given that y occurred is defined as follows. Let

$$Q(\bar{x}) \triangleq E[(x - \bar{x})(x - \bar{x})^T | y] \tag{B.34}$$

where $\bar{x}$ is some function of y. One can regard $Q(\bar{x})$ as a matrix-valued measure of the extent of fluctuations of x around $\bar{x} = \bar{x}(y)$. The following lemma shows how to choose $\bar{x}$ so as to 'minimize' $Q(\bar{x})$.

Lemma B.16
Consider the criterion $Q(\bar{x})$ and let $\hat{x}$ denote the conditional mean

$$\hat{x} = L[x|y] \tag{B.35}$$

Then $\hat{x}$ is the MVE of x in the sense that

$$Q(\hat{x}) \leq Q(\bar{x}) \quad \text{for all } \bar{x} \tag{B.36}$$

Proof. Since $E[\bar{x}|y] = \bar{x}$,

$$Q(\bar{x}) = E[xx^T + \bar{x}\bar{x}^T - \bar{x}x^T - x\bar{x}^T | y] = E[xx^T|y] + \bar{x}\bar{x}^T - \bar{x}\hat{x}^T - \hat{x}\bar{x}^T$$

$$= (\bar{x} - \hat{x})(\bar{x} - \hat{x})^T + E[(x - \hat{x})(x - \hat{x})^T | y] = (\bar{x} - \hat{x})(\bar{x} - \hat{x})^T + Q(\hat{x}) \geq Q(\hat{x})$$

with equality if and only if $\bar{x} = \hat{x}$. ∎

It follows from (B.36) that the conditional expectation $\hat{x}$ minimizes any scalar-valued monotonically nondecreasing function of Q, such as $\operatorname{tr} Q$ or $\det Q$. Moreover, it can be shown that under certain conditions on the distribution $p(x|y)$, $\hat{x}$ minimizes many other loss functions of the form $E[f(x - \bar{x})|y]$, for some scalar-valued function $f(\cdot)$ (Meditch, 1969). There are also, however, 'reasonable' loss functions which are not minimized by $\hat{x}$. To illustrate this fact, let $\dim x = 1$ and define

$$\tilde{x} = \arg \min_{\bar{x}} E[|x - \bar{x}| | y]$$

The loss function above can be written as

$$F(\bar{x}) \triangleq E[|x - \bar{x}| | y] = \int_{-\infty}^{\infty} |x - \bar{x}| p(x|y) dx$$

$$= \int_{-\infty}^{\bar{x}} (\bar{x} - x) p(x|y) dx + \int_{\bar{x}}^{\infty} (x - \bar{x}) p(x|y) dx$$

Now recall the following formula for the derivative of integrals whose limits and integrand depend on some parameter y:

$$\frac{d}{dy}\left\{\int_{u(y)}^{v(y)} f(x, y)dx\right\} = \int_{u(y)}^{v(y)} \frac{\partial f(x, y)}{\partial y} dx + v'(y)f(v(y), y) - u'(y)f(u(y), y)$$

which applies under weak conditions. Thus,

$$F'(\bar{x}) = \int_{-\infty}^{\bar{x}} p(x|y)dx - \int_{\bar{x}}^{\infty} p(x|y)dx$$

$$F''(\bar{x}) = 2p(\bar{x}|y)$$

from which it follows that $\bar{x}$ is uniquely defined by

$$\int_{-\infty}^{\bar{x}} p(x|y)dx = \int_{\bar{x}}^{\infty} p(x|y)dx$$

Thus, $\bar{x}$ is equal to the median of the conditional pdf $p(x|y)$. For nonsymmetric pdfs the median and the mean will in general differ.

B.6 Conditional Gaussian distributions

The following lemma characterizes conditional distributions for jointly Gaussian random variables.

Lemma B.17
Let x and y be two random Gaussian vectors, with

$$\begin{pmatrix} x \\ y \end{pmatrix} \sim \mathcal{N}\left[\begin{pmatrix} \bar{x} \\ \bar{y} \end{pmatrix}, R\right]$$

where

$$R = \begin{pmatrix} R_x & R_{xy} \\ R_{yx} & R_y \end{pmatrix} > 0$$

Then the conditional distribution of x given y, $p(x|y)$, is Gaussian with mean

$$\hat{x} = E[x|y] = \bar{x} + R_{xy}R_y^{-1}(y - \bar{y}) \tag{B.37}$$

and covariance matrix

$$\hat{P} = E[(x - \hat{x})(x - \hat{x})^T|y] = R_x - R_{xy}R_y^{-1}R_{yx} \tag{B.38}$$

Proof. By assumption

$$p(x, y) = \frac{1}{(2\pi)^{(nx+ny)/2}(\det R)^{1/2}} \\ \times \exp\left[-\frac{1}{2}((x-\bar{x})^T \; (y-\bar{y})^T)R^{-1}\begin{pmatrix} x - \bar{x} \\ y - \bar{y} \end{pmatrix}\right] \quad \text{(B.39)}$$

Consider the readily verified identity

$$\begin{pmatrix} I - R_{xy}R_y^{-1} \\ 0 & I \end{pmatrix} R \begin{pmatrix} I & 0 \\ -R_y^{-1}R_{yx} & I \end{pmatrix} = \begin{pmatrix} R_x - R_{xy}R_y^{-1}R_{yx} & 0 \\ 0 & R_y \end{pmatrix} \quad \text{(B.40)}$$

It follows from (B.40) that

$$\det R = \det R_y \det \hat{P} \quad \text{(B.41)}$$

and that

$$((x-\bar{x})^T \; (y-\bar{y})^T) R^{-1} \begin{pmatrix} x - \bar{x} \\ y - \bar{y} \end{pmatrix}$$
$$= ((x-\bar{x})^T \; (y-\bar{y})^T)\begin{pmatrix} I & 0 \\ -R_y^{-1}R_{yx} & I \end{pmatrix}$$
$$\times \begin{pmatrix} \hat{P}^{-1} & 0 \\ 0 & R_y^{-1} \end{pmatrix}\begin{pmatrix} I & -R_{xy}R_y^{-1} \\ 0 & I \end{pmatrix}\begin{pmatrix} x - \bar{x} \\ y - \bar{y} \end{pmatrix} \quad \text{(B.42)}$$
$$= ((x-\hat{x})^T \; (y-\bar{y})^T)\begin{pmatrix} \hat{P}^{-1} & 0 \\ 0 & R_y^{-1} \end{pmatrix}\begin{pmatrix} x - \hat{x} \\ y - \bar{y} \end{pmatrix}$$

Note that here (B.37) has been used as a definition of $\hat{x}$. Inserting (B.41) and (B.42) into (B.39) gives

$$p(x, y) = \frac{1}{(2\pi)^{ny/2}(\det R_y)^{1/2}} \exp\left[-\frac{1}{2}(y-\bar{y})^T R_y^{-1}(y-\bar{y})\right] \\ \times \frac{1}{(2\pi)^{nx/2}(\det \hat{P})^{1/2}} \exp\left[-\frac{1}{2}(x-\hat{x})^T \hat{P}^{-1}(x-\hat{x})\right] \quad \text{(B.43)}$$

Since the first factor in (B.43) is recognized as $p(y)$, it follows that the second must be $p(x|y)$. However, this factor is easily recognized as $\mathcal{N}(\hat{x}, \hat{P})$. This observation concludes the proof. ■

Corollary. Uncorrelated Gaussian random vectors are independent

Proof. Assume that x and y are uncorrelated. Then $R_{xy} = 0$. From the lemma $\hat{x} = \bar{x}$, $\hat{P} = R_x$ and from (B.43) $p(x, y) = p(y)p(x)$ which proves the independence. ■

It follows from (B.37) and Section B.5 that under the Gaussian hypothesis the MVE of x given that y has occurred, $\hat{x} = E[x|y]$, is an *affine* function of y. This is not necessarily true if x and y are not jointly Gaussian distributed. Thus, under the Gaussian hypothesis the *linear* MVE (or the BLUE; see Chapter 4) is identical to the *general* MVE. As a simple illustration of this fact the results of Complement C4.2 will be derived again using the formula for conditional expectation introduced above. For this purpose, reformulate the problem of Complement C4.2 to include it in the class of problems considered above. Let θ and y be two random vectors related by

$$\theta = \bar{\theta} + \varepsilon_1$$

$$y = \Phi\theta + \varepsilon_2$$

where $\bar{\theta}$ and Φ are given, and where $(\varepsilon_1^T \;\; \varepsilon_2^T)^T$ has zero mean and covariance matrix given by

$$\begin{pmatrix} P & 0 \\ 0 & S \end{pmatrix}$$

To determine the MVE $\hat{\theta} = E[\theta|y]$ of θ, assume further that θ and y are Gaussian distributed. Thus

$$\begin{pmatrix} \theta \\ y \end{pmatrix} \sim \mathcal{N}\left[\begin{pmatrix} \bar{\theta} \\ \Phi\bar{\theta} \end{pmatrix}, \begin{pmatrix} P & P\Phi^T \\ \Phi P & \Phi P\Phi^T + S \end{pmatrix} \right]$$

It follows from Lemma B.16 and (B.37) that the MVE of θ is given by

$$\hat{\theta} = E[\theta|y] = \bar{\theta} + P\Phi^T(\Phi P\Phi^T + S)^{-1}(y - \Phi\bar{\theta})$$

and its covariance matrix by

$$\hat{P} = P - P\Phi^T(S + \Phi P\Phi^T)^{-1}\Phi P$$

These are precisely the results (C4.2.4), (C4.2.5) obtained using the BLUE theory.

As a further application of the fact that the BLUE and MVE coincide under the Gaussian hypothesis, consider the following simple derivation of the expression of the BLUE for linear combinations of unknown parameters. Let θ denote the unknown parameter vector (which is regarded as a random vector) and let Y denote the vector of data from which information about θ is to be obtained. The aim is to determine the BLUE of $A\theta$, where A is a constant matrix. Since for any distribution of data

$$\text{MVE}(A\theta) = E[A\theta|Y] = AE[\theta|Y] = A\,\text{MVE}(\theta)$$

it follows immediately that

$$\text{BLUE}(A\theta) = A\,\text{BLUE}(\theta) \tag{B.44}$$

This property is used in the following section when deriving the Kalman–Bucy filter equations.

B.7 The Kalman–Bucy filter

Consider the following state space linear equations:

$$x_{k+1} = A_k x_k + w_k \tag{B.45a}$$

$$y_k = C_k x_k + v_k \tag{B.45b}$$

where A_k, C_k are (nonrandom) matrices of appropriate dimensions, and

$$Ew_k w_p^T = Q_k \delta_{k,p}$$
$$Ev_k v_p^T = R_k \delta_{k,p} \tag{B.45c}$$
$$Ew_k v_p^T = 0$$

Let $\hat{x}_k$ denote an estimate of x_k, with

$$E\hat{x}_k = x_k$$
$$E(\hat{x}_k - x_k)(\hat{x}_k - x_k)^T = P_k$$

Furthermore, let $\hat{x}_k$ depend linearly on the observations up to the time instant $k - 1$. Then, from (B.45c)

$$E(\hat{x}_k - x_k)v_k^T = 0$$

First consider the problem of determining the BLUE of x_k given the new measurement y_k. Denote the BLUE by $\tilde{x}_k$. This problem is exactly of the type treated in Complement C4.2 and reanalyzed at the end of Section B.6. Thus, it follows for example from (C4.2.4), or (B.37) that $\tilde{x}_k$ is given by

$$\tilde{x}_k = \hat{x}_k + P_k C_k^T (R_k + C_k P_k C_k^T)^{-1}(y_k - C_k \hat{x}_k)$$

and its covariance is

$$\tilde{P}_k = P_k - P_k C_k^T (R_k + C_k P_k C_k^T)^{-1} C_k P_k$$

Next consider the problem of determining the BLUE of x_{k+1}. Since w_k is uncorrelated with $\{y_k, y_{k-1}, \ldots\}$ and, therefore, with $(\tilde{x}_k - x_k)$, it readily follows from (B.44) and the state equation (B.45a) that the BLUE of x_{k+1} is given by

$$\hat{x}_{k+1} = A_k \tilde{x}_k$$

and the corresponding covariance matrix of the estimation errors by

$$P_{k+1} = A_k \tilde{P}_k A_k^T + Q_k$$

Thus, the BLUE $\hat{x}_k$ of x_k and its covariance matrix P_k obey the following recursive equations, called the Kalman–Bucy filter:

$$\hat{x}_{k+1} = A_k \hat{x}_k + A_k \Gamma_k (y_k - C_k \hat{x}_k)$$
$$\Gamma_k = P_k C_k^T (R_k + C_k P_k C_k^T)^{-1}$$
$$P_{k+1} = A_k (P_k - \Gamma_k C_k P_k) A_k^T + Q_k$$

Finally, recall from the discussion at the end of the previous section that under the Gaussian hypothesis the BLUE is identical to the general MVE.

B.8 Asymptotic covariance matrices for sample correlation and covariance estimates

Let $y(t)$ be a zero-mean stationary process, $r_k = Ey(t)y(t+k)$ and $\varrho_k = r_k/r_0$ denote its covariance and correlation at lag k, respectively, and let

$$\hat{r}_k = \frac{1}{N} \sum_{t=1}^{N-k} y(t)y(t+k) \quad (N = \text{number of data points})$$

$$\hat{\varrho}_k = \hat{r}_k/\hat{r}_0$$

be sample estimators of r_k and ϱ_k. Assume that r_k (and ϱ_k) converge exponentially to zero as $k \to \infty$ (which is not a restrictive assumption once the stationarity of $y(t)$ is accepted). Since many identification techniques (such as the related methods of correlation, least squares and instrumental variables) can be seen as maps {sample covariances or correlations} $\to$ {parameter estimates}, it is important to know the properties of the sample estimators $\{\hat{r}_k\}$ and $\{\hat{\varrho}_k\}$ (in particular their accuracy).

This section will derive the asymptotic variances–covariances

$$\sigma_{kp} \triangleq \lim_{N \to \infty} NE(\hat{r}_k - r_k)(\hat{r}_p - r_p) \quad 0 \leq k, p < \infty$$

From $\{\sigma_{kp}\}$ one can directly obtain the variances–covariances of the sample correlations

$$s_{kp} \triangleq \lim_{N \to \infty} NE(\hat{\varrho}_k - \varrho_k)(\hat{\varrho}_p - \varrho_p) \quad 0 \leq k, p < \infty$$

Indeed, since for large N

$$\hat{\varrho}_k - \varrho_k = \frac{\hat{r}_k}{\hat{r}_0} - \frac{r_k}{r_0} = \frac{\hat{r}_k - r_k}{r_0} - \frac{\hat{r}_k}{\hat{r}_0} \frac{\hat{r}_0 - r_0}{r_0} \approx \frac{\hat{r}_k - r_k}{r_0} - \varrho_k \frac{\hat{r}_0 - r_0}{r_0}$$

it follows that

$$s_{kp} = \frac{1}{r_0^2} (\sigma_{kp} - \varrho_p \sigma_{k0} - \varrho_k \sigma_{0p} + \varrho_k \varrho_p \sigma_{00}) \tag{B.46}$$

In the following paragraphs formulas will be derived for σ_{kp} and s_{kp}, firstly under the hypothesis that $y(t)$ is Gaussian distributed. Next the Gaussian hypothesis will be relaxed but $y(t)$ will be restricted to the class of linear processes. The formula for σ_{kp} will change slightly, while the formula for s_{kp} will remain unchanged. For ARMA processes (which are special cases of linear processes) compact formulas will be given for σ_{kp} and s_{kp}. Finally, some examples will show how the explicit expressions for $\{\sigma_{kp}\}$ and $\{s_{kp}\}$ can be used to assess the accuracy of $\{\hat{r}_k, \hat{\varrho}_k\}$ (which may be quite poor in some cases) and also to establish the accuracy properties of some estimation methods based on $\{\hat{r}_k\}$ or $\{\hat{\varrho}_k\}$.

Gaussian processes

Under the Gaussian assumption it holds by Lemma B.9 that

$$Ey(t)y(t+k)y(s)y(s+p) = r_k r_p + r_{t-s} r_{t-s+k-p} + r_{t-s-p} r_{t-s+k} \tag{B.47}$$

Section B.8   *Asymptotic covariance matrices*   571

Then, for N much larger than k and p,

$$NE(\hat{r}_k - r_k)(\hat{r}_p - r_p) \approx \frac{1}{N} \sum_{t=1}^{N-k} \sum_{s=1}^{N-p} E[y(t)y(t+k) - r_k][y(s)y(s+p) - r_p]$$

$$\approx \frac{1}{N} \sum_{t=1}^{N} \sum_{s=1}^{N} [r_{t-s}r_{t-s+k-p} + r_{t-s-p}r_{t-s+k}]$$

$$= \frac{1}{N} \sum_{\tau=-N}^{N} (N - |\tau|)(r_\tau r_{\tau+k-p} + r_{\tau-p} r_{\tau+k}) \quad (B.48)$$

where the neglected terms tend to zero as $N \to \infty$. Since r_τ converges exponentially to zero as $\tau \to \infty$, it follows that the term in (B.48) which contains $|\tau|$ tends to zero as $N \to \infty$ (cf. the calculations in Appendix A8.1). Thus,

$$\sigma_{kp} = \sum_{\tau=-\infty}^{\infty} (r_\tau r_{\tau+k-p} + r_{\tau-p} r_{\tau+k}) \quad (B.49)$$

Inserting (B.49) into (B.46) gives

$$s_{kp} = \sum_{\tau=-\infty}^{\infty} (\varrho_\tau \varrho_{\tau+k-p} + \varrho_{\tau-p} \varrho_{\tau+k} - 2\varrho_k \varrho_\tau \varrho_{\tau-p}$$

$$-2\varrho_p \varrho_\tau \varrho_{\tau+k} + 2\varrho_k \varrho_p \varrho_\tau^2) \quad (B.50)$$

The expressions (B.49) and (B.50) for σ_{kp} and s_{kp} are known as Bartlett's formulas.

Linear processes

Let $y(t)$ be the linear process

$$y(t) = \sum_{k=0}^{\infty} h_k e(t-k) \quad (B.51a)$$

where $\{e(t)\}$ is a sequence of independent random variables with

$$Ee(t) = 0 \quad\quad Ee(t)e(s) = \lambda^2 \delta_{t,s} \quad (B.51b)$$

Assume that h_k converges exponentially to zero as $k \to \infty$. Since

$$r_k = \sum_{i=0}^{\infty} \sum_{j=0}^{\infty} h_i h_j Ee(t-i)e(t+k-j) = \lambda^2 \sum_{i=0}^{\infty} h_i h_{i+k} \quad (B.52)$$

the assumption on $\{h_k\}$ guarantees that r_k converges exponentially to zero as $k \to \infty$. No assumption is made in the following about the distribution of $y(t)$.

A key result in establishing (B.49) (and (B.50)) was the identity (B.47) for Gaussian random variables. For non-Gaussian variables, (B.47) does not necessarily hold. In the present case,

$$\Delta \triangleq Ey(t)y(t+k)y(s)y(s+p)$$

$$= \sum_{i=0}^{\infty}\sum_{j=0}^{\infty}\sum_{l=0}^{\infty}\sum_{m=0}^{\infty} h_i h_j h_l h_m Ee(t-i)e(t+k-j)e(s-l)e(s+p-m) \quad \text{(B.53)}$$

It can be readily verified that

$$Ee(t-i)e(t+k-j)e(s-l)e(s+p-m)$$
$$= \lambda^4(\delta_{i,j-k}\delta_{l,m-p} + \delta_{t-i,s-l}\delta_{t+k-j,s+p-m} + \delta_{t-i,s+p-m}\delta_{t+k-j,s-l}) \quad \text{(B.54)}$$
$$+ (\mu - 3\lambda^4)\delta_{i,j-k}\delta_{l,m-p}\delta_{t-i,s-l}$$

where $\mu = Ee^4(t)$. Inserting (B.54) into (B.53),

$$\Delta = r_k r_p + r_{t-s}r_{t-s+k-p} + r_{t-s-p}r_{t-s+k} + (\mu - 3\lambda^4)\gamma_{s,t} \quad \text{(B.55a)}$$

where

$$\gamma_{s,t} = \sum_{i=0}^{\infty} h_i h_{i+k} h_{s-t+i} h_{s-t+i+p} \quad \text{(B.55b)}$$

and where $h_k = 0$ for $k < 0$ (by convention). Note that (B.55) reduces to (B.47) since for Gaussian variables $\mu = 3\lambda^4$. From (B.53)–(B.55) it follows similarly to (B.48), (B.49) that

$$\sigma_{kp} = \sum_{\tau=-\infty}^{\infty} (r_\tau r_{\tau+k-p} + r_{\tau-p}r_{\tau+k})$$
$$+ (\mu - 3\lambda^4) \lim_{N\to\infty} \frac{1}{N} \sum_{t=1}^{N}\sum_{s=1}^{N} \gamma_{s,t} \quad \text{(B.56)}$$

Next note that

$$\frac{1}{N}\sum_{t=1}^{N}\sum_{s=1}^{N} \gamma_{s,t} = \frac{1}{N}\sum_{\tau=-N}^{N} (N - |\tau|) \sum_{i=0}^{\infty} h_i h_{i+k} h_{\tau+i} h_{\tau+i+p} \quad \text{(B.57)}$$

By assumption there exist constants $0 \leq c < \infty$ and $0 \leq \alpha < 1$ such that

$$|h_k| \leq c\alpha^k \quad \text{for } k \geq 0$$

Thus

$$\sum_{\tau=-\infty}^{\infty} |\tau| \sum_{i=0}^{\infty} |h_i h_{i+k} h_{\tau+i} h_{\tau+i+p}|$$

$$= \sum_{\tau=-\infty}^{0} |\tau| \sum_{i=-\tau}^{\infty} |h_i h_{i+k} h_{\tau+i} h_{\tau+i+p}|$$

$$+ \sum_{\tau=1}^{\infty} |\tau| \sum_{i=0}^{\infty} |h_i h_{i+k} h_{\tau+i} h_{\tau+i+p}|$$

$$\leq \mathrm{const} \left[\sum_{\tau=-\infty}^{0} |\tau| \alpha^{-2\tau} + \sum_{\tau=1}^{\infty} |\tau| \alpha^{2\tau} \right]$$

$$= \mathrm{const} \sum_{\tau=-\infty}^{\infty} |\tau| \alpha^{2|\tau|} < \infty$$

This implies that the term in (B.57) which contains $|\tau|$, tends to zero as N tends to infinity. It follows that

$$\lim_{N \to \infty} \frac{1}{N} \sum_{t=1}^{N} \sum_{s=1}^{N} \gamma_{s,t} = \sum_{\tau=-\infty}^{\infty} \sum_{i=-\infty}^{\infty} h_i h_{i+k} h_{\tau+i} h_{\tau+i+p}$$
$$= \sum_{i=-\infty}^{\infty} \sum_{j=-\infty}^{\infty} (h_i h_{i+k})(h_j h_{j+p}) = \frac{1}{\lambda^4} r_k r_p \qquad (B.58)$$

Inserting (B.58) into (B.56),

$$\sigma_{kp} = \sum_{\tau=-\infty}^{\infty} (r_\tau r_{\tau+k-p} + r_{\tau-p} r_{\tau+k}) + (\mu/\lambda^4 - 3) r_k r_p \qquad (B.59)$$

which 'slightly' differs from the formula (B.49) obtained under the Gaussian hypothesis.

Next observe that the second term in (B.59) has no effect in the formula (B.46) for s_{kp}. Indeed,

$$r_k r_p - \varrho_p r_k r_0 - \varrho_k r_0 r_p + \varrho_k \varrho_p r_0^2 = 0$$

Thus the formula (B.50) for s_{kp} continues to hold in the case of possibly non-Gaussian linear processes.

ARMA processes

From the standpoint of applications, the expressions (B.49) or (B.59) for σ_{kp} are not very convenient due to the infinite sums which in practical computations must be truncated. For ARMA processes a more compact formula for σ_{kp} can be derived, which is more convenient than (B.59) for computations. Therefore, assume that $y(t)$ is an ARMA process given by (B.51), where

$$\sum_{k=0}^{\infty} h_k q^{-k} = \frac{C(q^{-1})}{A(q^{-1})} \qquad (B.60)$$

with $C(q^{-1})$ and $A(q^{-1})$ monic polynomials in q^{-1}. Define

$$\beta_k = \beta_{-k} = \sum_{\tau=-\infty}^{\infty} r_\tau r_{\tau+k}$$

Using $\{\beta_k\}$ one can write (B.59) as

$$\sigma_{kp} = \beta_{k-p} + \beta_{k+p} + (\mu/\lambda^4 - 3) r_k r_p \qquad (B.61)$$

Next note that

$$\sum_{k=-\infty}^{\infty} \beta_k z^{-k} = \sum_{\tau=-\infty}^{\infty} r_\tau \sum_{k=-\infty}^{\infty} r_{\tau+k} z^{-k} = \left(\sum_{\tau=-\infty}^{\infty} r_\tau z^\tau\right)\left(\sum_{p=-\infty}^{\infty} r_p z^{-p}\right) \quad \text{(B.62a)}$$
$$= (2\pi)^2 \phi^2(z)$$

where

$$\phi(z) \triangleq \frac{1}{2\pi} \sum_{\tau=-\infty}^{\infty} r_\tau z^{-\tau} = \frac{\lambda^2}{2\pi} \frac{C(z)C(z^{-1})}{A(z)A(z^{-1})} \quad \text{(B.62b)}$$

is the spectral density function of $y(t)$. It follows from (B.62) that $\{\beta_k\}$ are the covariances of the ARMA process

$$x(t) = \lambda \frac{C^2(q^{-1})}{A^2(q^{-1})} e(t)$$

Thus, to compute $\{\sigma_{kp}\}$ by (B.61) it is necessary to evaluate the covariances of the ARMA processes $y(t)$ and $x(t)$. Computation of ARMA covariances is a standard topic, and simple and efficient algorithms for doing this exist (see, for example, Complement C7.7). Note that no truncation or other kind of approximation is needed when using (B.61) to evaluate $\{\sigma_{kp}\}$.

The following two examples illustrate the way in which the theory developed in this section may be useful.

Example B.3 *Variance of $\hat{r}_k$ for a first-order AR process*
Let $y(t)$ be a Gaussian first-order autoregression

$$y(t) = \frac{1}{1 - aq^{-1}} e(t) \qquad |a| < 1$$

where to simplify the notation it is assumed that the variance of the white noise sequence $\{e(t)\}$ is given by

$$Ee^2(t) = (1 - a^2)$$

This in turn implies that the covariances of $y(t)$ are simply given by

$$r_k = a^{|k|} \qquad k = 0, \pm 1, \ldots$$

It is required to calculate the variance of $\hat{r}_k$, which is given by (cf. (B.61))

$$\sigma_{kk} = \beta_0 + \beta_{2k}$$

Thus $\{\beta_k\}$ must be evaluated. Now

$$\beta_k = Ex(t)x(t+k) = \frac{(1-a^2)^2}{2\pi i} \oint \frac{z^k}{(1 - az^{-1})^2(1 - az)^2} \frac{dz}{z}$$
$$= \frac{(1-a^2)^2}{2\pi i} \oint \frac{z^{k+1}}{(z-a)^2(1-az)^2} dz$$

Section B.8 — Asymptotic covariance matrices

$$= (1 - a^2)^2 \frac{(k+1)z^k(1-az) + 2az^{k+1}}{(1-az)^3}\bigg|_{z=a}$$

$$= a^k \frac{(k+1) - (k-1)a^2}{1-a^2}$$

$$= a^k\left(k + \frac{1+a^2}{1-a^2}\right)$$

which gives

$$\sigma_{kk} = (1 + a^{2k})\frac{1+a^2}{1-a^2} + 2ka^{2k}$$

It can be seen that for $|a|$ close to 1, σ_{kk} may take very large values. This implies that the sample covariances will have quite a poor accuracy for borderline stationary autoregressions. ∎

Example B.4 *Variance of $\hat{\varrho}_1$ for a first-order AR process*

Consider the same first-order autoregressive process as in the previous example. The least squares (or Yule–Walker) estimate of the parameter a (see, for example, Complement C8.1) is given by

$$\hat{a} = \frac{\hat{r}_1}{\hat{r}_0} = \hat{\varrho}_1$$

The asymptotic normalized variance of $\hat{a}$ follows from the general theory developed in Chapters 7 and 8. It is given by

$$\mathrm{var}(\hat{a}) = Ee^2(t)/r_0 = (1 - a^2) \tag{B.63}$$

The above expression for $\mathrm{var}(\hat{a})$ will now be derived using the results of this section. From (B.46),

$$\mathrm{var}(\hat{a}) = \mathrm{var}(\hat{\varrho}_1) = (\sigma_{11} - 2\varrho_1\sigma_{10} + \varrho_1^2\sigma_{00})/r_0^2 \tag{B.64}$$

where (see Example B.3)

$$r_0 = 1 \quad \varrho_1 = a$$

$$\sigma_{00} = 2\beta_0 = 2\frac{1+a^2}{1-a^2}$$

$$\sigma_{11} = \frac{1 + 4a^2 - a^4}{1 - a^2} \tag{B.65}$$

$$\sigma_{10} = 2\beta_1 = 2a\left(1 + \frac{1+a^2}{1-a^2}\right) = \frac{4a}{1-a^2}$$

Inserting the expressions (B.65) into (B.64),

$$\mathrm{var}(\hat{a}) = \frac{1 + 4a^2 - a^4 - 8a^2 + 2a^2(1+a^2)}{1-a^2} = 1 - a^2$$

which agrees with (B.63). It is interesting to note that while the estimates $\hat{r}_0$ and $\hat{r}_1$ have poor accuracy for $|a|$ close to 1, the estimate $\hat{a}$ obtained from $\hat{r}_0$ and $\hat{r}_1$ has quite a good accuracy. The reason is that the estimates $\hat{r}_0$ and $\hat{r}_1$ become highly correlated when $|a| \to 1$. ∎

B.9 Accuracy of Monte Carlo analysis

For many parametric identification methods the parameter estimates are asymptotically Gaussian distributed. In the text such a property was established for prediction error methods (see (7.59)), and for instrumental variable methods (see (8.29)). In this section it will be shown how the mean and the covariance matrix of a Gaussian distribution can be estimated and what accuracy these estimates will have.

Assume that m simulation runs have been made, for the same experimental condition, the same number of samples, but m independent realizations of the noise sequence. The resulting parameter estimates will be denoted by $\hat{\theta}^i$, $i = 1, \ldots, m$. According to the discussion above it can be assumed that $\hat{\theta}^i$ are independent and Gaussian distributed:

$$\hat{\theta}^i \sim \mathcal{N}(\theta_0, P/N) \quad i = 1, \ldots, m \tag{B.66}$$

where N denotes the number of samples. Natural estimates of θ_0 and P can be formed as

$$\bar{\theta} = \frac{1}{m} \sum_{i=1}^{m} \hat{\theta}^i \tag{B.67a}$$

$$\hat{P} = \alpha_m \sum_{i=1}^{m} N(\hat{\theta}^i - \bar{\theta})(\hat{\theta}^i - \bar{\theta})^T \tag{B.67b}$$

The estimate $\bar{\theta}$ is precisely the arithmetic mean. In (B.67b) α_m is a scale factor, which typically is taken as $1/m$ or $1/(m-1)$ (see below).

According to (B.66) it follows directly that

$$\bar{\theta} \sim \mathcal{N}\left(\theta_0, \frac{P}{mN}\right) \tag{B.68}$$

which also shows the mean and the covariance matrix of the estimate $\bar{\theta}$.

The analysis of $\hat{P}$ requires longer calculations. The mean value of $\hat{P}$ is straightforward to find:

$$E\hat{P} = \alpha_m N \sum_{i=1}^{m} E[\hat{\theta}^i \hat{\theta}^{iT} - \hat{\theta}^i \bar{\theta}^T - \bar{\theta}\hat{\theta}^{iT} + \bar{\theta}\bar{\theta}^T]$$

$$= \alpha_m N \sum_{i=1}^{m} (\theta_0 \theta_0^T + P/N) - \alpha_m N \sum_{i=1}^{m} E\hat{\theta}^i \frac{1}{m} \sum_{j=1}^{m} \hat{\theta}^{jT}$$

$$- \alpha_m N \sum_{i=1}^{m} E \frac{1}{m} \sum_{j=1}^{m} \hat{\theta}^j \hat{\theta}^{iT} + \alpha_m N m E \bar{\theta}\bar{\theta}^T$$

Section B.9 *Accuracy of Monte Carlo analysis* 577

$$= \alpha_m Nm(\theta_0\theta_0^T + P/N) - \alpha_m \frac{N}{m} \sum_{i=1}^{m} E\hat{\theta}^i \left(\hat{\theta}^{iT} + \sum_{\substack{j=1 \\ j \neq i}}^{m} \hat{\theta}^{jT} \right)$$

$$- \alpha_m \frac{N}{m} \sum_{i=1}^{m} E\left(\hat{\theta}^i + \sum_{\substack{j=1 \\ j \neq i}}^{m} \hat{\theta}^j \right) \hat{\theta}^{iT} + \alpha_m Nm(\theta_0\theta_0^T + P/(Nm))$$

$$= \alpha_m m(N\theta_0\theta_0^T + P) - 2\alpha_m N[\theta_0\theta_0^T + P/N + (m-1)\theta_0\theta_0^T]$$
$$+ \alpha_m m(N\theta_0\theta_0^T + P/m)$$

which gives

$$E\hat{P} = \alpha_m(m-1)P \tag{B.69}$$

The estimate $\hat{P}$ is hence unbiased if α_m is chosen as

$$\alpha_m = \frac{1}{m-1} \tag{B.70}$$

which is a classical result when estimating an unknown variance.

The mean square error of $\hat{P}$ will be derived using Lemma B.9. This mean square error will be analyzed componentwise. For this purpose the following expression will be calculated:

$$E[\hat{P}_{ij} - P_{ij}][\hat{P}_{kl} - P_{kl}] = E\hat{P}_{ij}\hat{P}_{kl} + P_{ij}P_{kl}\{1 - 2\alpha_m(m-1)\} \tag{B.71}$$

In particular, the choice $k = i$, $l = j$ will give the mean square error of the element $\hat{P}_{ij}$. To evaluate (B.71) one obviously has to determine $E\hat{P}_{ij}\hat{P}_{kl}$. From the definition (B.67b) and Lemma B.9,

$$E\hat{P}_{ij}\hat{P}_{kl} = \alpha_m^2 N^2 E \sum_{\mu=1}^{m} \sum_{\nu=1}^{m} (\hat{\theta}_i^\mu - \bar{\theta}_i)(\hat{\theta}_j^\mu - \bar{\theta}_j)(\hat{\theta}_k^\nu - \bar{\theta}_k)(\hat{\theta}_l^\nu - \bar{\theta}_l)$$

$$= \alpha_m^2 N^2 \sum_{\mu=1}^{m} \sum_{\nu=1}^{m} \{[E(\hat{\theta}_i^\mu - \bar{\theta}_i)(\hat{\theta}_j^\mu - \bar{\theta}_j)][E(\hat{\theta}_k^\nu - \bar{\theta}_k)(\hat{\theta}_l^\nu - \bar{\theta}_l)] \tag{B.72}$$

$$+ [E(\hat{\theta}_i^\mu - \bar{\theta}_i)(\hat{\theta}_k^\nu - \bar{\theta}_k)][E(\hat{\theta}_j^\mu - \bar{\theta}_j)(\hat{\theta}_l^\nu - \bar{\theta}_l)]$$
$$+ [E(\hat{\theta}_i^\mu - \bar{\theta}_i)(\hat{\theta}_l^\nu - \bar{\theta}_l)][E(\hat{\theta}_j^\mu - \bar{\theta}_j)(\hat{\theta}_k^\nu - \bar{\theta}_k)]\}$$

To proceed, first perform the following evaluation:

$$E(\hat{\theta}_i^\mu - \bar{\theta}_i)(\hat{\theta}_j^\nu - \bar{\theta}_j) = E\hat{\theta}_i^\mu \hat{\theta}_j^\nu - E\hat{\theta}_i^\mu \bar{\theta}_j - E\bar{\theta}_i \hat{\theta}_j^\nu + E\bar{\theta}_i \bar{\theta}_j$$

$$= \delta_{\mu\nu}(\theta_{0,i}\theta_{0,j} + P_{ij}/N) + (1 - \delta_{\mu\nu})\theta_{0,i}\theta_{0,j}$$

$$- E\hat{\theta}_i^\mu \frac{1}{m}\sum_{k=1}^{m} \hat{\theta}_j^k - \frac{1}{m} E \sum_{k=1}^{m} \hat{\theta}_i^k \hat{\theta}_j^\nu$$

$$+ (\theta_{0,i}\theta_{0,j} + P_{ij}/(Nm))$$

$$= \theta_{0,i}\theta_{0,j} + \delta_{\mu\nu}P_{ij}/N - \frac{1}{m}E\left[\hat{\theta}_i^\mu\left\{\hat{\theta}_j^\mu + \sum_{\substack{k=1 \\ k\neq\mu}}^m \hat{\theta}_j^k\right\}\right]$$

$$- \frac{1}{m}E\left[\left\{\hat{\theta}_i^\nu + \sum_{\substack{k=1 \\ k\neq\nu}}^m \hat{\theta}_i^k\right\}\hat{\theta}_j^\nu\right] + (\theta_{0,i}\theta_{0,j} + P_{ij}/Nm)$$

$$= \theta_{0,i}\theta_{0,j} + \delta_{\mu\nu}P_{ij}/N$$

$$- \frac{2}{m}[\{\theta_{0,i}\theta_{0,j} + P_{ij}/N\} + (m-1)\theta_{0,i}\theta_{0,j}]$$

$$+ (\theta_{0,i}\theta_{0,j} + P_{ij}/Nm)$$

$$= (P_{ij}/N)(\delta_{\mu\nu} - 1/m) \tag{B.73}$$

Now insert (B.73) into (B.72) to get

$$E\hat{P}_{ij}\hat{P}_{kl} = \alpha_m^2 \sum_{\mu=1}^m \sum_{\nu=1}^m \{P_{ij}(1-1/m)P_{kl}(1-1/m)$$

$$+ P_{ik}(\delta_{\mu\nu} - 1/m)P_{jl}(\delta_{\mu\nu} - 1/m) + P_{il}(\delta_{\mu\nu} - 1/m)P_{jk}(\delta_{\mu\nu} - 1/m)\}$$

$$= \alpha_m^2 P_{ij}P_{kl}(m-1)^2$$

$$+ \alpha_m^2(P_{ik}P_{jl} + P_{il}P_{jk})\sum_{\mu=1}^m\sum_{\nu=1}^m\left(\delta_{\mu\nu}^2 - \frac{2}{m}\delta_{\mu\nu} + \frac{1}{m^2}\right)$$

$$= \alpha_m^2 P_{ij}P_{kl}(m-1)^2 + \alpha_m^2(P_{ik}P_{jl} + P_{il}P_{jk})(m-1) \tag{B.74}$$

Inserting this result into (B.71) gives

$$E[\hat{P}_{ij} - P_{ij}][\hat{P}_{kl} - P_{kl}]$$
$$= P_{ij}P_{kl}\{1 - \alpha_m(m-1)\}^2 + (P_{ik}P_{jl} + P_{il}P_{jk})\alpha_m^2(m-1) \tag{B.75}$$

From (B.75) the following specialized results are obtained:

$$E[\hat{P}_{ij} - P_{ij}]^2 = P_{ij}^2[\{1 - \alpha_m(m-1)\}^2 + \alpha_m^2(m-1)]$$
$$+ P_{ii}P_{jj}\alpha_m^2(m-1) \tag{B.76a}$$

$$E[\hat{P}_{ii} - P_{ii}]^2 = P_{ii}^2[\{1 - \alpha_m(m-1)\}^2 + 2\alpha_m^2(m-1)] \tag{B.76b}$$

The relative precision of $\hat{P}_{ii}$ is hence given by

$$\beta \triangleq \frac{E[\hat{P}_{ii} - P_{ii}]^2}{P_{ii}^2} = \{1 - \alpha_m(m-1)\}^2 + 2\alpha_m^2(m-1) \tag{B.77}$$

The particular choice $\alpha_m = 1/(m-1)$ (see (B.70)) gives

$$\beta = \frac{2}{m-1} \tag{B.78}$$

The minimum value of β with respect to α_m,

$$\beta_{\min} = \min_{\alpha_m} \beta = \frac{2}{m+1} \tag{B.79a}$$

is obtained for

$$\alpha_m = \frac{1}{m+1} \tag{B.79b}$$

For large values of m the difference between (B.78) and (B.79a) is only marginal.

If m is large it is to be expected that approximately (cf. the central limit theorem)

$$\frac{\hat{P}_{ii} - P_{ii}}{P_{ii}} \sim \mathcal{N}(0, \beta) \tag{B.80}$$

In particular, it is found for the choice $\alpha_m = 1/m$ that with 95 percent probability

$$\left| \frac{\hat{P}_{ii} - P_{ii}}{P_{ii}} \right| \le 1.96 \sqrt{\beta} \approx \frac{1.96\sqrt{2}}{\sqrt{m}} \approx \frac{2.77}{\sqrt{m}} \tag{B.81}$$

Note that the exact distribution of $\hat{P}$ for finite m is Wishart (cf. Rao, 1973). Such a distribution can be seen as an extension of the χ^2 distribution to the multivariable case.

Bibliographical notes

For general results in probability theory, see Cramér (1946), Chung (1968), and for statistical inference, Kendall and Stuart (1961), Rao (1973), and Lindgren (1976). Lemma B.2 is adapted from Söderström (1975b). See also Hannan (1970) and Ljung (1985c) for further results on ergodicity (also called strong laws of large numbers) and for a thorough treatment of the central limit theorems.

Peterka (1981) presents a maximum *a posteriori* (or Bayesian) approach to identification and system parameter estimation.

The books Åström (1970), Anderson and Moore (1979), Meditch (1969) contain deep studies of stochastic dynamic systems. In particular, results such as those presented in Sections B.5–B.7 are discussed in detail. Original work on the Kalman filter appeared in Kalman (1960) and Kalman and Bucy (1961).

For the original derivation of the results on the asymptotic variances of sample covariances, see Bartlett (1946, 1966). For a further reading on this topic, consult Anderson (1971), Box and Jenkins (1976), Brillinger (1981) and Hannan (1970).

REFERENCES

Akaike, H. (1969). Fitting autoregressive models for prediction, *Ann. Inst. Statist. Math.*, Vol. 21, pp. 243–247.

Akaike, H. (1971). Information theory and an extension of the maximum likelihood principle, *2nd International Symposium on Information Theory, Tsahkadsor, Armenian SSR*. Also published in *Supplement to Problems of Control and Information Theory*, pp. 267–281, 1973.

Akaike, H. (1981). Modern development of statistical methods. In P. Eykhoff (ed.), *Trends and Progress in System Identification*. Pergamon Press, Oxford.

Alexander, T.S. (1986). *Adaptive Signal Processing*. Springer Verlag, Berlin.

Anděl, J., M.G. Perez and A.I. Negrao (1981). Estimating the dimension of a linear model, *Kybernetika*, Vol. 17, pp. 514–525.

Anderson, B.D.O. (1982). Exponential convergence and persistent excitation, *Proc. 21st IEEE Conference on Decision and Control, Orlando*.

Anderson, B.D.O. (1985). Identification of scalar errors-in-variables models with dynamics, *Automatica*, Vol. 21, pp. 709–716.

Anderson, B.D.O. and M. Deistler (1984). Identifiability in dynamic errors-in-variables models, *Journal of Time Series Analysis*, Vol. 5, pp. 1–13.

Anderson, B.D.O. and M.R. Gevers (1982). Identifiability of linear stochastic systems operating under linear feedback, *Automatica*, Vol. 18, pp. 195–213.

Anderson, B.D.O. and J.B. Moore (1979). *Optimal Filtering*. Prentice Hall, Inc., Englewood Cliffs.

Andersson, P. (1983). Adaptive forgetting through multiple models and adaptive control of car dynamics, *Report LIU-TEK LIC 1983:10*, Department of Electrical Engineering, Linköping University, Sweden.

Anderson, T.W. (1971). *Statistical Analysis of Time Series*. Wiley, New York.

Ansley, C.F. (1979). An algorithm for the exact likelihood of a mixed autoregressive-moving average process, *Biometrika*, Vol. 66, pp. 59–65.

Aoki, M. (1987). *State Space Modelling of Time Series*. Springer Verlag, Berlin.

Aris, R. (1978). *Mathematical Modelling Techniques*. Pitman, London.

Åström, K.J. (1968). Lectures on the identification problem – the least squares method, *Report 6806*, Division of Automatic Control, Lund Institute of Technology, Sweden.

Åström, K.J. (1969). On the choice of sampling rates in parametric identification of time series, *Information Sciences*, Vol. 1, pp. 273–278.

Åström, K.J. (1970). *Introduction to Stochastic Control Theory*. Academic Press, New York.

Åström, K.J. (1975). Lectures on system identification, chapter 3: Frequency response analysis, *Report 7504*, Department of Automatic Control, Lund Institute of Technology, Sweden.

Åström, K.J. (1980). Maximum likelihood and prediction error methods, *Automatica*, Vol. 16, pp. 551–574.

Åström, K.J. (1983a). Computer aided modelling, analysis and design of control systems – a perspective, *IEEE Control Systems Magazine*, Vol. 3, pp. 4–15.

Åström, K.J. (1983b). Theory and applications of adaptive control – a survey, *Automatica*, Vol. 19, pp. 471–486.

Åström, K.J. (1987). Adaptive feedback control, *Proc. IEEE*, Vol. 75, pp. 185–217.

Åström, K.J. and T. Bohlin (1965). Numerical identification of linear dynamic systems from normal operating records, *IFAC Symposium on Self-Adaptive Systems, Teddington, England*. Also in P.H. Hammond (ed.), *Theory of Self-Adaptive Control Systems*. Plenum, New York.

Åström, K.J., U. Borisson, L. Ljung and B. Wittenmark (1977). Theory and applications of self-tuning regulators, *Automatica*, Vol. 13, pp. 457–476.

Åström, K.J. and P. Eykhoff (1971). System identification – a survey, *Automatica*, Vol. 7, pp. 123–167.

Åström, K.J., P. Hagander and J. Sternby (1984). Zeros of sampled systems, *Automatica*, Vol. 20, pp. 31–38.

Åström, K.J. and T. Söderström (1974). Uniqueness of the maximum likelihood estimates of the parameters of an ARMA model, *IEEE Transactions on Automatic Control*, Vol. AC-19, pp. 769–773.

Åström, K.J. and B. Wittenmark (1971). Problems of identification and control, *Journal of Mathematical Analysis and Applications*, Vol. 34, pp. 90–113.

Åström, K.J. and B. Wittenmark (1989). *Adaptive Control*. Addison-Wesley, Reading, Mass.

Bai, E.W. and S.S. Sastry (1985). Persistency of excitation, sufficient richness and parameter convergence in discrete time adaptive control, *Systems and Control Letters*, Vol. 6, pp. 153–163.

Banks, H.T., J.M. Crowley and K. Kunisch (1983). Cubic spline approximation techniques for parameter estimation in distributed systems, *IEEE Transactions on Automatic Control*, Vol. AC-28, pp. 773–786.

Bartlett, M.S. (1946). On the theoretical specification and sampling properties of autocorrelated time-series, *Journal Royal Statistical Society*, Series B, Vol. 8, pp. 27–41.

Bartlett, M.S. (1966). *An Introduction to Stochastic Processes*. Cambridge University Press, London.

Basseville, M. and A. Benveniste (eds) (1986). *Detection of Abrupt Changes in Signals and Dynamical Systems*. (Lecture Notes in Control and Information Sciences no. 77), Springer Verlag, Berlin.

Bendat, J.S. and A.G. Piersol (1980). *Engineering Applications of Correlation and Spectral Analysis*. Wiley–Interscience, New York.

Benveniste, A. (1984). Design of one-step and multistep adaptive algorithms for the tracking of time-varying systems. *Rapport de Recherche no. 340*, IRISA/INRIA, Rennes. To appear in H.V. Poor (ed.), *Advances in Statistical Signal Processing*. JAI Press.

Benveniste, A. (1987). Design of adaptive algorithms for the tracking of time-varying systems, *International Journal of Adaptive Control and Signal Processing*, Vol. 1, pp. 3–29.

Benveniste, A. and C. Chaure (1981). AR and ARMA identification algorithms of Levinson type: an innovation approach, *IEEE Transactions on Automatic Control*, Vol. AC-26, pp. 1243–1261.

Benveniste, A. and J.J. Fuchs (1985). Single sample modal identification of a nonstationary stochastic process, *IEEE Transactions on Automatic Control*, Vol. AC-30, pp. 66–74.

Benveniste, A., M. Metivier and P. Priouret (1987). *Algorithmes Adaptatifs et Approximations Stochastiques*. Masson, Paris.

Benveniste, A. and G. Ruget (1982). A measure of the tracking capability of recursive stochastic algorithms with constant gains, *IEEE Transactions on Automatic Control*, Vol. AC-27, pp. 639–649.

Bergland, G.D. (1969). A guided tour of the fast Fourier transform, *IEEE Spectrum*, Vol. 6, pp. 41–52.

Bhansali, R.J. (1980). Autoregressive and window estimates of the inverse correlation function, *Biometrika*, Vol. 67, pp. 551–566.

Bhansali, R.J. and D.Y. Downham (1977). Some properties of the order of an autoregressive model selected by a generalization of Akaike's FPE criterion, *Biometrika*, Vol. 64, pp. 547–551.

Bierman, G.J. (1977). *Factorization Methods for Discrete Sequential Estimation*. Academic Press, New York.

Billings, S.A. (1980). Identification of nonlinear systems–a survey, *Proceedings of IEE*, part D, Vol. 127, pp. 272–285.

Billings, S.A. and W.S. Woon (1986). Correlation based model validity tests for non-linear models, *International Journal of Control*, Vol. 44, pp. 235–244.

Blundell, A.J. (1982). *Bond Graphs for Modelling Engineering Systems*. Ellis Horwood, Chichester.

Bohlin, T. (1970). Information pattern for linear discrete-time models with stochastic coefficients, *IEEE Transactions on Automatic Control*, Vol. AC-15, pp. 104–106.

Bohlin, T. (1971). On the problem of ambiguities in maximum likelihood identification, *Automatica*, Vol. 7, pp. 199–210.

Bohlin, T. (1976). Four cases of identification of changing systems. In R.K. Mehra and D.G. Lainiotis (eds), *System Identification: Advances and Case Studies*. Academic Press, New York.

Bohlin, T. (1978). Maximum-power validation of models without higher-order fitting, *Automatica*, Vol. 14, pp. 137–146.

Bohlin, T. (1987). Identification: practical aspects. In M. Singh (ed.), *Systems and Control Encyclopedia*. Pergamon, Oxford.

van den Boom, A.J.W. and R. Bollen (1984). The identification package SATER, *Proc. 9th IFAC World Congress, Budapest*.

van den Boom, A.J.W. and A.W.M. van den Enden (1974). The determination of the orders of process- and noise dynamics, *Automatica*, Vol. 10, pp. 244–256.

Box, G.E.P. and G.W. Jenkins (1976). *Time Series Analysis: forecasting and control* (2nd edn). Holden-Day, San Francisco.

Box, G.E.P. and D.A. Pierce (1970). Distribution of residual autocorrelations in autoregressive integrated moving average time series models, *J. Amer. Statist. Assoc.*, Vol. 65, pp. 1509–1526.

Brillinger, D.R. (1981). *Time Series: Data Analysis and Theory*. Holden-Day, San Francisco.

Burghes, D.N. and M.S. Borrie (1981). *Modelling with Differential Equations*. Ellis Horwood, Chichester.

Burghes, D.N., I. Huntley and J. McDonald (1982). *Applying Mathematics: A Course in Mathematical Modelling*. Ellis Horwood, Chichester.

Butash, T.C. and L.D. Davisson (1986). An overview of adaptive linear minimum mean square error prediction performance, *Proc. 25th IEEE Conference on Decision and Control, Athens*.

Caines, P.E. (1976). Prediction error identification methods for stationary stochastic processes, *IEEE Transactions on Automatic Control*, Vol. AC-21, pp. 500–505.

Caines, P.E. (1978). Stationary linear and nonlinear system identification and predictor set completeness, *IEEE Transactions on Automatic Control*, Vol. AC-23, pp. 583–594.

Caines, P.E. (1988). *Linear Stochastic Systems*. Wiley, New York.

Caines, P.E. and C.W. Chan (1975). Feedback between stationary stochastic processes, *IEEE Transactions on Automatic Control*, Vol. AC-20, pp. 498–508.

Caines, P.E. and C.W. Chan (1976). Estimation, identification and feedback. In R.K. Mehra and D.G. Lainiotis (eds), *System Identification – Advances and Case Studies*. Academic Press, New York.

Caines, P.E. and L. Ljung (1976). Prediction error estimators: Asymptotic normality and accuracy, *Proc. IEEE Conference on Decision and Control, Clearwater Beach*.

Carrol, R.J. and D. Ruppert (1984). Power transformations when fitting theoretical models to data, *Journal of the American Statistical Society*, Vol. 79, pp. 321–328.

Chatfield, C. (1979). Inverse autocorrelations, *Journal Royal Statistical Society*, Series A, Vol. 142, pp. 363–377.

Chavent, G. (1979). Identification of distributed parameter systems: about the output least square method, its implementation, and identifiability, *Proc. 5th IFAC Symposium on Identification and System Parameter Estimation, Darmstadt*.

Chen, H.F. (1985). *Recursive Estimation and Control for Stochastic Systems*. Wiley, New York.

Chen, H.F. and L. Guo (1986). Convergence of least-squares identification and adaptive control for stochastic systems, *International Journal of Control*, Vol. 44, pp. 1459–1476.

Chung, K.L. (1968). *A Course in Probability Theory*. Harcourt, Brace & World, New York.

Cioffi, J.M. (1987). Limited-precision effects in adaptive filtering, *IEEE Transactions on Circuits*

and *Systems*, Vol. CAS-34, pp. 821–833.
Čižek, V. (1986). *Discrete Fourier Transforms and Their Applications*. Adam Hilger, Bristol.
Clarke, D.W. (1967). Generalized least squares estimation of parameters of a dynamic model, *1st IFAC Symposium on Identification in Automatic Control Systems*, Prague.
Cleveland, W.S. (1972). The inverse autocorrelations of a time series and their applications, *Technometrics*, Vol. 14, pp. 277–298.
Cooley, J.W. and J.W. Tukey (1965). An algorithm for the machine computation of complex Fourier series, *Math of Computation*, Vol. 19, pp. 297–301.
Corrêa, G.O. and K. Glover (1984a). Pseudo-canonical forms, identifiable parametrizations and simple parameter estimation for linear multivariable systems: input–output models, *Automatica*, Vol. 20, pp. 429–442.
Corrêa, G.O. and K. Glover (1984b). Pseudo-canonical forms, identifiable parametrizations and simple parameter estimation for linear multivariable systems: parameter estimation, *Automatica*, Vol. 20, pp. 443–452.
Corrêa, G.O. and K. Glover (1987). Two-stage IV-based estimation and parametrization selection for linear multivariable identification, *International Journal of Control*, Vol. 46, pp. 377–401.
Cramér, H. (1946). *Mathematical Methods of Statistics*. Princeton University Press, Princeton.
Cybenko, G. (1980) The numerical stability of the Levinson-Durbin algorithm for Toeplitz systems of equations, *SIAM Journal on Sci. Stat. Comput.*, Vol. 1, pp. 303–319.
Cybenko, G. (1984). The numerical stability of the lattice algorithm for least squares linear prediction problems, *BIT*, Vol. 24, pp. 441–455.
Davies, W.D.T. (1970). *System Identification for Self-Adaptive Control*. Wiley–Interscience, London.
Davis, M.H.A. and R.B. Vinter (1985). *Stochastic Modelling and Control*. Chapman and Hall, London.
Davisson, L.D. (1965). The prediction error of stationary Gaussian time series of unknown covariance, *IEEE Transactions on Information Theory*, Vol. IT-11, pp. 527–532.
Deistler, M. (1986). Multivariate time series and linear dynamic systems, *Advances Stat. Analysis and Stat. Computing*, Vol. 1, pp. 51–85.
Delsarte, Ph. and Y.V. Genin (1986). The split Levinson algorithm, *IEEE Transactions on Acoustics, Speech and Signal Processing*, Vol. ASSP-34, pp. 470–478.
Demeure, C.J. and L.L. Scharf (1987). Linear statistical models for stationary sequences and related algorithms for Cholesky factorization of Toeplitz matrices, *IEEE Transactions on Acoustics, Speech and Signal Processing*, Vol. ASSP-35, pp. 29–42.
Dennis, J.E. Jr and R.B. Schnabel (1983). *Numerical Methods for Unconstrained Optimization and Nonlinear Equations*. Prentice Hall, Englewood Cliffs.
Dent, W.T. (1977). Computation of the exact likelihood function of an ARIMA process, *Journal Statist. Comp. and Simulation*, Vol. 5, pp. 193–206.
Dhrymes, Ph. J. (1978). *Introductory Econometrics*. Springer Verlag, New York.
Draper, N.R. and H. Smith (1981). *Applied Regression Analysis*. Wiley, New York.
Dugard, L. and I.D. Landau (1980). Recursive output error identification algorithms – theory and evaluation, *Automatica*, Vol. 16, pp. 443–462.
Dugré, J.P., L.L. Scharf and C. Gueguen (1986). Exact likelihood for stationary vector autoregressive moving average processes, *Stochastics*, Vol. 11, pp. 105–118.
Durbin, J. (1959). Efficient estimation of parameters in moving-average models, *Biometrika*, Vol. 46, pp. 306–316.
Durbin, J. (1960). The fitting of time series models, *Rev. Inst. Int. Statist.*, Vol. 28, pp. 233–244.
Egardt, B. (1979). *Stability of Adaptive Controllers* (Lecture Notes in Control and Information Sciences, no. 20). Springer Verlag, Berlin.
Eykhoff, P. (1974). *System Identification: Parameter and State Estimation*. Wiley, London.
Eykhoff, P. (ed.) (1981). *Trends and Progress in System Identification*. Pergamon, Oxford.
Fedorov, V.V. (1972). *Theory of Optimal Experiments*. Academic Press, New York.

Finigan, B.M. and I.H. Rowe (1974). Strongly consistent parameter estimation by the introduction of strong instrumental variables, *IEEE Transactions on Automatic Control*, Vol. AC-19, pp. 825–830.

Fortescue, T.R., L.S. Kershenbaum and B.E. Ydstie (1981). Implementation of self-tuning regulators with variable forgetting factors, *Automatica*, Vol. 17, pp. 831–835.

Freeman, T.G. (1985). Selecting the best linear transfer function model, *Automatica*, Vol. 21, pp. 361–370.

Friedlander, B. (1982). Lattice filters for adaptive processing, *Proc. IEEE*, Vol. 70, pp. 829–867.

Friedlander, B. (1983). A lattice algorithm for factoring the spectrum of a moving-average process, *IEEE Transactions on Automatic Control*, Vol. AC-28, pp. 1051–1055.

Friedlander, B. (1984). The overdetermined recursive instrumental variable method, *IEEE Transactions on Automatic Control*, Vol. AC-29, pp. 353–356.

Fuchs, J.-J. (1987). ARMA order estimation via matrix perturbation theory, *IEEE Transactions on Automatic Control*, Vol. AC-32, pp. 358–361.

Furht, B.P. (1973). New estimator for the identification of dynamic processes, *IBK Report*, Institut Boris Kidric Vinca, Belgrade, Yugoslavia.

Furuta, K., S. Hatakeyama and H. Kominami (1981). Structural identification and software package for linear multivariable systems, *Automatica*, Vol. 17, pp. 755–762.

Gardner, G., A.C. Harvey and G.D.A. Phillips (1980). An algorithm for exact maximum likelihood estimation of autoregressive moving average models by means of Kalman filtering, *Appl. Stat.*, Vol. 29, pp. 311–322.

K.F. Gauss (1809). *Teoria Motus Corporum Coelestium in Sectionibus Conicus Solem Ambientieum.* Reprinted translation: *Theory of the motion of the heavenly bodies moving about the sun in conic sections.* Dover, New York.

Gertler, J. and Cs. Bányász (1974). A recursive (on-line) maximum likelihood identification method, *IEEE Transactions on Automatic Control*, Vol. AC-19, pp. 816–820.

Gevers, M.R. (1986). ARMA models, their Kronecker indices and their McMillan degree, *International Journal of Control*, Vol. 43, pp. 1745–1761.

Gevers, M.R. and B.D.O. Anderson (1981). Representation of jointly stationary stochastic feedback processes, *International Journal of Control*, Vol. 33, pp. 777–809.

Gevers, M.R. and B.D.O. Anderson (1982). On jointly stationary feedback-free stochastic processes, *IEEE Transactions on Automatic Control*, Vol. AC-27, pp. 431–436.

Gevers, M. and L. Ljung (1986). Optimal experiment designs with respect to the intended model application, *Automatica*, Vol. 22, pp. 543–554.

Gevers, M. and V. Wertz (1984). Uniquely identifiable state-space and ARMA parametrizations for multivariable linear systems, *Automatica*, Vol. 20, pp. 333–347.

Gevers, M. and V. Wertz (1987a). Parametrization issues in system identification, *Proc. IFAC 10th World Congress, Munich*.

Gevers, M. and V. Wertz (1987b). Techniques for the selection of identifiable parametrizations for multivariable linear systems. In C.T. Leondes (ed.), *Control and Dynamic Systems*, Vol. 26: System Identification and Adaptive Control – Advances in Theory and Applications. Academic Press, New York.

Gill, P., W. Murray and M.H. Wright (1981). *Practical Optimization.* Academic Press, London.

Glover, K. (1984). All optimal Hankel-norm approximations of linear multivariable systems and their L^∞-error bounds, *International Journal of Control*, Vol. 39, pp. 1115–1193.

Glover, K. (1987). Identification: frequency-domain methods. In M. Singh (ed.), *Systems and Control Encyclopedia.* Pergamon, Oxford.

Gnedenko, B.V. (1963). *The Theory of Probability.* Chelsea, New York.

Godfrey, K.R. (1983). *Compartmental Models and Their Application.* Academic Press, New York.

Godfrey, K.R. and J.J. DiStefano, III (1985). Identifiability of model parameters, *Proc. 7th IFAC/IFORS Symposium on Identification and System Parameter Estimation, York.*

Godolphin, E.J. and J.M. Unwin (1983). Evaluation of the covariance matrix for the maximum likelihood estimator of a Gaussian autoregressive moving average process, *Biometrika*, Vol. 70, pp. 279–284.

Gohberg, I.C. and G. Heinig (1974). Inversion of finite Toeplitz matrices with entries being elements from a non-commutative algebra, *Rev. Roumaine Math. Pures Appl.*, Vol. XIX, pp. 623–665.

Golomb, S.W. (1967). *Shift Register Sequences*. Holden-Day, San Francisco.

Golub, G.H. and C.F. van Loan (1983) *Matrix Computations*. North Oxford Academic, Oxford.

Goodwin, G.C. (1987). Identification: experiment design. In M. Singh (ed.), *Systems and Control Encyclopedia*. Pergamon, Oxford.

Goodwin, G.C. and R.L. Payne (1973). Design and characterization of optimal test signals for linear single input–single output parameter estimation, *Proc. 3rd IFAC Symposium on Identification and System Parameter Estimation, the Hague*.

Goodwin, G.C. and R.L. Payne (1977). *Dynamic System Identification: Experiment Design and Data Analysis*. Academic Press, New York.

Goodwin, G.C., P.J. Ramadge and P.E. Caines (1980). Discrete time multi-variable adaptive control, *IEEE Transactions on Automatic Control*, Vol. AC-25, pp. 449–456.

Goodwin, G.C. and K.S. Sin (1984). *Adaptive Filtering, Prediction and Control*. Prentice Hall, Englewood Cliffs.

de Gooijer, J. and P. Stoica (1987). A min-max optimal instrumental variable estimation method for multivariable linear systems, *Technical Report No. 8712*, Faculty of Economics, University of Amsterdam. Also *International Journal of Control*, to appear 1989.

Granger, C.W.J. and P. Newbold (1977). *Forecasting Economic Time Series*. Academic Press, New York.

Graybill, F.A. (1983). *Matrices with Applications in Statistics* (2nd edn). Wadsworth International Group, Belmont, California.

Grenander, U. and M. Rosenblatt (1956). *Statistical Analysis of Stationary Time Series*. Almqvist och Wiksell, Stockholm, Sweden.

Grenander, U. and G. Szegö (1958). *Toeplitz Forms and Their Applications*. University of California Press, Berkeley, California.

Gueguen, C. and L.L. Scharf (1980). Exact maximum likelihood identification of ARMA models: a signal processing perspective, *Report ONR 36*, Department Electrical Engineering, Colorado State University, Fort Collins.

Guidorzi, R. (1975). Canonical structures in the identification of multivariable systems, *Automatica*, Vol. 11, pp. 361–374.

Guidorzi, R.P. (1981). Invariants and canonical forms for systems structural and parametric identification, *Automatica*, Vol. 17, pp. 117–133.

Gustavsson, I., L. Ljung and T. Söderström (1977). Identification of processes in closed loop: identifiability and accuracy aspects, *Automatica*, Vol. 13, pp. 59–75.

Gustavsson, I., L. Ljung and T. Söderström (1981). Choice and effect of different feedback configurations. In P. Eykhoff (ed.), *Trends and Progress in System Identification*. Pergamon, Oxford.

Haber, R. (1985). Nonlinearity test for dynamic processes, *Proc. 7th IFAC/IFORS Symposium on Identification and System Parameter Estimation, York*.

Haber, R. and L. Keviczky (1976). Identification of nonlinear dynamic systems, *Proc. 4th IFAC Symposium on Identification and System Parameter Estimation, Tbilisi*.

Hägglund, T. (1984). Adaptive control of systems subject to large parameter changes, *Proc. IFAC 9th World Congress, Budapest*.

Hajdasinski, A.K., P. Eykhoff, A.A.M. Damen and A.J.W. van den Boom (1982). The choice and use of different model sets for system identification, *Proc. 6th IFAC Symposium on Identification and System Parameter Estimation, Washington DC*.

Hannan, E.J. (1969). The identification of vector mixed autoregressive moving average systems, *Biometrika*, Vol. 56, pp. 222–225.

Hannan, E.J. (1970). *Multiple Time Series*. Wiley, New York.
Hannan, E.J. (1976). The identification and parameterization of ARMAX and state space forms, *Econometrica*, Vol. 44, pp. 713–723.
Hannan, E.J. (1980). The estimation of the order of an ARMA process, *Annals of Statistics*, Vol. 8, pp. 1071–1081.
Hannan, E.J. (1981). Estimating the dimension of a linear system, *Journal Multivar. Analysis*, Vol. 11, pp. 459–473.
Hannan, E.J. and M. Deistler (1988). *The Statistical Theory of Linear Systems*. Wiley, New York.
Hannan, E.J. and L. Kavalieris (1984). Multivariate linear time series models, *Advances in Applied Probability*, Vol. 16, pp. 492–561.
Hannan, E.J., P.R. Krishnaiah and M.M. Rao (eds) (1985). *Time Series in the Time Domain (Handbook of Statistics, Vol. 5)*. North-Holland, Amsterdam.
Hannan, E.J. and B.G. Quinn (1979). The determination of the order of an autoregression, *Journal Royal Statistical Society*, Series B. Vol. 41, pp. 190–195.
Hannan, E.J. and J. Rissanen (1982). Recursive estimation of mixed autoregressive-moving average order, *Biometrika*, Vol. 69, pp. 81–94.
Haykin, S. (1986). *Adaptive Filter Theory*. Prentice Hall, Englewood Cliffs.
Hill, S.D. (1985). Reduced gradient computation in prediction error identification, *IEEE Transactions on Automatic Control*, Vol. AC-30, pp. 776–778.
Ho, Y.C. (1963). On the stochastic approximation method and optimal filtering theory, *Journal of Mathematical Analysis and Applications*, Vol. 6, pp. 152–154.
Holst, J. (1977). Adaptive prediction and recursive estimation, *Report TFRT-1013*, Department of Automatic Control, Lund Institute of Technology, Sweden.
Honig, M.L. and D.G. Messerschmitt (1984). *Adaptive Filters: Structures, Algorithms and Applications*. Kluwer, Boston.
Hsia, T.C. (1977). *Identification: Least Squares Methods*. Lexington Books, Lexington, Mass.
Huber, P.J. (1973). Robust regression: asymptotics, conjectures and Monte Carlo, *Annals of Statistics*, Vol. 1, pp. 799–821.
Isermann, R. (1974). *Prozessidentifikation*. Springer Verlag, Berlin.
Isermann, R. (1980). Practical aspects of process identification, *Automatica*, Vol. 16, pp. 575–587.
Isermann, R. (1987). *Identifikation Dynamischer Systeme*. Springer Verlag, Berlin.
Isermann, R., U. Baur, W. Bamberger, P. Kneppo and H. Siebert (1974). Comparison of six on-line identification and parameter estimation methods, *Automatica*, Vol. 10, pp. 81–103.
Jakeman, A.J. and P.C. Young (1979). Refined instrumental variable methods of recursive time-series analysis. Part II: Multivariable systems, *International Journal of Control*, Vol. 29, pp. 621–644.
Jamshidi, M. and C.J. Herget (eds) (1985). *Computer-Aided Control Systems Engineering*. North-Holland, Amsterdam.
Janssen, P. (1987a). MFD models and time delays; some consequences for identification, *International Journal of Control*, Vol. 45, pp. 1179–1196.
Janssen, P. (1987b). General results on the McMillan degree and the Kronecker indices of ARMA- and MFD-models. *Report ER 87/08*, Measurement and Control Group, Faculty of Electrical Engineering, Eindhoven University of Technology, The Netherlands.
Janssen, P. (1988). On model parametrization and model structure selection for identification of MIMO systems. Doctoral dissertation, Eindhoven University of Technology, the Netherlands.
Janssen, P. and P. Stoica (1987). On the expectation of the product of four matrix-valued Gaussian random variables. *Technical Report 87-E-178*, Department of Electrical Engineering, Eindhoven University of Technology, The Netherlands. Also *IEEE Transactions on Automatic Control*, Vol. 33, pp. 867–870, 1988.
Jategaonkar, R.V., J.R. Raol and S. Balakrishna (1982). Determination of model order for dynamical system, *IEEE Transactions on Systems, Man and Cybernetics*, Vol. SMC-12, pp. 56–62.

Jenkins, G.M. and D.G. Watts (1969). *Spectral Analysis and Its Applications.* Holden-Day, San Francisco.
Jennrich, R.I. (1969). Asymptotic properties of non-linear least squares estimators, *Annals of Mathematical Statistics*, Vol. 40, pp. 633–643.
Jones, R.H. (1980). Maximum likelihood fitting of ARMA models to time series with missing observations, *Technometrics*, Vol. 22, pp. 389–395.
Kabaila, P. (1983). On output-error methods for system identification, *IEEE Transactions on Automatic Control*, Vol. AC-28, pp. 12–23.
Kailath, T. (1980). *Linear Systems.* Prentice Hall, Englewood Cliffs.
Kailath, T., A. Vieira and M. Morf (1978). Inverses of Toeplitz operators, innovations, and orthogonal polynomials, *SIAM Review*, Vol. 20, pp. 1006–1019.
Kalman, R.E. (1960). A new approach to linear filtering and prediction problems, *Transactions ASME, Journal of Basic Engineering*, Series D, Vol. 82, pp. 342–345.
Kalman, R.E. and R.S. Bucy (1961). New results in linear filtering and prediction theory, *Transactions ASME, Journal of Basic Engineering*, Series D, Vol. 83, pp. 95–108.
Karlsson, E. and M.H. Hayes (1986). ARMA modeling of time varying systems with lattice filters, *Proc. IEEE Conference on Acoustics, Speech and Signal Processing, Tokyo.*
Karlsson, E. and M.H. Hayes (1987). Least squares ARMA modeling of linear time-varying systems: lattice filter structures and fast RLS algorithms, *IEEE Transactions on Acoustics, Speech and Signal Processing*, Vol. ASSP-35, pp. 994–1014.
Kashyap, R.L. (1980). Inconsistency of the AIC rule for estimating the order of AR models, *IEEE Transactions on Automatic Control*, Vol. AC-25, pp. 996–998.
Kashyap, R.L. (1982). Optimal choice of AR and MA parts in autoregressive moving average models, *IEEE Transactions on Pattern Analysis and Machine Intelligence*, Vol. PAMI-4, pp. 99–103.
Kashyap, R.L. and A.R. Rao (1976). *Dynamic Stochastic Models from Empirical Data.* Academic Press, New York.
Kay, S.M. (1988). *Modern Spectral Estimation. Theory and Application.* Prentice-Hall, Englewood Cliffs.
Kendall, M.G. and S. Stuart (1961). *The Advanced Theory of Statistics*, Vol. II. Griffin, London.
Keviczky, L. and Cs. Bányász (1976). Some new results on multiple input-multiple output identification methods, *Proc. 4th IFAC Symposium on Identification and System Parameter Estimation, Tbilisi.*
Kubrusly, C.S. (1977). Distributed parameter system identification. A survey, *International Journal of Control*, Vol. 26, pp. 509–535.
Kučera, V. (1972). The discrete Riccati equation of optimal control, *Kybernetika*, Vol. 8, pp. 430–447.
Kučera, V. (1979). *Discrete Linear Control.* Wiley, Chichester.
Kumar, R. (1985). A fast algorithm for solving a Toeplitz system of equations, *IEEE Transactions on Acoustics, Speech and Signal Processing*, Vol. ASSP-33, pp. 254–267.
Kushner, H.J. and R. Kumar (1982). Convergence and rate of convergence of a recursive identification and adaptive control method which uses truncated estimators, *IEEE Transactions on Automatic Control*, Vol. AC-27, pp. 775–782.
Lai, T.L. and C.Z. Wei (1986). On the concept of excitation in least squares identification and adaptive control, *Stochastics*, Vol. 16, pp. 227–254.
Landau, I.D. (1979). *Adaptive Control. The Model Reference Approach.* Marcel Dekker, New York.
Lawson, C.L. and R.J. Hanson (1974). *Solving Least Squares Problems.* Prentice Hall, Englewood Cliffs.
Lee, D.T.L., B. Friedlander and M. Morf (1982). Recursive ladder algorithms for ARMA modelling, *IEEE Transactions on Automatic Control*, Vol. AC-29, pp. 753–764.
Lehmann, E.L. (1986). *Testing Statistical Hypotheses* (2nd edn). Wiley, New York.
Leondes, C.T. (ed.) (1987). *Control and Dynamic Systems*, Vol. 25–27: System Identification and Adaptive Control – Advances in Theory and Applications. Academic Press, New York.

Leontaritis, I.J. and S.A. Billings (1985). Input–output parametric models for non-linear systems, *International Journal of Control*, Vol. 41, pp. 303–344.

Leontaritis, I.J. and S.A. Billings (1987). Model selection and validation methods for non-linear systems, *International Journal of Control*, Vol. 45, pp. 311–341.

Levinson, N. (1947). The Wiener RMS (root-mean-square) error criterion in filter design and prediction, *J. Math. Phys.*, Vol. 25, pp. 261–278. Also in N. Wiener, *Extrapolation, Interpolation and Smoothing of Stationary Time Series*. Wiley, New York, 1949.

Lindgren, B.W. (1976). *Statistical Theory*. MacMillan, New York.

Ljung, G.M. and G.E.P. Box (1979). The likelihood function of stationary autoregressive-moving average models, *Biometrika*, Vol. 66, pp. 265–270.

Ljung, L. (1971). Characterization of the concept of 'persistently exciting' in the frequency domain, *Report 7119*, Division of Automatic Control, Lund Institute of Technology, Sweden.

Ljung, L. (1976). On the consistency of prediction error identification methods. In R.K. Mehra and D.G. Lainiotis (eds), *System Identification – Advances and Case Studies*. Academic Press, New York.

Ljung, L. (1977a). On positive real transfer functions and the convergence of some recursive schemes, *IEEE Transactions on Automatic Control*, Vol. AC-22, pp. 539–551.

Ljung, L. (1977b). Analysis of recursive stochastic algorithms, *IEEE Transactions on Automatic Control*, Vol. AC-22, pp. 551–575.

Ljung, L. (1977c). Some limit results for functionals of stochastic processes, *Report LiTH-ISY-I-0167*, Department of Electrical Engineering, Linköping University, Sweden.

Ljung, L. (1978). Convergence analysis of parametric identification methods, *IEEE Transactions on Automatic Control*, Vol. AC-23, pp. 770–783.

Ljung, L. (1980). Asymptotic gain and search direction for recursive identification algorithms, *Proc. 19th IEEE Conference on Decision and Control, Albuquerque*.

Ljung, L. (1981). Analysis of a general recursive prediction error identification algorithm, *Automatica*, Vol. 17, pp. 89–100.

Ljung, L. (1982). Recursive identification methods for off-line identification problems, *Proc. 6th IFAC Symposium on Identification and System Parameter Estimation, Washington DC*.

Ljung, L. (1984). Analysis of stochastic gradient algorithms for linear regression problems, *IEEE Transactions on Information Theory*, Vol. IT-30, pp. 151–160.

Ljung, L. (1985a). Asymptotic variance expressions for identified black-box transfer function models, *IEEE Transactions on Automatic Control*, Vol. AC-30, pp. 834–844.

Ljung, L. (1985b). Estimation of transfer functions, *Automatica*, Vol. 21, pp. 677–696.

Ljung, L. (1985c). A non-probabilistic framework for signal spectra, *Proc. 24th IEEE Conference on Decision and Control, Fort Lauderdale*.

Ljung, L. (1986). *System Identification Toolbox – User's Guide*. The Mathworks, Sherborn, Mass.

Ljung, L. (1987). *System Identification: Theory for the User*. Prentice Hall, Englewood Cliffs.

Ljung, L. and P.E. Caines (1979). Asymptotic normality of prediction error estimation for approximate system models, *Stochastics*, Vol. 3, pp. 29–46.

Ljung, L. and K. Glover (1981). Frequency domain versus time domain methods in system identification, *Automatica*, Vol. 17, pp. 71–86.

Ljung, L. and J. Rissanen (1976). On canonical forms, parameter identifiability and the concept of complexity, *Proc. 4th IFAC Symposium on Identification and System Parameter Estimation, Tbilisi*.

Ljung, L. and T. Söderström (1983). *Theory and Practice of Recursive Identification*. MIT Press, Cambridge, Mass.

Ljung, S. (1983). Fast algorithms for integral equations and least squares identification problems. Doctoral dissertation, Department of Electrical Engineering, Linköping University, Sweden.

Ljung S. and L. Ljung (1985). Error propagation properties of recursive least-squares adaptation algorithms, *Automatica*, Vol. 21, pp. 157–167.

Macchi, O. (ed.) (1984). *IEEE Transactions on Information Theory, Special Issue on Linear*

Adaptive Filtering, Vol. IT-30, no. 2.
McLeod, A.I. (1978). On the distribution of residual autocorrelations in Box–Jenkins models, *Journal Royal Statistical Society*, Series B, Vol. 40, pp. 296–302.
Marcus-Roberts, H. and M. Thompson (eds) (1976). *Life Science Models*. Springer Verlag, New York.
Mayne, D.Q. (1967). A method for estimating discrete time transfer functions, *Advances of Control, 2nd UKAC Control Convention, University of Bristol*.
Mayne, D.Q. and F. Firoozan (1982). Linear identification of ARMA processes, *Automatica*, Vol. 18, pp. 461–466.
Meditch, J.S. (1969). *Stochastic Optimal Linear Estimation and Control*. McGraw-Hill, New York.
Mehra, R.K. (1974). Optimal input signals for parameter estimation in dynamic systems – A survey and new results, *IEEE Transactions on Automatic Control*, Vol. AC-19, pp. 753–768.
Mehra, R.K. (1976). Synthesis of optimal inputs for multiinput-multioutput systems with process noise. In R.K. Mehra and D.G. Lainiotis (eds), *System Identification – Advances and Case Studies*. Academic Press, New York.
Mehra, R.K. (1979). Nonlinear system identification – selected survey and recent trends, *Proc. 5th IFAC Symposium on Identification and System Parameter Estimation, Darmstadt*.
Mehra, R.K. (1981). Choice of input signals. In P. Eykhoff (ed.), *Trends and Progress in System Identification*. Pergamon, Oxford.
Mehra, R.K. and D.G. Lainiotis (eds) (1976). *System Identification – Advances and Case Studies*. Academic Press, New York.
Mendel, J.M. (1973). *Discrete Techniques of Parameter Estimation: The Equation Error Formulation*. Marcel Dekker, New York.
Merchant, G.A. and T.W. Parks (1982). Efficient solution of a Toeplitz-plus-Hankel coefficient matrix system of equations, *IEEE Transactions on Acoustics, Speech and Signal Processing*, Vol. ASSP 30, pp 40–44.
Moore, J.B. and G. Ledwich (1980). Multivariable adaptive parameter and state estimators with convergence analysis, *Journal of the Australian Mathematical Society*, Vol. 21, pp. 176–197.
Morgan, B.J.T. (1984). *Elements of Simulation*. Chapman and Hall, London.
Moustakides, G. and A. Benveniste (1986). Detecting changes in the AR parameters of a nonstationary ARMA process, *IEEE Transactions on Information Theory*, Vol. IT-30, pp. 137–155.
Nehorai, A. and M. Morf (1984). Recursive identification algorithms for right matrix fraction description models, *IEEE Transactions on Automatic Control*, Vol. AC-29, pp. 1103–1106.
Nehorai, A. and M. Morf (1985). A unified derivation for fast estimation algorithms by the conjugate direction method, *Linear Algebra and Its Applications*, Vol. 72, pp. 119–143.
Nehorai, A. and P. Stoica (1988). Adaptive algorithms for constrained ARMA signals in the presence of noise, *IEEE Transactions on Acoustics, Speech and Signal Processing*, Vol. ASSP-36, pp. 1282–1291. Also *Proc. IEEE Conference on Acoustics, Speech and Signal Processing (ICASSP)*, Dallas.
Newbold, P. (1974). The exact likelihood function for a mixed autoregressive-moving average process, *Biometrika*, Vol. 61, pp. 423–426.
Ng, T.S., G.C. Goodwin and B.D.O. Anderson (1977). Identifiability of linear dynamic systems operating in closed-loop, *Automatica*, Vol. 13, pp. 477–485.
Nguyen, V.V. and E.F. Wood (1982). Review and unification of linear identifiability concepts, *SIAM Review*, Vol. 24, pp. 34–51.
Nicholson, H. (ed.) (1980). *Modelling of Dynamical Systems*, Vols 1 and 2. Peregrinus, Stevenage.
Norton, J.P. (1986). *An Introduction to Identification*. Academic Press, New York.
Ölçer, S., B. Egardt and M. Morf (1986). Convergence analysis of ladder algorithms for AR and ARMA models, *Automatica*, Vol. 22, pp. 345–354.
Oppenheim, A.V. and R.W. Schafer (1975). *Digital Signal Processing*. Prentice Hall, Englewood Cliffs.

Oppenheim, A.V. and A.S. Willsky (1983). *Signals and Systems*. Prentice Hall, Englewood Cliffs.

van Overbeek, A.J.M. and L. Ljung (1982). On-line structure selection for multivariable state-space models, *Automatica*, Vol. 18, pp. 529–543.

Panuska, V. (1968). A stochastic approximation method for identification of linear systems using adaptive filtering, *Joint Automatic Control Conference, Ann Arbor*.

Pazman, A. (1986). *Foundation of Optimum Experimental Design*. D. Reidel, Dordrecht.

Pearson, C.E. (ed.) (1974). *Handbook of Applied Mathematics*. Van Nostrand Reinhold, New York.

Peterka, V. (1975). A square root filter for real time multivariate regression, *Kybernetika*, Vol. 11, pp. 53–67.

Peterka, V. (1981). Bayesian approach to system identification. In P. Eykhoff (ed.), *Trends and Progress in System Identification*. Pergamon, Oxford.

Peterka, V. and K. Šmuk (1969). On-line estimation of dynamic parameters from input–output data, *4th IFAC Congress, Warsaw*.

Peterson, W.W. and E.J. Weldon, Jr (1972). *Error-Correcting Codes* (2nd edn). MIT Press, Cambridge, Mass.

Polis, M.P. (1982). The distributed system parameter identification problem: A survey of recent results, *Proc. 3rd IFAC Symposium on Control of Distributed Parameter Systems, Toulouse*.

Polis, M.P. and R.E. Goodson (1976). Parameter identification in distributed systems: a synthesizing overview, *Proceedings IEEE*, Vol. 64, pp. 45–61.

Polyak, B.T. and Ya. Z. Tsypkin (1980). Robust identification, *Automatica*, Vol. 16, pp. 53–63.

Porat, B. and B. Friedlander (1985). Asymptotic accuracy of ARMA parameter estimation methods based on sample covariances, *Proc. 7th IFAC/IFORS Symposium on Identification and System Parameter Estimation, York*.

Porat, B. and B. Friedlander (1986). Bounds on the accuracy of Gaussian ARMA parameter estimation methods based on sample covariances, *IEEE Transactions on Automatic Control*, Vol. AC-31, pp. 579–582.

Porat, B., B. Friedlander and M. Morf (1982). Square root covariance ladder algorithms, *IEEE Transactions on Automatic Control*, Vol. AC-27, pp. 813–829.

Potter, J.E. (1963). New statistical formulas, *Memo 40*, Instrumentation Laboratory, MIT, Cambridge, Mass.

Priestley, M.B. (1982). *Spectral Analysis and Time Series*. Academic Press, London.

Rabiner, L.R. and B. Gold (1975). *Theory and Applications of Digital Signal Processing*. Prentice Hall, Englewood Cliffs.

Rake, H. (1980). Step response and frequency response methods, *Automatica*, Vol. 16, pp. 519–526.

Rake, H. (1987). Identification: transient- and frequency-response methods. In M. Singh (ed.), *Systems and Control Encyclopedia*. Pergamon, Oxford.

Rao, C.R. (1973). *Linear Statistical Inference and Its Applications*. Wiley, New York.

Ratkowsky, D.A. (1983). *Nonlinear Regression Modelling*. Marcel Dekker, New York.

Reiersøl, O. (1941). Confluence analysis by means of lag moments and other methods of confluence analysis, *Econometrica*, Vol. 9, pp. 1–23.

Rissanen, J. (1974). Basis of invariants and canonical forms for linear dynamic systems, *Automatica*, Vol. 10, pp. 175–182.

Rissanen, J. (1978). Modeling by shortest data description, *Automatica*, Vol. 14, pp. 465–471.

Rissanen, J. (1979). Shortest data description and consistency of order estimates in ARMA processes, *International Symposium on Systems Optimization and Analysis* (eds A. Bensoussan and J.L. Lions), pp. 92–98.

Rissanen, J. (1982). Estimation of structure by minimum description length, *Circuits, Systems and Signal Processing*, Vol. 1, pp. 395–406.

Rogers, G.S. (1980). *Matrix Derivatives*. Marcel Dekker, New York.

Rowe, I.H. (1970). A bootstrap method for the statistical estimation of model parameters,

International Journal of Control, Vol. 12, pp. 721-738.
Rubinstein, R.Y. (1981). *Simulation and the Monte Carlo Method.* Wiley-Interscience, New York.
Samson, C. (1982). A unified treatment of fast algorithms for identification, *International Journal of Control*, Vol. 35, pp. 909-934.
Saridis, G.N. (1974). Comparison of six on-line identification algorithms, *Automatica*, Vol. 10, pp. 69-79.
Schmid, Ch. and H. Unbehauen (1979). Identification and CAD of adaptive systems using the KEDDC package, *Proc. 5th IFAC Symposium on Identification and System Parameter Estimation, Darmstadt.*
Schwarz, G. (1978). Estimating the dimension of a model, *Annals of Statistics*, Vol. 6, pp. 461-464.
Schwarze, G. (1964). Algorithmische Bestimmung der Ordnung und Zeitkonstanten bei P-, I- und D-Gliedern mit zwei unterschiedlichen Zeitkonstanten und Verzögerung bis 6. Ordnung, *Messen, Steuren, Regeln*, vol. 7, pp. 10-19.
Seborg, D.E., T.F. Edgar and S.L. Shah (1986). Adaptive control strategies for process control: a survey, *AIChE Journal*, Vol. 32, pp. 881-913.
Sharman, K.C. and T.S. Durrani (1988). Annealing algorithms for adaptive array processing. *8th IFAC/IFORS Symposium on Identification and System Parameter Estimation, Beijing.*
Shibata, R. (1976) Selection of the order of an autoregressive model by Akaike's information criterion, *Biometrika*, Vol. 63, pp. 117-126.
Shibata, R. (1980). Asymptotically efficient selection of the order of the model for estimating parameters of a linear process, *Annals of Statistics*, Vol. 8, pp. 147-164.
Sin, K.S. and G.C. Goodwin (1980). Checkable conditions for identifiability of linear systems operating in closed loop, *IEEE Transactions on Automatic Control*, Vol. AC-25, pp. 722-729.
Sinha, N.K. and S. Puthenpura (1985). Choice of the sampling interval for the identification of continuous-time systems from samples of input/output data, *Proceedings IEE*, Part D, vol. 132, pp. 263-267.
Söderström, T. (1973). An on-line algorithm for approximate maximum likelihood identification of linear dynamic systems, *Report 7308*, Department of Automatic Control, Lund Institute of Technology, Sweden.
Söderström, T. (1974). Convergence properties of the generalized least squares identification method, *Automatica*, Vol. 10, pp. 617-626.
Söderström, T. (1975a). Comments on 'Order assumption and singularity of information matrix for pulse transfer function models', *IEEE Transactions on Automatic Control*, Vol. AC-20, pp. 445-447.
Söderström, T. (1975b). Ergodicity results for sample covariances, *Problems of Control and Information Theory*, Vol. 4, pp. 131-138.
Söderström, T. (1975c). Test of pole-zero cancellation in estimated models, *Automatica*, Vol. 11, pp. 537-541.
Söderström, T. (1977). On model structure testing in system identification, *International Journal of Control*, Vol. 26, pp. 1-18.
Söderström, T. (1981). Identification of stochastic linear systems in presence of input noise, *Automatica*, Vol. 17, pp. 713-725.
Söderström, T. (1987). Identification: model structure determination. In M. .gh (ed.), *Systems and Control Encyclopedia.* Pergamon, Oxford.
Söderström, T., I. Gustavsson and L. Ljung (1975). Identifiability conditions for linear systems operating in closed loop, *International Journal of Control*, vol. 21, pp. 243-255.
Söderström, T., L. Ljung and I. Gustavsson (1974). On the accuracy of identification and the design of identification experiments, *Report 7428*, Department of Automatic Control, Lund Institute of Technology, Sweden.
Söderström, T., L. Ljung and I. Gustavsson (1978). A theoretical analysis of recursive identi-

fication methods, *Automatica*, Vol. 14, pp. 231–244.

Söderström, T. and P. Stoica (1980). On criterion selection in prediction error identification of least squares models, *Bul. Inst. Politeh. Buc.*, Series Electro., Vol. 40, pp. 63–68.

Söderström, T. and P. Stoica (1981a). On criterion selection and noise model parametrization for prediction error identification methods, *International Journal of Control*, Vol. 34, pp. 801–811.

Söderström, T. and P. Stoica (1981b). On the stability of dynamic models obtained by least-squares identification, *IEEE Transactions on Automatic Control*, Vol. AC-26, pp. 575–577.

Söderström, T. and P. Stoica (1981c). Comparison of some instrumental variable methods: consistency and accuracy aspects, *Automatica*, Vol. 17, pp. 101–115.

Söderström, T. and P. Stoica (1982). Some properties of the output error method, *Automatica*, Vol. 18, pp. 93–99.

Söderström, T. and P. Stoica (1983). *Instrumental Variable Methods for System Identification*. (Lecture Notes in Control and Information Sciences, no. 57). Springer Verlag, Berlin.

Söderström, T. and P. Stoica (1984). On the generic consistency of instrumental variable estimates, *Proc. 9th IFAC World Congress, Budapest*.

Söderström, T. and P. Stoica (1988). On some system identification techniques for adaptive filtering, *IEEE Transactions on Circuits and Systems*, Vol. CAS-35, pp. 457–461.

Söderström, T., P. Stoica and E. Trulsson (1987). Instrumental variable methods for closed loop systems, *Proc. 10th IFAC Congress, Munich*.

Solbrand, G., A. Ahlén and L. Ljung (1985). Recursive methods for off-line estimation, *International Journal of Control*, Vol. 41, pp. 177–191.

Solo, V. (1979). The convergence of AML, *IEEE Transactions on Automatic Control*, Vol. AC-24, pp. 958–963.

Solo, V. (1980). Some aspects of recursive parameter estimation, *International Journal of Control*, Vol. 32, pp. 395–410.

de Souza, C.E., M.R. Gevers and G.C. Goodwin (1986). Riccati equations in optimal filtering of nonstabilizable systems having singular state transition matrices, *IEEE Transactions on Automatic Control*, Vol. AC-31, pp. 831–838.

Steiglitz, K. and L.E. McBride (1965). A technique for the identification of linear systems, *IEEE Transactions on Automatic Control*, Vol. AC-10, pp. 461–464.

Stewart, G.W. (1973) *Introduction to Matrix Computations*. Academic Press, New York.

Stoica, P. (1976). The repeated least squares identification method, *Journal A*, Vol. 17, pp. 151–156.

Stoica, P. (1981a). On multivariate persistently exciting signals, *Bul. Inst. Politehnic Bucuresti*, Vol. 43, pp. 59–64.

Stoica, P. (1981b). On a procedure for testing the orders of time series, *IEEE Transactions on Automatic Control*, Vol. AC-26, pp. 572–573.

Stoica, P. (1981c). On a procedure for structural identification, *International Journal of Control*, Vol. 33, pp. 1177–1181.

Stoica, P. (1983). Generalized Yule–Walker equations and testing the orders of multivariate time-series, *International Journal of Control*, Vol. 37, pp. 1159–1166.

Stoica, P., P. Eykhoff, P. Janssen and T. Söderström (1986). Model structure selection by cross-validation, *International Journal of Control*, Vol. 43, pp. 1841–1878.

Stoica, P., B. Friedlander and T. Söderström (1985a). The parsimony principle for a class of model structures, *IEEE Transactions on Automatic Control*, Vol. AC-30, pp. 597–600.

Stoica, P., B. Friedlander and T. Söderström (1985b). Optimal instrumental variable multistep algorithms for estimation of the AR parameters of an ARMA process, *Proc. 24th IEEE Conference on Decision and Control, Fort Lauderdale*. An extended version appears in *International Journal of Control*, vol. 45, pp. 2083–2107, 1987.

Stoica, P., B. Friedlander and T. Söderström (1985c). An approximate maximum likelihood approach to ARMA spectral estimation, *Proc. 24th IEEE Conference on Decision and Control, Fort Lauderdale*. An extended version appears in *International Journal of Control*, Vol. 45,

pp. 1281–1310, 1987.
Stoica, P., B. Friedlander and T. Söderström (1986). Asymptotic properties of high-order Yule–Walker estimates of frequencies of multiple sinusoids, *Proc. IEEE Conference on Acoustics, Speech and Signal Processing, ICASSP '86, Tokyo.*
Stoica, P., J. Holst and T. Söderström (1982). Eigenvalue location of certain matrices arising in convergence analysis problems, *Automatica*, Vol. 18, pp. 487–489.
Stoica, P. and R. Moses (1987). On the unit circle problem: the Schur–Cohn procedure revisited. *Technical report SAMPL-87-06*, Department of Electrical Engineering, The Ohio State University, Columbus.
Stoica, P., R. Moses, B. Friedlander and T. Söderström (1986). Maximum likelihood estimation of the parameters of multiple sinusoids from noisy measurements. *3rd ASSP Workshop on Spectrum Estimation and Modelling, Boston.* An expanded version appears in *IEEE Transaction on Acoustics, Speech and Signal Processing,* Vol. ASSP-37, pp. 378–392, 1989.
Stoica, P. and A. Nehorai (1987a). A non-iterative optimal min-max instrumental variable method for system identification, *Technical Report No. 8704*, Department of Electrical Engineering, Yale University, New Haven. Also *International Journal of Control,* Vol. 47, pp. 1759–1769, 1988.
Stoica, P. and A. Nehorai (1987b). On multistep prediction error methods. *Technical Report No. 8714*, Department of Electrical Engineering, Yale University, New Haven. Also to appear in *Journal of Forecasting*.
Stoica, P. and A. Nehorai (1987c). On the uniqueness of prediction error models for systems with noisy input-output data, *Automatica*, Vol. 23, pp. 541–543.
Stoica, P. and A. Nehorai (1988). On the asymptotic distribution of exponentially weighted prediction error estimators, *IEEE Transactions on Acoustics, Speech and Signal Processing,* Vol. ASSP-36, pp. 136–139.
Stoica, P., A. Nehorai and S.M. Kay (1987). Statistical analysis of the least squares autoregressive estimator in the presence of noise, *IEEE Transactions on Acoustics, Speech and Signal Processing,* Vol. ASSP-35, pp. 1273–1281.
Stoica, P. and T. Söderström (1981a). Asymptotic behaviour of some bootstrap estimators, *International Journal of Control,* Vol. 33, pp. 433–454.
Stoica, P. and T. Söderström (1981b). The Steiglitz–McBride algorithm revisited: convergence analysis and accuracy aspects, *IEEE Transactions on Automatic Control,* Vol. AC-26, pp. 712–717.
Stoica, P. and T. Söderström (1982a). Uniqueness of maximum likelihood estimates of ARMA model parameters – an elementary proof, *IEEE Transactions on Automatic Control,* Vol. AC-27, pp. 736–738.
Stoica, P. and T. Söderström (1982b). A useful input parameterization for optimal experiment design, *IEEE Transactions on Automatic Control,* Vol. AC-27, pp. 986–989.
Stoica, P. and T. Söderström (1982c). Comments on the Wong and Polak minimax approach to accuracy optimization of instrumental variable methods, *IEEE Transactions on Automatic Control,* Vol. AC-27, pp. 1138–1139.
Stoica, P. and T. Söderström (1982d). Instrumental variable methods for identification of Hammerstein systems, *International Journal of Control,* Vol. 35, pp. 459–476.
Stoica, P. and T. Söderström (1982e). On nonsingular information matrices and local identifiability, *International Journal of Control,* Vol. 36, pp. 323–329.
Stoica, P. and T. Söderström (1982f). On the parsimony principle, *International Journal of Control,* Vol. 36, pp. 409–418.
Stoica, P. and T. Söderström (1982g). Uniqueness of prediction error estimates of multivariable moving average models, *Automatica*, Vol. 18, pp. 617–620.
Stoica, P. and T. Söderström (1983a). Optimal instrumental-variable methods for the identification of multivariable linear systems, *Automatica*, Vol. 19, pp. 425–429.
Stoica, P. and T. Söderström (1983b). Optimal instrumental variable estimation and approximate

implementation, *IEEE Transactions on Automatic Control*, Vol. AC-28, pp. 757–772.

Stoica, P. and T. Söderström (1984). Uniqueness of estimated k-step prediction models of ARMA processes, *Systems and Control Letters*, Vol. 4, pp. 325–331.

Stoica, P. and T. Söderström (1985). Optimization with respect to covariance sequence parameters, *Automatica*, Vol. 21, pp. 671–675.

Stoica, P. and T. Söderström (1987). On reparametrization of loss functions used in estimation and the invariance principle. *Report UPTEC 87113R*, Department of Technology, Uppsala University, Sweden.

Stoica, P., T. Söderström, A. Ahlén and G. Solbrand (1984). On the asymptotic accuracy of pseudo-linear regression algorithms, *International Journal of Control*, Vol. 39, pp. 115–126.

Stoica, P., T. Söderström, A. Ahlén and G. Solbrand (1985). On the convergence of pseudo-linear regression algorithms, *International Journal of Control*, Vol. 41, pp. 1429–1444.

Stoica, P., T. Söderström and B. Friedlander (1985). Optimal instrumental variable estimates of the AR parameters of an ARMA process, *IEEE Transactions on Automatic Control*, Vol. AC-30, pp. 1066–1074.

Strang, G. (1976). *Linear Algebra and Its Applications*. Academic Press, New York.

Strejc, V. (1980). Least squares parameter estimation, *Automatica*, Vol. 16, pp. 535–550.

Treichler, J.R., C.R. Johnson Jr and M.G. Larimore (1987). *Theory and Design of Adaptive Filters*. Wiley, New York.

Trench, W.F. (1964). An algorithm for the inversion of finite Toeplitz matrices, *J. SIAM*, Vol. 12, pp. 512–522.

Tsypkin, Ya. (1987). *Grundlagen der Informationellen Theorie der Identifikation*. VEB Verlag Technik, Berlin.

Tyssø, A. (1982). CYPROS, an interactive program system for modeling, identification and control of industrial and nontechnical processes, *Proc. American Control Conference, Arlington, VA*.

Unbehauen, H. and B. Göhring (1974). Tests for determining model order in parameter estimation, *Automatica*, Vol. 10, pp. 233–244.

Unbehauen, H. and G.P. Rao (1987). *Identification of Continuous Systems*. North-Holland, Amsterdam.

Verbruggen, H.B. (1975). Pseudo random binary sequences, *Journal A*, Vol. 16, pp. 205–207.

Vieira, A. and T. Kailath (1977). On another approach to the Schur–Cohn criterion, *IEEE Transactions on Circuits and Systems*, Vol. CAS-24, pp. 218–220.

Wahlberg, B. (1985). Connections between system identification and model reduction, *Report LiTH-ISY-I-0746*, Department of Electrical Engineering, Linköping University, Sweden.

Wahlberg, B. (1986). On model reduction in system identification, *Proc. American Control Conference, Seattle*.

Wahlberg, B. (1987). On the identification and approximation of linear systems. Doctoral dissertation, no. 163, Department of Electrical Engineering, Linköping University, Sweden.

Wahlberg, B. and L. Ljung (1986). Design variables for bias distribution in transfer function estimation, *IEEE Transactions on Automatic Control*, Vol. AC-31, pp. 134–144.

Walter, E. (1982). *Identifiability of State Space Models*. (Lecture Notes in Biomathematics no. 46). Springer Verlag, Berlin.

Walter, E. (ed.) (1987). *Identifiability of Parametric Models*. Pergamon, Oxford.

Weisberg, S. (1985). *Applied Linear Regression* (2nd edn). Wiley, New York.

Wellstead, P.E. (1978). An instrumental product moment test for model order estimation, *Automatica*, Vol. 14, pp. 89–91.

Wellstead, P.E. (1979). *Introduction to Physical System Modelling*. Academic Press, London.

Wellstead, P.E. (1981). Non-parametric methods of system identification, *Automatica*, Vol. 17, pp. 55–69.

Wellstead, P.E. and R.A. Rojas (1982). Instrumental product moment model-order testing: extensions and applications, *International Journal of Control*, Vol. 35, pp. 1013–1027.

Werner, H.J. (1985). More on BLU estimation in regression models with possibly singular covariances, *Linear Algebra and Its Applications*, Vol. 64, pp. 207–214.

Wetherill, G.B., P. Duncombe, K. Kenward, J. Köllerström, S.R. Paul and B.J. Vowden (1986). *Regression Analysis with Applications*. Chapman and Hall, London.

Whittle, P. (1953). The analysis of multiple stationary time series, *Journal Royal Statistical Society*, Vol. 15, pp. 125–139.

Whittle, P. (1963). On the fitting of multivariate autoregressions and the approximate canonical factorization of a spectral density matrix, *Biometrika*, Vol. 50, pp. 129–134.

Widrow, B. and S.D. Stearns (1985). *Adaptive Signal Processing*. Prentice Hall, Englewood Cliffs.

Wieslander, J. (1976). IDPAC – User's guide, *Report TFRT-3099*, Department of Automatic Control, Lund Institute of Technology, Sweden.

Wiggins, R.A. and E.A. Robinson (1966). Recursive solution to the multichannel filtering problem, *Journal Geophysical Research*, Vol. 70, pp. 1885–1891.

Willsky, A. (1976). A survey of several failure detection methods, *Automatica*, Vol. 12, pp. 601–611.

Wilson, G.T. (1969). Factorization of the covariance generating function of a pure moving average process, *SIAM Journal Numerical Analysis*, Vol. 6, pp. 1–8.

Wong, K.Y. and E. Polak (1967). Identification of linear discrete time systems using the instrumental variable approach, *IEEE Transactions on Automatic Control*, Vol. AC-12, pp. 707–718.

Young, P.C. (1965). On a weighted steepest descent method of process parameter estimation, *Report*, Cambridge University, Engineering Laboratory, Cambridge.

Young, P.C. (1968). The use of linear regression and related procedures for the identification of dynamic processes, *Proc. 7th IEEE Symposium on Adaptive Processes, UCLA, Los Angeles*.

Young, P.C. (1970). An instrumental variable method for real-time identification of a noisy process, *Automatica*, Vol. 6, pp. 271–287.

Young, P.C. (1976). Some observations on instrumental variable methods of time series analysis, *International Journal of Control*, Vol. 23, pp. 593–612.

Young, P. (1981). Parameter estimation for continuous-time models: a survey, *Automatica*, Vol. 17, pp. 23–39.

Young, P. (1982). A computer program for general recursive time-series analysis, *Proc. 6th IFAC Symposium on Identification and System Parameter Estimation, Washington DC*.

Young, P. (1984). *Recursive Estimation and Time-Series Analysis*. Springer Verlag, Berlin.

Young, P.C. and A.J. Jakeman (1979). Refined instrumental variable methods of recursive time-series analysis. Part I: Single input, single output systems, *International Journal of Control*, Vol. 29, pp. 1–30.

Young, P.C., A.J. Jakeman and R. McMurtrie (1980). An instrumental variable method for model order identification, *Automatica*, Vol. 16, pp. 281–294.

Zadeh, L.A. (1962). From circuit theory to systems theory, *Proc. IRE*, Vol. 50, pp. 856–865.

Zadeh, L.A. and E. Polak (1969). *System Theory*. McGraw-Hill, New York.

Zarrop, M.B. (1979a). A Chebyshev system approach to optimal input design, *IEEE Transactions on Automatic Control*, Vol. AC-24, pp. 687–698.

Zarrop, M.B. (1979b). *Optimal Experiment Design for Dynamic System Identification*. (Lecture Notes in Control and Information Sciences no. 21). Springer Verlag, Berlin.

Zarrop, M. (1983). Variable forgetting factors in parameter estimation, *Automatica*, Vol. 19, pp. 295–298.

van Zee, G.A. and O.H. Bosgra (1982). Gradient computation in prediction error identification of linear discrete-time systems, *IEEE Transactions on Automatic Control*, Vol. AC-27, pp. 738–739.

Zohar, S. (1969). Toeplitz matrix inversion: the algorithm of W.F. Trench, *J. ACM*, Vol. 16, pp. 592–601.

Zohar, S. (1974). The solution of a Toeplitz set of linear equations, *J. ACM*, Vol. 21, pp. 272–278.

Zygmund, A. (1968). *Trigonometric Series* (rev. edn). Cambridge University Press, Cambridge.

ANSWERS AND FURTHER HINTS TO THE PROBLEMS

2.1 $\text{mse}(\hat{\theta}_1) = \dfrac{N+1}{N^2} < \text{mse}(\hat{\theta}_2) = \dfrac{3}{N}$

2.2 $\text{var}(\hat{\alpha}) = \dfrac{6\lambda^2}{N(N+1)(2N+1)} \approx \dfrac{3\lambda^2}{N^3}$

2.3 $E\hat{\mu} = \mu \quad \text{var}(\hat{\mu}) = \dfrac{\sigma}{N}$

$E\hat{\sigma}_1 = \dfrac{N-1}{N}\sigma \quad E\hat{\sigma}_2 = \sigma$

$\text{var}(\hat{\sigma}_1) = 2\dfrac{N-1}{N^2}\sigma^2 \quad \text{mse}(\hat{\sigma}_1) = \dfrac{2N-1}{N^2}\sigma^2$

$\text{var}(\hat{\sigma}_2) = \text{mse}(\hat{\sigma}_2) = \dfrac{2}{N-1}\sigma^2 > \text{mse}(\hat{\sigma}_1)$

2.7 *Hint.* Show that $V''(\theta)$ is positive definite.

3.4 $r_y(k) = \dfrac{\lambda^2}{1-a^2}(-a)^k$

3.8 $\text{var}[\hat{h}(k)] = \dfrac{1}{N}\dfrac{b^2}{1-a^2}[1 + (2k-3)(1-a^2)a^{2k-2}] \quad k \geq 1$

3.9 *Hint.* Show that

$U_N(\omega) = \begin{cases} Na/2i & \omega = \tilde{\omega} \\ 0 & \omega = 2\pi j/N, j \neq n \end{cases}$

3.10 *Hint.* Show that

$\hat{H}(e^{i\omega}) \approx \dfrac{1}{a}Y_N(\omega)[1 - e^{-i\omega}], \quad U(\omega) = \dfrac{a}{1 - e^{-i\omega}}$

4.1 (a) Case (i) $\hat{a} = \dfrac{1}{N(N-1)}[2(2N+1)S_0 - 6S_1]$

$\hat{b} = \dfrac{6}{N(N-1)(N+1)}[-(N+1)S_0 + 2S_1]$

Case (ii) $\hat{a} = \dfrac{1}{2N+1}S_0$

$\hat{b} = \dfrac{3}{N(N+1)(2N+1)}S_1$

(b) Case (i) $\hat{b} = \dfrac{6}{N(N+1)(2N+1)}\left[-\dfrac{N+1}{2}S_0 + S_1\right]$

Case (ii) $\hat{b} = \dfrac{3}{N(N+1)(2N+1)} S_1$

(c) $\text{var}[s(1)] = \text{var}[s(N)] = \dfrac{2\lambda^2(2N-1)}{N(N+1)}$

$\text{var}[s(t)]$ is minimized for $t = (N+1)/2$; $\text{var}[s((N+1)/2)] = \dfrac{\lambda^2}{N}$

(d) $\varrho = -\left(\dfrac{3(N+1)}{2(2N+1)}\right)^{1/2} \approx -\dfrac{\sqrt{3}}{2}$

(e) $N = 3$: $(\hat{\theta} - \theta)^T \begin{pmatrix} 3 & 6 \\ 6 & 14 \end{pmatrix}(\hat{\theta} - \theta) = 0.2$

$N = 8$: $(\hat{\theta} - \theta)^T \begin{pmatrix} 8 & 36 \\ 36 & 204 \end{pmatrix}(\hat{\theta} - \theta) = 0.2$

4.2 In $\mathcal{M}_1$ $\text{var}(\hat{b}) = \dfrac{12}{N(N^2-1)}$

In $\mathcal{M}_2$ $\text{var}(\hat{b}) = \dfrac{2}{(N-1)^2}$

4.4 $\hat{\theta} = (\Phi_1^T \Phi_1)^{-1} \Phi_1^T y_1 \quad \hat{y}_2 = \Phi_2 \hat{\theta}$

Hint. First write

$Y - \Phi\theta = \begin{pmatrix} y_1 \\ 0 \end{pmatrix} - \begin{pmatrix} \Phi_1 & 0 \\ \Phi_2 & -I \end{pmatrix} \begin{pmatrix} \theta \\ y_2 \end{pmatrix}$

4.5 Hint. Use $\sum_{t=1}^{N} t^k = O(N^{k+1}/(k+1))$, N large

4.6 Hint. $\Phi^T \Phi = \dfrac{N}{2} I \quad \hat{\theta} = \dfrac{2}{N} \sum_{t=1}^{N} \varphi(t) y(t)$

4.7 The set of minimum points = the set of solutions to the normal equations. This set is given by

$\begin{pmatrix} \alpha \\ \sum_{t=1}^{N} y(t)/N - \alpha \end{pmatrix}$ α an arbitrary parameter

The pseudoinverse of Φ is

$\Phi^\dagger = \dfrac{1}{2N} \begin{pmatrix} 1 & \cdots & 1 \\ 1 & \cdots & 1 \end{pmatrix}$

Minimum norm solution = $\dfrac{1}{2N} \sum_{t=1}^{N} y(t) \begin{pmatrix} 1 \\ 1 \end{pmatrix}$

4.8 $|u(t)| = \beta$, $t = 1, \ldots, N$

4.9 $x_1 = B(B^T B)^{-1} d \quad x_2 = x_1$

$x_3 = B(\delta I + B^T B)^{-1} d \to x_1$ as $\delta \to 0$

4.10 Hint. Take $F = (\Phi^T R^{-1} \Phi)^{-1} \Phi^T \Phi$

4.12 *Hint.* Show that $NP_{LS} = \sum_{\tau=-N}^{N} r(\tau) - \frac{1}{N}\sum_{\tau=-N}^{N}|\tau|r(\tau)$

and that $\left(\frac{1}{N}\sum_{\tau=-N}^{N} r(\tau)\right)\Phi^T R^{-1}\Phi \to 1$ as $N \to \infty$.

5.2 *Hint.* Use the result (A3.1.10) with $H(q^{-1}) = 1 - 2\cos\bar{\omega}q^{-1} + q^{-2}$.

5.3 *Hint.* $C_p = \sum_{j=1}^{2}\frac{a_j^2}{4}(\delta_{k_j,p} + \delta_{M-k_j,p})$

5.4 $|\varrho_1| \leq 1,\ 2\varrho_1^2 - 1 \leq \varrho_2 \leq 1$

5.5 The contour of the area for an MA(2) process is given by the curves

$$\varrho_1 + \varrho_2 = -\frac{1}{2}$$

$$\varrho_1 - \varrho_2 = \frac{1}{2}$$

$$\varrho_1^2 = 4\varrho_2(1 - 2\varrho_2)$$

For an AR(2) process: $|\varrho_1| \leq 1,\ 2\varrho_1^2 - 1 \leq \varrho_2 \leq 1$

5.6 (a) $r(\tau) = a^2(1 - 2\alpha)^{|\tau|}$

Hint. Use the relation $Eu = E[E(u/y)]$

(b) $\phi(\omega) = \frac{a^2}{2\pi}\frac{1 - (1 - 2\alpha)^2}{1 + (1 - 2\alpha)^2 - 2(1 - 2\alpha)\cos\omega}$

5.7 (a) $r(\tau) = 1 - \frac{2\tau}{n},\ \tau = 0,\ldots,n-1;\ r(\tau) = -r(\tau - n)$

(c) $u(t)$ is pe of order n.

5.8 *Hint.* Let $y(t) = C(q^{-1})e(t)$ and set $\beta_j = |c_j|$ and first show

$$|r_k| \leq \sum_{j=0}^{n-k}\beta_j\beta_{j+k}$$

$$\sum_{k=1}^{n}|r_k| \leq \frac{1}{2}(\beta_0 \ldots \beta_n)\begin{pmatrix} 0 & 1 & \ldots & 1 \\ 1 & & & \\ \vdots & & \ddots & 1 \\ 1 & \ldots & 1 & 0 \end{pmatrix}\begin{pmatrix}\beta_0 \\ \vdots \\ \beta_n\end{pmatrix}$$

Use the result $x^T Q x \leq \lambda_{\max}(Q)x^T x$ which holds for a symmetric matrix Q. When Q has the structure $Q = ee^T - I$, where $e = (1\ \ldots\ 1)$, its eigenvalues can be calculated exactly by using Corollary 1 of Lemma A.5.

5.9 *Hint.* Use Lemma A.1 to invert the $(M|M)$ matrix R_u.

6.1 The boundary of the area is given by the lines:

$a_2 = 1$ $-2 \leq a_1 \leq 2$

$a_2 = -1 - a_1$ $-2 \leq a_1 \leq 0$

$a_2 = -1 + a_1$ $0 \leq a_1 \leq 2$

6.2 $y(t) - 0.8y(t-1) = e(t) - 0.5e(t-1)\quad Ee^2(t) = 0.8$

6.3 $\phi_y(\omega) = \dfrac{5\pi}{1.25 - \cos\omega}$

6.4 $B^T = C = (1 \ 0 \ \ldots \ 0)$, $x(t) = (y(t) \ \ldots \ y(t+1-n))^T$

6.5 (a) A_s, B_s coprime, $\min(na - na_s, nb - nb_s) = 0$, $nc \geq nc_s$.
(b) Many examples can be constructed where $\min(na - na_s, nb - nb_s) > 0$, $nc \geq nc_s$.

6.6 (a) The sign of a_{12} cannot be determined.
(b) No single parameter can be determined, only the combinations ba_{12}, $a_{11} + a_{22}$, $a_{11}a_{22} - a_{12}^2$.
(c) All parameters can be uniquely determined.

6.7 $G(q^{-1}; K) = \dfrac{Khq^{-1}}{1 - q^{-1}}$ $H(q^{-1}; K) = \dfrac{1 + (2 - \sqrt{3})q^{-1}}{(1 - q^{-1})^2}$

$\Lambda(K) = K^2 r h^3 (2 + \sqrt{3})/6$

Hint. Intermediate results are

$$R_1 = rh \begin{pmatrix} K^2 h^2/3 & Kh/2 \\ Kh/2 & 1 \end{pmatrix} \quad P = rh \begin{pmatrix} K^2 h^2 \left(\dfrac{1}{3} + \dfrac{1}{\sqrt{12}}\right) & Kh\left(\dfrac{1}{2} + \dfrac{1}{\sqrt{12}}\right) \\ Kh\left(\dfrac{1}{2} + \dfrac{1}{\sqrt{12}}\right) & 1 + \dfrac{1}{\sqrt{12}} \end{pmatrix}$$

7.1 $\hat{y}(t+1|t) = \dfrac{0.5}{1 + 0.5q^{-1}} y(t)$ $\Gamma[\hat{y}(t+1|t) - y(t+1)]^2 = 4$

Hint. Compute $\phi_y(\omega)$ and make a spectral factorization to get the model $y(t) = v(t) + 0.5v(t-1)$, $Ev(t)v(s) = 4\delta_{t,s}$.

7.3 (b) $(1 - q^{-1})y(t) = (1 - aq^{-1})e(t)$

7.4 $\dfrac{\partial \varepsilon(t, \theta)}{\partial b_i} = -\dfrac{D(q^{-1})}{C(q^{-1})F(q^{-1})} u(t - i)$

$\dfrac{\partial \varepsilon(t, \theta)}{\partial c_i} = -\dfrac{1}{C(q^{-1})} \varepsilon(t - i)$

$\dfrac{\partial \varepsilon(t, \theta)}{\partial d_i} = \dfrac{1}{D(q^{-1})} \varepsilon(t - i)$

$\dfrac{\partial \varepsilon(t, \theta)}{\partial f_i} = \dfrac{D(q^{-1})B(q^{-1})}{C(q^{-1})F^2(q^{-1})} u(t - i)$

7.7 (a) Hint. Show that $V(x^{(k+1)}) - V(x^{(k)}) = -\alpha g_k^T S g_k + O(\alpha^2)$ with $g_k = V'(x^{(k)})^T$.

7.8 Hint. Show first that $\det(Q + \Delta Q) - \det Q = \text{tr}[(\det Q)Q^{-1}\Delta Q] + O(\|\Delta Q\|^2)$

7.9 $L_1(q^{-1}; \theta) = C(\theta)[qI - A(\theta) + K(\theta)C(\theta)]^{-1}K(\theta)$
$L_2(q^{-1}; \theta) = C(\theta)[qI - A(\theta) + K(\theta)C(\theta)]^{-1}B(\theta)$
Hint for (b). Use Example 6.5.

7.10 $\dfrac{\partial^2 V}{\partial \theta_i \partial \theta_j} = 2 \sum_{t=1}^{N} \dfrac{\partial \varepsilon(t, \theta)}{\partial \theta_i} \dfrac{\partial \varepsilon(t, \theta)}{\partial \theta_j} + 2 \sum_{t=1}^{n} \varepsilon(t, \theta) \dfrac{\partial^2 \varepsilon(t, \theta)}{\partial \theta_i \partial \theta_j}$

$\dfrac{\partial \varepsilon(t, \theta)}{\partial a_i} = \dfrac{1}{C(q^{-1})} y(t - i)$ $\dfrac{\partial \varepsilon(t, \theta)}{\partial b_i} = \dfrac{-1}{C(q^{-1})} u(t - i)$

$\dfrac{\partial \varepsilon(t, \theta)}{\partial c_i} = \dfrac{-1}{C(q^{-1})} \varepsilon(t - i, \theta)$

$\dfrac{\partial^2 \varepsilon(t, \theta)}{\partial a_i \partial a_j} = 0$ $\dfrac{\partial^2 \varepsilon(t, \theta)}{\partial a_i \partial b_j} = 0$ $\dfrac{\partial^2 \varepsilon(t, \theta)}{\partial b_i \partial b_j} = 0$

$$\frac{\partial^2 \varepsilon(t, \theta)}{\partial a_i \partial c_j} = \frac{-1}{C^2(q^{-1})} y(t - i - j) \quad \frac{\partial^2 \varepsilon(t, \theta)}{\partial b_i \partial c_j} = \frac{1}{C^2(q^{-1})} u(t - i - j)$$

$$\frac{\partial^2 \varepsilon(t, \theta)}{\partial c_i \partial c_j} = \frac{2}{C^2(q^{-1})} \varepsilon(t - i - j, \theta)$$

7.11 $\hat{y}(t + k|t) = \dfrac{(-a)^{k-1}(c - a)}{1 + cq^{-1}} y(t)$

where $|c| < 1$ is uniquely given by

$$\lambda^2(1 + cz)(1 + cz^{-1}) \equiv \lambda_e^2 + \lambda_\varepsilon^2(1 + az)(1 + az^{-1})$$

If the processes are not Gaussian this is still the best linear predictor (in the mean square sense).

7.12 *Hint.* Show that $\hat{\theta} - \theta_0 \approx [E\psi^T(t, \theta_0)\psi(t, \theta_0)]^{-1} \left(\dfrac{1}{N} \sum_{t=1}^{N} \psi(t, \theta_0)\varepsilon(t, \theta_0) \right)$

7.13 The asymptotic criteria minimized by the LS and OE methods are:

$$V_{LS} = \int_{-\pi}^{\pi} |A(e^{i\omega})|^2 \left| G(e^{i\omega}) - \frac{B(e^{i\omega})}{A(e^{i\omega})} \right|^2 \phi_u(\omega) d\omega$$

$$+ \int_{-\pi}^{\pi} |A(e^{i\omega})|^2 \phi_e(\omega) d\omega$$

$$V_{OE} = \int_{-\pi}^{\pi} \left| G(e^{i\omega}) - \frac{B(e^{i\omega})}{A(e^{i\omega})} \right|^2 \phi_u(\omega) d\omega + \text{constant}$$

7.14 (a) $NP_1 = \begin{pmatrix} 1 - a_0^2 & 0 \\ 0 & \dfrac{\lambda^2}{\sigma^2(1 + a_0^2)} \end{pmatrix}$

$NP_2 = \begin{pmatrix} 1 - a_0^2 & 0 & b_0(1 - a_0^2) \\ 0 & \lambda^2/\sigma^2 & 0 \\ b_0(1 - a_0^2) & 0 & b_0^2(1 - a_0^2) + \lambda^2/\sigma^2 \end{pmatrix}$

(c) *Hint.* Use the fact that $\begin{pmatrix} F^T P_2^{-1} F & F^T Q F \\ F^T Q F & F^T Q P_2 Q F \end{pmatrix} \geq 0$

and Lemma A.3

7.17 (a) *Hint.* As an intermediate step, show that

$$\frac{1}{N} \sum_{\tau=-N}^{N} |\tau| r_\tau r_{\tau+k} \to 0 \text{ as } N \to \infty.$$

(b) $a^* = a$, $P = 1 - a^2$

7.18 (a) Apply a PEM to the model structure given in part (b). Alternatively, note that $y(t)$ is an ARMA process, $A_0(q^{-1})y(t) = C_0(q^{-1})w(t)$, for some C_0, and use a PEM for ARMA estimation.

(b) Let $C(q^{-1})$ and λ_w^2 with C stable be uniquely determined from θ by

$$C(z)C(z^{-1})\lambda_w^2 \equiv A(z)A(z^{-1})\lambda_e^2 + \lambda_v^2$$

Then $\varepsilon(t, \theta) = [A(q^{-1})/C(q^{-1})]y(t)$. Now the vector $(a_1 \ldots a_n \ \beta\lambda_v^2 \ \beta\lambda_e^2)'$ and θ give the same $V(\theta)$ for all β. Remedy: Set $\lambda_e^2 = 1$ during the optimization and compute the correct β-value from $\min V(\theta)$. As an alternative set $\tilde{\theta} = (a_1 \ldots a_n \rho^2)^T \rho^2 = \lambda_v^2/\lambda_e^2$ and $\bar{\theta} = (\tilde{\theta} \ \lambda_w^2)$. Minimize V w.r.t. $\tilde{\theta}$ and set $\lambda_w^2 = \min V/N$, giving $\bar{\theta}$. Finally θ can be computed by a 1-to-1 transformation form $\bar{\theta}$.

7.20 *Hint.* Set $P(q^{-1}) = \begin{pmatrix} \tilde{A}(q^{-1}) & \tilde{B}(q^{-1}) \\ \bar{A}(q^{-1}) & \bar{B}(q^{-1}) \end{pmatrix}$, $x(t) = \begin{pmatrix} y(t) \\ u(t) \end{pmatrix}$

and consider the problem of $E\|P(q^{-1})x(t)\|^2 = \min$. Show that $\tilde{A} = \hat{A}$, $\tilde{B} = \hat{B}$, $\bar{A} = I$, $\bar{B} = 0$.

7.22 *Hint.* The stationary solutions to (i) must satisfy

$$E\left[A(q^{-1})\left\{\frac{1}{A(q^{-1})}y(t)\right\} - B(q^{-1})\left\{\frac{1}{A(q^{-1})}u(t)\right\}\right]$$

$$\times \begin{pmatrix} \frac{1}{A(q^{-1})} y(t-i) \\ \frac{1}{A(q^{-1})} u(t-i) \end{pmatrix} = 0 \quad i = 1, \ldots, n$$

8.2 For the extended IV estimate the weighting matrix changes from Q to $T^T Q T$.

8.3 (a) $R = \begin{pmatrix} 0 & 1 & 0 \\ -b_1 & 0 & 1 \\ -(b_2 & ab_1) & 0 & 0 \end{pmatrix}$

(b) R is singular $\Leftrightarrow ab_1 = b_2 \Leftrightarrow 1 + az, b_1z + b_2z^2$ are not coprime. When $ab_1 = b_2$, R will be singular for all inputs. If $b_2 \neq ab_1$ the matrix R is nonsingular for pe inputs.

8.4 $n\tau \geq nb$, with $n\tau = nb$ being preferable.

8.6 $\hat{\theta}_{LS} = \begin{pmatrix} -\sigma^2/(2\sigma^2 + \lambda^2) \\ 1 \end{pmatrix}$ which corresponds to a stable model

$\hat{\theta}_{IV} = \begin{pmatrix} -1 \\ 1 \end{pmatrix}$ which corresponds to a model with a pole on the stability boundary

8.7 $P_{IV} = \frac{\lambda^2}{b^2\sigma^2}\begin{pmatrix} 1 + c^2 & -bc \\ -bc & b^2(1 + c^2) \end{pmatrix}$

8.8 (a) $P_{IV}^{opt} = \frac{\lambda^2}{b^2\sigma^2}\begin{pmatrix} (1-a^2)(1-ac)^2 & -bc(1-a^2)(1-ac) \\ -bc(1-a^2)(1-ac) & b^2(1-a^2c^2) \end{pmatrix}$

(c) *Hint.* $\tilde{P} = \frac{\lambda^4(c-a)^2(1-a^2)}{b^2\sigma^2[b^2\sigma^2 + \lambda^2(c-a)^2]} \begin{pmatrix} 1-ac \\ -bc \end{pmatrix}(1-ac \quad -bc)$

8.9 *Hint.* Show that $\mathcal{N}(EZ(t)Z^T(t)) \subset \mathcal{N}(R^T)$, and that this implies rank $R \leq$ rank $EZ(t)Z^T(t)$.

8.11 Set $r_k = Ey(t)y(t-k)$ and $\tilde{r}_k = E[C(q^{-1})y(t)][C(q^{-1})y(t-k)]$. Then

$P_1 = \frac{\tilde{r}_0}{r_1^2} \quad P_2 = \frac{(1+a^2)\tilde{r}_0 - 2a\tilde{r}_1}{r_1^2(1+a^2)^2}$

and $P_1 < P_2$ for example if $ac < 0$.

9.2 $\text{var}(\hat{b}) = \left[\sum_{t=1}^{N} \lambda^{2(N-t)} u^2(t)\right] / \left[\sum_{t=1}^{N} \lambda^{N-t} u^2(t)\right]^2$

9.3 (a) $V(t) = \frac{(\hat{\theta}_0 - \theta_0)^2 + P_0^2\lambda^2 t}{(1 + P_0 t)^2}$

9.4 (a) $P^{-1}(t) = \lambda P^{-1}(t-1) + \varphi(t)\varphi^T(t)$
$[P^{-1}(t)\hat{\theta}(t)] = \lambda[P^{-1}(t-1)\hat{\theta}(t-1)] + \varphi(t)y(t)$
$$\hat{\theta}(t) = \left[\lambda^t P^{-1}(0) + \sum_{s=1}^{t} \lambda^{t-s}\varphi(s)\varphi^T(s)\right]^{-1}\left[\lambda^t P^{-1}(0)\hat{\theta}(0) + \sum_{s=1}^{t} \lambda^{t-s}\varphi(s)y(s)\right]$$

9.6 (b) The set D_g is given by
$-1 < c_1 + c_2 < 1, \; -1 < c_1 - c_2 < 1$

9.12 (a) *Hint.* First show the following. Let h be an arbitrary vector and let $\phi(\omega)$ be the spectral density of $h^T B^T \varphi(t)$. Then
$$h^T \tilde{M} h = \int_{-\pi}^{\pi} \operatorname{Re} \frac{1}{C(e^{i\omega})} \phi(\omega) d\omega \geq 0.$$

(b) The contour of the set D_l is given by
$c_1 + c_2 = -1, \; c_2 = -1 + c_1, \; 8(c_2 - 0.5)^2 = 2 - c_1^2$

9.13 $L(\theta_0) = \dfrac{1-a^2}{(c-a)} \begin{pmatrix} \dfrac{a}{1-a^2} & \dfrac{-c}{1-ac} \\ \dfrac{c}{1-a^2} & \dfrac{a+ac^2-2c}{(1-a^2)(1-ac)} \end{pmatrix}$

with eigenvalues $\lambda_1 = -1$, $\lambda_2 = -\dfrac{1}{1-ac}$.

9.15 $\hat{u}(t) = \operatorname{sign}[\psi^T(t)r(t)]$ where
$\varphi^T(t+1) = (u(t) \; \psi^T(t)), \; P(t) = \begin{pmatrix} \varrho(t) & r^T(t) \\ r(t) & R(t) \end{pmatrix}$

Hint. Use Corollary 1 of Lemma A.5.

9.16 *Hint.* $\sigma_N = \dfrac{1}{N^{2\alpha}}[\hat{\theta}(0) - \theta]^2 + \dfrac{\lambda^2}{N^{2\alpha}} \sum_{k=1}^{N} k^{2\alpha-2}$

10.1 (a) $N \operatorname{var}(\hat{a}) = \lambda^2 \dfrac{(1-a^2-2abf)\sigma^2 + f^2\lambda^2}{\sigma^2(b^2\sigma^2 + \lambda^2)}$ $N \operatorname{var}(\hat{b}) = \dfrac{\lambda^2}{\sigma^2}$

(b) $N \operatorname{var}(\hat{f}) = \dfrac{\lambda^2(1-\alpha^2)}{b^2(b^2\sigma^2 + \lambda^2)}$

10.2 (a) $\operatorname{var}(\hat{b}) = \dfrac{\lambda^2(1-b^2k^2)}{\sigma^2 + k^2\lambda^2}$

(b) $\bar{a} = bk$, $\bar{b} = b$. Solving these equation with LS will give $\hat{b}_2$ and with a weighted LS will give $\hat{b}_3$.

(c) $\operatorname{var}(\hat{b}_1) = \dfrac{\lambda^2}{\sigma^2} > \operatorname{var}(\hat{b})$

$\operatorname{var}(\hat{b}_2) = \dfrac{\lambda^2}{(1+k^2)^2}\left[\dfrac{k^2(1-b^2k^2)}{b^2\sigma^2 + \lambda^2} + \dfrac{1}{\sigma^2}\right] > \operatorname{var}(\hat{b})$

$\operatorname{var}(\hat{b}_3) = \operatorname{var}(\hat{b})$

10.3 The system is identifiable in case (c) but not in cases (a), (b).

10.4 $\hat{g} = \dfrac{\sigma_\varepsilon^2}{\sigma_\varepsilon^2 + \sigma_e^2} g$

10.5 (a) *Hint.* The identifiability identity can be written
$$AB_0 + BC_0 - A_0B \equiv B_0C$$
(b) $a = \dfrac{\beta}{b_0}(a_0 - c_0)$. It implies minimum variance control and $Ey^2(t) = \lambda^2$.
Hint. Evaluate (9.55b).

10.6 $A = \dfrac{\lambda}{b}[2\delta(1 + a^2) + 2a^2]^{1/2}$

$\omega = \arccos\left[\dfrac{a}{\delta(1 - a^2) - a^2}\right]$

10.7 (a) *Hint.* The information matrix is nonnegative definite.
(b) *Hint.* Choose the regulator gains k_1 and k_2 such that $a + bk_1 = -a - bk_2$.

10.8 (b) *Hint.* The signal $\dfrac{1}{1 + aq^{-1}}u(t)$ can be taken as an AR(1) process with appropriate parameters.

11.3 $EW_N = \lambda^2 + \dfrac{\lambda^2}{N}\left[2 + (\alpha^2 - 1)\dfrac{2b^2\sigma^2 + \lambda^2}{b^2\sigma^2 + \lambda^2}\right]$

11.4 $Ey^2(t) = \lambda^2 + \lambda^2 \dfrac{(c_0 - a_0 + b_0k)^2}{1 - (a_0 - b_0k)^2}\quad k = \dfrac{a - c}{b}$

$Ey^2(t) \geq \lambda^2$ with equality for all models satisfying
$(a - c)b_0 = (a_0 - c_0)b$

11.5 (b) *Hint.* For a first-order model
$$W_N = \lambda^2(1 + a_0^2) + \dfrac{\lambda^2(a_0^2 - a^2)^2}{1 - a_0^2}$$
For a second-order model
$$W_N = \lambda^2(1 + a_0^2) + \dfrac{\lambda^2}{1 - a_0^2}[(a_0^2 - a_1^2 + a_2)^2 + (a_1a_2)^2 - 2a_0(a_0^2 - a_1^2 + a_2)(-a_1a_2)]$$

11.6 (a) $E\varepsilon^2(t) = \lambda^2\left[1 + c^{2n+2}\dfrac{1 - c^2}{1 - c^{2n+2}}\right]$

(b) Case I: 5%, 1.19%, 0.29%. Case II: 50%, 33.3%, 25%

11.7 (a) $N\,\text{var}(\hat{a}) = 1 - a_0^2$, $N\,\text{var}(\hat{a}_1) = 1$.
(b) *Hint.* Use Lemma A.2.
(c) *Hint.* Choose the assessment criterion
$$W_\alpha(\theta) = \dfrac{N}{\lambda^2}(\alpha^T\ 0)(\theta - \theta_0)(\theta - \theta_0)^T\begin{pmatrix}\alpha\\0\end{pmatrix}$$
with α being an arbitrary vector.

11.8 (a) $P_1 = \begin{pmatrix}\dfrac{1}{\lambda^2 + b^2\sigma^2} & 0 \\ 0 & \dfrac{1}{\sigma^2}\end{pmatrix}\quad P_2 = \dfrac{1}{\sigma^2}\begin{pmatrix}1 & 0 & 0\\0 & 1 & 0\\0 & 0 & 1\end{pmatrix}$

(b) $A_{\alpha_1} = \lambda^2 + \dfrac{\lambda^2}{N}\left[\dfrac{\lambda^2 + s^2b^2}{\lambda^2 + \sigma^2b^2} + \dfrac{s^2}{\sigma^2}\right]$

$A_{\alpha_2} = \lambda^2 + \dfrac{\lambda^2}{N}\dfrac{3s^2}{\sigma^2}$

604 *Answers and further hints*

11.9 $Ey = m$ $\text{var}(y) = 2m + 4 \sum_{j=1}^{m} (m - j) \frac{r_u^2(j)}{r_u^2(0)}$

The test quantity (11.18) has mean m and variance $2m$.

11.10 *Hint.* Use

$$\left.\frac{\partial^2 W_N}{\partial \theta^2}\right|_{\theta=\theta_0} = 2(\det \Lambda) E\psi(t)\Lambda^{-1}\psi^T(t)$$

See (A7.1.4) and Problem 7.8.

12.1 $y(t) = \frac{b}{1 + a}[1 - (-a)^t]$

12.3 (a) $\begin{pmatrix} \hat{a} \\ \hat{b} \end{pmatrix} = \begin{pmatrix} a_0 \\ b_0 \end{pmatrix} + \frac{(1 + a_0)m_y - b_0 m_u}{\frac{b_0^2 \sigma^2 + \lambda^2}{1 - a_0^2}(\sigma^2 + m_u^2) + m_y^2 \sigma^2} \begin{pmatrix} -\sigma^2 m_y \\ \frac{b_0^2 \sigma^2 + \lambda^2}{1 - a_0^2} m_u \end{pmatrix}$

are consistent $\Leftrightarrow m_y = \frac{b_0}{1 + a_0} m_u.$

$b_0/(1 + a_0)$ is the static gain.

(b) $\hat{a} = a_0$, $\hat{b} = b_0$, $\hat{c} = (1 + a_0)m_y - b_0 m_u$

12.4 Set $r = (b^2\sigma^2 + \lambda^2)/(1 - a^2)$

(a) $P = \frac{\lambda^2}{\sigma^2 r} \begin{pmatrix} \sigma^2 & 0 & \sigma^2 m \\ 0 & r & 0 \\ \sigma^2 m & 0 & \sigma^2(m^2 + r) \end{pmatrix}$

(b) $P = \begin{pmatrix} \lambda^2/r & 0 \\ 0 & \lambda^2/\sigma^2 \end{pmatrix}$

(c) $\begin{pmatrix} \hat{a} \\ \hat{b} \end{pmatrix} = \begin{pmatrix} a \\ b \end{pmatrix} + \frac{\lambda^2(1 - a)}{b^2\sigma^2(3 + a) + 4\lambda^2} \begin{pmatrix} 2 \\ -b \end{pmatrix}$

(d) $P = \frac{\lambda^2}{b^2\sigma^2} \begin{pmatrix} 2 & -b \\ -b & 2b^2 \end{pmatrix}$

(e) $P_{(a)} = P_{(b)} < P_{(d)}$

12.6 $\text{var}[y^{(1)}(t) - y^{(2)}(t)] = a^{2t}\left[\frac{\lambda^2}{1 - a^2} + \text{var}(x(0))\right]$

12.7 (a) $N \text{var}(\hat{a} - \hat{c}) = 1 - \hat{a}_0^2\hat{c}_0^2$; $N(\hat{a} - \hat{c})^2/(1 - \hat{a}^2\hat{c}^2) \xrightarrow{\text{dist}} \chi^2(1)$
(b) F-test; whiteness test for $\hat{r}_y(\tau)$
(c) 10^4

12.10 $\frac{\partial V}{\partial x(1)} = 2\sum_{t=1}^{N} \varepsilon(t, \theta') \frac{\partial \varepsilon(t, \theta')}{\partial x(1)}$

where $\frac{\partial \varepsilon(t, \theta')}{\partial x(1)}$ can be recursively computed as

$\frac{\partial \varepsilon(t, \theta')}{\partial x(1)} = \frac{\partial \varepsilon(t - 1, \theta')}{\partial x(1)} F(\theta), \quad \frac{\partial \varepsilon(1, \theta')}{\partial x(1)} = C$

12.11 $\hat{x}(1) = -\frac{1 - c^2}{1 - c^{2N}} \sum_{t=1}^{N} (-c)^{t-1} \sum_{s=1}^{t} (-c)^{t-s} y(s)$

$W = \frac{1 - c^2}{1 - c^{2N}} \lambda^2 \to 0$, as $N \to \infty$

12.12 $\mathscr{X}_1$ gives
$$N \operatorname{var}(\hat{S}) = \frac{\lambda^2}{(1+a_0)^2}\left[S^2 \frac{1-a_0^2}{b_0^2\sigma^2 + \lambda^2} + \frac{1}{\sigma^2}\right] = 1050$$
$\mathscr{X}_2$ gives
$$N \operatorname{var}(\hat{S}) = \frac{\lambda^2}{\sigma^2}\frac{1}{(1+a_0)^2} = 100$$
The variance is always smaller for $\mathscr{X}_2$.

12.14 $\operatorname{var}(\hat{\alpha}) = \dfrac{\alpha^2}{N}\dfrac{e^{2\alpha h}-1}{(\alpha h)^2}$

Case (a): $\operatorname{var}(\hat{\alpha})$ is minimal for $h_0 = 0.797/\alpha$

12.15 (a) $a = -e^{-\alpha h}$, $b = \dfrac{\beta}{\alpha}(1-e^{-\alpha h})$

(b) $\hat{a} = -e^{-\alpha h}$, $\hat{b} = \dfrac{\beta}{\alpha - \gamma}(e^{-\gamma h} - e^{-\alpha h}) \to b$ when $\gamma \to 0$

AUTHOR INDEX

Ahlén A., 226, 358, 509
Alexander, T.S., 357
Akaike, H., 232, 444, 456
Anděl, J., 456
Anderson, B.D.O., 120, 149, 158, 169, 171, 174, 177, 226, 258, 259, 412, 579
Anderson, T.W., 129, 200, 292, 579
Andersson, P., 358
Ansley, C.F., 250
Aoki, M., 129
Aris, R., 8
Åström, K.J., 8, 54, 83, 129, 149, 158, 159, 171, 226, 231, 248, 325, 357, 418, 473, 492, 508, 509, 510, 579

Bai, E.W., 120
Balakrishna, S., 456
Bamberger, W., 357
Banks, H.T., 7
Bányasz, Cs., 238, 358
Bartlett, M.S., 92, 579
Basseville, M., 357, 457, 509
Baur, U., 357
Bendat, J.S., 55
Benveniste, A., 285, 357, 358, 457, 509
Bergland, G.D., 55
Bhansali, R.J., 291, 457
Bierman, G.J., 358
Billings, S.A., 7, 171, 456
Blundell, A.J., 8
Bohlin, T., 129, 171, 226, 325, 358, 456, 492, 509
Bollen, R., 510
van den Boom, A.J.W., 171, 456, 510
Bosgra, O.H., 226
Borisson, U., 357
Borrie, M.S., 8
Box, G.E.P., 226, 250, 438, 456, 461, 464, 509, 579
Brillinger, D.R., 45, 46, 55, 129, 579
Bucy, R.S., 579
Burghes, D.N., 8
Butash, T.C., 456

Caines, P.E., 8, 226, 228, 357, 500, 509
Carrol, R.J., 171
Chan, C.W., 500, 509
Chatfield, C., 291

Chaure, C., 358
Chavent, G., 7
Chen, H.F., 357, 358
Chung, K.L., 551, 579
Cioffi, J.M., 358
Čížek, V., 130
Clarke, D.W., 226, 238
Cleveland, W.S., 291
Cooley, J.W., 55
Corrêa, G.O., 171
Cramér, H., 210, 267, 551, 579
Crowley, J.M., 7
Cybenko, G., 295, 358

Damen, A.A.M., 171
Davies, W.D.T., 54, 97, 129, 142, 143
Davis, M.H.A., 8
Davisson, L.D., 444, 456
Deistler, M., 8, 171, 259
Delsartre, Ph., 295
Demeure, C.J., 294
Dennis, J.E., Jr., 226, 259
Dent, W.T., 250
Dhrymes, Ph.J., 83
DiStefano, J.J. III, 171
Downham, D.Y., 457
Draper, N.R., 83
Dugard, L., 357
Dugré, J.P., 250
Duncombe, P., 83
Durbin, J., 292, 293
Durrani, T.S., 495

Edgar, T.F., 357
Egardt, B., 357, 358
van den Enden, A.W.M., 456
Eykhoff, P., 8, 54, 97, 171, 449, 456, 500

Fedorov, V.V., 412
Finigan, B.M., 284
Firoozan, F., 226
Fortesque, T.R., 358
Freeman, T.G., 456
Friedlander, B., 90, 92, 95, 135, 252, 259, 285, 289, 290, 295, 310, 311, 358, 360, 364, 374, 456, 495
Fuchs, J.J., 285, 437, 456
Furht, B.P., 358
Furuta, K., 510

Gardner, G., 250
Gauss, K.F., 60

Genin, Y.V., 295
Gertler, J., 358
Gevers, M., 171, 177, 412
Gill, P., 226, 259
Glover, K., 7, 54, 171, 226
Gnedenko, B.V., 548
Godfrey, K.R., 171
Godolphin, E.J., 254
Gohberg, I.C., 254
Göhring, B., 456
Gold, B., 129
Golomb, S.W., 138
Golub, G.H., 546
Goodson, R.E., 7
Goodwin, G.C., 8, 120, 177, 238, 297, 357, 358, 412, 465, 509, 545
deGooijer, J., 305
Granger, C.W.J., 509
Graybill, F.A., 546
Grenander, U., 83, 135, 243
Gueguen, C., 250
Guidorzi, R., 171
Guo, L., 358
Gustavsson, I., 31, 357, 412, 416, 420, 465

Haber, R., 7, 499
Hagander, P., 473
Hagglund, T., 358
Hajdasinski, A.K., 171
Hannan, E.J., 8, 55, 129, 171, 182, 203, 457, 579
Hanson, R.J., 211, 537
Harvey, A.C., 250
Hatakeyama, S., 510
Hayes, M.H., 358
Haykin, S., 357, 358
Heinig, G., 254
Herget, C.J., 501
Hill, S.D., 226
Ho, Y.C., 325
Holst, J., 232, 358
Honig, M.L., 357, 358
Hsia, T.C., 8
Huber, P.J., 509
Huntley, I., 8

Isermann, R., 8, 357, 412, 509

Jakeman, A.J., 284, 456
Jamshidi, M., 501
Janssen, P., 171, 449, 456, 500, 509, 556
Jategaonkar, R.V., 456

Jenkins, G.M., 54, 226, 438, 464, 509, 579
Jennrich, R.I., 83
Johnson, C.R., 357
Jones, R.H., 296

Kabaila, P., 226, 242
Kailath, T., 159, 171, 182, 183, 254, 298, 316, 511
Kalman, R.E., 579
Karlsson, E., 358
Kashyap, R.L., 8, 156, 182, 456, 457
Kavalieris, L., 171
Kay, S.M., 55, 223
Kendall, M.G., 579
Kenward, K., 83
Kershenbaum, L.S., 358
Keviczky, L., 7, 238
Kneppo, P., 357
Köllerstrom, J., 83
Kominami, H., 510
Krishnaiah, P.R., 8
Kubrusly, C.S., 7
Kučera, V., 171, 177, 181, 182
Kumar, R., 298, 358
Kunisch, K., 7
Kushner, H.J., 358

Lai, T.L., 120
Lainiotis, D.G., 8
Landau, I.D., 357
Larimore, M.G., 357
Lawson, C.L., 211, 537
Ledwich, G., 358
Lee, D.T.L., 358
Lehmann, E.L., 83, 456
Leondes, C.T., 8
Leontaritis, I.J., 171, 456
Levinson, N., 293
Lindgren, B.W., 579
Ljung, G.M., 250
Ljung, L., 8, 31, 45, 55, 57, 122, 129, 162, 171, 203, 226, 344, 350, 357, 358, 412, 416, 420, 465, 501, 509, 550, 579
Ljung, S., 358
van Loan, C.F., 546

Macchi, O., 357
Marcus-Roberts, H., 8
Mayne, D.Q., 226, 284
McBride, L.E., 225
McDonald, J., 8
McLeod, A.I., 458
McMurtrie, R., 456
Meditch, J.S., 565, 579
Mehra, R.K., 7, 8, 412, 509
Mendel, J.M., 8
Merchant, G.A., 182
Messerschmitt, D.G., 357, 358
Metivier, M., 358
Moore, J.B., 149, 158, 169, 171, 174, 177, 226, 258, 358, 579

Morf, M., 171, 254, 294, 358, 374
Morgan, B.J.T., 97
Moses, R., 297, 298, 495
Moustakides, G., 457
Murray, W., 226, 259

Negrao, A.I., 456
Nehorai, A., 171, 223, 231, 232, 259, 294, 305, 351
Newbold, P., 250, 509
Ng, T.S., 412
Nguyen, V.V., 171
Nicholson, H., 8
Norton, J.P., 8

Ölcer, S., 358
Oppenheim, A.V., 129, 130
van Overbeek, A.J.M., 171

Panuska, V., 358
Parks, T.W., 182
Paul, B., 83
Payne, R.L., 8, 238, 297, 412, 465, 509, 545
Pazman, A., 412
Pearson, C.E., 173, 267, 295, 546
Perez, M.G., 456
Peterka, V., 358, 546, 560, 579
Peterson, W.W., 138
Phillips, G.D.A., 250
Picree, D.A., 456, 461
Piersol, A.G., 55
Polak, E., 138, 284, 304, 358
Polis, M.P., 7
Polyak, B.T., 509
Porat, B., 92, 95, 358, 374
Potter, J.E., 358
Priestley, M.B., 55
Priouret, P., 358
Puthenpura, S., 509

Quinn, B.G., 457

Rabiner, L.R., 129
Rake, H., 54
Ramadge, P.J., 357
Rao, A.R., 8, 156, 182
Rao, C.R., 83, 89, 579
Rao, G.P., 8
Rao, M.M., 8
Raol, J.R., 456
Ratkowsky, D.A., 83
Reiersøl, O., 284
Rissanen, J., 171, 457
Robinson, E.A., 311
Rogers, G.S., 83
Rojas, R.A., 456
Rosenblatt, M., 83, 243
Rowe, I.H., 284
Rubinstein, R.Y., 97
Ruget, G., 358
Ruppert, D., 171

Samson, C., 358
Saridis, G.N., 357
Sastry, S., 120
Schafer, R.W., 129, 130
Scharf, L.L., 250, 294
Schmid, Ch., 510
Schnabel, R.B., 226, 259
Schwarz, G., 457
Schwarze, G., 54
Seborg, D.E., 357
Shah, S.L., 357
Sharman, K.C., 495
Shibata, R., 449, 456
Siebert, H., 357
Sin, K.S., 120, 177, 357, 358, 412
Sinha, N.K., 509
Smith, H., 83
Šmuk, K., 546
Söderström, T., 8, 31, 90, 92, 95, 135, 162, 170, 171, 182, 183, 224, 225, 226, 229, 231, 234, 239, 247, 248, 259, 265, 268, 276, 284, 285, 289, 290, 294, 296, 297, 303, 304, 310, 344, 350, 357, 358, 412, 416, 420, 433, 437, 449, 456, 464, 465, 594, 500, 509, 579
Solbrand, G., 226, 358, 509
Solo, V., 358
de Sousa, C.E., 177
Stearns, S.D., 247, 349, 357
Steiglitz, K., 225
Sternby, J., 473
Stewart, G.W., 546
Stoica, P., 8, 90, 92, 95, 122, 123, 129, 135, 170, 171, 182, 183, 223, 224, 225, 226, 229, 231, 232, 234, 239, 247, 248, 259, 265, 268, 276, 284, 285, 289, 290, 294, 296, 297, 298, 303, 304, 305, 310, 351, 358, 412, 437, 449, 456, 464, 495, 500, 509, 556
Strang, G., 546
Strejc, V., 546
Stuart, S., 579
Szegö, G., 135

Thompson, M., 8
Treichler, J.R., 357
Trench, W.F., 298
Trulsson, E., 285, 412
Tsypkin, Ya.Z., 509
Tukey, J.W., 55
Tyssø, A., 510

Unbehauen, H., 8, 456, 510
Unwin, J.M., 254

Verbruggen, H.B., 129
Vieria, A., 254, 298
Vinter, R.B., 8
Vowden, B.J., 83

Wahlberg, B., 7, 226, 509
Walter, E., 171
Watts, D.G., 54
Wei, C.Z., 120
Weisberg, S., 83
Weldon, E.J. Jr., 138
Wellstead, P.E., 8, 55, 456
Werner, H.J., 88
Wertz, V., 171
Wetherill, G.B., 83

Whittle, P., 223, 311
Widrow, B., 247, 349, 357
Wieslander, J., 429, 510
Wiggins, R.A., 311
Willsky, A., 129, 509
Wilson, G.T., 171, 181, 182
Wittenmark, B., 325, 357
Wong, K.Y., 284, 304, 358
Wood, E.F., 171
Woon, W.S., 456

Wright, M.H., 226, 259
Ydstie, B.E., 358
Young, P.C., 8, 226, 284, 285, 354, 358, 456, 509, 510
Zadeh, L.A., 31, 138
Zarrop, M.B., 351, 412, 509
van Zee, G.A., 226
Zohar, S., 298
Zygmund, A., 357

SUBJECT INDEX

Accuracy: of closed loop systems indentification, 401; of correlation analysis, 58; of frequency analysis, 41; of instrumental variable estimates, 268; of Monte Carlo analysis, 576; of noise variance estimate, 224; of prediction error estimate, 205
adaptive systems; 120, 320
Akaike's information criterion (AIC), 442, 445, 446
aliasing, 472
a posteriori probability density function, 559
approximate ML method, 333
approximation for RPEM, 329
approximation models, 7, 21, 28, 204, 209, 220, 228
ARARX model, 154, 169, 494
ARMA covariance evaluation, 251
ARMAX model, 149, 152, 162, 192, 196, 213, 215, 219, 222, 282, 331, 333, 346, 409, 416, 433, 478, 482, 489, 491, 495, 505
assessment criterion, 438, 464
asymptotic distribution, 205, 240, 268, 276, 285, 424, 427, 428, 441, 458, 461, 547, 550
asymptotic estimates, 203
asymptotically best consistent estimate, 91
autocorrelation test, 423, 457
autoregressive (AR) model, 127, 128, 151, 169, 199, 223, 253, 289, 292, 310, 361, 373, 446, 574, 575
autoregressive integrated moving average (ARIMA) model, 151, 180, 461, 479, 481, 507
autoregressive moving average (ARMA) model, 97, 103, 122, 125, 151, 163, 177, 207, 215, 229, 247, 249, 257, 289, 298, 355, 494, 505, 573

backward prediction, 312, 368
balance equation, 4
Bartlett's formula, 571
basic assumptions, 202, 264, 383
Bayes estimation, 560
best linear unbiased estimate (BLUE), 67, 82, 83, 88, 567, 569
bias, 18, 562
binary polynomial, 140
BLUE for singular residual covariance, 88
BLUE under linear constraints, 83

canonical parametrization, 155, 156
central limit theorem, 550
changes of sign test, 428
chi-two (χ^2) distribution, 73, 441, 461, 556
Cholesky factorization, 252, 259, 292, 314, 516
clock period, 113
closed loop operation, 25, 29, 381, 453, 500
comparison of model structures, 440

computational aspects, 74, 86, 181, 211, 231, 274, 277, 292, 298, 310, 350, 361, 373
computer packages, 501
condition number, 135, 534
conditional expectation, 565, 566
conditional Gaussian distribution, 566
conditional mean, 565, 566
conditional probability density function, 559
consistency, 19, 121, 186, 203, 221, 265, 266, 283, 346, 449, 505
consistent structure selection rules, 449
constraint for optimal experiment design, 402
continuous time models, 7, 158
controlled autoregressive integrated moving average (CARIMA, ARIMAX) model, 479
controlled autoregressive (ARX) model, 151, 185, 207, 233, 373, 452, 496
controller form state-space equation, 492
convergence analysis, 345, 346
convergence in distribution, 547, 551
convergence in mean square, 547, 548, 549
convergence in probability, 547, 551
convergence with probability one, 547, 548, 549
correlation analysis, 12, 42, 58
correlation coefficient, 78
covariance function, 42, 55, 100, 129, 143, 424, 457
covariance matching property, 317
covariance matrix, 65, 135, 183, 189, 205, 222, 226, 240, 270, 285, 286, 292, 303, 305, 514, 552, 560, 570; positive definiteness, 295, 316
Cramér-Rao lower bound, 66, 210, 223, 242, 560
criteria with complexity terms, 442, 444
criterion: for optimal experiment, 401; for parameter estimation, 62, 188, 190, 228, 324, 328, 342, 373, 497; with complexity term for model validation, 442, 444
cross-checking, 500
cross-correlation test, 426, 455, 457
cross-variance function, 57, 426
cross-spectral density, 57
cross-validation, 500

dead time, 33, 487
decaying transients, 504
delta operator, 477
determinant ratio test, 437
diagonal form model, 155, 165
difference operator, 477
differential equation, 343
direct identification, 389
direct sum, 518
discrete Fourier transform (DFT), 43, 45, 102, 130
distributed parameter systems, 7
drift, 180, 474, 481, 507
drifting disturbance, 180, 474, 481, 507
dynamic system, 1

Empirical transfer function estimate, 45
equation error, 62, 186
equation error method, 186
ergodic process, 102, 120, 547
errors in variables, 257
estimation of mean values, 479
exact likelihood function, 199, 249
experimental condition, 9, 28, 29, 401, 468
exponential smoothing, 217
extended least squares (ELS), 333

F distribution, 74, 558
F-test, 74, 441, 444, 446
fast LS lattice/ladder algorithms, 361, 373
fault detection, 320
feedback, 25, 28, 29, 381, 500
filtered prediction error, 473
final prediction error (FPE), 443, 445, 446
finite impulse response (FIR), 43, 151
finite state system, 138
Fisher information matrix, 210, 410, 433, 562
forgetting factor, 324, 340, 349, 374
Fourier harmonic decomposition, 80
Fourier transform, 43
frequency analysis, 37
frequency domain aspects, 7, 469, 470
frequency resolution, 47
full polynomial form model, 154, 165, 182, 233, 260, 264, 277, 373

Gain sequence for recursive algorithm, 342
Gauss–Newton algorithm, 213, 217, 218
Gaussian distribution, 198, 205, 210, 230, 240, 242, 250, 268, 276, 285, 424, 427, 428, 458, 497, 550, 551, 552, 566, 576
Gaussian process, 570
general linear model, 148, 188, 194, 388
generalized least squares (GLS) method, 154, 198, 236, 431
generic consistency, 266, 271
geometrical interpretation: Householder transformation, 527; least squares estimate, 64
Given's transformation, 527
global convergence, 352
global minimum points, 434
Gohberg–Heinig–Semencul formula, 256
gradient computation, 213, 217

Hammerstein model, 160
hierarchical model structures, 441, 462, 464
Householder transformation, 527
hypothesis testing, 74, 423, 441

identifiability, 19, 167, 204, 382, 388, 395, 396, 416
identification-method, 9, 29, 492
IDPAC, 429, 501
improved frequency analysis, 39
impulse, 23, 37, 121
impulse response, 23, 37, 54
indirect GLS method, 238
indirect identification, 395, 407
indirect prediction error method, 220, 408
information matrix, 210, 410, 433, 562
initial estimates for iterative schemes, 214

initial values:
    for computation of prediction errors, 490, 505;
    for recursive algorithms, 326, 335
innovation, 158, 397, 501
input–output covariance matrix, 133, 183
input signal, 28, 96, 468
input signal amplitude, 469
instrumental product moment matrix, 437
instrumental variable method: asymptotic distribution, 268, 285;
    basic IV method, 261;
    closed loop operation, 386;
    comparison with PEM, 276, 286;
    computational aspects, 274, 277;
    consistency, 265;
    description, 260;
    extended IV method, 262, 305, 358;
    generic consistency, 266, 271;
    linear transformation of instruments, 281;
    min-max optimal IV method, 303;
    model structure determination, 437;
    multistep algorithm, 276;
    optimal IV method, 272, 274, 286;
    optimally weighted extended IV method, 305;
    recursive algorithms, 327, 358;
    underparametrized models, 269
instruments, 262
integrated models, 481
inverse covariance function, 291
inverse Yule–Walker equations, 291
irreducible polynomial, 140

joint input–output identification, 396, 412, 500

Kalman(-Bucy) filter, 176, 196, 325, 568
Kalman gain, 157, 217
Kronecker product, 131, 233, 466, 542

lag window, 46
lattice filter, 314, 361, 373
lattice structure, 313
least mean square (LMS), 349
least squares (LS) method, 12, 60, 184, 185, 191, 198, 233, 261, 321, 361, 373, 427, 452
Lebesgue measure, 266
Levinson–Durbin algorithm (LDA), 254, 292, 298, 310
likelihood function, 198, 560
linear difference equations, 133, 152
linear filtering, 55
linear in the parameters model, 148
linear process, 571
linear regression, 28, 60, 151, 185, 207, 233, 373, 562, 564
linear systems of equations, 518, 520, 534
linear transformation of IV, 281
linearity, 499
local convergence, 344, 354, 355
local minima, 244, 247, 493
loss function, 14, 62, 189, 228, 324, 328, 342, 373, 496, 497
low-pass filtering, 112
Lyapunov equation, 128
Lyapunov function, 175, 346

Subject index 611

Markov estimate, 67, 79, 82, 90
mathematical modelling, 4
MATLAB, 501
matrices: Cholesky factorization, 252, 259, 292, 314, 516;
  condition number, 135, 534;
  determinant, 514;
  eigenvalues, 348;
  generalized inverse, 520;
  Hankel matrix, 181;
  idempotent matrix, 463, 467, 537, 557;
  inversion lemma, 175, 323, 330, 511;
  Kronecker product, 131, 233, 466, 542;
  Moore–Penrose generalized inverse, 520;
  norm, 532;
  nullspace, 518;
  orthogonal complement, 518;
  orthogonal matrix, 522, 527, 528, 533, 539;
  orthogonal projector, 522, 538;
  orthogonal triangularization, 75, 278, 527;
  QR method, 75, 527;
  partitioned matrix, 511;
  positive definite, 120, 183, 295, 297, 316, 512, 515, 516, 544;
  pseudoinverse, 63, 89, 518, 520, 524;
  range, 518, 538;
  rank, 512, 537, 541;
  singular values, 518, 524, 533;
  Sylvester matrix, 134, 246, 249, 307, 540;
  Toeplitz matrix, 181, 250, 292, 294, 298, 311;
  trace, 66;
  unitary matrix, 131;
  Vandermonde matrix, 131;
  vectorization operator, 233, 543
matrix fraction description (MFD), 182
matrix inversion lemma, 175, 323, 330, 511
maximum aposteriori (MAP) estimation, 559
maximum length PRBS, 138, 143
maximum likelihood (ML) estimation, 198, 249, 256, 497, 559
mean value, 100, 479
minimum variance estimation (MVE), 565, 567
minimum variance regulator, 404, 409, 418
model, 3, 146
model approximation, 7, 21, 28, 204, 209, 220
model classification, 146
model dimension determination, 71
model order, 71, 152, 422
model output, 15, 423
model parametrization, 146
model structure, 9, 29, 146, 148, 188, 260, 275, 278, 441, 482
model structure determination, 71, 422, 440, 482
model validation, 29, 422, 499
model verification, 499
modeling, 1, 146
modified spectral analysis, 385
modulo-two addition, 137
moment generating function, 553
Monte Carlo analysis, 269, 446, 576
moving average (MA) model, 127, 128, 151, 258, 291, 494
multistep optimal IVM, 276

multistep prediction, 229, 453
multivariable system, 154, 155, 157, 165, 201, 208, 226, 228, 233, 264, 277, 373, 384

nested model structures, 87, 278
Newton–Raphson algorithm, 212, 217, 218; for spectral factorization, 181
noise-free part of regressor vector, 265
noisy input data, 256, 281
nonlinear models, 7, 160
nonlinear regression, 91
nonparametric methods, 10, 28, 32
nonparametric models, 9
nonstationary process, 180, 474, 479, 481, 507
nonzero mean data, 473, 503
normal equations, 75, 79
null hypothesis, 423
numerical accuracy, 432

on-line robustification, 498
open loop operation, 264, 403
optimal accuracy, 209, 272
optimal experimental condition, 401, 411, 471, 544
optimal IV method, 272, 274
optimal loss function, 497
optimal prediction, 192, 216, 229
optimization algorithms, 212
ordinary differential equation approach, 343
orthogonal projection, 64, 522, 538
orthogonal triangularization, 75, 278, 527
oscillator, 34
outliers, 495, 497
output error method (OEM), 198, 239, 409
output error model structure, 153, 459, 494
overdetermined linear equations, 62, 262, 278
overparametrization, 162, 433, 459

parameter identifiability, 19, 167, 204, 388, 416, 432, 452
parameter vector, 9, 14, 60, 148
parametric method, 12, 28
parametric models, 9, 60
Parseval's formula, 52
parsimony principle, 85, 161, 438, 464
partial correlation coefficient, 313
periodic signals, 129
periodogram, 45
persistent excitation, 28, 117, 133, 184, 202, 264, 383, 408, 434, 468
pole-zero cancellation, 433
polynomial trend, 60, 77, 79, 479
polynomial trend model determination, 79
portmanteau statistics, 461
positive realness condition, 348, 352, 355
practical aspects, 348, 468
precomputations, 473
prediction, 21, 192, 229
prediction error, 22, 188, 362
prediction error method:
  asymptotic distribution, 205, 240;
  asymptotic estimates, 203;
  basis for parsimony principle, 464;
  closed loop operation, 387, 416;
  comparison with IV, 276, 286;

612 Subject index

computational aspects, 211;
consistency, 186, 203, 221;
description, 188;
indirect algorithm, 220;
optimal accuracy, 209;
recursive algorithm, 328, 345;
relation to LS, GLS and OE methods, 198, 233, 236, 239;
relation to ML method, 198;
statistical efficiency, 209;
underparametrized models, 210, 220
prediction error variance, 439
predictor, 188
prefiltering, 262, 472
probability density function (pdf), 198, 547, 559
probability of level change, 115
product moment matrix, 437
program package, 501
pseudocanonical parametrization, 155, 265
pseudoinverse, 63, 89, 518, 520, 524
pseudolinear regression (PLR), 333, 334, 346, 352, 355
pseudo random binary sequence (PRBS), 15, 97, 102, 109, 124, 128, 137

quasilinearization, 217
QR method, 75, 527

random walk, 180, 507
random wave, 127
rank test, 437
rational approximation of weighted sequences, 245
rational spectral density, 173
real-time identification, 324, 354
reciprocal polynomial, 151
recursive extended instrumental variables, 358
recursive identification, 29, 77, 320
recursive instrument variables (RIV), 327, 334
recursive least squares (RLS), 321, 354, 354, 498
recursive prediction error method (RPEM), 328, 334, 345, 353, 354
reflection coefficient, 302, 313, 318, 366, 378
regressor, 60
regularization, 81
relaxation algorithm, 236
residual, 62, 423, 429
Riccati equation, 157, 174, 177, 196
robust RLS algorithm, 498
robustness, 495
root-N consistent estimate, 91
Rouché's theorem, 295

sample correlation, 570
sample covariance matrix, 189, 270, 570
sampling, 158, 472, 507, 508
sampling interval, 472, 473, 507, 508
Schur–Cohn procedure, 298
search direction, 349
sensitivity: to noise, 41;
to rounding errors, 76, 532
shift between different regulators, 382, 388, 404
shift register, 137
significance level for hypothesis testing, 74, 425, 445

single input single output (SISO) model, 147, 152
singular residual covariance, 88
singular value decomposition, 518, 522, 533
sinusoidal signal, 98, 104, 125, 126
sliding window RLS, 354
Slutzky's lemma, 551
software, 501
spectral analysis, 43, 383
spectral characteristics, 100, 127, 129
spectral density, 55, 102, 117, 122, 130, 383
spectral factorization, 98, 149, 157, 172, 229, 258, 478
square root algorithm, 350
square wave, 127
stability of estimated AR models, 295, 316
stability of LS models, 224
state space models, 157, 166, 174, 178, 179, 196, 568
statistical efficiency, 210, 562
Stearns' conjecture, 247
Steiglitz–McBride method, 225
step function, 19, 97, 121
step length, 342
step response, 11, 32, 469
stochastic approximation, 349, 356
stochastic differential equation, 158, 507
stochastic gradient algorithm, 349, 356
strictly positive real condition, 348, 352, 355
structural index, 156, 161, 422
system, 9, 28
system identifiability, 167, 382, 388, 395
system identification, 1, 4, 5

test on covariance functions, 423, 426, 457
theoretical analysis, 17, 186, 202, 264, 334
time delay, 33, 487
time invariance, 500
transient analysis, 10, 32
trend, 60, 77, 79, 479
truncated weighted function, 43, 61, 117

unbiased estimate, 18, 65, 67, 560
underparametrization, 162, 204, 209, 220, 269
unimodality, 244, 247, 493
uniqueness properties, 161, 182, 183
updating linear regression models, 86
updating the prediction error function, 353

validation, 29, 422, 499

Weierstrass theorem, 173
weighted least squares, 262, 324, 373, 498
weighting function, 12, 42, 54, 61, 128, 245, 247, 284, 411
weighting matrix, 262, 278
white noise, 10, 65, 109, 121, 148, 187, 245, 423
Whittle–Wiggins–Robinson algorithm, 310
Whittle's formula, 223
Wiener–Hopf equation, 42
Wiener process, 158

Yule–Walker equations, 288, 298, 311, 575